Introduction to Control Systems

THIRD EDITION

ARUN K GHOSH

Formerly

Head, Instrumentation Centre, University of Kalyani

Principal, Murshidabad College of Engineering and Technology, Berhampore

Principal, Bengal College of Engineering and Technology, Durgapur

RUMI GHOSH

Operatica Solutions LLC

San Jose, CA, USA

PHI Learning Private Limited

Delhi-110092

2026

In fond memory of ***Shri Asoke K. Ghosh*** *(October 1942 – February 2024), Founder Chairman and Managing Director of PHI Learning, whose vision endlessly inspires.*

The Legacy Continues....

Published by Pushpita Ghosh, PHI Learning Private Limited, Rimjhim House, 111, Patparganj Industrial Estate, Delhi-110092 and Printed by Syndicate Binders, A-20, Hosiery Complex, Noida, Phase-II Extension, Noida-201305 (N.C.R. Delhi).

₹1295.00

INTRODUCTION TO CONTROL SYSTEMS, Third Edition
Arun K Ghosh and Rumi Ghosh

ISBN-978-81-19364-18-3 (Print Book)
ISBN-978-81-19364-19-0 (e-Book)

To

the fond memory of our mother and grandmother

Manorama

Contents

Preface

The second edition of this book was published more than a decade ago. During this period, a lot of change has taken place in the method of examination, especially in competitive exams and therefore, in the teaching pattern in undergraduate engineering colleges. Previously, the emphasis was on free-response long-answer type questions. Now, short-answer MCQs, demanding a knowledge of the concepts, are the trend. Therefore, we thought it worthwhile to revise almost every chapter and discuss intricate concepts through short-answer type examples, collecting them mostly from the GATE tests of last few years.

To make the book more student-friendly, we dropped one chapter—Preliminary Mathematics, and one appendix—Matlab Primer. Preliminary Mathematics was dropped because we have discussed necessary advanced mathematics in brief wherever needed. Matlab Primer is removed because it is no longer the only computer-based resource to study control systems. Other more powerful free languages and tools, such as Python, have emerged.

Two friends—Professor Adhish Chakraborty and Professor Arun Mondal of GNIT—gave us feedback on what students expect from a textbook on control systems. We profoundly thank them for that.

Way back in 2014, Professor Aneesh Kumar K T, Electrical Engineering Department, Mar Baselios College of Engineering & Technology, Trivandrum pointed out two errors in the Review Questions of the *Compensators* chapter. We are grateful to him for taking the trouble to do so. We will be happy to receive more feedback from anybody at arunkghosh_bect@yahoo.com.

We would also like to thank Professor Sukanya Dutta of Supreme Knowledge Foundation, WB, Professor Babar Ali of the Future Institute of Technology, WB and Professor Cevdet Islek of NTNU of Norway for their interest in this publication.

Headings of some topics have been oloured gray. This indicates that these topics may be skipped for a quick reading of the text.

Arun K Ghosh
Rumi Ghosh

1

Elementary Concepts

1.1 PREAMBLE

Initially, the concept of control engineering was introduced for electrical and mechanical processes as a theoretical tool to help control them. But today it has developed to such an extent that its field of application extends not only to every branch of engineering but also to such diverse fields as economics or biology. In our everyday life, we encounter so many applications of controls, like control of temperature in a refrigerator, maintenance of water levels in toilet flush cisterns or steering a motor car's course. Some of these controls are automatic, though many are manual. The biological world presents an amazing amount of automatic control. Our body is an excellent example. To mention only a few elementary such controls we can say that the body normally maintains a steady temperature throughout the summer and the winter when external temperature varies widely; our eyes automatically limit light to the retina by squinting in high sun or opening in full in darkness. There are many, many more highly complicated control processes running in our body and man has so far been able to unveil only a few them. Our primary objective of this study, of course, will concern engineering applications.

1.2 WHAT IS CONTROL?

Suppose we want to build a liquid-filled bath which should be maintained at a fixed temperature. We need this bath to do some experiments, or produce something through a chemical reaction.

For this purpose, we need to measure the temperature of the bath, compare it with a given set-point and heat or cool the bath accordingly. This is what is called the temperature control of the bath.

Thus, we need a temperature *controller* which manipulates the heating or cooling (may be called the *manipulated variable*) by comparing the temperature with a set-point (called the *reference input*). This manipulated variable is then fed to the bath (called the *process*) resulting to a controlled temperature (*controlled variable*). The bath temperature may be disturbed by loss of heat to or gain of heat from the ambience, which may be called *disturbance*.

This entire business is shown diagrammatically in Figure 1.1.

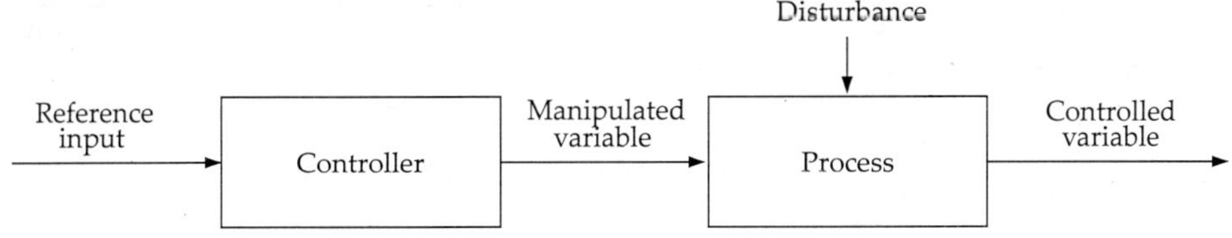

Figure 1.1: Generalized presentation of temperature control of a bath.

So, the components of a control system, as we learned from this example, can be described in general terms as given in the Table 1.1.

Table 1.1: Generalized definitions related to a control system

Term	*What it means*
Reference (or command) input	The desired value with which the output is compared.
Controller	Which limits the deviations of the output of the process within a specified range through (i) a measurement of the output, (ii) its comparison with a desired value and then (iii) by application of a manipulated variable.
Manipulated variable	The quantity or condition that the controller introduces in order to manipulate the value of the controlled variable so that it reaches a desired value.
Process	A progressively continuing operation—natural, artificial or voluntary—that consists of a series of control actions or movements systematically directed to achieve a particular result or end.
Disturbance	The unwanted input to the system that adversely affects the value of the output or response.
Controlled variable	The quantity or condition which is generally measured and controlled through some manoeuvre. Normally, it is the output or response of the process.

With this background, we now look into two important concepts in control systems, namely *open-loop* and *closed-loop* systems.

1.3 OPEN-LOOP SYSTEM

In an open-loop system the actuating device controls the process directly. Which means the output is *not* compared with a reference and given as a feedback to the system. Obviously, in such systems the *output has no direct effect on the control action.* Figure 1.2 represents an open-loop system where the actual arrangement is shown in Figure 1.2a, and the corresponding block diagram, in Figure 1.2b.

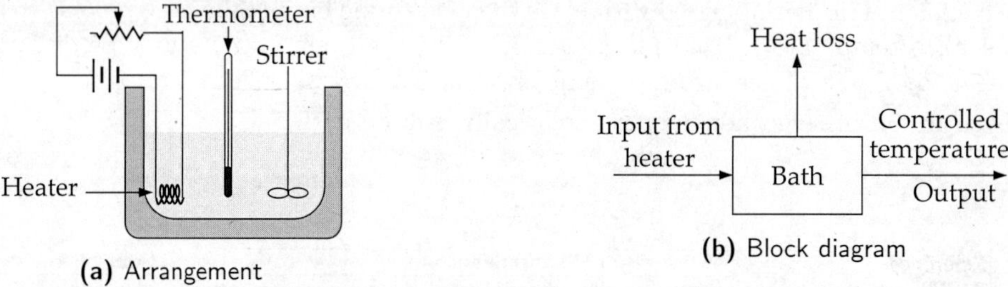

(a) Arrangement

(b) Block diagram

Figure 1.2: Open-loop control system.

Consider our old example of temperature control of a bath. Here, we may dip an electric heater into the liquid and control the current through it in such a way that the heat gained by the bath and the heat lost from it are equal so that the liquid maintains a steady temperature [Figure 1.2a]. This is an example of an open-loop control.

In real life, open loop systems can be found at many places. Consider, for example the traffic lights on the road. In the intersection of two roads, three lights—green, amber and red—either allow the traffic to flow, or warn that the signal will change soon, or stop the flow in that direction altogether. These controls are done with the help of preset timers which operate irrespective of the traffic situation at a particular instant. This means that there is *no feedback* to the lights from the traffic situation on the road. So, this is an example of an open-loop control. A similar open-loop control works in a washing machine where the machine has no automatic feedback from its own system about how dirty the washables are and therefore how long to rinse them.

1.4 CLOSED-LOOP SYSTEM

The temperature control in a liquid bath, where the heater is switched ON or OFF with the help of a contact thermometer and relay combination [Figure 1.3a], is an example of a closed-loop control.

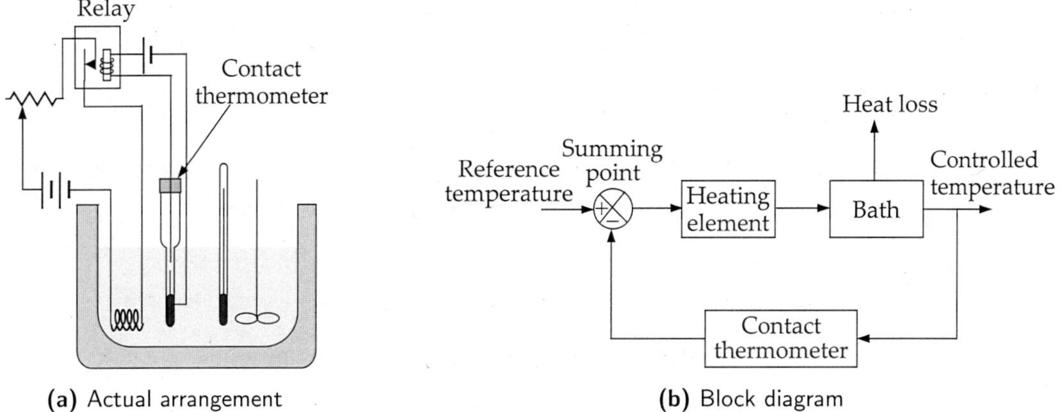

(a) Actual arrangement (b) Block diagram

Figure 1.3: Feedback control of temperature of a bath.

Here the bath temperature (i.e. output) is continuously measured and compared with the desired value (reference input) through a feedback of the output. The difference between the actual output and the reference input, called the *error signal*, is used to automatically actuate a control element such that the output maintains the desired value. Figure 1.3b gives a block diagram presentation of the closed-loop control. The *summing point* indicated in the diagram is basically the *error detector*.

Similar automatic temperature controls are maintained in electric irons, geysers, refrigerators or air-conditioners. So, they are also examples of closed-loop controls. The keyword for diagnosing closed-loop control is *feedback*. If the physical quantity that has to be controlled is *sensed by the system itself* and the feedback generates a control, it is a closed-loop one.

Some electronic calculators, for example, automatically switch off if they are not in use for some fixed time. Is it an example of open-loop control? The answer is, no. Because, the calculator has a sensing and feedback mechanism to tell the circuitry whether it is in use or not.

In this context we want to mention that a closed-loop system with one block in the forward path, one block in the feedback path, one take-off point and one summing-point, as shown in Figure 1.4, is called a *canonical*[1] closed-loop system.

[1]Meaning *simple or standard form.*

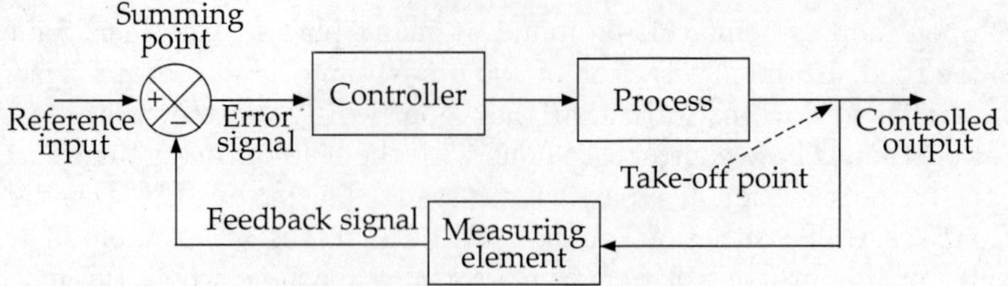

Figure 1.4: Closed-loop control.

Table 1.2 lists the differences between open-loop and closed-loop control systems.

Table 1.2: Distinction between open-loop and closed-loop control systems

Open-Loop	*Closed-Loop*
1. Simplest and economical	1. Somewhat complicated and therefore costly
2. Less stability problem	2. May become unstable and generate oscillations
3. Less accurate control	3. More accurate control
4. Affected by disturbances	4. Almost independent of disturbances
5. Convenient when the output is difficult to measure	5. Cannot be installed if the output is not amenable to measurement

A more involved comparison is provided in Table 5.1 at page 110 in Chapter 5.

REVIEW QUESTIONS

1.1 (i) What do you mean by feedback control system?
(ii) Distinguish between an open-loop and a closed-loop system.

1.2 Choose the correct answer:

(i) In an open-loop control system
(A) the output is measured and compared with a reference input
(B) the output has no control on the input
(C) the control action is actuated by a feedback signal
(D) none of the above statements is correct

(ii) In a closed-loop control system
(A) the output is measured and compared with a reference input to produce an error signal
(B) the output has no control over the input
(C) the output signal is added to the input signal
(D) none of the above statements is correct

(iii) Automatic traffic lights on roads are examples of
(A) closed-loop system
(B) open-loop system

(C) both open-loop and closed-loop system
(D) none of the above

1.3 Indicate which category—open-loop or closed-loop—the following systems belong to:

(i) A bicycle dynamo generating electricity for the head lamp
(ii) An automatic electric iron
(iii) A sprinkler watering a lawn
(iv) A microwave oven
(v) A household refrigerator
(vi) A household geyser
(vii) A self-operated household pump lifting water to the tank and maintaining water level automatically
(viii) An electronic calculator which turns off if not in use for 6 minutes
(ix) The household electric energy meter
(x) The speed regulation in electric fans

Answers to Selected Questions

1.2. (i) (B) (ii) (A) (iii) (B)

1.3. (i) open-loop (ii) closed-loop (iii) open-loop (iv) open-loop (v) closed-loop
(vi) closed-loop (vii) closed-loop (viii) closed-loop (ix) open-loop (x) open-loop

2

Control Systems and Transfer Functions

Does a static system—meaning a system not changing with time—need be controlled? The answer is 'no'. Only a dynamic system has to be controlled. And, mathematically, a dynamic system is represented by differential equations. So, differential equations need be solved to study control systems. Which is not easy always.

Laplace transform helps convert differential equations to algebraic equations, though in the s-space[1]. If we need solutions in the familiar t-space, we can perform *Inverse Laplace transform* to revert back there.

The generalized input-output relation of a dynamic system can be written as

$$\begin{aligned} a_n\frac{d^n c(t)}{dt^n} + a_{n-1}\frac{d^{n-1}c(t)}{dt^{n-1}} + \cdots + a_1\frac{dc(t)}{dt} + a_0 c(t) \\ = b_m\frac{d^m r(t)}{dt^m} + b_{m-1}\frac{d^{m-1}r(t)}{dt^{m-1}} + \cdots + b_1\frac{dr(t)}{dt} + b_0 r(t) \end{aligned}$$

where a's and b's are system parameters which are constants for a linear time-invariant (LTI) system (defined in Section 2.3). Taking Laplace transform of both sides and assuming all initial conditions $= 0$, we get

$$(a_n s^n + a_{n-1}s^{n-1} + \cdots + a_1 s + a_0)C(s) = (b_m s^m + b_{m-1}s^{m-1} + \cdots + b_1 s + b_0)R(s)$$

A convenient way of representing a control system is to define it by its transfer function.

2.1 TRANSFER FUNCTION

The transfer function of a system is defined as the ratio of the relaxed (meaning, initial conditions $= 0$) Laplace transforms of the output and the input. That is

$$\text{Transfer function} = \left.\frac{\text{Laplace transform of the output}}{\text{Laplace transform of the input}}\right|_{\text{initial conditions}=0}$$

Symbolically, if $r(t)$ = input (reference) and $c(t)$ = output (controlled), respectively and if $\mathscr{L}\{r(t)\}$[2] $= R(s)$ and $\mathscr{L}\{c(t)\} = C(s)$, then the transfer function is given by

$$\begin{aligned} \text{Transfer function} &= \left.\frac{C(s)}{R(s)}\right|_{\text{initial conditions}=0} \\ &= \frac{b_m s^m + b_{m-1}s^{m-1} + \cdots + b_1 s + b_0}{a_n s^n + a_{n-1}s^{n-1} + \cdots + a_1 s + a_0} \end{aligned} \tag{2.1}$$

Diagrammatically, the t-space and s-space representations are given in Figures 2.1a and 2.1b, respectively.

---

[1]A brief introduction to Laplace transform is given in Appendix *Laplace Transform* at page 515.

[2]We will use $\mathscr{L}\{f(t)\}$ to indicate Laplace transform of $f(t)$. The Laplace-transformed quantity will be written with a capital letter, such as $\mathscr{L}\{f(t)\} = F(s)$.

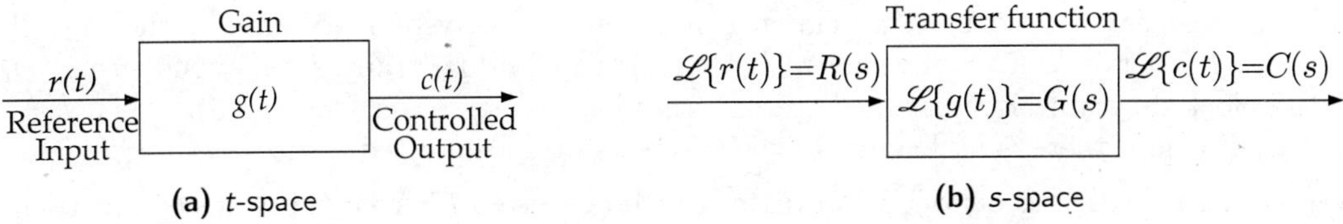

(a) t-space

(b) s-space

Figure 2.1: Representation of input, output and gain.

EXAMPLE 2.1. The relationship between the input $x(t)$ and output $y(t)$ of a system is given by

$$\frac{d^2y}{dt^2} = x(t-2) + \frac{d^2x}{dt^2}$$

The transfer function of the system is

(A) $1 + \dfrac{e^{-2s}}{s^3}$ (B) $1 + \dfrac{e^{2s}}{s^3}$ (C) $1 + s^2e^{-2s}$ (D) $1 + s^2e^{2s}$

Solution Laplace transform of the given equation yields

$$s^2Y(s) = \left(\frac{e^{-2s}}{s} + s^2\right)X(s) \qquad \text{[using time-shifting property]}$$

$$\Rightarrow \qquad \text{Transfer function} = \frac{Y(s)}{X(s)} = \frac{\dfrac{e^{-2s}}{s} + s^2}{s^2} = 1 + \frac{e^{-2s}}{s^3}$$

Ans. (A)

EXAMPLE 2.2. The transfer function of the system described by

$$\frac{d^2y}{dt^2} + \frac{dy}{dt} = \frac{du}{dt} + 2u$$

with u as input and y as output is

(A) $\dfrac{(s+2)}{(s^2+s)}$ (B) $\dfrac{(s+1)}{(s^2+s)}$ (C) $\dfrac{2}{(s^2+s)}$ (D) $\dfrac{2s}{(s^2+s)}$

Solution Taking Laplace transform of both sides, we get

$$(s^2+s)Y(s) = (s+2)U(s)$$

Therefore,

$$\frac{Y(s)}{U(s)} = \frac{(s+2)}{(s^2+s)}$$

Ans. (A)

Now that the transfer function has been defined, we will check a few of its properties.

Properties of Transfer Functions

- Transfer function offers a symbolic picture of the dynamic characteristics of a linear time-invariant system (defined in Section 2.3).

- If the transfer functions of individual components of a system are known, the overall transfer function can be determined by reducing individual transfer functions either by the block diagram or signal flow graph technique (explained in Chapter 3). This is, of course, true if there is no loading of the system by the connected devices.
- Poles and zeros (see below) of a system can be determined from the transfer function. Position of poles and zeros plays a pivotal role in the overall behaviour of a system.
- The transfer function, with the knowledge of feedback factor, helps us find out the characteristic equation, which, in turn, gives us an insight into the stability of the system.

Poles and Zeros

Our formulation of transfer function, as given in Eq. (2.1), can be factored and rewritten as

$$\text{Transfer function, } G(s) \equiv K\frac{(s-b_1')(s-b_2')\cdots(s-b_m')}{(s-a_1')(s-a_2')\cdots(s-a_n')} \tag{2.2}$$

where, K is called the *gain factor*. Equation (2.2) helps us define poles and zeros of a transfer function as follows.

Pole	*Zero*
1. Values of s for which $G(s) \to \infty$ are called poles. From Eq. (2.2) it is clear that $a_1', a_2' \ldots a_n'$are such values.	1. Values of s for which $G(s) \to 0$ are called zeros. From Eq. (2.2) it is clear that $b_1', b_2' \ldots b_m'$ are such values.
2. If a value is not repeated, it is called a *simple pole*. In case it is repeated, it is called *multiple pole.*	2. If a value is not repeated, it is called a *simple zero*. In case it is repeated, it is called *multiple zero.*

Poles and zeros are plotted in a complex plane treating $s = \sigma + \jmath\omega$. It is customary to indicate poles by crosses ($\times$) and zeros by small circles ($\circ$). A typical plot is shown in Figure 2.2.

EXAMPLE 2.3. Find the poles and zeros of the transfer function

$$G(s) = \frac{(s^2+3s+2)(s^2-2s+2)}{(s^2+7s+12)(s+3)(s^2+9)}$$

Solution The transfer function can be rewritten as

$$G(s) = \frac{(s+1)(s+2)(s-1+\jmath 1)(s-1-\jmath 1)}{(s+3)^2(s+4)(s+\jmath 3)(s-\jmath 3)} \tag{i}$$

From Eq. (i) we find poles are at $s = -3,\ -4,\ -\jmath 3$ and $+\jmath 3$, where the one at $s = -4$ is simple but the other at $s = -3$ is multiple, having a multiplicity of 2.

Zeros are at $s = -1,\ -2,\ 1+\jmath 1$ and $1-\jmath 1$. The pole-zero plot is shown in Figure 2.2.

Note: Complex poles occur in conjugate pairs. Why? Because the transfer function is real.

EXAMPLE 2.4. A system has poles at $s = 4,\ -2$ and zero at $s = -3$. Sketch the pole-zero plot and find the transfer function if the gain factor is 4.

Solution The plot is given in Figure 2.3.
The transfer function is given by

$$G(s) = \frac{4(s+3)}{(s+2)(s-4)}$$

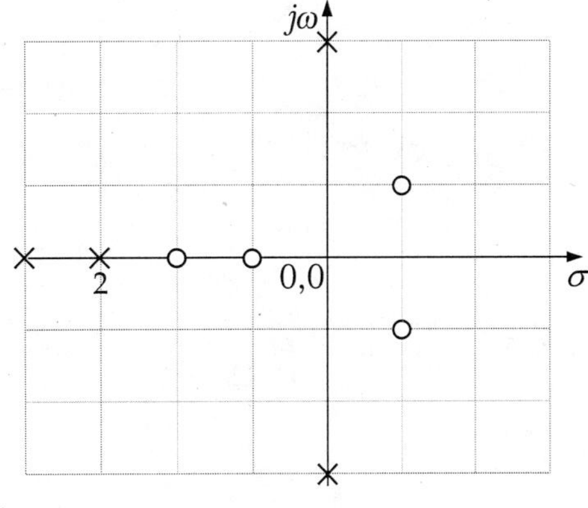

Figure 2.2: Pole-zero plot (Example 2.3). The number 2 at pole $s = -3$ indicates that there are two poles at the same location.

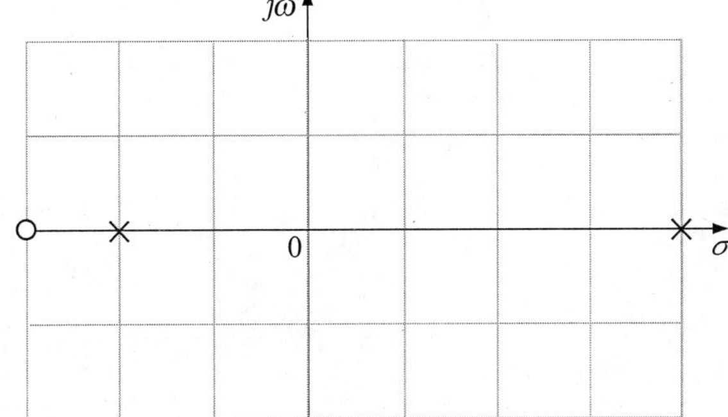

Figure 2.3: Pole-zero plot (Example 2.4).

EXAMPLE 2.5. Draw poles and zeros for $V(s) = \dfrac{s+1}{s^2+7s+12}$ and evaluate $v(t)$.

Solution The denominator of the function can be factored to give it the form of

$$V(s) = \frac{(s+1)}{(s+3)(s+4)} \equiv \frac{A}{s+3} + \frac{B}{s+4}$$

Then by applying rules of partial fractions[3],

$$A = (s+3)V(s)|_{s=-3} = \left.\frac{s+1}{s+4}\right|_{s=-3} = -2$$

$$B = (s+4)V(s)|_{s=-4} = \left.\frac{s+1}{s+3}\right|_{s=-4} = 3$$

Therefore, $$V(s) = \frac{3}{s+4} - \frac{2}{s+3} \qquad \text{(i)}$$

Taking inverse Laplace transform of Eq. (i), we get

$$v(t) = 3e^{-4t} - 2e^{-3t}$$

The pole-zero plot is given in Figure 2.4.

EXAMPLE 2.6. Find poles and zeros for the circuit given in Figure 2.5a.

[3]see Appendix *Rules of Partial Fractions* at page 520.

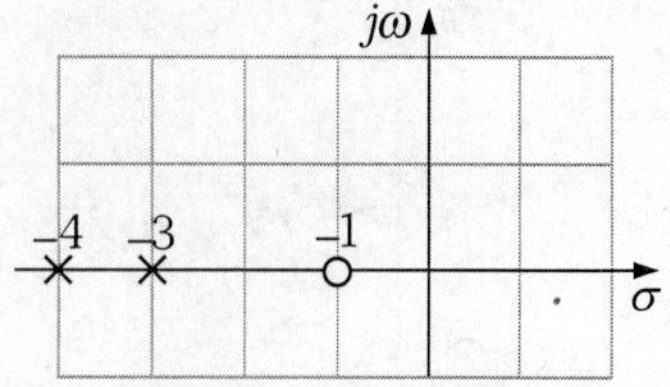

Figure 2.4: Pole-zero plot (Example 2.5).

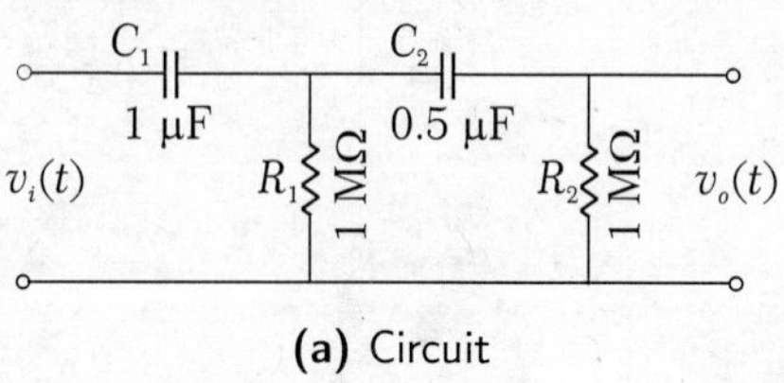

(a) Circuit

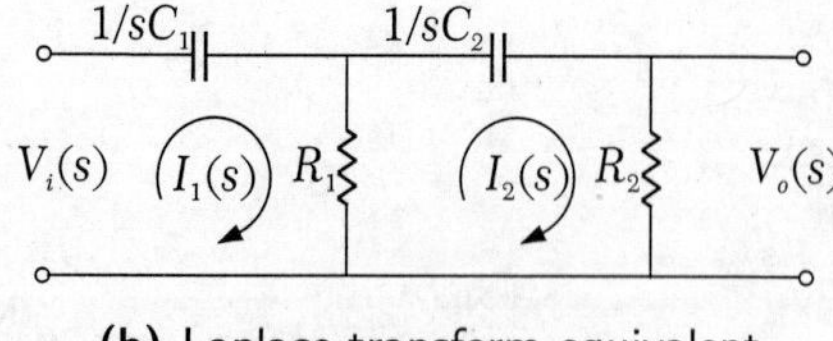

(b) Laplace transform equivalent

Figure 2.5: Example 2.6.

Solution Applying KVL to the Laplace transform of the circuit [see Figure 2.5b], we get

$$V_i(s) = \left(\frac{1}{sC_1} + R_1\right) I_1(s) - R_1 I_2(s) \tag{i}$$

$$0 = \left(\frac{1}{sC_2} + R_2 + R_1\right) I_2(s) - R_1 I_1(s) \tag{ii}$$

$$V_o(s) = R_2 I_2(s) \tag{iii}$$

Eliminating $I_1(s)$ from Eq. (i) with the help of Eq. (ii), we get

$$\begin{aligned} V_i(s) &= \left[\left(\frac{1}{sC_1} + R_1\right)\left(\frac{1}{sR_1C_2} + \frac{R_2}{R_1} + 1\right) - R_1\right] I_2(s) \\ &= \left(\frac{1}{s^2R_1C_1C_2} + \frac{1}{sC_2} + \frac{R_2}{sC_1R_1} + R_2 + \frac{1}{sC_1}\right) I_2(s) \end{aligned} \tag{iv}$$

Putting $R_1 = R_2 = 1\ M\Omega$, $C_1 = 1\ \mu F$ and $C_2 = 0.5\ \mu F$, we get from Eq. (iv)

$$\begin{aligned} V_i(s) &= \left(\frac{10^6}{0.5s^2} + \frac{10^6}{0.5s} + \frac{10^6}{s} + 10^6 + \frac{10^6}{s}\right) I_2(s) \\ &= 10^6\left(\frac{2}{s^2} + \frac{4}{s} + 1\right) I_2(s) \end{aligned} \tag{v}$$

From Eqs. (iii) and (v), we get

$$\frac{V_o(s)}{V_i(s)} = \frac{s^2}{s^2 + 4s + 2} = \frac{s^2}{(s + 0.586)(s + 3.414)} \tag{vi}$$

From Eq. (vi) we find that there are two zeros at $s = 0$ and poles are at $s = -0.586$ and -3.414.

Note: While drawing Laplace equivalent circuits the following Laplace domain expressions for R, L and C are used:

Element	*Voltage in time-domain*	*Voltage in Laplace-domain*	*Laplace-domain behaviour*
Resistance R	$Ri(t)$	$RI(s)$	R
Inductance L	$L\dfrac{di(t)}{dt}$	$sLI(s)$	sL
Capacitance C	$\dfrac{1}{C}\int i(t)dt$	$\dfrac{1}{sC}I(s)$	$\dfrac{1}{sC}$

EXAMPLE 2.7. The transfer function of a system possesses a zero at $s = -4$ and poles at $s = 0$ (twice), $\jmath 2$, $-\jmath 2$, -2 and -3. If the value of the transfer function is 0.1 at $s = 1$, find the gain factor of the system.

Solution The transfer function of the system is given by

$$G(s) = \frac{K(s+4)}{s^2(s-\jmath 2)(s+\jmath 2)(s+2)(s+3)} = \frac{K(s+4)}{s^2(s^2+4)(s^2+5s+6)}$$

Now, from the given condition

$$G(s) = 0.1 = \left.\frac{K(s+4)}{s^2(s^2+4)(s^2+5s+6)}\right|_{s=1} = \frac{5K}{(5)(12)}$$

Therefore, $$K = 1.2$$

Type and Order of Transfer Functions

The definitions are as follows:

Type: Indicates the number of *poles at the origin.* If there are n poles at $s = 0$, then type $= n$.

Order: Is determined by the highest power of s in the *denominator polynomial.* If the highest power of the denominator polynomial is m, then order $= m$.

Let us consider an example.

EXAMPLE 2.8. A transfer function is given by

$$G(s) = \frac{100(s+5)(s+50)}{s^4(s+10)(s^2+3s+10)}$$

What are its type and order?

Solution The transfer function indicates 4 poles at the origin and if the denominator of the transfer function is expressed in a polynomial form, the highest power of s will be 7. Therefore, its type is 4, and the order is 7.

Kinds of Transfer Functions

Transfer functions are considered to be of three kinds, namely:

1. Strictly proper
2. Proper
3. Improper

Suppose in a transfer function, the degree of the polynomial of the numerator $N(s) = m$ and that of the polynomial of the denominator $D(s) = n$. This means that there are m zeros and n poles of the transfer function. The transfer function may belong to one of the following three categories depending on the relative values of m and n:

$m < n$: This means that the number of poles exceeds that of zeros. For example, if $G(s)$ denotes the transfer function where $N(s)$ and $D(s)$ are numerator and denominator polynomials, i.e.

$$G(s) = \frac{N(s)}{D(s)} = \frac{s^3 + n_1 s^2 + n_2 s + n_3}{s^4 + d_1 s^3 + d_2 s^2 + d_3 s + d_4}$$

it is a *strictly proper transfer function*, because

$$\deg[N(s)] = 3 < \deg[D(s)] = 4$$

A strictly proper transfer function approaches zero when the s approaches infinity, i.e.

$$\lim_{s \to \infty} G(s) \to 0$$

All physical control systems conform to this property. Also, integral of the real part of the strictly proper transfer function is zero.

$m = n$: In this case, for $s \to \infty$, $G(s) \to K$. This situation may arise in some physical systems. Which is why such transfer functions are called *proper transfer functions.*

$m > n$: This implies, for $s \to \infty$, $G(s) \to \infty$. A physical system having an infinite transfer function is meaningless and therefore, such transfer function is called the *improper transfer function.*

Characteristic Equation

The denominator polynomial of *the closed-loop transfer function* is called the *characteristic polynomial.* The characteristic polynomial equated to zero is called the *characteristic equation.*

If $G(s)$ = forward path transfer function and $H(s)$ = feedback path transfer function of a closed loop control system (CLTF), then the closed-loop transfer function is given by[4]

$$\text{CLTF} = \frac{G(s)}{1 + G(s)H(s)}$$

The characteristic equation is given by

$$1 + G(s)H(s) = 0$$

If Eq. (2.1) represents a closed-loop transfer function , then

$$a_n s^n + a_{n-1} s^{n-1} + \cdots + a_1 s + a_0 = 0$$

is its characteristic equation. Obviously, the roots of the characteristic equation are poles of the closed-loop transfer function.

EXAMPLE 2.9. If the open-loop transfer function of negative feedback control system is

$$G(s)H(s) = \frac{K}{(s+1)^3}$$

then the gain K for a closed-loop pole to be at $-\dfrac{1}{2} + \jmath\dfrac{\sqrt{3}}{2}$ is

[4]see Eq. (3.6) at page 33.

(A) 1 (B) 2 (C) 3 (D) 4

Solution From the characteristic equation we get

$$1 + G(s)H(s) = 1 + \frac{K}{(s+1)^3} = 0$$

or,

$$K = -(s+1)^3$$

Therefore,

$$K\Big|_{s=-\frac{1}{2}+\jmath\frac{\sqrt{3}}{2}} = \left|\frac{1}{2} + \jmath\frac{\sqrt{3}}{2}\right|^3 = \left|\sqrt{\frac{1}{4}+\frac{3}{4}}\right|^3 \qquad [\because K \text{ is real}]$$

$$= 1$$

Ans. (A)

Minimum Phase System

If neither a pole nor a zero of a transfer function lies on the right half of the s-plane, the transfer function is called the *minimum phase transfer function* and the system, *minimum phase system.* Conversely, transfer functions having poles or zeros in the right half of the s-plane are called *non-minimum phase transfer functions* and the systems, *non-minimum phase systems.* For example, a system having transfer function as

$$G(s) = \frac{1 + s\tau_1}{1 + s\tau_2}$$

is a minimum phase system, while the one having a transfer function of

$$G(s) = \frac{1 - s\tau_1}{1 + s\tau_2}$$

is a non-minimum phase system. The corresponding pole-zero diagrams are shown in Figures 2.6a and 2.6b.

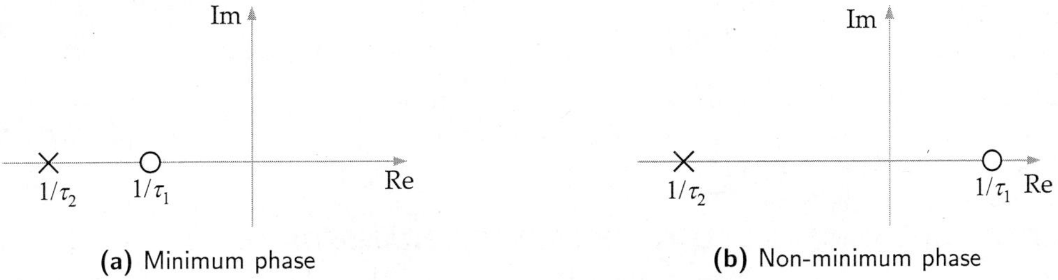

(a) Minimum phase **(b)** Non-minimum phase

Figure 2.6: Pole-zero plots of systems.

We will have occasions to discuss the reasons for such nomenclature in Sections 6.7 and 12.9 at pages 163 and 332, respectively.

EXAMPLE 2.10. Of the following transfer functions, the function that has a minimum phase characteristic is

(A) $\frac{1+s}{1-s}$ (B) $\frac{1-s}{1+s}$ (C) $\frac{e^{-3}}{s}$ (D) $\frac{s}{(s+1)(s+2)}$

Solution Of the four choices, the first two are ruled out because they possess pole/zero in the right-hand plane (RHP). For the remaining two, the phase factors are calculated as follows:

$$\text{(C)}\ \frac{e^{-3}}{s} \equiv \frac{e^{-3}}{\jmath\omega} = \left|\frac{e^{-3}}{\omega}\right| \angle -90^\circ$$

$$\text{(D)}\ \frac{s}{(s+1)(s+2)} \equiv \frac{\jmath\omega}{(\jmath\omega+1)(\jmath\omega+2)} = \left|\frac{\omega}{\sqrt{(\omega^2+1)(\omega^2+4)}}\right| \angle\left(90^\circ - \tan^{-1}\omega - \tan^{-1}\frac{\omega}{2}\right)$$

Obviously, the answer is (C) which has a fixed negative phase with no variation.

EXAMPLE 2.11. The transfer function of a passive circuit has its poles and zeros on

(A) left and right halves respectively of the s-plane
(B) right and left halves respectively of the s-plane
(C) right half of the s-plane
(D) left half of the s-plane

Solution The answer is (D) as can be seen from Examples 2.29, 2.30 and 2.31.

EXAMPLE 2.12. An analog LTI system has a negative phase-shift which varies linearly with frequency. The magnitude response is frequency independent. The system transfer function is

(A) $\dfrac{1-s\tau}{1+s\tau}$ (B) $e^{-s\tau}$ (C) $\dfrac{1-s^2\tau^2}{1+s^2\tau^2}$ (D) $\dfrac{1-s\tau}{1+s\tau}e^{-s\tau}$

Solution To figure out the phase-shift, we need to convert each transfer function to frequency domain by substituting $s \equiv \jmath\omega$. Then we get

$$\text{(A)}\ \frac{1-\jmath\omega\tau}{1+\jmath\omega\tau} \equiv \frac{\sqrt{1+(-\omega\tau)^2}\angle -\tan^{-1}\omega\tau}{\sqrt{1+(\omega\tau)^2}\angle\tan^{-1}\omega\tau} = |1|\angle -2\tan^{-1}\omega\tau$$

$$\text{(B)}\ e^{-\jmath\omega\tau} = \cos\omega\tau - \jmath\sin\omega\tau \equiv \sqrt{(\cos\omega\tau)^2 + (-\sin\omega\tau)^2}\angle -\tan^{-1}\tan\omega\tau = |1|\angle -\omega\tau$$

$$\text{(C)}\ \frac{1+\omega^2\tau^2}{1-\omega^2\tau^2} \equiv \left|\frac{1+\omega^2\tau^2}{1-\omega^2\tau^2}\right|\angle 0^\circ$$

$$\text{(D)}\ \frac{1-\jmath\omega\tau}{1+\jmath\omega\tau}e^{-\jmath\omega\tau} \equiv |1|\angle -2\tan^{-1}\omega\tau - \omega\tau$$

So, of the four choices, only (B) satisfies the given conditions.

Systems are usually checked by giving some standard inputs to them. Next we check Laplace transform of those standard inputs.

2.2 STANDARD INPUTS AND THEIR LAPLACE TRANSFORMS

Unit Step Input

Mathematically, the unit step input, $u(t)$, is expressed as

$$u(t) = \begin{cases} 0 & \text{if } t \le 0 \\ 1 & \text{if } t > 0 \end{cases}$$

Graphically, the input looks like that in Figure 2.7a.

The Laplace transform of the unit step input can be worked out as follows:

$$\mathscr{L}\{u(t)\} = \int_{t=0}^{\infty} u(t)e^{-st}dt = \int_0^{\infty} e^{-st}dt = -\frac{1}{s}\left[e^{-st}\right]_0^{\infty} = \frac{1}{s}$$

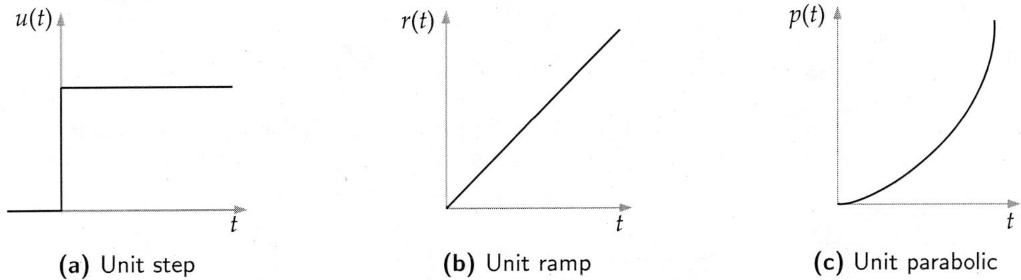

(a) Unit step (b) Unit ramp (c) Unit parabolic

Figure 2.7: Graphical presentation of standard inputs.

Unit Ramp Input

The unit ramp input, $r(t)$, is represented as

$$r(t) = t$$

Graphically, it looks like Figure 2.7b. Its Laplace transform can be worked out as follows:

$$\mathscr{L}\{r(t)\} = \int_{t=0}^{\infty} te^{-st}dt = \left[-\frac{te^{-st}}{s} + \int \frac{e^{-st}}{s}dt\right]_0^{\infty} = \left[-\frac{te^{-st}}{s}\right]_0^{\infty} - \left[\frac{e^{-st}}{s^2}\right]_0^{\infty} = \frac{1}{s^2}$$

Unit Parabolic Input

The mathematical form of the parabolic input, $p(t)$, is:

$$p(t) = \frac{t^2}{2}$$

Its graphical form is shown in Figure 2.7c. The Laplace transform is

$$\mathscr{L}\{p(t)\} = \frac{1}{2}\int_{t=0}^{\infty} t^2e^{-st}dt = \frac{1}{2}\left[-\frac{t^2}{s}e^{-st}\right]_0^{\infty} + \frac{1}{s}\int_0^{\infty} te^{-st}dt = 0 + \frac{1}{s}\left(\frac{1}{s^2}\right) = \frac{1}{s^3}$$

Unit Impulse Input

This input is defined in terms of the Dirac delta function which is given by

$$\delta(t) = \begin{cases} \lim\limits_{\varepsilon \to 0} \dfrac{1}{\varepsilon} & \text{for } 0 \le t \le \varepsilon \\ 0 & \text{for } t > \varepsilon \end{cases}$$

Graphically, the δ-function may be presented by Figure 2.8. The delta function can also be written as

$$\delta(t) = \frac{du(t)}{dt}$$

where $u(t)$ is the unit step function.

Thus, the Laplace transform of the impulse input works out as

$$\mathscr{L}\{\delta(t)\} = \mathscr{L}\left\{\frac{du(t)}{dt}\right\} = s \cdot \frac{1}{s} = 1 \tag{2.3}$$

Table 2.1 summarizes what we discussed about the test signals.

Note: The Laplace transform of each successive signal is the integral of the preceding one.

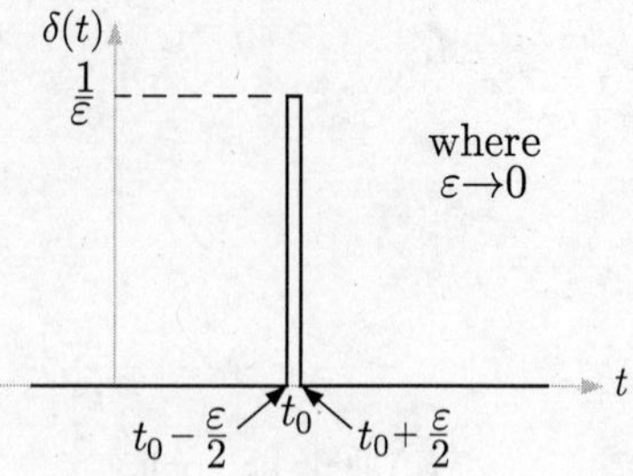

Figure 2.8: Schematic presentation of unit impulse input.

Table 2.1: Standard test signals

Signal	$r(t)$	$R(s)$
Unit impulse	$\delta(t) = \begin{cases} \lim\limits_{\varepsilon \to 0} \frac{1}{\varepsilon} & \text{for } 0 \le t \le \varepsilon \\ 0 & \text{for } t > \varepsilon \end{cases}$	$\mathscr{L}\{\delta(t)\} = 1$
Unit step	$u(t) = \begin{cases} 0 & \text{for } t \le 0 \\ 1 & \text{for } t > 0 \end{cases}$	$\mathscr{L}\{u(t)\} = \frac{1}{s}$
Unit ramp	$r(t) = 1$	$\mathscr{L}\{r(t)\} = \frac{1}{s^2}$
Unit parabolic	$p(t) = \frac{t^2}{2}$	$\mathscr{L}\{p(t)\} = \frac{1}{s^3}$

Impulse response and transfer function. We have seen in Eq. (2.3) that the Laplace transform of the impulse input is 1. We know that the transfer function is defined in Eq. (2.1) as

$$G(s) = \frac{C(s)}{R(s)}$$

which means that the Laplace transform of the response (or output) of a system is given by

$$C(s) = G(s)R(s)$$

For an impulse input, $R(s)$ is equal

So, the response of an impulse input equals the transfer function, i.e.,

$$C(s) = G(s) \tag{2.4}$$

This is an important result. Let us work out a couple of examples so as to bear it in mind.

EXAMPLE 2.13. The impulse response of a system is e^{-5t}. Determine its transfer function.

Solution Here, $c(t) = e^{-5t}$. Its Laplace transform gives

$$C(s) = \frac{1}{s+5} = G(s)$$

EXAMPLE 2.14. The unit impulse response of a system is $e^{-t}(1-\cos 3t)$. Determine its transfer function.

Solution Here, $c(t) = e^{-t}(1 - \cos 3t) = e^{-t} - e^{-t}\cos 3t$. Therefore,

$$\mathscr{L}\{c(t)\} = \mathscr{L}\{e^{-t}\} - \mathscr{L}\{e^{-t}\cos 3t\} = \frac{1}{s+1} - \frac{s+1}{(s+1)^2+9}$$

$$= \frac{9}{(s+1)(s^2+2s+10)} = G(s)$$

EXAMPLE 2.15. Find the impulse response of the system, the transfer function of which is given by

$$G(s) = \frac{s}{s^2 + 4s + 2}$$

Solution The transfer function can be written as

$$G(s) = \frac{s}{(s + 0.586)(s + 3.414)} \equiv \frac{A}{s + 0.586} + \frac{B}{s + 3.414}$$

Now,

$$A = (s + 0.586)C(s)|_{s=-0.586} = \left.\frac{s}{s + 3.414}\right|_{s=-0.586} = -0.21$$

$$B = (s + 3.414)C(s)|_{s=-3.414} = \left.\frac{s}{s + 0.586}\right|_{s=-3.414} = 1.21$$

From Eq. (2.4) $$C(s) = G(s) = \frac{1.21}{s + 3.414} - \frac{0.21}{s + 0.586}$$

On taking inverse Laplace transform of the above equation, we get

$$c(t) = 1.21e^{-3.414t} - 0.21e^{-0.586t}$$

EXAMPLE 2.16. Consider an LTI system with impulse response $h(t) = e^{-5t}u(t)$. If the output of the system is $y(t) = e^{-3t}u(t) - e^{-5t}u(t)$, then the input, $x(t)$, is given by

(A) $e^{-3t}u(t)$ (B) $2e^{-3t}u(t)$ (C) $e^{-5t}u(t)$ (D) $2e^{-5t}u(t)$

Solution Given,

$$h(t) = e^{-5t}u(t)$$

Therefore, $$\mathscr{L}\{h(t)\} = H(s) = \frac{1}{s + 5}$$

$$y(t) = e^{-3t} - e^{-5t}$$

Therefore, $$\mathscr{L}\{y(t)\} = Y(s) = \frac{1}{s + 3} - \frac{1}{s + 5} = \frac{2}{(s + 3)(s + 5)}$$

So, $$X(s) = \frac{Y(s)}{H(s)} = \frac{2\cancel{(s + 5)}}{(s + 3)\cancel{(s + 5)}} = \frac{2}{s + 3}$$

$$\mathscr{L}^{-1}\{X(s)\} = x(t) = 2e^{-3t}u(t)$$

Ans. (B)

EXAMPLE 2.17. Let a causal[5] LTI system be characterized by the following differential equation, with initial rest condition

$$\frac{d^2y(t)}{dt^2} + 7\frac{dy(t)}{dt} + 10y(t) = 4x(t) + 5\frac{dx(t)}{dt}$$

where, $x(t)$ and $y(t)$ are input and output, respectively. The impulse response of the system is [$u(t)$ is the unit step function]

(A) $2e^{-2t}u(t) - 7e^{-5t}u(t)$ (B) $-2e^{-2t}u(t) + 7e^{-5t}u(t)$
(C) $7e^{-2t}u(t) - 2e^{-5t}u(t)$ (D) $-7e^{-2t}u(t) + 2e^{-5t}u(t)$

[5]see Section *Causal and Anti-causal Systems* at page 22 for definition.

Solution Taking Laplace transform of the given equation, with initial rest condition, we get

$$(s^2 + 7s + 10)Y(s) = (4 + 5s)X(s)$$

Therefore,
$$\frac{Y(s)}{X(s)} = \frac{5s+4}{(s+2)(s+5)}$$

Since,
$$\begin{aligned}\text{impulse response} &= \mathscr{L}^{-1}\{\text{transfer function}\}\\ &= \mathscr{L}^{-1}\left\{\frac{5s+4}{(s+2)(s+5)}\right\}\\ &= \mathscr{L}^{-1}\left\{-\frac{2}{s+2} + \frac{7}{s+5}\right\}\\ &= -2e^{-2t}u(t) + 7e^{-5t}u(t)\end{aligned}$$

Ans. (B)

EXAMPLE 2.18. The impulse response of a system is $h(t) = t\,u(t)$. For an input $u(t-1)$, the output is

(A) $\frac{t^2}{2}u(t)$ (B) $\frac{t(t-1)}{2}u(t-1)$

(C) $\frac{(t-1)^2}{2}u(t-1)$ (D) $\frac{t^2-1}{2}u(t-1)$

Solution

Given,
$$h(t) = t\,u(t)$$

Therefore,
$$\mathscr{L}\{h(t)\} \equiv H(s) = \frac{1}{s^2} = G(s) \qquad [\because \text{impulse response}]$$

Input
$$r(t) = u(t-1)$$

$$\mathscr{L}\{r(t)\} = R(s) = \frac{e^{-s}}{s}$$

$$C(s) = G(s)R(s) = \frac{1}{s^2}\cdot\frac{e^{-s}}{s}$$

So, output
$$c(t) = \mathscr{L}^{-1}\{C(s)\} = \frac{(t-1)^2}{2}u(t-1)$$
[see line 5 of Table A.1, apply time-shifting]

Ans. (C)

2.3 CONTROL SYSTEM CATEGORIES

On the basis of certain criteria, control systems can be divided into the following categories:

1. Linear and non-linear systems
2. Time-varying and time-invariant systems
3. Causal and anti-causal systems
4. Deterministic and stochastic systems
5. SISO and MIMO

We consider them in that order and see what the criteria are.

Linear and Non-linear Systems

A system is called linear if it satisfies both (i) additivity and (ii) homogeneity properties. If it does not satisfy any one of them, it is called non-linear. Now, the question is what are additivity and homogeneity properties?

Additivity property. If x and y belong to the domain of the function f, additivity demands

$$f(x+y) = f(x) + f(y)$$

Let us now examine a few functions and see if they satisfy the additivity criterion.

EXAMPLE 2.19. Let a function be defined by

$$u(y) = \frac{d^2y(t)}{dt^2} + 8\frac{dy(t)}{dt} + y(t) \tag{i}$$

Does it satisfy the additivity property?

Solution Let us check what happens to the function if we substitute $x + y$ for y in Eq. (i).

$$\begin{aligned} u(x+y) &= \frac{d^2}{dt^2}[x(t) + y(t)] + 8\frac{d}{dt}[x(t) + y(t)] + [x(t) + y(t)] \\ &= \left[\frac{d^2x(t)}{dt^2} + 8\frac{dx(t)}{dt} + x(t)\right] + \left[\frac{d^2y(t)}{dt^2} + 8\frac{dy(t)}{dt} + y(t)\right] \\ &= u(x) + u(y) \end{aligned}$$

Therefore, $u(y)$ is an additive function.

EXAMPLE 2.20. Let a function be defined by

$$f(x) = x^2 \tag{i}$$

Does it satisfy additivity criterion?

Solution As before, we substitute $x + y$ for x in Eq. (i). Then,

$$f(x+y) = (x+y)^2 \neq x^2 + y^2$$

Thus, $f(x)$ is not an additive function.

EXAMPLE 2.21. Let

$$u(y) = y\frac{d^2y}{dt^2} + 7\frac{dy}{dt} \tag{i}$$

Is it an additive function?

Solution On substituting $x + y$ for y in Eq. (i), we get

$$u(x+y) = (x+y)\frac{d^2}{dt^2}(x+y) + 7\frac{d}{dt}(x+y) \neq \left[x\frac{d^2x}{dt^2} + 7\frac{dx}{dt}\right] + \left[y\frac{d^2y}{dt^2} + 7\frac{dy}{dt}\right]$$

Thus, $u(y)$ is not an additive function.

Homogeneity property. The homogeneity property demands that for any y belonging to the domain of the function f, if α is a scalar constant, then

$$f(\alpha y) = \alpha f(y)$$

Let us now examine whether the functions, that we tested for additivity, satisfy the homogeneity criterion.

EXAMPLE 2.22. Is the function defined by

$$u(y) = \frac{d^2y(t)}{dt^2} + 8\frac{dy(t)}{dt} + y(t) \tag{i}$$

homogeneous?

Solution We substitute αy for y in Eq. (i) and observe

$$\frac{d^2(\alpha y)}{dt^2}+8\frac{d(\alpha y)}{dt}+\alpha y=\alpha\left[\frac{d^2y}{dt^2}+8\frac{dy}{dt}+y\right]$$

Therefore, $u(y)$ defined by Eq. (i) is a homogeneous function.

EXAMPLE 2.23. Does the function defined by

$$f(x)=x^2 \tag{i}$$

satisfy the homogeneity test?

Solution We see the result of substitution of x by αx in Eq. (i) that

$$f(\alpha x)=(\alpha x)^2\neq\alpha x^2$$

Therefore, $f(x)$ defined in Eq. (i) does not stand the homogeneity test.

EXAMPLE 2.24. What about the homogeneity of the function defined in the following equation?

$$u(y)=y\frac{d^2y}{dt^2}+7\frac{dy}{dt} \tag{i}$$

Solution The same test of substitution of y by αy yields

$$(\alpha y)\frac{d^2(\alpha y)}{dt^2}+7\frac{d(\alpha y)}{dt}=\alpha^2y\frac{d^2y}{dt^2}+7\alpha\frac{dy}{dt}\neq\alpha\left[y\frac{d^2y}{dt^2}+7\frac{dy}{dt}\right]$$

Hence, the function $u(y)$ defined in Eq. (i) is not homogeneous.

Considering all the three functions cited above, we arrive at the conclusion that only the one given in Example 2.22 is linear because it satisfies both additivity and homogeneity properties while the functions given in Examples 2.23 and 2.24 do not, and hence are non-linear.

EXAMPLE 2.25. The response of a system to a unit impulse is $y(t)=e^{-(t-1)}$. Which one of the following is the correct statement about the system?

(A) The system is non-linear.
(B) The system is unstable in open-loop.
(C) The steady-state gain of the system for a unit step input is 0.2.
(D) The steady-state gain of the system for a unit step input is 1.0.

Solution We observe that for $t\to\infty$, $y(t)=0$. Therefore, the options (B), (C) and (D) are not true. To test the linearity, let us check if $y(t)$ is additive.

$$\begin{aligned}y(t+T)&=e^{-5(t+T-1)}=e^{-5(t-1)}\cdot e^{-5T}\\&\neq y(t)+y(T)\end{aligned}$$

Ans. (A)

Time-varying and Time-invariant Systems

If a system has a time-dependent input function, $x(t)$, and a time-dependent output function, $y(t)$, the system will be considered time-invariant if a time-delay on the input, $x(t-\Delta t)$ directly equates to a time-delay of the output function $y(t-\Delta t)$. The block-diagram [Figure 2.9]

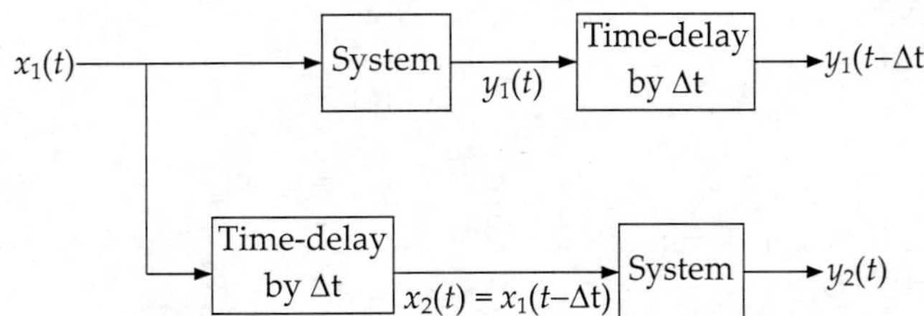

Figure 2.9: Block-diagram illustrating time-invariant system.

illustrates the time-invariance for a deterministic continuous-time single-input single-output system. The system is time-invariant iff[6] $y_2(t) = y_1(t - \Delta t)$ for all time t, for all real constant Δt and for all input $x_1(t)$.

In practical terms, if the parameters of a control system vary with time, it is a *time-varying* system. If they do not vary with time, the system is *time-invariant.*

A simple rocket is a time-varying system because while it is moving in its trajectory, one parameter of its mathematical modelling—mass—is continuously changing with time because of burning of its fuel. However, a simple pendulum of mass m, executing a simple harmonic motion about its mean position of rest, is a time-invariant system, and therefore, its trajectory can be described by a simple equation

$$m\frac{d^2x}{dt^2} = -\alpha x$$

where its parameters—m and α—are independent of time.

An LR circuit, within its current tolerance, is a time-invariant system. It is described by

$$e = L\frac{di}{dt} + Ri$$

where L and R are constants. But if the current is high, L and R become $L(t)$ and $R(t)$ because they get hot with time and their values change thereby. Obviously then, the LR circuit is not time-invariant.

EXAMPLE 2.26. If the input $x(t)$ and output $y(t)$ of a system are related as $y(t) = \max\{0,\ x(t)\}$, then the system is

(A) linear and time-variant
(B) linear and time-invariant
(C) non-linear and time-variant
(D) non-linear and time-invariant

Solution Given, $y(t) = \max\{0,\ x(t)\}$. Let us define

$$y_1(t) = \max\{0,\ x_1(t)\} \qquad y_2(t) = \max\{0,\ x_2(t)\}$$
$$y_3(t) = \max\{0,\ [x_1(t) + x_2(t)]\} \qquad y_4(t) = \max\{0,\ \alpha x(t)\}$$
$$y_5(t) = \alpha \max\{0,\ x(t)\}$$

To check if the system is linear, we examine the homogeneity and additivity property of the relation. Since, $y_4(t) = y_5(t)$, it is homogeneous. To check additivity let, $x_1(t) = 1$ and $x_2(t) = -1$. Then,

$$y_1(t) = \max\{0,\ 1\} = 1 \qquad y_2(t) = \max\{0,\ -1\} = 0$$
$$y_3(t) = \max\{0,\ 0\} = 0 \qquad y_3(t) \neq y_1(t) + y_2(t)$$

[6] meaning *if and only if*

Hence, the system is non-linear. Now, we check the time-invariance. If the input is $x_1(t-\Delta t)$, then

$$y_1(t) = \max\{0,\ x_1(t-\Delta t\}$$
$$y(t-\Delta t) = \max\{0,\ x_1(t-\Delta t\} = y_1(t)$$

So, it is time-invariant.
Ans. (D)

Causal and Anti-causal Systems

In a causal system, the output of the system does not depend on future inputs, but only on past input. This means that for the system to be causal, it should be independent from the future values only. This implies that past dependency will cause no problem for the system from becoming causal.

Causal systems are practically or physically realizable systems. Let us consider some examples.

EXAMPLE 2.27. If the input, $x(t)$, and output, $y(t)$, of a system are related as follows, are they causal?

(a) $y(t) = Ax(t)$, where A is a constant.
(b) $y(t) = Ax(t-1)$
(c) $y(t) = x(t) + Ax(t+1)$

Solution

(a) This is a causal system. Here, the signal is only dependent on the present values of $x(t)$. For example if we substitute $t = 5$, the output will show for that instant of time only. Therefore, as the output has no dependence on the future value of the input, it is a causal system.
(b) This is also a causal system. Here, the system depends on past values. For instance, if we substitute $t = 5$, the output will by $Ax(4)$, which is dependent on a previous value rather than the present input $x(5)$. By no means, it depends upon future values. Therefore, this system is also a causal system.
(c) This is not a causal system, rather it can be called an anti-causal system. Here, the system has two parts. The part $x(t)$, as we have discussed earlier, depends only upon the present values of $x(t)$. So, there is no issue with it. However, if we take the case of $x(t+1)$, it clearly depends on the future values of $x(t)$ because if we put $t = 2$, the output will contain $Ax(3)$ which is a future value in comparison to the present input $x(2)$. Hence, it is not causal.

EXAMPLE 2.28. A system with input $x(t)$ and output $y(t)$ is defined by the input-output relation

$$y(t) = \int_{-\infty}^{-2t} x(\tau)d\tau$$

The system will be

(A) causal, time-invariant and unstable
(B) causal, time-invariant and stable
(C) non-causal, time-invariant and unstable

(D) non-causal, time-variant and unstable

Solution Causality: Since $y(t)$ depends on $x(-2t)$, it is non-causal.
Time-variance:

$$y(t) = \int_{-\infty}^{-2t} x(\tau - \tau_0) \neq y(t - \tau_0)$$

So, it is time-variant.
Stability:
Output $y(t)$ is unbounded for a bounded input. For example,

if $$x(\tau) = e^{-\tau} \qquad \text{(bounded)}$$

$$y(t) = \int_{-\infty}^{-2t} e^{-\tau} d\tau = \left[\frac{e^{-\tau}}{-1}\right]_{-\infty}^{-2t} \qquad \text{(unbounded)}$$

Ans. (D)

Deterministic and Stochastic Control Systems

If a control system produces a predictable and repetitive output when it is subjected to known input and disturbance, it is called a *deterministic* control system. A swinging simple pendulum, for example, is a deterministic system, because, at a given place, it will oscillate with the same time-period for a small displacement of its bob.

When the output of a control system becomes statistically predictable, it is called a *stochastic* system. A cluster of molecules of a gas in an enclosure, for example, forms a stochastic system, because for a given temperature of the enclosure, all the molecules will not acquire the same velocity. Some molecules will have very high velocities, some very low and most of them will acquire a certain velocity that is called the *most probable velocity.* In other words, the velocity of molecules is statistically predictable. There are many more examples of stochastic systems. In fact, most of the natural systems are stochastic systems. Take, for example, the temperature at a certain place on the earth's surface. At a certain date of the year, the temperature has a statistically predictable value, called the normal temperature, although energy supply and energy reception systems are the same comprising the sun and the earth.

SISO and MIMO

SISO is the acronym for Single Input Single Output system while MIMO for Multiple Input Multiple Output system. The terms are self-explanatory and therefore, require no elaborate discussion. We may mention here, we will be considering *mostly* SISO systems in our studies.

We already considered a few examples to see how to calculate transfer functions of systems. Here, we consider a few more.

EXAMPLE 2.29. Find the transfer function of the simple LCR circuit given in Figure 2.10a.

Solution From the Laplace equivalent circuit [Figure 2.10b], we get

$$E_i(s) = \left[Ls + R + \frac{1}{Cs}\right] I(s) \qquad \text{(iii)}$$

$$E_o(s) = \frac{1}{Cs} I(s) \qquad \text{(iv)}$$

(a) *LCR* circuit (b) Laplace equivalent

Figure 2.10: Example 2.29.

Dividing Eq. (iv) by Eq. (iii), we get after a little manipulation

$$\frac{E_o(s)}{E_i(s)} = G(s) = \frac{1}{s^2LC + sRC + 1} = \frac{1}{\dfrac{s^2}{\omega_n^2} + s\tau + 1}$$

where, $\omega_n = \dfrac{1}{\sqrt{LC}}$ and $\tau = RC$.

EXAMPLE 2.30. For the electrical network shown in Figure 2.11a, determine the transfer function $V_o(s)/V_i(s)$:

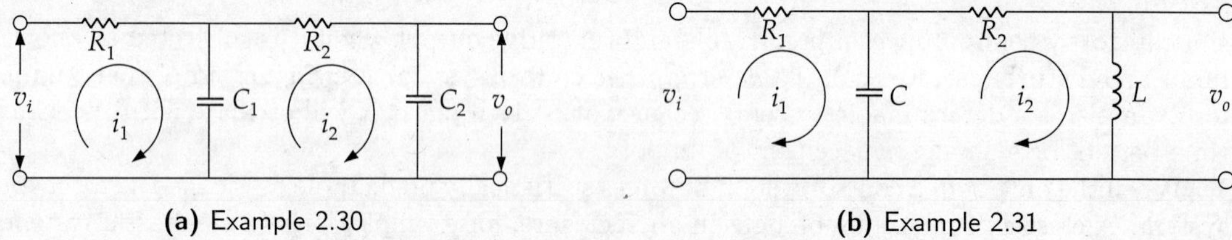

(a) Example 2.30 (b) Example 2.31

Figure 2.11: Electrical networks.

Solution The Laplace equivalent of the given circuit can be obtained by substituting $C_1 \equiv 1/sC_1$, $C_2 \equiv 1/sC_2$, $i_1 \equiv I_1(s)$, $i_2 \equiv I_2(s)$, $v_i \equiv V_i(s)$ and $v_o \equiv V_o(s)$. Hence,

$$\left.\begin{aligned} V_i(s) &= R_1 I_1(s) + \frac{I_1(s) - I_2(s)}{sC_1} \\ 0 &= R_2 I_2(s) + \frac{I_2(s)}{sC_2} + \frac{I_2(s) - I_1(s)}{sC_1} \\ V_o(s) &= \frac{I_2(s)}{sC_2} \end{aligned}\right\} \qquad \text{(i)}$$

Eliminating $I_1(s)$ and $I_2(s)$ from Eqs. (i), we get after a little algebraic manipulation

$$\frac{V_o(s)}{V_i(s)} = \frac{1}{s^2 R_1 R_2 C_1 C_2 + s(R_1C_1 + R_1C_2 + R_2C_2) + 1}$$

EXAMPLE 2.31. Find the transfer function for the network given in Figure 2.11b.

Solution Applying KVL to the Laplace transform of the circuit as before, we get

$$\left.\begin{aligned} V_i(s) &= \left(R_1 + \frac{1}{sC}\right) I_1(s) - \frac{1}{sC} I_2(s) \\ 0 &= \left(R_2 + sL + \frac{1}{sC}\right) I_2(s) - \frac{1}{sC} I_1(s) \\ V_o(s) &= sLI_2(s) \end{aligned}\right\} \qquad \text{(i)}$$

Eliminating $I_1(s)$ and $I_2(s)$ from Eqs. (i), we get after a little algebraic manipulation

$$\frac{V_o(s)}{V_i(s)} = \frac{sL}{s^2R_1LC + s(R_1R_2C + L) + R_1 + R_2}$$

EXAMPLE 2.32. Find the transfer function of a system described by

$$\frac{d^2y}{dt^2} + 3\frac{dy}{dt} + 4y = 2\frac{d^2x}{dt^2} + 6x$$

Solution Taking Laplace transform of the equation, we get

$$(s^2 + 3s + 4)Y(s) = (2s^2 + 6)X(s)$$

Therefore, the transfer function is

$$\frac{Y(s)}{X(s)} = \frac{2s^2 + 6}{s^2 + 3s + 4}$$

EXAMPLE 2.33. Find the transfer function of a system given by

$$3\frac{d^2c(t)}{dt^2} + 5\frac{dc(t)}{dt} + c(t) = r(t) + 3r(t-2)$$

Solution The second function on the RHS of the given equation has a different time base. So, we have to apply the second shifting property when we take Laplace transform of the equation. Thus, the Laplace transform of the equation is

$$(3s^2 + 5s + 1)C(s) = [1 + 3e^{-2s}]R(s)$$

Therefore,

$$\frac{C(s)}{R(s)} = \frac{1 + 3e^{-2s}}{3s^2 + 5s + 1}$$

EXAMPLE 2.34. Determine V_2/V_1 of the circuit[7] given in Figure 2.12a.

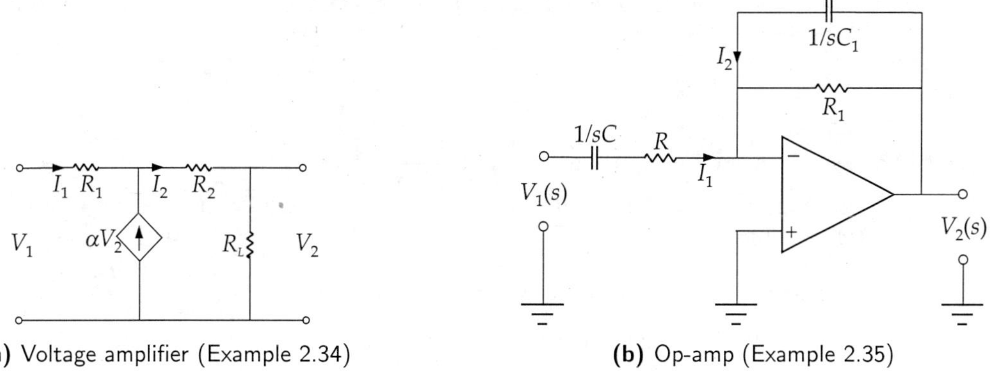

(a) Voltage amplifier (Example 2.34) **(b)** Op-amp (Example 2.35)

Figure 2.12: Circuits.

[7]This ratio is *not* the transfer function. It is rather the *amplification factor*.

Solution Using KVL, we get

$$V_1 = R_1I_1 + (R_2 + R_L)I_2$$

$$I_2 = I_1 + \alpha V_2 \quad \text{(i)}$$

$$V_2 = I_2R_L \quad \text{(ii)}$$

Therefore,

$$\begin{aligned} V_1 &= R_1(I_2 - \alpha V_2) + (R_2 + R_L)I_2 && \text{[using Eq. (i)]} \\ &= (R_1 + R_2 + R_L)I_2 - \alpha R_1V_2 && \text{[using Eq. (ii)]} \\ &= \left[\frac{R_1 + R_2}{R_L} + 1 - \alpha R_1\right] V_2 \end{aligned}$$

Thus,

$$\frac{V_2}{V_1} = \frac{R_L}{R_1 + R_2 + R_L - \alpha R_1R_L} = \frac{R_L}{(1 - \alpha R_L)R_1 + R_2 + R_L}$$

EXAMPLE 2.35. Find $V_2(s)/V_1(s)$ of the circuit given in Figure 2.12b where the op-amp is ideal.

Solution Since the op-amp is ideal, $I_1 = -I_2$. This gives

$$\frac{V_1(s)}{R + \dfrac{1}{sC}} = -\frac{V_2(s)}{\left[\dfrac{R_1}{sC_1}\right] \div \left[R_1 + \dfrac{1}{sC_1}\right]} = -\frac{V_2(s)}{\dfrac{R_1}{sC_1R_1 + 1}}$$

or,

$$\frac{V_2(s)}{V_1(s)} = -\frac{\dfrac{R_1}{sC_1R_1 + 1}}{R + \dfrac{1}{sC}} = -\frac{sCR_1}{(sC_1R_1 + 1)(sCR + 1)}$$

REVIEW QUESTIONS

2.1 Choose the correct answer:

(i) A system has the transfer function of $\dfrac{1-s}{1+s}$. Its gain at 1 rad/s is

(A) 1 (B) 0 (C) −1 (D) 0.5

(ii) For the characteristic equation $s(s+4)(s+7) + k(s+1)(s+5) = 0$, the roots of the system at $k \to \infty$ will be at

(A) -1, -5 and ∞ (B) -4, -7 and ∞
(C) -4, ∞ and $-\infty$ (D) 0, -1 and -4

(iii) A system has the transfer function of $\dfrac{1-s}{1+s}$. It is known as

(A) minimum phase system (B) non-minimum phase system
(C) low-pass system (D) high-pass system

(iv) The transfer function of a system is given by $G(s) = \dfrac{10(s+1)(s+10)}{s^4(s+5)(s^23s+10)}$. The type and order of the system are respectively

(A) 4 and 7 (B) 4 and 9 (C) 7 and 5 (D) 5 and 7

(v) The type number of a transfer function denotes the number of

(A) poles at the origin (B) zeros at the origin
(C) poles at infinity (D) none of these

(vi) Given that $G(s) = \dfrac{K}{s^2(s+2)(s+3)}$. The type and order of the system are

(A) 3 and 3 (B) 2 and 4 (C) 3 and 1 (D) 3 and 0

(vii) A system has a transfer function $\dfrac{1-s}{1+s}$. It is known as

(A) low-pass system (B) high-pass system
(C) all pass system (D) minimum phase system

(viii) The system that can be represented by a transfer function is a

(A) nonlinear time dependent system
(B) nonlinear time varying system
(C) linear time varying system
(D) linear time independent system

(ix) The closed-loop gain of the system in the given figure is

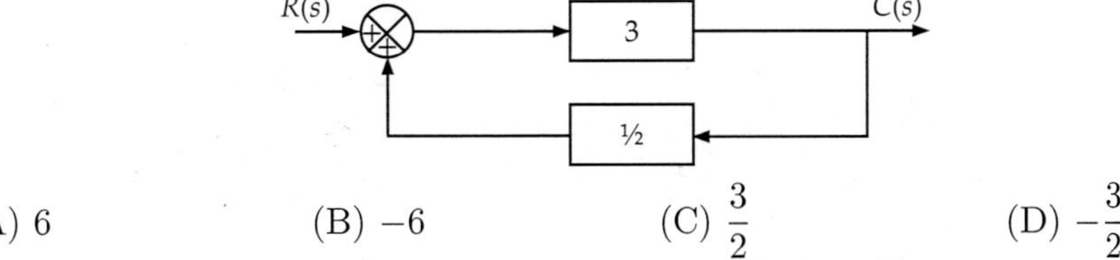

(A) 6 (B) −6 (C) $\dfrac{3}{2}$ (D) $-\dfrac{3}{2}$

(x) The transfer function of a simple RC integrator circuit shown in the figure is

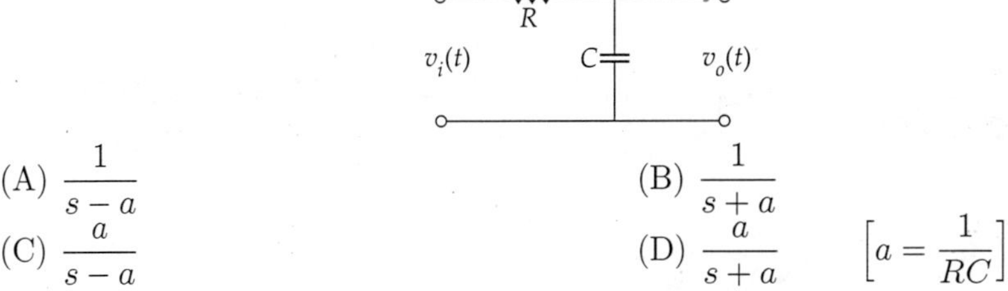

(A) $\dfrac{1}{s-a}$ (B) $\dfrac{1}{s+a}$
(C) $\dfrac{a}{s-a}$ (D) $\dfrac{a}{s+a}$ $\left[a = \dfrac{1}{RC}\right]$

(xi) For linear time-invariant systems, that are Bounded Input Bounded stable, which one of the following statements is TRUE?

(A) The impulse response will be integral, but may not be absolutely integrable.
(B) The unit impulse response will have finite support.
(C) The unit step response will be absolutely integrable.
(D) The unit step response will be bounded.

2.2 What is a linear time-invariant system?

2.3 Consider the system given in Figure 2.13, where

$$G(s) = \frac{s+25}{s^4+10s^3+35s^2+50s+24}$$

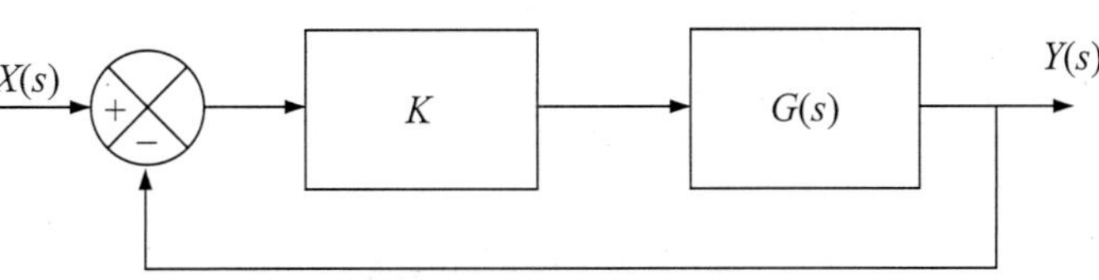

Figure 2.13: Question 2.3.

(a) If one pole is at $s = -1$, what are other open-loop poles and zeros?

(b) What is the characteristic equation of the system?

2.4 The dynamical behaviours of a few systems are represented by the following differential equations. Find which ones of them are linear time-invariant systems.

(A) $M\dfrac{d^2x}{dt^2} + D\dfrac{dx}{dt} + Kx + f(x)$ (B) $L\dfrac{di}{dt} + Ri + \dfrac{\int i\,dt}{C} = e(i)$

(C) $M(t)\dfrac{d^2x}{dt^2} + K(t)x = f(t,x)$ (D) $Li\dfrac{di}{dt} + Ri^2 = P(i)$

2.5 A system transfer function is given by

$$G(s) = \frac{10(s+1)(s+3)}{s(s+2)(s+1-\jmath 2)(s+1+\jmath 2)}$$

Plot poles and zeros in the s-plane and determine the value of the transfer function at $s = 1$.

2.6 The pole-zero plot of a system is given by Figure 2.14. If $G(s) = 10$ for $s = 1$, find the transfer function of the system.

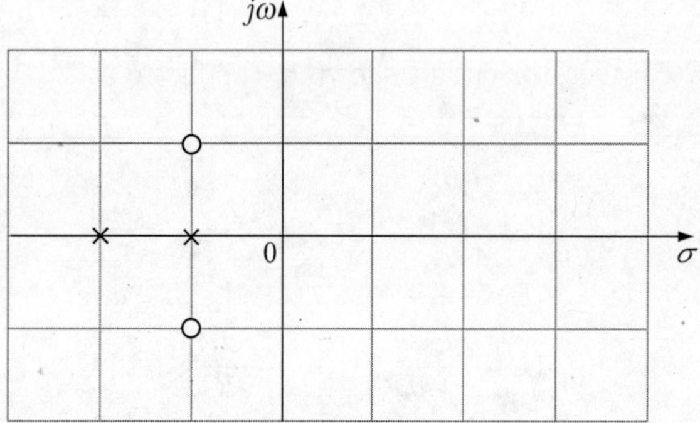

Figure 2.14: Question 2.6.

2.7 The unit impulse response of a system is $e^{-it}\sin\omega t$. Find the transfer function of the system.

2.8 The transfer function of system is given by

$$G(s) = \frac{Y(s)}{X(s)} = \frac{6}{s^2+5s+2}$$

Find the differential equation that describes the dynamical behaviour of the system.

2.9 What is the difference between order and type of control system? Discuss with examples.

2.10 A system has an open-loop transfer function of $\dfrac{K}{s^3(Ts+1)}$. What are the type and order of the system?

2.11 Find the transfer function of the network shown in Figure 2.15a.

2.12 Obtain the transfer function of the circuit shown in Figure 2.15b.

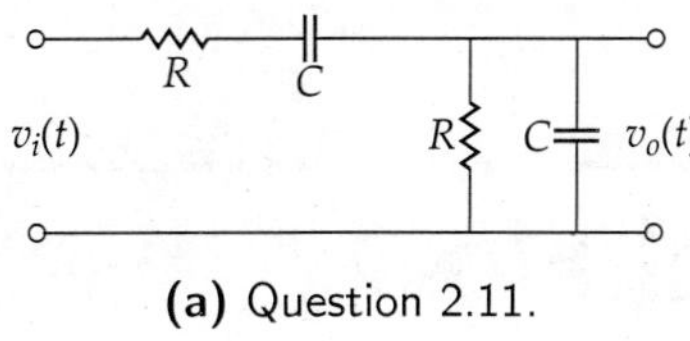

(a) Question 2.11.

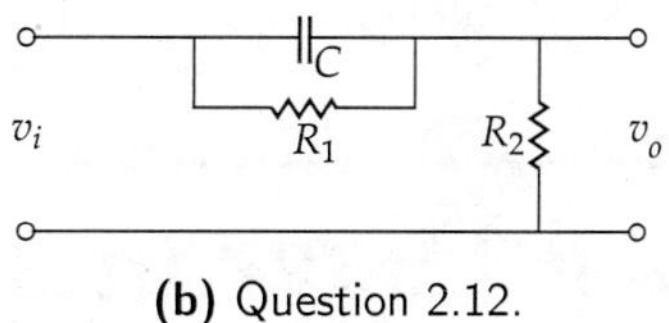

(b) Question 2.12.

Figure 2.15: Circuits.

Answers to Selected Questions

2.1. (i) (A) (ii) (A) (iii) (B) (iv) (A) (v) (A) (vi) (B) (vii) (C) (viii) (D) (ix) (B) (x) (D) (xi) (D)

2.3. (a) Poles at $s = -2,\ -3,\ -4$. Zero at $s = -25$
(b) Characteristic equation is $s^4 + 10s^3 + 35s^2 + (50 + K)s + (24 + 25K) = 0$

2.4. (A) and (B) 2.5. 3.33 2.6. $\dfrac{12(s^2+2s+2)}{(s+1)(s+2)}$ 2.7. $\dfrac{\omega}{s^2+2s+(1+\omega^2)}$

2.8. $\dfrac{d^2y}{dt^2} + 5\dfrac{dy}{dt} + 2y = 6x$ 2.10. Type 3, order 4 2.11. $\dfrac{sR_1R_2C + R_2}{R_1 + R_2 + sR_1R_2C}$

2.12. $\dfrac{sRC}{s^2R^2C^2 + 2sRC + R + 1}$

3

Block Diagrams and Signal Flow Graphs

3.1 BLOCK DIAGRAMS

We have seen, transfer functions can be represented by blocks in a block diagram to give a visual presentation of interrelationship between different constituents of the control system. A block diagram can, therefore, be termed as a shorthand pictorial presentation of the cause and effect relationship between the output and input of a control system (see Figure 3.1a).

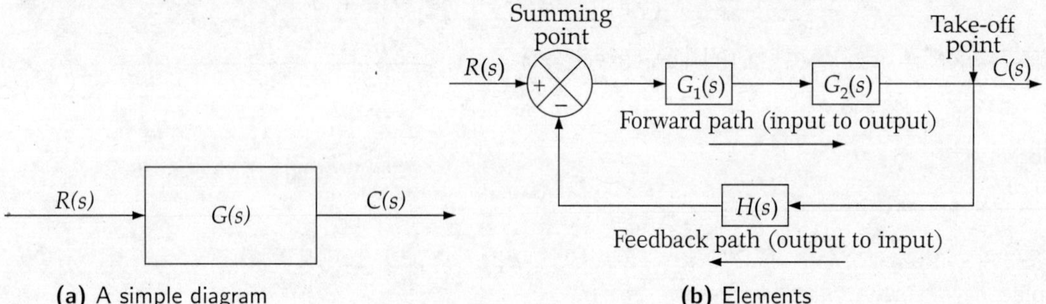

(a) A simple diagram

(b) Elements

Figure 3.1: Block diagram.

The block diagram technique has been developed not only to show the interrelationship between constituents of a control system but also to do algebraic manipulations through their reduction procedure. Before we proceed to do that, let us know the names of different parts/segments of a block diagram used in describing control systems.

Elements of a Block Diagram

Summing point. Represented by a circle with inner[1] indications of signs [see Figure 3.1b], these points show the addition or subtraction of signals from different sources.

Take-off point. [2]If the output of a signal branches off from one point, the point of branching off is called the *take-off point* [see Figure 3.1b].

Forward and feedback paths. Forward and feedback paths are indicated in a block diagram given in Figure 3.1b. A forward path is indicated by an arrow pointing towards right while a left pointing arrow indicates a feedback path.

Table 3.1 lists different symbols that are normally used in a block diagram.

[1]Some authors prefer *outer* indication of signs.

[2]Also called *Pick-off* point.

Table 3.1: List of symbols normally used in block diagrams

Symbol	*Meaning*	*Symbol*	*Meaning*
$R(s)$	Laplace-transformed reference signal	$C(s)$	Laplace-transformed controlled output
$G(s)$	Transfer function of the forward path	$H(s)$	Transfer function of the feedback path
$E(s)$	Laplace-transformed error signal	$B(s)$	Laplace-transformed feedback signal

Drawing Block Diagrams

We are familiar with drawing block diagrams of simple control systems—like open-loop and closed-loop systems—right from the beginning. Here we will consider how to draw block diagrams if electrical circuits are given or dynamical equations are provided. A few examples will help us understand the procedure.

EXAMPLE 3.1. Draw the block diagram for the electrical circuit given in Figure 3.13a.

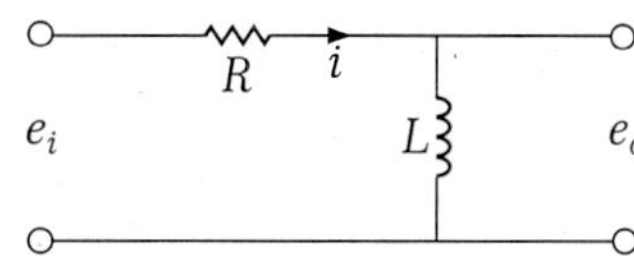

Figure 3.2: Electrical circuit.

Solution Using KCL and KVL, we can write the following equations:

$$i = \frac{e_i - e_o}{R} \qquad e_o = L\frac{di}{dt}$$

Laplace transform of these equations, assuming all initial conditions = 0, are

$$I(s) = \frac{E_i(s) - E_o(s)}{R} \quad \cdots \text{(i)} \qquad E_o(s) = sLI(s) \quad \cdots \text{(ii)}$$

The two block diagrams given in Figures 3.3a and 3.3b represent the two equations individually. These diagrams can be merged as in Figure 3.3c to represent the given circuit.

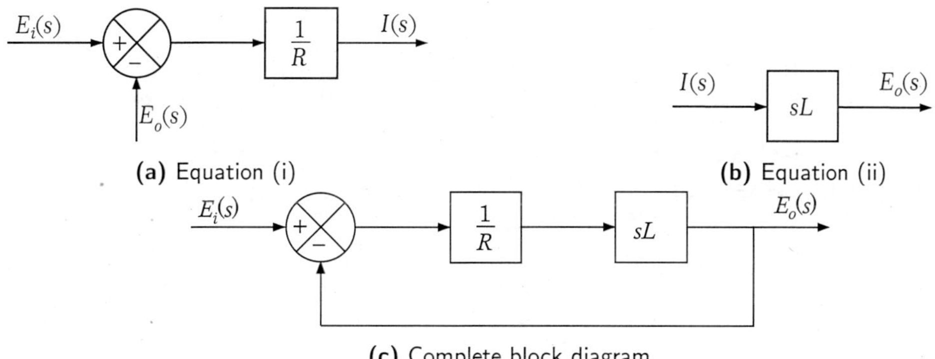

(a) Equation (i)

(b) Equation (ii)

(c) Complete block diagram

Figure 3.3: Example 3.1.

EXAMPLE 3.2. Draw the block diagram for the circuit given in Figure 3.4.

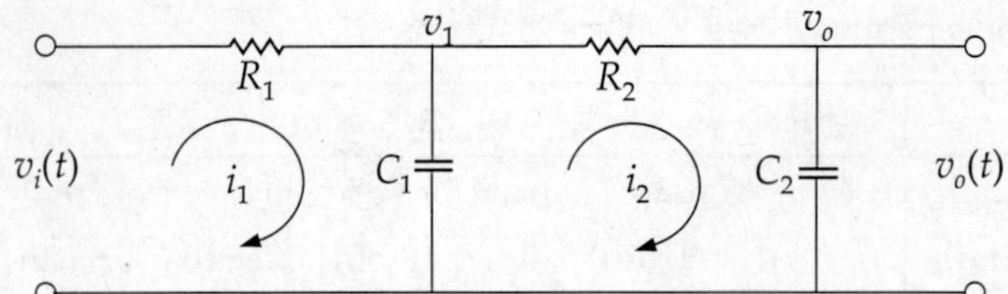

Figure 3.4: Circuit (Example 3.2).

Solution Applying KVL and KCL to the circuit, we get

$$i_1 = \frac{v_i - v_1}{R_1} \qquad v_1 = \frac{\int (i_1 - i_2)dt}{C_1}$$

$$i_2 = \frac{v_1 - v_o}{R_2} \qquad v_o = \frac{\int i_2 \, dt}{C_2}$$

Their Laplace transforms, under relaxed conditions, yield

$$I_1(s) = \frac{1}{R_1}[V_i(s) - V_1(s)] \tag{i}$$

$$V_1(s) = \frac{1}{sC_1}[I_1(s) - I_2(s)] \tag{ii}$$

$$I_2(s) = \frac{1}{R_2}[V_1(s) - V_o(s)] \tag{iii}$$

$$V_o(s) = \frac{1}{sC_2}I_2(s) \tag{iv}$$

Equation (i) can be converted to a block diagram as follows. We note, $V_i(s)$ is the input corresponding to $v_i(t)$ of the given circuit. So, putting this in the input of a summing point we draw Figure 3.5a to represent Eq. (i).

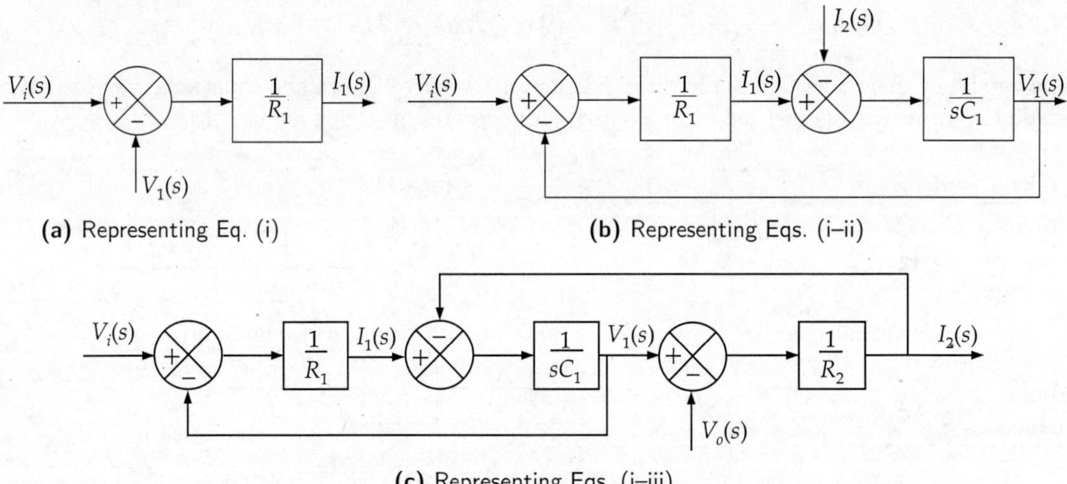

(a) Representing Eq. (i)

(b) Representing Eqs. (i–ii)

(c) Representing Eqs. (i–iii)

Figure 3.5: Parts of the block diagram.

Equation (ii) modifies the diagram to its form given in Figure 3.5b. Equation (iii)) extends the diagram to what is presented in Figure 3.5c.

The last equation, Eq. (iv), completes the diagram to look like what is given in Figure 3.6.

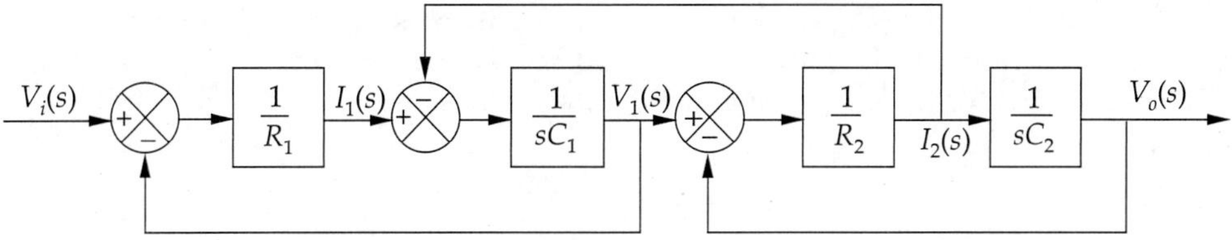

Figure 3.6: Complete block diagram representing Eqs. (i–iv).

Reduction of a Closed-Loop Block Diagram

Figure 3.7 represents the block diagram of a closed-loop control system.

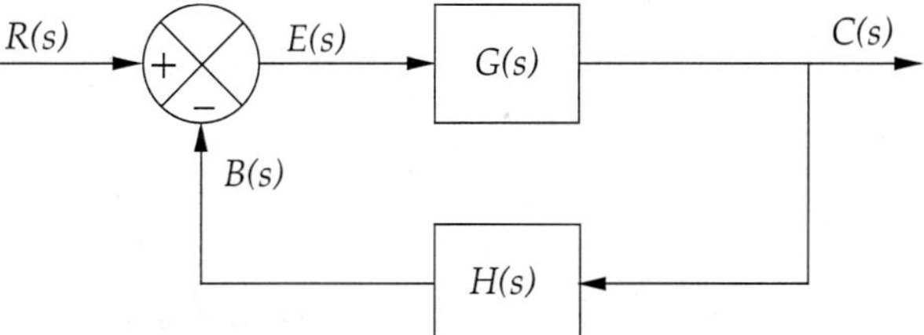

Figure 3.7: Block diagram of a closed-loop control system.

Different quantities, all Laplace transformed, that are used are defined in Table 3.1. From Figure 3.7, we get

$$C(s) = E(s)G(s) \tag{3.1}$$

$$B(s) = C(s)H(s) \tag{3.2}$$

$$E(s) = R(s) - B(s) \tag{3.3}$$

Substituting Eq. (3.3) in Eq. (3.1), we get

$$C(s) = [R(s) - B(s)]G(s) \tag{3.4}$$

Plugging in the value of $B(s)$ from Eq. (3.2) in Eq. (3.4), we get

$$C(s) = R(s)G(s) - C(s)H(s)G(s) \tag{3.5}$$

Equation (3.5), on rearrangement, yields the closed-loop transfer function as

$$\boxed{T(s) = \frac{C(s)}{R(s)} = \frac{G(s)}{1 + G(s)H(s)}} \tag{3.6}$$

Equation (3.6) is an important result which indicates that the closed-loop of Figure 3.7 can be reduced to give the diagram of Figure 3.8.

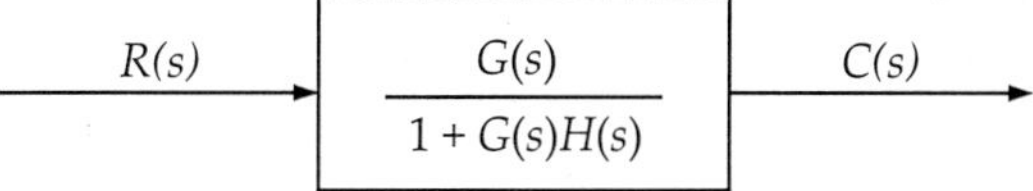

Figure 3.8: Reduced form of closed-loop of Figure 3.7.

Note: The quantity $G(s)H(s)$ appearing in the denominator of Eq. (3.6) is often called the *loop transfer function* or *open-loop transfer function.*

EXAMPLE 3.3. As shown in Figure 3.9, a negative feedback system has an amplifier of gain 100 with $\pm 10\%$ tolerance in the forward path, and an attenuator of value $\frac{9}{100}$ in the feedback path. The overall gain of the system is approximately

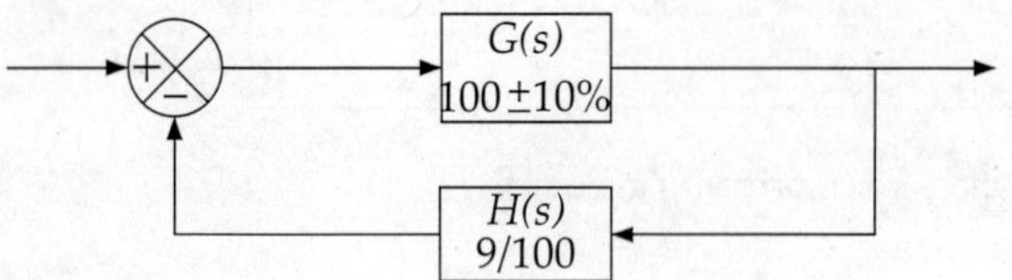

Figure 3.9: Example 3.3.

(A) $10 \pm 1\%$ (B) $10 \pm 2\%$ (C) $10 \pm 5\%$ (D) $10 \pm 10\%$

Solution Given, $G(s) = 100 \pm 10\%, H(s) = \frac{9}{100}$. We know, overall gain is

$$\frac{C(s)}{R(s)} = \frac{G(s)}{1 + G(s)H(s)}$$

When $$G(s) = 100 + 10 = 110$$

$$\frac{C(s)}{R(s)} = \frac{110}{1 + 110 \times \frac{9}{100}} \cong 10.1$$

When $$G(s) = 100 - 10 = 90$$

$$\frac{C(s)}{R(s)} = \frac{90}{1 + 90 \times \frac{9}{100}} \cong 9.9$$

Hence, the variation in overall gain $= 10 \pm 0.1 = 10 \pm 1\%$.
Ans. (A)

Reduction Rules

The block diagram presentation of a given system may be reduced to a simplified block diagram with fewer blocks than the original diagram. This is done by applying block diagram reduction rules which are presented in Table 3.2.

It is better to follow a general procedure for block diagram reduction. The steps of the procedure are:

Step 1: Reduce cascade blocks.
Step 2: Reduce parallel blocks.
Step 3: Eliminate feedback loops.
Step 4: Move summing points to right and take-off points to left.
Step 5: Repeat steps 1 to 4 until a simple form is obtained.

However, these steps cannot be followed in all cases in the given order. Reduction of block diagrams requires a little bit of practice when one develops a skill as to what step has to be taken at what time.

Table 3.2: Block diagram reduction rules

Transformation	*Original diagram*	*Equivalent diagram*
Combining blocks in cascade	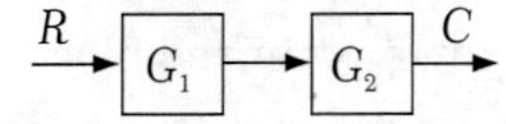 $R \to G_1 \to G_2 \to C$	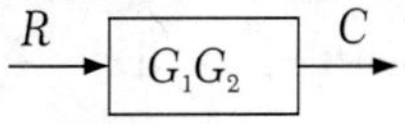 $R \to G_1G_2 \to C$
Combining blocks in parallel	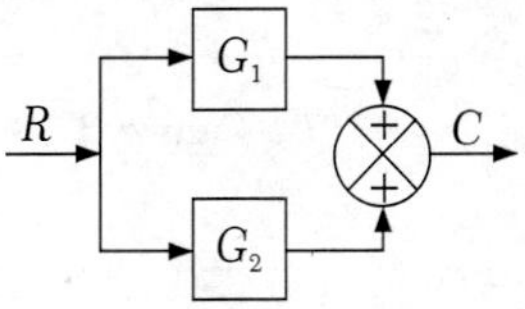 $R \to G_1, G_2 \to (+,+) \to C$	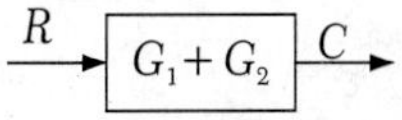 $R \to G_1 + G_2 \to C$
Moving a summing point beyond a block	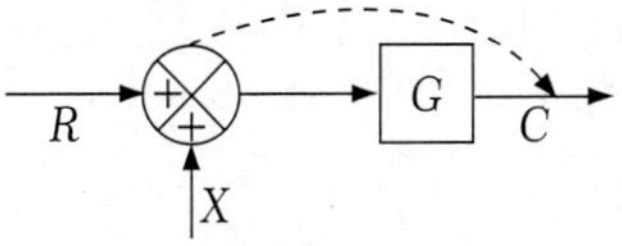 $C = (R + X)G$	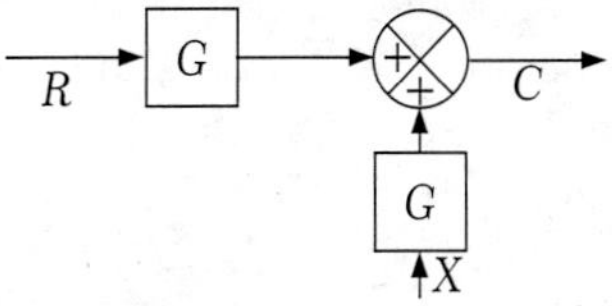 $C = RG + XG$
Moving a summing point behind a block	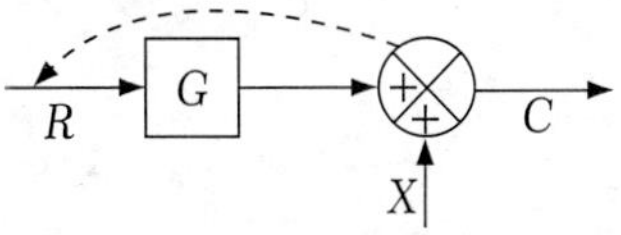 $R \to G \to (+,+) \to C$, X	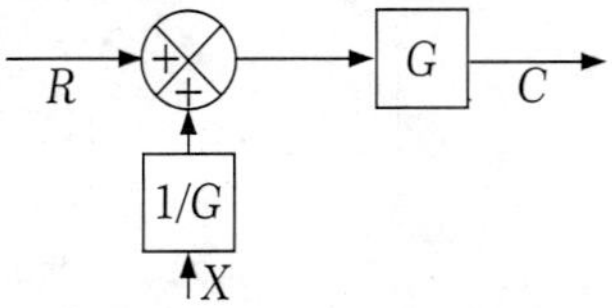 $R \to (+,+) \to G \to C$, $X \to 1/G$
Eliminating a feedback loop	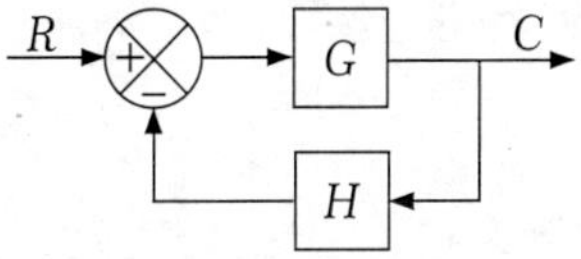 $R \to (+,-) \to G \to C$, H	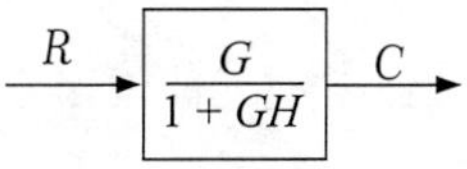 $R \to \dfrac{G}{1+GH} \to C$
Moving a take-off point beyond a block	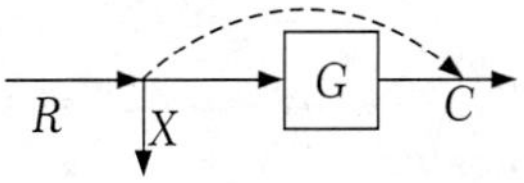 $X = R \quad C = RG$	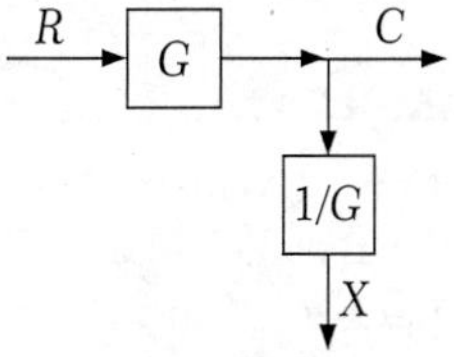 $X = R \quad C = RG$
Moving a take-off point behind a block	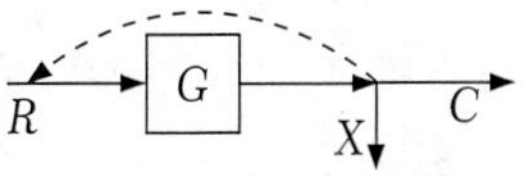 $X = C = RG$	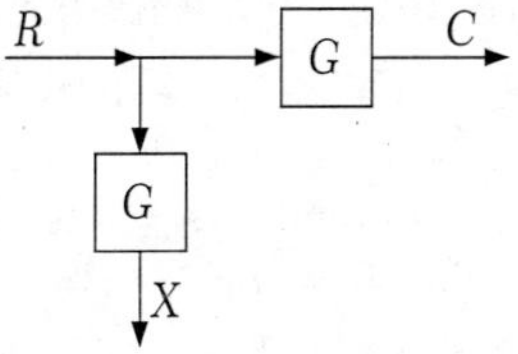 $X = C = RG$

Multi-input system

In case a multi-input system has to be reduced, the following steps should be followed (see Example 3.6):

Step 1: Put all but one inputs to zero. Find the resultant output.

Step 2: Repeat Step 1 until all inputs are covered.

Step 3: Find the resultant output by superposition.

Having this background, we will now consider a few block diagram reductions.

EXAMPLE 3.4. Determine the transfer function of the block diagram given in Figure 3.10.

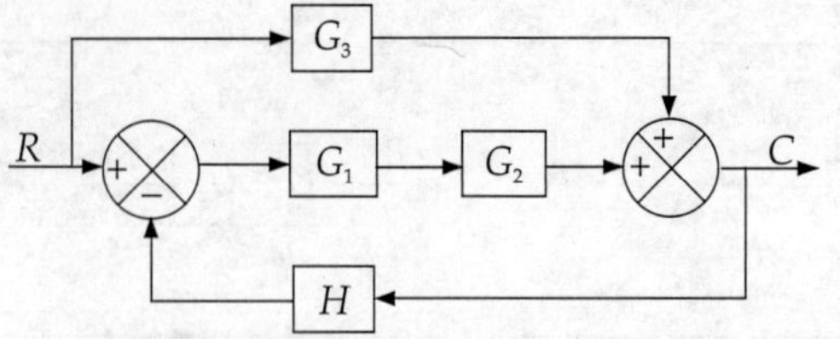

Figure 3.10: Block diagram (Example 3.4).

Solution First we combine cascaded and parallel blocks. As a result, we get the feedback block of Figure 3.11a. Next, we reduce this feedback loop to arrive at the final reduced form as given in Figure 3.11b.

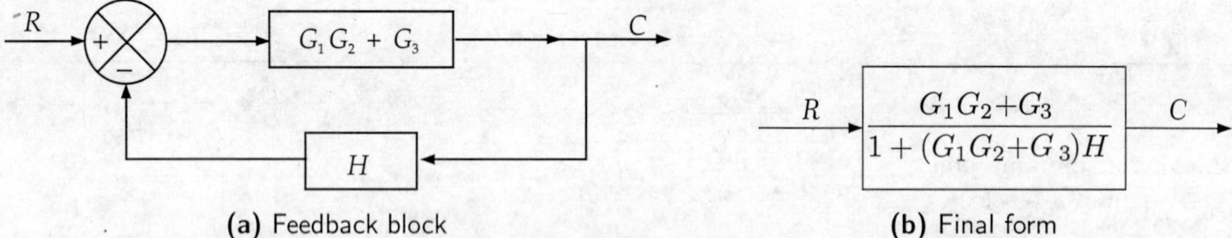

(a) Feedback block

(b) Final form

Figure 3.11: Example 3.4.

The transfer function is given by

$$\frac{C(s)}{R(s)} = \frac{G_1G_2 + G_3}{1 + (G_1G_2 + G_3)H}$$

EXAMPLE 3.5. Reduce the block diagram of Figure 3.12a and obtain the transfer function.

Solution By inspection we find that after the first summing point there is a feedback loop and then a parallel block. These are independent. So, we take the following steps: (i) reduce the feedback loop and the parallel block, (ii) reduce the resulting feedback loop on the left, and (iii) multiply the resultant by G_6. The successive steps are shown in ①, ② and ③ in Figure 3.12b.

EXAMPLE 3.6. Find the output of the system shown in Figure 3.13a.

Solution This is a multi-input system. So, initially, we put $X = 0$ and reduce the resultant diagram. Successive stages of block diagram reduction are shown in Figure 3.13. The similar reduction with $R = 0$ is shown in Figure 3.14.

The final output, as discussed before, is obtained by superposition of the two individual outputs. Thus,

$$C = C_1 + C_2 = \frac{G_1G_2G_3R + G_3X}{1 + G_3H_1 + G_1G_2G_3H_2}$$

EXAMPLE 3.7. Find $C(s)/R(s)$ of the multiple-loop feedback control system shown in Figure 3.15 by the block diagram reduction technique.

Solution By inspection, we observe that if the take-off point after G_3 can be moved to that after G_4, the diagram will result in 3 feedback loops. Figure 3.15a shows how the point is moved to the desired place.

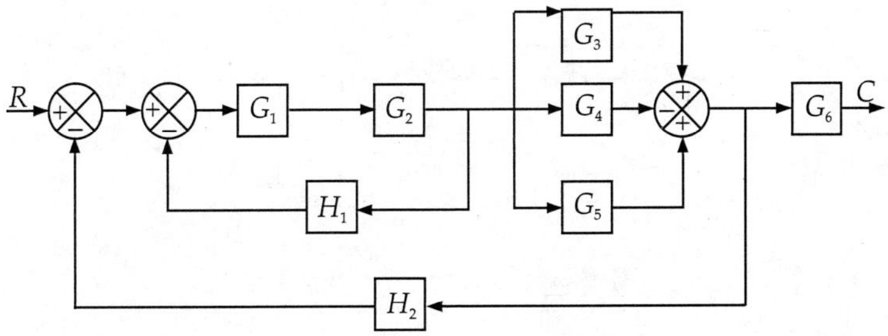

R
G1
G2
G3
G4
G5
G6
C
H1
H2

(a) Block diagram

R
G1G2 / 1+G1G2H1
G3 − G4 + G5
G6
C
H2
1
G1G2(G3 − G4 + G5) / 1+G1G2H1
1+H2 [G1G2(G3 − G4 + G5) / 1+G1G2H1]
2
G1G2G6(G3 − G4 + G5) / 1+G1G2H1+ H2 G1G2(G3 − G4 + G5)
3

(b) Successive stages of block diagram reduction

Figure 3.12: Example 3.5.

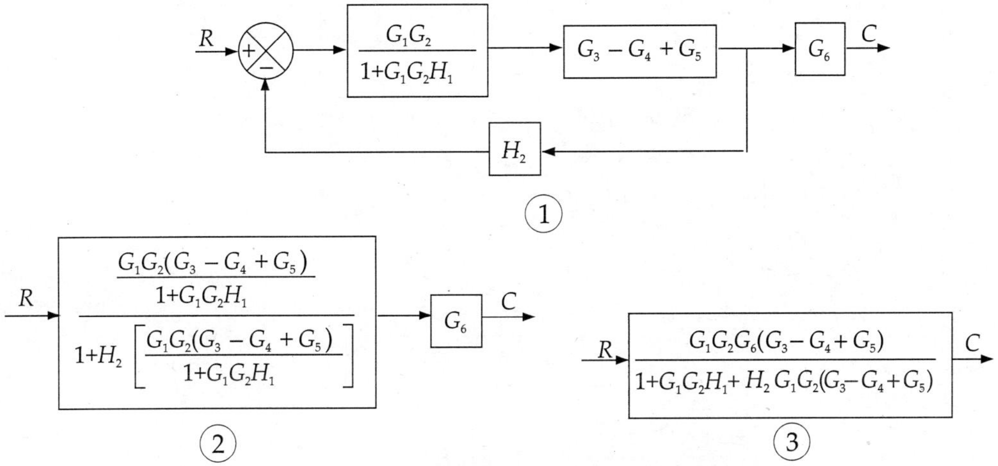

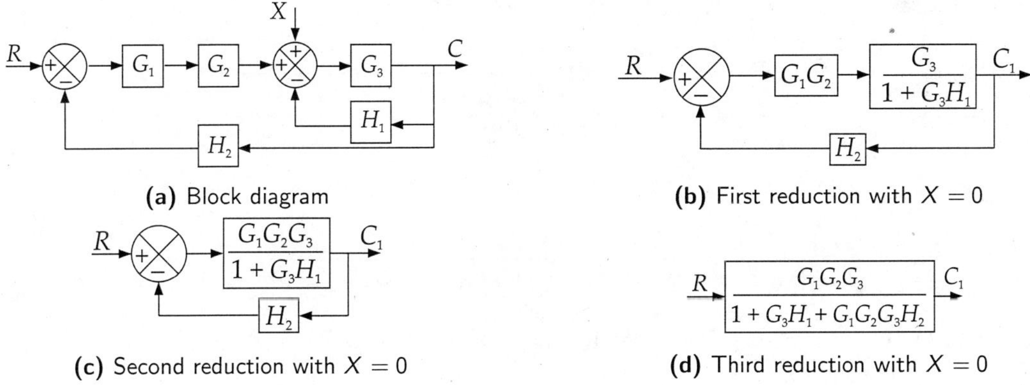

X
R
G1
G2
G3
C
H1
H2

(a) Block diagram

R
G1G2
G3 / 1 + G3H1
C1
H2

(b) First reduction with $X = 0$

R
G1G2G3 / 1 + G3H1
C1
H2

(c) Second reduction with $X = 0$

R
G1G2G3 / 1 + G3H1 + G1G2G3H2
C1

(d) Third reduction with $X = 0$

Figure 3.13: Example 3.6.

Now we can reduce the feedback loop involving H_3 when the diagram assumes the form of Figure 3.15b.

Next, the feedback loops involving H_2/G_4 and H_1 are reduced in succession. Corresponding diagrams are shown in Figures 3.15c and 3.15d.

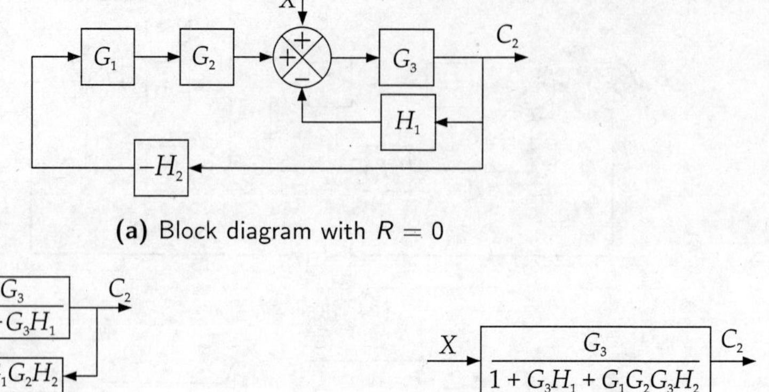

(a) Block diagram with $R = 0$

(b) First reduction

(c) Second reduction

Figure 3.14: Reduction with $R = 0$.

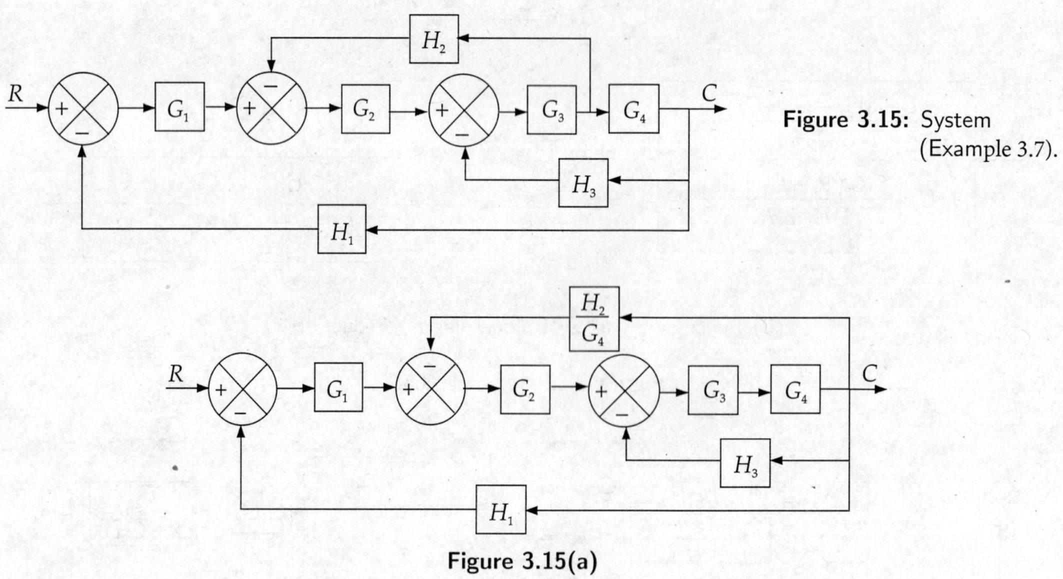

Figure 3.15: System (Example 3.7).

Figure 3.15(a)

Hence,

$$\frac{C(s)}{R(s)} = \frac{G_1G_2G_3G_4}{1 + G_3G_4H_3 + G_2G_3H_2 + G_1G_2G_3G_4H_1}$$

EXAMPLE 3.8. Convert the system given in Figure 3.16 into a unity feedback system and find the characteristic equation.

Solution First we reduce the parallel blocks to get Figure 3.17a.

The next cascade reduction is shown in Figure 3.17b. Now, we have to convert it to a unity feedback system. Let the present forward transfer function be G. We need to find G' such that

$$\frac{G}{1+2G} = \frac{G'}{1+G'} \quad \Rightarrow \quad G' = \frac{G}{1+G}$$

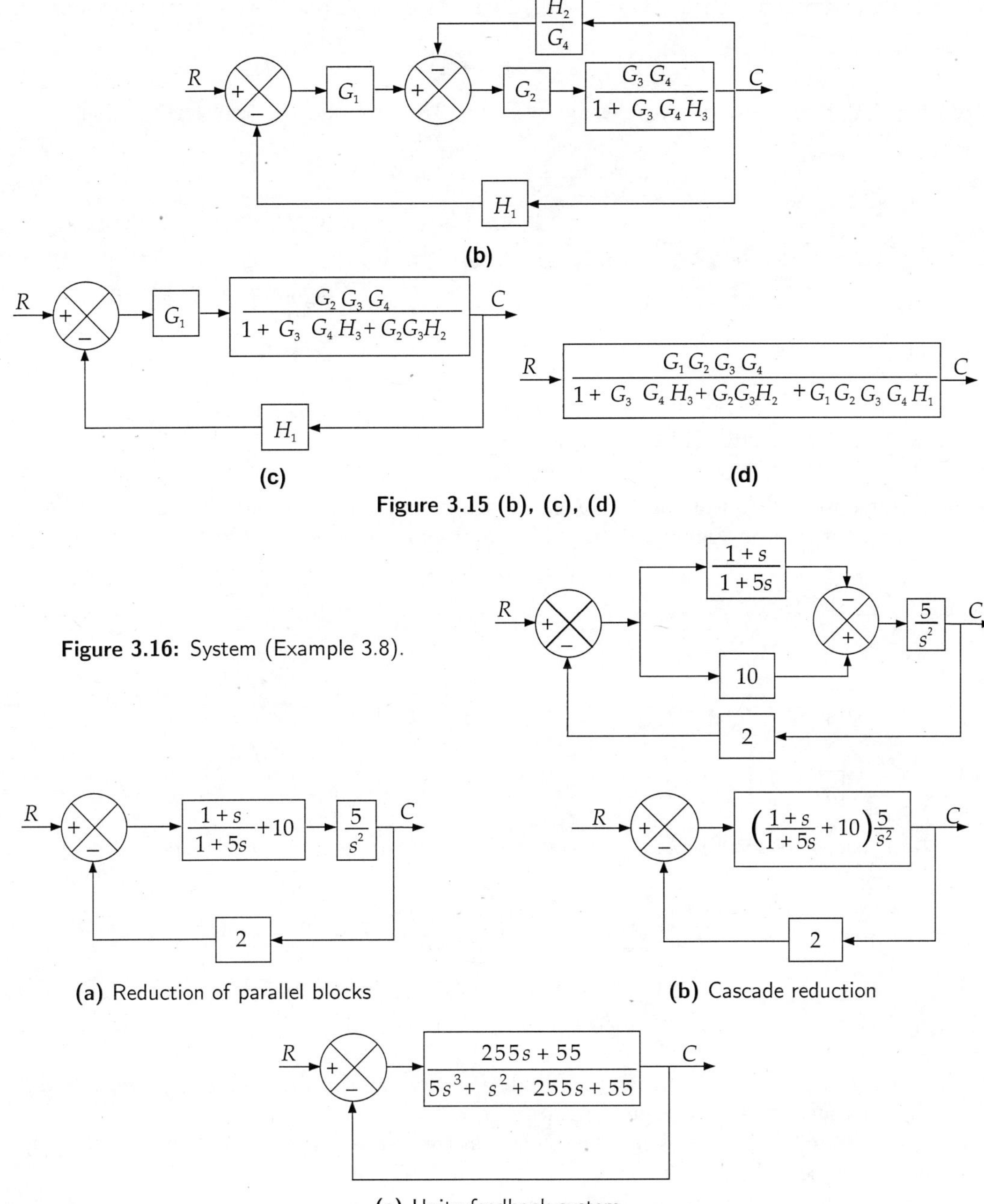

Figure 3.15 (b), (c), (d)

Figure 3.16: System (Example 3.8).

(a) Reduction of parallel blocks

(b) Cascade reduction

(c) Unity feedback system

Figure 3.17: Example 3.8.

Therefore,

$$G' = \frac{\left(\dfrac{1+s}{1+5s} + 10\right)\dfrac{5}{s^2}}{1 + \left(\dfrac{1+s}{1+5s} + 10\right)\dfrac{5}{s^2}} = \frac{55 + 255s}{5s^3 + s^2 + 255s + 55}$$

So, the required unity feedback system is as shown in Figure 3.17c. The characteristic equation is, therefore,

$$5s^3 + s^2 + 510s + 110 = 0$$

EXAMPLE 3.9. Reduce the block diagram of Figure 3.18 and find the transfer function.

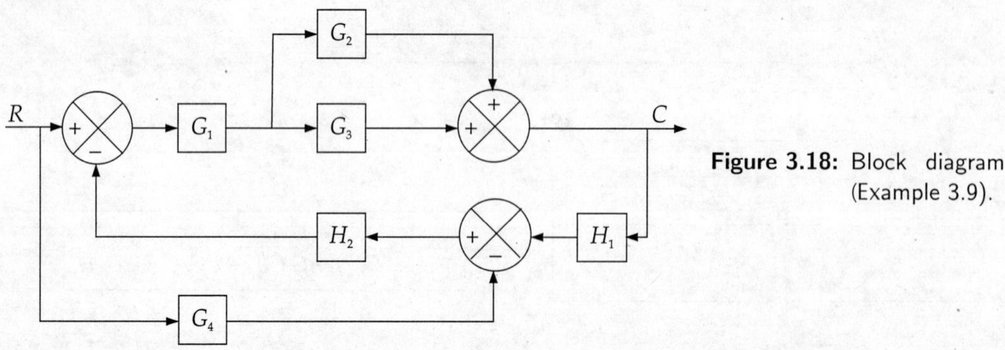

Figure 3.18: Block diagram (Example 3.9).

Solution First we reduce the parallel blocks and combine the resultant with the series block. The resulting diagram is Figure 3.19a. Now we observe that one forward path through G_4

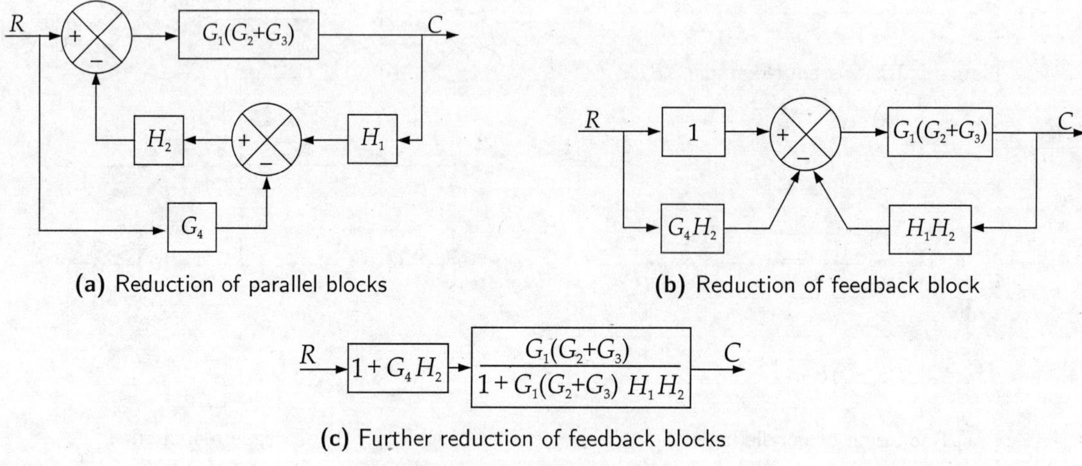

(a) Reduction of parallel blocks

(b) Reduction of feedback block

(c) Further reduction of feedback blocks

Figure 3.19: Example 3.9.

and H_2, and a feedback path through H_1 and H_2 merge at the first summing point. Shown explicitly, the situation is as in Figure 3.19b.

In this diagram, we have two parallel blocks on the left and a feedback loop on the right. On reduction, these yield Figure 3.19c.

Therefore, the transfer function is given by

$$\begin{aligned}\frac{C(s)}{R(s)} &= (1 + G_4H_2)\left[\frac{G_1(G_2 + G_3)}{1 + G_1(G_2 + G_3)H_1H_2}\right] \\ &= \frac{G_1G_2 + G_1G_3 + G_1G_2G_4H_2 + G_1G_3G_4H_2}{1 + G_1G_2H_1H_2 + G_1G_3H_1H_2}\end{aligned}$$

EXAMPLE 3.10. Using the block diagram reduction technique find the transfer function for the block diagram given in Figure 3.20.

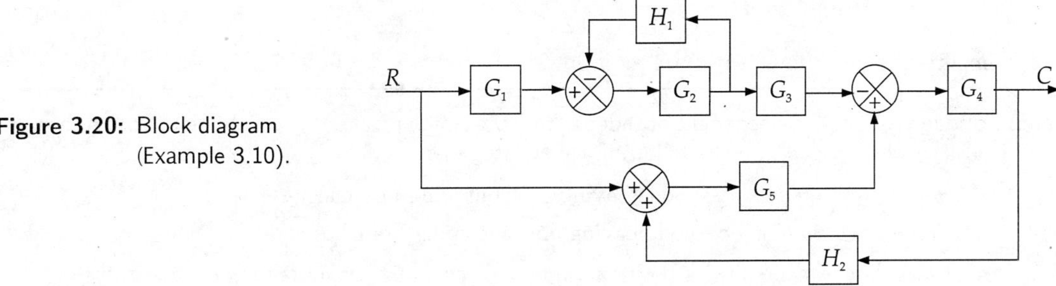

Figure 3.20: Block diagram (Example 3.10).

Solution First we reduce the feedback loop involving H_1. The diagram then assumes the form of Figure 3.21a. We observe that one forward path through G_5 and a feedback path H_2G_5 merge at the first line summing point. The situation is as shown in Figure 3.21b.

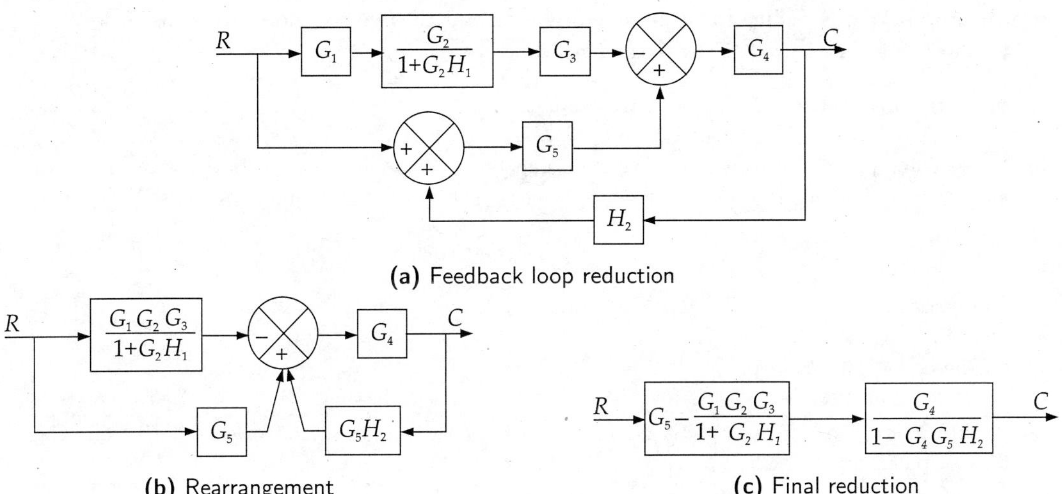

Figure 3.21: Example 3.10.

The reduction, given in Figure 3.21c, follows from this diagram. Therefore, the transfer function is given by

$$\frac{C(s)}{R(s)} = \frac{G_4G_5 + G_2G_4G_5H_1 - G_1G_2G_3G_4}{1 - G_4G_5H_2 + G_2H_1 - G_2G_4G_5H_1H_2}$$

3.2 SIGNAL FLOW GRAPHS

Introduced by SJ Mason[3] about seventy years ago, signal flow graphs are more elegant and powerful for dealing with more complex interrelationships where block diagram technique becomes cumbersome. However, it can be utilized for linear systems only.

A signal flow graph (often called *flow graph*) represents a system by directed line segments originating from or terminating at points, called *nodes*. Before we discuss how to construct signal flow graphs, let us consider its constituent elements.

[3]Samuel Jefferson Mason (1921–1974) was an American electronics engineer. Mason's doctoral dissertation

Elements of Signal Flow Graph

The elements of signal flow graph are as follows:

Node: A node represents a dependent or independent variable. $x_1,\ x_2, \ldots x_9$ in Figure 3.22 are nodes. Nodes can be divided into the following three categories:

1. *Source (or Input) node*—node having only outgoing branches (x_1).
2. *Sink (or Output) node*—node having only incoming branches (x_7).
3. *Chain node*—node having both outgoing and incoming branches ($x_3,\ x_4,\ x_5,\ x_6$).

Branch: Lines joining nodes are branches. An arrow on the branch indicates the direction of flow of the signal and the number written on it indicates the *gain* of the branch.
In Figure 3.22, $x_3 = Ax_2$, where A is the gain of the branch.

Forward path: All paths leading to output and originating at the input are forward paths. If we denote forward paths in Figure 3.22 by p_1, and p_2, they are:

$p_1 \ : \ x_1x_2x_3x_4x_5x_6x_7$ $\qquad$ $p_2 \ : \ x_1x_2x_3x_5x_6x_7$

Loop or Feedback path: A path originating from and terminating at the same node, without crossing any other node twice is a loop. There are 4 loops in Figure 3.22, namely

$L_1 \ : \ x_3x_4x_3$ $\qquad$ $L_2 \ : \ x_4x_5x_6x_4$

$L_3 \ : \ x_5x_5$ $\qquad$ $L_4 \ : \ x_3x_5x_6x_4x_3$

Of these, L_3 is a self-loop.

Forward path gain: Product of all gains associated with a forward path is the forward path gain. Gains for two paths are:

$p_1 \ : \ ACEH$ $\qquad$ $p_2 \ : \ AGH$

Loop gain: Product of gains associated with a loop is the loop gain. Gains for four loops are:

$L_1 \ : \ BC$ $\qquad$ $L_2 \ : \ -EHD$

$L_3 \ : \ F$ $\qquad$ $L_4 \ : \ -GHDB$

Non-touching loops: Loops having no common nodes are non-touching loops. L_1 and L_3 is a pair of non-touching loops.

They are pictorially presented in Figure 3.22 as well.

The transfer function of a system can be determined from signal flow graph by using *Mason's Gain Formula.*

was on signal-flow graphs and he is often credited with inventing them. However, there is a controversy and it is said that the concept of a signal-flow graph was originally worked out in 1942 by Shannon [Claude Shannon, in full Claude Elwood Shannon (1916–2001), American mathematician and electrical engineer who laid the theoretical foundations for digital circuits and information theory, a mathematical communication model.] in dealing with analog computers.

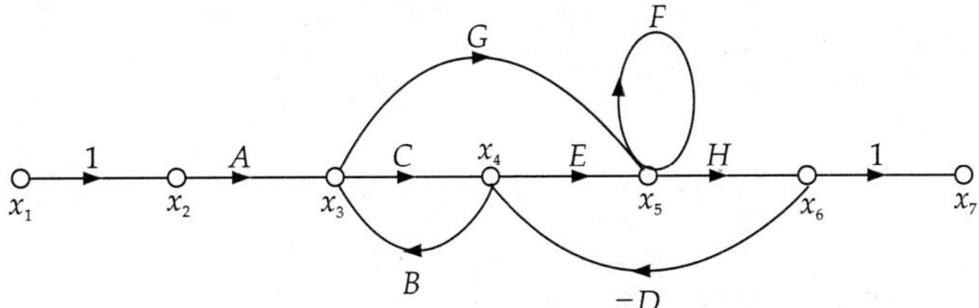

Figure 3.22: A signal flow graph showing different elements.

Mason's Gain Formula

Mason's gain formula is given by

$$\text{Transfer function} = \frac{C(s)}{R(s)} = \frac{\sum_{i=1}^{N} p_i \Delta_i}{\Delta} \tag{3.7}$$

where N = total number of forward paths
p_i = gain of the i-th forward path
Δ = determinant of the graph
Δ_i = path-factor for the i-th path

The determinant of the graph Δ and the path-factor for the i-th path Δ_i are defined as follows:

$$\begin{aligned}\Delta = 1 &- \Sigma(\text{all individual loop gains}) \\ &+ \Sigma(\text{gain products of all possible combinations of } \textit{two} \text{ non-touching loops}) \\ &- \Sigma(\text{gain products of all possible combinations of } \textit{three} \text{ non-touching loops}) + \cdots\end{aligned}$$

$$\Delta_i = 1 - (\text{part of } \Delta \text{ that is non-touching to the path } p_i)$$

We can find out the transfer function of signal flow graph of Figure 3.22 from the following data:

$$N = 2$$
$$p_1 = ACEH \qquad \Delta_1 = 1 \text{ (since all loops are touching } p_1)$$
$$p_2 = AGH \qquad \Delta_2 = 1 \text{ (since all loops are touching } p_2)$$
$$\Delta = 1 - (BC - EHD + F - GHDB) + BCF$$

The transfer function is given by

$$\frac{C(s)}{R(s)} = \frac{ACEH + AGH}{1 - (BC - EHD + F - GHDB) + BCF}$$

Now we work out a few examples to familiarize ourselves with the procedure.

EXAMPLE 3.11. Draw a signal flow graph for the following algebraic equations:

$$x_2 = ax_1 + bx_3 \tag{i}$$
$$x_3 = cx_2 + dx_4 \tag{ii}$$
$$x_4 = ex_2 + fx_3 + gx_4 \tag{iii}$$
$$x_5 = hx_2 + kx_4 \tag{iv}$$

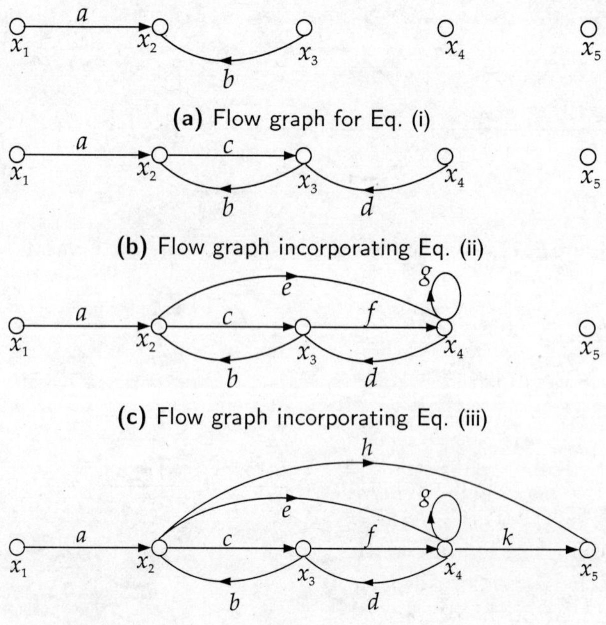

(a) Flow graph for Eq. (i)

(b) Flow graph incorporating Eq. (ii)

(c) Flow graph incorporating Eq. (iii)

(d) Complete flow graph incorporating Eq. (iv)

Figure 3.23: Flow graph construction from equations.

Solution We note that there are five variables, $x_1,\ x_2, \ldots x_5$, in total in the four equations. So, first we plot five nodes and then draw the flow graph for Eq. (i) as shown in Figure 3.23a.

Incorporating flow lines according to Eqs. (ii), (iii), as shown in Figures 3.23b and 3.23c, finally the complete signal flow graph is drawn as shown in Figure 3.23d.

EXAMPLE 3.12. For the signal flow graph shown in the following figure, the transfer function is $\dfrac{C(s)}{R(s)} =$ ______

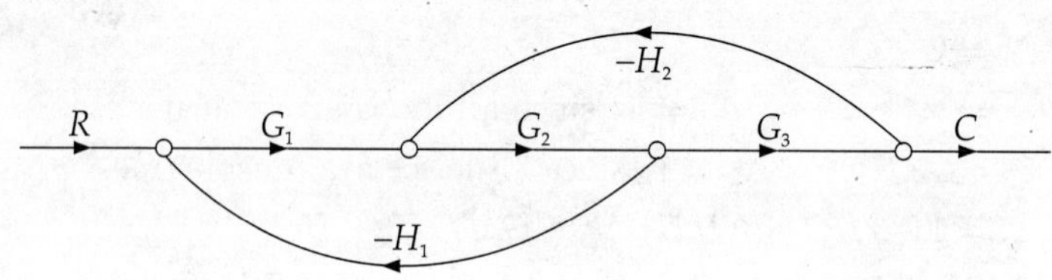

Solution Forward path is only one whereas there are two loops. There is no non-touching loop. So, we have

$$p_1 = G_1G_2G_3$$

$$L_1 = -G_1G_2H_1 \qquad L_2 = -G_2G_3H_2$$

$$\Delta = 1 - (-G_1G_2H_1 - G_2G_3H_2) \qquad \Delta_1 = 1$$

Therefore, from Mason's gain formula, we have

$$\frac{C(s)}{R(s)} = \frac{G_1G_2G_3}{1 + G_1G_2H_1 + G_2G_3H_2}$$

EXAMPLE 3.13. The transfer function between y_2 and y_1 in the following figure is

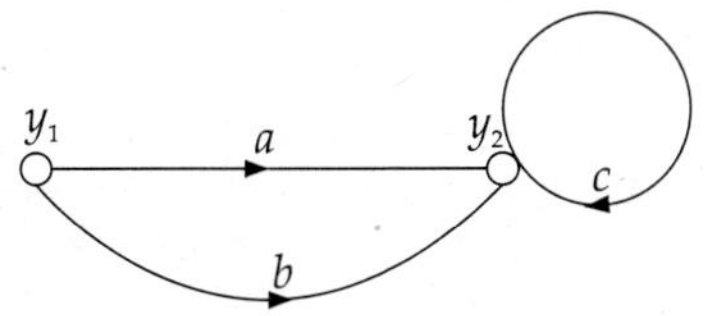

(A) $a+b$ (B) $(a+b)c$ (C) $\dfrac{a+b}{1-c}$ (D) $\dfrac{a+b}{1+c}$

Solution There are two forward paths and one loop. So, we have

$$p_1 = a \qquad p_2 = b$$
$$L_1 = c$$
$$\Delta = 1 - c \qquad \Delta_1 = 1 = \Delta_2$$

Therefore
$$\frac{C(s)}{R(s)} = \frac{a+b}{1-c}$$

Ans. (C)

EXAMPLE 3.14. The transfer function of the open-loop system $G(s)$ which is represented by the signal flow graph shown in the figure below is

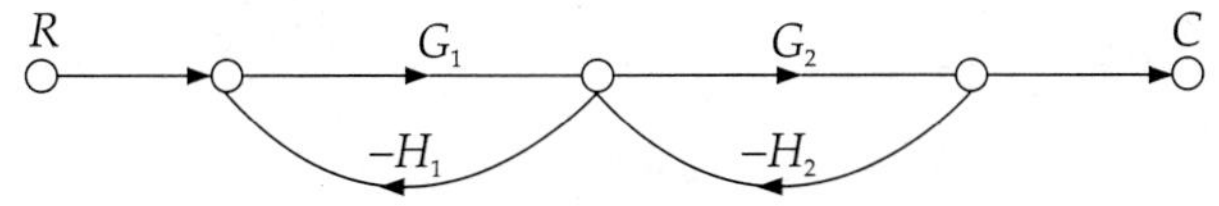

(A) $\dfrac{G_1G_2}{1+G_1H_1+G_2H_2}$ (B) $\dfrac{G_1G_2}{1+G_1H_1G_2H_2}$

(C) $\dfrac{G_1G_2}{(1+G_1H_1)(1+G_2H_2)}$ (D) $\dfrac{G_1G_2}{1+(G_1+H_1)(G_2+H_2)}$

Solution There exists one forward path and two loops. So, we have

$$p_1 = G_1G_2$$
$$L_1 = -G_1H_1 \qquad L_2 = -G_2H_2$$
$$\Delta = 1 + G_1H_1 + G_2H_2 \qquad \Delta_1 = 1$$
$$\therefore \quad \frac{C}{R} = \frac{G_1G_2}{1+G_1H_1+G_2H_2}$$

Ans. (A)

EXAMPLE 3.15. The signal flow graph representation of a control system is shown in the following figure.

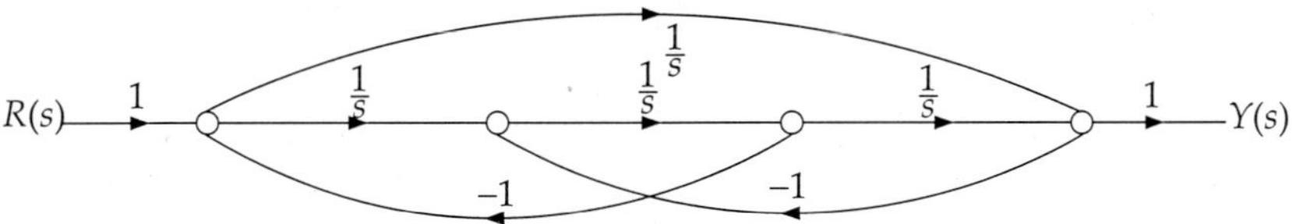

The transfer function $\dfrac{Y(s)}{R(s)}$ is computed as

(A) $\dfrac{1}{s}$ (B) $\dfrac{s^2+1}{s(s^2+2)}$ (C) $\dfrac{s(s^2+1)}{s^2+2}$ (D) $1-\dfrac{1}{s}$

Solution There are two forward paths and two loops. Therefore, we have

$$p_1 = \frac{1}{s^3} \qquad p_2 = \frac{1}{s}$$

$$L_1 = -\frac{1}{s^2} \qquad L_2 = -\frac{1}{s^2}$$

$$\Delta = 1 + \frac{2}{s^2} \qquad \Delta_1 = 1 = \Delta_2$$

Thus,

$$\frac{Y(s)}{R(s)} = \frac{\frac{1}{s^3} + \frac{1}{s}}{1 + \frac{2}{s^2}} = \frac{1 + s^2}{s(s^2 + 2)}$$

Ans. (B)

EXAMPLE 3.16. A filter is represented by the signal flow graph shown in the following figure.

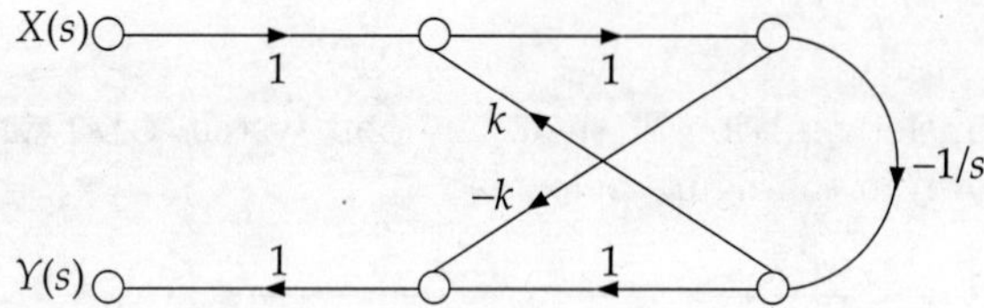

Its input is $x(t)$ and output is $y(t)$. The transfer function of the filter is

(A) $\frac{-(1+ks)}{s+k}$ (B) $\frac{(1+ks)}{s+k}$ (C) $\frac{-(1-ks)}{s+k}$ (D) $\frac{(1-ks)}{s+k}$

Solution The signal flow graph contains two forward paths and one loop. Therefore, we have

$$p_1 = -\frac{1}{s} \qquad p_2 = -k$$

$$L_1 = -\frac{k}{s}$$

$$\Delta = 1 + \frac{k}{s} \qquad \Delta_1 = 1 = \Delta_2$$

Thus,

$$\frac{Y(s)}{X(s)} = \frac{-\frac{1}{s} - k}{1 + \frac{k}{s}} = \frac{-(1+ks)}{s+k}$$

Ans. (A)

EXAMPLE 3.17. Which of the options is an equivalent representation of the signal flow graph shown here?

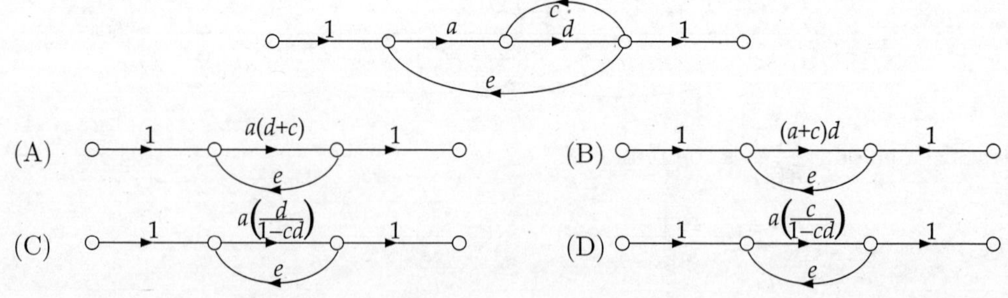

Solution The transfer function corresponding to the given signal flow graph is

$$\text{Transfer function} = \frac{ad}{1 - ade - dc}$$

The transfer function corresponding to the option (C) is

$$\text{Transfer function} = \frac{a\left[\dfrac{d}{1-cd}\right]}{1 - \dfrac{ade}{1-cd}} = \frac{ad}{1 - ade - dc}$$

Ans. (C)

EXAMPLE 3.18. In the system whose signal flow graph is shown in the following figure, $U_1(s)$ and $U_2(s)$ are inputs. The transfer function $\dfrac{Y(s)}{U_1(s)}$ is

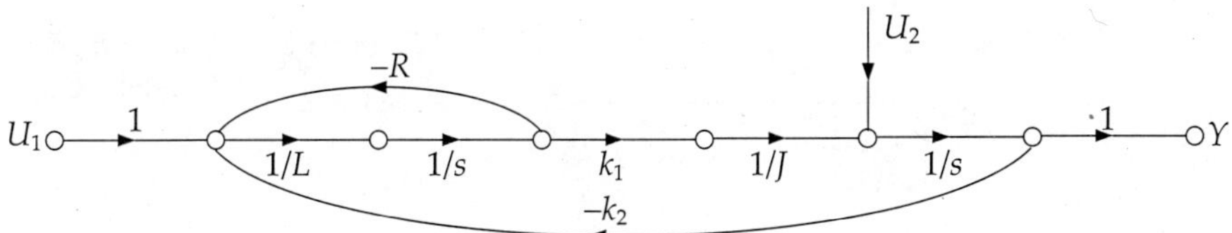

(A) $\dfrac{k_1}{JLs^2 + JRs + k_1k_2}$ (B) $\dfrac{k_1}{JLs^2 - JRs - k_1k_2}$

(C) $\dfrac{k_1 - U_2(R + sL)}{JLs^2 + (JR - U_2L)s + k_1k_2 - U_2R}$ (D) $\dfrac{k_1 - U_2(sL - R)}{JLs^2 - (JR + U_2L)s - k_1k_2 + U_2R}$

Solution We note that the transfer function $\dfrac{Y(s)}{U_1(s)}$ needs be determined. Therefore, the input U_2 does not come into the picture. Hence,

$$\frac{Y(s)}{U_1(s)} = \frac{\dfrac{k_1}{JLs^2}[1]}{1 - \left[-\dfrac{R}{Ls} - \dfrac{k_1k_2}{JLs^2}\right]} = \frac{k_1}{JLs^2 - JRs - k_1k_2}$$

Ans. (B)

EXAMPLE 3.19. For the signal flow graph shown in the figure, which one of the following expressions is equal to the transfer function $\left.\dfrac{Y(s)}{X_2(s)}\right|_{X_1(s)=0}$?

(A) $\dfrac{G_1}{1 + G_2(1 + G_1)}$ (B) $\dfrac{G_2}{1 + G_1(1 + G_2)}$

(C) $\dfrac{G_1}{1 + G_1G_2}$ (D) $\dfrac{G_2}{1 + G_1G_2}$

Solution Required $\left.\dfrac{Y(s)}{X_2(s)}\right|_{X_1(s)=0}$. So, we put $X_1(s) = 0$. Then

Forward path gain $\quad p_1 = G_2 \quad \Delta_1 = 1$

Loop gains $\quad L_1 = -G_1 \quad L_2 = -G_1G_2$

$$\Delta = 1 - (L_1 + L_2) \qquad \text{[no non-touching loop-pair]}$$
$$= 1 + G_1(1 + G_2)$$

Transfer function $\quad \dfrac{Y(s)}{X_2(s)} = \dfrac{G_2}{1 + G_1(1 + G_2)}$

Ans. (B)

EXAMPLE 3.20. Find the transfer function for the system given by signal flow graph of Figure 3.24.

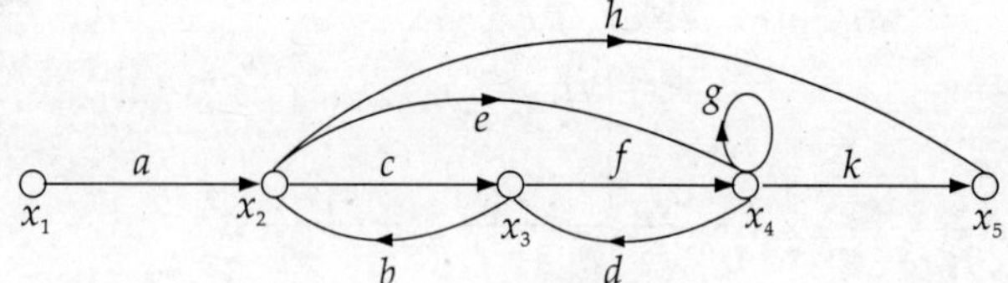

Figure 3.24: Signal flow graph (Example 3.20).

Solution The number of forward paths, $N = 3$. Their gains are

$$p_1 = acfk \qquad p_2 = aek \qquad p_3 = ah$$

Loops and their gains are

$$L_1 = bc \qquad L_2 = df \qquad L_3 = g \qquad L_4 = bed$$

Non-touching loop pair is $(L_1,\ L_3)$. Therefore,

$$\Delta = 1 - (bc + df + g + bed) + bc(g)$$

All loops are touching p_1 and p_2. Therefore, $\Delta_1 = \Delta_2 = 1$. L_2 and L_3 do not touch p_3. Therefore, $\Delta_3 = 1 - df - g$. Thus, the transfer function is given by

$$\text{Transfer function} = \frac{p_1\Delta_1 p_2\Delta_2 + p_3\Delta_3}{\Delta} = \frac{acfk + aek + ah(1 - df - g)}{1 - bc - df - g - bed + bcg}$$

EXAMPLE 3.21. Find the transfer function for the system given by signal flow graph of Figure 3.25.

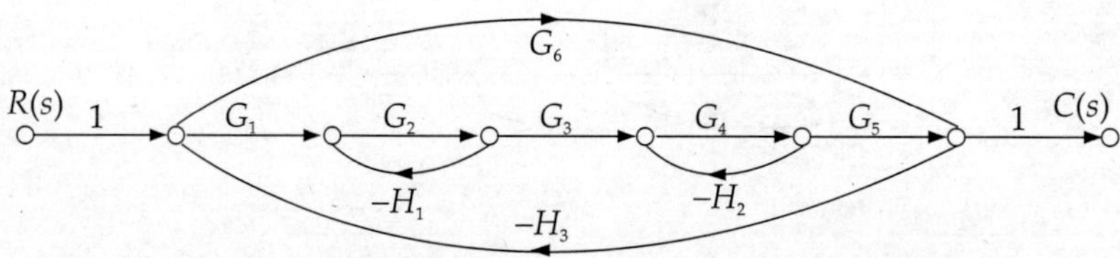

Figure 3.25: Signal flow graph (Example 3.21).

Solution Here forward paths, their gains and number are

$$p_1 = G_1G_2G_3G_4G_5 \qquad p_2 = G_6 \qquad N = 2$$

Loops and their gains are

$$L_1 = -G_2H_1 \qquad L_2 = -G_4H_2$$
$$L_3 = -G_6H_3 \qquad L_4 = -G_1G_2G_3G_4G_5H_3$$

The first three loops are non-touching. Possible combinations of *two* non-touching loops and the corresponding loop-gains are

$$L_1L_2 = G_2G_4H_1H_2 \qquad L_1L_3 = G_2G_6H_1H_3 \qquad L_2L_3 = G_4G_6H_2H_3$$

Possible combination of *three* non-touching loops and the corresponding loop-gain are

$$L_1L_2L_3 = -G_2G_4G_6H_1H_2H_3$$

Therefore,

$$\begin{aligned}
\Delta &= 1 + (G_2H_1 + G_4H_2 + G_6H_3 + G_1G_2G_3G_4G_5H_3) \\
&\quad + (G_2G_4H_1H_2 + G_2G_6H_1H_3 + G_4G_6H_2H_3) + (G_2G_4G_6H_1H_2H_3) \\
\Delta_1 &= 1 - 0 \quad (\because \text{ all loops touch } p_1) \\
\Delta_2 &= 1 + G_2H_1 + G_4H_2 + G_2G_4H_1H_2
\end{aligned}$$

Thus,

$$\text{Transfer function} = \frac{G_1G_2G_3G_4G_5 + G_6(1 + G_2H_1 + G_4H_2 + G_2G_4H_1H_2)}{\begin{aligned}&[1 + (G_2H_1 + G_4H_2 + G_6H_3 + G_1G_2G_3G_4G_5H_3) \\ &+ (G_2G_4H_1H_2 + G_2G_6H_1H_3 + G_4G_6H_2H_3) \\ &+ (G_2G_4G_6H_1H_2H_3)]\end{aligned}}$$

EXAMPLE 3.22. Find the transfer function of the network given in Figure 3.26a using signal flow graph and Mason's gain formula:

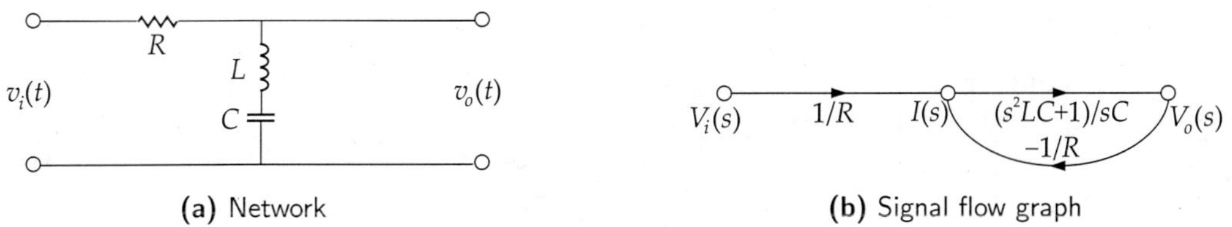

(a) Network **(b)** Signal flow graph

Figure 3.26: Example 3.22.

Solution From the equivalent network in s-space, we get

$$I(s) = \frac{V_i(s) - V_o(s)}{R} \qquad V_o(s) = \left(sL + \frac{1}{sC}\right) I(s)$$

Taking $V_i(s)$, $I(s)$ and $V_o(s)$ as three nodes, we draw the signal flow graph as shown in Figure 3.26b.

There is only one forward path (p_1) and loop (L_1) here. Their gains are

$$p_1 = \frac{s^2LC+1}{sRC} \qquad L_1 = -\frac{s^2LC+1}{sRC}$$

Therefore

$$\Delta = 1 + \frac{s^2LC+1}{sRC} \qquad \Delta_1 = 1 \quad (\because \text{ no non-touching loop exists})$$

$$\text{Transfer function} = \frac{V_o(s)}{V_i(s)} = \frac{\dfrac{s^2LC+1}{sRC}}{1+\dfrac{s^2LC+1}{sRC}} = \frac{s^2LC+1}{s^2LC+1+sRC}$$

EXAMPLE 3.23. Find the transfer function of the *RC*-network given in Figure 3.27a by the signal flow graph method.

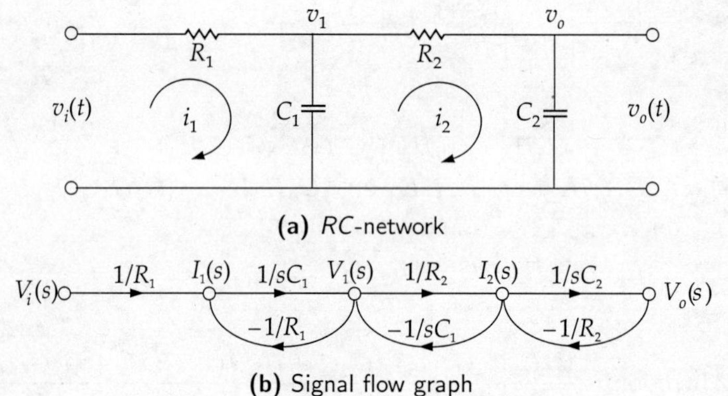

Figure 3.27: Example 3.23.

Solution From the *s*-space conversion of the network, we get

$$I_1(s) = \frac{1}{R_1}[V_i(s) - V_1(s)] \qquad V_1(s) = \frac{1}{sC_1}[I_1(s) - I_2(s)]$$

$$I_2(s) = \frac{1}{R_2}[V_1(s) - V_o(s)] \qquad V_o(s) = \frac{1}{sC_2}I_2(s)$$

The corresponding signal flow graph is given in Figure 3.27b.

The forward path and gain are

$$p_1 = \frac{1}{s^2R_1R_2C_1C_2}$$

Loops and gains are

$$L_1 = -\frac{1}{sR_1C_1} \qquad L_2 = -\frac{1}{sR_2C_1} \qquad L_3 = -\frac{1}{sR_2C_2}$$

All loops are touching the forward path. Therefore, $\Delta_1 = 1$.

$$\begin{aligned}\Delta &= 1 - (L_1 + L_2 + L_3) + L_1L_3 \\ &= 1 + \frac{1}{sR_1C_1} + \frac{1}{sR_2C_1} + \frac{1}{sR_2C_2} + \frac{1}{s^2R_1R_2C_1C_2}\end{aligned}$$

Thus, the transfer function is given by

$$\frac{V_o(s)}{V_i(s)} = \frac{p_1\Delta_1}{\Delta} = \frac{1}{s^2R_1R_2C_1C_2 + sR_2C_2 + sR_1C_2 + sR_1C_1 + 1}$$

EXAMPLE 3.24. Draw the signal flow graph from the given differential equation

$$3\frac{d^3y}{dx^3} + 11\frac{d^2y}{dx^2} + \frac{5}{2}\frac{dy}{dx} + 16y = x$$

Solution The Laplace transform of the given equation, assuming all initial conditions = 0, is

$$3s^3Y(s) + 11s^2Y(s) + \frac{5}{2}sY(s) + 16Y(s) = X(s)$$

Therefore, transfer function $= \dfrac{Y(s)}{X(s)} = \dfrac{1}{3s^3 + 11s^2 + \frac{5}{2}s + 16} = \dfrac{\frac{1}{3s^3}}{1 - \left(-\frac{11}{3s} - \frac{5}{6s^2} - \frac{16}{3s^3}\right)}$

$$\equiv \frac{p_1\Delta_1}{1-(L_1+L_2+L_3)}$$

where $p_1 = \dfrac{1}{3s^3}$ $\qquad \Delta_1 = 1$

$L_1 = \dfrac{1}{s}\left(-\dfrac{11}{3}\right) \qquad L_2 = \dfrac{1}{s^2}\left(-\dfrac{5}{6}\right) \qquad L_3 = \dfrac{1}{s^3}\left(-\dfrac{16}{3}\right)$

The signal flow graph, drawn accordingly, is presented in Figure 3.28.

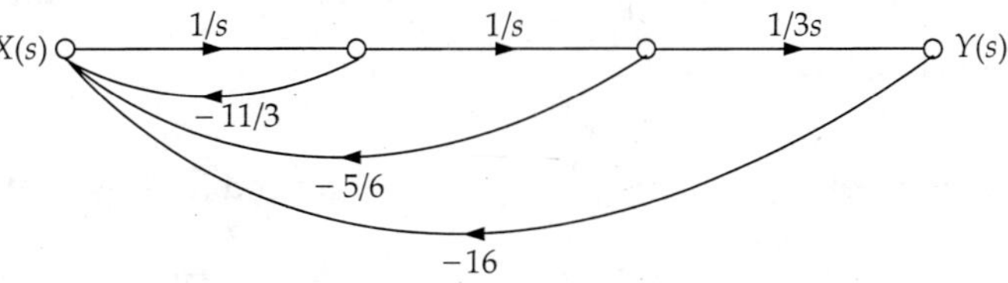

Figure 3.28: Signal flow graph (Example 3.24).

EXAMPLE 3.25. Find the overall transfer function through a signal flow graph and Mason's gain formula for the following equations

$x_2 = x_1 - H_3x_5 \qquad x_3 = G_1x_2 - H_1x_4 \qquad x_4 = G_2x_3 - H_2x_5$

$x_5 = G_3x_4 - G_5x_6 \qquad x_6 = -H_4x_6 + G_4x_2$

Solution The signal flow graph is shown in Figure 3.29.

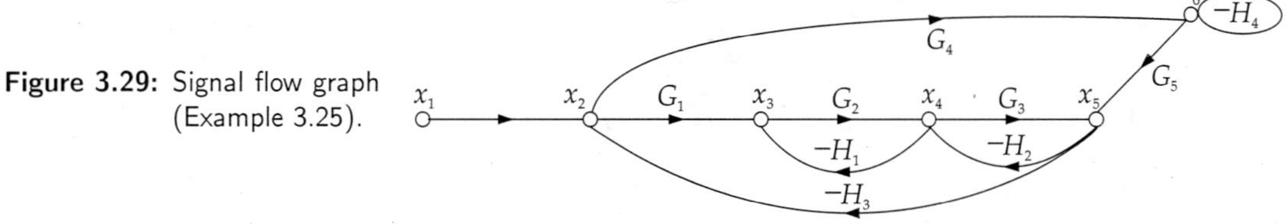

Figure 3.29: Signal flow graph (Example 3.25).

We may note here that though x_1 is the input, the output is x_5. Hence, forward paths and gains are

$$p_1 = G_1G_2G_3 \qquad p_2 = G_4G_5$$

Loops and gains are

$$L_1 = -G_2H_1 \qquad L_2 = -G_3H_2 \qquad L_3 = -H_4 \qquad L_4 = -G_4G_5H_3$$
$$L_5 = -G_1G_2G_3H_3$$

Now,

$$\begin{aligned}\Delta &= 1-(L_1+L_2+L_3+L_4+L_5)+(L_1L_3+L_1L_4+L_2L_3+L_3L_5)\\ &= 1+(G_2H_1+G_3H_2+H_4+G_4G_5H_3+G_1G_2G_3H_3)\\ &\quad +(G_2H_1H_4+G_2G_4G_5H_1H_3+G_3H_2H_4+G_1G_2G_3H_3H_4)\\ &= (1+H_4)(1+G_2H_1+G_3H_2+G_1G_2G_3H_3)+(1+G_2H_1)G_4G_5H_3\\ \Delta_1 &= 1+H_4\\ \Delta_2 &= 1+G_2H_1\end{aligned}$$

Therefore,

$$\frac{x_5}{x_1} = \frac{G_1G_2G_3(1+H_4)+G_4G_5(1+G_2H_1)}{(1+H_4)(1+G_2H_1+G_3H_2+G_1G_2G_3H_3)+(1+G_2H_1)G_4G_5H_3}$$

EXAMPLE 3.26. The signal flow graph of a system is given by Figure 3.30. Find out the overall gain of the system.

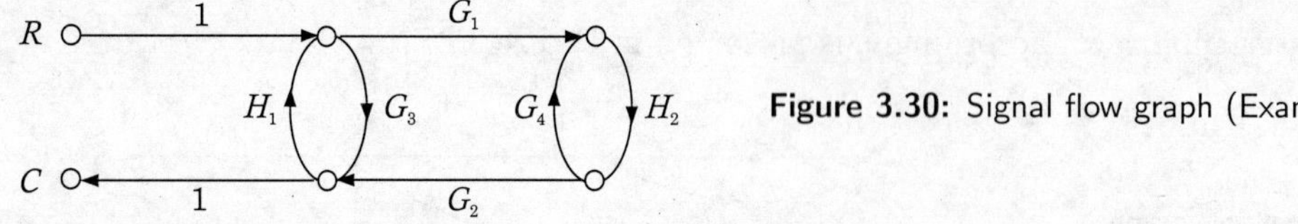

Figure 3.30: Signal flow graph (Example 3.26).

Solution Here, forward paths, their gain-products and path factors are

$$p_1 = G_1H_2G_2 \qquad \Delta_1 = 1 \qquad p_2 = G_3 \qquad \Delta_2 = 1 - G_4H_2$$

Loops and their gains are

$$L_1 = G_3H_1 \qquad L_2 = G_4H_2 \qquad L_3 = G_1H_2G_2H_1$$

There exists one pair of non-touching loops, namely L_1 and L_2. Therefore, the determinant of the graph is

$$\Delta = 1 - G_3H_1 - G_4H_2 - G_1G_2H_1H_2 + G_3G_4H_1H_2$$

So, the overall gain, i.e. the transfer function is

$$\frac{C(s)}{R(s)} = \frac{G_1G_2H_2 + G_3(1-G_4H_2)}{1 - G_3H_1 - G_4H_2 - G_1G_2H_1H_2 + G_3G_4H_1H_2}$$

EXAMPLE 3.27. Find the transfer function of the system represented by the signal flow graph shown in Figure 3.31.

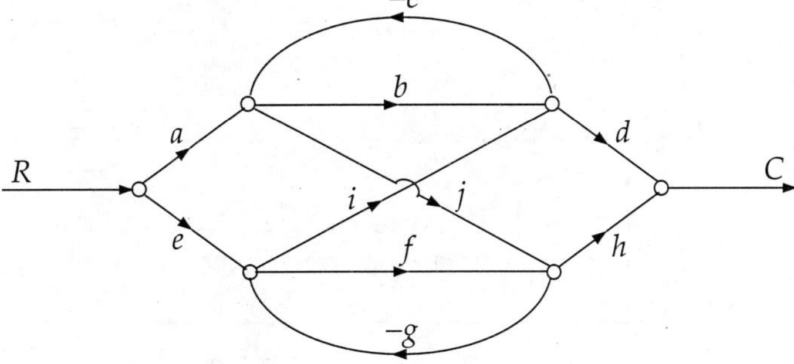

Figure 3.31: Signal flow graph (Example 3.27).

Solution Forward path-gains and their path-factors are

$$p_1 = abd \quad \Delta_1 = 1 + fg \quad p_2 = ajh \quad \Delta_2 = 1 \quad p_3 = efh \quad \Delta_3 = 1 + bc$$
$$p_4 = eid \quad \Delta_4 = 1 \quad p_5 = -ajgid \quad \Delta_5 = 1 \quad p_6 = -eicjh \quad \Delta_6 = 1$$

Loops and their gains are

$$L_1 = -bc \qquad L_2 = -fg \qquad L_3 = cjgi$$

Non-touching loops are L_1 and L_2. Therefore, the determinant of the graph is

$$\Delta = 1 + bc + fg - cgij + bcfg$$

Using Mason's gain formula, the required transfer function is given by

$$\frac{C(s)}{R(s)} = \frac{abd(1 + fg) + ahj + efh(1 + bc) + edi - adgij - cehij}{1 + bc + fg - cgij + bcfg}$$

Conversion of Block Diagrams to Signal Flow Graphs

Notwithstanding that block diagram is an useful tool to show the relationship between constituent units of a control system, sometimes it becomes cumbersome to reduce it to yield the transfer function which is needed for further manipulation. Signal flow graphs, in that respect, pose little problem. Therefore, sometimes one has to convert a block diagram to a signal flow graph to obtain the required transfer function. We will now consider a few examples to show how to make such conversions.

EXAMPLE 3.28. Draw the signal flow graph and determine C/R for the block diagram shown in Figure 3.32a.

Solution We observe that the block diagram consists of 4 summing points and 3 take-off points. So, to draw the signal flow graph, we place $4 + 3 = 7$ nodes, name them $x1 \cdots x7$ for our convenience and draw links between them as shown in Figure 3.32b.
Forward paths and loops are

$$p_1 = x_1x_2x_3x_4x_5x_6x_7 \qquad p_2 = x_1x_2x_3x_4x_6x_7$$
$$L_1 = x_2x_3x_4x_5x_2 \qquad L_2 = x_3x_4x_5x_6x_7x_3 \qquad L_3 = x_3x_4x_6x_7x_3$$
$$L_4 = x_1x_2x_3x_4x_5x_6x_7x_1 \qquad L_5 = x_1x_2x_3x_4x_6x_7x_1$$

Forward path gains are

$$p_1 = G_1G_2G_3 \qquad p_2 = -G_1G_4$$

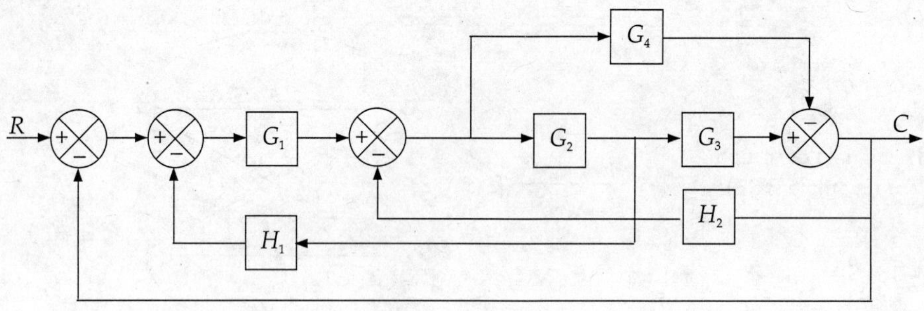

(a) Block diagram

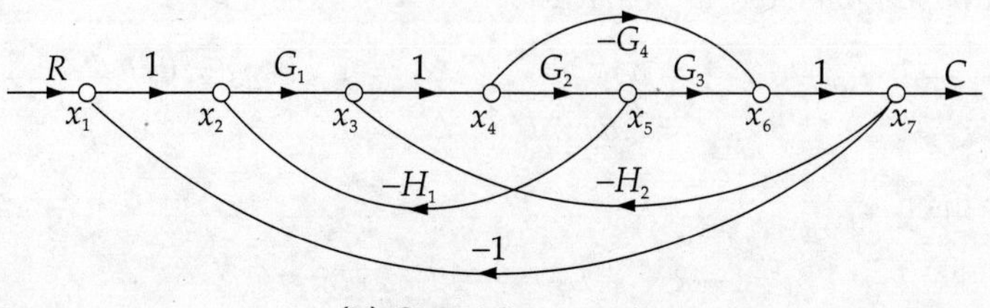

(b) Signal flow graph

Figure 3.32: Example 3.28.

Loop gains are

$$L_1 = -G_1G_2H_1 \qquad L_2 = -G_2G_3H_2 \qquad L_3 = G_4H_2 \qquad L_4 = -G_1G_2G_3$$

$$L_5 = G_1G_4$$

There is no non-touching loop and all loops touch the two forward paths. Therefore,

$$\Delta_1 = 1 \qquad \Delta_2 = 1$$

$$\begin{aligned}\Delta &= 1 - (L_1 + L_2 + L_3 + L_4 + L_5)\\ &= 1 + G_1G_2H_1 + G_2G_3H_2 - G_4H_2 + G_1G_2G_3 - G_1G_4\end{aligned}$$

Therefore, $\quad$ transfer function $= \dfrac{C}{R} = \dfrac{p_1\Delta_1 + p_2\Delta_2}{\Delta}$

$$= \frac{G_1G_2G_3 - G_1G_4}{1 + G_1G_2H_1 + G_2G_3H_2 - G_4H_2 + G_1G_2G_3 - G_1G_4}$$

EXAMPLE 3.29. Find the overall transfer function using Mason's gain formula of the system whose block diagram is shown in Figure 3.33a.

Solution The signal flow graph of the block diagram is given in Figure 3.33b. Here, forward paths and gains are

$$p_1 = G_1G_2G_3 \qquad p_2 = G_4$$

Loops and gains are

$$L_1 = -G_1G_2H_3 \qquad L_2 = -G_2G_3H_2 \qquad L_3 = -G_1G_2G_3H_1 \qquad L_4 = -G_4H_1$$

The non-touching loop-pair and its gain are

$$L_2L_4 = G_2G_3G_4H_1H_2$$

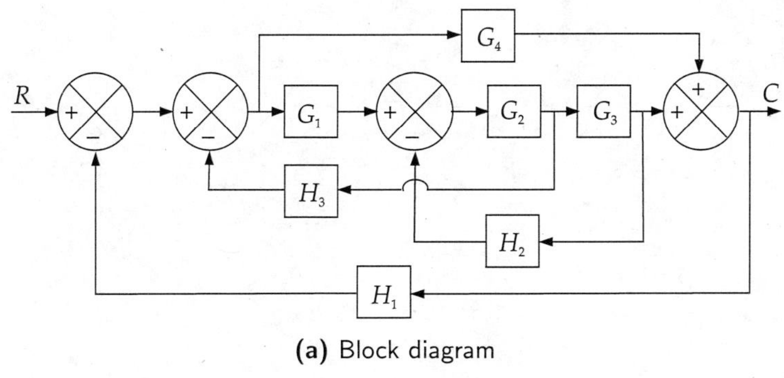

(a) Block diagram

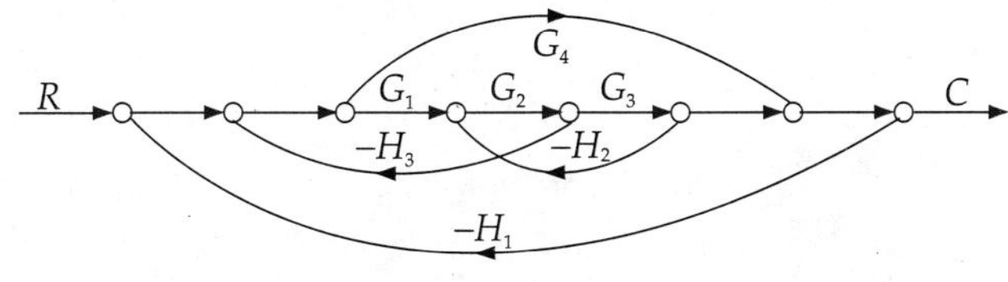

(b) Signal flow graph

Figure 3.33: Example 3.29.

All loops and pairs touch p_1, but L_2 does not touch p_2. Therefore,

$$\Delta_1 = 1 \qquad \Delta_2 = 1 + G_2G_3H_2$$

$$\Delta = 1 + G_1G_2H_3 + G_2G_3H_2 + G_1G_2G_3H_1 + G_4H_1 + G_2G_3G_4H_1H_2$$

Thus, the transfer function is given by

$$\frac{C(s)}{R(s)} = \frac{G_1G_2G_3 + G_2G_3G_4H_2 + G_4}{1 + G_1G_2H_3 + G_2G_3H_2 + G_1G_2G_3H_1 + G_4H_1 + G_2G_3G_4H_1H_2}$$

EXAMPLE 3.30. Find $C(s)/V(s)$ of the block diagram given in Figure 3.34a by signal flow graph and Mason's gain formula.

Solution The signal flow graph is given in Figure 3.34b.

We note that this is a multi-input system. So, when considering the transfer function between $C(s)$ and $V(s)$, we put $R(s) = 0$. Then, the only forward path, its gain product and path factor are

$$p_1 = (-1)G_2 = -G_2 \qquad \Delta_1 = 1$$

Loops and their gains are

$$L_1 = -G_1G_2H_1 \qquad L_2 = -G_1G_2$$

There is no non-touching loop pair. So, the determinant of the graph is

$$\Delta = 1 + G_1G_2 + G_1G_2H_1$$

The required transfer function is, therefore, given by

$$\frac{C(s)}{V(s)} = \frac{G_2}{1 + G_1G_2 + G_1G_2H_1}$$

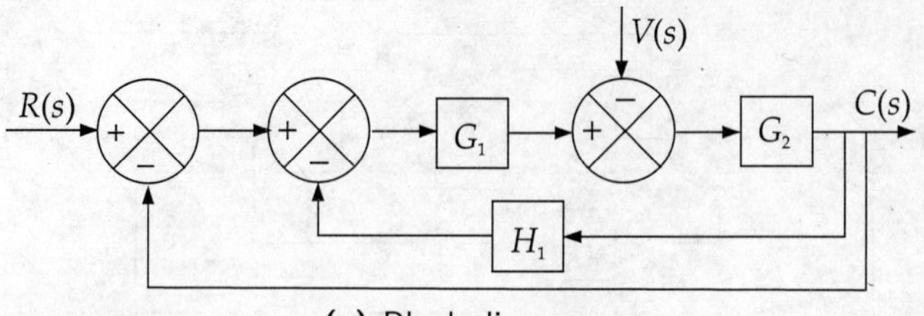

(a) Block diagram

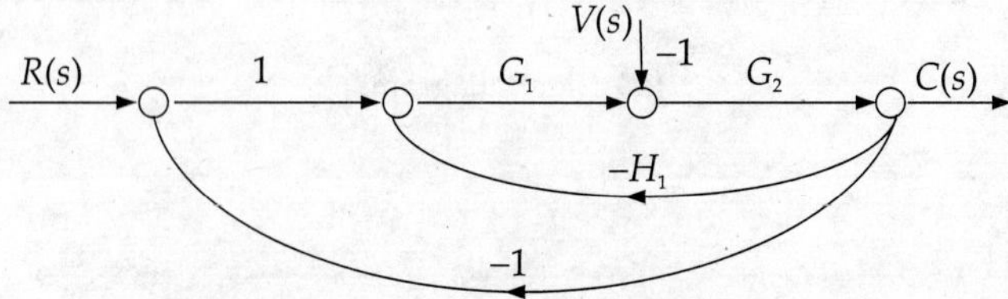

(b) Signal flow graph

Figure 3.34: Example 3.30.

EXAMPLE 3.31. The system shown in the figure below

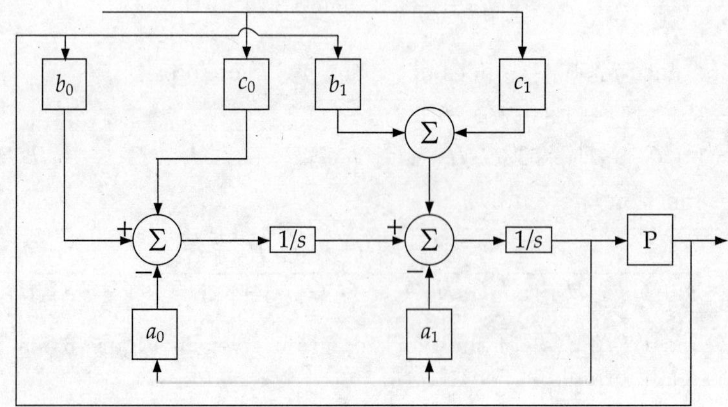

Block diagram 1

can be reduced in the form

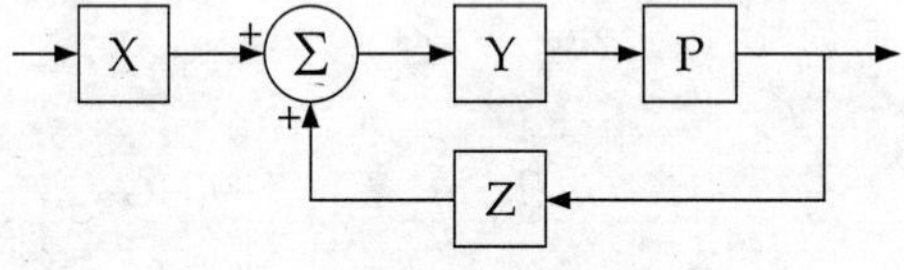

Block diagram 2

with

(A) $X = c_0 s + c_1$, $Y = \dfrac{1}{(s^2 + a_0 s + a_1)}$, $Z = b_0 s + b_1$

(B) $X = 1$, $Y = \dfrac{(c_0 s + c_1)}{(s^2 + a_0 s + a_1)}$, $Z = b_0 s + b_1$

(C) $X = c_1 s + c_0$, $Y = \dfrac{(b_1 s + b_0)}{(s^2 + a_1 s + a_0)}$, $Z = 1$

(D) $X = c_1 s + c_0$, $Y = \dfrac{1}{(s^2 + a_1 s + a_0)}$, $Z = b_1 s + b_0$

Solution We convert the Block diagram 1 to the following signal flow graph

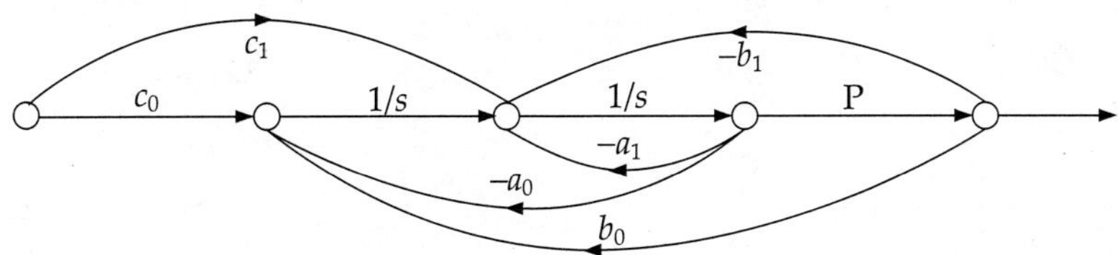

Here, forward path and loop gains are

$$p_1 = \frac{c_0 P}{s^2} \qquad p_2 = \frac{c_1 P}{s} \qquad L_1 = -\frac{a_1}{s} \qquad L_2 = -\frac{a_0}{s^2} \qquad L_3 = \frac{P b_0}{s^2} \qquad L_4 = \frac{P b_1}{s}$$

The determinant of the graph and path factors are

$$\Delta = 1 + \frac{a_1}{s} + \frac{a_0}{s^2} - \frac{P b_0}{s^2} - \frac{P b_1}{s}$$

$$\Delta_1 = \Delta_2 = 1$$

The transfer function from this graph is

$$\frac{C(s)}{R(s)} = \frac{\dfrac{c_0 P}{s^2} + \dfrac{c_1 P}{s}}{1 + \dfrac{a_1}{s} + \dfrac{a_0}{s^2} - \dfrac{P b_0}{s^2} - \dfrac{P b_1}{s}}$$

$$= \frac{(c_0 + c_1 s)P}{(s^2 + a_1 s + a_0) - P(b_0 + s b_1)}$$

$$= \frac{\dfrac{(c_0 + c_1 s)P}{(s^2 + a_1 s + a_0)}}{1 - \dfrac{P(b_0 + s b_1)}{(s^2 + a_1 s + a_0)}} \qquad \text{(i)}$$

Now, the transfer function from the Block diagram 2 is

$$\frac{C(s)}{R(s)} = \frac{XYP}{1 - YPZ} \qquad \text{(ii)}$$

Comparing corresponding terms of Eqs. (i) and (ii), we get

$$X = (c_0 + c_1 s) \qquad Y = \frac{1}{(s^2 + a_1 s + a_0)} \qquad Z = b_0 + s b_1$$

Ans. (D)

EXAMPLE 3.32. For the block diagram shown in Figure 3.35a, the transfer function $Y(s)/R(s)$ is

(A) $\dfrac{2s + 3}{s + 1}$ (B) $\dfrac{3s + 2}{s - 1}$ (C) $\dfrac{s + 1}{3s + 2}$ (D) $\dfrac{3s + 2}{s + 1}$

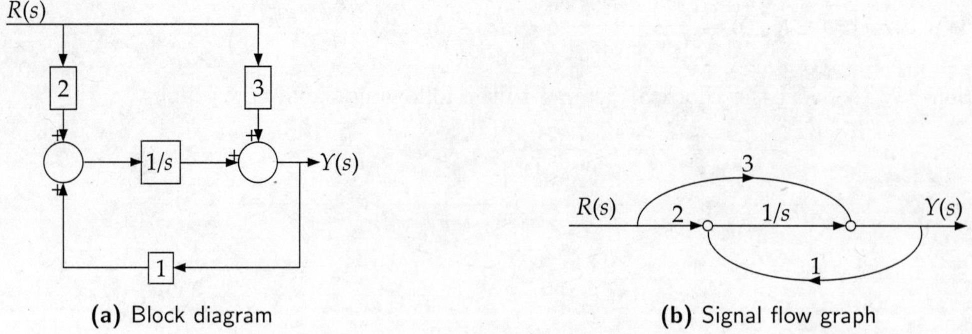

(a) Block diagram **(b)** Signal flow graph

Figure 3.35: Example 3.32.

Solution The corresponding signal flow graph is shown in Figure 3.35b. From this, we observe

$$\text{Forward paths} \quad p_1 = 3, \quad p_2 = \frac{2}{s}$$

$$\text{Loop} \quad L_1 = 1/s$$

$$\text{Therefore,} \quad \frac{Y(s)}{R(s)} = \frac{3 + (2/s)}{1 - (1/s)} = \frac{3s + 2}{s - 1}$$

Ans. (B)

REVIEW QUESTIONS

3.1 The block diagram of a control system is given in the following figure.

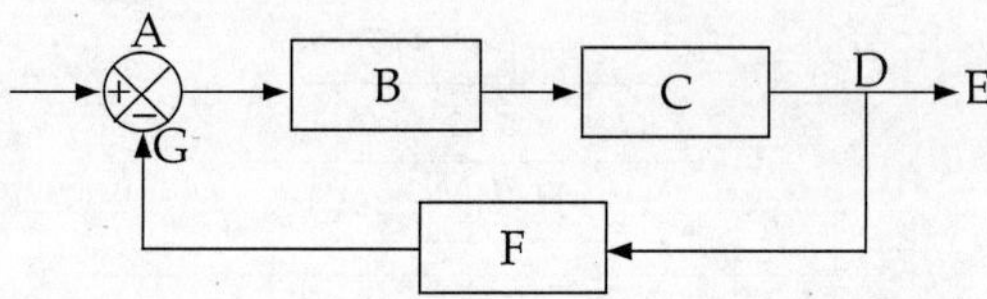

Match the following

Column I	*Column II*
A	i. Forward path transfer function
B	ii. Feedback signal
C	iii. Summing point
D	iv. Feedback transfer function
E	v. Control element
F	vi. Take-off point
G	vii. Controlled output

3.2 Choose the correct answer

(i) The signal flow graph of a system is shown in the following figure. The number of forward paths and the number of pairs of non-touching loops are respectively

(A) 3, 1 (B) 3, 2 (C) 4, 2 (D) 2, 4

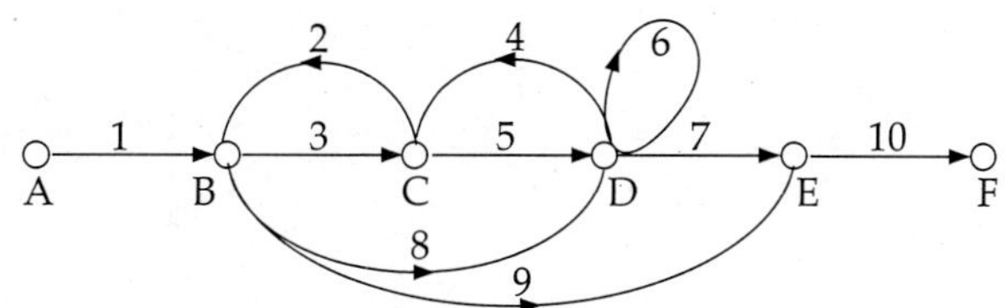

(ii) Signal flow graph is used to obtain the

(A) stability of a system
(B) transfer function of a system
(C) controllability of a system
(D) observability of a system

(iii) In a signal flow graph, loops are said to be non-touching if they

(A) do not touch alternate nodes
(B) do not touch any node
(C) do not possess any common node
(D) possess at least one common node

(iv) Signal flow graph approach is applicable to

(A) linear system only
(B) non-linear system only
(C) both linear and non-linear systems
(D) none of these

(v) The signal flow graph of a system is shown in the following figure. $U(s)$ is the input and $C(s)$ is the output.

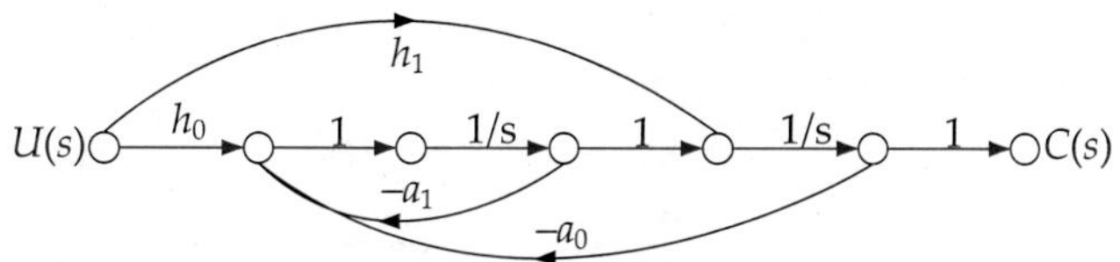

Assuming, $h_1 = b_1$ and $h_0 = b_0 - b_1 a_1$, then input-output transfer function $G(s) = \frac{C(s)}{U(s)}$ of the system is given by

(A) $G(s) = \dfrac{b_0 s + b_1}{s^2 + a_0 s + a_1}$
(B) $G(s) = \dfrac{a_1 s + a_0}{s^2 + b_1 s + b_0}$
(C) $G(s) = \dfrac{b_1 s + b_0}{s^2 + a_1 s + a_0}$
(D) $G(s) = \dfrac{a_0 s + a_1}{s^2 + b_0 s + b_1}$

(vi) The signal flow graph for the system is shown in the following figure. The transfer function $\frac{Y(s)}{U(s)}$ for this system is

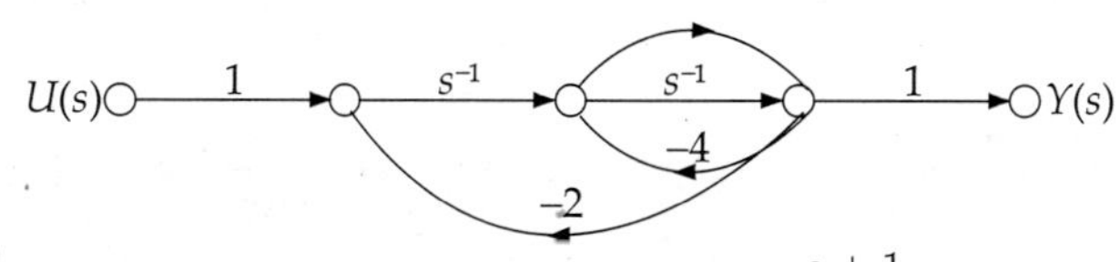

(A) $\dfrac{s+1}{5s^2+6s+2}$
(B) $\dfrac{s+1}{s^2+6s+2}$
(C) $\dfrac{s+1}{s^2+4s+2}$
(D) $\dfrac{1}{5s^2+6s+2}$

3.3 (a) What do you mean by the block diagram of a system?

(b) What is Mason's gain formula?

(c) Obtain the overall transfer function of the block diagram shown in the following figure by block diagram reduction technique.

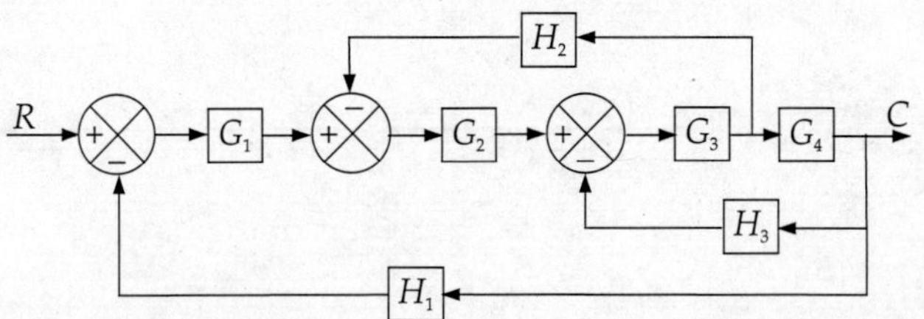

(d) Convert the block diagram into its signal flow graph and find transfer function using the gain formula.

3.4 Calculate the transfer function of the closed-loop system shown in the following figure.

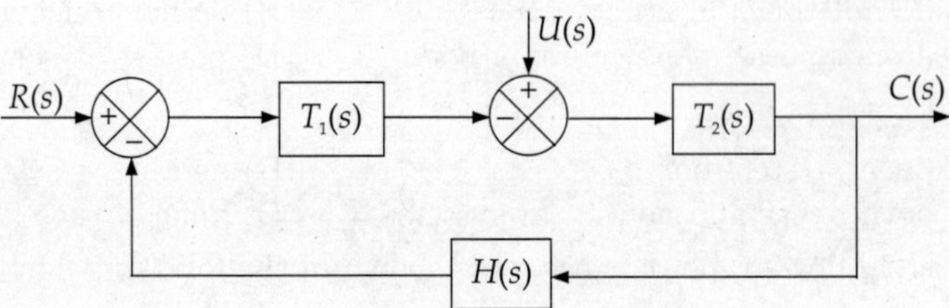

3.5 Find the transfer functions of the systems represented by signal flow graphs shown in the following figures:

(a)

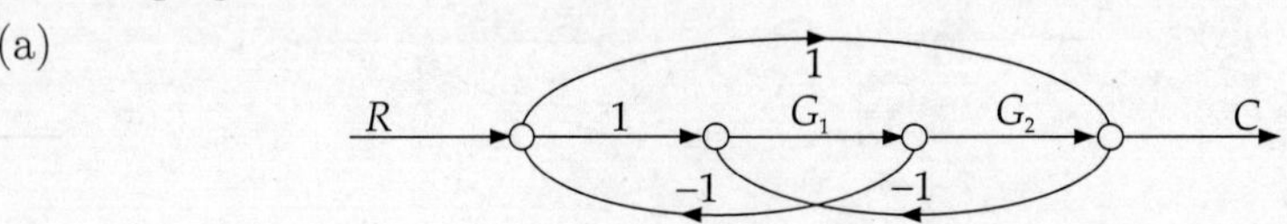

(b)

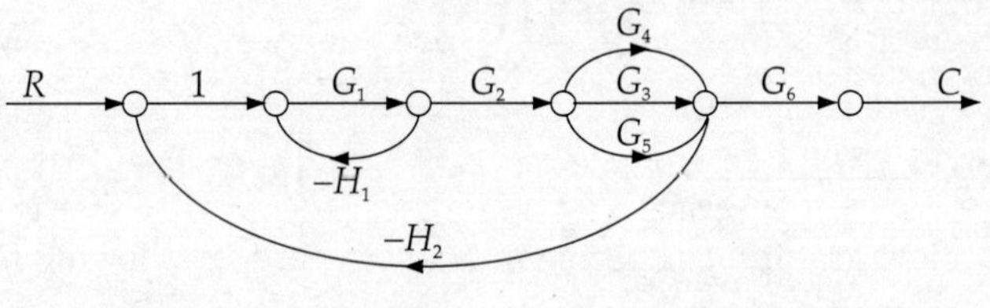

3.6 Simplify the block diagram of the following figure using block diagram reduction techniques.

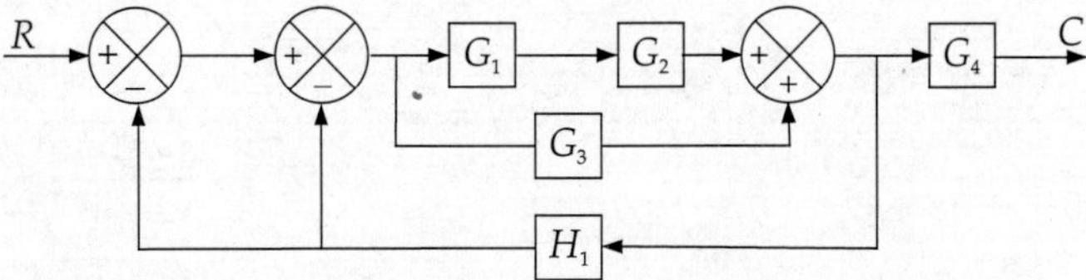

3.7 Find C/R for the signal flow graph shown in the following figure.

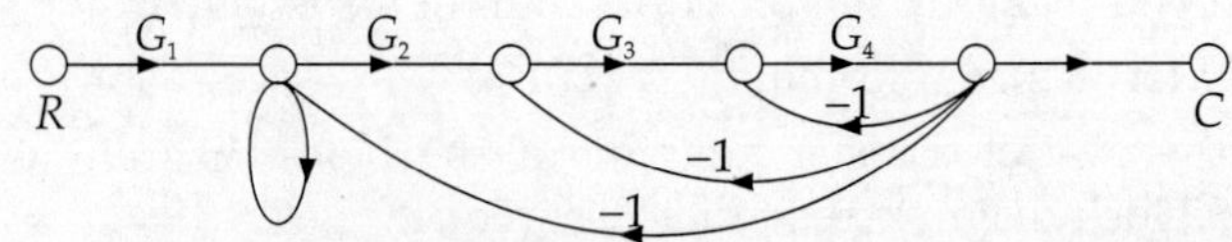

[*Hints:* Assume gain 1 for the segments where gains are not mentioned.]

3.8 Find the transfer function of the signal flow graph given in the following figure.

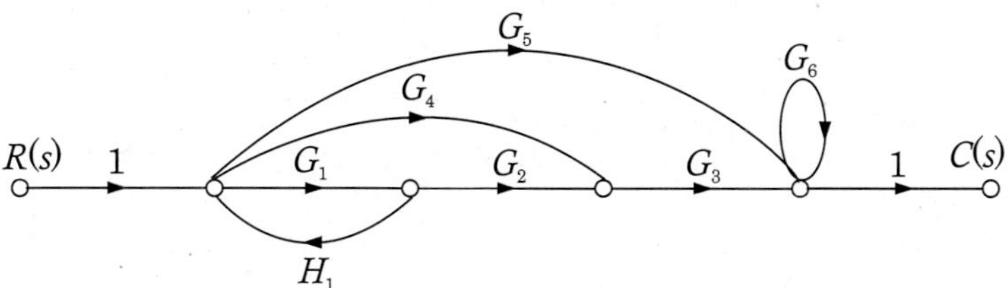

3.9 Draw the signal flow graph of the system of Figure 3.15 (Example 3.7 at page 38) and find $C(s)/R(s)$ using Mason's gain formula.

3.10 With the help of the block diagram reduction algebra find out the transfer function of the system shown in the following figure.

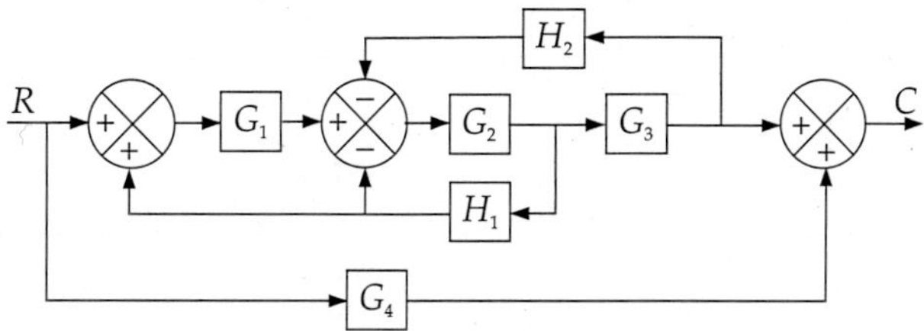

3.11 Reduce the complete block diagram obtained in Example 3.2 at page 33 by block diagram reduction technique.

3.12 Using block diagram reduction techniques, find the closed-loop transfer function $C(s)/R(s)$ whose block diagram is given in the following figure.

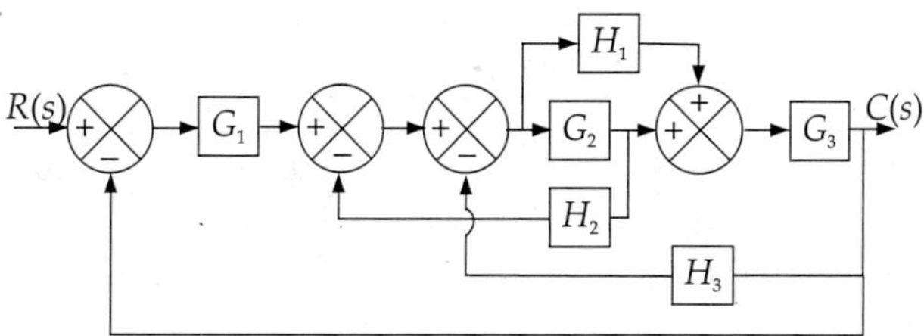

3.13 Find out the closed loop transfer function for the system given in the following figure by the block diagram reduction method.

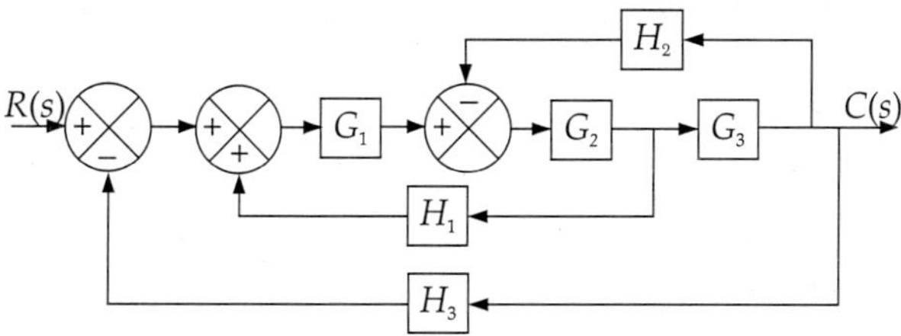

3.14 Find the transfer function for the system whose block diagram representation is shown in the following figure.

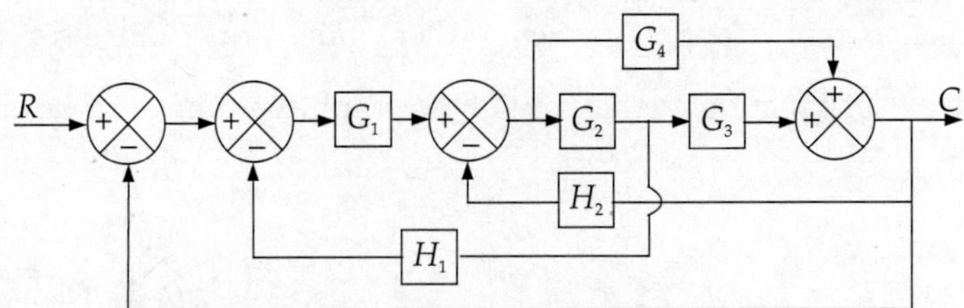

3.15 Using block diagram reduction technique find the transfer function of the block diagram shown in the following figure.

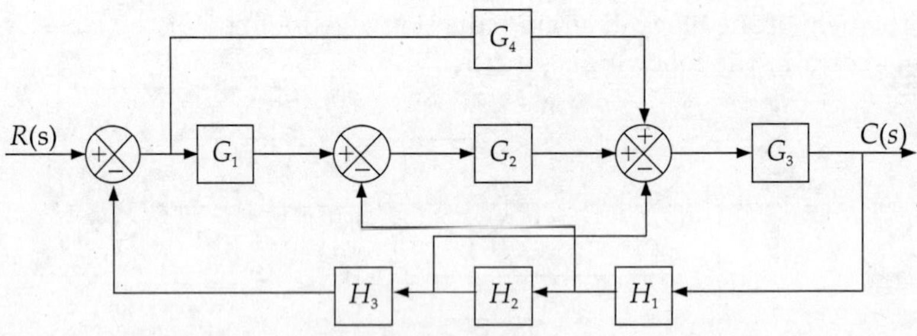

3.16 Consider the following set of equations:

$$x_2 = t_{12}x_1 + t_{32}x_2$$
$$x_3 = t_{23}x_3 + t_{43}x_4$$
$$x_4 = t_{24}x_2 + t_{34}x_3 + t_{44}x_4$$
$$x_5 = t_{45}x_4$$

(a) First construct the signal flow graphs for individual equations.
(b) Then construct the signal flow graph as a whole.
(c) With the help of Mason's gain formula determine the overall gain of the system.

Answers to Selected Questions

3.1. A→(iii) B→(v) C→(i) D→(vi) E→(vii) F→(iv) G→(ii)

3.2. (i) (A) (ii) (B) (iii) (C) (iv) (A) (v) (C) (vi) (A)

3.3. (d) $\dfrac{C}{R} = \dfrac{G_1G_2G_3G_4}{1 + G_2G_3H_2 + G_3G_4H_3 + G_1G_2G_3G_4H_1}$ 3.4. $\dfrac{C}{R} = \dfrac{UT_2 - RT_1T_2}{1 + T_1T_2H}$

3.5. (a) $\dfrac{C}{R} = 1$ (b) $\dfrac{C}{R} = \dfrac{G_1G_2G_6(G_3 + G_4 + G_5)}{1 + G_1H_1 + G_1G_2G_3H_2 + G_1G_2G_4H_2 + G_1G_2G_5H_2}$

3.6. $\dfrac{C}{R} = \dfrac{G_1G_2 + G_3}{1 + 2(G_1G_2 + G_3)H_1}$ 3.7. $\dfrac{C}{R} = G_1$

3.8. $\dfrac{C(s)}{R(s)} = 1 - G_1H_1 - G_6 + G_1G_6H_1$

3.9. $\dfrac{C(s)}{R(s)} = \dfrac{G_1G_2G_3G_4}{1 + G_2G_3H_2 + G_3G_4H_3 + G_1G_2G_3G_4H_1}$

3.10. $\dfrac{C(s)}{R(s)} = \dfrac{G_4 + G_2G_4H_1 + G_2G_3G_4H_2 - G_1G_2G_4H_1 + G_1G_2G_3}{1 + G_2H_1 + G_2G_3H_2 - G_1G_2H_1}$

3.11. $\dfrac{V_o(s)}{V_i(s)} = \dfrac{1}{s^2R_1R_2C_1C_2 + s(R_1C_1 + R_2C_2 + R_1C_2) + 1}$

3.12. $\dfrac{C(s)}{R(s)} = \dfrac{G_1G_2G_3 + G_1G_3H_1}{1 + G_2H_2 + H_1H_2 + G_2G_3H_3 + G_3H_1H_3 + G_1G_2G_3 + G_1G_3H_1}$

3.13. $\dfrac{C(s)}{R(s)} = \dfrac{G_1G_2G_3}{1 + G_2G_3H_2 - G_1G_2H_1 + G_1G_2G_3H_3}$

3.14. $\dfrac{C}{R} = \dfrac{G_1G_2G_3 + G_1G_4}{1 + G_1G_2G_3 + G_1G_4 + G_2G_3H_2 + G_4H_2 + G_1G_2H_1}$

3.15. $\dfrac{C(s)}{R(s)} = \dfrac{G_3(G_1G_2 + G_4)}{1 + G_2G_3H_1 + (G_1G_2 + G_4)G_3H_1H_2H_3}$

3.16(c). $\dfrac{C(s)}{R(s)} = \dfrac{t_{12}t_{23}t_{34}t_{45} + t_{12}t_{24}t_{45}}{1 - t_{23}t_{32} - t_{34}t_{43} - t_{44}}$

4

Systems and System Components

In Chapter 1, we gave the definition of a system. If a system has to be controlled to our desire, we have to find out its transfer function and to find its transfer function we have to build its mathematical model with its controllable parameters. So, in this chapter we will deal with the methods of building mathematical models of systems and then a few components that are of importance for controlling systems.

4.1 MATHEMATICAL MODELLING OF PHYSICAL SYSTEMS

Mathematical modelling of a system generally means describing its dynamical behaviour through differential equations. Such equations are obtained by utilizing the physical laws associated with the process. This approach applies equally to mechanical, electrical, fluid and thermodynamic systems. In mechanical systems, two kinds of motion—translational and rotational—take place. We recapitulate relevant mathematics of these motions from the modelling standpoint.

Translational Systems

If the motion takes place along a straight line, we call it a translational system. When a translation takes place, it may face three kinds of opposing forces originating from (i) inertia, (ii) friction or damping and (iii) elasticity or spring action. Let us consider them in some detail.

Inertial force. Newton's law states that when a force is applied to a body at rest or of uniform motion in a straight line, it generates an acceleration in it. But the mass of the body tends to remain in its original state which is called *inertia*. This means that if the body was at rest before the application of the force, it has a tendency to remain at rest. Thus, inertia is an opposing force which is given by

$$F_M(t) = M\frac{d^2x}{dt^2} = M\frac{dv}{dt}$$

where, M, x and v indicate mass, displacement and velocity of the body, respectively.

It can be diagrammatically represented as in Figure 4.1a.

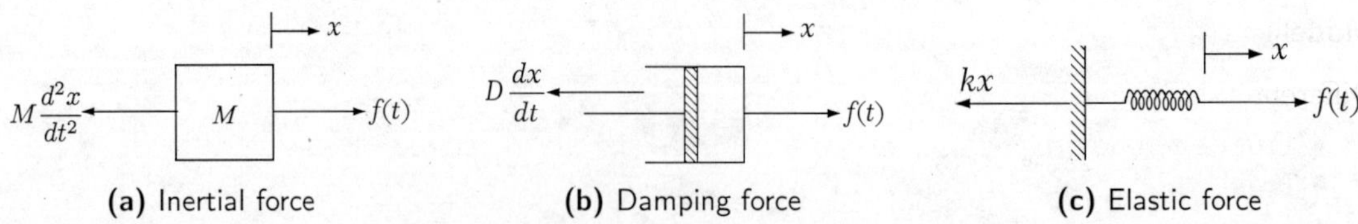

(a) Inertial force (b) Damping force (c) Elastic force

Figure 4.1: Schematic representation of forces.

Frictional or damping force. Since a body does not move in absolute vacuum, it experiences a frictional force from its surroundings, be it from the solid on which it moves or from the fluid which it wades through. This force is called *viscous friction* in general because it is proportional to the velocity v with which the body moves at a particular instant. So, the damping force F_D is given by

$$F_D(t) = Dv = D\frac{dx}{dt}$$

where D is the *damping coefficient.*

The damping force is represented by what is called a *dashpot* as shown in Figure 4.1b.

Damping forces not necessarily come into play owing to natural causes. Sometimes they are deliberately introduced to prevent or subdue oscillation in some systems, or increase stability in some others. Imagine riding a bike having frictionless tyres on frozen ice!

Elastic or spring force. When a body, such as a spring, suffers elongation, a restoring force, originating from the elasticity of the material, tries to put it back to its original shape. This force of restoration F_k, proportional to the displacement of the body, is given by

$$F_k = kx$$

where k is called the *spring constant* or *stiffness.* Its diagrammatic form is given in Figure 4.1c.

Rotational Systems

The angular displacement of a body about a fixed axis generates its rotational motion. Like force in translational motion, a rotation of a body at rest is caused by an externally applied torque which is resisted by three kinds of opposing torques. They are—(i) inertial torque, (ii) damping torque and (iii) spring torque.

Inertial torque. Like mass in translational motion, moment of inertia J stores the inertia of the existing state in rotational motion. Thus, the inertial torque is written as

$$T_I(t) = J\frac{d\omega}{dt} = J\frac{d^2\theta}{dt^2}$$

where θ and ω are angular displacement and angular velocity, respectively.

Damping torque. Damping torque, which is proportional to the angular velocity of the body, is given by

$$T_D(t) = D\omega = D\frac{d\theta}{dt}$$

Spring torque. The spring torque is proportional to the angular displacement of the body and is given by

$$T_k(t) = k\theta$$

Comparable parameters of these two kinds of motion are given in Table 4.1.

Modelling Mechanical Systems

Differential equations, describing mechanical systems, can be set up in two ways:

- From free body diagrams
- From nodal representations

Each method has its own advantages. While free body diagrams can be set up intuitively, nodal representations are helpful in drawing equivalent electrical networks. We consider both the methods and solve a few examples.

Translation		*Rotation*	
Parameter	*Symbol*	*Parameter*	*Symbol*
Mass	M	Inertia	J
Force	f	Torque	T
Velocity	v	Angular velocity	ω
Displacement	x	Angular displacement	θ

Table 4.1: Comparable parameters of translational and rotational motion

Free body diagram method

In the free body diagram method, the force balance at equilibrium condition of each associated mass (i.e. body) is considered and the equation is set up for each. To find out the transfer function of the whole system, intermediate variables are eliminated.

EXAMPLE 4.1. Deduce the equation governing the motion of the mass M in Figure 4.2. Find out also the transfer function.

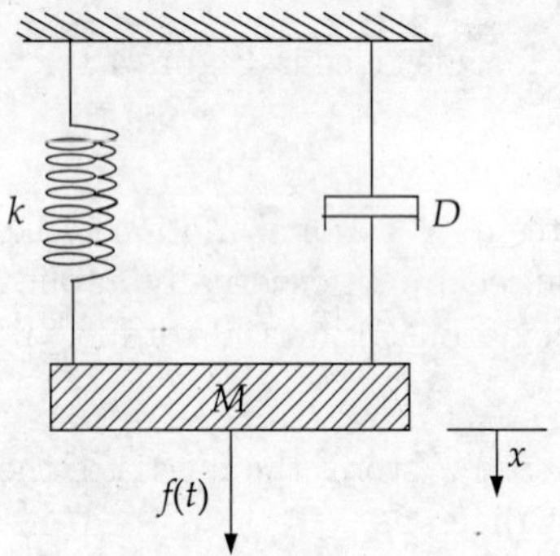

Figure 4.2: Mechanical system (Example 4.1).

Solution When the force f acting on the mass M produces a displacement x, three opposing forces act on it. So, as discussed earlier, the differential equation governing the motion is given by the force balance condition

$$f = M\frac{d^2x}{dt^2} + D\frac{dx}{dt} + kx \tag{i}$$

Taking Laplace transform of Eq. (i) with all initial conditions = 0 (i.e. system is relaxed)[1] and rearranging terms, we get the transfer function as

$$F(s) = s^2MX(s) + sDX(s) + kX(s)$$

or,

$$\frac{X(s)}{F(s)} = \frac{1}{s^2M + sD + k}$$

EXAMPLE 4.2. Find the transfer function of the system involving two bodies M_1 and M_2 given in Figure 4.3.

Solution We start from the mass M_1 on which the external force $f(t)$ acts. Treating it as a free body, its force balance is shown in Figure 4.4a.

While drawing the figure, we need to keep it in mind that the inertial force causes the displacement x_1 of the mass M_1, but viscous friction and spring forces cause the differential

[1]Unless stated otherwise, Laplace transforms will be made with all initial conditions = 0.

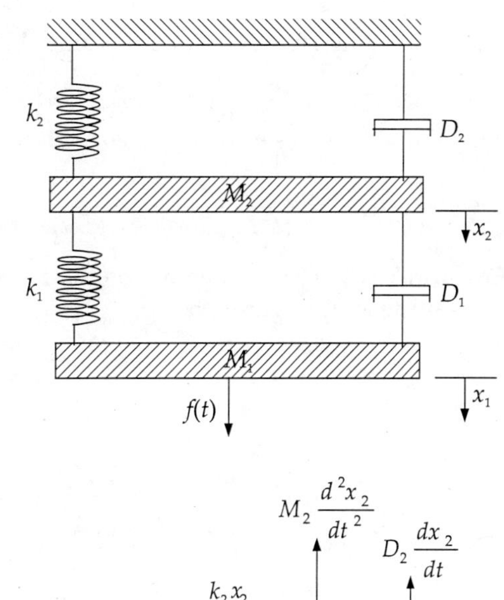

Figure 4.3: Two-body mechanical system (Example 4.2).

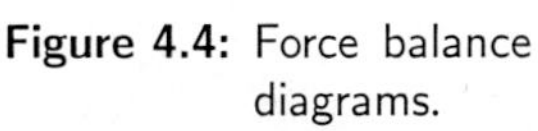

Figure 4.4: Force balance diagrams.

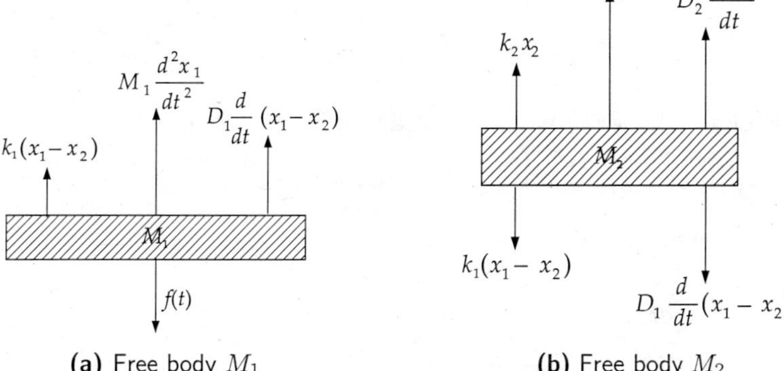

(a) Free body M_1 **(b)** Free body M_2

displacement $(x_1 - x_2)$ between M_1 and M_2. The corresponding differential equation and its Laplace transform are given by Eqs. (i) and (ii), respectively.

$$M_1 \frac{d^2x_1}{dt^2} + D_1 \frac{d}{dt}(x_1 - x_2) + k_1(x_1 - x_2) = f(t) \tag{i}$$

$$s^2 M_1 X_1(s) + sD_1[X_1(s) - X_2(s)] + k[X_1(s) - X_2(s)] = F(s) \tag{ii}$$

On rearranging Eq. (ii), we get

$$X_1(s) = \frac{F(s) + sD_1 X_2(s) + k_1 X_2(s)}{s^2 M_1 + sD_1 + k_1} \tag{iii}$$

Next, taking M_2 as a free body, its force balance diagram is drawn (Figure 4.4b). While drawing this diagram, we kept it in mind that the damping and spring force which were reactions (opposing forces) on M_1, work as action (i.e. forward force) on M_2 in keeping with Newton's 3rd law of motion. From Figure 4.4b, the following differential equation and its Laplace transform can be written as

$$M_2 \frac{d^2x_2}{dt^2} + D_2 \frac{dx_2}{dt} + k_2 x_2 = D_1 \frac{d}{dt}(x_1 - x_2) + k_1(x_1 - x_2) \tag{iv}$$

$$(s^2 M_2 + sD_2 + k_2) X_2(s) = (sD_1 + k_1)[X_1(s) - X_2(s)] \tag{v}$$

On rearranging Eq. (v) and eliminating $X_1(s)$ with the help of Eq. (iii), we get

$$[\{s^2M_2 + s(D_1 + D_2) + (k_1 + k_2)\}(s^2M_1 + sD_1 + k_1) - (sD_1 + k_1)^2]X_2(s) = F(s)(sD_1 + k_1)$$

Therefore, the transfer function is given by

$$\frac{X_2(s)}{F(s)} = \frac{sD_1 + k_1}{\{s^2M_2 + s(D_1 + D_2) + (k_1 + k_2)\}(s^2M_1 + sD_1 + k_1) - (sD_1 + k_1)^2}$$

EXAMPLE 4.3. Determine the differential equations describing the dynamics of the mechanical system shown in Figure 4.5.

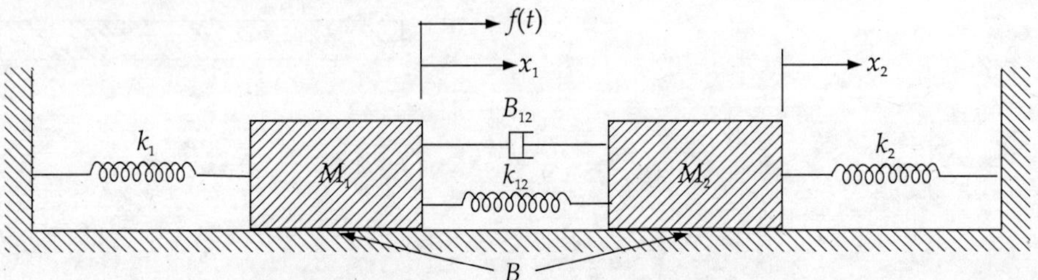

Figure 4.5: Mechanical system (Example 4.3).

Solution The force balance diagrams of free bodies M_1 and M_2 are shown in Figures 4.6a and 4.6b, respectively.

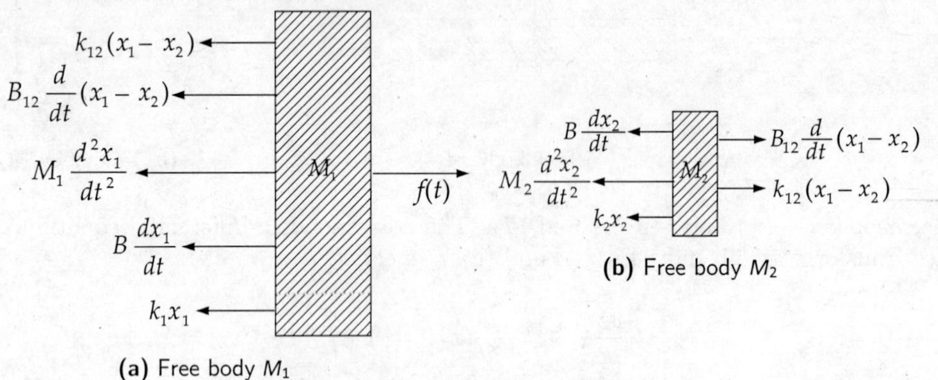

Figure 4.6: Force balance diagrams.

The corresponding differential equations are given by

$$M_1\frac{d^2x_1}{dt^2} + B_{12}\frac{d}{dt}(x_1 - x_2) + B\frac{dx_1}{dt} + k_{12}(x_1 - x_2) + k_1x_1 = f(t) \quad \text{(i)}$$

$$M_2\frac{d^2x_2}{dt^2} + B\frac{dx_2}{dt} + k_2x_2 = B_{12}\frac{d}{dt}(x_1 - x_2) + k_{12}(x_1 - x_2) \quad \text{(ii)}$$

The system dynamics is described by Eqs. (i) and (ii).

Nodal representation method

In the nodal representation method, a network diagram for the mechanical system is first drawn following the given steps.

Step 1: A reference line is drawn. It embodies fixed point(s) of the system.

Step 2: Nodes[2] corresponding to displacements, are plotted.

Step 3: Masses are connected between nodes and the reference line.

Step 4: Viscous frictions and springs associated with each mass are connected between nodes and reference line, parallel to the respective mass.

Step 5: Viscous frictions and springs between two displacements are connected between the two respective nodes.

Step 6: External applied force is connected between the relevant node and the reference line.

Step 7: For each node, force balance equations are written.

The procedure will be clear from the following examples, especially from Example 4.4, where we show it step by step explicitly.

EXAMPLE 4.4. Find, by the nodal representation method, the transfer function for the mechanical system shown in Figure 4.3.

Solution The step by step procedure of drawing a nodal presentation is given as follows:

Step 1. Corresponding to the fixed point at the top, from where the system is suspended, we draw a reference line at the bottom.

Step 2. Corresponding to two displacements x_1 and x_2, two nodes are plotted.

Step 3. M_1 and M_2 are connected between x_1, x_2 and the reference line.

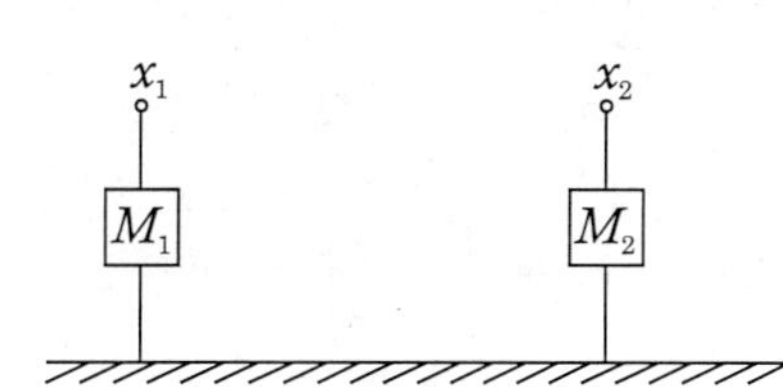

[2]Means 'labelled points in a tree diagram at which subordinate points branch off'. The literal meaning of the word is 'knot'.

Step 4. D_2 and k_2 are associated with the mass M_2. So, they are connected between x_2 and the reference line, parallel to M_2.

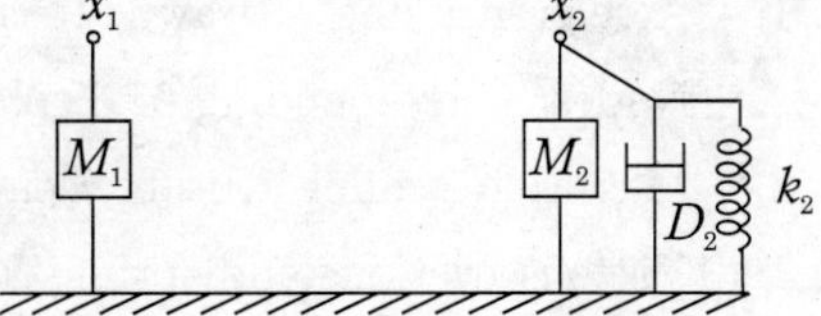

Step 5. D_1 and k_1 act between x_1 and x_2. So, they are connected accordingly.

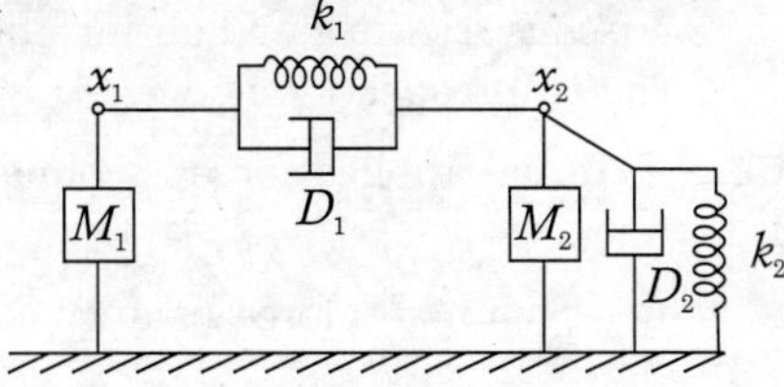

Step 6. $f(t)$ is connected between x_1 and the reference line.

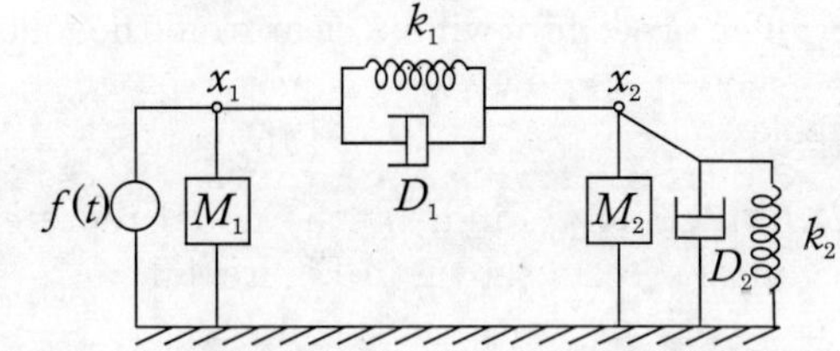

The *Step 6* gives the complete diagram.

Once the diagram is drawn, it is easy to set up differential equations balancing left and right side of each node separately. The force balance equation for node at x_1 is given by

$$f(t) = M_1 \frac{d^2x_1}{dt^2} + D_1 \frac{d}{dt}(x_1 - x_2) + k_1(x_1 - x_2) \tag{i}$$

That at node x_2 is given by

$$D_1 \frac{d}{dt}(x_1 - x_2) + k_1(x_1 - x_2) = M_2 \frac{d^2x_2}{dt^2} + D_2 \frac{dx_2}{dt} + k_2 x_2 \tag{ii}$$

We observe that Eqs. (i) and (ii) are identical to Eqs. (i) and (iv) respectively of Example 4.2. The rest of the procedure of finding out the transfer function i s the same as given i n Example 4.2.

EXAMPLE 4.5. Find the differential equations for Figure 4.5 by the nodal representation method.

Solution The relevant nodal diagram is shown in Figure 4.7. The force balance equation at node x_1 is

$$f(t) = M_1 \frac{d^2x_1}{dt^2} + B\frac{dx_1}{dt} + k_1x_1 + B_{12}\frac{d}{dt}(x_1 - x_2) + k_{12}(x_1 - x_2) \tag{i}$$

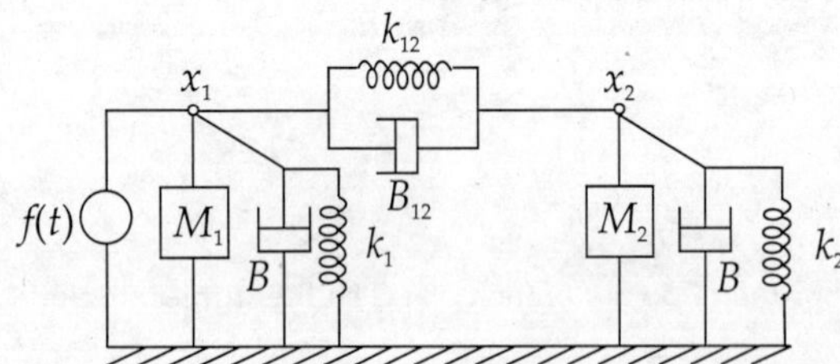

Figure 4.7: Nodal diagram (Example 4.5).

That at node x_2 is

$$B_{12}\frac{d}{dt}(x_1 - x_2) + k_{12}(x_1 - x_2) = M_2\frac{d^2x_2}{dt^2} + B\frac{dx_2}{dt} + k_2x_2 \quad \text{(ii)}$$

We note that Eqs. (i) and (ii) are identical to Eqs. (i) and (ii) respectively of Example 4.3.

Analogous Systems

In electrical systems, voltage V, current i and magnetic flux linkage Φ are considered three important parameters. Table 4.2 shows how these parameters can be interrelated through passive components of circuits, namely R, L and C.

Table 4.2: Interrelationships of V (voltage), i (current) and Φ (magnetic flux linkage)

Component	*Relation between*		
	V–i	i–V	i–Φ
R	$V = iR$	$i = \frac{V}{R}$	$i = \frac{1}{R}\frac{d\Phi}{dt}$
L	$V = L\frac{di}{dt}$	$i = \frac{1}{L}\int V\,dt$	$i = \frac{\Phi}{L}$
C	$V = \frac{1}{C}\int i\,dt$	$i = C\frac{dV}{dt}$	$i = C\frac{d^2\Phi}{dt^2}$

We observe that similar differential equations for dynamic systems can be written for mechanical and electrical systems as follows:

Mechanical		*Electrical*	
Translation:	$M\frac{d^2x}{dt^2} + D\frac{dx}{dt} + kx = f$	Voltage:	$L\frac{d^2Q}{dt^2} + R\frac{dQ}{dt} + \frac{Q}{C} = V$
Rotation:	$J\frac{d^2\theta}{dt^2} + D\frac{d\theta}{dt} + k\theta = T$	Current:	$C\frac{d^2\Phi}{dt^2} + \frac{1}{R}\frac{d\Phi}{dt} + \frac{\Phi}{L} = i$

Comparing these equations, we can draw an analogy between different physical quantities as shown in Table 4.3.

Table 4.3: Analogous quantities

Translation	*Rotation*	*Electrical analogy*	
		Force-voltage	*Force-current*
f	T	V	i
M	J	L	C
D	D	R	$1/R$
k	k	$1/C$	$1/L$
x	θ	Q	Φ

We will now see how these analogies can be exploited to frame differential equations for dynamic systems.

EXAMPLE 4.6. By using force-current $(f - i)$ analogy, draw the electrical analogy of the mechanical system given in Figure 4.8a.

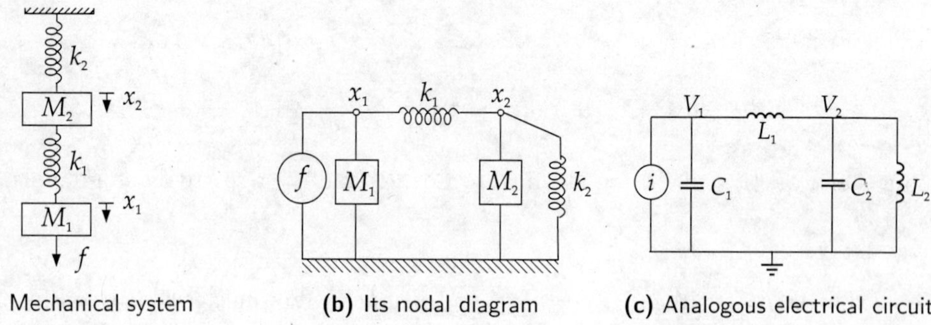

(a) Mechanical system **(b)** Its nodal diagram **(c)** Analogous electrical circuit

Figure 4.8: Example 4.6.

Solution The nodal diagram for the system is given in Figure 4.8b. Force-balance equations at x_1 and x_2 are

$$f = M_1\frac{d^2x_1}{dt^2} + k_1(x_1 - x_2) \tag{i}$$

$$k_1(x_1 - x_2) - M_2\frac{d^2x_2}{dt^2} + k_2x_2 \tag{ii}$$

To arrive at their electrical analogs, we substitute $f \to i$, $M \to C$, $k \to 1/L$ and $x \to \Phi$ in Eqs. (i) and (ii) to obtain

$$i = C_1\frac{d^2\Phi_1}{dt^2} + \frac{1}{L_1}(\Phi_1 - \Phi_2) = C_1\frac{dV_1}{dt} + \frac{1}{L_1}\Big(\int V_1\,dt - \int V_2\,dt\Big) \tag{iii}$$

$$\frac{1}{L_1}\Big(\int V_1\,dt - \int V_2\,dt\Big) = C_2\frac{d^2\Phi_2}{dt^2} + \frac{\Phi_2}{L_2} = C_2\frac{dV_2}{dt} + \frac{\int V_2 dt}{L_2} \tag{iv}$$

To draw the analogous electrical circuit,

1. We plot two nodes, V_1 and V_2 and an earth line.
2. Then the left and right loops (see Figure 4.8c) are drawn from Eqs. (iii) and (iv), respectively.

Note: To satisfy Eqs. (iii) and (iv), we have to draw L_1 and L_2, and not $1/L_1$ and $1/L_2$, in place of k_1 and k_2. So, as told earlier, that the electrical analog *looks similar to the nodal diagram* in the f–i analogy, but not exactly identical to it as dictated by the analogy.

EXAMPLE 4.7. Find the electrical analogous circuits of the mechanical system shown in Figure 4.9a.

Solution The nodal diagram for the system is given in Figure 4.9b. Force-balance equations at nodes x_1 and x_2 are

$$f = M_1\frac{d^2x_1}{dt^2} + D\frac{d}{dt}(x_1 - x_2) \tag{i}$$

$$D\frac{d}{dt}(x_1 - x_2) = M_2\frac{d^2x_2}{dt^2} + k_2x_2 \tag{ii}$$

For the f–V analogy equations, we put $f \to V$, $M \to L$, $D \to R$, $k \to 1/C$ and $x \to Q$ to

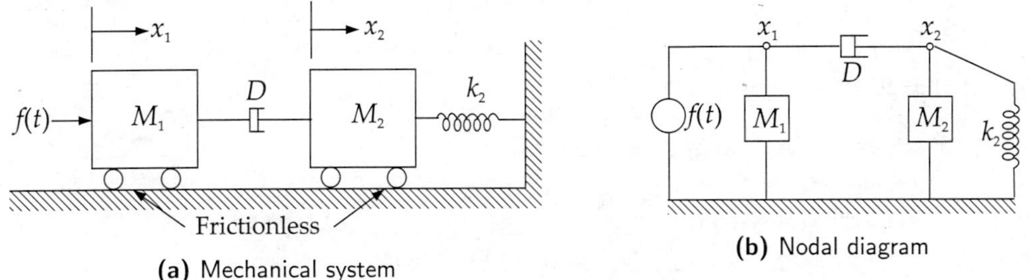

(a) Mechanical system

(b) Nodal diagram

Figure 4.9: Example 4.7.

get from Eqs. (i) and (ii)

$$V = L_1 \frac{d^2 Q_1}{dt^2} + R\frac{d}{dt}(Q_1 - Q_2) = L_1 \frac{di_1}{dt} + R(i_1 - i_2) \tag{iii}$$

$$R(i_1 - i_2) = L_2 \frac{d^2 Q_2}{dt^2} + \frac{Q_2}{C_2} = L_2 \frac{di_2}{dt} + \frac{\int i_2\, dt}{C_2} \tag{iv}$$

Equations (iii) and (iv) are represented by Figure 4.10a.

(a) f–V analog

(b) f–i analog

Figure 4.10: Electrical analogs.

Note: While M_1 and M_2 are connected *in parallel* in the nodal diagram, their analogous quantities L_1 and L_2 are connected *in series* in the f–V analog diagram. Similar series-parallel inversion occurs for $D \to R$ and $k \to C$.

Next, we consider the f–i analogy. Here, we substitute $f \to i$, $M \to C$, $D \to 1/R$ and $k \to 1/L$ in Eqs. (i) and (ii) to get the following equations

$$i = C_1 \frac{d^2 \Phi_1}{dt^2} + \frac{1}{R}\frac{d}{dt}(\Phi_1 - \Phi_2) = C_1 \frac{dV_1}{dt} + \frac{1}{R}(V_1 - V_2) \tag{v}$$

$$\frac{1}{R}(V_1 - V_2) = C_2 \frac{d^2 \Phi_2}{dt^2} + \frac{\Phi_2}{L_2} = C_2 \frac{dV_2}{dt} + \frac{\int V_2\, dt}{L_2} \tag{vi}$$

Equations (v) and (vi) are represented by Figure 4.10b. As in Example 4.6, we note the similarity between the nodal diagram and the f–i analogous diagram.

EXAMPLE 4.8. Set up differential equations for the rotational mechanical system shown in Figure 4.11a and draw the electrical analog from T–i analogy.

Solution The nodal representation diagram of the rotational system is given in Figure 4.11b.

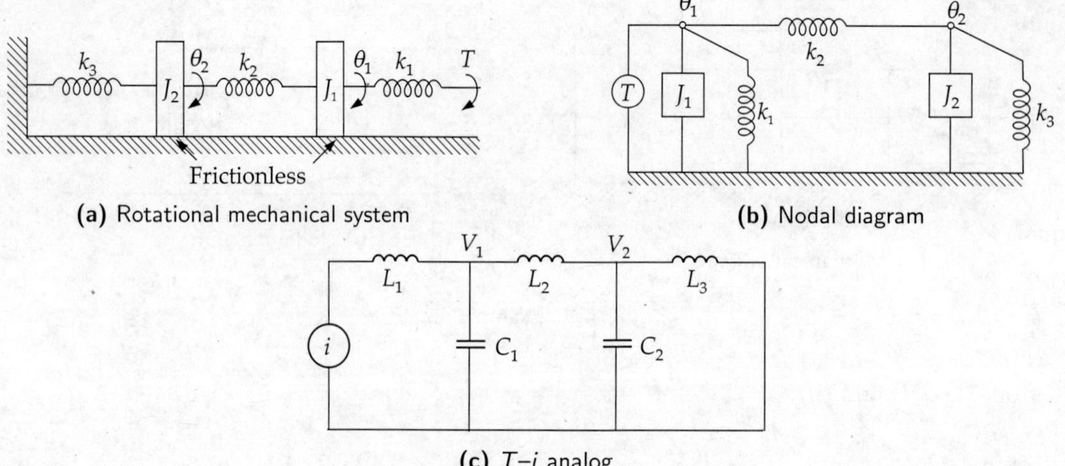

Figure 4.11: Example 4.8.

Torque-balance equations corresponding to nodes θ_1 and θ_2 are as follows:

$$T = J_1 \frac{d^2\theta_1}{dt^2} + k_1\theta_1 + k_2(\theta_1 - \theta_2) \qquad \text{(i)}$$

$$k_2(\theta_1 - \theta_2) = J_2 \frac{d^2\theta_2}{dt^2} + k_3\theta_2 \qquad \text{(ii)}$$

Substituting $T \to i$, $J \to C$, $k \to 1/L$ and $\theta \to \Phi$ from T–i analogy, we get from Eqs. (i) and (ii)

$$i = C_1 \frac{d^2\Phi_1}{dt^2} + \frac{\Phi_1}{L_1} = \frac{\Phi_1 - \Phi_2}{L_2} = C_1 \frac{dV_1}{dt} + \frac{\int V_1\, dt}{L_1} + \frac{\int (V_1 - V_2) dt}{L_2} \qquad \text{(iii)}$$

$$\frac{\int (V_1 - V_2) dt}{L_2} = C_2 \frac{d^2\Phi_2}{dt^2} + \frac{\Phi_2}{L_3} = C_2 \frac{dV_2}{dt} + \frac{\int V_2\, dt}{L_3} \qquad \text{(iv)}$$

Figure 4.11c represents Eqs. (iii) and (iv).

EXAMPLE 4.9. A rotational mechanical system is shown in Figure 4.12a. Determine system equations and hence find electrically analogous systems from (a) torque-current and (b) torque-voltage analogies.

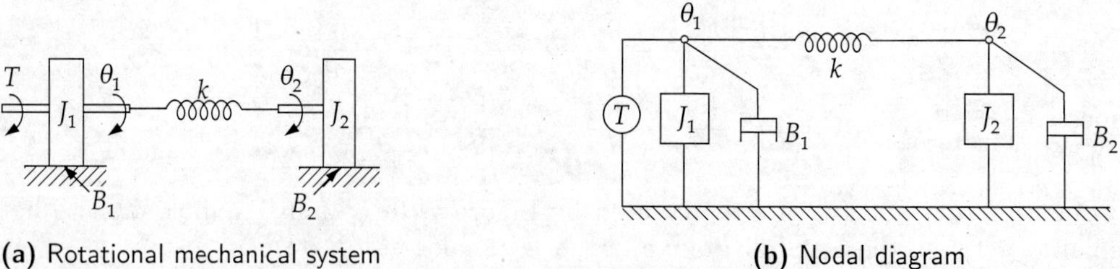

Figure 4.12: Example 4.9.

Solution The nodal representation diagram of the system is given in Figure 4.12b. Torque-balance equations corresponding to nodes q_1 and q_2 are given by

$$T = J_1 \frac{d^2\theta_1}{dt^2} + B_1 \frac{d\theta_1}{dt} + k(\theta_1 - \theta_2) \tag{i}$$

$$k(\theta_1 - \theta_2) = J_2 \frac{d^2\theta_2}{dt^2} + B_2 \frac{d\theta_2}{dt} \tag{ii}$$

Equations (i) and (ii) are the required system equations.

(a) To find electrically analogous circuit from the T–i analogy, we substitute $T \to i$, $J \to C$, $B \to 1/R$, $k \to 1/L$ and $q \to F$ in Eqs. (i) and (ii) to get

$$i = C_1 \frac{d^2\Phi_1}{dt^2} + \frac{1}{R_1}\frac{d\Phi_1}{dt} + \frac{\Phi_1 - \Phi_2}{L} = C_1 \frac{dV_1}{dt} + \frac{V_1}{R_1} + \frac{\int (V_1 - V_2)\, dt}{L} \tag{iii}$$

$$\frac{\int (V_1 - V_2)\, dt}{L} = C_2 \frac{dV_2}{dt} + \frac{V_2}{R_2} \tag{iv}$$

Equations (iii) and (iv) are represented by the circuit given in Figure 4.13a.

Figure 4.13: Analogous circuits.

(b) To draw the T–V analogous circuit, we first substitute $T \to V$, $J \to L$, $B \to R$, $k \to 1/C$ and $q \to Q$ in Eqs. (i) and (ii) to obtain

$$V = L_1 \frac{d^2Q_1}{dt^2} + R_1 \frac{dQ_1}{dt} + \frac{Q_1 - Q_2}{C} = L_1 \frac{di_1}{dt} + R_1 i_1 + \frac{\int (i_1 - i_2)\, dt}{C} \tag{v}$$

$$\frac{\int (i_1 - i_2)\, dt}{C} = L_2 \frac{di_2}{dt} + R_2 i_2 \tag{vi}$$

When translated to an electrical network, Eqs. (v) and (vi) yield what is shown in Figure 4.13b.

Mechanical Coupling

In mechanical systems, parts may be connected through gears. We consider a connection between two toothed wheels A and B of radii r_A and r_B having number of teeth n_A and n_B respectively (Figure 4.14a).

The wheels touch at the point T. Let, at any instant, the angular displacements of A and B be θ_A and θ_B. Assuming no friction loss, work done by A will equal that done by B. So, if T_A and T_B be the torques acting on them,

$$T_A\theta_A = T_B\theta_B$$

or,

$$\frac{T_A}{T_B} = \frac{\theta_B}{\theta_A} \tag{4.1}$$

(a) Gear train

(b) Motor driving a load

Figure 4.14: Mechanical coupling.

Secondly, the linear distances covered by A and B will be equal. This condition gives us

$$r_A\theta_A = r_B\theta_B$$

or,

$$\frac{r_A}{r_B} = \frac{\theta_B}{\theta_A} \tag{4.2}$$

Thirdly, because the size of teeth of both the wheels must match, their numbers must be proportional to their radii. That gives

$$\frac{r_A}{r_B} = \frac{n_A}{n_B} \tag{4.3}$$

Again, from Eq. (4.2) we get

$$\theta_B = \frac{r_A}{r_B}\theta_A \tag{4.4}$$

Therefore,

$$\frac{d\theta_B}{dt} = \frac{r_A}{r_B}\frac{d\theta_A}{dt}$$

$\Rightarrow$

$$\frac{\omega_B}{\omega_A} = \frac{r_A}{r_B} \tag{4.5}$$

where, ω_A and ω_B are angular velocities of A and B. Combining Eqs. (4.1) to (4.3) and (4.5), we get

$$\frac{T_A}{T_B} = \frac{r_A}{r_B} = \frac{n_A}{n_B} = \frac{\theta_B}{\theta_A} = \frac{\omega_B}{\omega_A} \tag{4.6}$$

Now, consider a gear system as shown in Figure 4.14b, where a load is being driven by a motor through a gear train.

If T_M = torque generated by the motor, J_A = moment of inertia of the motor shaft, D_A = viscous friction of the motor shaft and T_A = transmitted torque to the load side, we can set up the torque-balance equation for the motor shaft as

$$T_M = J_A\frac{d^2\theta_A}{dt^2} + D_A\frac{d\theta_A}{dt} + T_A \tag{4.7}$$

A similar equation for the load side can be written as

$$T_B = J_B\frac{d^2\theta_B}{dt^2} + D_B\frac{d\theta_A}{dt} + T_L \tag{4.8}$$

where J_B, D_B correspond to load inertia and viscous friction, and T_L, T_B are torques developed at the load and B. Assuming no friction loss at the gear transmission, we have from Eq. (4.6)

$$T_B = \frac{n_B}{n_A}$$

Substituting this value in Eq. (4.8), we get on rearrangement

$$T_A = \left(\frac{n_A}{n_B}\right)\left(J_B\frac{d^2\theta_B}{dt^2} + D_B\frac{d\theta_A}{dt} + T_L\right) \tag{4.9}$$

Using the relation $\dfrac{\theta_B}{\theta_A} = \dfrac{n_A}{n_B}$ of Eq. (4.6), we get from Eq. (4.9)

$$T_A = \left(\frac{n_A}{n_B}\right)^2\left(J_B\frac{d^2\theta_B}{dt^2} + D_B\frac{d\theta_A}{dt}\right) + \left(\frac{n_A}{n_B}\right)T_L$$

Substituting the value of T_A in Eq. (4.7), we get

$$\begin{aligned} T_M &= \left[J_A + \left(\frac{n_A}{n_B}\right)^2 J_B\right]\frac{d^2\theta_A}{dt^2} + \left[D_A + \left(\frac{n_A}{n_B}\right)^2 D_B\right]\frac{d\theta_A}{dt} + \left(\frac{n_A}{n_B}\right)T_L \\ &\equiv J'_A\frac{d^2\theta_A}{dt^2} + D'_A\frac{d\theta_A}{dt} + T'_L \end{aligned}$$

where

$$\begin{aligned} J'_A &= J_A + \left(\frac{n_A}{n_B}\right)^2 J_B \\ D'_A &= D_A + \left(\frac{n_A}{n_B}\right)^2 D_B \\ T'_L &= \left(\frac{n_A}{n_B}\right)T_L \end{aligned} \tag{4.10}$$

The relations given in Eq. (4.10) pertain to the primary of the gears. To arrive at similar relations for the secondary of the gears, we multiply both sides of Eq. (4.7) by (n_B/n_A) to get

$$\left(\frac{n_B}{n_A}\right)T_M = \left(\frac{n_B}{n_A}\right)J_A\frac{d^2\theta_A}{dt^2} + \left(\frac{n_B}{n_A}\right)D_A\frac{d\theta_A}{dt} + \left(\frac{n_B}{n_A}\right)T_A \tag{4.11}$$

Using $\dfrac{\theta_A}{\theta_B} = \dfrac{n_B}{n_A} = \dfrac{T_B}{T_A}$ of Eq. (4.6), we get from Eq. (4.11)

$$\left(\frac{n_B}{n_A}\right)T_M = \left(\frac{n_B}{n_A}\right)^2 J_A\frac{d^2\theta_B}{dt^2} + \left(\frac{n_B}{n_A}\right)^2 D_A\frac{d\theta_B}{dt} + T_B \tag{4.12}$$

Substituting the value of T_B from Eq. (4.8) into Eq. (4.12), we get on rearranging

$$\left(\frac{n_B}{n_A}\right)T_M = \left[J_B + \left(\frac{n_B}{n_A}\right)^2 J_A\right]\frac{d^2\theta_B}{dt^2} + \left[D_B + \left(\frac{n_B}{n_A}\right)^2 D_A\right]\frac{d\theta_B}{dt} + T_L$$

or,

$$T'_M = J'_B\frac{d^2\theta_B}{dt^2} + D'_B\frac{d\theta_B}{dt} + T_L$$

where

$$\begin{aligned} J'_B &= J_B + \left(\frac{n_B}{n_A}\right)^2 J_A \\ D'_B &= D_B + \left(\frac{n_B}{n_A}\right)^2 D_A \\ T'_M &= \left(\frac{n_B}{n_A}\right) T_M \end{aligned} \tag{4.13}$$

The three relations in Eq. (4.13) are our desired relations for the secondary of the gears. We will find application of these relations later.

4.2 CONTROL SYSTEM COMPONENTS

A general feedback control system is shown in Figure 4.15 in the block diagram form.

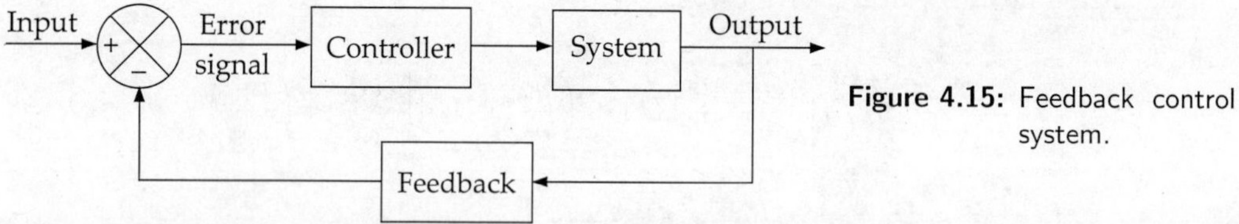

Figure 4.15: Feedback control system.

From the diagram, it may be seen that a general feedback control system consists of four components, namely

1. System to be controlled (or Plant)
2. Feedback element (or Transducer)
3. Error signal generator
4. Control signal generator

In the following sections we will consider a few representative devices for the first three of these components. Control signal generators are mostly electronic devices, the discussion of which is beyond the scope of this text.

System to be Controlled

Servomotor. Servomotors are used in control systems to generate a high acceleration from a stationary condition. Depending on their input or control, these motors can be classified as shown in Table 4.4.

Current type	*Motor type*
dc	Armature-controlled Field-controlled
ac	

Table 4.4: Servomotor classification

dc servomotor. First we consider dc servomotors as control system components. As already told, they are of two types.

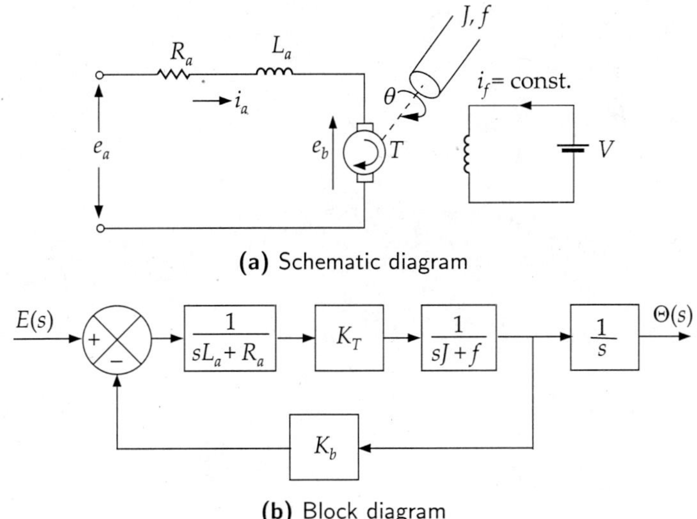

(a) Schematic diagram

(b) Block diagram

Figure 4.16: Armature-controlled dc motor.

1. Armature-controlled dc servomotor. The schematic diagram of an armature-controlled dc motor is shown in Figure 4.16a.

Here, R_a, L_a, i_a = armature resistance, inductance and current, respectively
i_f = field current
e_a, e_b = applied voltage and back emf, respectively
J, f = moment of inertia and coefficient of friction of the motor shaft and load
T, θ = torque generated by the motor and angular deflection of the motor shaft
V = voltage applied in the field coil

Air-gap flux Φ is proportional to the field current. Therefore,

$$\Phi = k_f i_f \tag{4.14}$$

where k_f is a constant. The torque is proportional to the flux and armature current. Therefore,

$$T = k_t \Phi i_a = k_t k_f i_f i_a \qquad \text{[using Eq. (4.14)]}$$
$$= K_T i_a \tag{4.15}$$

where $K_T = k_t k_f i_f$, i_f being constant. The back emf is proportional to the rotational speed. Hence,

$$e_b = K_b \frac{d\theta}{dt} \tag{4.16}$$

The Laplace transform of Eq. (4.16) is

$$E_b(s) = K_b s \Theta(s) \tag{4.17}$$

Applying KVL to the armature circuit, we get

$$e_a = i_a R_a + L_a \frac{di_a}{dt} + e_b$$

From its Laplace transform we get

$$E_a(s) - E_b(s) = (sL_a + R_a) I_a(s) \tag{4.18}$$

The load-torque equation is

$$J\frac{d^2\theta}{dt^2} + f\frac{d\theta}{dt} = T = K_T i_a \qquad \text{[using Eq. (4.15)]} \tag{4.19}$$

The Laplace transform of Eq. (4.19) is

$$(s^2 J + sf)\Theta(s) = K_T I_a(s) \tag{4.20}$$

Eliminating $I_a(s)$ from Eq. (4.20) with the help of Eq. (4.18), we get

$$\begin{aligned}(s^2 J + sf)\Theta(s) &= \left[\frac{E_a(s) - E_b(s)}{sL_a + R_a}\right] K_T \\ &= \left[\frac{E_a(s) - sK_b\Theta(s)}{sL_a + R_a}\right] K_T \qquad \text{[using Eq. (4.17)]}\end{aligned} \tag{4.21}$$

On rearranging Eq. (4.21), we obtain

$$[(s^2 J + sf)(sL_a + R_a) + sK_b K_T]\Theta(s) = K_T E_a(s)$$

which, on further rearrangement, yields the transfer function

$$G(s) = \frac{\Theta(s)}{E_a(s)} = \frac{K_T}{s[(sJ + f)(sL_a + R_a) + K_b K_T]} \tag{4.22}$$

To draw a block diagram of the system, we observe that the transfer function can be written as

$$G(s) = \frac{\dfrac{K_T}{(sJ + f)(sL_a + R_a)}}{1 + \dfrac{K_T K_b}{(sJ + f)(sL_a + R_a)}} \cdot \frac{1}{s}$$

It is now clear that the block diagram should look like Figure 4.16b. Obviously, it is a *closed-loop system.*

If the armature inductance L_a is small, we get from Eq. (4.22)

$$\begin{aligned}G(s) &= \frac{\Theta(s)}{E_a(s)} = \frac{K_T}{s[(sJ + f)R_a + K_b K_T]} = \frac{K_T/R_a}{s^2 J + s\left(f + \dfrac{K_b K_t}{R_a}\right)} \equiv \frac{K_T/R_a}{s^2 J + sf'} \\ &\equiv \frac{K}{s(s\tau + 1)}\end{aligned}$$

where $f' = f + \dfrac{K_b K_t}{R_a}$ = effective coefficient of friction

$K = \dfrac{K_T}{R_a f'}$ = motor gain

$\tau = \dfrac{1}{f'}$ = motor time-constant

The expression for f' shows that the back emf effectively increases the friction of the motor.

2. Field-controlled dc motor. A schematic diagram of the field-controlled dc motor is shown in Figure 4.17a.

(a) Schematic diagram

(b) Block diagram

Figure 4.17: Field-controlled dc motor.

Here $R_f,\ L_f$ = field winding resistance and inductance
$i_a,\ i_f$ = armature current (constant) and field current
$e_b,\ e_f$ = back emf and field voltage
$J,\ f$ = moment of inertia and coefficient of friction of the motor shaft and load
$T,\ \theta$ = torque generated by the motor and angular displacement of the motor shaft

In contrast to the armature-controlled motor, here the armature current is constant while the field current is variable. Therefore, Eq. (4.15) can be rewritten as

$$T = K_T' i_f$$

Applying KVL to the field circuit, we get

$$L_f \frac{di_f}{dt} + R_f i_f = e_f \tag{4.23}$$

The Laplace transform of Eq. (4.23), with initial conditions = 0, is

$$(sL_f + R_f)I_f(s) = E_f(s) \tag{4.24}$$

The load torque equation and its Laplace transform are

$$J\frac{d^2\theta}{dt^2} + f\frac{d\theta}{dt} = T = K_T' i_f$$

$$(s^2 J + sf)\Theta(s) = K_t' I_f(s) \tag{4.25}$$

Eliminating $I_f(s)$ from Eq. (4.24) with the help of Eq. (4.25), we get on rearranging

$$\frac{\Theta(s)}{E_f(s)} = \frac{K_T'}{s(sJ + f)(sL_f + R_f)} = \frac{K_T'/(R_f f)}{s\left(s\dfrac{J}{f} + 1\right)\left(s\dfrac{L_f}{R_f} + 1\right)}$$

or,

$$G(s) = \frac{K'}{s(s\tau_m + 1)(s\tau_f + 1)}$$

where $K' = \dfrac{K_T'}{R_f f}$ = motor gain

$\tau_m = \dfrac{J}{f}$ = motor time-constant

$\tau_f = \dfrac{L_f}{R_f}$ = field circuit time-constant

The block diagram for the field-controlled servomotor is given in Figure 4.17b. It may be observed from the block diagram that the field-controlled servomotor is an *open-loop system*.

The ac servomotor. An ac servomotor is basically a 2-phase induction motor with low X/R ratio (X = reactance), thus making the torque-speed characteristic nearly linear. The comparison of torque-speed characteristics between an ordinary two-phase induction motor and a servomotor is shown in Figure 4.18a.

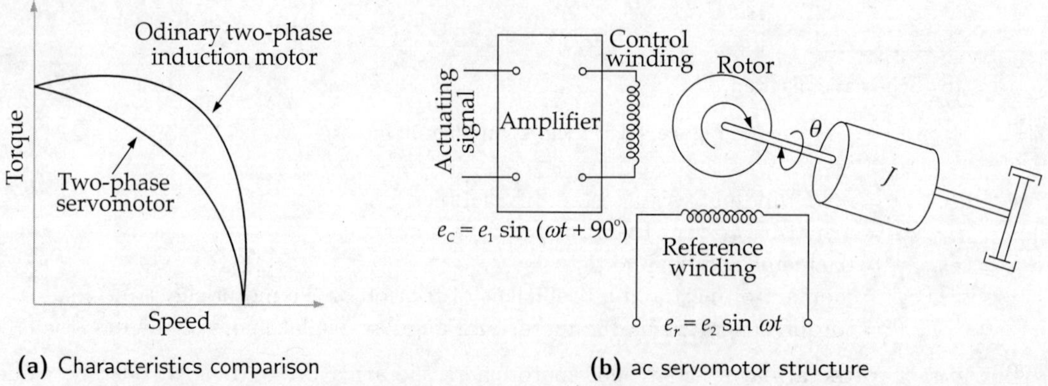

(a) Characteristics comparison (b) ac servomotor structure

Figure 4.18: ac servomotor: torque-speed characteristics and structure.

An ac servomotor is used for low-power, high acceleration applications. A schematic diagram of an ac servomotor is shown in Figure 4.18b.

The two-phase motor has two windings at right angles to each other. While the reference winding is given a steady voltage, the amplitude of the control winding voltage varies according to the actuating signal. But the latter voltage always remains 90° out of phase with the former (see Figure 4.19a). This can be achieved by feeding voltage to the control winding through a capacitor.

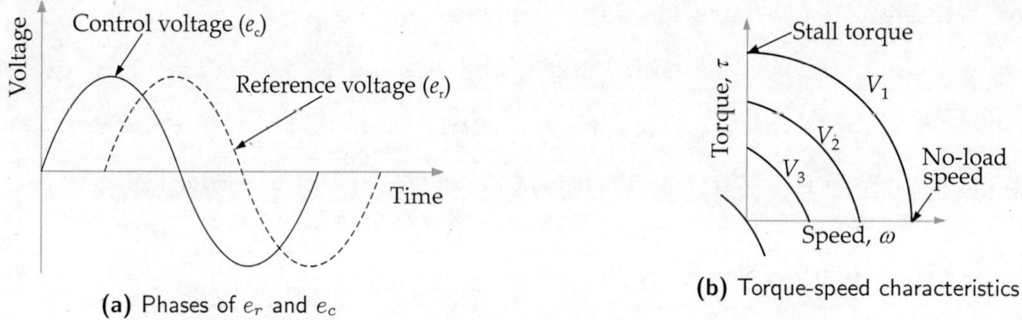

(a) Phases of e_r and e_c (b) Torque-speed characteristics

Figure 4.19: Emf phases and torque-speed of ac servomotor.

This variation in phase of voltage between the two mutually perpendicular windings generates a rotational motion of the rotor owing to the rotation of the resultant magnetic flux (see Figure 4.20).

The torque-speed characteristics, as a function of the control voltage, are shown in Figure 4.19b. The torque corresponding to $\omega = 0$ (i.e. no speed) is called *stall torque*, while the speed corresponding to $\tau = 0$ (i.e. no torque) is called *no-load speed.*

Assuming a linear relationship between torque and speed, we can write

$$\tau = m\omega + c \tag{4.26}$$

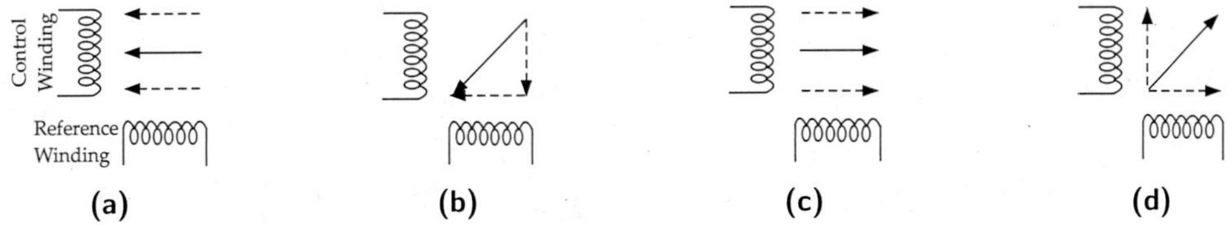

Figure 4.20: Rotation of the resultant magnetic flux. Individual magnetic vector:- - ->; Resultant magnetic vector: —>. a, b, c, d show different phases of rotation.

At $\omega = 0$, let $\tau = \tau_0$. Then

$$\tau_0 = c = Ke_c \tag{4.27}$$

where, e_c is the control voltage and K is a constant. So

$$K = \frac{\tau_0}{e_c}$$

Again, at $\tau = 0$, let $\omega = \omega_0$. Therefore,

$$m = -\frac{c}{\omega_0} = -\frac{\tau_0}{\omega_0}$$

Thus, Eq. (4.26) can be rewritten as

$$\tau = -\frac{\tau_0}{\omega_0}\omega + Ke_c \tag{4.28}$$

Its Laplace transform yields

$$T(s) = -D\Omega(s) + KE_c(s) \tag{4.29}$$

where, $D\left(= \frac{\tau_0}{\omega_0}\right)$ is called the *damping factor*[3]. Also, $\omega = \frac{d\theta}{dt}$. So, its Laplace transform gives $\Omega(s) = s\Theta(s)$. Substituting this value in Eq. (4.29), we get

$$T(s) = -sD\Theta(s) + KE_c(s) \tag{4.30}$$

From mechanical considerations, we obtain

$$\tau = J\frac{d^2\theta}{dt^2} + D\frac{d\theta}{dt} \tag{4.31}$$

The Laplace transform of Eq. (4.31) gives

$$T(s) = (s^2J + sD)\Theta(s) \tag{4.32}$$

From Eqs. (4.30) and (4.32), we get after a little rearrangement

$$s(sJ + 2D)\Theta(s) = KE_c(s)$$

$$\frac{\Theta(s)}{E_c(s)} = \frac{K}{s(sJ + 2D)} = \frac{K/(2D)}{s\left(s\frac{J}{2D} + 1\right)} \tag{4.33}$$

or,

$$G(s) = \frac{K_m}{s(s\tau_t + 1)} \tag{4.34}$$

[3]aka coefficient of viscous friction.

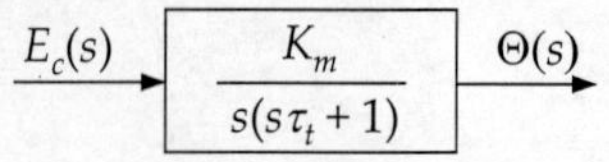

$$E_c(s) \rightarrow \boxed{\frac{K_m}{s(s\tau_t+1)}} \rightarrow \Theta(s)$$

Figure 4.21: Block diagram of an ac servomotor.

where, $K_m = \dfrac{K}{2D}$ = motor gain constant and $\tau_t = \dfrac{J}{2D}$ = motor time-constant.

The block diagram of an ac servomotor is given in Figure 4.21. We note that this is also an *open-loop system* like the dc field-controlled servomotor.

EXAMPLE 4.10. An ac servomotor has its both windings excited with 115 V ac. Its stall torque is 2 lb-ft and the coefficient of viscous friction = 0.2 lb-ft-s.

(a) Determine its no load speed.

(b) If it is connected to a load of 0.8 lb-ft (the coefficient of viscous friction = 0.05 lb-ft-s) through a gear pass with a ratio of 4, calculate the speed at which the motor will run.

Solution

(a) From Eq. (4.28) $\tau = -D\omega + Ke_c$. For $\tau = 0$, $\omega = \omega_0$. Therefore, no-load speed is

$$\omega_0 = \frac{Ke_c}{D} = \frac{\tau_0}{D} = \frac{2}{0.2} = 10 \text{ rad/s}$$

(b) When N = gear ratio and D_L = coefficient of viscous friction due to load, the total viscous friction D_T is [see Eq. (4.10)]

$$D_T = D + N^2 D_L = 0.2 + 4^2 \times 0.05 = 1.0 \text{ lb-ft-s}$$

Therefore from $\tau_L = Ke_c - D_T\omega = \tau_0 - D_T\omega$, we get

$$\omega = \frac{\tau_0 - \tau_L}{D_T} = \frac{2 - 0.8}{1} = 1.2 \text{ rad/s}$$

Stepper motor. Stepper motors move in steps in response to input pulses, i.e. they are actuators in incremental motion control systems. They are widely used in digitally controlled machines, such as, computer printers, disc- and tape-drives, etc. Generally, such motors are of two types:

- Variable reluctance type
- Permanent magnet type

We will consider only the first type here.

In a variable reluctance stepper motor, the stator and rotor are made of one or more stacks. Stators are fixed to a common frame while the rotors have a common shaft [see Figure 4.22a]. Both stators and rotors have a toothed structure [see Figure 4.22b]. Stators are excited by pulses while rotors are unexcited.

If n_t is the number of teeth of the rotor and θ is the angle of misalignment between the rotor and stator teeth, then there are two positions

(i) $\theta = 0°$

(ii) $\theta = \pm\dfrac{360°}{2n_t} = \pm\dfrac{180°}{n_t}$

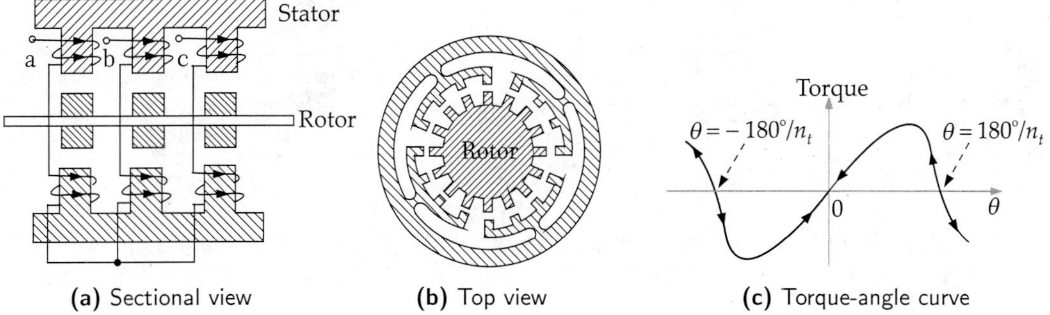

(a) Sectional view **(b)** Top view **(c)** Torque-angle curve

Figure 4.22: Stepper motor.

when no torque will be acting on the rotor. Of these two positions, the first constitutes a stable equilibrium, i.e. a small disturbance from here will bring the rotor back to this position. The second position, on the other hand, is unstable meaning thereby that a small disturbance form here makes the rotor move away from this position (see Figure 4.22c for the torque vs. angle curve). As a consequence, the rotor locks into $\theta = 0°$ position.

The construction of the motor is such that rotor teeth of all stacks are perfectly aligned, but stator teeth of stacks are displaced by an angle of

$$\phi = \frac{360°}{Nn_t}$$

where N is the number of stacks.

Consider a 3-stack 24-teeth motor where the stack x is aligned with the rotor. Then the angle of displacement of stacks y and z is

$$\phi = \frac{360°}{3 \times 24} = 5°$$

Now, if the y stator is excited by a pulse, the rotor will move 5° in one direction while a pulse excitation of the z stator will make the rotor move by the same angle in the opposite direction. Thus, motions corresponding to pulse trains of $yzxyz$ and $zyxzy$ will be opposite to each other. Care should be taken to time the pulses so that the rotor finds enough time to lock itself to each step rather than skipping any.

Feedback Element: Tachometer

Tachometers[4] are used to measure rotational speed. These basically miniature low voltage generators are often used as feedback elements. They can be of two types—dc and ac.

The dc tachometer. In an ordinary generator, the rotor has inertia. But a tachometer rotor should have minimum inertia. To achieve this end, moving coil tachometers use iron-less rotors. The rotor along with the armature winding is shaped as a cup and is hung between the permanent magnet poles and the inner iron structure like a cantilever. Figure 4.23 gives a diagrammatic presentation of a dc tachometer.

The generated voltage is given by the familiar Faraday relation

$$e(t) \propto \frac{d\Phi}{dt}$$

[4]The Greek word *takhos* means speed.

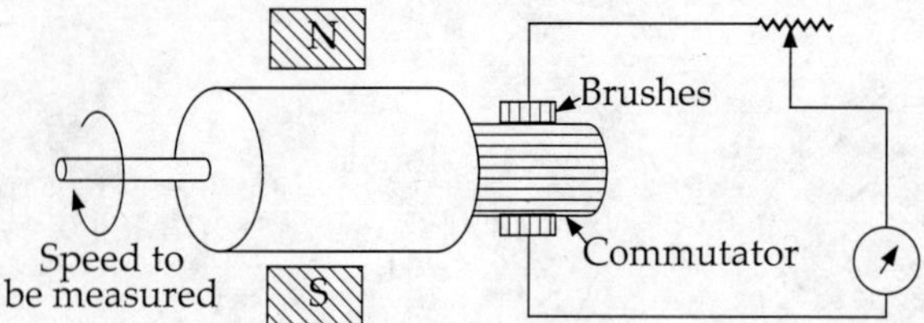

Figure 4.23: Schematic diagram of a dc tachometer.

where Φ is the flux linkage. Since the change in flux linkage varies with the angular velocity $\omega(t)$ of the rotor

$$e(t) = K\omega(t) \tag{4.35}$$

Taking Laplace transform of Eq. (4.35), we get the following expression for the transfer function of the dc tachometer

$$\frac{E(s)}{\Omega(s)} = K$$

The unit of K is V/krpm. Typically, K value is $\sim$ 6 V/krpm. As in dc generators, the ac output is converted to dc by a commutator.

The ac tachometer. Structurally, an ac tachometer resembles a two-phase induction motor. It consists of two stator windings arranged in space quadrature along with a rotor which is just an aluminium cup that rotates in an air-gap that contains magnetic flux. Figure 4.24 gives a diagrammatic representation of the ac tachometer.

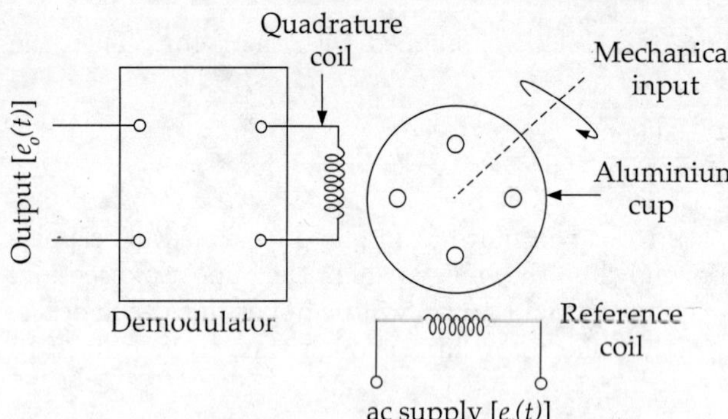

Figure 4.24: Schematic diagram of an ac tachometer.

A sinusoidal voltage, given by

$$e_r(t) = E_r \sin \omega_c t$$

is applied to the reference winding. No emf is induced in the output winding when the rotor is stationary. However, as the rotor turns, a voltage of frequency ω_c is induced in the output winding. The amplitude of this voltage is proportional to the speed of rotation. In case the direction of rotation changes, the output voltage suffers a 180° phase shift.

To filter out the carrier frequency, a phase-sensitive demodulator can be used in the output which has the form of

$$e_o(t) = e(t) \sin \omega_c t$$

The forms of the modulated signal and the final filtered output are shown in Figures 4.25a and 4.25c.

The transfer function of the ac tachometer, therefore, has the same form as that of a dc tachometer.

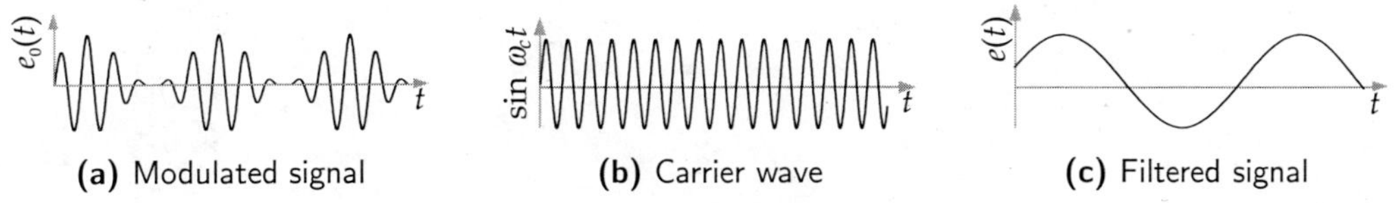

(a) Modulated signal **(b)** Carrier wave **(c)** Filtered signal

Figure 4.25: $e_o(t)$ and $e(t)$ of ac tachometer.

EXAMPLE 4.11. A tachometer has a gain of 0.05 V/(rad/s). Determine the following:

(a) The output when the shaft-speed is 20 rad/s.

(b) The shaft-speed in rad/s when the output is 1.8 V.

Solution

(a) $e(t) = K\omega(t) = 0.05 \times 20 \text{ V} = 1 \text{ V}$

(b) $\omega(t) = \dfrac{e(t)}{K} = \dfrac{1.8}{0.05} = 36 \text{ rad/s}$

EXAMPLE 4.12. Figure 4.26a represents a closed-loop control system for regulating the speed of a field-controlled dc motor. Determine the reference voltage if the speed is to be maintained at 150 rpm.

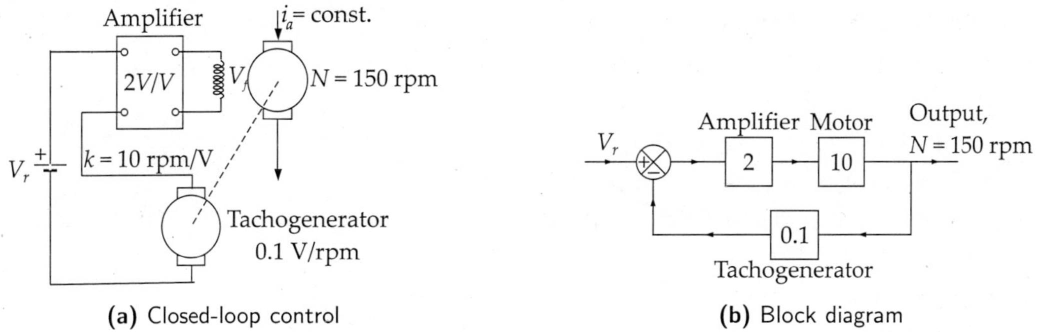

(a) Closed-loop control **(b)** Block diagram

Figure 4.26: Example 4.12.

Solution The block diagram of the system looks like that given in Figure 4.26b. From the block diagram, we get

$$\frac{N}{V_r} = \frac{2 \times 10}{1 + 2 \times 10 \times 0.1} = \frac{20}{3}$$

Therefore,

$$V_r = \frac{3N}{20} = \frac{3 \times 160}{20} = 22.5 \text{ V}$$

Error Signal Generator

The error signal generator can be of many kinds. We consider here only two simple such devices, namely, potentiometer and synchro.

Potentiometer. The potentiometer i s a simple device that converts liner or angular displacement into. voltage. .A fixed voltage is applied between two terminals of a fixed resistor in the

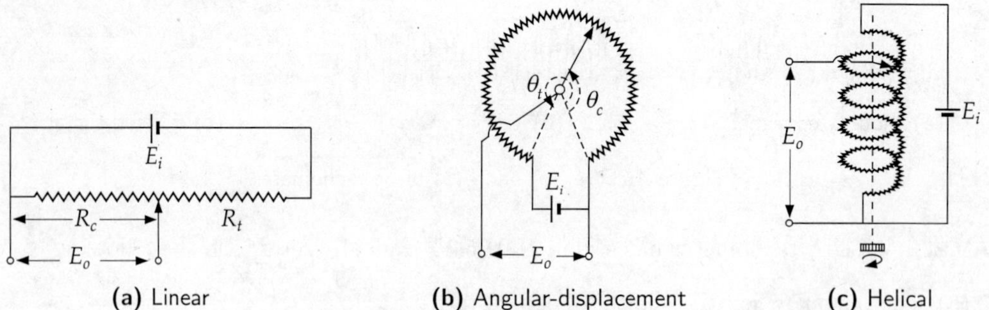

(a) Linear **(b)** Angular-displacement **(c)** Helical

Figure 4.27: Potentiometers.

form of a wound wire. A third terminal, called *wiper*, moves over this resistance and generates a variable voltage (see Figure 4.27). It is easy to see from the Figures 4.27a and 4.27b that the following relations hold good for linear and angular potentiometers:

Linear: $$\frac{R_c}{R_t} = \frac{E_o}{E_i}$$

Angular: $$\frac{\theta_c}{\theta_t} = \frac{E_o}{E_i}$$

where the subscripts c, t, o and i indicate control, total, output and input, respectively. For an N-turn angular potentiometer, $\theta_t = 2\pi N$.

EXAMPLE 4.13. A 3-turn, 100 kΩ potentiometer with 1% linearity uses a 30 V supply. Find the following:

(a) The potentiometer constant in V/turn.
(b) The range of voltages at mid-point setting.
(c) The voltage at the mid-point when the potentiometer is loaded with 500 kΩ (assume, the potentiometer is perfectly linear).

Solution

(a) The potentiometer constant K_P is given by

$$K_P = \frac{30}{3} = 10 \text{ V/turn}$$

(b) For the mid-point setting, the resistance is 50 kΩ. Therefore,

$$\text{Voltage at the mid-point } = 10 \times 1.5 \text{ (turn)} = 15 \text{ V}$$

Given, linearity = 1%. Therefore,

$$\text{Voltage error } = \pm 30 \times 0.01 \text{ V} = \pm 0.3 \text{ V}$$

Thus, $$\text{Voltage range } = 15 \pm 0.3 \text{ V} = 14.7 \text{ to } 15.3 \text{ V}$$

(c) We get from Figure 4.28a

$$R = 50 \text{ k}\Omega \parallel 500 \text{ k}\Omega = \frac{50 \times 500}{50 + 500} = 45.45 \text{ k}\Omega$$

By voltage division,

$$\text{Voltage } = \frac{30}{50 + 45.45} \times 45.45 \text{ V} = 14.3 \text{ V}$$

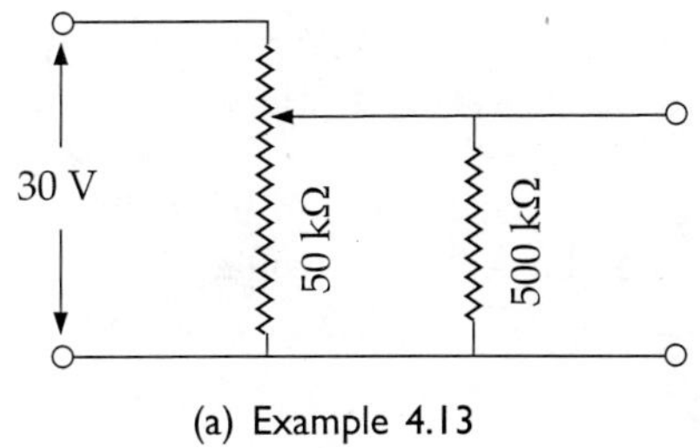

(a) Example 4.13

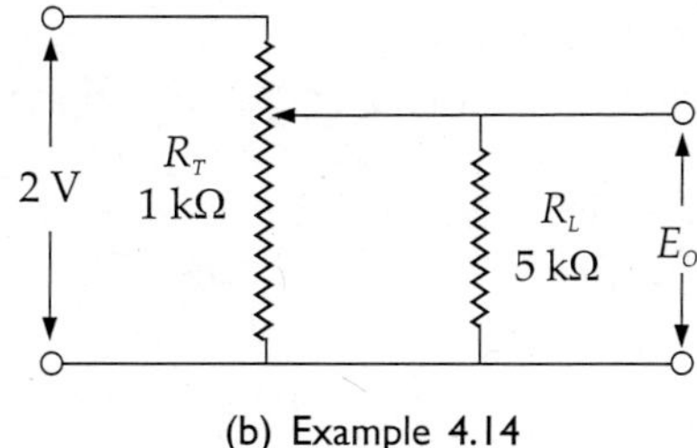

(b) Example 4.14

Figure 4.28: Potentiometers.

EXAMPLE 4.14. (a) Calculate the output voltage E_o if the excitation voltage is 2 V and the setting ratio is 0.4 (see Figure 4.28b).

(b) Also calculate the loading error.

(c) If the potentiometer is a multi-turn unit having a shaft rotation of 1080°, find the positional error associated with loading.

Solution (a) Let α = setting ratio. The resistance due to the load

$$R = (1-\alpha)R_T + (\alpha R_T \parallel R_L) = 0.6 + \frac{(0.4)(1)(5)}{(0.4)(1)+5} = 0.97 \text{ k}\Omega$$

Therefore
$$E_o = \frac{(\alpha R_T \parallel R_L)}{R} E_i = \frac{0.37}{0.97} \times 2 = 0.763 \text{ V}$$

(b) The error due to loading is given by

$$\text{Error } = \alpha E_i - E_o = (0.4)(2) - 0.763 = 0.037 \text{ V}$$

(c) Positional error $= \dfrac{\text{maximum angle of rotation}}{\text{applied voltage}} \times \text{error} = \dfrac{1080°}{2} \times 0.037 = 20°.$

EXAMPLE 4.15. A helical 5-turn potentiometer has a resistance of 50 kΩ and 10,000 winding turns. Find the linearity when the measured resistance at its mid-point is 25.5 kΩ. Find the resolution of the potentiometer in V/turn and potentiometer constant in V/rad if 100 V is applied to it.

Solution The mid-point should have a resistance of 25 kΩ. But the measured value is 25.5 kΩ. So, deviation is 0.5 kΩ. Therefore,

$$\text{Linearity} = \frac{\text{Deviation}}{\text{Desired value}} \times 100 = \frac{0.5}{25} \times 100 = 2\%$$

$$\text{Resolution} = \frac{100}{10000} = 10 \text{ mV/turn}$$

$$\text{Potentiometer constant} = \frac{E_i}{2\pi N} = \frac{100}{2\pi \times 5} = 3.183 \text{ V/rad}$$

Synchro. Basically electromechanical devices, based on variation of reluctance, synchros mea-sure angles or perform functions related to angle measurement such as remote control of angle or computation of rectangular component of vectors. They are widely used as detectors owing to their ruggedness in construction and reliability of performance. Though they find various applications, two of their uses (i) as transmitter and (ii) as control transformer are of interest to us here.

Synchro transmitter. A synchro transmitter consists of a wound rotor on a dumbbell shaped magnetic substance and a wound stator which are arranged concentrically so that the motion of the rotor produces a variable mutual inductance between the two windings [Figure 4.29a]. The rotor is laminated, has a single-phase winding, and is connected to supply lines through precision slip-rings, while the stator winding is three-phase, the phases being displaced by 120° in space. Let the rotor of the synchro transmitter be excited by an ac voltage e_r given by

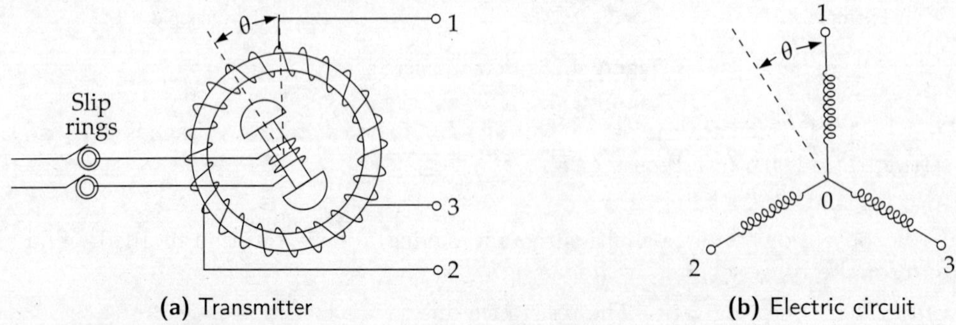

Figure 4.29: Schematic diagram of a synchro transmitter and its electric circuit.

$$e_r = V_r \sin \omega t$$

At the *electric zero*, i.e. when the angle θ between the rotor and stator winding 1 equals 0°, the voltage induced between the terminal 1 and the neutral 0 [see Figure 4.29b] is given by

$$e_{10}(t) = AV_r \sin \omega t$$

where A is a constant. Then the voltages induced between terminals (2 and 0) and (3 and 0) are

$$e_{20}(t) = AV_r \sin \omega t \cos 120° = -\frac{1}{2}AV_r \sin \omega t$$

$$e_{30}(t) = AV_r \sin \omega t \cos 240° = -\frac{1}{2}AV_r \sin \omega t$$

Consequently, the voltages developed between the terminals will be

$$e_{12}(t) = e_{10}(t) - e_{20}(t) = \frac{3}{2}AV_r \sin \omega t \tag{4.36}$$

$$e_{23}(t) = e_{20}(t) - e_{30}(t) = 0 \tag{4.37}$$

$$e_{31}(t) = e_{30}(t) - e_{10}(t) = -\frac{3}{2}AV_r \sin \omega t \tag{4.38}$$

Voltages, shown by Eqs. (4.36) to (4.38), demonstrate that despite having its three-phase construction, only single-phase voltages develop in the terminals.

Now, we consider the case when the rotor subtends an angle θ with respect to the electric zero configuration. Then voltages developed in each stator coil with respect to the neutral 0 are

$$e_{10}(t) = AV_r \sin \omega t \cos \theta$$

$$e_{20}(t) = AV_r \sin \omega t \cos(120° - \theta)$$

$$e_{30}(t) = AV_r \sin \omega t \cos(240° - \theta)$$

Therefore, the voltages developed between terminals are

$$\begin{aligned}
e_{12}(t) &= e_{10}(t) - e_{20}(t) = AV_r \sin\omega t[\cos\theta - \cos(120^\circ - \theta)] \\
&= AV_r \sin\omega t \left[\cos\theta + \left(\frac{1}{2}\cos\theta - \frac{\sqrt{3}}{2}\sin\theta\right)\right] \\
&= AV_r \sin\omega t \left[\frac{3}{2}\cos\theta - \frac{\sqrt{3}}{2}\sin\theta\right] \\
&= \sqrt{3}AV_r \sin\omega t \sin(\theta + 120^\circ) \\
e_{23}(t) &= e_{20}(t) - e_{30}(t) = AV_r \sin\omega t[\cos(120^\circ - \theta) - \cos(240^\circ - \theta)] \\
&= AV_r \sin\omega t \left[-\frac{1}{2}\cos\theta + \frac{\sqrt{3}}{2}\sin\theta + \frac{1}{2}\cos\theta + \frac{\sqrt{3}}{2}\sin\theta\right] \\
&= \sqrt{3}AV_r \sin\omega t \sin\theta \\
e_{31}(t) &= e_{30}(t) - e_{10}(t) = AV_r \sin\omega t[\cos(240^\circ - \theta) - \cos\theta] \\
&= AV_r \sin\omega t \left[\left(-\frac{1}{2}\cos\theta - \frac{\sqrt{3}}{2}\sin\theta\right) - \cos\theta\right] \\
&= AV_r \sin\omega t \left[-\frac{3}{2}\cos\theta - \frac{\sqrt{3}}{2}\sin\theta\right] \\
&= \sqrt{3}AV_r \sin\omega t \sin(\theta + 240^\circ)
\end{aligned}$$

Figure 4.30 is a plot of these voltages vs. the rotor position. We observe, for each angular position of the rotor, the set of voltages generated by the three terminals is unique. So, a measurement of this set of voltages determines the angular position of the rotor.

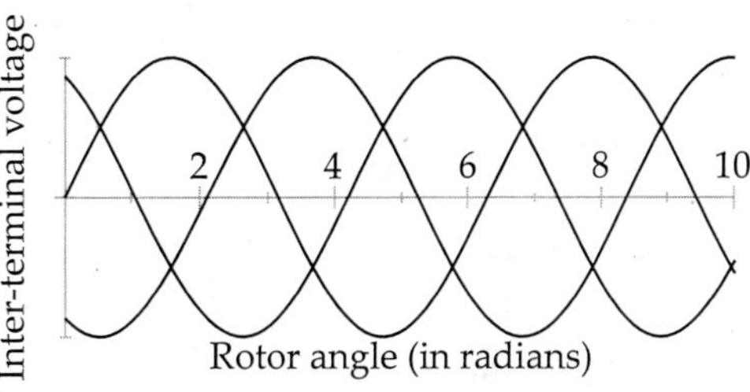

Figure 4.30: Inter-terminal voltages vs. rotor angular position plot for the synchro transmitter.

Synchro control transformer. A synchro control transformer is used to detect an error be-tween two shaft positions and convert it to an electric signal. Since a comparison between two is involved, synchro control transformer is used in conjunction with a synchro transmitter. Their constructions are mostly similar except that the control transformer rotor is of cylindrical shape [see Figure 4.31a]. This is necessary to maintain a uniform flux in the air-gap when the rotor rotates. The uniform air-gap flux, in turn, provides a uniform impedance for the rotor output terminals which are often connected to a control device through an amplifier. The idea is that the amplifier should see a constant input impedance.

In the error detection operation, the output of stator windings of the sychro transmitter is connected to that of the synchro control transformer, thus ensuring that the same current flows through the circuits [see Figure 4.31b]. The voltage induced between rotor terminals of the control transformer is

$$e_0(t) = K_c V_r \sin\omega t \cos\phi \tag{4.39}$$

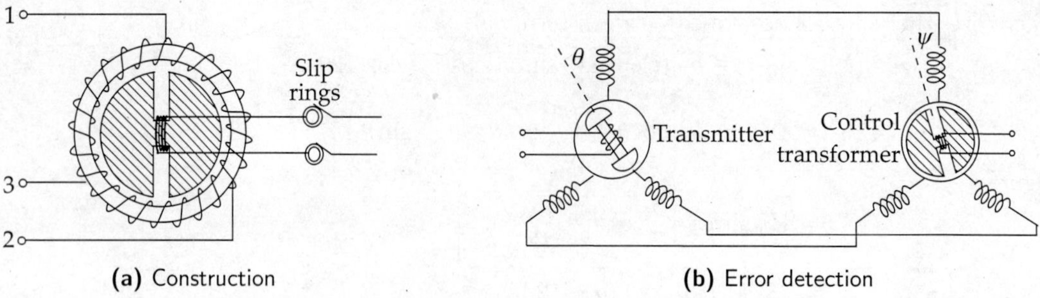

(a) Construction (b) Error detection

Figure 4.31: Synchro control transformer.

where e_0 = error voltage, K_c = sensitivity of the control transformer and ϕ = angle between the two rotors.

When $\phi = 90°$, the error voltage is zero. This position is known as the *electrical zero* for the control transformer. Consider the case when θ is the transmitter rotor angle and ψ is the control transformer rotor angle. Then,

$$\phi = 90° - (\theta - \psi)$$

Substituting this value in Eq. (4.39), the error voltage is

$$e_0(t) = K_c V_r \sin \omega t \sin(\theta - \psi) \tag{4.40}$$

Equation (4.40) shows that the error voltage generated by the rotor of the control transformer is not linear with respect to the relative angular positions of the two shafts because it is a sine function of the same. However, since a sine function remains almost linear within 0° to 15° angle variation, for small relative angular displacements between the two shafts

$$e_0(t) = K_c V_r (\theta - \psi) \sin \omega t$$

This situation suits well as far as error detection is concerned.

4.3 A FEW ELEMENTARY SYSTEMS

Now that we have discussed mathematical modelling of physical systems and control system components, we may discuss a few elementary control systems, namely

1. Servomechanism
2. Temperature control
3. Liquid level control

Servomechanism

By servomechanism or simply *servo*, we mean control of position—be it linear or angular—and its derivative, speed. The former is known as *positional servomechanism* and the latter, *rate servomechanism*.

Positional servomechanism. A simple position control system is shown in Figure 4.32. Here, the input indicates the desired position of the load. The input and feedback constitute a bridge in which if the input is changed, an error voltage e_E appears at the amplifier input. The amplifier output drives the motor to make the feedback pointer assume an identical position on the potentiometer in order that $e_E = 0$.

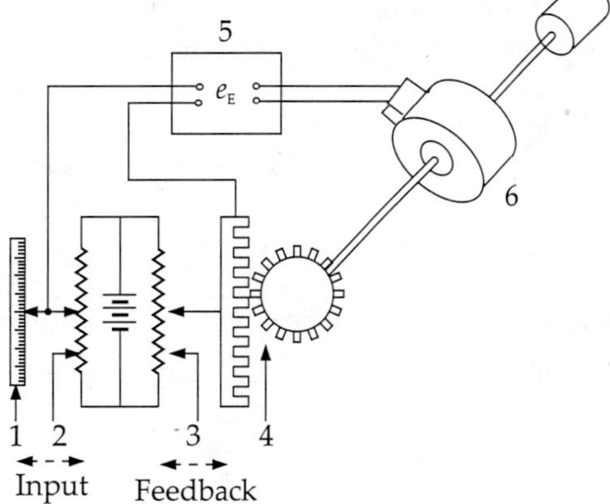

Figure 4.32: Position control system. 1: scale, 2,3: potentiometer, 4: gear train, 5: amplifier, 6: servomotor.

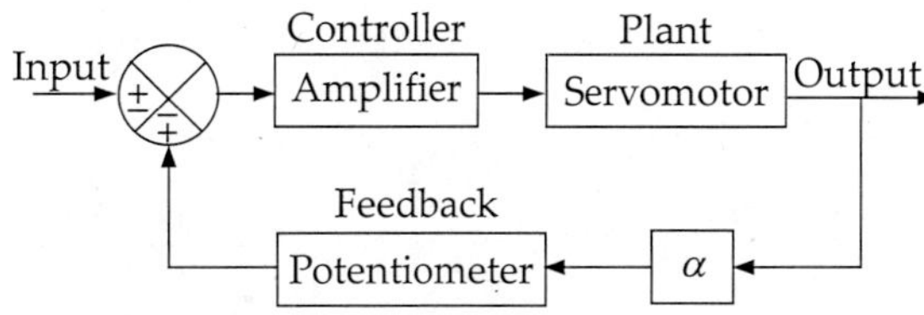

Figure 4.33: Block diagram of position control system.

In the block diagram form it looks like that shown in Figure 4.33.

Let K_p, K_a, K_m = potentiometer constant, amplifier gain and motor gain

$J_m,\ D_m$ = moment of inertia and viscous friction of the motor shaft and the load

$n_m,\ n_p$ = number of teeth of the motor gear and the potentiometer

$J_f,\ D_f$ = moment of inertia and viscous friction of the feedback gear train

$\theta_r(t),\ \theta_m(t)$ = angular position of the reference and the motor shaft

$e_r(t)$ = input reference voltage corresponding to the marked position of the load $= K_p\theta_r(t)$

$e_f(t)$ = feedback voltage corresponding to feedback gear position $= K_p\theta_m(t)$

Then, the error signal

$$e(t) = e_r(t) - e_f(t) = K_p\theta_r(t) - K_p\theta_m(t) \tag{4.41}$$

The Laplace transform of Eq. (4.41) is

$$E(s) = K_p[\Theta_r(s) - \Theta_m(s)]$$

The voltage input $e_c(t)$ to the motor is

$$e_c(t) = K_a e(t) = K_a K_p[\theta_r(t) - \theta_m(t)] \tag{4.42}$$

The Laplace transform of Eq. (4.42) is

$$E_c(s) = K_a K_p[\Theta_r(s) - \Theta_m(s)] \tag{4.43}$$

The gear train ratio α to the feedback is

$$\alpha = \frac{n_m}{n_p}$$

We saw in Eq. (4.10) that the reflected inertia and viscous friction of the motor due to feedback gear train can be written as

$$J = J_m + \alpha^2 J_f$$
$$D = D_m + \alpha^2 D_f$$

Also, we know from Eq. (4.33)

$$\frac{\Theta_m(s)}{E_c(s)} = \frac{K_m}{s(sJ + 2D)} \tag{4.44}$$

Combining Eqs. (4.43) and (4.44), we get

$$\begin{aligned}\Theta_m(s) &= \frac{K_m}{s(sJ + 2D)} E_c(s) = \frac{K_m K_a K_p}{s(sJ + 2D)} [\Theta_r(s) - \Theta_m(s)] \\ &= \frac{K\Theta_r(s)}{s(sJ + 2D)} - \frac{K\Theta_m(s)}{s(sJ + 2D)}\end{aligned} \tag{4.45}$$

where $K = K_m K_a K_p$. On rearranging terms in Eq. (4.45), we get

$$\Theta_m(s)\left[1 + \frac{K}{s(sJ + 2D)}\right] = \frac{K\Theta_r(s)}{s(sJ + 2D)}$$

Thus, the transfer function of the system is given by

$$G(s) = \frac{\Theta_m(s)}{\Theta_r(s)} = \frac{K}{s(sJ + 2D) + K} = \frac{K}{s^2 J + 2sD + K}$$

Rate servomechanism. Here, the derivative of position i.e. speed is to be held constant. So, the controlled variable here is the speed of rotation of a motor and hence the feedback element may be a tachometer. The relevant block diagram is given in Figure 4.34.

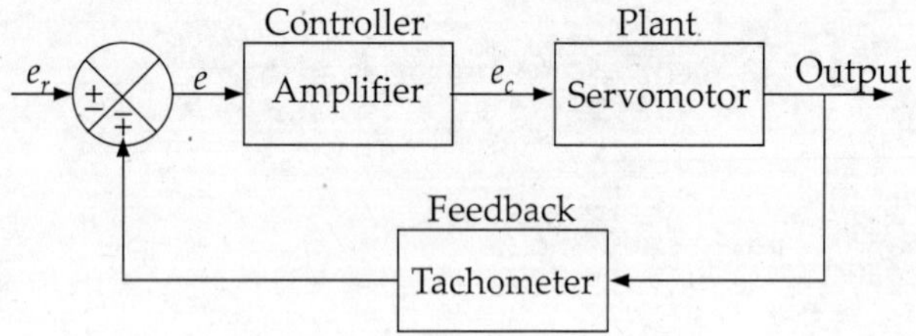

Figure 4.34: Block diagram of rate servomechanism.

For the dc tachometer, we know [see Eq. (4.35)]

$$e_f(t) = K_t \omega_t(t)$$

where K_t is the tachometer constant and ω_t, its angular velocity. So, the error signal is

$$e(t) = e_r(t) - e_f(t) = e_r(t) - K_t \omega_t(t)$$

The voltage input to the motor is

$$e_c(t) = K_a e(t) = K_a[e_r(t) - K_t \omega_t(t)] = K_a \left[e_r(t) - K_t \frac{d\theta}{dt}\right] \tag{4.46}$$

If this voltage is fed to an ac servomotor at rest, the generated stall torque τ_0 is given by [see Eq. (4.27)]

$$\tau_0 = K e_c$$

where K is a constant. Also, from mechanical considerations

$$J\frac{d^2\theta}{dt^2} + D\frac{d\theta}{dt} = \tau_0 = Ke_c$$

or,

$$\left(\frac{1}{K}\right)\frac{d^2\theta}{dt^2} + \left(\frac{D}{K}\right)\frac{d\theta}{dt} = e_c \tag{4.47}$$

Substituting the value of e_c from Eq. (4.46) in Eq. (4.47), we get

$$\left(\frac{1}{K}\right)\frac{d^2\theta}{dt^2} + \left(\frac{D}{K}\right)\frac{d\theta}{dt} = K_a\left[e_r(t) - K_t\frac{d\theta}{dt}\right]$$

or,

$$e_r = \left(\frac{1}{KK_a}\right)\frac{d^2\theta}{dt^2} + \left(\frac{D}{KK_a} + K_t\right)\frac{d\theta}{dt}$$

At constant speed, $\dfrac{d^2\theta}{dt^2} = 0$. Therefore,

$$\frac{d\theta}{dt} = \frac{e_r}{\dfrac{D}{KK_a} + K_t} = \frac{e_rKK_a}{D + KK_aK_t} = \frac{e_r\left(\dfrac{K}{D}\right)K_a}{1 + \left(\dfrac{K}{D}\right)K_aK_t} = \frac{e_rK_mK_t}{1 + K_mK_aK_t}$$

where $K/D = K_m$ is called the *velocity gain constant* of the motor.

Thermal System: Temperature Control

Let us consider a temperature control system in which there is an equal inflow and outflow of liquid and a steady temperature of the liquid in the bath is maintained by the arrangement shown in Figure 4.35.

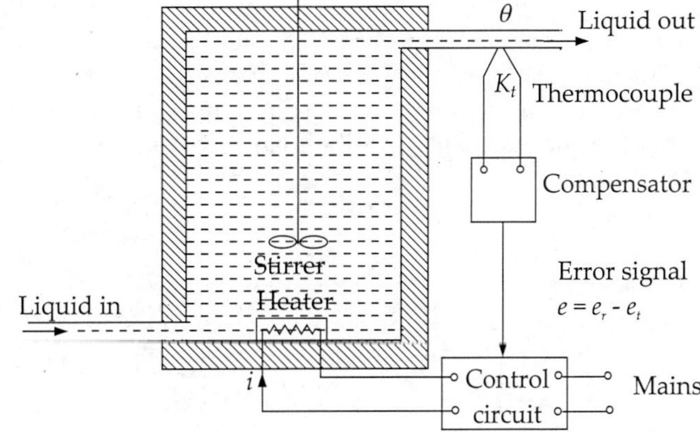

Figure 4.35: Temperature controller.

Before building a mathematical model of the system, we need to make the following simplifying assumptions:

1. The bath is perfectly insulated i.e. there is no heat loss either by conduction or by radiation from the bath.
2. The stirrer thoroughly mixes the liquid so that the temperature within the bath is uniform everywhere.

3. The control circuit is a zero-order instrument, i.e.

$$i = K_c \tag{4.48}$$

where K_c is a constant.

Now, if

$m,\ \theta,\ c,\ \rho$ = mass, temperature, specific heat and density of the liquid in the tank
θ_i = temperature of the liquid flowing in
R = resistance of the heating element
V = volume of liquid flowing per second

then we have from the condition of heat balance

$$i^2 R = mc\frac{d\theta}{dt} + V\rho c(\theta - \theta_i) \tag{4.49}$$

If the present values of the variables are connected to their zero-subscripted steady-state values as $e = e_0 + \Delta e$, $\theta = \theta_0 + \Delta\theta$ and $\theta_l = \theta_{i0} + \Delta\theta_i$ then, substituting Eq. (4.48) in Eq. (4.49) and utilizing these conditions, we have

$$K_c^2(e_0 + \Delta e)R = mc\frac{d\Delta\theta}{dt} + V\rho c(\theta_0 + \Delta\theta - \theta_{i0} - \Delta\theta_i) \tag{4.50}$$

Equating all incremental values to zero, we have from Eq. (4.50) the steady-state equation as

$$K_c^2 e_0^2 R = V\rho c(\theta_0 - \theta_{i0}) \tag{4.51}$$

Subtracting Eq. (4.51) from Eq. (4.50) and neglecting the term containing $(\Delta e)^2$, we have

$$2K_c^2 e_0\, \Delta e\, R = mc\frac{d\Delta\theta}{dt} + \frac{1}{R_t}(\Delta\theta - \Delta\theta_i)$$

or,

$$K\Delta e = \tau\frac{d\Delta\theta}{dt} + (\Delta\theta - \Delta\theta_i) \tag{4.52}$$

where $R_t = (1/V\rho c)$, $K = 2K_c^2 e_0\, \Delta e\, RR_t$ and $\tau = R_t mc$. R_t can be called the *thermal resistance of the heat transfer process* and τ, the time-constant.

We note that the thermocouple converts temperature to an emf e_t, which is fed to the comparator. The comparator subtracts this emf from a reference emf e_r and generates an error emf. Therefore, we can write the following equations for the incremental changes:

$$\Delta e_t = K_t \Delta\theta \tag{4.53}$$

$$\Delta e = \Delta e_r - \Delta e_t \tag{4.54}$$

The Laplace transforms of Eqs. (4.52), (4.53) and (4.54) yield

$$K\Delta E(s) = \tau s\, \Delta\Theta(s) + \Delta\Theta(s) - \Delta\Theta_i(s)$$

$$\Rightarrow \quad \Delta\Theta(s) = \frac{K\Delta E(s)}{\tau s + 1} + \frac{\Delta\Theta_i(s)}{\tau s + 1} \tag{4.55}$$

$$\Delta E_t(s) = K_t \Delta\Theta(s) \tag{4.56}$$

$$\Delta E(s) = \Delta E_r(s) - \Delta E_t(s) \tag{4.57}$$

Equations (4.55), (4.56) and (4.57) help us draw the block diagram of Figure 4.36 for the system. Here, the forward-path transfer function and the disturbance transfer function are given by

$$G(s) = \frac{K}{\tau s + 1}, \qquad D(s) = \frac{\Delta\Theta_i(s)}{\tau s + 1}$$

Figure 4.36: Block diagram of temperature control system.

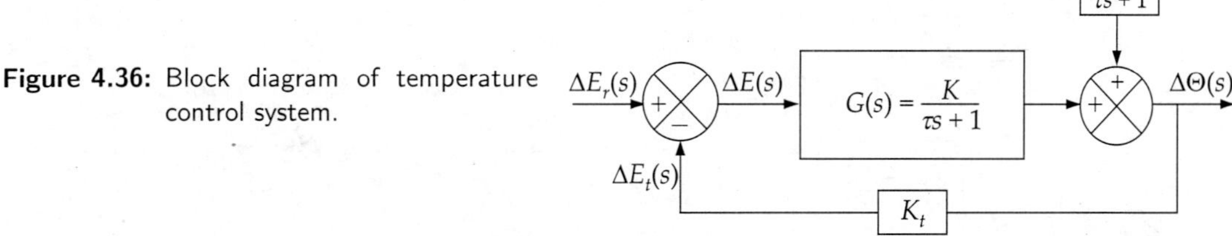

The disturbance is caused by the change in temperature of the inflowing liquid.

If there is no change in temperature of the inflowing liquid, the disturbance signal is zero. Then, we get from Figure 4.36

$$\Delta\Theta(s) = \Delta E(s)G(s) = [\Delta E_r(s) - \Delta E_t(s)]G(s) = \frac{K}{\tau s+1}\cdot\Delta E_r(s) - \frac{KK_t}{\tau s+1}\cdot\Delta\Theta(s)$$

or, $$\Delta\Theta(s)\left[1+\frac{KK_t}{\tau s+1}\right] = \frac{K}{\tau s+1}\cdot\Delta E_r(s)$$

or, $$\frac{\Delta\Theta(s)}{\Delta E_r(s)} = \frac{K}{\tau s+1+KK_t}$$

If the reference signal is changed by a step, the corresponding change in the temperature of the tank liquid can be obtained from

$$\Delta\Theta(s) = \frac{1}{s}\cdot\frac{K}{\tau s+1+KK_t} \tag{4.58}$$

The steady state value of $\theta(t)$ can be obtained by taking inverse Laplace transform of Eq. (4.58) and then putting $t\to\infty$. Alternatively, it can be derived straightaway from Eq. (4.58) using the final value theorem[5] as follows:

$$\Delta\theta(t)_{ss} = \lim_{s\to 0} s\Delta\Theta(s) = \lim_{s\to 0}\left[s\cdot\frac{1}{s}\cdot\frac{K}{\tau s+1+KK_t}\right] = \frac{K}{1+KK_t}$$

We note that the feedback loop reduces the temperature change by a factor of $1/(1+KK_t)$ if there is an unwanted change in the reference voltage.

It can be found by the same method that any fluctuation in the temperature of the inflowing liquid cannot disturb the temperature of the liquid in the tank much because of the reduction by the same factor. So, by controlling the loop-gain, the influence of unwanted changes can be kept under control.

Liquid System: Level Control

Control of liquid levels is an important feature in many industrial processes. We consider a simple system, shown in Figure 4.37a, where the level is being maintained by a controller which actuates a pump to supply liquid at a rate of Q_i m^3/s. Let Q_o be the liquid outflow rate and h, the liquid head in the tank. Of course, a steady liquid-level can be maintained by making $Q_i = Q_o$. However, there may be fluctuations in the supply rate due to many factors and, therefore, a level control measure is necessary.

The well-known Poiseuille's equation[6] states that for a laminar flow of liquid through a

[5]See Appendix *Laplace Transform* Section A at page 518.

[6]See, for example, *The General Properties of Matter* by CJ Smith, Edward Arnold Co.

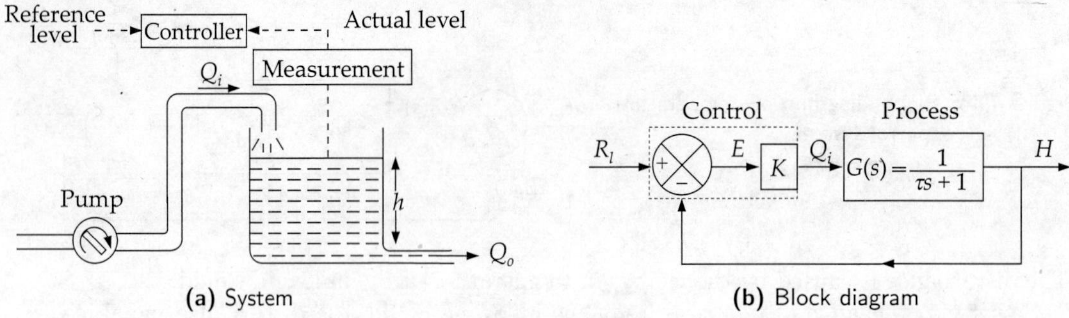

Figure 4.37: Liquid level control.

horizontal pipe

$$\Delta p = \frac{8\mu l}{\pi a^4} Q_o = R' Q_o$$

where μ = viscosity of the fluid flowing through the pipe
$l,\ a$ = length and inner radius of the pipe
Q_o = rate of fluid flowing out of the pipe
Δp = pressure head that causes the flow

$R' = 8\mu l/(\pi a^4)$ is called the *fluid resistance*. Since, $\Delta p = h\rho g$ where g is the acceleration due to gravity and h is the height of the liquid in the tank, we can write

$$h = RQ_o$$

where $R = R'/(\rho g)$.

The rate of storage of liquid in the tank is the difference between the inflow rate and outflow rate of the liquid. Hence, if C is the cross-sectional area of the tank, then

$$C\frac{dh}{dt} = Q_i - Q_o = Q_i - \frac{h}{R}$$

or,

$$RC\frac{dh}{dt} + h = RQ_i \tag{4.59}$$

Also, if $r_l,\ h$ = reference and actual level of liquid in the tank
e = error in liquid level = $r_l - h$
K = gain of the controller

then we have

$$RQ_i = Ke \tag{4.60}$$

The Laplace transforms of Eqs. (4.59) and (4.60) give

$$(\tau s + 1)H(s) = KE(s) \qquad \text{where} \qquad \tau = RC$$

The system may be represented by the block diagram of Figure 4.37b. The block diagram gives the following relations between the error and the reference signals:

$$\frac{H(s)}{R_l(s)} = \frac{KG(s)}{1+KG(s)} = \frac{K}{\tau s + K + 1}$$

$$E(s) = R_l(s) - H(s) = R_l(s)\left(1 - \frac{K}{\tau s + K + 1}\right)$$

$$= R_l(s) \cdot \frac{1+\tau s}{\tau s + K + 1}$$

or,

$$\frac{E(s)}{R_l(s)} = \frac{1+\tau s}{\tau s + K + 1}$$

For a step change in the reference signal,

$$\Delta E(s) = \frac{1}{s} \cdot \frac{\tau s + 1}{\tau s + K + 1}$$

the steady-state value of which is given by

$$\Delta e(t)_{ss} = \lim_{s\to 0} s\,\Delta E(s) = \lim_{s\to 0} \frac{\tau s + 1}{\tau s + 1 + K} = \frac{1}{1+K} \tag{4.61}$$

Equation (4.61) suggests that to decrease the error in the level of the liquid, one has to increase K. This, in effect, means to use a high power pump. However, one has to consider the cost of the control at the same time and so the question of optimization arises.

4.4 TRANSPORTATION LAG IN SYSTEMS

We know that transfer functions represent the ratio of Laplace transforms of output and input in linear time-invariant systems. Generally, they are ratios of two polynomials in the s-space. When they are derived, apart from assuming that all initial conditions are set to zero, there is another *tacit assumption that there is no time delay for the error signal generation.* But, in many cases this assumption is not valid. Even computer controlled systems have time delays, or transportation lags, because the computer takes some time to perform numerical operations before it produces an error signal. For mechanical, pneumatic or chemical systems, the time delay is no doubt greater.

Suppose a signal generated at a distance d travels with a velocity v to reach the controller. If T be the corresponding time lag, it is given by

$$T = \frac{d}{v}$$

Because of this time delay, the output lags behind the controller by T seconds. Thus, the monitored feedback signal $b(t)$ and output $c(t)$ are related as

$$b(t) = c(t - T) \tag{4.62}$$

The Laplace transform of Eq. (4.62) yields

$$B(s) = e^{-sT}C(s) \tag{4.63}$$

Equation (4.63) gives the transfer function between the feedback signal and the output as

$$\frac{B(s)}{C(s)} = e^{-sT} \tag{4.64}$$

The function e^{-sT}, called the *transportation lag* or *input delay* or *dead time*, is very useful in handling control problems. We may recall, we encountered this function in Example 2.33 at page 25. We will consider other examples in Example 9.15 at page 209 and Example 10.27 at page 251.

REVIEW QUESTIONS

4.1 Choose the correct answers:

(i) In position control system the device used for rate feedback voltage is called

(A) potentiometer (B) synchrotransmitter
(C) tachogenerator (D) servomotor

(ii) In the force-voltage analogy, mass is analogous to

(A) charge (B) current (C) inductance (D) resistance

(iii) A rotational system is described by the differential equation

$$2\frac{d\omega}{dt} + \omega = 10$$

where the right hand side represents the constant torque acting on the system and ω represents the angular velocity. The solution for ω is

(A) $10(1 - e^{-t/2})$ (B) $10e^{-t/2}$ (C) $10(1 + e^{-t/2})$ (D) $5(1 - e^{-t/2})$

(iv) An ac servomotor is basically a

(A) universal motor (B) single-phase induction motor
(C) two-phase induction motor (D) three-phase induction motor

(v) In force-voltage analogous system, displacement is equivalent to

(A) current (B) flux (C) charge (D) inductance

4.2 Derive the transfer function of an armature-controlled dc motor.

4.3 Draw the typical torque-speed characteristics of a two-phase induction motor. How is this achieved and why? Show that the transfer function of a two-phase induction motor can be written in the form

$$T(s) = \frac{\Theta_m(s)}{V_z(s)} = \frac{\text{Motor shaft displacement}}{\text{Control phase voltage}} = \frac{K_m}{s(1 + s\tau_m)}$$

What are the expressions for K_m and τ_m and what are they called?

4.4 (a) Determine the transfer function of a dc servomotor in the field controlled mode.

(b) Derive the transfer function of a two-phase, 50 Hz servomotor having the following specifications:

Starting torque	: 0.16 N-m
Inertia of motor	: 1×10^{-5} kg-m^2
Supply voltage	: 115 V
No-load angular velocity	: 304 rad/s

Assume a straight line characteristics and the viscous friction is zero.

4.5 Consider the mechanical system shown in Figure 4.38a. $F(t)$ is the external force applied to the system and $y(t)$ is the displacement of the mass element M from the equilibrium position. Draw the free-body diagram of the mass element and write down the ordinary differential equation that describes the system dynamics.

4.6 For the mechanical system shown in Figure 4.38b, write the differential equations of performance with equivalent mechanical network.

Figure 4.38: Mechanical systems of Questions 4.5 and 4.6.

4.7 Derive the system equations and find the value of $X_2(s)/F(s)$ for the system shown in the following figure.

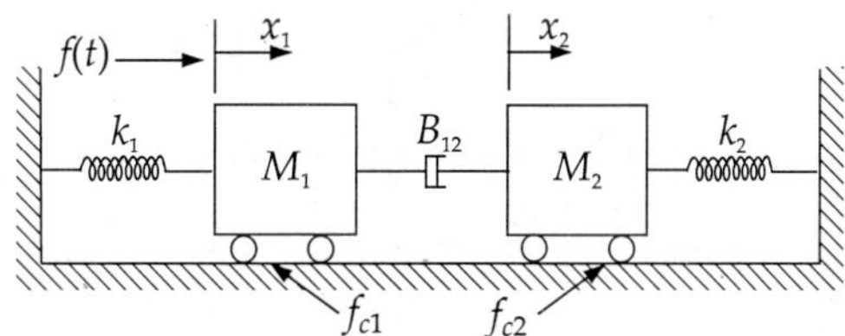

4.8 For the mechanical system shown in Figure 4.39a, (a) draw the equivalent mechanical network, (b) write performance equations and (c) draw the analogous electrical network.

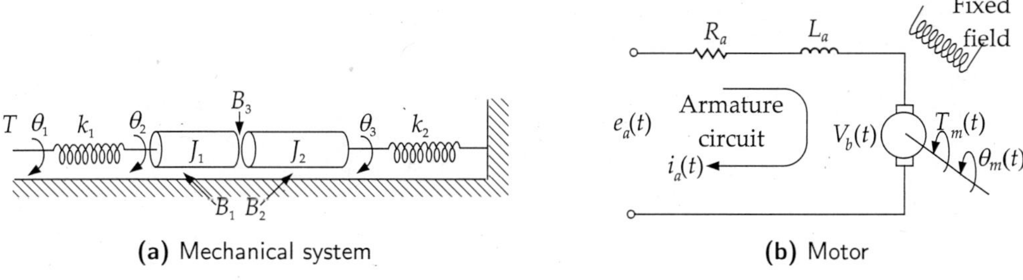

Figure 4.39: Questions 4.8 and 4.9.

4.9 (a) Is there any special constructional features for the motors used in servo system?

(b) Find the transfer function of the motor given in Figure 4.39b.

(c) What is the order of the transfer function?

(d) Sketch the torque-speed curve of the motor.

4.10 Write short notes on the following:

(a) The ac tachogenerator
(b) Tachometer
(c) Synchro transmitter control transformer
(d) Synchro pair
(e) Position and speed control of an automobile
(f) Liquid level controller in two-tank process
(g) A temperature control system.

Answers to Selected Questions

4.1. (i) (C) (ii) (C) (iii) (A) (iv) (C) (v) (C) 4.4. (b) $\dfrac{\Theta(s)}{E_c(s)} = \dfrac{139.1}{s(s+105.3)}$

4.6. $M_1\dfrac{d^2x}{dt^2} + B_2\dfrac{dx}{dt} + B_1\dfrac{d}{dt}(x - x_1) = F(t)$ $\quad M_2\dfrac{d^2x_1}{dt^2} + kx_1 = B_1\dfrac{d}{dt}(x - x_1)$

4.7. $f(t) = M_1\dfrac{d^2x_1}{dt^2} + f_{c1}\dfrac{dx_1}{dt} + k_1x_1 + B_{12}\dfrac{d}{dt}(x_1 - x_2)$

$$B_{12}\frac{d}{dt}(x_1 - x_2) = M_2\frac{d^2x_2}{dt^2} + f_{c2}\frac{dx_2}{dt} + k_2x_2$$

$$\frac{X_2(s)}{F(s)} = \frac{sB_{12}}{\left[s^2M_1 + s(f_{c1} + B_{12}) + k_1\right]\left[s^2M_2 + s(f_{c2} + B_{12}) + k_2\right] + s^2B_{12}^2}$$

4.8. (b) $J_1\dfrac{d^2\theta_1}{dt^2} + B_1\dfrac{d\theta_1}{dt} + B_3\dfrac{d}{dt}(\theta_1 - \theta_2) + k_1\theta_1 = T$

$$J_2\frac{d^2\theta_1}{dt^2} + B_2\frac{d\theta_1}{dt} + k_2\theta_2 = B_3\frac{d}{dt}(\theta_1 - \theta_2)$$

(c) Analogous (f–i analogy) electrical equations are

$$L_1\frac{di_1}{dt} + R_1i_1 + R_3(i_1 - i_2) + \frac{\int i_1 dt}{C_1} = V$$

$$L_2\frac{di_2}{dt} + R_2i_2 + \frac{\int i_2 dt}{C_2} = R_3(i_1 - i_2)$$

Circuit may be drawn from these equations.

5

Feedback Control Characteristics

5.1 INTRODUCTION

At the outset, we would like to mention that the feedback can be of two types, namely

- Degenerative or negative feedback
- Regenerative or positive feedback

Unless otherwise specified, normally the word 'feedback' in control systems means the degenerative feedback.

In Chapter 4, we have seen how to obtain mathematical models of components of control systems. Also, in Chapter 1 we discussed feedback control systems. In this chapter, we will study the following:

1. Transfer functions of open-loop and closed-loop systems
2. The effect of parameter variation in open-loop and closed-loop systems
3. The effects of feedback on
   (a) Sensitivity
   (b) Time-constant
   (c) Overall gain
   (d) Stability
   (e) Disturbance
4. Steady-state error of a control system.

5.2 TRANSFER FUNCTION

Open-loop System

The block diagram and signal flow graph of an open-loop control system are shown in Figures 5.1a and 5.1b.

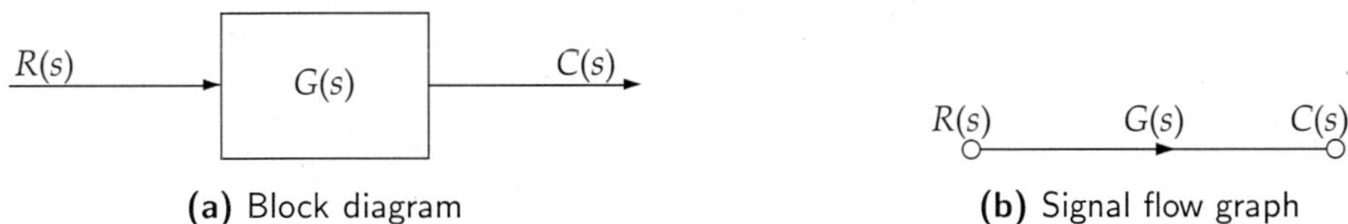

(a) Block diagram **(b)** Signal flow graph

Figure 5.1: Open-loop system.

The output of the open-loop system is

$$C(s) = G(s)R(s)$$

Therefore, though $G(s)$ should be called the open-loop transfer function, the convention is to define the open-loop transfer function (OLTF) as the signal from the feedback path. Thus,

$$\text{OLTF} = G(s)H(s)$$

Unless otherwise specified, the value of $H(s)$ is usually assumed to be 1.

Closed-loop System

The block diagram and signal flow graph of a closed-loop system are shown in Figures 5.2a and 5.2b, respectively.

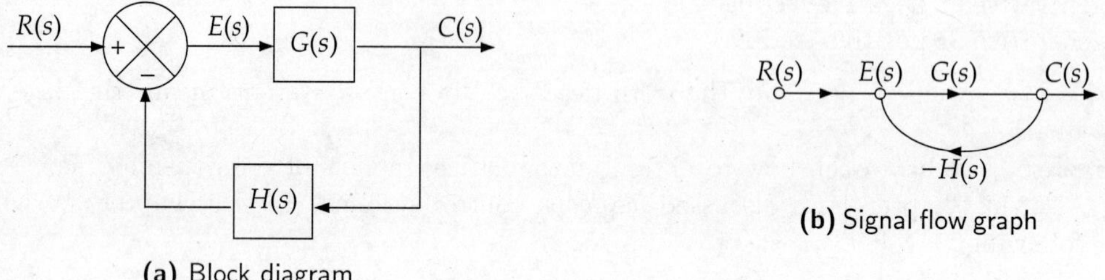

(a) Block diagram

(b) Signal flow graph

Figure 5.2: Closed-loop system.

The basic difference between open-loop and closed-loop systems is the generation of *error signal*, $E(s)$, in the latter. The output of the closed-loop system is, therefore,

$$C(s) = E(s)G(s) = [R(s) - H(s)]G(s)$$

The closed-loop transfer function (CLTF) is thus,

$$\text{CLTF} = \frac{C(s)}{R(s)} = \frac{G(s)}{1 + G(s)H(s)} \tag{5.1}$$

5.3 EFFECTS OF FEEDBACK

We know that feedback exists only for closed-loop systems. For open-loop systems, there is no feedback. However, for the sake of comparison about the effect of feedback, we have to consider the corresponding situations in open-loop systems as well.

On Parameter Variation

Open-loop system. Let, $\Delta G(s)$ = change in $G(s)$ due to parameter variation, and $\Delta C(s)$ = corresponding change in the output. Then,

$$\begin{aligned} C(s) + \Delta C(s) &= [G(s) + \Delta G(s)]R(s) = G(s)R(s) + \Delta G(s)R(s) \\ &= C(s) + \Delta G(s)R(s) \end{aligned}$$

Therefore,

$$\Delta C(s) = \Delta G(s)R(s) \tag{5.2}$$

Closed-loop system. From Eq. (5.1),

$$C(s) = \left[\frac{G(s)}{1+G(s)H(s)}\right]R(s)$$

Therefore,

$$\begin{aligned} C(s)+\Delta C(s) &= \left[\frac{G(s)+\Delta G(s)}{1+[G(s)+\Delta G(s)]H(s)}\right]R(s) \\ &= \left[\frac{G(s)+\Delta G(s)}{1+G(s)H(s)}\right]R(s) \qquad \text{[neglecting } \Delta G(s)H(s)] \\ &= \frac{G(s)}{1+G(s)H(s)}R(s) + \frac{\Delta G(s)}{1+G(s)H(s)}R(s) \\ &= C(s) + \frac{\Delta G(s)}{1+G(s)H(s)}R(s) \end{aligned}$$

Thus,

$$\Delta C(s) = \frac{\Delta G(s)}{1+G(s)H(s)}R(s) \tag{5.3}$$

Comparing Eqs. (5.2) and (5.3), we observe that the closed-loop variation in output is $1/[1+G(s)H(s)]$ times that of the open-loop system. Usually, $G(s)H(s)$ is high which means that the feedback lowers the variation in output with respect to parameter variation.

On Sensitivity

If K is a parameter of the system and T is the transfer function of the system, then sensitivity of T with respect to K is defined as

$$S_K^T = \frac{\partial T/T}{\partial K/K}$$

Open-loop system. Here, $T = G$. Hence,

$$S_G^T = \frac{\partial G/G}{\partial G/G} = 1$$

Hence, the sensitivity of an open-loop system to variation of G is 100%.

Closed-loop system. Here,

$$T = \frac{G}{1+GH}$$

$$\frac{\partial T}{\partial G} = \frac{1}{1+GH} - \frac{GH}{(1+GH)^2} \tag{5.4a}$$

$$\frac{\partial T}{\partial H} = -\frac{G^2}{(1+GH)^2} \tag{5.4b}$$

Therefore,

$$\begin{aligned} S_G^T &= \frac{\partial T}{\partial G}\cdot\frac{G}{T} \\ &= \left[\frac{1}{1+GH} - \frac{GH}{(1+GH)^2}\right](1+GH) \qquad \text{[using Eq. (5.4a)]} \\ &= \frac{1}{1+GH} \qquad (5.5) \\ &< 1 \qquad \text{[since } GH \text{ is high]} \quad (5.6) \end{aligned}$$

Thus, sensitivity for a closed-loop system is reduced by a factor. That essentially means that the closed-loop system is much less sensitive to G variation.

Now,
$$\begin{aligned} S_H^T &= \frac{\partial T}{\partial H} \cdot \frac{H}{T} \\ &= -\frac{G^2}{(1+GH)^2} \cdot \frac{H(1+GH)}{G} && \text{[using Eq. (5.4b)]} \\ &= -\frac{GH}{1+GH} && (5.7) \\ &= -1 && [\because\ GH \gg 1]\ (5.8) \end{aligned}$$

So, a feedback loop is highly sensitive to H variation.

Note: (a) Since a closed-loop system is highly sensitive to H variation, it is necessary to use robust feedback elements that are *not* affected by environmental factors.

(b) G is normally made of power elements while H comprises measuring/monitoring elements that operate at low power levels. So, it is easier and less costly to maintain the constancy of H.

EXAMPLE 5.1. Feedback control systems are (choose the correct answer):

(A) Insensitive to both forward- and feedback-path parameter changes

(B) Less sensitive to feedback-path parameter changes than to forward-path parameter changes

(C) Less sensitive to forward-path parameter changes than to feedback-path parameter changes

(D) Equally sensitive to forward- and feedback-path parameter changes

Solution From Eqs. (5.6) and (5.8), we conclude that feedback control systems are less sensitive to forward-path parameter changes than to feedback-path parameter changes.
Ans. (C)

On Time-constant

Let us consider an elementary system where the process is of first order and the negative feedback is of zero order[1].

Open-loop system. Here,

$$G(s) = \frac{C(s)}{R(s)} = \frac{K}{1+sT} \tag{5.9}$$

For a unit impulse input, $R(s) = 1$. Therefore,

$$C(s) = \frac{K}{T} \cdot \frac{1}{s+(1/T)}$$

Thus,
$$c(t) = \frac{K}{T} \exp[-(t/T)] \tag{5.10}$$

[1]Means order of the transfer function of the feedback element.

Closed-loop system. Here,

$$T(s) = \frac{G(s)}{1 + G(s)H(s)}$$

If $H(s) = K'$, then

$$T(s) = \frac{\dfrac{K}{1+sT}}{1 + \dfrac{KK'}{1+sT}} = \frac{K/T}{s + \dfrac{1+KK'}{T}} \tag{5.11}$$

Therefore,

$$c(t) = \frac{K}{T}\exp[-(1/T')] \tag{5.12}$$

where

$$T' = \frac{T}{1+KK'}$$

Now, $T' < T$, because $KK' \gg 1$. Therefore, a comparison of Eq. (5.12) with Eq. (5.10) reveals that a closed-loop system has a lower time-constant, or faster response, than an open-loop system.

On Overall Gain

In the case of an open-loop system,

$$\text{Overall gain} = G(s)$$

while, for a closed-loop system

$$\text{Overall gain} = \frac{G(s)}{1 + G(s)H(s)}$$

Since $G(s)H(s) > 1$, the overall gain of a closed-loop system is less than that of an open-loop system possessing the same process gain.

On Stability

Consider a first order system with zero order feedback.

Open-loop system. From Eq. (5.9)

$$G(s) = \frac{K/T}{s + \dfrac{1}{T}}$$

Therefore, the pole is at $-\dfrac{1}{T}$.

Closed-loop system. From Eq. (5.11)

$$T(s) = \frac{K/T}{s + [(1 + KK'/T)]}$$

Therefore, the pole is at $-\dfrac{1+KK'}{T}$.

The poles are plotted in Figure 5.3a. The figure shows that for increasing values of K', the location of the pole shifts more towards the left in the s-plane. This means that, with proper feedback, the stability of a system may be increased. However, we will see later that, in general, open-loop systems, which are simpler in construction, are more stable than closed-loop systems which are rather complex in design.

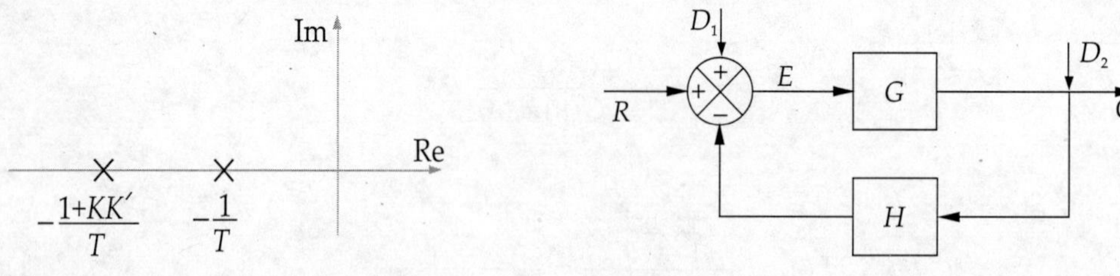

(a) Pole plot for open- and closed-loop **(b)** Closed-loop with disturbances

Figure 5.3: Stability and disturbance.

On Disturbance

Open-loop system. Here, if the disturbance is the on the input side, it gets multiplied by G while, if it is on the output side, it remains as such. So, in an open-loop system, there is no question of controlling the disturbance.

Closed-loop system. Consider Figure 5.3b where D_1 and D_2 indicate disturbances.

Here,
$$C \pm D_2 = (E \pm D_1)G = (R - CH \pm D_1)G$$

or,
$$C\left(\frac{1}{G} + H\right) = R \pm D_1 \mp \frac{D_2}{G}$$

or,
$$C = \frac{G}{1+GH}R \pm \frac{G}{1+GH}D_1 \mp \frac{1}{1+GH}D_2$$
$$= \frac{R}{H} \pm \frac{D_1}{H} \mp \frac{D_2}{GH} \qquad [\text{for } GH \gg 1]$$

This result shows that the value of D_2 gets considerably reduced, whereas that of D_1 is not much reduced. Thus, a closed-loop system considerably reduces noise (or disturbance) at process output, but not so effectively at the process input.

EXAMPLE 5.2. A feedback system with high gain K, is shown in Figure 5.4. The closed-loop

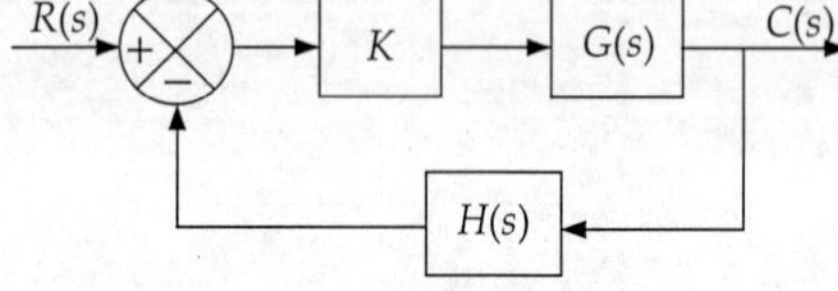

Figure 5.4: Example 5.2.

transfer function is (choose the correct answer)

(A) sensitive to perturbations in $G(s)$ and $H(s)$
(B) sensitive to perturbations in $G(s)$ but not perturbations in $H(s)$
(C) sensitive to perturbations in $H(s)$ but not to perturbations in $G(s)$
(D) insensitive to perturbations in $G(s)$ and $H(s)$

Solution The closed-loop transfer function is given by

$$T = \frac{KG}{1+KGH} \tag{i}$$

Determination of the sensitivity of T to variations in G:

$$S_G^T = \frac{\partial T}{\partial G} \cdot \frac{G}{T}$$

Now,
$$\frac{\partial T}{\partial G} = \frac{K(1+KGH) - K^2GH}{(1+KGH)^2}$$

$$= \frac{K}{(1+KGH)^2} \tag{ii}$$

$$\frac{G}{T} = \frac{1+KGH}{K} \quad \text{[from Eq. (i)]} \tag{iii}$$

Therefore,
$$S_G^T = \frac{K}{(1+KGH)^2} \cdot \frac{1+KGH}{K} \quad \text{[from Eqs. (ii) and (iii)]}$$

$$= \frac{1}{1+KGH} \cong \frac{1}{KGH}$$

$$\ll 1 \tag{iv}$$

Determination of the sensitivity of T to variations in H:

$$S_H^T = \frac{\partial T}{\partial H} \cdot \frac{H}{T}$$

Now,
$$\frac{\partial T}{\partial H} = -\frac{KG \cdot KG}{(1+KGH)^2} \tag{v}$$

$$\frac{H}{T} = H \cdot \frac{1+KGH}{KG} \quad \text{[from Eq. (i)]} \tag{vi}$$

Therefore,
$$S_H^T = -\frac{K^2G^2}{(1+KGH)^2} \cdot \frac{H(1+KGH)}{KG} \quad \text{[from Eqs. (v) and (vi)]}$$

$$= -\frac{KGH}{(1+KGH} \cong -\frac{KGH}{KGH}$$

$$= -1 \tag{vii}$$

From Eqs. (iv) and (vii) we observe that T is sensitive to perturbations in H but not to perturbations in G.

Ans. (C)

On Steady-state Error

The error $e(t)$ is defined as

$$e(t) = r(t) - c(t)$$

Therefore,
$$E(s) = R(s) - C(s)$$

The steady-state error is defined as

$$e_{ss} = e(t)|_{t\to\infty}$$

which on applying the final value theorem[2] turns out to be

$$e_{ss} = \lim_{s\to 0} sE(s) \tag{5.13}$$

Now, let us examine how the steady-state errors for a step input to an open-loop and closed-loop systems compare.

[2]See Appendix *Laplace Transform* Eq. (A.8) at page 518.

Open-loop system. Here,

$$E(s) = R(s) - R(s)G(s) = [1 - G(s)]R(s) \tag{5.14}$$

$R(s)$ being equal to $\frac{1}{s}$, we have from Eqs. (5.13) and (5.14)

$$e_{ss} = \lim_{s \to 0}[1 - G(s)] = 1 - G(0)$$

Closed-loop system. Here, for a unity feedback system,

$$E(s) = R(s) - R(s)\frac{G(s)}{1 + G(s)} = \frac{1}{1 + G(s)}R(s) \tag{5.15}$$

Therefore, for a step input, we have from Eqs. (5.13) and (5.15)

$$e_{ss} = \lim_{s \to 0} \frac{1}{1 + G(s)} = \frac{1}{1 + G(0)}$$

Since $G(0)$ is high, the value of steady-state error is much less in a closed-loop system. In other words, a closed-loop system gives much better accuracy than an open-loop system.

Table 5.1 summarizes the differences between an open-loop and a closed-loop system.

Table 5.1: Distinction between an open-loop and a closed-loop system

Property	*Comments*
Transfer function and gain	The open-loop gain is higher because $G(s)$ is normally high.
Parameter variation	Variation in output with parameter variation is lower in a closed-loop.
Sensitivity	Closed-loop is much less sensitive to G variation, while open-loop sensitivity is 100%.
Response	Since a closed-loop system has a lower time-constant, its response is faster.
Stability	For the closed-loop system, the location of the pole may be shifted to a more stable position by adjusting K'.
Disturbance	The question of control of disturbance does not arise for open-loop systems while noise or disturbance is considerably reduced at the output of closed-loop systems.
Steady-state error	Error is much less in closed-loop systems.

EXAMPLE 5.3. Calculate the sensitivity of the system given in Figure 5.5 with respect to (a) the forward path transfer gain and (b) feedback path transfer function at $\omega = 1.3$ rad/s.

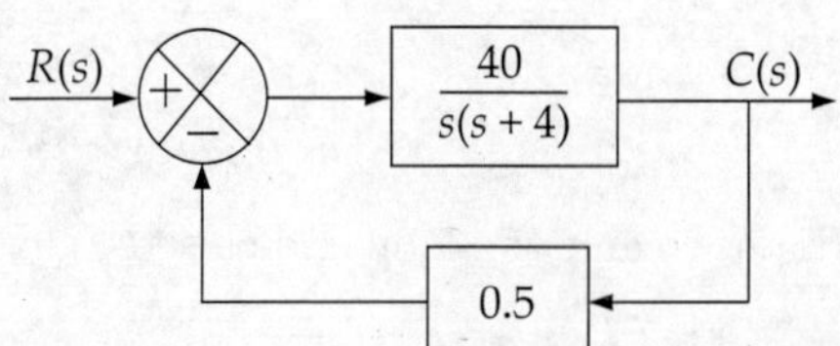

Figure 5.5: System (Example 5.3).

Solution Given: $G = \dfrac{40}{s(s+4)}$ and $H = 0.5$

(a) $S_G^T = \dfrac{1}{1+GH} = \dfrac{s^2+4s}{s^2+4s+20}$

$$S_G^T\Big|_{\jmath\omega} = \left.\frac{-\omega^2+4\jmath\omega}{-\omega^2+4\jmath\omega+20}\right|_{\omega=1.3} = \frac{-1.69+\jmath 5.2}{-1.69+\jmath 5.2+20} = \frac{-1.69+\jmath 5.2}{18.31+\jmath 5.2}$$

Therefore,
$$|S_G^T| = \sqrt{\frac{(-1.69)^2+(5.2)^2}{(18.31)^2+(5.2)^2}} = 0.287$$

(b) $S_H^T = \dfrac{-GH}{1+GH} = -\dfrac{20}{s^2+4s+20}$

$$S_H^T\Big|_{\jmath\omega} = -\left.\frac{20}{-\omega^2+4\jmath\omega+20}\right|_{\omega=1.3} = -\frac{20}{18.31+\jmath 5.2}$$

Therefore,
$$|S_H^T| = \frac{20}{\sqrt{(18.31)^2+(5.2)^2}} = 1.051$$

EXAMPLE 5.4. $G(s) = \dfrac{\omega_n^2}{s^2+2\zeta\omega_n s+\omega_n^2}$ where $\omega_n = 8\pi/T$, $T = 6.28$ s and $\zeta = 0.3$. Calculate the open-loop and closed-loop sensitivities for changes in A and H where A is the gain and H is the feedback path transfer function.

Solution $\omega_n = \dfrac{8\pi}{6.28} = 4\,\text{rad/s}$. $G(s) = \dfrac{4^2}{s^2+2(0.3)(4)+4^2} = \dfrac{16}{s^2+2.4s+16}$

For open-loop sensitivity,

$$T(s) = AG(s)$$

Hence,
$$\frac{\partial T}{\partial A} = G(s) \quad \text{and} \quad \frac{A}{T} = \frac{1}{G(s)}$$

Therefore,
$$S_A^T = \frac{\partial T}{\partial A}\cdot\frac{A}{T} = 1$$

There is no question of having S_H^T here.
For closed-loop sensitivity

$$T(s) = \frac{AG(s)}{1+AG(s)H} = \frac{16A}{s^2+2.4s+16(AH+1)}$$

$$\frac{\partial T}{\partial A} = \frac{16A}{s^2+2.4s+16(AH+1)}\left[1-\frac{16AH}{s^2+2.4s+16(AH+1)}\right]$$

$$S_A^T = 1-\frac{16AH}{s^2+2.4s+16(AH+1)} = \frac{s^2+2.4s+16}{s^2+2.4s+16(AH+1)}$$

$$\frac{\partial T}{\partial H} = -\frac{(16A)^2}{[s^2+2.4s+16(AH+1)]^2}$$

$$S_H^T = \frac{\partial T}{\partial H}\cdot\frac{H}{T} = -\frac{(16A)^2}{[s^2+2.4s+16(AH+1)]^2}\cdot\frac{H}{T}$$

$$= -\frac{16AH}{s^2+2.4s+16(AH+1)}$$

EXAMPLE 5.5. Refer to the control system given in Figure 5.6.

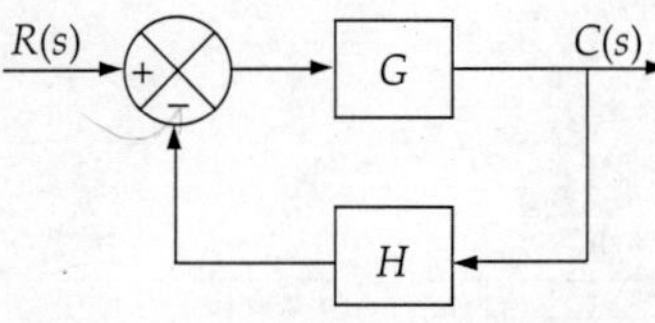

Figure 5.6: Control system (Example 5.5).

$G = 210$ and $H = 0.12$.
Find (a) reference voltage E_r when $E_o = 240\,\text{V}$, and (b) percentage change in the generator terminal when the forward path gain has been reduced by 12%.

Solution $T(s) = \dfrac{E_o(s)}{E_r(s)} = \dfrac{G(s)}{1+G(s)H(s)} = \dfrac{210}{1+(210)(0.12)} = 8.015$

(a) $E_r = \dfrac{E_o}{T} = \dfrac{240}{8.015} = 29.24\text{ V}$

(b) $S_G^T = \dfrac{\partial T}{\partial G} \cdot \dfrac{G}{T} = \left[\dfrac{1}{1+GH} - \dfrac{GH}{(1+GH)^2}\right] \cdot (1+GH) = \dfrac{1}{1+GH} = \dfrac{1}{26.2}$

Now, $$S_G^T = \frac{\partial T/T}{\partial G/G} = \frac{\%\text{ change in } T}{\%\text{ change in } G} = \frac{1}{26.2}$$

Therefore, $$\%\text{ change in } T = \frac{12}{26.2} = 0.458$$

5.4 REGENERATIVE FEEDBACK

As already mentioned, in a regenerative feedback system, the fed back signal is added to the reference signal rather than the conventional practice of subtracting it. This is mostly not desired in a control system.

For example, if one desires to control the temperature of a room at 25°C, the control system continuously measures the room temperature, subtracts the feedback signal from the reference, i.e. 25°C, and if the result is negative, which means the room temperature is higher, the cooling mechanism is switched on. So, in a negative feedback system, a negative error signal means cooling and a positive error signal means heating is required. Now, if the feedback is positive which means the signal is added to the reference, the result is always positive and hence, the cooling mechanism is never switched on.

For similar requirements, the degenerative or negative feedback is sometimes resorted to. From the mathematical viewpoint, a positive feedback has the transfer function

$$\frac{C(s)}{R(s)} = \frac{G(s)}{1 - G(s)H(s)}$$

The negative sign in the denominator implies that if $G(s)H(s) = 1$ the system gives an infinite output for a finite input.

However, in some special cases the regenerative or positive feedback is needed to increase the loop gain of the feedback system. Consider the feedback system as depicted in the block diagram of Figure 5.7.

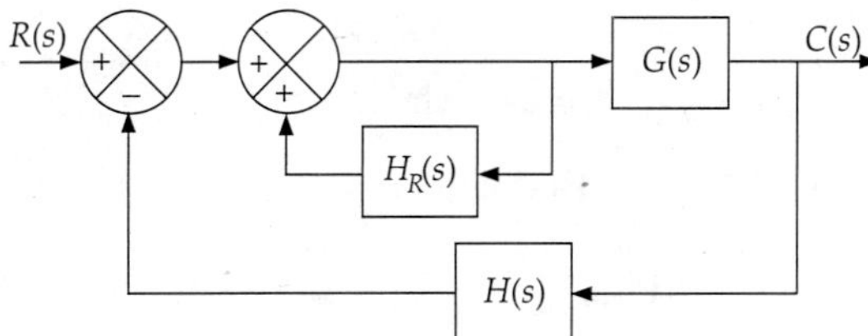

Figure 5.7: Use of regenerative feedback.

Here, the inner loop reduces to $\dfrac{1}{1-H_R(s)}$, making the total transfer function

$$\frac{C(s)}{R(s)} = \frac{\dfrac{G(s)}{1-H_R(s)}}{1+\dfrac{G(s)H(s)}{1-H_R(s)}} = \frac{G(s)}{1-H_R(s)+G(s)H(s)} \tag{5.16}$$

In this system, if $H_R(s) \approx 1$, Eq. (5.16) reduces to

$$\frac{C(s)}{R(s)} \approx \frac{1}{H(s)} \tag{5.17}$$

Equation (5.17) shows that by a proper choice of the regenerative feedback $H_R(s)$, the closed-loop transfer function can be made insensitive to $G(s)$.

REVIEW QUESTIONS

5.1 Choose the correct answer

(i) The output of a feedback control system should be a function of

(A) input (B) feedback signal
(C) reference and output (D) all of above

(ii) In a closed-loop control system the error signal is

(A) the difference between the measured value and the reference value
(B) the sum of the measured value and the reference value
(C) ratio of the measured value and the reference value
(D) none of the above

(iii) Consider the following characteristics:

1. Sensitivity is 100
2. Disturbance can be controlled
3. Steady-state error is much less
4. Gain is higher

Which of these characteristics are true for a closed-loop system?

(A) 1 and 2 (B) 2 and 3 (C) 3 and 4 (D) 1 and 4

(iv) Which of the characteristics stated in (C) above are true for an open-loop system?

(A) 1 and 2 (B) 2 and 3 (C) 3 and 4 (D) 1 and 4

(v) The effect of negative feedback is to

(A) reduce the overall gain
(B) increase the sensitivity of parameter variation in forward path

(C) slow the dynamic response

(D) decrease the bandwidth

5.2 (a) Draw the block diagram of a closed-loop control system and indicate the following on it:

(i) plant

(ii) command input

(iii) controlled output

(iv) actuating signal

(v) feedback element and control element

(b) Mention the important features of closed-loop control systems.

5.3 (a) Define sensitivity of a system variable T with respect to another parameter X.

(b) Find the sensitivity of the overall transfer function of the closed-loop system shown in Figure 5.8 with respect to (i) forward path transfer function and (ii) feedback path transfer function at the frequency 1.4 rad/s.

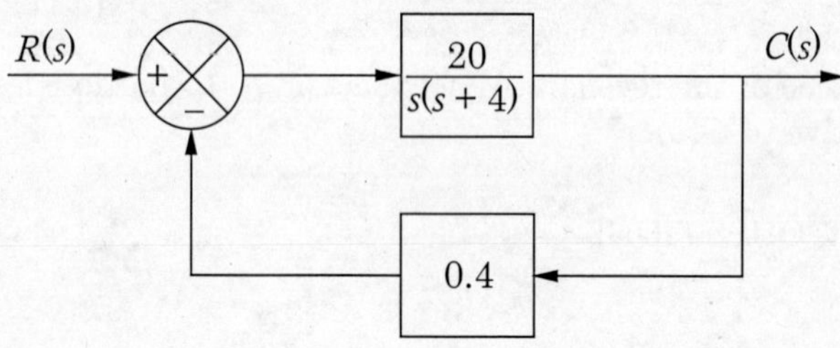

Figure 5.8: System [Question 5.3(b)].

5.4 (a) What is regenerative feedback? What is its application?

(b) Show that the use of negative feedback in a control system reduces the sensitivity of the system to parameter variation and disturbance signal.

Answers to Selected Questions

5.1. (i) (A) (ii) (A) (iii) (B) (iv) (D) (v) (A)

5.3. (b)(i) $S_G^T = 0.72$, (ii) $S_H^T = -0.97$

6

Time-domain Performance of Control Systems

6.1 INPUT SIGNALS

Control systems are basically dynamic systems. So, studying their behaviour with time is important. Indeed, the study of variation of output of a control system is important from many perspectives including its stability. However, the actual input to the system is usually unknown and, therefore, some standard test input signals are used in the study. We discussed such signals in Chapter 2 (Section 2.2) and their Laplace transforms and the results were summarized in Table 2.1 at page 16.

6.2 RESPONSES

Once an input is given to a system, it will produce an output or response. The response can be

(i) Transient
(ii) Steady-state

The transient response is that part of the response which approaches to zero as time tends to infinity. The part of the response that remains after the transients die out, is called the steady-state response. The parameter, which is important in steady state, is the steady-state error, e_{ss}. It is defined as

$$e_{ss} = \lim_{t\to\infty} e(t) = \lim_{t\to\infty} [r(t) - c(t)] \tag{6.1}$$

where, $e(t)$ = error, $r(t)$ = reference input and $c(t)$ = output.

6.3 TIME RESPONSE ANALYSIS

First Order System

If the dynamic relation between the reference input $r(t)$ and output $c(t)$ of a system is of the form of

$$a_1 \frac{dc(t)}{dt} + a_0 c(t) = b_0 r(t)$$

then it can be written as

$$\tau \frac{dc(t)}{dt} + c(t) = Kr(t) \tag{6.2}$$

where $\tau = a_1/a_0$ is the *time constant* and $K = b_0/a_0$ is the *static sensitivity* of the system. Taking Laplace transform of Eq. (6.2) and assuming all initial conditions = 0, we have

$$(s\tau + 1)C(s) = KR(s) \tag{6.3}$$

Equation (6.3) yields the transfer function

$$\boxed{G(s) \equiv \frac{C(s)}{R(s)} = \frac{K}{s\tau + 1}}$$

The order of the transfer function is 1. So, such systems are called *first order systems.* The block diagram and signal flow graph of a first order system are given in Figures 6.1a and 6.1b, respectively.

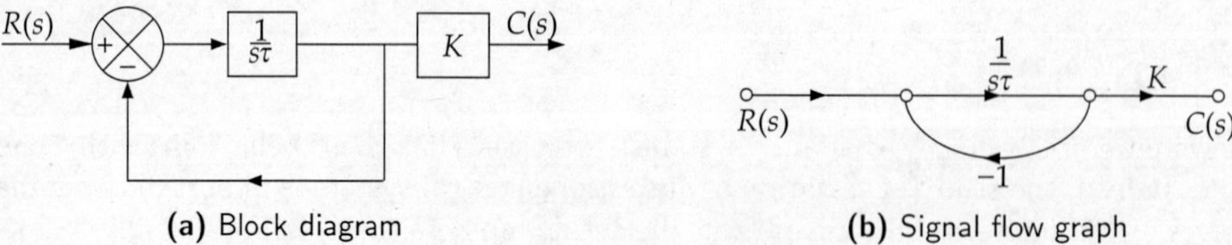

(a) Block diagram **(b)** Signal flow graph

Figure 6.1: First order system.

Now, we check the response of the system for different standard inputs. For convenience, let us assume that $K = 1$ so that the transfer function is

$$G(s) = \frac{1}{s\tau + 1}$$

Unit impulse input. Since, for this input $R(s) = 1$, we have

$$C(s) = \frac{1}{s\tau + 1} = \frac{1}{\tau} \cdot \frac{1}{s + \frac{1}{\tau}}$$

In time-domain,

$$\mathscr{L}^{-1}\{C(s)\} = c(t) = \frac{1}{\tau} e^{-(t/\tau)}$$

Therefore, as $t \to 0 \quad c(t) \to \frac{1}{\tau}$

and as $t \to \infty \quad c(t) \to 0$

The time-response curve is given in Figure 6.2.

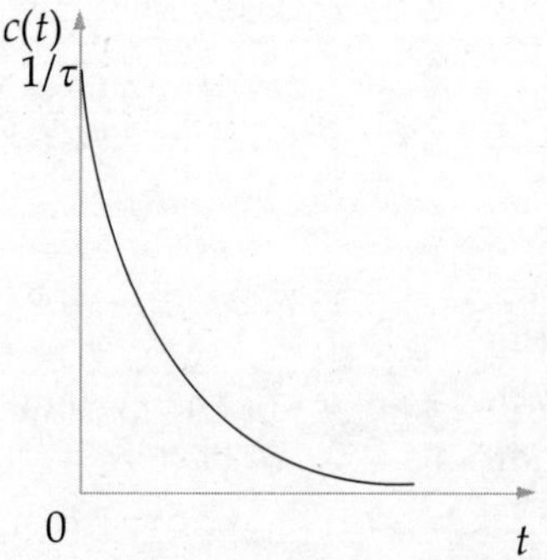

Figure 6.2: Time-response curve for unit impulse input to a first order system.

Unit step input. In the case of unit step input, $R(s) = 1/s$. Therefore,

$$C(s) = \frac{1}{s} \cdot \frac{1}{s\tau + 1} = \frac{1}{s} - \frac{1}{s + \frac{1}{\tau}} \tag{6.4}$$

In t-space
$$\mathscr{L}^{-1}\{C(s)\} = c(t) = 1 - e^{-t/\tau} \tag{6.5}$$

$$\text{Initial slope of the curve} = \left.\frac{dc(t)}{dt}\right|_{t=0} = \left.\frac{1}{\tau}e^{-t/\tau}\right|_{t=0} = \frac{1}{\tau} \tag{6.6}$$

It may be observed from Eq. (6.5) that

Transfer function
$$G(s) = \frac{C(s)}{R(s)} = \frac{1}{s\tau + 1} \tag{6.7}$$

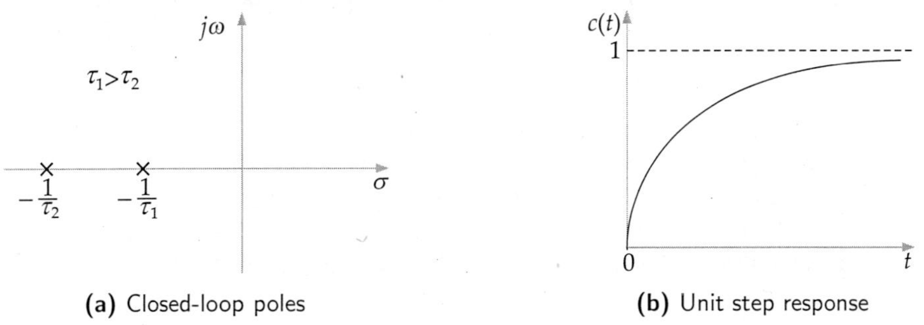

(a) Closed-loop poles **(b)** Unit step response

Figure 6.3: First order system.

Poles are obtained by solving the characteristic equation $s\tau + 1 = 0$, yielding $s = -1/\tau$.

Note:
- Equation (6.6) shows that if the value of τ is high, the initial slope is low and vicè versa.
- Thus, the lower the τ, the faster the response.
- Also, the lower the τ, the higher the stability. This is apparent from a plot of its poles in Figure 6.3a, where it is shown that the pole corresponding to τ_2 is farther away from the Im-axis, indicating that it offers higher stability.

The steady-state error is given as

$$e_{ss} = \lim_{t\to\infty} [r(t) - c(t)] = 1 - 1 = 0$$

An interesting result may be obtained from Table 6.1 where values of the output are given against t/τ. It may be seen that the step response of the first order system attains 99.3% of the final value for $t = 5\tau$. The time-response curve is shown in Figure 6.3b.

Table 6.1: Response with time for unit step input

t/τ	1	2	3	4	5
$c(t)$ in %	63.2	86.0	95.0	98.2	99.3

EXAMPLE 6.1. A first order process has a static gain of 1, a time constant of 1 s and a transport delay of 0.1 s. Its transfer function is

(A) $\dfrac{e^{-s}}{1 + 0.1s}$ (B) $\dfrac{e^{-10s}}{1 + s}$ (C) $\dfrac{e^{-0.1s}}{1 + s}$ (D) $\dfrac{e^{-0.1s}}{1 + 0.1s}$

Solution Since this is a first order system with a time delay, let the transfer function be

$$G(s) = \frac{Ke^{-s\tau_d}}{1+\tau s}$$

where terms have their usual significance. Given, $K = 1$, $\tau = 1\,\text{s}, \tau_d = 0.1\,\text{s}$. So the transfer function is

$$G(s) = \frac{e^{-0.1s}}{1+s}$$

Ans. (C)

EXAMPLE 6.2. A unit step is applied at $t = 0$ to a first order system without time delay. The response has a value of 1.264 units at $t = 10$ min and 2 units at steady state. What is the transfer function of the system?

Solution Since it is a first order system, let the transfer function be given by

$$G(s) = \frac{K}{1+\tau s} \tag{i}$$

We know that for a step input of step K, the output is given by

$$c(t) = K\left(1 - e^{-t/\tau}\right)$$

However, since the steady state output is 2 units, $K = 2$ and the output is given by

$$c(t) = 2\left(1 - e^{-t/\tau}\right) \tag{ii}$$

Now, from the given condition at $t = 10$ min $= 600$ s, and $c(t) = 1.264$, we get from Eq. (ii)

$$1.264 = 2\left(1 - e^{-600/\tau}\right)$$

or, $$1 - e^{-600/\tau} = \frac{1.264}{2} = 0.632$$

or, $$e^{-600/\tau} = 1 - 0.632 = 0.368$$

or, $$\frac{-600}{\tau} = \ln 0.368 = -1$$

Therefore, $$\tau = 600\ \text{s}$$

Thus, we get the required transfer function from Eq. (i) as

$$G(s) = \frac{2}{1+600s}$$

EXAMPLE 6.3. A system with transfer function of $1/(\tau s + 1)$, subjected to a step input takes 10 seconds to reach 50% of the step height. The value of τ is

(A) 6.9 s (B) 10 s (C) 14.4 s (D) 20 s

Solution For an input of step A, the response of a first order system is given by

$$c(t) = A\left(1 - e^{-t/\tau}\right)$$

Given, when $t = 10$ s, $c(t) = A/2$. So, we have

$$\frac{A}{2} = A\left(1 - e^{-10/\tau}\right)$$

$$\Rightarrow \qquad e^{-10/\tau} = 1 - \frac{1}{2} = \frac{1}{2}$$

$$\Rightarrow \qquad -\frac{10}{\tau} = \ln 0.5 = -0.6931$$

$$\Rightarrow \qquad \tau = \frac{10}{0.6931} = 14.4 \text{ s}$$

Ans. (C)

EXAMPLE 6.4. In the feedback scheme shown in Figure 6.4, the time-constant of the closed-loop system will be

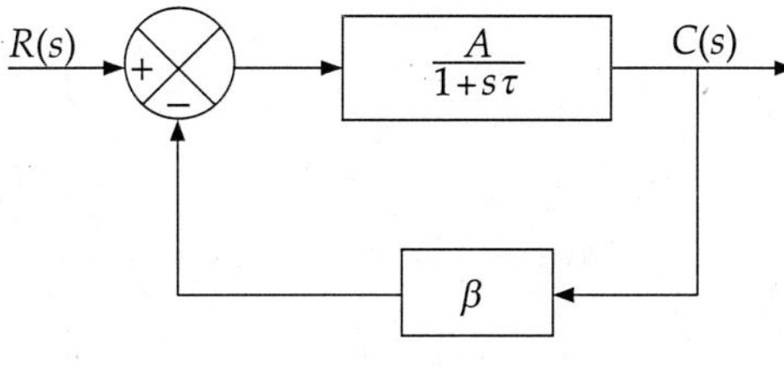

Figure 6.4: Feedback scheme (Example 6.4).

(A) $A\beta\tau$ (B) $1 + A\beta\tau$ (C) τ (D) $\dfrac{\tau}{1 + A\beta}$

Solution The closed-loop transfer function is

$$\frac{C(s)}{R(s)} = \frac{\dfrac{A}{1 + s\tau}}{1 + \dfrac{A\beta}{1 + s\tau}} = \frac{A}{s\tau + 1 + A\beta} = \frac{A/(1 + A\beta)}{s \cdot \dfrac{\tau}{1 + A\beta} + 1} \equiv \frac{K}{s\tau' + 1}$$

So, the time constant, $\tau' = \dfrac{\tau}{1 + A\beta}$.

Ans. (D)

EXAMPLE 6.5. For the system $2/(s + 1)$, the approximate time taken for a step response to reach 98% of its final value is

(A) 1 s (B) 2 s (C) 4 s (D) 8 s

Solution Given system $G(s) = \dfrac{C(s)}{R(s)} = \dfrac{2}{s + 1}$, step input $R(s) = \dfrac{1}{s}$. Therefore,

$$\text{Output} \quad C(s) = G(s)R(s) = \frac{2}{s + 1} \cdot \frac{1}{s}$$

$$= \frac{2}{s} - \frac{2}{s + 1} \qquad \text{(i)}$$

Taking inverse Laplace transform of Eq. (i), we get

$$c(t) = 2\left(1 - e^{-t}\right) = 2 \times 0.98$$

$$\Rightarrow \qquad 1 - e^{-t} = 0.98$$

$$\Rightarrow \qquad e^{-t} = 0.02$$

$$\Rightarrow \qquad -t = \ln 0.02 = -3.91 \text{ s}$$

Ans. (C)

Unit ramp input. For the unit ramp input, $R(s) = \dfrac{1}{s^2}$. Therefore,

$$C(s) = \frac{1}{s^2} \cdot \frac{1}{s\tau + 1} \equiv \frac{A}{s^2} + \frac{B}{s} + \frac{C}{s\tau + 1} \text{ (say)}$$

Then[1],

$$A = \left[s^2 C(s)\right]_{s=0} = \left[\frac{1}{s\tau + 1}\right]_{s=0} = 1$$

$$B = \left[\frac{d}{ds}\left\{s^2 C(s)\right\}\right]_{s=0} = \left[\frac{d}{ds}\left(\frac{1}{s\tau + 1}\right)\right]_{s=0} = -\left.\frac{\tau}{(s\tau + 1)^2}\right|_{s=0} = -\tau$$

$$C = \left[(s\tau + 1)C(s)\right]_{s=-\frac{1}{\tau}} = \left.\frac{1}{s^2}\right|_{s=-\frac{1}{\tau}} = \tau^2$$

Thus,

$$C(s) = \frac{1}{s^2} - \frac{\tau}{s} + \frac{\tau^2}{s\tau + 1} = \frac{1}{s^2} - \frac{\tau}{s} + \frac{\tau}{s + \dfrac{1}{\tau}} \tag{6.8}$$

which, on inverse Laplace transform, yields

$$c(t) = t - \tau + \tau e^{-t/\tau} = t - \tau\left(1 - e^{-t/\tau}\right) \tag{6.9}$$

The time-response curve is given in Figure 6.5. The steady-state error is

$$e_{\text{ss}} = \lim_{t \to \infty}\left[r(t) - c(t)\right] = t - (t - \tau) = \tau$$

from Eq. (6.9) and $r(t) = t$.

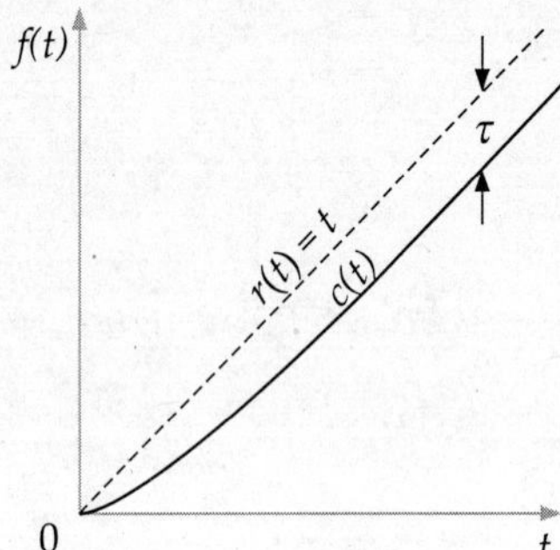

Figure 6.5: Time-response curve for the unit ramp input to a first order system. - - -, input and ———, output.

Note: For the input: $\dfrac{dr(t)}{dt} = u(t)$ and $\left.\dfrac{du(t)}{dt}\right|_{t=0} = \delta(t)$. The same interrelation also holds good for the corresponding responses. This means that the unit step response equals the first differential of the ramp response and so on.

Second Order Systems

If the dynamic input-output relationship of a system is represented by a differential equation of the form of

$$a_2 \frac{d^2 c(t)}{dt^2} + a_1 \frac{dc(t)}{dt} + a_0 c(t) = b_0 r(t) \tag{6.10}$$

[1]See Appendix *Rules of Partial Fractions*, Example B.3 at page 521 for the method.

it is called a second order system because we will presently see that the transfer function of the system has an order of 2.

Equation (6.10) can be written as

$$\frac{1}{\omega_n^2}\frac{d^2c(t)}{dt^2}+\frac{2\zeta}{\omega_n}\frac{dc(t)}{dt}+c(t)=Kr(t) \tag{6.11}$$

where,
$$\frac{1}{\omega_n^2}=\frac{a_2}{a_0},\quad \frac{2\zeta}{\omega_n}=\frac{a_1}{a_0},\quad K=\frac{b_0}{a_0}$$

ω_n, ζ and K are called *natural frequency of oscillation*, *damping ratio* and *static sensitivity* (or *gain*), respectively. Taking Laplace transform of Eq. (6.11) and assuming it to be a relaxed system (i.e., all initial conditions = 0), we get

$$\left(\frac{s^2}{\omega_n^2}+\frac{2\zeta s}{\omega_n}+1\right)C(s)=KR(s) \tag{6.12}$$

On rearranging Eq. (6.12), we get the transfer function for the second order system as

$$\boxed{G(s)\equiv\frac{C(s)}{R(s)}=\frac{K}{\dfrac{s^2}{\omega_n^2}+\dfrac{2\zeta s}{\omega_n}+1}=\frac{K\omega_n^2}{s^2+2\zeta\omega_n s+\omega_n^2}} \tag{6.13}$$

This can also be written in the form of

$$\frac{C(s)}{R(s)}=\frac{K}{s(s\tau+1)+K}=\frac{\dfrac{K}{\tau}}{s^2+\dfrac{s}{\tau}+\dfrac{K}{\tau}} \tag{6.14}$$

where

$$\begin{aligned}\frac{K}{\tau}=\omega_n^2 \quad &\Rightarrow \quad \omega_n=\sqrt{\frac{K}{\tau}}\\ 2\zeta\omega_n=\frac{1}{\tau} \quad &\Rightarrow \quad \zeta=\frac{1}{2\omega_n\tau}\end{aligned} \tag{6.15}$$

The results given in Eq. (6.15) are important because they indicate how the undamped natural frequency and therefore, damping ratio behave with the gain. With the increase in gain, the natural frequency increases which, in turn, reduces the damping ratio.

The block diagram and signal flow graph representations of a second order system are given in Figures 6.6a and 6.6b, respectively. From Eq. (6.13), we see that the characteristic equation

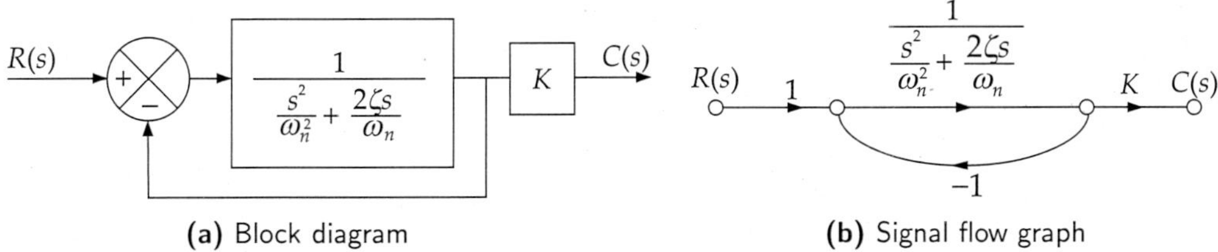

(a) Block diagram **(b)** Signal flow graph

Figure 6.6: Second order system representation.

for the system is

$$\boxed{s^2 + 2\zeta\omega_n s + \omega_n^2 = 0}$$

Therefore, its roots are given by

$$s_1,\ s_2 = -\zeta\omega_n \pm \omega_n\sqrt{\zeta^2 - 1} \tag{6.16}$$

Depending on the value of ζ, three distinct cases present themselves from Eq. (6.16) as shown in Table 6.2.

Case	*Damping ratio*	*System nature*
1	$\zeta > 1$	Overdamped
2	$\zeta = 1$	Critically damped
3	$\zeta < 1$	Underdamped

Table 6.2: Three distinct cases of second order systems

EXAMPLE 6.6. Match the transfer functions of the second-order systems with the nature of the system given below.

Transfer functions	*Nature of system*
P. $\dfrac{15}{s^2 + 5s + 15}$	I.Overdamped
Q. $\dfrac{25}{s^2 + 10s + 25}$	II.Critically damped
R. $\dfrac{35}{s^2 + 18s + 35}$	III.Underdamped

(A) P→I, Q→II, R→III (B) P→II, Q→I, R→III
(C) P→III, Q→II, R→I (D) P→III, Q→I, R→II

Solution Considering the first transfer function, we get

$$\frac{15}{s^2 + 5s + 15} \equiv \frac{\omega_{n1}^2}{s^2 + 2\zeta\omega_{n1}s + \omega_{n1}^2}$$

$$\omega_{n1} = \sqrt{15} = 3.872\,\text{rad/s}$$

$$2\zeta_1\omega_{n1} = 5$$

Therefore, $$\zeta_1 = \frac{5}{2 \times 3.872} = 0.64 \qquad \text{[underdamped]}$$

From the second transfer function, we get

$$\frac{25}{s^2 + 10s + 25} \equiv \frac{\omega_{n2}^2}{s^2 + 2\zeta\omega_{n2}s + \omega_{n2}^2}$$

$$\omega_{n2}^2 = \sqrt{25} = 5\,\text{rad/s}$$

$$2\zeta_2\omega_{n2} = 10$$

Therefore, $$\zeta_2 = \frac{10}{2 \times 5} = 1 \qquad \text{[critically damped]}$$

Ans. (C)

We will consider only the unit step input, i.e. $R(s) = 1/s$, for the second order system.

Underdamped ($0 < \zeta < 1$) system. For the second order underdamped system, we have

$$C(s) = \frac{\omega_n^2}{s^2 + 2\zeta\omega_n s + \omega_n^2} \cdot R(s) = \frac{\omega_n^2}{s^2 + 2\zeta\omega_n s + \omega_n^2} \cdot \frac{1}{s}$$

$$= \frac{\omega_n^2}{s(s + \zeta\omega_n - \jmath\omega_n\sqrt{1-\zeta^2})(s + \zeta\omega_n + \jmath\omega_n\sqrt{1-\zeta^2})} \tag{6.17}$$

$$\equiv \frac{A}{s} + \frac{B}{(s + \zeta\omega_n - \jmath\omega_n\sqrt{1-\zeta^2})} + \frac{C}{(s + \zeta\omega_n + \jmath\omega_n\sqrt{1-\zeta^2})} \tag{6.18}$$

Substituting the values of A, B and C [as obtained in Eqs. (B.4), (B.5) and (B.6) in Appendix *Rules of Partial Fractions*] in Eq. (6.18), we get

$$C(s) = \frac{1}{s} - \left(\frac{\zeta}{2\jmath\sqrt{1-\zeta^2}} + \frac{1}{2}\right)\frac{1}{s + \zeta\omega_n - \jmath\omega_n\sqrt{1-\zeta^2}} + \left(\frac{\zeta}{2\jmath\sqrt{1-\zeta^2}} - \frac{1}{2}\right)\frac{1}{s + \zeta\omega_n + \jmath\omega_n\sqrt{1-\zeta^2}} \tag{6.19}$$

The inverse Laplace transform of Eq. (6.19) yields

$$c(t) = 1 - \left(\frac{\zeta}{2\jmath\sqrt{1-\zeta^2}} + \frac{1}{2}\right)e^{-\left(\zeta - \jmath\sqrt{1-\zeta^2}\right)\omega_n t} + \left(\frac{\zeta}{2\jmath\sqrt{1-\zeta^2}} - \frac{1}{2}\right)e^{-\left(\zeta + \jmath\sqrt{1-\zeta^2}\right)\omega_n t} \tag{6.20}$$

Equation (6.20) yields on simplification

$$c(t) = 1 - e^{-\zeta\omega_n t}\left[\frac{\zeta}{\sqrt{1-\zeta^2}}\sin\omega_d t + \cos\omega_d t\right] \tag{6.21}$$

where, $\omega_d = \omega_n\sqrt{1-\zeta^2}$. Thus,

$$\boxed{c(t) = 1 - \frac{e^{-\zeta\omega_n t}}{\sqrt{1-\zeta^2}}\sin(\omega_d t + \phi)} \tag{6.22}$$

where $\tan\phi = \dfrac{\sqrt{1-\zeta^2}}{\zeta}$.

The plot of Eq. (6.22) shows a sine wave decaying in an exponential envelope (see Figure 6.7).

Note:
1. Equation (6.17) consists of a real pole at the origin and two complex poles.
2. The exponentially decaying amplitude of the response is caused by the real pole whereas the sinusoidal oscillation is caused by the complex poles.
3. The time constant of the exponential decay is actually reciprocal of the real part of the system pole.

Damping ratio. In this context, let us see what is the physical definition of the damping ratio ζ. Often it is defined as the ratio of the exponential decay frequency to the natural frequency of the system. That is

$$\zeta = \frac{\zeta\omega_n}{\omega_n} = \frac{\text{Exponential decay frequency}}{\text{Natural frequency (rad/s)}} \tag{6.23}$$

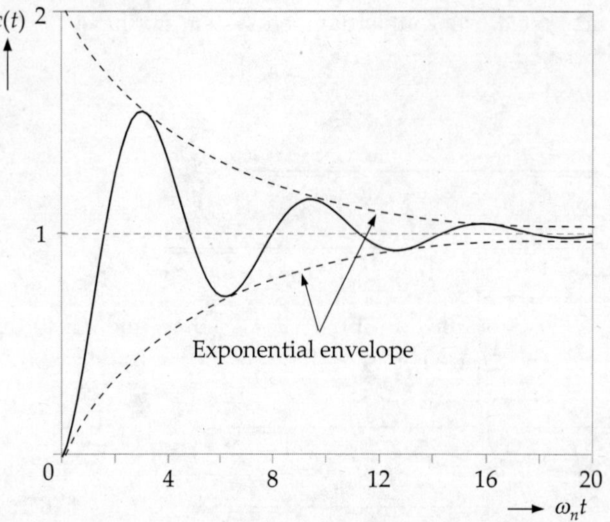

Figure 6.7: Plot of Eq. (6.22).

Also,

$$\text{Exponential decay frequency} = \frac{1}{\text{Exponential time constant}}$$

$$\text{Natural frequency (rad/s)} = \frac{2\pi}{\text{Natural period (s)}}$$

Therefore, from Eq. (6.23) we have

$$\zeta = \frac{1}{2\pi} \cdot \frac{\text{Natural period (s)}}{\text{Exponential time constant}}$$

The exponential decay frequency σ_D is thus defined as

$$\sigma_D = \zeta\omega_n$$

Critically damped ($\zeta = 1$) system. For the critically damped second order system,

$$C(s) = \frac{\omega_n^2}{s(s+\omega_n)^2} = \frac{1}{s} - \frac{\omega_n}{(s+\omega_n)^2} - \frac{1}{s+\omega_n} \tag{6.24}$$

Therefore,

$$c(t) = 1 - \omega_n t e^{-\omega_n t} - e^{-\omega_n t} = 1 - (1+\omega_n t)e^{-\omega_n t} \tag{6.25}$$

Overdamped ($\zeta > 1$) system. In the case of overdamped second order system, the response works out to be

$$c(t) = 1 - \frac{\zeta + \sqrt{\zeta^2 - 1}}{2\sqrt{\zeta^2 - 1}} e^{\left(-\zeta + \sqrt{\zeta^2 - 1}\right)\omega_n t} + \frac{\zeta - \sqrt{\zeta^2 - 1}}{2\sqrt{\zeta^2 - 1}} e^{\left(-\zeta - \sqrt{\zeta^2 - 1}\right)\omega_n t} \tag{6.26}$$

The response curves for critically damped and over-damped systems are given in Figure 6.8.

Note: It is apparent from the curves that

1. The higher the value of ζ, the more the rise time and therefore, the settling time (see Figure 6.7).
2. A critically-damped second order system attains the step input value at the time constant value of nearly 8 (see Figure 6.8) while this value was nearly 5 for the first order system (see Table 6.1).

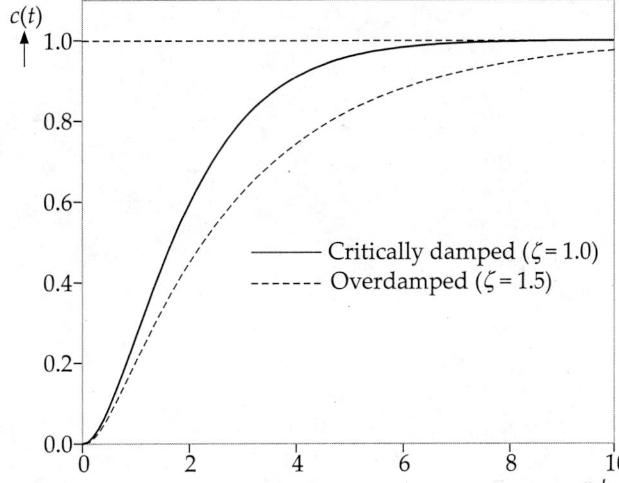

Figure 6.8: Response curves for critically damped and overdamped systems.

Performance Parameters for Response

The performance parameters for response are defined in Table 6.3.

Table 6.3: Performance parameters for response

Parameter	*Definition*
Delay time, t_d	The time required for the step response of the system to reach 50% of the final value.
Rise time, t_r	The time required for the step response to reach the final output for the first time in underdamped systems or grow from 10% to 90% of the value in overdamped systems.
Peak time, t_p	The time required by the step response of the system to reach the first peak which is the maximum value of the output.
Maximum overshoot, M_p	The maximum amount by which the output deviates from the input. It is generally expressed as percent of the input.
Settling time, t_s	The time the system takes to settle within a given per cent of the final value.

These parameters are indicated in Figure 6.9 which shows the step response of a second order system. Now, let us discuss these parameters one by one.

Delay time. According to the definition, $c(t)|_{t=t_d} = \frac{1}{2}$. So, from Eqs. (6.22), (6.25) and (6.26), we get

$$\begin{aligned} \frac{e^{-\zeta x}}{\sqrt{1-\zeta^2}}\sin(\sqrt{1-\zeta^2}x+\phi) &= \frac{1}{2} \qquad \text{for} \quad 0<\zeta<1 \\ (1+x)e^{-x} &= \frac{1}{2} \qquad \text{for} \quad \zeta=1 \\ \frac{\zeta+\sqrt{\zeta^2-1}}{2\sqrt{\zeta^2-1}}e^{(-\zeta+x\sqrt{\zeta^2-1})} + \frac{\zeta-\sqrt{\zeta^2-1}}{2\sqrt{\zeta^2-1}}e^{(-\zeta-x\sqrt{\zeta^2-1})} &= \frac{1}{2} \qquad \text{for} \quad \zeta>1 \end{aligned} \tag{6.27}$$

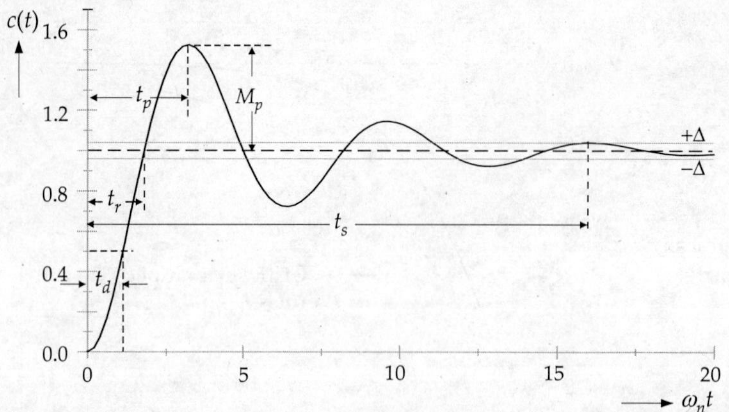

Figure 6.9: Graphical presentation of performance parameters of an underdamped second order system. - - - input, — output. $\pm\Delta$ lines indicate 2% tolerance band.

where $x = \omega_n t_d$. Solutions to Eq. (6.27) can be obtained for different values of ζ and plotted to draw a graph of ζ vs. x. A linear approximation to this graph is given by the relation

$$x = 1 + 0.7\zeta \tag{6.28}$$

Equation (6.28) yields the approximate expression for t_d as

$$\boxed{t_d = \frac{1 + 0.7\zeta}{\omega_n}} \tag{6.29}$$

Rise time. We consider the underdamped system for simplicity. If t_r = time required to attain the steady-state value, we note that the sine factor of Eq. (6.22) should have a zero value to make $c(t) = 1$. Thus, from Eq. (6.22) we get

$$\sin(\omega_d t_r + \phi) = \sin N\pi \tag{6.30}$$

For the first time, i.e., for $N = 1$, we have from Eq. (6.30)

$$\omega_n t_r \sqrt{1-\zeta^2} + \phi = \pi$$

or,

$$\boxed{t_r = \frac{\pi - \phi}{\omega_n \sqrt{1-\zeta^2}}} \tag{6.31}$$

where

$$\phi = \tan^{-1} \frac{\sqrt{1-\zeta^2}}{\zeta}$$

Peak time. If t_p is the time required to reach the first peak, we can find an expression for it from Eq. (6.22) using the condition that $c(t)$ will be maximum here. Thus,

$$\left.\frac{dc(t)}{dt}\right|_{t=t_p} = 0$$

$$\Rightarrow \quad -\frac{e^{\zeta\omega_n t_p}}{\sqrt{1-\zeta^2}}\omega_d \cos(\omega_d t_p + \phi) + \frac{\zeta\omega_n e^{\zeta\omega_n t_p}}{\sqrt{1-\zeta^2}}\omega_d \sin(\omega_d t_p + \phi) = 0$$

$$\Rightarrow \quad \omega_d \cos(\omega_d t_p + \phi) - \zeta\omega_n \sin(\omega_d t_p + \phi) = 0$$

$$\Rightarrow \quad \sqrt{1-\zeta^2}\cos[(\omega_n\sqrt{1-\zeta^2})t_p + \phi] - \zeta \sin[(\omega_n\sqrt{1-\zeta^2})t_p + \phi] = 0$$

$$\Rightarrow \quad \sin\phi \cos[(\omega_n\sqrt{1-\zeta^2})t_p + \phi] - \cos\phi \sin[(\omega_n\sqrt{1-\zeta^2})t_p + \phi] = 0$$

$$\Rightarrow \quad \sin[(\omega_n\sqrt{1-\zeta^2})t_p + \phi - \phi] = 0 = \sin N\pi \tag{6.32}$$

For the first peak, $N = 1$. Therefore, from Eq. (6.32)

$$\boxed{t_p = \frac{\pi}{\omega_n\sqrt{1-\zeta^2}}} \tag{6.33}$$

Maximum overshoot. Since it is the overshoot value at the peak time,

$$M_p = c(t)\big|_{t=t_p} - 1 = -\frac{e^{\zeta\omega_n t_p}}{\sqrt{1-\zeta^2}}\sin[(\omega_n\sqrt{1-\zeta^2})t_p + \phi] \tag{6.34}$$

Substituting t_p value from Eq. (6.33) in Eq. (6.34), after a little algebraic manipulation, we get

$$\boxed{M_p = e^{-\frac{\pi\zeta}{\sqrt{1-\zeta^2}}}} \tag{6.35}$$

Maximum per cent overshoot can be obtained by multiplying right-hand side of Eq. (6.35) by 100. It may be observed that M_p is a function of ζ only. Its behaviour with ζ is shown in Figure 6.10.

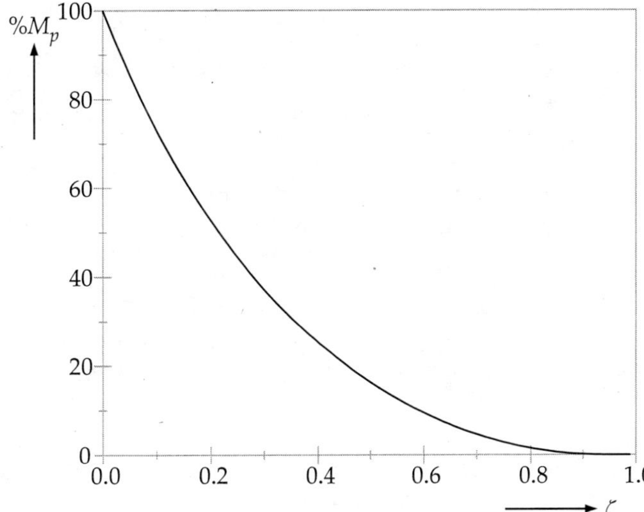

Figure 6.10: ζ vs. $\%M_p$ plot.

EXAMPLE 6.7. A unity feedback system has open-loop transfer function $G(s) = 25/\{s(s+6)\}$. The peak overshoot in the step-input response of the system is approximately equal to

(A) 5% (B) 10% (C) 15% (D) 20%

Solution Given, $G(s) = 25/\{s(s+6)\}$ and $H(s) = 1$. Therefore, the characteristic equation of the system is

$$1 + G(s)H(s) = 1 + \frac{25}{s^2+6s} = s^2 + 6s + 25 = 0 \tag{i}$$

Comparing Eq. (i) with the standard characteristic equation of a second order system, $s^2 + 2\zeta\omega_n s + \omega_n^2$, we get

$$\omega_n^2 = 25 \quad \Rightarrow \quad \omega_n = 5$$
$$2\zeta\omega_n = 6 \quad \Rightarrow \quad \zeta = 0.6$$

$$\%M_p = \left\{ e^{-\dfrac{\pi\zeta}{\sqrt{1-\zeta^2}}} \right\} \times 100 = 9.478 \cong 10$$

Ans. (B)

Settling time. If t_s is the time for settling within 2% of the final value, then

$$\boxed{t_s = \frac{4}{\zeta\omega_n}} \tag{6.36}$$

This approximate relation can be arrived at from the following considerations. Suppose we want that the output settles within a tolerance band of $\pm\Delta$. That means the amplitude of the oscillating sine factor should equal Δ. Therefore,

$$\frac{e^{\zeta\omega_n t_p}}{\sqrt{1-\zeta^2}} = \Delta \tag{6.37}$$

Taking logarithm of both sides of Eq. (6.37) and rearranging, we get

$$-\zeta\omega_n t_s = \ln(\Delta\sqrt{1-\zeta^2}) = \ln\Delta - \frac{\zeta^2}{2}$$

or,

$$t_s = \frac{\zeta}{2\omega_n} - \frac{\ln\Delta}{\zeta\omega_n} \tag{6.38}$$

Since $\zeta < 1$, the second term on the right hand side of Eq. (6.38) dominates and so we may write it as

$$t_s \approx -\frac{\ln\Delta}{\zeta\omega_n} \tag{6.39}$$

Substituting $\Delta = 0.02$ in Eq. (6.39), we get the approximate relation of Eq. (6.36).

Steady-state error. From the definition of error

$$e(t) = r(t) - c(t) = \frac{e^{-\zeta\omega_n t}}{\sqrt{1-\zeta^2}} \sin(\omega_d t + \phi)$$

Therefore,

$$e_{ss} = e(t)\big|_{t\to\infty} = 0$$

With this background on the response of systems, we will work out a few problems and then discuss how steady-state error can be calculated without having recourse to calculation of actual response as a function of time.

EXAMPLE 6.8. A measurement system with input $x(t)$ and output $y(t)$ is described by the differential equation

$$3\frac{dy}{dx} + 5y = 8x$$

The static sensitivity of the system is

(A) 0.60 (B) 1.60 (C) 1.67 (D) 2.67

Solution The given equation can be written as

$$\frac{3}{5}\frac{dy}{dx} + y = \frac{8}{5}x$$

Comparing it with Eq. (6.2) at page 115, the static sensitivity is found to be $\frac{8}{5} = 1.6$.

Ans. (B)

EXAMPLE 6.9. A linear time-invariant system initially at rest, when subjected to a step input, gives a response $y(t) = te^{-t},\ t > 0$. The transfer function of the system is

(A) $\dfrac{1}{(s+1)^2}$ (B) $\dfrac{1}{s(s+1^2}$ (C) $\dfrac{s}{(s+1)^2}$ (D) $\dfrac{1}{s(s+1)}$

Solution Given response is $y(t) = te^{-t}$ and input $r(t) = u(t)$. So,

$$\mathscr{L}\{y(t)\} = Y(s) = \frac{1}{(s+1)^2}$$

$$\mathscr{L}\{r(t)\} = R(s) = \frac{1}{s}$$

Therefore,

$$\frac{Y(s)}{R(s)} = \frac{s}{(s+1)^2}$$

Ans. (C)

EXAMPLE 6.10. A stable real linear time-invariant system with single pole p, has a transfer function $H(s) = (s^2+100)/(s-p)$ with a dc gain of 5. The smallest positive frequency, in rad/s, at unity gain is closed to

(A) 8.84 (B) 11.08 (C) 78.13 (D) 122.87

Solution We need to figure out the value of p first. For that, we note that the given transfer function has a dc gain of 5, i.e. at $\omega = 0$ the magnitude of the transfer function is 5. This means that

$$|G(s)| = \left|\frac{s^2+100}{s-p}\right|_{\omega=0} \equiv \left|\frac{(\jmath\omega)^2+100}{\jmath\omega-p}\right|_{\omega=0} = 5 \tag{i}$$

From Eq. (i), we get $p = -20$. Now, we are required to find the minimum frequency at which the gain is 1. This gives the condition

$$\left|\frac{(\jmath\omega)^2+100}{\jmath\omega+20}\right| = 1$$

or,

$$\frac{100-\omega^2}{\sqrt{\omega^2+20^2}} = 1$$

or,

$$\omega^4 - 201\omega^2 + 9600 = 0$$

or,

$$x^2 - 201x + 9600 = 0 \qquad [\text{substituting } x = \omega^2] \tag{ii}$$

Solving Eq. (ii) for the minimum value of x yields $\omega \cong 8.84$.
Ans. (A)

EXAMPLE 6.11. A unity feedback system has an open-loop transfer function $G(s) = K/[s(s+3]$. The value of K that yields a damping ratio of 0.5 for the closed-loop system is

(A) 1 (B) 3 (C) 5 (D) 9

Solution The characteristic equation is given by

$$s(s+3) + K = 0$$

So, we have by comparison with a standard second order system

$$\omega_n^2 = K \qquad \Rightarrow \omega_n = \sqrt{K}$$

$$2\zeta\omega_n = 3 \qquad \Rightarrow \zeta = \frac{3}{2\omega_n} = \frac{3}{2\sqrt{K}} = 0.5 \quad \text{(given)}$$

$$\text{or,}\quad \sqrt{K} = 3 \qquad \Rightarrow K = 9$$

Ans. (D)

EXAMPLE 6.12. If a first order system and its time response to a unit step input are as shown in Figures 6.11a and 6.11b, respectively, then the gain K is

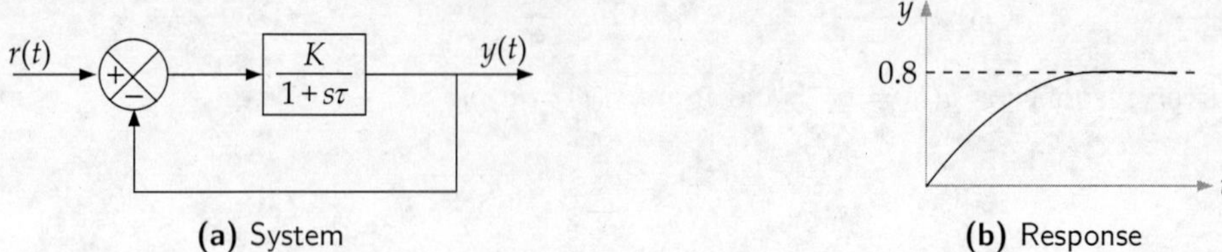

(a) System **(b)** Response

Figure 6.11: Example 6.12.

(A) 0.25 (B) 0.8 (C) 1 (D) 4

Solution Let us calculate the response as follows:

$$\frac{Y(s)}{R(s)} = \frac{K}{s\tau + K + 1}$$

$$\Rightarrow \qquad Y(s) = \frac{1}{s} \cdot \frac{K}{s\tau + K + 1} \qquad \left[\because R(s) = \frac{1}{s}\right]$$

Therefore,

$$y_{ss} = \lim_{s\to 0} sY(s) = \lim_{s\to 0} s \cdot \frac{1}{s} \cdot \frac{K}{s\tau + K + 1} = \frac{K}{K+1} \qquad \text{(i)}$$

From Eq. (i) and the response curve, we get

$$\frac{K}{K+1} = 0.8 \qquad \Rightarrow \quad 1 + \frac{1}{K} = \frac{1}{0.8} \qquad \Rightarrow \quad K = \frac{0.8}{0.2} = 4$$

Ans. (D)

EXAMPLE 6.13. The unit step response of a system with the transfer function $G(s) = (1 - 2s)/(1 + s)$ is given by which of the following waveforms?

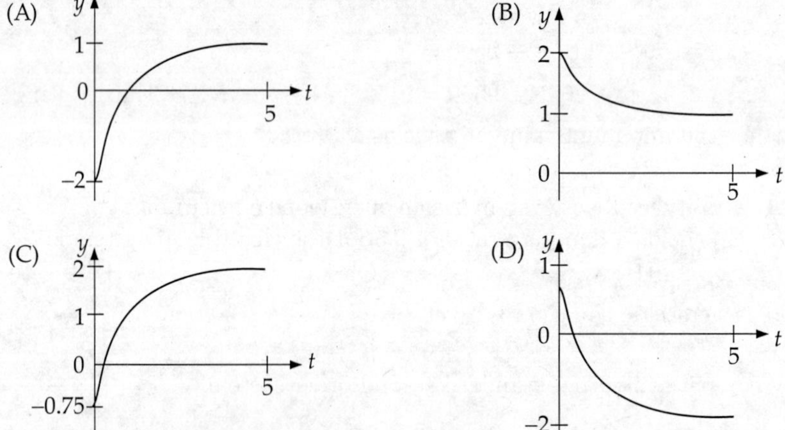

Solution From the given transfer function we get its step response as

$$Y(s) = G(s)U(s) = \frac{1-2s}{1+s} \cdot \frac{1}{s} \equiv \frac{A}{s} + \frac{B}{s+1} \qquad \text{(i)}$$

Now,

$$A = \left.\frac{1-2s}{s+1}\right|_{s=0} = 1 \qquad B = \left.\frac{1-2s}{s}\right|_{s=-1} = -3$$

Substituting these values in Eq. (i) and taking its inverse Laplace transform, we get

$$y(t) = \left(1 - 3e^{-t}\right)u(t) \tag{ii}$$

Equation (ii) indicates that

$$\lim_{t=0} y(t) = -2 \qquad \text{and} \qquad \lim_{t\to\infty} y(t) = 1$$

which tallies with the curve (A).
Ans. (A)

EXAMPLE 6.14. A unity feedback system has open-loop transfer function $G(s) = 100/[s(s+p)]$. The time at which the response to a unit step input reaches its peak is $\pi/8$ seconds.

(i) The damping coefficient for the closed-loop system is
(A) 0.4 (B) 0.6 (C) 0.8 (D) 1

(ii) The value of p is
(A) 6 (B) 12 (C) 14 (D) 16

Solution Since, $H(s) = 1$, the closed-loop transfer function is given by

$$\frac{C(s)}{R(s)} = \frac{100}{s^2 + ps + 100}$$

Therefore, comparing it with the standard second-order transfer function we have

$$\omega_n^2 = 100 \qquad \Rightarrow \omega_n = 10$$

$$2\zeta\omega_n = p \qquad \Rightarrow \zeta = \frac{p}{20}$$

(i) Now

$$t_p = \frac{\pi}{\omega_n\sqrt{1-\zeta^2}} = \frac{\pi}{8} \text{ (given)} \quad \Rightarrow \quad \omega_n\sqrt{1-\zeta^2} = 8 \quad \Rightarrow \quad \sqrt{1-\zeta^2} = \frac{8}{10} = 0.8$$

$$\Rightarrow \quad \zeta^2 = 1 - 0.64 = 0.36 \qquad \therefore \quad \zeta = 0.6$$

Ans. (B)

(ii) $p = 2\zeta\omega_n = 2 \times 0.6 \times 10 = 12$
Ans. (B)

EXAMPLE 6.15. A second-order real system has the following properties:

(a) the damping ratio $\zeta = 0.5$ and undamped natural frequency $\omega_n = 10$ rad/s.
(b) the steady state value of the output, to a unit step input, is 1.02.

The transfer function of the system is

(A) $\dfrac{1.02}{s^2 + 5s + 100}$ (B) $\dfrac{102}{s^2 + 10s + 100}$ (C) $\dfrac{100}{s^2 + 10s + 100}$ (D) $\dfrac{102}{s^2 + 5s + 100}$

Solution The standard second-order transfer function is

$$G(s) = K\frac{\omega_n^2}{s^2 + 2\zeta\omega_n s + \omega_n^2}$$

Given, $\zeta = 0.5$ and $\omega_n = 10$. Therefore,

$$G(s) = K\frac{100}{s^2 + 10s + 100}$$

Now, the steady state output is 1.02. This means that

$$c(\infty) = \lim_{s\to 0} s \cdot \frac{1}{s}\left[\frac{100}{s^2 + 10s + 100}\right]K = 1.02$$

$$\Rightarrow \qquad K = 1.02$$

Therefore, $$G(s) = \frac{1.02 \times 100}{s^2 + 10s + 100} = \frac{102}{s^2 + 10s + 100}$$

Ans. (B)

EXAMPLE 6.16. The unit step response of a unity feedback system with open-loop transfer function $G(s) = K/[(s + 1)(s + 2)]$ is shown in Figure 6.12.

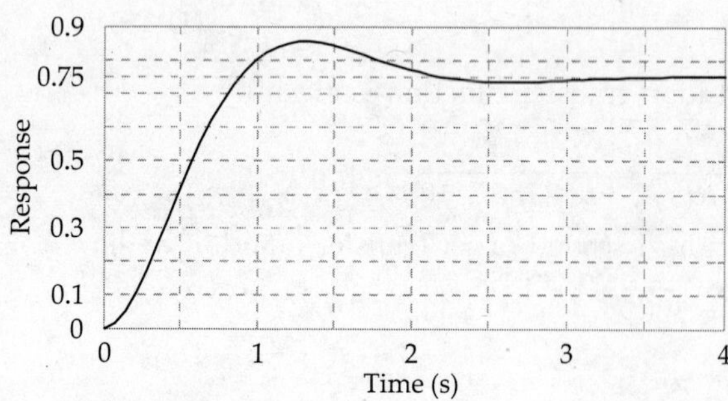

Figure 6.12: Example 6.16.

The value of K is

(A) 0.5 (B) 2 (C) 4 (D) 6

Solution From the given forward transfer function and unity feedback, the closed-loop transfer function is obtained as

$$\frac{C(s)}{R(s)} = \frac{K}{(s + 1)(s + 2) + K} = \frac{K}{s^2 + 3s + 2 + K}$$

or, $$C(s) = \frac{K}{s^2 + 3s + 2 + K} \cdot \frac{1}{s} \qquad \left[\because R(s) = \frac{1}{s}\right]$$

We get from the given diagram $c_{ss} = 0.75$. So,

$$\lim_{s\to 0} s \cdot C(s) = \lim_{s\to 0} \frac{K}{s^2 + 3s + 2 + K} = 0.75$$

$$\Rightarrow \qquad \frac{K}{2 + K} = \frac{3}{4}$$

$$\Rightarrow \qquad K = 6$$

Ans. (D)

EXAMPLE 6.17. For a unity feedback system with open-loop gain of $G(s) = \dfrac{9}{s(s+3)}$, the closed-loop system will have a damping ratio of

(A) 0.33 (B) 0.5 (C) 1.0 (D) 3

Solution The closed-loop transfer function is given by

$$\frac{C(s)}{R(s)} = \frac{\dfrac{9}{s(s+3)}}{1+\dfrac{9}{s(s+3)}} = \frac{9}{s^2+3s+9}$$

Comparing the closed-loop transfer function to that of a standard second order system, we have

$$\omega_n^2 = 9 \qquad \Rightarrow \omega_n = 3$$

$$2\zeta\omega_n = 3 \qquad \Rightarrow \zeta = \frac{3}{2\omega_n} = \frac{3}{2\times 3} = 0.5$$

Ans. (B)

EXAMPLE 6.18. The transfer function of a control system is given by $\dfrac{C(s)}{R(s)} = \dfrac{25}{s^2+6s+25}$. The first maximum value of the response occurs at a time $t_{\max}$ given by

(A) $\dfrac{\pi}{8}$ (B) $\dfrac{\pi}{4}$ (C) $\dfrac{\pi}{2}$ (D) π

Solution Comparing the closed-loop transfer function to that of a standard second order system, we have

$$\omega_n^2 = 25 \qquad \Rightarrow \omega_n = 5$$

$$2\zeta\omega_n = 6 \qquad \Rightarrow \zeta = \frac{6}{2\omega_n} = \frac{6}{2\times 5} = 0.6$$

$$t_{\max} = \frac{\pi}{5\sqrt{1-0.6^2}} = \frac{\pi}{4}$$

Ans. (B)

EXAMPLE 6.19. The impulse response of a system S_1 is given by $x_1(t) = 4e^{-2t}$. The step response of a system S_2 is given by $x_2(t) = 2\left(1-e^{-3t}\right)$.

(a) Find the transfer functions of the systems S_1 and S_2.

(b) The two systems are cascaded together without any interaction. Find the unit step response of the cascaded system.

(c) Sketch the unit ramp response of the cascaded system.

Solution (a) Taking Laplace transform of the response of S_1, we get

$$X_1(s) = \frac{4}{s+2}$$

Transfer function of S_1

$$G_1(s) = \frac{4}{s+2} \qquad [\because Y_1(s) = 1]$$

Taking Laplace transform of the response of S_2, we get

$$X_2(s) = 2\left(\frac{1}{s} - \frac{1}{s+3}\right) = \frac{6}{s(s+3)}$$

$$Y_2(s) = \frac{1}{s}$$

Transfer function of S_2 $$G_2(s) = \frac{6}{s(s+3)} \cdot s = \frac{6}{s+3}$$

(b) The transfer function of the cascaded system is

$$G(s) = G_1(s)G_2(s) = \frac{24}{(s+2)(s+3)}$$

The Laplace transform of unit ramp is $R(s) = \dfrac{1}{s^2}$. Therefore,

$$C(s) = \frac{24}{(s+2)(s+3)} \cdot \frac{1}{s^2} \equiv \frac{A}{s^2} + \frac{B}{s} + \frac{C}{s+2} + \frac{D}{s+3}$$

$$A = \left.\frac{24}{(s+2)(s+3)}\right|_{s=0} = 4$$

$$B = \frac{d}{ds}\left[s^2 C(s)\right]_{s=0} = \frac{d}{ds}\left[\frac{24}{(s+2)(s+3)}\right] = -\left.\frac{24(2s+5)}{(s+2)^2(s+3)^2}\right|_{s=0} = -\frac{10}{3}$$

$$C = \left.\frac{24}{s^2(s+3)}\right|_{s=-2} = 6$$

$$D = \left.\frac{24}{s^2(s+2)}\right|_{s=-3} = -\frac{8}{3}$$

$$C(s) = \frac{4}{s^2} - \frac{10}{3}s + \frac{6}{s+2} - \frac{8}{3}e^{-3t}$$

Therefore, $$c(t) = 4t - \frac{10}{3}u(t) + 6e^{-2t} - \frac{8}{3}e^{-3t}$$

(c) The plot of the response is as given in Figure 6.13.

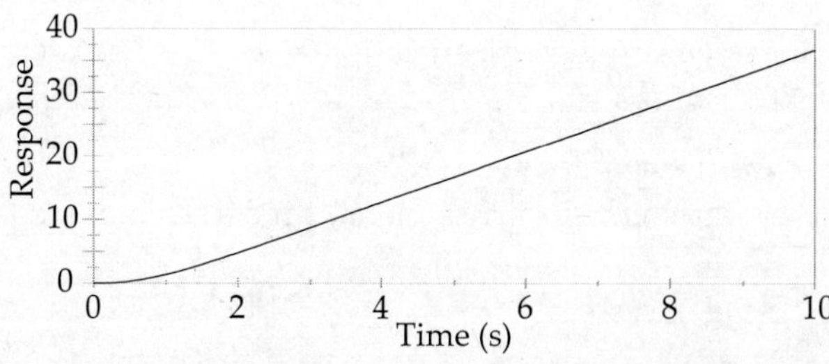

Figure 6.13: Response (Example 6.19).

EXAMPLE 6.20. For a unity feedback system having

$$G(s) = \frac{25}{s(s+10)}$$

find (a) ω_n, (b) ζ, (c) ω_d, (d) t_p and (e) M_p.

Solution Since it is a unity feedback system, its overall transfer function is given by

$$T(s) = \frac{G(s)}{1+G(s)} = \frac{25}{s(s+10)+25} = \frac{25}{s^2+10s+25} \equiv \frac{\omega_n^2}{s^2+2\zeta\omega_n s+\omega_n^2}$$

Comparing the denominator of the given transfer function with that of the standard second order system, we get

(a) $\omega_n^2 = 25$. Therefore, $\omega_n = 5$ rad/s

(b) $2\zeta\omega_n = 10$. Therefore, $\zeta = \dfrac{5}{\omega_n} = 1$ (critically damped)

(c) $\omega_d = \omega_n\sqrt{1-\zeta^2} = 0$

(d) $t_p = \dfrac{\pi}{\omega_n\sqrt{1-\zeta^2}} \to \infty$ (no peak)

(e) $M_p = 0$, since it has no peak

EXAMPLE 6.21. The open-loop transfer function of a unity feedback system is given by

$$G(s) = \frac{K}{s(1+sT)}$$

where T and K are positive constants. By what factor should the gain K be reduced such that

(a) the peak overshoot of the system to unit step input is reduced from 75% to 25%, and

(b) the damping ratio increases from 0.1 to 0.6?

Solution The overall transfer function of the system is given by

$$T(s) = \frac{G(s)}{1+G(s)} = \frac{K}{s(1+sT)+K} = \frac{K/T}{s^2+\dfrac{s}{T}+\dfrac{K}{T}} \equiv \frac{\omega_n^2}{s^2+2\zeta\omega_n s+\omega_n^2}$$

By comparison, we get $\omega_n = \sqrt{K/T}$ and $\zeta = \dfrac{1}{2\sqrt{KT}}$.

(a) From the given condition and Eq. (6.35), we get

$$(M_p)_1 = 100e^{-\frac{\pi\zeta_1}{\sqrt{1-\zeta_1^2}}} = 75 \quad \text{(i)}$$

$$(M_p)_2 = 100e^{-\frac{\pi\zeta_2}{\sqrt{1-\zeta_2^2}}} = 25 \quad \text{(ii)}$$

On taking logarithms of both sides, we get

$$-\frac{\pi\zeta_1}{\sqrt{1-\zeta_1^2}} = \ln\frac{75}{100} = \ln 3 - \ln 4 \quad \text{[from Eq. (i)]}$$

or,

$$\frac{\pi\zeta_1}{\sqrt{1-\zeta_1^2}} = \frac{\ln 4 - \ln 3}{\pi} = 0.0916 \quad \text{(iii)}$$

$$\frac{\pi\zeta_2}{\sqrt{1-\zeta_2^2}} = \frac{\ln 4 - \ln 1}{\pi} = 0.4413 \quad \text{[from Eq. (ii)] (iv)}$$

Substituting $\zeta_1 = \dfrac{1}{2\sqrt{K_1 T}}$ and $\zeta_2 = \dfrac{1}{2\sqrt{K_2 T}}$, we get from Eqs. (iii) and (iv)

$$\frac{1}{\sqrt{4K_1T-1}} = 0.0916 \qquad \text{or,} \qquad K_1T = 30.045 \tag{v}$$

$$\frac{1}{\sqrt{4K_2T-1}} = 0.4413 \qquad \text{or,} \qquad K_2T = 1.534 \tag{vi}$$

Thus from Eqs. (v) and (vi), we have

$$\frac{K_1}{K_2} = \frac{30.045}{1.534} \tag{vii}$$

Equation (vii) suggests that K should be reduced nearly 20 times to lower the peak overshoot from 75% to 25%.

(b) Here,

$$\frac{\zeta_1}{\zeta_2} = \sqrt{\frac{K_2T}{K_1T}} \qquad \text{or,} \qquad \frac{K_1}{K_2} = \frac{\zeta_2^2}{\zeta_1^2} = \frac{(0.6)^2}{(0.1)^2} = 36 \tag{viii}$$

The import of Eq. (viii) is that K has to decrease by 1/36 of its present value to increase the damping ratio from 0.1 to 0.6.

EXAMPLE 6.22. Find the time response of the following system for input $r(t) = \sin t$:

$$T(s) = \frac{1}{(s+1)(s^2+1)}$$

Solution

$$r(t) = \sin t \qquad \Rightarrow R(s) = \frac{1}{s^2+1}$$

Therefore,

$$C(s) = \frac{1}{(s+1)(s^2+1)^2} \equiv \frac{A}{s+1} + \frac{Bs+C}{(s^2+1)^2} + \frac{Ds+E}{s^2+1} \quad \text{(say)}$$

Equating the numerators of both sides, we get from comparing the coefficients of s^6, s^5, s^4, s^3, s^2, s and the constants

$$A = \frac{1}{4},\ B = -\frac{1}{2},\ C = \frac{1}{2},\ D = -\frac{1}{4} \text{ and } E = \frac{1}{4}$$

Thus,

$$\begin{aligned} C(s) &= \frac{1}{4}\cdot\frac{1}{s+1} - \frac{1}{2}\cdot\frac{s-1}{(s^2+1)^2} - \frac{1}{4}\cdot\frac{s-1}{s^2+1} \\ &= \frac{1}{4}\cdot\frac{1}{s+1} - \frac{1}{2}\cdot\frac{s-1}{(s^2+1)^2} + \frac{1}{2}\cdot\frac{1}{(s^2+1)^2} - \frac{1}{4}\cdot\frac{s}{s^2+1} + \frac{1}{4}\cdot\frac{1}{s^2+1} \end{aligned}$$

On performing inverse Laplace transform, we get

$$\begin{aligned} c(t) &= \frac{1}{4}e^{-t} - \frac{1}{2}\frac{d}{dt}(t\sin t) + \frac{1}{2}t\sin t - \frac{1}{4}\cos t + \frac{1}{4}\sin t \\ &= \frac{1}{4}e^{-t} - \frac{1}{2}\sin t - \frac{1}{2}t\cos t + \frac{1}{2}t\sin t - \frac{1}{4}\cos t + \frac{1}{4}\sin t \\ &= \frac{1}{4}e^{-t} - \frac{1}{2\sqrt{2}}\sin\left(t + \frac{\pi}{4}\right) + \frac{1}{\sqrt{2}}t\sin\left(t - \frac{\pi}{4}\right) \end{aligned}$$

EXAMPLE 6.23. A step input is applied to a unity feedback system having the following open-loop transfer function:

$$G(s) = \frac{15}{s(s+3)}$$

Find (a) the closed-loop transfer function, (b) ω_n, (c) ζ and (d) ω_d.

Solution (a) The closed-loop transfer function $= \dfrac{G(s)}{1+G(s)} = \dfrac{15}{s(s+3)+15} = \dfrac{15}{s^2+3s+15}$

(b) Comparing the denominator of closed-loop transfer function with that of a standard second order transfer function, we get

$$\omega_n^2 = 15 \qquad \text{or,} \qquad \omega_n = \sqrt{15} = 3.873 \text{ rad/s}$$

(c) From the same comparison, we also get

$$2\zeta\omega_n = 3 \qquad \text{or,} \qquad \zeta = \frac{3}{2\omega_n} = \frac{3}{2\sqrt{15}} = 0.387$$

(d) $\omega_d = \omega_n\sqrt{1-\zeta^2} = 3.571$ rad/s

EXAMPLE 6.24. A second order control system, having $\zeta = 0.4$ and $\omega_n = 5$ rad/s, is subjected to a step input. Determine (a) closed-loop transfer function, (b) t_r, (c) t_p, (d) t_s and (e) M_p.

Solution (a) The standard closed-loop transfer function for a second order control system is given by

$$\frac{C(s)}{R(s)} = \frac{\omega_n^2}{s^2 + 2\zeta\omega_n s + \omega_n^2}$$

Substituting the values of ζ and ω_n, we get

$$\frac{C(s)}{R(s)} = \frac{25}{s^2+4s+25}$$

(b) $t_r = \dfrac{\pi - \cos^{-1}\zeta}{\omega_n\sqrt{1-\zeta^2}} = \dfrac{\pi - \cos^{-1}0.4}{5\sqrt{1-0.16}} = 0.433$ s

(c) $t_p = \dfrac{\pi}{\omega_n\sqrt{1-\zeta^2}} = \dfrac{\pi}{5\sqrt{1-0.16}} = 0.686$ s

(d) $t_s = \dfrac{4}{\zeta\omega_n} = \dfrac{4}{(0.4)(0.5)} = 2$ s

(e) $M_p = 100\exp\left(-\dfrac{\pi\zeta}{\sqrt{1-\zeta^2}}\right) = 100\exp\left(-\dfrac{0.4\pi}{\sqrt{1-0.16}}\right) = 25.4\%$

EXAMPLE 6.25. A unity feedback control system has the following open-loop transfer function:

$$G(s) = \frac{4s+1}{4s^2}$$

Find expressions for its time response when it is subjected to (a) unit impulse input and (b) unit step input.

Solution The closed-loop transfer function of the system is given by

$$T(s) = \frac{C(s)}{R(s)} = \frac{4s+1}{4s^2+4s+1} = \frac{4s+1}{(2s+1)^2}$$

(a) For a unit impulse input, $R(s) = 1$.

Therefore, $$C(s) = \frac{4s+1}{(2s+1)^2} = \frac{s+\frac{1}{4}}{\left(s+\frac{1}{2}\right)^2} \equiv \frac{A}{\left(s+\frac{1}{2}\right)^2} + \frac{B}{s+\frac{1}{2}} \text{ (say)}$$

Then, $$A = \left(s+\frac{1}{2}\right)^2 C(s)\bigg|_{s=-\frac{1}{2}} = \left(s+\frac{1}{2}\right)_{s=-\frac{1}{2}} = -\frac{1}{4}$$

$$B = \frac{d}{ds}\left\{\left(s+\frac{1}{2}\right)^2 C(s)\right\}_{s=-\frac{1}{2}} = 1$$

Thus, $$C(s) = -\frac{1}{4\left(s+\frac{1}{2}\right)^2} + \frac{1}{s+\frac{1}{2}}$$

The inverse Laplace transform of the above equation yields

$$c(t) = e^{-\frac{t}{2}} - \frac{1}{4}te^{-\frac{t}{2}} = \left[1 - \frac{t}{4}\right] e^{-\frac{t}{2}}$$

(b) For a unit step input, $R(s) = \dfrac{1}{s}$. Therefore,

$$C(s) = \frac{s+\frac{1}{4}}{s\left(s+\frac{1}{2}\right)^2} \equiv \frac{A}{s} + \frac{B}{\left(s+\frac{1}{2}\right)^2} + \frac{C}{s+\frac{1}{2}} \text{ (say)}$$

$$A = sC(s)|_{s=0} = \frac{s+\frac{1}{4}}{\left(s+\frac{1}{2}\right)^2}\bigg|_{s=0} = 1$$

$$B = \left(s+\frac{1}{2}\right)^2 C(s)\bigg|_{s=-\frac{1}{2}} = \frac{s+\frac{1}{4}}{s}\bigg|_{s=-\frac{1}{2}} = \frac{1}{2}$$

$$C = \frac{d}{ds}\left\{\left(s+\frac{1}{2}\right)^2 C(s)\right\}\bigg|_{s=-\frac{1}{2}} = \frac{d}{ds}\left\{\frac{s+\frac{1}{4}}{s}\right\}\bigg|_{s=-\frac{1}{2}} = \left[\frac{1}{s} - \frac{s+\frac{1}{4}}{s^2}\right]_{s=-\frac{1}{2}} = -1$$

Thus, $$C(s) = \frac{1}{s} + \frac{1}{2\left(s+\frac{1}{2}\right)^2} - \frac{1}{s+\frac{1}{2}}$$

Its inverse Laplace transform yields

$$c(t) = 1 + \frac{1}{2}te^{-\frac{t}{2}} - e^{-\frac{t}{2}} = 1 - \left(1 - \frac{t}{2}\right) e^{-\frac{t}{2}}$$

EXAMPLE 6.26. Determine the transfer function of the control system given in Figure 6.14a and derive an expression for its output if the input is a step having a magnitude of 2 units.

Solution The block diagram can be reduced as shown in Figure 6.14b. The reduced block diagram yields the closed-loop transfer function of the system as

$$T(s) = \frac{C(s)}{R(s)} = \frac{4}{s(s+4)+4+4(s+2)} = \frac{4}{s^2+8s+12} = \frac{4}{(s+2)(s+6)}$$

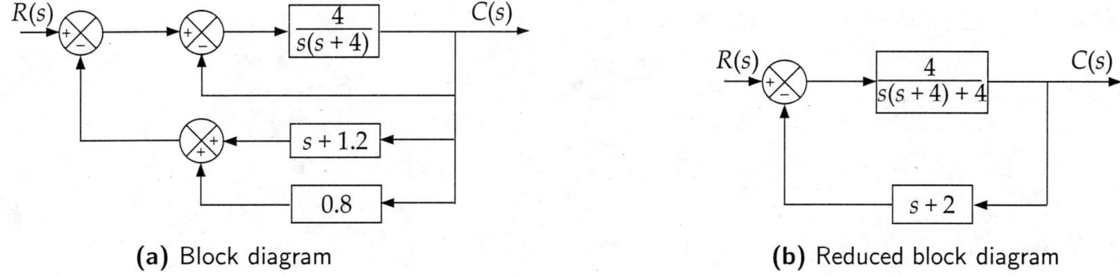

(a) Block diagram **(b)** Reduced block diagram

Figure 6.14: Example 6.26.

Given, $R(s) = \dfrac{2}{s}$. Therefore,

$$C(s) = \frac{8}{s(s+2)(s+6)} \equiv \frac{A}{s} + \frac{B}{s+2} + \frac{C}{s+6} \text{ (say)}$$

Then,

$$A = sC(s)|_{s=0} = \left.\frac{8}{(s+2)(s+6)}\right|_{s=0} = \frac{2}{3}$$

$$B = (s+2)C(s)|_{s=-2} = \left.\frac{8}{s(s+6)}\right|_{s=-2} = -1$$

$$C = (s+6)C(s)|_{s=-6} = \left.\frac{8}{s(s+2)}\right|_{s=-6} = \frac{1}{3}$$

Thus,

$$C(s) = \frac{2}{3}\cdot\frac{1}{s} - \frac{1}{s+2} + \frac{1}{3}\cdot\frac{1}{s+6}$$

Its inverse Laplace transform yields

$$c(t) = \frac{2}{3} - e^{-2t} + \frac{1}{3}e^{-6t}$$

EXAMPLE 6.27. The measurement conducted on a servomechanism shows the error response to be

$$e(t) = 1.66\ e^{-8t}\ \sin(6t + 37^\circ)$$

where the input is a sudden unit displacement. Determine the natural frequency, damping ratio and damped angular frequency of the system.

Solution Given, $r(t) = 1$. Therefore,

$$e(t) = r(t) - c(t) = 1 - c(t)$$

Since the error is sinusoidal, it is a second order underdamped system where, for a step input,

$$c(t) = 1 - \frac{e^{-\zeta\omega_n t}}{\sqrt{1-\zeta^2}}\ \sin\left(\omega_d t + \tan^{-1}\frac{\sqrt{1-\zeta^2}}{\zeta}\right)$$

Thus,

$$e(t) = \frac{e^{-\zeta\omega_n t}}{\sqrt{1-\zeta^2}}\ \sin\left(\omega_d t + \tan^{-1}\frac{\sqrt{1-\zeta^2}}{\zeta}\right) = 1.66\ e^{-8t}\ \sin(6t + 37^\circ) \qquad \text{(i)}$$

Comparing the two sides of Eq. (i), we get

$$\tan^{-1}\frac{\sqrt{1-\zeta^2}}{\zeta} = 37^\circ \quad \Rightarrow \quad \frac{\sqrt{1-\zeta^2}}{\zeta} = \tan 37^\circ = 0.7536 \quad \Rightarrow \quad \zeta = 0.7986$$

$$\omega_d = 6 \text{ rad/s} \quad \Rightarrow \quad \omega_n = \frac{6}{\sqrt{1-\zeta^2}} = 9.969 \text{ rad/s}$$

EXAMPLE 6.28. In the block diagram shown in Figure 6.15

$$G(s) = A/s^2 \qquad H(s) = ms + n$$

For $A = 10$, determine the values of m and n for a step input with a time constant 0.1 s which gives a peak over-shoot of 30%.

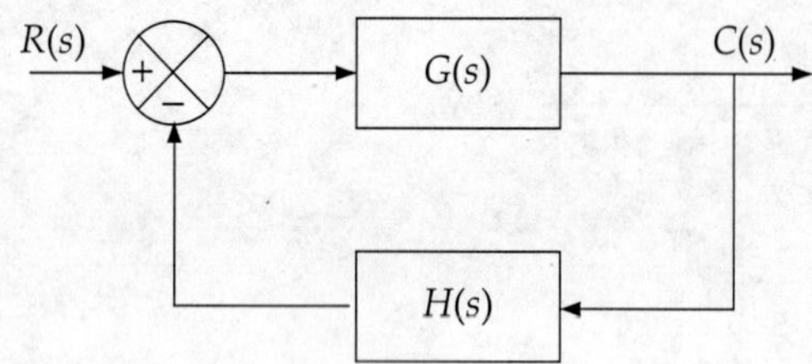

Figure 6.15: Block diagram (Example 6.28).

Solution The overall transfer function of the system is given by

$$T(s) = \frac{C(s)}{R(s)} = \frac{G(s)}{1+G(s)} = \frac{A}{s^2 + Ams + An} \equiv \frac{\omega_n^2}{s^2 + 2\zeta\omega_n s + \omega_n^2}$$

Comparing the denominators of the transfer function of the system and that of the standard second order transfer function, we get

$$2\zeta\omega_n = Am = 10m \qquad \omega_n^2 = An = 10n \tag{i}$$

From Eq. (i), we find

$$\zeta = \frac{10m}{2\omega_n} = \frac{10m}{2\sqrt{10n}} \tag{ii}$$

Now, Time constant, $\dfrac{1}{\zeta\omega_n} = \dfrac{2}{10m} = \dfrac{1}{5m} = 0.1$ s (given)

Therefore, $m = 2$

Substituting this value in Eq. (ii), we get

$$\zeta = \frac{10 \times 2}{2\sqrt{10m}} = \frac{10}{\sqrt{10n}} \tag{iii}$$

Again, Peak overshoot, $\%M_p = 100 \exp\left(-\dfrac{\pi\zeta}{\sqrt{1-\zeta^2}}\right) = 30$ (given)

$$\Rightarrow \quad -\frac{\pi\zeta}{\sqrt{1-\zeta^2}} = \ln(0.3) = -1.204 \quad \Rightarrow \quad \pi^2\zeta^2 = (1.204)^2(1-\zeta^2)$$

$$\Rightarrow \quad \zeta = \sqrt{\frac{(1.204)^2}{\pi^2 + (1.204)^2}} = 0.3579$$

Substituting the value of ζ in Eq. (iii), we get

$$\sqrt{\frac{10}{n}} = 0.3579 \qquad \therefore \quad n = \frac{10}{(0.3579)^2} = 78.07$$

EXAMPLE 6.29. The system and its response curve are given in Figures 6.16a and 6.16b. Determine the values of M, B and K.

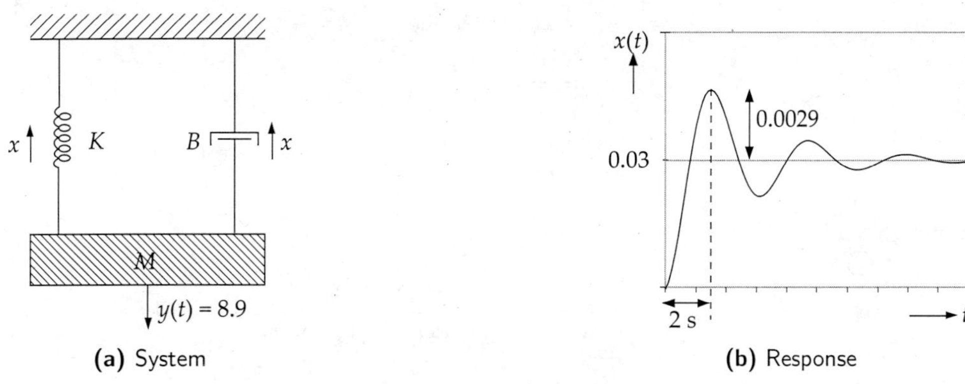

(a) System **(b)** Response

Figure 6.16: Example 6.29.

Solution The differential equation for the system can be written as

$$M\frac{d^2x(t)}{dt^2} + B\frac{dx(t)}{dt} + Kx(t) = y(t)$$

Its Laplace transform, for all initial conditions = 0, is

$$(s^2M + sB + K)X(s) = Y(s)$$

Therefore, the transfer function of the system is

$$T(s) = \frac{X(s)}{Y(s)} = \frac{1}{s^2M + sB + K} = \frac{1/M}{s^2 + s\dfrac{B}{M} + \dfrac{K}{M}} \tag{i}$$

This yields the characteristic equation

$$s^2 + s\frac{B}{M} + \frac{K}{M} = 0 \equiv s^2 + 2\zeta\omega_n s + \omega_n^2$$

By comparison, we get

$$\omega_n = \sqrt{K/M}, \qquad 2\zeta\omega_n = \frac{B}{M}$$

Therefore,

$$\zeta = \frac{B}{2\omega_n M} = \frac{B}{2\sqrt{KM}} \tag{ii}$$

Now, considering the peak overshoot

$$M_p = \exp\left(-\frac{\pi\zeta}{\sqrt{1-\zeta^2}}\right) = \frac{0.0029}{0.03} = 0.0967 \qquad \text{[from the response curve]}$$

$$\Rightarrow \qquad \frac{\pi\zeta}{\sqrt{1-\zeta^2}} = -\ln 0.0967 = 2.3365$$

$$\Rightarrow \qquad \zeta = \frac{2.3365}{\sqrt{\pi^2 + 2.3365^2}} = 0.597 \tag{iii}$$

From Figure 6.16a (a) and Eq. (i),

$$X(s) = \frac{8.9}{s(Ms^2 + Bs + K)}$$

So, $$x(\infty) = 0.03 = \lim_{s\to 0} sX(s) = \lim_{s\to 0} \frac{8.9}{Ms^2 + Bs + K} = \frac{8.9}{K}$$

Therefore, $$K = \frac{8.9}{0.03} = 296.67 \text{ N/m} \qquad \text{(iv)}$$

Again, $$t_p = \frac{\pi}{\omega_n\sqrt{1-\zeta^2}} = 2 \qquad \text{[from response]}$$

$$\Rightarrow \qquad \omega_n = \frac{\pi}{2\sqrt{1-\zeta^2}} = \frac{\pi}{2\sqrt{1-0.597^2}} = 1.958\,\text{rad/s} \qquad \text{(v)}$$

Combining results of Eqs. (iv) and (v), we get

$$\omega_n = \sqrt{\frac{K}{M}} = 1.958 \qquad \Rightarrow \qquad M = \frac{K}{\omega_n^2} = \frac{296.67}{(1.958)^2} = 77.383 \text{ kg} \qquad \text{(vi)}$$

And Eq. (ii), together with Eq. (vi) yields

$$\zeta = \frac{B}{2\sqrt{KM}}$$

or, $$B = 2\zeta\sqrt{KM} = 2(0.597)\sqrt{(296.67)(77.383)} = 180.91 \text{ N-s/m}$$

EXAMPLE 6.30. In the closed-loop system shown in Figure 6.17, determine the value of K and a such that the per cent peak overshoot due to the unit step input does not exceed 10. Settling time should be equal to 1 s.

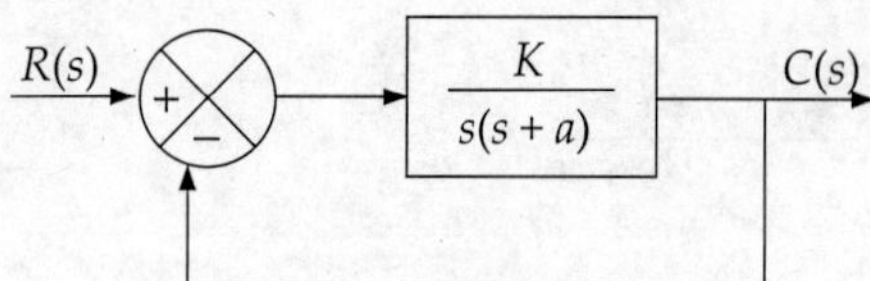

Figure 6.17: Closed-loop system (Example 6.30).

Solution The overall transfer function of the system is

$$T(s) = \frac{C(s)}{R(s)} = \frac{K}{s(s+a)+K} = \frac{K}{s^2 + as + K}$$

Comparing the characteristic equation, $s^2 + as + K = 0$, with that of a standard second order system, we get

$$\omega_n = \sqrt{K}, \qquad \zeta = \frac{a}{2\omega_n} = \frac{a}{2\sqrt{K}} \qquad \text{(i)}$$

Now, $$t_s = \frac{4}{\zeta\omega_n} = 1\,s \qquad \text{(given)}$$

Therefore, $$\zeta\omega_n = 4 = \frac{a}{2}$$

or, $$a = 8 \qquad \text{(ii)}$$

The peak overshoot condition gives

$$\exp\left(-\frac{\pi\zeta}{\sqrt{1-\zeta^2}}\right) = 0.1 \qquad \Rightarrow \qquad \frac{\pi\zeta}{\sqrt{1-\zeta^2}} = 2.303 \tag{iii}$$

Substituting the values of ζ from Eq. (i) in Eq. (iii), we get

$$\frac{\frac{\pi a}{2\sqrt{K}}}{\sqrt{1-\frac{a^2}{4K}}} = 2.303 \qquad \Rightarrow \qquad \frac{\pi a}{\sqrt{4K-a^2}} = 2.303$$

Putting the value of a from Eq. (ii) in this equation, we get

$$\frac{\pi(8)}{\sqrt{4K-64}} = 2.303 \qquad \Rightarrow \qquad \frac{64\pi^2}{2.303^2} = 4K - 64 \qquad \Rightarrow \qquad K = \frac{1}{4}\left[\frac{64\pi^2}{2.303^2} + 64\right] = 45.774$$

EXAMPLE 6.31. Determine the value of controller gains K_p and K_1 for the system shown in Figure 6.18 so that its damping ratio and natural frequency of oscillation become 1 and 4 Hz respectively.

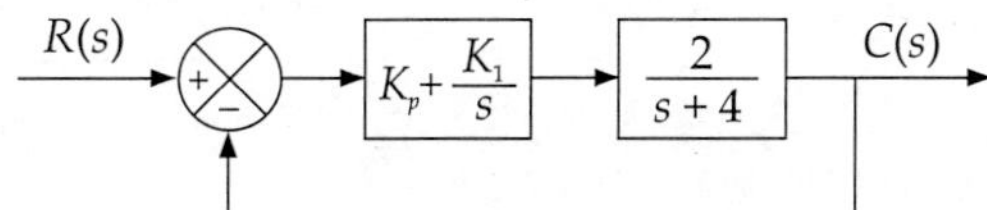

Figure 6.18: Control system (Example 6.31).

Solution Here,

$$G(s) = \frac{2(K_1 + K_p s)}{s(s+4)}$$

$$T(s) = \frac{G(s)}{1+G(s)} = \frac{2(K_1 + K_p s)}{s(s+4) + 2(K_1 + K_p s)} = \frac{2(K_1 + K_p s)}{s^2(2K_p + 4s + 2K_1)}$$

Comparing the characteristic equation, $s^2(2K_p + 4s + 2K_1) = 0$ with that of a standard second order system, we get

$$\omega_n^2 = 2K_1$$

Given $\nu_n = 4$. Then,

$$\omega_n = 2\pi\nu_n = 8\pi$$

$$\Rightarrow \qquad K_1 = \frac{\omega_n^2}{2} = \frac{64\pi^2}{2} = 315.83$$

and,

$$2\zeta\omega_n = 2K_p + 4$$

or,

$$K_p = \frac{2\zeta\omega_n - 4}{2} = \zeta\omega_n - 2 = 8\pi - 2 = 23.13 \qquad [\because \zeta = 1 \text{ (given)}]$$

Performance parameters from location of poles. Suppose a pair of dominant poles of a system are given by $p_1 = -a + \jmath b$ and $p_2 = -a - \jmath b$ as shown in Figure 6.19.
Then, the damping ratio is given by

$$\zeta = \cos\phi = \frac{a}{\sqrt{a^2+b^2}} \tag{6.40}$$

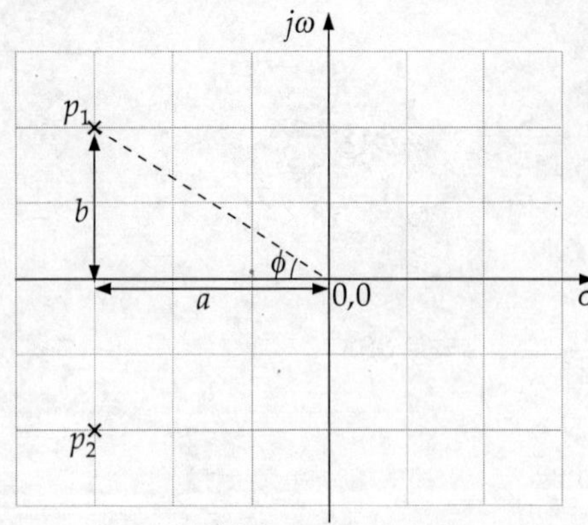

Figure 6.19: Dominant pole locations.

The natural frequency of oscillation is obtained from the radial distance of the pole from the origin as

$$\omega_n = \sqrt{a^2 + b^2} \tag{6.41}$$

The damped oscillation frequency ω_d and exponential decay frequency σ_D can be found from Eq. (6.42) as

$$\omega_d \equiv \omega_n \sqrt{1 - \zeta^2} = b \tag{6.42}$$

$$\sigma_D \equiv \zeta \omega_n = a \tag{6.43}$$

EXAMPLE 6.32. The dominant poles of a servosystem are located at $s = (-2 \pm \jmath 2)$. The damping ratio of the system is

(A) 1 (B) 0.8 (C) 0.707 (D) 0.6

Solution From Eq. (6.40) we get

$$\zeta = \cos\phi = \frac{2}{\sqrt{2^2 + 2^2}} = \frac{1}{\sqrt{2}} = 0.707$$

Ans. (C)

Alternatively

Let the transfer function be

$$G(s) = \frac{K}{(s + 2 - \jmath 2)(s + 2 + \jmath 2)} = \frac{K}{s^2 + 4s + 8}$$

Then

$$\omega_n^2 = 8 \quad \Rightarrow \quad \omega_n = \sqrt{8}$$

$$2\zeta\omega_n = 4 \quad \Rightarrow \quad \zeta = \frac{4}{2\sqrt{8}} = 0.707$$

6.4 ANALYSIS OF STEADY-STATE ERROR

We have already defined steady-state error in Eq. (6.1) at page 115 and calculated the same for first and second order systems in time-domain by a method that involves the inverse Laplace transform of responses. But this error can be calculated directly from the Laplace transformed values of the response. We will discuss the procedure in this section.

From Eq. (6.1), we know that

$$e_{ss}(t) = \lim_{t \to \infty} e(t) \tag{6.44}$$

Applying the final value theorem[2] to Eq. (6.44), we get

$$e_{ss}(t) = \lim_{s \to 0} sE(s) \tag{6.45}$$

Consider a canonical feedback system given in Figure 6.20.

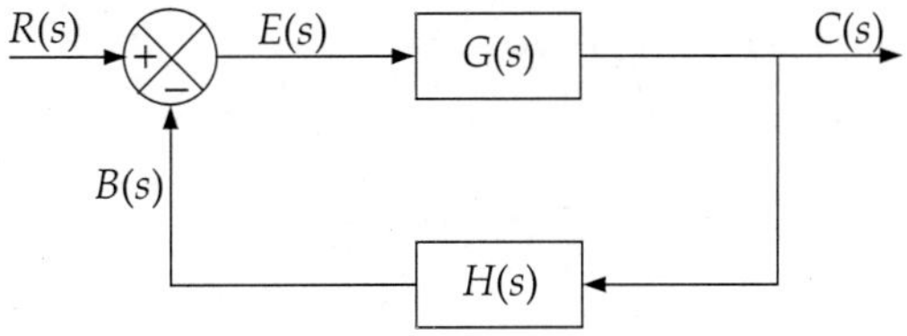

Figure 6.20: Canonical feedback control system.

The overall transfer function for this system is given by

$$T(s) = \frac{C(s)}{R(s)} = \frac{G(s)}{1 + G(s)H(s)} \tag{6.46}$$

From Figure 6.20, we get

$$E(s) = R(s) - B(s) = R(s) - C(s)H(s) = R(s) - E(s)G(s)H(s) \tag{6.47}$$

This yields

$$E(s) = \frac{R(s)}{1 + G(s)H(s)} \tag{6.48}$$

Therefore

$$\boxed{e_{ss} = \lim_{s \to 0} sE(s) = \lim_{s \to 0} \frac{sR(s)}{1 + G(s)H(s)}} \tag{6.49}$$

Note: From Eq. (6.49) we find that

(i) the numerator is related to the input

(ii) the denominator, which is the characteristic polynomial, relates to the type of the system.

Thus, the steady-state error of a system depends on:

- Input to the system
- Type of the system

Before we discuss the effects of input and type of system on the steady-state error, let us define what are called *static error coefficients* or *static error constants.*

Static Error Constants

Often called *figures of merit* of a control system, static error constants are three in number, namely

1. Static position error constant, K_p
2. Static velocity error constant, K_v

[2] See Appendix *Laplace Transform*, Section A at page 515.

3. Static acceleration error constant, K_a

They are defined in such a way that the higher the value of the constant, the less the value of e_{ss}.

Note: Each error constant variable is the derivative of its preceding one, i.e., velocity is the derivative of position and acceleration is the derivative of velocity. These originate from their definitions.

Input Types and Error Constants

Step input. Here, $R(s) = \frac{1}{s}$

Therefore,
$$e_{ss} = \lim_{s\to 0}\left[\frac{s}{1+G(s)H(s)}\cdot\frac{1}{s}\right] = \lim_{s\to 0}\frac{1}{1+G(s)H(s)} = \frac{1}{1+\lim\limits_{s\to 0}G(s)H(s)}$$
$$\equiv \frac{1}{1+K_p} \tag{6.50}$$

where

$$\boxed{K_p = \lim_{s\to 0} G(s)H(s)} \tag{6.51}$$

K_p is called the *static position error constant.*

Ramp input. Here, $R(s) = \frac{1}{s^2}$

Therefore,
$$e_{ss} = \lim_{s\to 0}\left[\frac{s}{1+G(s)H(s)}\cdot\frac{1}{s^2}\right] = \lim_{s\to 0}\frac{1}{s[1+G(s)H(s)]} = \frac{1}{\lim\limits_{s\to 0}sG(s)H(s)}$$
$$\equiv \frac{1}{K_v} \tag{6.52}$$

where

$$\boxed{K_v = \lim_{s\to 0} sG(s)H(s)} \tag{6.53}$$

K_v is called the *static velocity error constant.*

Note: There is an s in its definition. This indicates a differentiation for once in consonance with the fact that velocity is the first derivative of the position variable.

Parabolic input. Here, $R(s) = \frac{1}{s^3}$

Therefore,
$$e_{ss} = \lim_{s\to 0}\left[\frac{s}{1+G(s)H(s)}\cdot\frac{1}{s^3}\right] = \lim_{s\to 0}\frac{1}{s^2[1+G(s)H(s)]} = \frac{1}{\lim\limits_{s\to 0}s^2G(s)H(s)}$$
$$\equiv \frac{1}{K_a} \tag{6.54}$$

where

$$\boxed{K_a = \lim_{s\to 0} s^2G(s)H(s)} \tag{6.55}$$

K_a is called the *static acceleration constant.* Predictably, its definition contains s^2 which indicates a second derivative of the position variable.

System Types and Steady-State Errors

We know that the *type* of a system is determined by the *number of poles at the origin* (i.e. at $s = 0$) existing in the open-loop transfer function, $G(s)H(s)$, of a system. The general form of $G(s)H(s)$ is

$$G(s)H(s) = \frac{K(1+sT_1)(1+sT_2)\cdots(1+sT_m)}{s^n(1+sT_a)(1+sT_b)\cdots(1+sT_n)} \tag{6.56}$$

In Eq. (6.56), the index n indicates the number of poles at the origin. Therefore, it is a type n system.

EXAMPLE 6.33. Identify the type of the system for which

$$G(s) = \frac{K(1+3s)}{s^2} \text{ and } H(s) = \frac{1+5s}{s^2+6s+7}$$

Solution Here,

$$G(s)H(s) = \frac{K(1+3s)(1+5s)}{s^2(s^2+6s+7)}$$

Since, the number of poles at the origin is 2 (indicated by s^2 in the denominator), type = 2.

EXAMPLE 6.34. Identify the type of the system for which

$$G(s) = \frac{4}{s^2+6s+7} \text{ and } H(s) = s+3$$

Solution Here,

$$G(s)H(s) = \frac{4(s+3)}{s^2+6s+7}$$

It has no poles at the origin. Therefore, type = 0.

With this background, we see how the static error constants, and for that matter, steady-state errors, are related to types of systems.

Type 0. Here,

$$G(s)H(s) = \frac{K(1+sT_1)(1+sT_2)\cdots(1+sT_m)}{(1+sT_a)(1+sT_b)\cdots(1+sT_n)}$$

<u>Step input</u> $$K_p = \lim_{s\to 0} G(s)H(s) = \frac{K\cdot 1\cdot 1\cdots 1}{1\cdot 1\cdots 1} = K$$

Therefore, $$e_{\text{ss}} = \frac{1}{1+K_p} = \frac{1}{1+K}$$

<u>Ramp input</u> $$K_v = \lim_{s\to 0} sG(s)H(s) = \lim_{s\to 0} \frac{sK\cdot 1\cdot 1\cdots 1}{1\cdot 1\cdots 1} = 0$$

Therefore, $$e_{\text{ss}} = \frac{1}{K_v} = \infty$$

<u>Parabolic input</u> $$K_a = \lim_{s^2\to 0} s^2G(s)H(s) = \lim_{s\to 0} \frac{s^2K\cdot 1\cdot 1\cdots 1}{1\cdot 1\cdots 1} = 0$$

Therefore, $$e_{\text{ss}} = \frac{1}{K_a} = \infty$$

Type 1. Here,

$$G(s)H(s) = \frac{K(1+sT_1)(1+sT_2)\cdots(1+sT_m)}{s(1+sT_a)(1+sT_b)\cdots(1+sT_n)}$$

Step input
$$K_p = \lim_{s\to 0} G(s)H(s) = \lim_{s\to 0} \frac{K\cdot 1\cdot 1\cdots 1}{s\cdot 1\cdot 1\cdots 1} = \infty$$

Therefore,
$$e_{ss} = \frac{1}{1+K_p} = \frac{1}{1+\infty} = 0$$

Ramp input
$$K_v = \lim_{s\to 0} sG(s)H(s) = \lim_{s\to 0} \frac{Ks\cdot 1\cdot 1\cdots 1}{s\cdot 1\cdot 1\cdots 1} = K$$

Therefore,
$$e_{ss} = \frac{1}{K_v} = \frac{1}{K}$$

Parabolic input
$$K_a = \lim_{s^2\to 0} s^2G(s)H(s) = \lim_{s\to 0} \frac{Ks^2\cdot 1\cdot 1\cdots 1}{s\cdot 1\cdot 1\cdots 1} = 0$$

Therefore,
$$e_{ss} = \frac{1}{K_a} = \frac{1}{0} = \infty$$

Type 2. Here,

$$G(s)H(s) = \frac{K(1+sT_1)(1+sT_2)\cdots(1+sT_m)}{s^2(1+sT_a)(1+sT_b)\cdots(1+sT_n)} \tag{6.57}$$

Step input
$$K_p = \lim_{s\to 0} G(s)H(s) = \infty$$

Therefore,
$$e_{ss} = \frac{1}{1+K_p} = 0$$

Ramp input
$$K_v = \lim_{s\to 0} sG(s)H(s) = \infty$$

Therefore,
$$e_{ss} = \frac{1}{K_v} = 0$$

Parabolic input
$$K_a = \lim_{s^2\to 0} s^2G(s)H(s) = K$$

Therefore,
$$e_{ss} = \frac{1}{K_a} = \frac{1}{K}$$

The results, so far obtained, are summarized in Table 6.4. The following points emerge from the table:

(a) Ramp and parabolic inputs should not be applied to type 0 system.
(b) Parabolic input should not be applied to type 1 system.
(c) For types ≥ 3, $e_{ss} = 0$ for all inputs.

Two more points which are not apparent from Table 6.4, are as follows:

- e_{ss} for a linear combination of inputs can be superimposed.
- The method fails for sinusoidal inputs, because the final value theorem cannot be applied to that case.

Table 6.4: Steady-state error for different types and different inputs

Type	Step input $e_{ss} = \dfrac{1}{1+K_p}$	Ramp input $e_{ss} = \dfrac{1}{K_v}$	Parabolic input $e_{ss} = \dfrac{1}{K_a}$
0	$\dfrac{1}{1+K}$	∞	∞
1	0	$\dfrac{1}{K}$	∞
2	0	0	$\dfrac{1}{K}$
3	0	0	0

(*Steady-state error for* spans the Step, Ramp and Parabolic input columns.)

EXAMPLE 6.35. A unity feedback system has open-loop transfer function $G(s)$. The steady-state error is zero for

(A) step input and type-1 $G(s)$ (B) ramp input and type-1 $G(s)$
(C) step input and type-0 $G(s)$ (D) ramp input and type-0 $G(s)$

Solution The Table 6.4 shows that the answer is (A). However, it can be worked out as follows.

We know, the steady-state error for a system with open-loop transfer function $G(s)$ and feedback $H(s)$ is given by [see Eq. (6.49)]

$$e_{ss} = \lim_{s\to 0} \frac{sR(s)}{1 + G(s)H(s)}$$

Here $H(s) = 1$. So, we have

$$e_{ss} = \lim_{s\to 0} \frac{sR(s)}{1 + G(s)} \qquad \text{(i)}$$

First, we consider inputs. The Laplace transform of the step input has s in the denominator while that of the ramp input has s^2 in the denominator. So, after cancelling the s in the numerator, a ramp input leaves s in the denominator which blows off e_{ss} as can be seen from Eq. (i). So, ramp inputs are ruled out. Now, our Eq. (i) turns out to be

$$e_{ss} = \frac{1}{1 + \lim\limits_{s\to 0} G(s)}$$

From this equation, we find that e_{ss} can be zero if the type of $G(s)$ can produce s on the numerator. The type-0 $G(s)$ contains no s in the denominator, while a type-1 does. Therefore, we need a step input and type-1 $G(s)$.

Ans. (A)

EXAMPLE 6.36. The system $G(s) = \dfrac{0.8}{s^2 + s - 2}$ is subjected to a step input. The system output $y(t)$ as $t \to \infty$ is

(A) 0.8 (B) 0.4 (C) -0.4 (D) unbounded

Solution Applying the final value theorem, we get

$$c(t)\big|_{t\to\infty} = \lim_{s\to 0} s \cdot \frac{1}{s} \cdot \frac{0.8}{s^2 + s - 2} = -0.4$$

Ans. (C)

EXAMPLE 6.37. A single-input single-output system with c as output and r as input, is described by

$$\frac{d^2c}{dt^2} + 2\frac{dc}{dt} + 10y = 5\frac{dr}{dt} - 3r$$

For the above system find an input $u(t)$, with zero initial condition, that produces the same output as with no input and with initial conditions: $\dfrac{dc(0^-)}{dt} = -4$, $c(0^-) = 1$.

Solution With no input, the Laplace transform of the given system is given by

$$[s^2C(s) - sc(0) - c'(0)] + 2[sC(s) - c(0)] + 10C(s) = 0$$

or,
$$[s^2 + 2s + 10]C(s) - sc(0) - c'(0) - 2c(0) = 0$$

or,
$$[s^2 + 2s + 10]C(s) - [s \times 1 + (-4) + 2 \times 1] = 0 \qquad [\because c'(0) = -4,\ c(0) = 1]$$

or,
$$\frac{s-2}{s^2 + 2s + 10} = C(s) \qquad \text{(i)}$$

With input, and zero initial conditions, the Laplace transform of the given system yields

$$[s^2C(s) - sc(0) - c'(0)] + 2[sC(s) - c(0)] + 10C(s) = 5sR(s) - 3R(s)$$

$$[s^2 + 2s + 10]C(s) = [5s - 3]R(s) \qquad \text{[since initial conditions = 0]}$$

or,
$$C(s) = \frac{5s-3}{s^2 + 2s + 10}R(s) \qquad \text{(ii)}$$

Given that both the outputs are the same, we have from Eqs. (i) and (ii)

$$\frac{s-2}{s^2 + 2s + 10} = \frac{5s-3}{s^2 + 2s + 10}R(s)$$

or,
$$R(s) = \frac{s-2}{5s-3} = \frac{1}{5}\left[\frac{5s-10}{5s-3}\right]$$

$$= \frac{1}{5}\left[1 - \frac{7}{5s-3}\right] = \frac{1}{5}\left[1 - \frac{\frac{7}{5}}{s - \frac{3}{5}}\right]$$

Therefore,
$$\mathscr{L}^{-1}\{R(s)\} = r(t) = \frac{1}{5}\left[\delta(t) - \frac{7}{5}e^{\frac{3}{5}t}u(t)\right]$$

Hence, the required input is

$$r(t) = \frac{1}{5}\delta(t) - \frac{7}{25}e^{\frac{3}{5}t}u(t)$$

EXAMPLE 6.38. A unity feedback closed-loop second order system has a transfer function of

$$\frac{81}{s^2 + 0.6s + 9}$$

and is excited by a step input of 10 units. The steady-state error of the output is

(A) 10 (B) 0.0 (C) 1.0 (D) 0.1

Solution Given: $G(s)H(s) = \dfrac{81}{s^2 + 0.6s + 9}$ and $R(s) = \dfrac{10}{s}$. Therefore,

$$e_{ss} = \lim_{s\to 0} s \cdot \frac{10}{s} \cdot \frac{1}{1 + \dfrac{81}{s^2 + 0.6s + 9}} = \frac{10}{1+9} = 1$$

Ans. (C)

EXAMPLE 6.39. For a unit step input, a system with a closed-loop transfer function of

$$G(s) = \frac{20}{s^2 + 2s + 5}$$

has a steady-state output of

(A) 10 (B) 5 (C) 2 (D) 4

Solution Given: $R(s) = \frac{1}{s}$. Applying the final value theorem, the stead-state output is given by

$$c(t)\big|_{ss} = \lim_{s\to 0} s \cdot C(s) = \lim_{s\to 0} s \cdot G(s)R(s) = \lim_{s\to 0} s \cdot \frac{20}{s^2 + 2s + 5} \cdot \frac{1}{s} = \frac{20}{5} = 4$$

Ans. (D)

EXAMPLE 6.40. The loop transfer function of a system is given by

$$G(s)H(s) = \frac{10e^{-0.1s}}{s(s+5)(s^2 + 4s + 16)}$$

When the system is subjected to a parabolic input of $t^2/2$, the steady-state error is

(A) zero (B) 0.1 (C) infinity (D) $e^{-0.1}/0.1$

Solution The Laplace transform of the input is $R(s) = 1/s^3$. Therefore,

$$e_{ss} = \lim_{s\to 0} \frac{sR(s)}{1 + G(s)H(s)} = \lim_{s\to 0} \frac{s}{s^3} \cdot \frac{1}{1 + \dfrac{10e^{-0.1s}}{s(s+5)(s^2 + 4s + 16)}}$$

$$= \lim_{s\to 0} \frac{1}{s^2} \cdot \frac{(s+5)(s^2 + 4s + 16)}{s(s+5)(s^2 + 4s + 16) + 10e^{-0.1s}} = \infty$$

Ans. (C)

EXAMPLE 6.41. The steady-state error due to a unit step disturbance input $D(s)$ in Figure 6.21a is

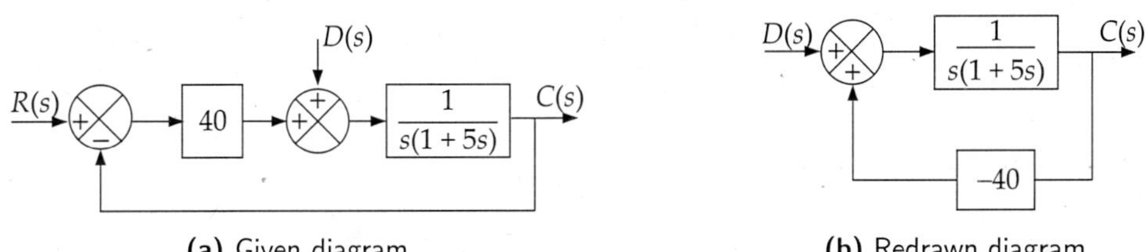

(a) Given diagram **(b)** Redrawn diagram

Figure 6.21: System (Example 6.41).

(A) 0 (B) 0.012 (C) 0.021 (D) 0.025

Solution To find the influence of $D(s) = \frac{1}{s}$, we put $R(s) = 0$ and redraw the block diagram as shown in Figure 6.21b. Then

$$G(s)H(s) = \frac{-40}{s(1+5s)} \quad \Rightarrow \quad 1 + G(s)H(s) = \frac{s(1+5s) - 40}{s(1+5s)}$$

$$e_{ss} = \lim_{s\to 0} \frac{sD(s)}{1 + G(s)H(s)} = \lim_{s\to 0} s \cdot \frac{1}{s} \cdot \frac{s(1+5s)}{s(1+5s) - 40} = 0$$

Ans. (A)

EXAMPLE 6.42. The dynamic response of a measurement system is given by the differential equation

$$4\frac{d^2y}{dt^2} + 640\frac{dy}{dt} + 40000y = 80000x$$

where y is the output and x is the input.

(a) Determine

(i) the natural frequency

(ii) the damping ratio, and

(iii) the static sensitivity of the system

(b) If it is used to measure a point of frequency of 10 Hz, what would be the ratio of the indicated amplitude to the input amplitude?

Solution We write the given differential equation as

$$\frac{4}{40000}\frac{d^2y}{dt^2} + \frac{640}{40000}\frac{dy}{dt} + y = \frac{80000}{40000}$$

Comparing it with Eq. (6.11) at page 121, we get

(a)(i) $\dfrac{1}{\omega_n^2} = \dfrac{4}{40000} \quad \Rightarrow \quad \omega_n^2 = 10000 \quad \Rightarrow \quad \omega_n = 100 \text{ rad/s}$

(ii) $\dfrac{2\zeta}{\omega_n} = \dfrac{640}{40000} \quad \Rightarrow \quad \zeta = \dfrac{640}{40000} \times \dfrac{100}{2} = 0.8$

(iii) $K = \dfrac{80000}{40000} = 2$

(b) The Laplace transform of the given equation yields

$$(4s^2 + 640s + 40000)Y(s) = 80000X(s)$$

Therefore, $\dfrac{Y(s)}{X(s)} = \dfrac{20000}{s^2 + 160s + 10000} \Rightarrow \dfrac{y(\jmath\omega)}{x(\jmath\omega)} = \dfrac{20000}{-\omega^2 + \jmath 160\omega + 10000}$

Thus, at the given frequency 10 Hz

$$\left|\frac{y(\jmath\omega)}{x(\jmath\omega)}\right|_{\omega=2\pi\times 10} = \left.\frac{20000}{\sqrt{\left(10000 - \omega^2\right)^2 + (160\omega)^2}}\right|_{\omega=20\pi}$$

$$= \frac{20000}{\sqrt{[10000 - (20\pi)^2]^2 + (160 \times 20\pi)^2}} = 1.7044$$

EXAMPLE 6.43. The open-loop transfer function of a unity feedback system is given by

$$G(s) = \frac{100}{s^2(s+4)(s^2+5s+25)}$$

Find static error coefficients and the steady-state error of the system when it is subjected to an input of $r(t) = 2 + 4t + 2t^2$.

Solution Since unity feedback, $H(s) = 1$. The static error coefficients are

$$K_p = \lim_{s\to 0} G(s)H(s) = \lim_{s\to 0} \frac{100}{s^2(s+4)(s^2+5s+25)} = \infty$$

$$K_v = \lim_{s\to 0} sG(s)H(s) = \lim_{s\to 0} \frac{100}{s(s+4)(s^2+5s+25)} = \infty$$

$$K_a = \lim_{s\to 0} s^2G(s)H(s) = \lim_{s\to 0} \frac{100}{(s+4)(s^2+5s+25)} = \frac{100}{(4)(25)} = 1$$

Taking the Laplace transform of the input, we get

$$R(s) = \frac{2}{s} + \frac{4}{s^2} + \frac{4}{s^3}$$

We observe that this is a combination of step, ramp and parabolic inputs. Hence the steady-state error is given by

$$e_{\text{ss}} = \frac{2}{1+K_p} + \frac{4}{K_v} + \frac{4}{K_a}$$

Substituting the values obtained for the error coefficients, we get

$$e_{\text{ss}} = \frac{2}{1+\infty} + \frac{4}{\infty} + \frac{4}{1} = 4$$

EXAMPLE 6.44. A servomechanism, the block diagram of which is given in Figure 6.22, is designed to keep a radar antenna pointed at a flying aeroplane. If the aeroplane is flying with a velocity of 600 km/h, at a range of 2 km, and the maximum tracking error is to be within 0.1°, determine the required velocity error coefficient.

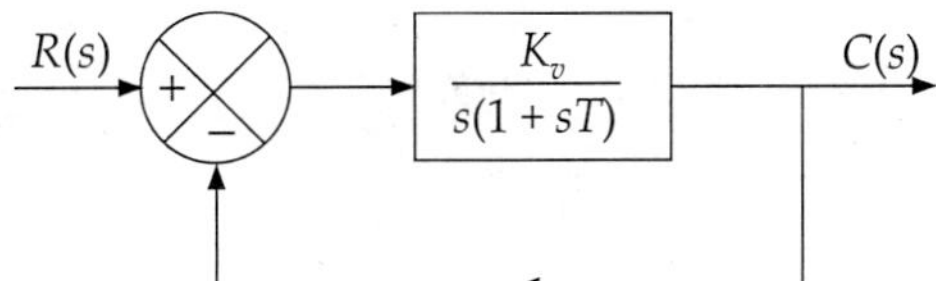

Figure 6.22: Block diagram of servomechanism (Example 6.44).

Solution Given, velocity = 600 km/h = 166.67 m/s, radial distance = 2 km = 2000 m. Therefore, the angular velocity is

$$\omega = \frac{v}{r} = \frac{166.67}{2000} = 0.083 \text{ rad/s} = 4.756 \text{ deg/s}$$

The input to the tracking system is the angle swept in time t s. So,

$$r(t) = \omega t = 4.756t \quad \Rightarrow \quad R(s) = \frac{4.756}{s^2}$$

The open-loop transfer function of the system is

$$G(s)H(s) = \frac{K_v}{s(1+sT)} \qquad \therefore \quad 1 + G(s)H(s) = \frac{s(1+sT)+K_v}{s(1+sT)}$$

Thus, the steady-state error is given by

$$e_{\text{ss}} = \lim_{s\to 0} \frac{R(s)}{1+G(s)H(s)} = \lim_{s\to 0} \frac{4.756}{s^2} \cdot \frac{s(1+sT)}{s(1+sT)+K_v} = \frac{4.756}{K_v}$$

To maintain steady-state tracking error at 0.1°,

$$\frac{4.756}{K_v} = 0.1 \qquad \Rightarrow \quad K_v = 47.56/\text{s}$$

EXAMPLE 6.45. Find type, error coefficients and e_{ss} for the system, shown in Figure 6.23a, for the inputs (a) $r(t) = 12$, (b) $r(t) = 6t$ and (c) $r(t) = 12 + 6t + \frac{1}{2}t^2$.

$$G_1(s) = \frac{25}{(s+5)(s+10)} \qquad H_1(s) = 5s$$

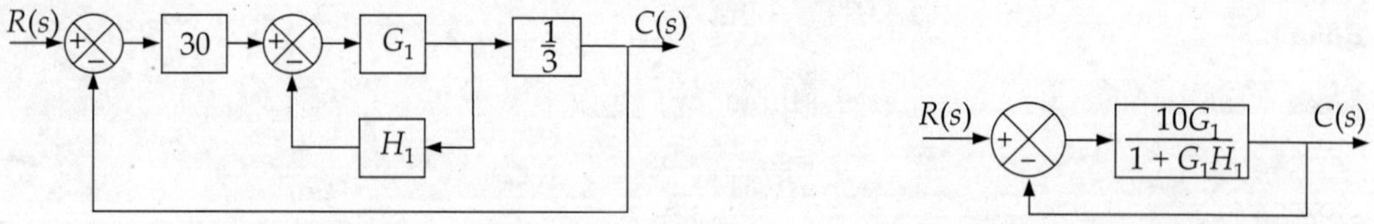

(a) Given block diagram

(b) Reduced block diagram

Figure 6.23: Example 6.45.

Solution The block diagram can be reduced to give Figure 6.23b. Therefore,

$$G(s) = \frac{10G_1}{1+G_1H_1} = \frac{250}{(s+5)(s+10)+125s} = \frac{250}{s^2+140s+50} \qquad H(s) = 1$$

So, the open-loop transfer function and closed-loop transfer function of the system are

$$\text{Open-loop transfer function} = G(s)H(s) = \frac{250}{s^2+140s+50}$$

$$\text{Closed-loop transfer function} = \frac{G(s)}{1+G(s)H(s)} = \frac{\dfrac{250}{s^2+140s+50}}{1+\dfrac{250}{s^2+140s+50}} = \frac{250}{s^2+140s+300}$$

The closed-loop transfer function has no poles at the origin. Therefore, it is a type 0 system. For a type 0 system, only the static position error coefficient, K_p, exists, its value being

$$K_p = \lim_{s\to 0} G(s)H(s) = \lim_{s\to 0} \frac{250}{s^2+140s+50} = 5$$

(a) Here, $R(s) = \frac{12}{s}$. Therefore,

$$e_{ss} = \lim_{s\to 0} \frac{sR(s)}{1+G(s)H(s)} = \frac{12}{1+K_p} = 2$$

(b) Here, $R(s) = \frac{6}{s^2}$. Therefore,

$$e_{ss} = \lim_{s\to 0} \frac{sR(s)}{1+G(s)H(s)} = \lim_{s\to 0} \frac{6}{s[1+G(s)H(s)]} = \infty$$

(c) Here, $r(t) = 12 + 6t + \frac{1}{2}t^2$. Its Laplace transform in relaxed condition is $R(s) = \frac{12}{s} + \frac{6}{s^2} + \frac{1}{2s^2}$. Errors for individual inputs can be superimposed and we have already seen that e_{ss} for the second part is ∞. So, without calculating it for the third part we can say that for this input also $e_{ss} = \infty$.

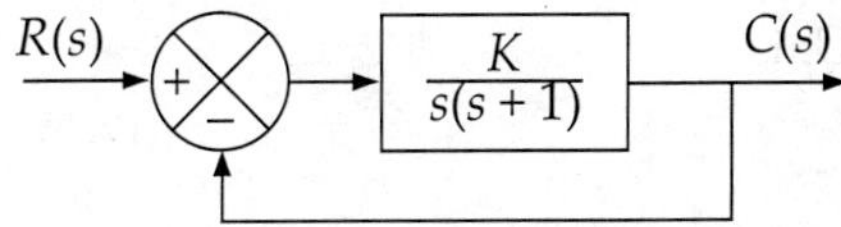

Figure 6.24: Control system (Example 6.46).

EXAMPLE 6.46. Determine the range of values of K of the system shown in Figure 6.24 so that $e_{ss} < 0.004$ when $r(t) = 0.2t$.

Solution The open-loop transfer function for the system is

$$G(s)H(s) = \frac{K}{s(s+1)} \qquad \text{which gives} \quad 1 + G(s)H(s) = \frac{s(s+1)+K}{s(s+1)}$$

Now, $r(t) = 0.2t$ which gives $R(s) = \dfrac{0.2}{s^2}$ Thus,

$$e_{ss} = \lim_{s\to 0} \frac{sR(s)}{1+G(s)H(s)} = \lim_{s\to 0}\left[\frac{0.2}{s^2} \cdot \frac{s^2(s+1)}{s^2+s+K}\right] = \lim_{s\to 0} \frac{0.2(s+1)}{s^2+s+K} = \frac{0.2}{K}$$

Given, $e_{ss} < 0.004$. So,

$$\frac{0.2}{K} < 0.004 \qquad \text{which means,} \quad K > \frac{0.2}{0.004} \qquad \text{or,} \quad K > 50$$

EXAMPLE 6.47. The transfer function of a system is $1/(1+sT)$. The input to the system is $tu(t)$. The output would track the system. Find the value of the steady-state error.

Solution Given: $G(s) = \dfrac{1}{1+sT}$, and since the output would track the system, $H(s) = 1$.

Also, $$r(t) = tu(t) = t \quad \text{for} \quad t > 0 \qquad \Rightarrow \quad R(s) = \frac{1}{s^2}$$

Therefore, $$e_{ss} = \lim_{s\to 0} \frac{sR(s)}{1+G(s)H(s)} = \lim_{s\to 0} \frac{s(1+sT)\cdot \dfrac{1}{s^2}}{2+sT} = \infty$$

Alternative method

We can arrive at the same result from the following study in the t-space. For this case, the closed-loop transfer function is given by

$$\frac{C(s)}{R(s)} = \frac{G(s)}{1+G(s)H(s)} = \frac{1}{2+sT}$$

Therefore,

$$C(s) = \frac{1}{2+sT} \cdot \frac{1}{s^2} = \frac{1/T}{s^2\left(s+\frac{2}{T}\right)} \equiv \frac{P}{s^2} + \frac{Q}{s} + \frac{R}{s+\frac{2}{T}}$$

where P, Q and R are constants. The constants are evaluated as follows:

$$P = s^2C(s)\big|_{s=0} = \left.\frac{1/T}{s+\frac{2}{T}}\right|_{s=0} = \frac{1}{2}$$

$$Q = \frac{d}{ds}\left[s^2C(s)\right]_{s=0} = \frac{1}{T}\cdot\frac{d}{ds}\left.\frac{1}{s+\frac{2}{T}}\right|_{s=0} = -\frac{1}{T}\cdot\left.\frac{1}{\left(s+\frac{2}{T}\right)^2}\right|_{s=0} = -\frac{T}{4}$$

$$R = \left.\left(s+\frac{2}{T}\right)C(s)\right|_{s=-\frac{2}{T}} = \left.\frac{1}{s^2T}\right|_{s=-\frac{2}{T}} = \frac{T}{4}$$

Thus,

$$C(s) = \frac{1}{2s^2} - \frac{T}{4s} + \frac{T}{4\left(s + \frac{2}{T}\right)}$$

which on inverse Laplace transform yields

$$c(t) = \frac{t}{2} - \frac{T}{4} + \frac{T}{4} \cdot e^{-\frac{2t}{T}}$$

Therefore, the steady-state error is given by

$$e_{ss} = \lim_{t\to\infty} [r(t) - c(t)] = \lim_{t\to\infty} \left[t - \left(\frac{t}{2} - \frac{T}{4} + \frac{T}{4} \cdot e^{-\frac{2t}{T}} \right) \right] = \lim_{t\to\infty} \left[\frac{t}{2} + \frac{T}{4} \left(1 - e^{-\frac{2t}{T}} \right) \right] = \infty$$

Note: The alternative method obviously demonstrates the power of the final value theorem as well as the convenience offered by it.

EXAMPLE 6.48. A unit ramp input is applied to the system shown in Figure 6.25. The steady-state error in its output is

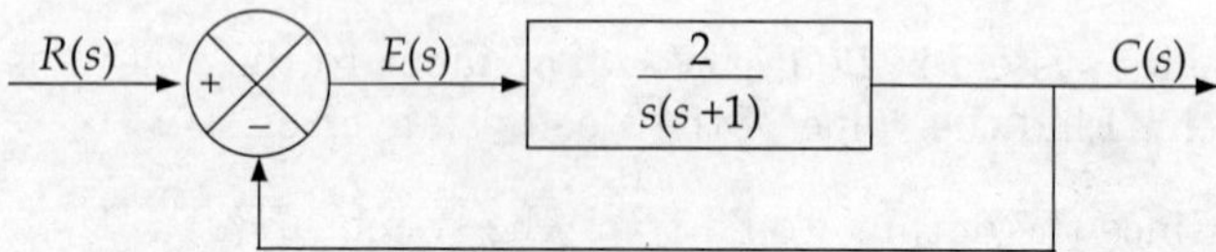

Figure 6.25: System (Example 6.48).

(A) 0 (B) 0.5 (C) 1 (D) 2

Solution Given: $R(s) = 1/s^2$ and $G(s)H(s) = 2/[s(s+1)]$. Therefore,

$$e_{ss} = \lim_{s\to 0} sE(s) = \lim_{s\to 0} \frac{sR(s)}{1 + G(s)H(s)} = \lim_{s\to 0} s \cdot \frac{1}{s^2} \cdot \frac{s(s+1)}{s(s+1)+2} = \frac{1}{2} = 0.5$$

Ans. (B)

EXAMPLE 6.49. For a system having

$$G(s)H(s) = \frac{K(s+4)}{s(s^3 + 5s^2 + 6s)}$$

find the (a) type of the system, (b) static error constants and (c) error due to input $\frac{A}{2}t^2$.

Solution (a) $G(s)H(s) = \frac{K(s+4)}{s(s^3 + 5s^2 + 6s)} = \frac{K(s+4)}{s^2(s^2 + 5s + 6)}$. Therefore, type = 2.

(b) Since it is a type 2 system, $K_p = K_v = \infty$.

$$K_a = \lim_{s\to 0} s^2 G(s)H(s) = \lim_{s\to 0} \frac{K(s+4)}{s^2 + 5s + 6} = \frac{2K}{3}$$

(c) The input is parabolic. Its Laplace transform is $R(s) = \frac{A}{s^3}$. Therefore,

$$e_{ss} = \frac{A}{K_a} = \frac{3A}{2K}$$

Dynamic Error Coefficients

In the previous section we discussed how to evaluate the steady-state error utilizing the final value theorem. The method suffers from the following limitations:

- It allows determination of error coefficients for only three kinds of input.
- Also, for those permitted kinds of input, it does not show how the error varies with time.

These disadvantages can be overcome by following the method of finding generalized error coefficients that are often termed as *dynamic error coefficients.*

The error function, as we found in Eq. (6.48), is given by

$$\frac{E(s)}{R(s)} = \frac{1}{1+G(s)H(s)}$$

The denominator on the RHS of the equation is a function of s and it can be written as a power series in s as follows

$$\frac{E(s)}{R(s)} = \frac{1}{K_1} + \frac{1}{K_2}s + \frac{1}{K_3}s^2 + \cdots$$

$$\Rightarrow \qquad E(s) = \frac{1}{K_1}R(s) + \frac{1}{K_2}sR(s) + \frac{1}{K_3}s^2R(s) + \cdots \tag{6.58}$$

The inverse Laplace transform of Eq. (6.58) yields

$$e(t) = \frac{1}{K_1}r(t) + \frac{1}{K_2}\frac{d}{dt}r(t) + \frac{1}{K_3}\frac{d^2}{dt^2}r(t) + \cdots \tag{6.59}$$

Therefore, if the input function is known, dynamic error coefficients can be figured out through a comparison between Eq. (6.59) and the polynomial obtained from the inverse Laplace transform[3]. The following examples will make the procedure clear.

EXAMPLE 6.50. The open-loop transfer function of a servo system with unity feedback is given by

$$G(s) = \frac{500}{s(1+0.1s)}$$

(a) Evaluate the error series of the system.

(b) Determine the steady-state error of the system for the input $r(t) = 1 + 2t + t^2$.

Solution Since it is a unity feedback system, we have

$$\frac{E(s)}{R(s)} = \frac{1}{1+G(s)} = \frac{s+0.1s^2}{500+s+0.1s^2}$$

The long division of the RHS of the equation yields

$$\begin{array}{rllll}
 & +0.002s & +0.000196s^2 & -0.000000792s^3 & \\
\hline
500+s+0.1s^2\,\big) & +s & +0.1s^2 & & \\
 & +s & +0.002s^2 & +0.0002s^3 & \\
\hline
 & & +0.098s^2 & -0.0002s^3 & \\
 & & +0.098s^2 & +0.000196s^3 & +0.0000196s^4 \\
\hline
 & & & -0.000396s^3 & -0.0000196s^4 \\
 & & & -0.000396s^3 & -0.000000792s^4 \\
\hline
\end{array}$$

[3] Some authors define error coefficients as $C_i = 1/K_i$.

Thus,

$$\frac{E(s)}{R(s)} = 0.002s + 0.000196s^2 - 0.000000792s^3 + \cdots$$

$$\Rightarrow \quad E(s) = 0.002sR(s) + 0.000196s^2R(s) - 0.000000792s^3R(s) + \cdots \tag{i}$$

$$\equiv \frac{1}{K_1}R(s) + \frac{1}{K_2}sR(s) + \frac{1}{K_3}s^2R(s) + \frac{1}{K_4}s^3R(s) + \cdots \tag{ii}$$

The inverse Laplace transform of Eq. (i) yields

$$e(t) = 0.002\frac{dr(t)}{dt} + 0.000196\frac{d^2r(t)}{dt^2} - 0.000000792\frac{d^3r(t)}{dt^3} + \cdots \tag{iii}$$

We are given, $r(t) = 1 + 2t + t^2$. Therefore,

$$\frac{dr(t)}{dt} = 2 + 2t, \qquad \frac{d^2r(t)}{dt^2} = 2, \qquad \frac{d^3r(t)}{dt^3} = 0 \tag{iv}$$

Substituting results of Eqs. (iv) in Eq. (iii), we get the required error series as

$$e(t) = 0.002(2 + 2t) + 0.0000196(2) - 0.000000792(0) + \cdots = 0.004392 + 0.004t$$

(a) The dynamic error coefficients can be calculated by comparing Eq. (i) and Eq. (ii) as follows:

$$K_1 = \frac{1}{0} = \infty \qquad K_2 = \frac{1}{0.002} = 500$$

$$K_3 = \frac{1}{0.000196} = 5102.04 \qquad K_4 = -\frac{1}{0.000000792} = -1262626.26$$

(b) The steady-state error in time-domain is

$$\lim_{t\to\infty} e(t) = \lim_{t\to\infty} (0.004392 + 0.004t) = \infty$$

EXAMPLE 6.51. The open-loop transfer function of a unity feedback system is given by

$$G(s) = \frac{50}{s(s+10)}$$

Find out dynamic error coefficients and the error series when the input is $r(t) = 1 + 2t + t^2$.

Solution It being a unity feedback system, we have

$$\frac{E(s)}{R(s)} = \frac{10s + s^2}{50 + 10s + s^2}$$

Note: Since we want to find out a polynomial in ascending powers of s, we have to arrange the numerator and denominator in that order, before performing the long division.

By long division (not shown), we get

$$E(s) = (0)R(s) + 0.2sR(s) - 0.02s^2R(s) + (0)s^3R(s) + 0.004s^4R(s) + \cdots \tag{i}$$

$$\equiv \frac{1}{K_1}R(s) + \frac{1}{K_2}sR(s) + \frac{1}{K_3}s^2R(s) + \frac{1}{K_4}s^3R(s) + \frac{1}{K_5}s^4R(s) + \cdots \tag{ii}$$

Comparing the coefficients of Eq. (i) and Eq. (ii), we get the dynamic error coefficients as follows:

$$K_1 = \frac{1}{0} = \infty \qquad K_2 = \frac{1}{0.2} = 5 \qquad K_3 = -\frac{1}{0.02} = -50$$

$$K_4 = \frac{1}{0} = \infty \qquad K_5 = \frac{1}{0.004} = 250$$

Now, the inverse Laplace transform of Eq. (i) yields

$$e(t) = (0)r(t) + 0.2\frac{dr(t)}{dt} - 0.02\frac{d^2r(t)}{dt^2} + (0)\frac{d^3r(t)}{dt^3} + 0.004\frac{d^4r(t)}{dt^4} + \cdots$$

The input is $r(t) = 1 + 2t + t^2$. So,

$$\frac{dr(t)}{dt} = 2 + 2t, \qquad \frac{d^2r(t)}{dt^2} = 2$$

Therefore, the error series is given by

$$e(t) = 0.2(2 + 2t) - 0.02(2) = 0.36 + 0.4t$$

6.5 OTHER ERROR PARAMETERS

We now define three other important error parameters, namely,

1. Integral Square Error (ISE)
2. Integral Absolute Error (IAE)
3. Integral Time-weighted Absolute Error (ITAE)

All the measures require a fixed experiment to be performed on the system by fixing a set-point or disturbance change and the integrals are evaluated over a fixed time period, in theory from $t = 0$ to $t = \infty$, but usually until a time long enough for the responses to settle.

Integral Square Error (ISE)

Unlike the steady-state error, the integral square error is a measure of the system error through the entire time period, i.e. from $t = 0$ to $t = \infty$. It is defined as

$$\text{ISE} = \int_0^\infty e^2(t)\, dt \tag{6.60}$$

Advantages. ISE integrates the square of the error over time. Since the square of a large error will be much bigger, ISE will put more weight to large errors than smaller ones. Control systems specified to minimize ISE will tend to eliminate large errors quickly, but will tolerate small errors persisting for a long period of time.

Disadvantage. Often this leads to a fast response, but with a considerable low amplitude, oscillation.

Alternative definition. Equation (6.60) gives the magnitude of the ISE from the t-space representation $e(t)$ of the error function. It would be convenient if we could find it out from the s-space representation $E(s)$ of the error function.

A real integration theorem states that

$$\int_0^\infty f(t)dt = \lim_{t\to\infty}\int_0^t f(t)dt$$

Applying the final value theorem, we get from this equation

$$\lim_{t\to\infty}\int_0^t f(t)dt = \lim_{s\to 0} s\mathscr{L}\left\{\int_0^t f(t)dt\right\} = \lim_{s\to 0} \cancel{s}\frac{F(s)}{\cancel{s}}$$

or,

$$\int_0^\infty f(t)dt = \lim_{s\to 0} F(s) \tag{6.61}$$

Combining Eqs (6.60) and (6.61), we get

$$\text{ISE} = \lim_{s\to 0} E^2(s) \tag{6.62}$$

EXAMPLE 6.52. A unity feedback position control system has a forward path transfer function $G(s) = K/s$. For unit step input, compute the value of K that minimizes ISE.

Solution Given: $G(s) = K/s$, $H(s) = 1$. Also, for a unit step input $R(s) = 1/s$. Therefore,

$$T(s) = \frac{C(s)}{R(s)} = \frac{G(s)}{1+G(s)} = \frac{\dfrac{K}{s}}{1+\dfrac{K}{s}} = \frac{K}{s+K}$$

Thus,

$$C(s) = \frac{K}{s(s+K)} = \frac{1}{s} - \frac{1}{s+K} \tag{i}$$

The inverse Laplace transform of Eq. (i) yields

$$c(t) = 1 - e^{-Kt} \qquad \Rightarrow \quad e(t) = 1 - c(t) = e^{-Kt}$$

Therefore,

$$\text{ISE} = \int_0^\infty e^{-2Kt}\, dt = -\left.\frac{e^{-2Kt}}{2K}\right|_0^\infty = \frac{1}{2K}$$

This result indicates that ISE $\to 0$ as $K \to \infty$.

EXAMPLE 6.53. A unity feedback system has a forward path transfer function

$$G(s) = \frac{10}{s(s+5)}$$

It is subjected to a unit ramp input. The integral of the squared error (ISE) is

(A) zero (B) infinity (C) always negative (D) 0.5

Solution We know

$$E(s) = \frac{R(s)}{1 + G(s)H(s)} = \frac{1}{s^2} \cdot \frac{s^2 + 5s}{s^2 + 5s + 10} = \frac{s + 5}{s(s^2 + 5s + 10)} \qquad \left[\because R(s) = \frac{1}{s^2}\right]$$

So, from Eq. (6.62)

$$\text{ISE} = \lim_{s \to 0} \left[\frac{s + 5}{s(s^2 + 5s + 10)}\right]^2 = \infty$$

Ans. (B)

EXAMPLE 6.54. The forward path transfer function of a unity feedback system is given by

$$G(s) = \frac{(1 + 5s)(1 + 10s)(1 + 2s)}{(1 + s)(1 + 8s)(1 + 20s)}$$

If $e(t)$ is the error to a unit impulse input, the value of the performance index

$$J = \int_0^\infty e(t)dt$$

is equal to

(A) zero (B) infinity (C) −12 (D) 0.5

Solution Given $R(s) = 1,\ H(s) = 1$. Therefore,

$$E(s) = \frac{1}{1 + G(s)} = \frac{(1 + s)(1 + 8s)(1 + 20s)}{(1 + s)(1 + 8s)(1 + 20s) + (1 + 5s)(1 + 10s)(1 + 2s)}$$

From Eq. (6.61), we have

$$J = \int_0^\infty e(t)dt = \lim_{s \to 0} E(s) = \lim_{s \to 0} \frac{(1 + s)(1 + 8s)(1 + 20s)}{(1 + s)(1 + 8s)(1 + 20s) + (1 + 5s)(1 + 10s)(1 + 2s)} = 0.5$$

Ans. (D)

Note: This is neither ISE nor IAE. This is a new parameter defined by the examiner.

Integral Absolute Error (IAE)

The integral absolute error is a measure of average error over a certain time period, T, and is defined as

$$\text{IAE} = \int_0^T |e(t)|\ dt$$

If it is a second order underdamped system, the IAE looks like what is presented in Figure 6.26.

Advantage. IAE integrates the absolute error over time. It does not add weight to any of the errors in a system's response.

Disadvantage. It tends to produce slower response than ISE optimal systems, but usually with less sustained oscillation.

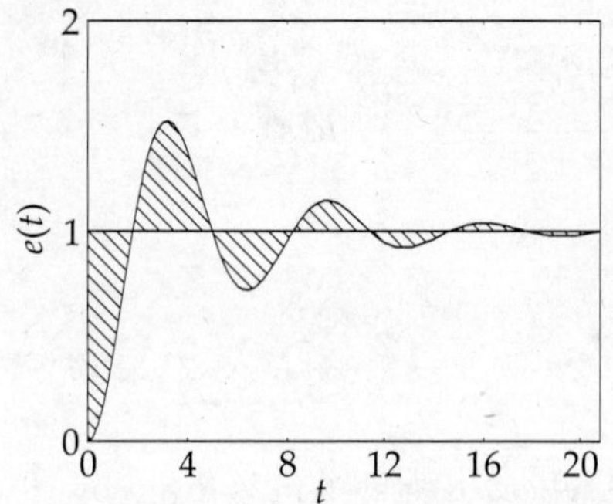

Figure 6.26: Shaded area represents the integral absolute error of an underdamped second order system.

Integral Time-weighted Absolute Error (ITAE)

ITAE integrates the absolute error multiplied by the time over time and is defined as

$$\text{IAE} = \int_0^T t|e(t)|\, dt$$

Advantages. This method of error estimation essentially weights errors which exist after a long time much more heavily than those at the start of the response. ITAE tuning produces systems which settle much more quickly than the other two tuning methods.

Disadvantage. ITAE tuning produces systems with sluggish initial response (necessary to avoid sustained oscillation).

6.6 EFFECTS OF ADDITION OF POLES AND ZEROS ON RESPONSE OF SYSTEMS

It is interesting to study the effects of addition of poles and zeros on the response of systems. While doing that, we will restrict ourselves to

(a) Closed-loop transfer functions only.
(b) Second order underdamped systems and step response only.
(c) Poles and zeros will be located on the left half plane of the s-space.

Addition of Poles

Let the closed-loop transfer function be

$$T(s) = \frac{1}{s^2 + 0.6s + 1}$$

Obviously it is a second order system with $\omega_n = 1$ and $\zeta = 0.3$. We add a pole to make it

$$T(s) = \frac{1}{s^2 + 0.6s + 1} \cdot \frac{1}{\tau_p s + 1} \tag{6.63}$$

Figure 6.27a shows step responses for the system represented by Eq. (6.63) for $\tau_p = 0$, 1, and 2. Here we observe that as τ_p increases, rise time increases, but peak overshoot decreases. We add that, though observed in a particular case, the conclusion is general.

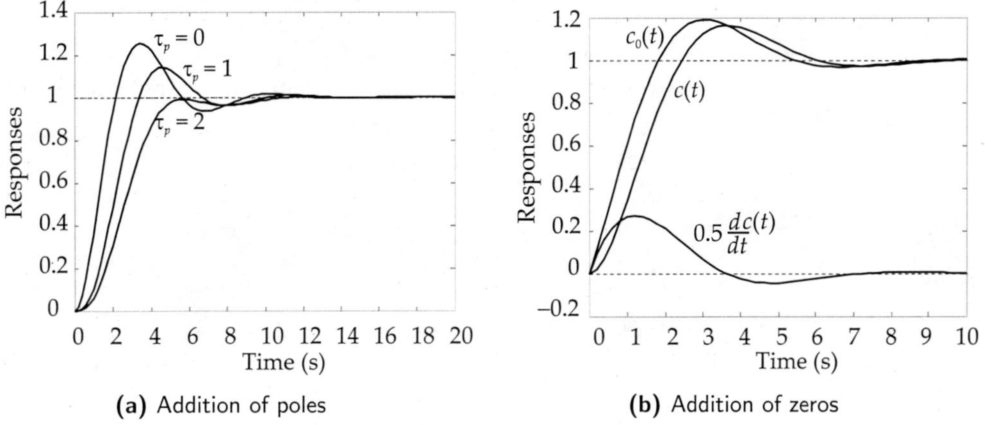

(a) Addition of poles

(b) Addition of zeros

Figure 6.27: Effect of addition of poles and zeros to closed-loop transfer function.

Addition of Zeros

Let the closed-loop transfer function be

$$T(s) = \frac{1}{s^2 + s + 1} \tag{6.64}$$

We add a zero to it at $s = -1/\tau_z$ so that

$$T(s) = \frac{1 + \tau_z s}{s^2 + s + 1} \tag{6.65}$$

Equation (6.65) can be written as

$$T(s) = \frac{1}{s^2 + s + 1} + \tau_z s \cdot \frac{1}{s^2 + s + 1} \tag{6.66}$$

So, if $c(t)$ be the step response of Eq. (6.64), that of Eq. (6.66) is

$$c_0(t) = c(t) + \tau_z \frac{d}{dt} c(t)$$

Figure 6.27b shows the plot of $c(t)$, $\tau_z \frac{d}{dt} c(t)$ and $c_0(t)$. It is clear from the figure that the zero term decreases the rise time but increases the peak overshoot. The effects of addition of poles and zeros are summarized in Table 6.5.

Table 6.5: Effects of addition of poles and zeros

As the *pole* nears origin	Rise time (t_r) ⟶ Increases (↑) Peak overshoot (M_p) ⟶ Decreases (↓)
As the *zero* nears origin	Rise time (t_r) ⟶ Decreases (↓) Peak overshoot (M_p) ⟶ Increases (↑)

6.7 MINIMUM AND NON-MINIMUM PHASE SYSTEMS

In Section 2.1 at page 13, we had talked about minimum and non-minimum phase systems. Here, we want to see the difference of response that these systems offer.

We recall that minimum-phase systems have only positive terms in their transfer functions. If the system is of non-minimum phase, the sweep in phase area is larger. Let us compare the unit step responses of minimum and non-minimum phase systems, the transfer functions of which are given by

$$\frac{C(s)}{R(s)} = G_{\text{min}}(s) = \frac{s+1}{s^2+s+1}$$

$$\frac{C(s)}{R(s)} = G_{\text{non-min}}(s) = \frac{-s+1}{s^2+s+1}$$

We present the unit step response of the systems in Figures 6.28a and 6.28b, respectively. The difference in behaviour is apparent from Figure 6.28b. The non-minimum phase system

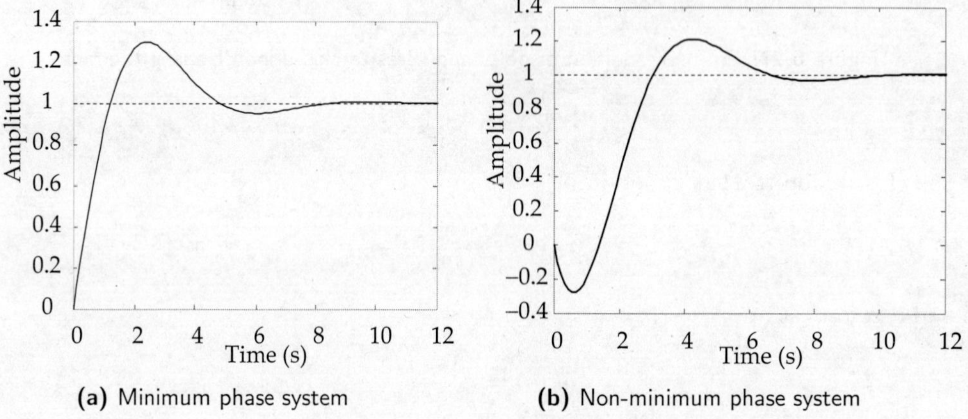

(a) Minimum phase system

(b) Non-minimum phase system

Figure 6.28: Unit step response of minimum and non-minimum phase systems.

produces an initial response that is going in the opposite direction to that of the input. This is typical for such systems.

REVIEW QUESTIONS

6.1 Choose the correct answer

(i) The characteristic equation of a system is $s^2 + 2s + 2 = 0$. The system is

(A) critically damped (B) underdamped
(C) overdamped (D) none of these

(ii) For a control system to be operated in overdamped condition, the value of the damping ratio should be

(A) 0 (B) > 1 (C) 1 (D) < 1

(iii) The steady-state error for a type 3 system for a unit step input is

(A) zero (B) infinity (C) one (D) three

(iv) The transfer function of a system is $\frac{10}{1+s}$. The steady-state error due to step input is

(A) 10 (B) 0 (C) $\frac{1}{11}$ (D) ∞

(v) The second order system defined by transfer function

$$\frac{C(s)}{R(s)} = \frac{25}{s^2 + 5s + 25}$$

is given a unit step input. The time taken for the output to settle within ±2% of the final value is

(A) 1.2 s (B) 1.6 s (C) 2 s (D) 0.4 s

(vi) Addition of a pole to the closed-loop transfer function

(A) increases rise time (B) decreases rise time
(C) increases overshoot (D) has no effect

(vii) The measure of average error over a specified time period is called

(A) integral absolute error (B) integral square error
(C) steady-state error (D) none of these

(viii) A second order system has a damping ratio of 0.9. The system is

(A) underdamped (B) overdamped
(C) critically damped (D) insufficient information

(ix) A system has a single pole at the origin. Its impulse response will be

(A) constant (B) ramp
(C) decaying exponential (D) oscillatory

(x) For a type 1 system, the steady-state error due to step input is

(A) $\dfrac{1}{1+K_p}$ (B) $\dfrac{1}{K_p}$
(C) 0 (D) ∞

(xi) The settling time of a second order underdamped system for settling within ±2% of the final value is given by

(A) $\dfrac{4}{\zeta\omega_n}$ (B) $\dfrac{3}{\zeta\omega_n}$ (C) $\dfrac{4\zeta}{\omega_n}$ (D) $\dfrac{2}{\zeta\omega_n}$

(xii) For a unit step input, a system with a closed-loop transfer function of $20/(s^2+2s+5)$ has a steady-state output of

(A) 10 (B) 5 (C) 2 (D) 4

(xiii) For a second order system given by

$$2\frac{d^2y}{dt^2} + 4\frac{dy}{dt} + 8y = 8x$$

the damping ratio is

(A) 0.1 (B) 0.25 (C) 0.333 (D) 0.5

(xiv) The unit step response of a particular control system is $c(t) = 1 - 10e^{-t}$. The transfer function is

(A) $\dfrac{10}{s+1}$ (B) $\dfrac{s-9}{s+1}$
(C) $\dfrac{1-9s}{s+1}$ (D) $\dfrac{1-9s}{s(s+1)}$

(xv) The steady-state error for a type 2 system subjected to a unit ramp input is

(A) 2 (B) 1 (C) 0 (D) ∞

(xvi) Addition of a zero to the closed-loop transfer function

(A) increases rise time (B) decreases rise time

(C) increases overshoot (D) has no effect

(xvii) Assuming zero initial condition, the response $y(t)$ of the system given below to a unit step input $u(t)$ is

$U(s)$ → $1/s$ → $Y(s)$

(A) $u(t)$ (B) $tu(t)$ (C) $\frac{t^2}{2}u(t)$ (D) $e^{-1}u(t)$

(xviii) The impulse response of a system is $h(t) = tu(t)$. For an input $u(t-1)$, the output is

(A) $\frac{t^2}{2}u(t)$ (B) $\frac{t(t-1)}{2}u(t-1)$

(C) $\frac{(t-1)^2}{2}u(t-1)$ (D) $\frac{t^2-1}{2}u(t-1)$

6.2 Consider the following block diagram:

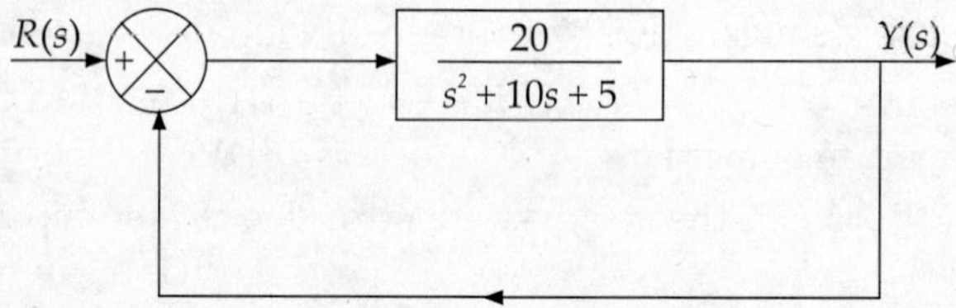

(a) Find the closed-loop transfer function $T(s) = Y(s)/R(s)$.

(b) Find the closed-loop unit step response.

(c) Find the final value of $y(t)$.

6.3 (a) Starting from the cold condition, a first order system when subjected to a step input, has a temperature rise of 25°C after an hour and 37.5°C after 2 hours. Calculate the final steady temperature rise and the thermal time constant.

(b) Sketch the time-domain response of a typical underdamped second order system to a step input. On this sketch, indicate the following:

(i) Maximum peak overshoot, M_p

(ii) Rise time, t_r

(iii) Delay time, t_d

(iv) Settling time, t_s

(v) Steady-state error, e_{ss}

6.4 (a) A second order differential equation is given by

$$\frac{d^2x}{dt^2} + 5\frac{dx}{dt} + 7x = 7y$$

Find the following:

(i) Undamped natural frequency

(ii) Damping ratio

(iii) Damped natural frequency

(iv) Damping coefficient

(v) Time constant

(b) Explain the unit impulse response of a first order system.

6.5 Consider a unity feedback system with forward transfer function given by

$$G(s) = \frac{1}{(s+1)(s+2)}$$

The steady-state error in the output of the system for a unit step input is ——(up to 2 decimal places).

6.6 The unit step response $y(t)$ of a unity feedback system with open-loop transfer function

$$G(s)H(s) = \frac{K}{(s+1)^2}(s+2)$$

is shown in the following figure. The value of K is ——(up to 2 decimal places).

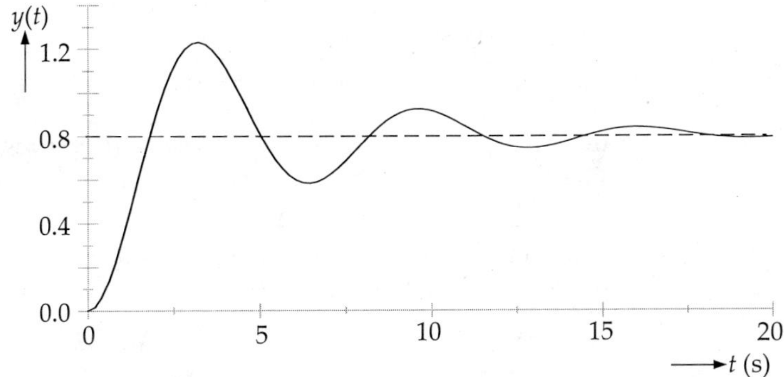

6.7 The closed-loop transfer function of a system is

$$T(s) = \frac{4}{s^2 + 0.4s + 4}$$

The steady-state error due to unit step input is ——.

6.8 Find the expression for % peak overshoot of a second order underdamped system when subjected to a unit step input.

6.9 Consider the unity feedback system with $G(s) = \dfrac{K}{s(s+4)}$. If $K_v = 12$, find the percent overshoot and peak time.

6.10 (a) Find the error coefficient for the following system as shown in in the following figure.

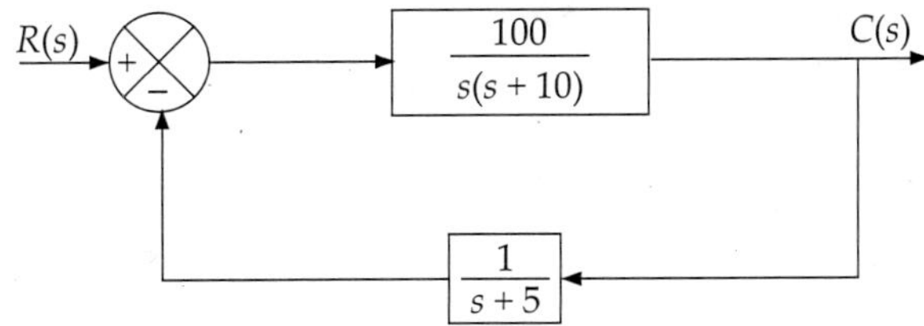

(b) What do you mean by transient response and steady state response?

6.11 What is steady-state error? Suppose unit ramp input is given to type 0, type 1 and type 2 systems. Compare the steady-state error for these three cases.

6.12 (a) Find out the time response analysis of a unity feedback underdamped control system having open loop gain of $\dfrac{\omega_n^2}{s^2 + 2\zeta\omega_n s}$ for unit step input.

(b) Define and calculate the value of peak time and settling time for the above time response equation.

6.13 Derive an expression for step response of an underdamped second order system. Sketch the response indicating all time-domain specifications.

6.14 Find the steady-state error of the unity feedback system having $G(s) = \dfrac{10}{0.1s+1}$ in following the input $r(t) = 1 + tu(t)$.

6.15 (a) Find out the response of a typical underdamped second order closed-loop system with unity feedback having unit step input. Draw the response curve.

(b) A second order servo system has poles at $-1 \pm \jmath 2$ and zero at $-1 \pm \jmath 0$. Its steady-state output for a unit step input is $\frac{1}{2}$. Determine the transfer function.

6.16 (a) Sketch the unit step response of a second order system indicating peak overshoot, rise time and settling time.

(b) A temperature sensing device can be modelled as a first order system with a time constant of 6 s. It is suddenly subjected to a step input of 30°C to 150°C. What temperature will be indicated in 10 s after the process has started?

6.17 (a) Define the following:

(i) Rise time (ii) Delay time
(iii) Settling time (iv) Overshoot

(b) The open-loop transfer function of a unity feedback control system is given by

$$G(s) = \frac{A}{s(1+sT)}$$

(i) By what factor should the amplifier gain A be multiplied so that the damping ratio is increased from a value of 0.2 to 0.6?

(ii) By what factor should the amplifier gain be multiplied so that the overshoot of the unit step response is reduced from 80% to 20%?

6.18 (a) Find out the response of a typical underdamped second order closed-loop system with unity feedback having step input.

(b) For a unity feedback closed loop control system the open-loop transfer function is given by

$$G(s) = \frac{25}{s(s+5)}$$

Calculate the value of rise time (t_r) and maximum overshoot (M_p) when the system is subjected to a unit step input.

6.19 Evaluate the static error constants for a unity feedback system having a forward path transfer function $G(s) = 50/s(s+10)$. Estimate steady-state errors of the systems for the input $r(t)$ given by $r(t) = 1 + 2t + t^2$.

Answers to Selected Questions

6.1. (i) (B) (ii) (B) (iii) (A) (iv) (C) (v) (B) (vi) (A) (vii) (A) (viii) (A) (ix) (A) (x) (C) (xi) (A) (xii) (D) (xiii) (D) (xiv) (B) (xv) (C) (xvi) (B) (xvii) (B) (xviii) (C)

6.2. (a) $20/(s^2 + 10s + 25)$ (B) $y(t) = 0.8 - 4(t + 0.2)e^{-5t}$ (c) $y_{ss} = 0.8$

6.3. (a) 3.79 h 6.4. (a) (i) 2.65 rad/s (ii) 0.95 (iii) 0.87 rad/s (iv) 0.71 (v) 0.4 s

6.5. 0.66 6.6. 8

6.7. 0

6.9. 38.6%, 0.47 s 6.10. (A) $K_1 = \infty,\ K_2 = 2,\ K_3 = -10,\ K_4 = -66.67$ 6.14. ∞

6.15. (b) $\dfrac{2(s+1)}{s^2 + 2s + 5}$ 6.16. (b) 127.3°C 6.17. (b) (i) 0.11 (ii) 0.024

6.18. (b) 0.484 s, 16.3% 6.19. $K_p = \infty$, $K_v = 50$, $K_a = 0$. e_{ss} for the given input is ∞.

7

Basics of Controllers

7.1 INTRODUCTION

We know that an automatic control system can be of two kinds—open-loop and closed-loop—depending on whether the control element receives a feedback from the output or not. The block diagrams of Figures 7.1a and 7.1b give a visual presentation of their difference.

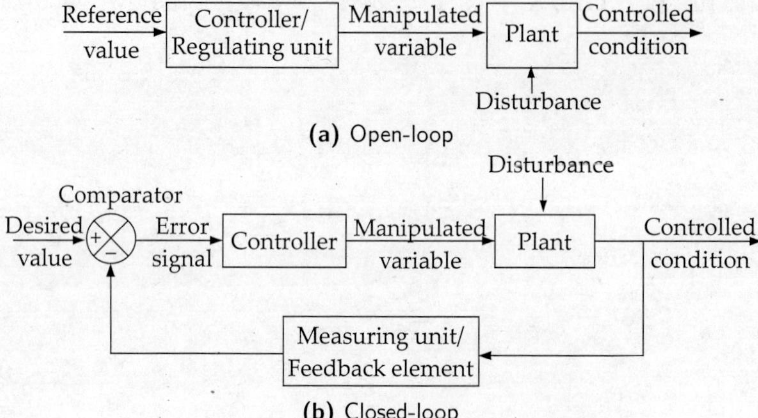

Figure 7.1: Automatic control systems.

From these diagrams we see that controllers constitute an integral element in either control system. In an automatic control system, the controlled value of the variable is continuously compared with its desired value and the deviation is automatically rectified. This procedure can be accomplished via four basic modes:

1. Two-position or ON-OFF control
2. Proportional control
3. Integral control
4. Derivative control

Except the first one, various combinations of other three basic modes may be employed to suit the plant characteristics. So, controllers can be divided into following six categories depending on their mode of control:

1. Proportional (P) controller
2. Integral (I) controller
3. Derivative (D) controller
4. Proportional plus integral (PI) controller
5. Proportional plus derivative (PD) controller
6. Proportional plus integral plus derivative (PID) controller

In the following sections we will consider these controllers and evaluate their performance.

7.2 TWO-POSITION (ON-OFF) CONTROLLER

Two-position controllers are generally electric devices such as relays, solenoid operated valves, etc. Schematically, their action may be represented as in Figure 7.2a.

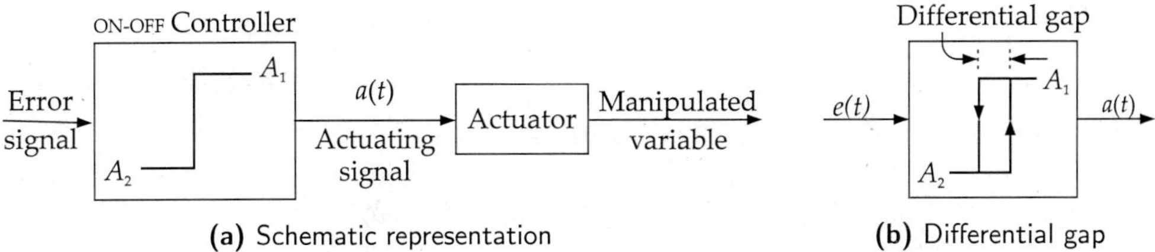

(a) Schematic representation (b) Differential gap

Figure 7.2: Two-position ON-OFF controller.

If $a(t)$ is the actuating signal, then mathematically the action can be described as

$$a(t) = \begin{cases} A_1 \text{ (Max or ON position)} & = a_1 e(t) > 0 \\ A_2 \text{ (Min or OFF position)} & = a_2 e(t) < 0 \end{cases}$$

The range through which the error signal must pass before switching takes place is called the *differential gap* (see Figure 7.2b).

The advantages and disadvantages of this type of control are listed in Table 7.1.

Table 7.1: Advantages and disadvantages of ON-OFF controller

Advantages	*Disadvantages*
1. Simple, economical.	1. Transient conditions and steady-state accuracy are not much improved.
2. Used in domestic appliances as well as in industry where rough control is necessary.	2. Controlled output oscillates around the set point.

7.3 PROPORTIONAL (P) CONTROLLER

In a proportional controller, the actuating signal is proportional to the error signal. Mathematically, if $a(t)$ is the actuating signal, then

$$a(t) \propto e(t) = K_p e(t) \tag{7.1}$$

where K_p is called the *controller gain*. Though generally its value is greater than 1, it can be less than 1 in some applications. Since in most of the cases it amplifies the error signal, it facilitates detection of small deviations and their eventual remedy by the actuator.

The Laplace transform of Eq. (7.1) is

$$A(s) = K_p E(s)$$

Accuracy

Consider, for example, the second order system given in Figure 7.3a.

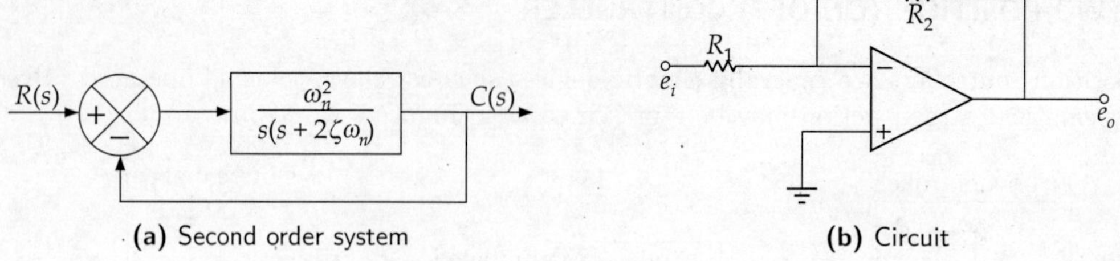

Figure 7.3: Proportional controller.

Here, with the addition of a proportional controller the open-loop transfer function becomes

$$G(s)H(s) = \frac{K_p\omega_n^2}{s(s+2\zeta\omega_n)}$$

Therefore,

$$e_{ss} = \lim_{s\to 0}\frac{sR(s)}{1+G(s)H(s)} = \lim_{s\to 0}\frac{sR(s)}{1+\dfrac{K_p\omega_n^2}{s(s+2\zeta\omega_n)}} = \lim_{s\to 0}\frac{s^2(s+2\zeta\omega_n)}{s(s+2\zeta\omega_n)+K_p\omega_n^2}\cdot R(s) \tag{7.2}$$

Step input. For a step input, $R(s) = 1/s$. Therefore, from Eq. (7.2)

$$e_{ss} = \lim_{s\to 0}\frac{s^2(s+2\zeta\omega_n)}{s(s+2\zeta\omega_n)+K_p\omega_n^2}\cdot\frac{1}{s} = \lim_{s\to 0}\frac{s(s+2\zeta\omega_n)}{s(s+2\zeta\omega_n)+K_p\omega_n^2} = 0$$

Ramp input. In this case, $R(s) = 1/s^2$. So, from Eq. (7.2)

$$e_{ss} = \lim_{s\to 0}\frac{s^2(s+2\zeta\omega_n)}{s(s+2\zeta\omega_n)+K_p\omega_n^2}\cdot\frac{1}{s^2} = \frac{2\zeta}{K_p\omega_n} \tag{7.3}$$

Parabolic input. In the case of parabolic input, $R(s) = 1/s^3$. Therefore, from Eq. (7.2)

$$e_{ss} = \lim_{s\to 0}\frac{s^2(s+2\zeta\omega_n)}{s(s+2\zeta\omega_n)+K_p\omega_n^2}\cdot\frac{1}{s^3} = \infty$$

The advantages and disadvantages of a proportional controller are listed in Table 7.2.

Realization

A simple op-amp realization of the proportional controller is indicated in Figure 7.3b. The input-output relationship for such a circuit is given by

$$\left|\frac{e_o}{e_i}\right| = \frac{R_2}{R_1}$$

EXAMPLE 7.1. The open-loop transfer function of a dc motor is given as

$$\frac{\omega(s)}{V_a(s)} = \frac{10}{1+10s}$$

When connected in feedback as shown in the following figure, the approximate value of K_a that will reduce the time-constant of the closed-loop system by one hundred times as compared to that of the open-loop system is

Table 7.2: Advantages and disadvantages of a proportional controller

Advantages	*Disadvantages*
1. This type of controller increases the forward path gain. So, the response becomes faster for overdamped systems as a consequence. For underdamped systems, however, it increases peak overshoot though reducing the rise time.	1. Causes offset i.e. it may cause a sustained deviation from the desired value.
2. e_{ss} is reduced. Also, as K_p increases, e_{ss} decreases. However, an increase of K_p may make the system unstable if K_p is too high.	2. It increases the maximum overshoot of the system because it lowers ζ.
3. It is a stable control as long as K_p is not high.	

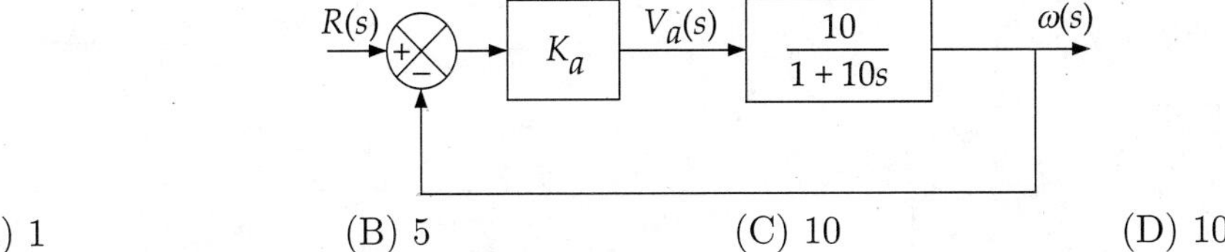

(A) 1 (B) 5 (C) 10 (D) 100

Solution The open-loop transfer function is given by

$$\left.\frac{\omega(s)}{V_a(s)}\right|_{\text{open-loop}} = \frac{10}{1+10s} \equiv \frac{K}{1+s\tau_{\text{open-loop}}} \tag{i}$$

By comparison, we get from Eq. (i) $\tau_{\text{open-loop}} = 10$. We need to make $\tau_{\text{closed-loop}} = 100\tau_{\text{open-loop}}$ by introducing the controller of gain K_a and closing the loop. The closed-loop transfer function is given by

$$\left.\frac{\omega(s)}{V_a(s)}\right|_{\text{closed-loop}} = \frac{K_a\left[\dfrac{10}{1+10s}\right]}{1+K_a\left[\dfrac{10}{1+10s}\right]} = \frac{10K_a}{10s+\left(10K_a+1\right)}$$

$$= \frac{\dfrac{10K_a}{10K_a+1}}{1+\left[\dfrac{10}{10K_a+1}\right]s} \equiv \frac{K}{1+s\tau_{\text{closed-loop}}} \tag{ii}$$

Equation (ii) suggests that $\tau_{\text{closed-loop}} = 10/\left(10K_a+1\right)$. From the given condition $\tau_{\text{open-loop}} = 100 \times \tau_{\text{closed-loop}}$, we get

$$10 = 100 \times \left[\frac{10}{10K_a+1}\right]$$

On simplification,

$$K_a = 9.9 \cong 10$$

Ans. (C)

7.4 INTEGRAL (I) CONTROLLER

In an integral controller, the actuating action can be mathematically put as

$$a(t) \propto \int e(t)\ dt = K_i \int e(t)\ dt \tag{7.4}$$

where K_i, a constant, is the *controller gain.*

The Laplace transform of Eq. (7.4) is

$$A(s) = \frac{K_i}{s} E(s) \tag{7.5}$$

Realization

A simple op-amp realization of the integral controller is shown in Figure 7.4a. This is nothing but an integrator circuit for which the output-input relation is given by

$$e_o = \frac{1}{RC} \int |e_i| \, dt \tag{7.6}$$

The Laplace transform of Eq. (7.6) yields the transfer function of the controller as

$$G(s) = \frac{E_o(s)}{E_i(s)} = \frac{1}{s} \cdot \frac{1}{RC}$$

which has the form of Eq. (7.5).

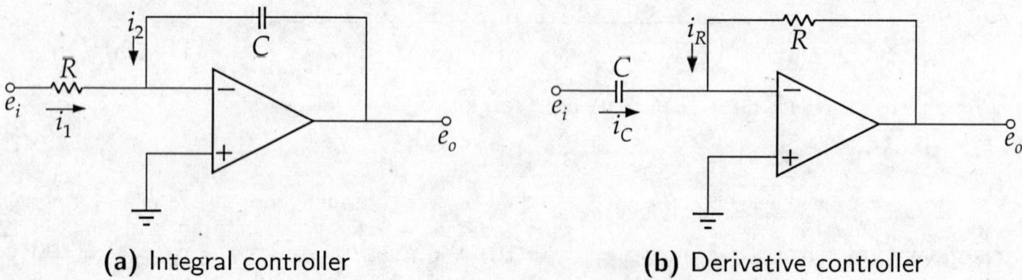

Figure 7.4: Op-amp realization of controllers.

7.5 DERIVATIVE (D) CONTROLLER

The derivative controller produces an actuating signal from the error signal as

$$a(t) \propto \frac{de(t)}{dt} = K_d \frac{de(t)}{dt} \tag{7.7}$$

where K_d is a constant. The Laplace transform of Eq. (7.7) is

$$A(s) = K_d s E(s)$$

Realization

The simple op-amp realization (Figure 7.4b) of the derivative controller is nothing else but a differentiator for which the output is related to the input as

$$e_o = RC \frac{de_i}{dt} \tag{7.8}$$

From the Laplace transform of Eq. (7.8), we get the transfer function as

$$G(s) = \frac{E_o(s)}{E_i(s)} = sRC$$

Advantages and disadvantages of derivative controllers are given in Table 7.3.

The disadvantages of a derivative controller far outweigh the only one advantage it offers. This is why *it is never used alone.*

Table 7.3: Advantages and disadvantages of derivative controllers

Advantages	*Disadvantages*
1. Also called the *rate controller*, it improves the transient response of the system.	1. It amplifies the noise signal.
	2. It may cause saturation effect in the actuator.
	3. It does not improve e_{ss} because of generation of a zero in the transfer function.

7.6 PROPORTIONAL PLUS INTEGRAL (PI) CONTROLLER

In the case of a PI controller

$$a(t) = K_p e(t) + K_i \int e(t)\, dt \tag{7.9}$$

$$\Rightarrow \qquad A(s) = \left(1 + \frac{K_i}{s}\right) E(s)$$

if $K_p = 1$.

Control Action

The actuating action of a PI controller for error input of a unit step can be worked out from Eq. (7.9) as

$$a(t) = K_p u(t) + K_i \int u(t)\, dt = K_p + K_i t$$

Thus, a step error signal will produce a ramp actuating signal as shown in Figure 7.5a.

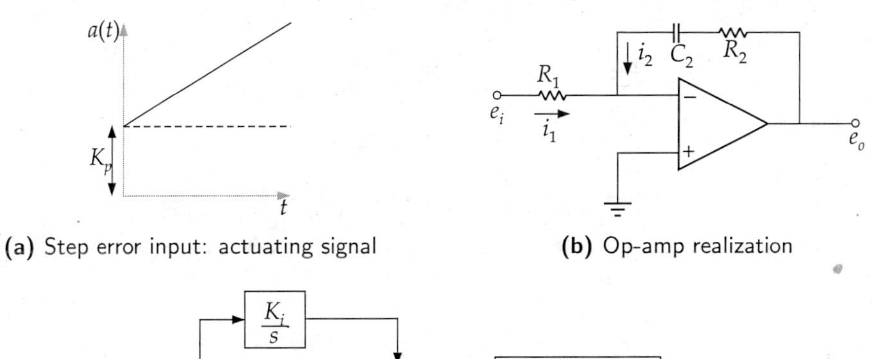

(a) Step error input: actuating signal

(b) Op-amp realization

(c) Attached to a second order plant

Figure 7.5: PI controller.

Stability

To check the stability of a PI controller, let us consider a second order closed-loop system to which it is attached (see Figure 7.5c).

The overall transfer function of the system is given by

$$T(s) = \frac{C(s)}{R(s)} = \frac{\left(1+\dfrac{K_i}{s}\right)\left[\dfrac{\omega_n^2}{s(s+2\zeta\omega_n)}\right]}{1+\left(1+\dfrac{K_i}{s}\right)\left[\dfrac{\omega_n^2}{s(s+2\zeta\omega_n)}\right]} = \frac{(s+K_i)\omega_n^2}{s^3+2\zeta\omega_n s^2+(s+K_i)\omega_n^2}$$

The characteristic equation is

$$s^3+2\zeta\omega_n s^2+(s+K_i)\omega_n^2 = 0 \tag{7.10}$$

It possesses three roots. If $K_i > 2\zeta\omega_n$, then two roots have positive real parts that will make the system unstable. However if $K_i < 2\zeta\omega_n$, then all the three roots have negative real parts indicating a stable system[1].

Now we discuss the steady-state error or, in other words, the accuracy that a PI controller offers.

Accuracy

The steady-state error for the system with PI controller is

$$e_{ss} = \lim_{s\to 0} sE(s) = \lim_{s\to 0} \frac{sR(s)}{1+G(s)H(s)} = \lim_{s\to 0} \frac{s^3(s+2\zeta\omega_n)R(s)}{s^3+2\zeta\omega_n s^2+\omega_n^2 s++K_i\omega_n^2}$$

Step input. In the case of step input,

$$e_{ss} = \lim_{s\to 0} \frac{s^3(s+2\zeta\omega_n)}{s^3+2\zeta\omega_n s^2+\omega_n^2 s++K_i\omega_n^2}\cdot\frac{1}{s} = 0$$

Ramp input. In the case of ramp input,

$$e_{ss} = \lim_{s\to 0} \frac{s^3(s+2\zeta\omega_n)}{s^3+2\zeta\omega_n s^2+\omega_n^2 s++K_i\omega_n^2}\cdot\frac{1}{s^2} = 0$$

Parabolic input. Here,

$$e_{ss} = \lim_{s\to 0} \frac{s^3(s+2\zeta\omega_n)}{s^3+2\zeta\omega_n s^2+\omega_n^2 s++K_i\omega_n^2}\cdot\frac{1}{s^3} = \frac{2\zeta}{K_i\omega_n}$$

Table 7.4 reveals that the accuracy of a PI controller is superior to that of a P controller.

Advantages

The advantages of a PI controller are as follows:

1. e_{ss} is reduced, which means the accuracy improves.
2. The order and type of the system increase by 1. That improves damping and reduces overshoot.
3. The noise is filtered.
4. The offset is eliminated.

[1]For a discussion on the location of poles vis-à-vis the stability of a system, see Section 8.3 at page 190. The condition can be worked out using Routh's rule.

Table 7.4: Comparison between P and PI controllers

Input	Steady-state error	
	P ($K_p = 1$)	*PI*
Step	0	0
Ramp	$\frac{2\zeta}{\omega_n}$	0
Parabolic	∞	$\frac{2\zeta}{K_i \omega_n}$

Realization

The simple op-amp realization of a PI controller is given in Figure 7.5b. For the circuit given in Figure 7.5b

$$i_1 = \frac{e_i}{R_1} = |i_2|$$

Therefore,

$$e_o = \frac{1}{C_2}\int |i_2| \; dt + i_2 R_2 = \frac{1}{R_1 C_2}\int |e_i| \; dt + \frac{R_2}{R_1} e_i \tag{7.11}$$

The Laplace transform of Eq. (7.11) yields

$$\frac{E_o(s)}{E_i(s)} = \frac{R_2}{R_1} + \frac{1}{R_1 C_2 s} \equiv K_p + \frac{K_i}{s}$$

7.7 PROPORTIONAL PLUS DERIVATIVE (PD) CONTROLLER

In a PD controller

$$a(t) = K_p e(t) + K_d \frac{de(t)}{dt}$$

Its Laplace transform produces

$$\frac{A(s)}{E(s)} = 1 + K_d s$$

Control Action

If the error signal is a unit step, then

$$a(t) = K_p u(t) + K_d \frac{du(t)}{dt} = K_p + K_d \delta(t)$$

which indicates that the actuating signal is an impulse. However, if the error signal is a ramp, it is easy to see that the actuating signal is also a ramp (see Figure 7.6a).

Transient Response

Figure 7.6c gives the block diagram of the PD-controlled second order process. Its transfer function is

$$T(s) = \frac{C(s)}{R(s)} = \frac{(1 + K_d s)\omega_n^2}{s^2 + (2\zeta\omega_n + K_d \omega_n^2)s + \omega_n^2}$$

The characteristic equation

$$s^2 + (2\zeta\omega_n + K_d \omega_n^2)s + \omega_n^2 = 0$$

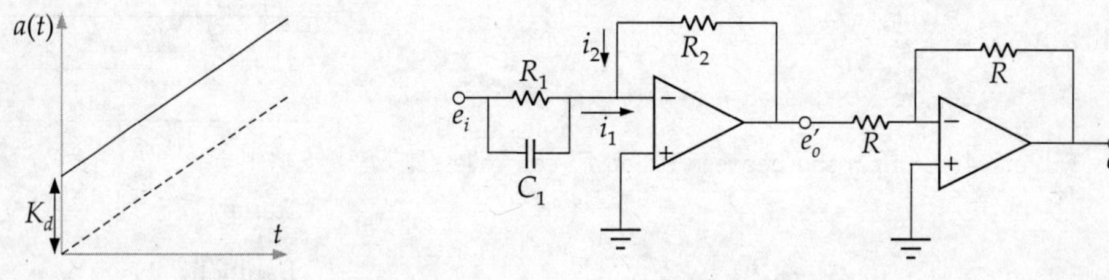

(a) Actuating signal for ramp error signal **(b)** Op-amp realization

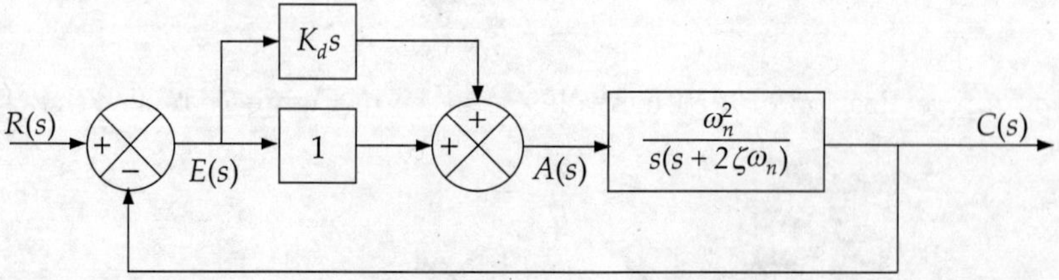

(c) PD-controlled second order process

Figure 7.6: PD controller.

can be rewritten as

$$s^2 + 2\left(\zeta + \frac{K_d\omega_n}{2}\right)\omega_n s + \omega_n^2 = s^2 + 2\zeta'\omega_n s + \omega_n^2 = 0$$

where

$$\zeta' = \zeta + \frac{K_d\omega_n}{2} \tag{7.12}$$

Equation (7.12) shows that the effective damping of the system increases by inducting a PD controller. As a result, the maximum overshoot is reduced (see Figure 6.10 at page 127 for ζ vs. M_p graph).

Accuracy

Here,

$$e_{\text{ss}} = \lim_{s\to 0}\frac{sR(s)}{1+G(s)H(s)} = \lim_{s\to 0}\frac{s^2(s+2\zeta\omega_n)R(s)}{s^2+(2\zeta\omega_n+K_d\omega_n^2)s+\omega_n^2} \tag{7.13}$$

Step input. In the case of step input,

$$e_{\text{ss}} = \lim_{s\to 0}\frac{s^2(s+2\zeta\omega_n)}{s^2+(2\zeta\omega_n+K_d\omega_n^2)s+\omega_n^2}\cdot\frac{1}{s} = 0$$

Ramp input. In the case of ramp input,

$$e_{\text{ss}} = \lim_{s\to 0}\frac{s^2(s+2\zeta\omega_n)}{s^2+(2\zeta\omega_n+K_d\omega_n^2)s+\omega_n^2}\cdot\frac{1}{s^2} = \frac{2\zeta}{\omega_n} \tag{7.14}$$

Parabolic input. In the case of parabolic input,

$$e_{\text{ss}} = \lim_{s\to 0}\frac{s^2(s+2\zeta\omega_n)}{s^2+(2\zeta\omega_n+K_d\omega_n^2)s+\omega_n^2}\cdot\frac{1}{s^3} = \infty$$

Comparing Eq. (7.13) with Eq. (7.2), we find that the steady-state error is nearly the same for both P and PD controllers. But, PD controller scores higher over the transient response because it increases the damping ratio and thus reduces the peak overshoot.

Realization

An op-amp realization of the PD controller circuit is shown in Figure 7.6b. From the circuit we get $i_1 = -i_2 \quad e_o = -e'_o$. Applying KVL to the Laplace-transformed circuit, we obtain

$$E_i(s) = \frac{R_1 \cdot \dfrac{1}{sC_1}}{R_1 + \dfrac{1}{sC_1}} \cdot I_1(s) \qquad \Rightarrow \quad I_1(s) = \frac{sR_1C_1 + 1}{R_1} \cdot E_i(s) \tag{7.15}$$

$$E'_o = E_o(s) = R_2 I_2(s) = R_2|I_1(s)| = R_2 \left|\frac{sR_1C_1+1}{R_1}\right| \cdot E_i(s) = \left|R_2C_1 s + \frac{R_2}{R_1}\right| \cdot E_i(s)$$

[using Eq. (7.15)]

Thus,
$$\frac{E_o(s)}{E_i(s)} = \frac{R_2}{R_1} + R_2C_1 s \equiv K_p + K_d s$$

7.8 PROPORTIONAL PLUS INTEGRAL PLUS DERIVATIVE (PID) CONTROLLER

As the name suggests, in a PID controller the actuating signal has the following form:

$$a(t) = K_p e(t) + K_i \int e(t)\ dt + K_d \frac{de(t)}{dt}$$

The block diagram of the PID controller action is shown in Figure 7.7.

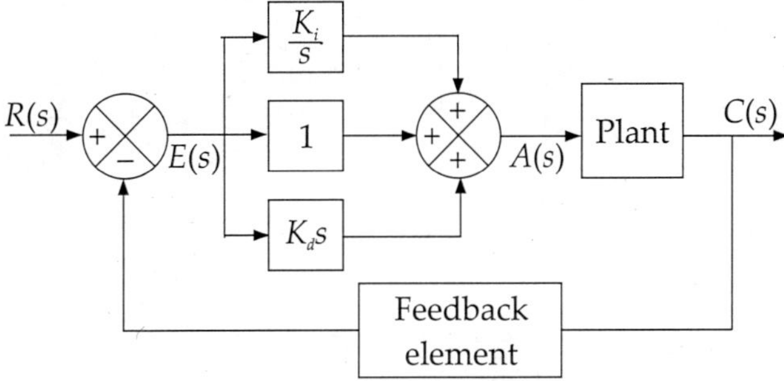

Figure 7.7: PID control action.

Without making a detailed analysis, we may observe that the PD control improves the transient part and the PI control improves the steady-state part. Therefore, from the principle of superposition of errors, we may conclude that a combination of PI and PD control improves the overall control.

The op-amp realization of PID controller is given in Figure 7.8. Table 7.5 presents a summary of salient features of three basic modes of control to help select a controller for a particular process.

Actions of P, PI, PD and PID Controllers

To observe visually the actions of P, PI, PD and PID controllers, let us consider a system, the open-loop transfer function of which is given by

$$G(s) = \frac{1}{s^2 + 10s + 20} \tag{7.16}$$

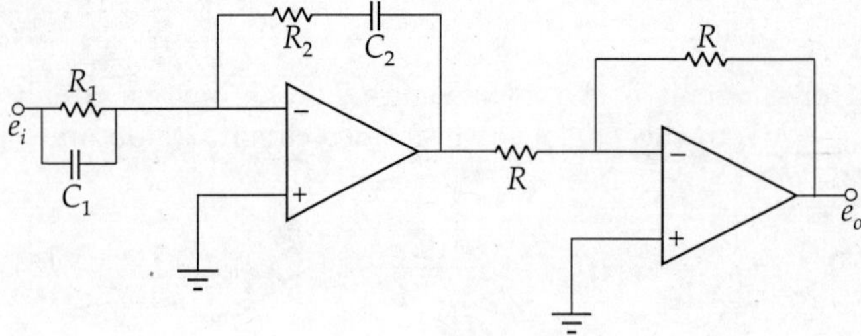

Figure 7.8: PID controller circuit.

Table 7.5: Selection of control action

Proportional action is chosen if	*Integral action is chosen if*	*Derivative action is chosen if*
1. Load changes are small.	1. Offset is not permitted.	1. Transient response parameters such as rise time, maximum overshoot etc. are to be improved upon.
2. Offset can be tolerated.	2. High degree of accuracy is required.	2. Plant load changes have to be tracked by the controller.
3. Error signal, which is small, needs amplification.		

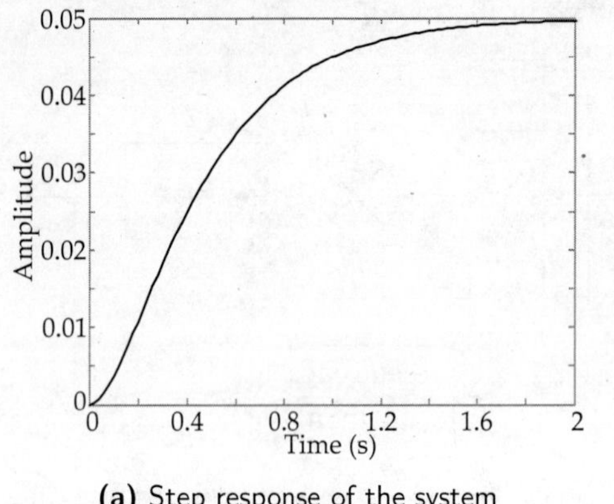

(a) Step response of the system

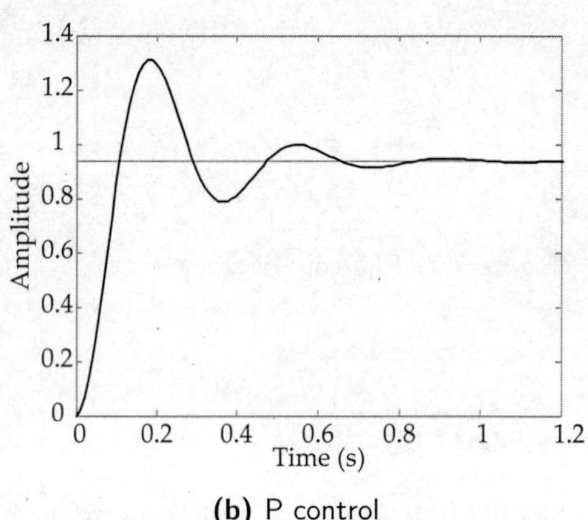

(b) P control

Figure 7.9: System response without any control and with P control.

Its response to a unit step input is shown in Figure 7.9a. We observe that the plant produces an output of 0.05 when the input is 1. So, the steady-state error is 0.95 which is pretty high.

P control. We introduce a proportional controller to reduce the rise time as well as steady-state error. With $K_p = 300$, our closed-loop transfer function becomes

$$G(s) = \frac{K_p}{s^2 + 10s + (20 + K_p)} = \frac{300}{s^2 + 10s + 320} \tag{7.17}$$

The step response of Eq. (7.17) is shown in Figure 7.9b. It shows that both rise time and steady-state error situations have improved with the introduction of a proportional controller.

PI control. Now we introduce an integral controller with $K_i = 50$, say. We know, this introduces a factor $\left[K_p + \left(K_i/s\right)\right]$ in the numerator of the open-loop transfer function. So, the closed-loop transfer function becomes

$$\frac{C(s)}{R(s)} = \frac{K_p s + K_i}{s^3 + 10s^2 + \left(K_p + 20\right)s + K_i} = \frac{300s + 50}{s^3 + 10s^2 + 320s + 50}$$

The corresponding step response is shown in Figure 7.10a. Evidently,

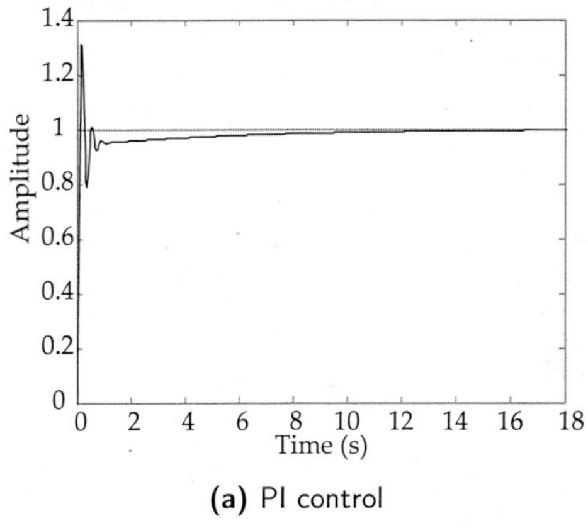

(a) PI control

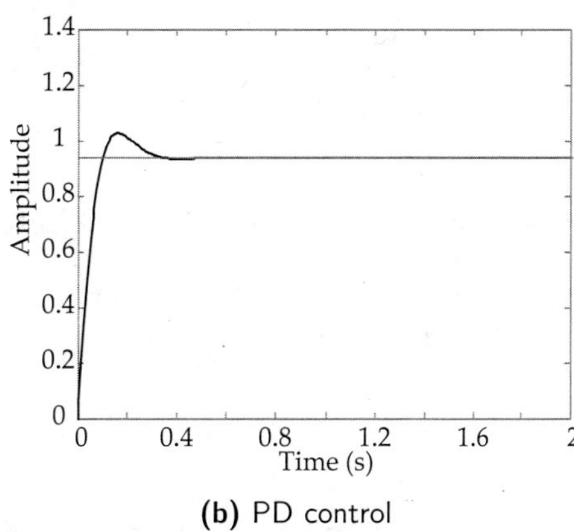

(b) PD control

Figure 7.10: PI and PD control.

- The rise time and steady-state error are reduced
- The overshoot and settling time have increased

with the addition of integral control.

PD control. Next, we remove the integral controller and introduce a derivative controller with $K_d = 15$. So the closed-loop transfer function is now

$$G(s) = \frac{K_d s + K_p}{s^2 + \left(10 + K_d\right)s + \left(20 + K_p\right)} = \frac{15s + 300}{s^2 + 25s + 320} \tag{7.18}$$

The step response of Eq. (7.18) is displayed in Figure 7.10b. It shows that with the introduction of derivative action

- Both overshoot and settling time have improved
- But the steady-state error is considerable

PID control. With PID control, the closed-loop transfer function of the system assumes the form of

$$\frac{C(s)}{R(s)} = \frac{K_d s^2 + K_p s + K_i}{s^3 + \left(10 + K_d\right)s^2 + \left(20 + K_p s\right) + K_i}$$

Now, let us check the step response. The result is shown in Figure 7.11. The combined action of PID control has produced

- A low system response time
- A low settling time
- No overshoot
- No steady-state error

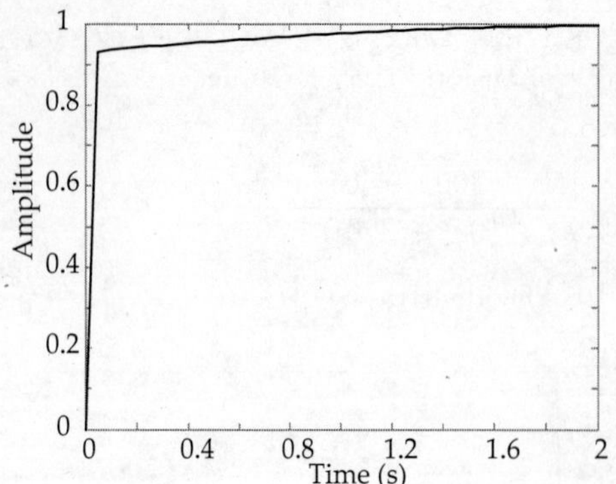

Figure 7.11: PID controlled step response.

7.9 RATE FEEDBACK CONTROLLER

Also called *derivative feedback controller* and *output derivative controller*, this controller incorporates derivative action in the feedback path as shown in Figure 7.12. Here,

$$a(t) = e(t) - K_D \frac{dc(t)}{dt}$$

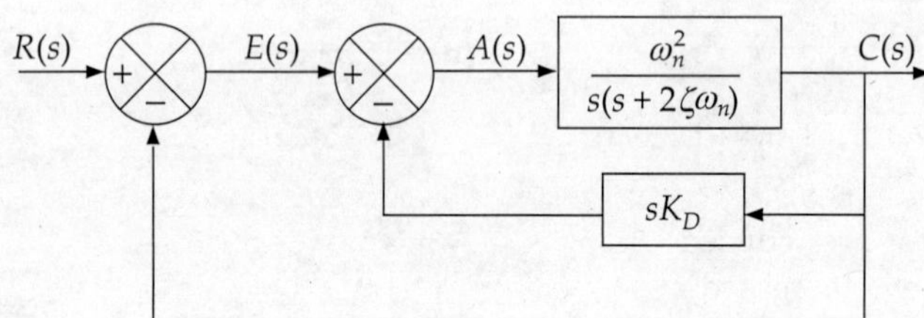

Figure 7.12: Rate feedback control action.

Thus, the forward path gain is

$$G(s) = \frac{C(s)}{R(s)} = \frac{\dfrac{\omega_n^2}{s(s+2\zeta\omega_n)}}{1 + \dfrac{sK_D\omega_n^2}{s(s+2\zeta\omega_n)}} = \frac{\omega_n^2}{s^2 + (2\zeta\omega_n + K_D\omega_n^2)s}$$

The overall transfer function is

$$\frac{C(s)}{R(s)} = \frac{G(s)}{1+G(s)} = \frac{\omega_n^2}{s^2 + (2\zeta\omega_n + K_D\omega_n^2)s + \omega_n^2}$$

The characteristic equation is

$$s^2 + (2\zeta\omega_n + K_D\omega_n^2)s + \omega_n^2 = 0 \tag{7.19}$$

Equation (7.19) can be written in the standard form of the characteristic equation of a second order system as

$$s^2 + 2\zeta'\omega_n s + \omega_n^2 = 0$$

where

$$\zeta' = \frac{2\zeta\omega_n + K_D\omega_n^2}{2\omega_n} = \zeta + \frac{K_D\omega_n}{2} \tag{7.20}$$

It is clear from Eq. (7.20) that the damping ratio is increased. Consequently,

(a) the maximum overshoot is decreased[2], and

[2]See Figure 6.10 at page 127.

(b) the rise-time is increased[3].

Now, let us check the accuracy of the rate feedback controller.

Accuracy

The steady-state error for the rate feedback controller is given by

$$e_{ss} = \lim_{s\to 0} \frac{sR(s)}{1+G(s)H(s)} = \lim_{s\to 0} \frac{s[s^2+(2\zeta\omega_n + K_D\omega_n^2)s]R(s)}{s^2+(2\zeta\omega_n+K_D\omega_n^2)s+\omega_n^2} = \lim_{s\to 0} \frac{s^2(s+2\zeta\omega_n+K_D\omega_n^2)R(s)}{s^2+(2\zeta\omega_n+K_D\omega_n^2)s+\omega_n^2}$$

Step input. In the case of step input,

$$e_{ss} = \lim_{s\to 0} \frac{s^2(s+2\zeta\omega_n+K_D\omega_n^2)}{s^2+(2\zeta\omega_n+K_D\omega_n^2)s+\omega_n^2}\cdot\frac{1}{s} = 0$$

Ramp input. Here,

$$e_{ss} = \lim_{s\to 0} \frac{s^2(s+2\zeta\omega_n+K_D\omega_n^2)}{s^2+(2\zeta\omega_n+K_D\omega_n^2)s+\omega_n^2}\cdot\frac{1}{s^2} = \frac{2\zeta\omega_n+K_D\omega_n^2}{\omega_n^2} = \frac{2\zeta}{\omega_n}+K_D \quad (7.21)$$

Parabolic input. In the case of parabolic input,

$$e_{ss} = \lim_{s\to 0} \frac{s^2(s+2\zeta\omega_n+K_D\omega_n^2)}{s^2+(2\zeta\omega_n+K_D\omega_n^2)s+\omega_n^2}\cdot\frac{1}{s^3} = \infty$$

A comparison between the PD controller and the rate feedback controller, both of which employ the derivative action, will be of interest here.

Comparison between PD and Rate Feedback Controllers

1. Damping ratio increases in both the cases. As a consequence, peak overshoot and settling time decrease while rise time increases in both the cases.
2. The PD controller introduces an additional zero at $s = -1/K_d$ in the closed loop transfer function while the rate feedback controller does not. We have seen in Section 6.4 that the addition of a zero reduces rise time but increases the peak overshoot. So, even if $K_d = K_D$, the responses will not be the same because of this zero factor.
3. The type of the system does not change in both the cases, which implies that the pattern of the steady-state error will remain the same for both the cases. However, by comparing Eq. (7.14) with Eq. (7.21), we find that the ramp input error is higher for the rate feedback controller.

EXAMPLE 7.2. In the system shown in Figure 7.13, the damping ratio is to be made 0.8 using derivative control. Determine the value of K_d.

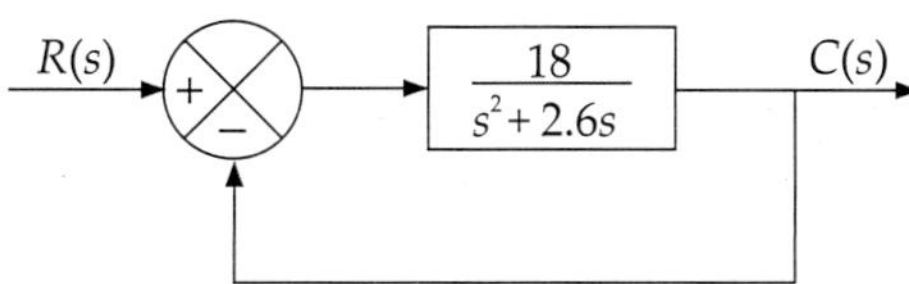

Figure 7.13: System (Example 7.2).

[3] See Eq. (7.14).

Solution Here,

$$T(s) = \frac{C(s)}{R(s)} = \frac{\dfrac{18}{s^2+2.6s}}{1+\dfrac{18}{s^2+2.6s}} = \frac{18}{s^2+2.6s+18}$$

Therefore the characteristic equation is

$$s^2 + 2.6s + 18 = 0 \equiv s^2 + 2\zeta\omega_n s + \omega_n^2$$

By comparison,

$$\omega_n^2 = 18 \quad \Rightarrow \quad \omega_n = 4.243 \text{ rad/s}$$

$$2\zeta\omega_n = 2.6 \quad \Rightarrow \quad \zeta = \frac{2.6}{(2)(4.243)} = 0.3$$

Now, with the derivative control, ζ' has to be 0.8. So, from Eq. (7.20)

$$K_d = (\zeta' - \zeta)\frac{2}{\omega_n} = (0.8 - 0.3)\frac{2}{4.243} = 0.24$$

EXAMPLE 7.3. In the system shown in Figure 7.14, determine the derivative feedback constant K_D which will increase the damping factor of the system to 0.6. What is the steady-state error to unit ramp input with this setting of the derivative feedback constant?

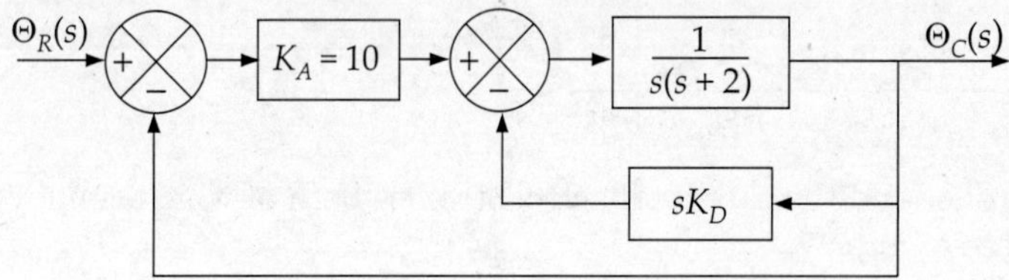

Figure 7.14: System (Example 7.3).

Solution Given, $G_1(s) = \dfrac{1}{s(s+2)}$ and $H_1(s) = sK_0$. The transfer function of the inner feedback loop is

$$G(s) = \frac{G_1(s)}{1 + G_1(s)H_1(s)} = \frac{1}{s(s+2) + sK_D}$$

And the overall transfer function is

$$T(s) = \frac{\Theta_C(s)}{\Theta_R(s(} = \frac{K_AG(s)}{1+K_AG(s)} = \frac{K_A}{s(s+2)+sK_D+K_A} = \frac{10}{s^2+(2+K_D)s+10}$$

The characteristic equation is

$$s^2 + (2+K_D)s + 10 = 0 \equiv s^2 + 2\zeta\omega_n s + \omega_n^2$$

By comparison,

$$\omega_n^2 = 10 \quad \Rightarrow \quad \omega_n = 3.162 \text{ rad/s}$$

$$2\zeta\omega_n = 2 + K_D \quad \Rightarrow \quad K_D = 2\zeta\omega_n - 2 = 2(0.6)(3.162) - 2 = 1.8$$

Therefore,

$$e_{\text{ss}} = \lim_{s\to 0} \frac{s\Theta_R(s)}{1+K_AG(s)} = \lim_{s\to 0} \frac{s \cdot \dfrac{1}{s^2}}{1+\dfrac{10}{s^2+2s+1.8s}} = \lim_{s\to 0} \frac{s+3.8}{s^2+3.8s+10} = 0.38$$

EXAMPLE 7.4. Figure 7.15 shows the block diagram of a position control system.

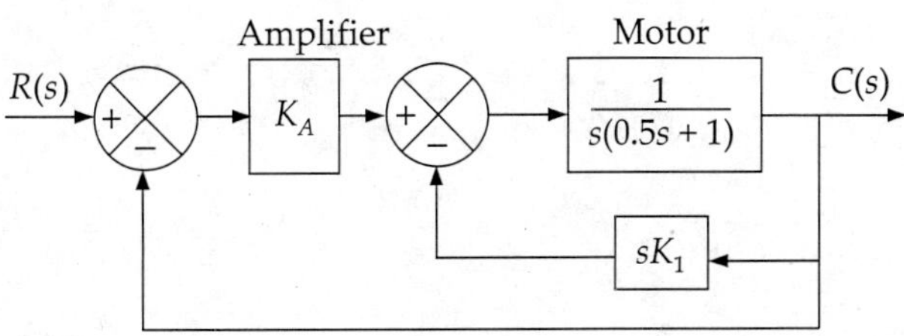

Figure 7.15: Position control system (Example 7.4).

(a) In the absence of the derivative feedback, i.e. $K_1 = 0$, determine the damping ratio of the system for amplifier gain $K_A = 5$.

(b) Find suitable values of parameters K_A and K_1 so that the damping ratio of the system is increased to 0.7 without affecting the steady state error as obtained in part (a).

Solution

(a) Here,

$$T(s) = \frac{C(s)}{R(s)} = \frac{\dfrac{s}{s(0.5s+1)}}{1 + \dfrac{s}{s(0.5s+1)}} = \frac{10}{s^2 + 2s + 10}$$

Therefore, the characteristic equation is

$$s^2 + 2s + 10 = 0 \equiv s^2 + 2\zeta\omega_n s + \omega_n^2$$

By comparison,

$$\omega_n^2 = 10 \quad \Rightarrow \quad \omega_n = 3.162 \text{ rad/s}$$

$$2\zeta\omega_n = 2 \quad \Rightarrow \quad \zeta = \frac{1}{3.162} = 0.316$$

(b) The steady-state error (for a ramp input) for part (a) is given by

$$e_{ss} = \lim_{s\to 0} \frac{sR(s)}{1 + \dfrac{5}{s(0.5s+1)}} = \lim_{s\to 0} \frac{0.5s+1}{s(0.5s+1)+5} = 0.2$$

Now, the transfer function for the inner loop is

$$G_1(s) = \frac{\dfrac{1}{s(0.5s+1)}}{1 + \dfrac{sK_1}{s(0.5s+1)}} = \frac{1}{s(0.5s+1) + sK_1}$$

So, the overall transfer function is

$$T(s) = \frac{K_A G_1(s)}{1 + K_A G_1(s)} = \frac{\dfrac{K_A}{s(0.5s+1)+sK_1}}{1 + \dfrac{K_A}{s(0.5s+1)+sK_1}} = \frac{K_A}{0.5s^2 + (K_1+1)s + K_A}$$

$$= \frac{K_A/0.5}{s^2 + \dfrac{K_1+1}{0.5} + \dfrac{K_A}{0.5}}$$

The characteristic equation is

$$s^2 + \frac{K_1+1}{0.5} + \frac{K_A}{0.5} = 0 \equiv s^2 + 2\zeta\omega_n s + \omega_n^2$$

By comparison,

$$\omega_{n'}^2 = \frac{K_A}{0.5} \quad \Rightarrow \quad \omega_{n'} = \sqrt{\frac{K_A}{0.5}}$$

$$2\zeta\omega_{n'} = \frac{K_1+1}{0.5}. \quad \Rightarrow \quad \zeta' = \frac{1}{2\omega_n'} = \frac{(K_1+1)}{0.5\times 2}\sqrt{\frac{0.5}{K_A}} = 0.7 \text{ (given)}$$

Therefore,

$$\frac{K_1+1}{\sqrt{K_A}} = \frac{0.7}{\sqrt{0.5}} \tag{i}$$

Now,

$$e_{ss} = \lim_{s\to 0} \frac{s \times \frac{1}{s^2}}{1 + \frac{K_A}{s(0.5s+1)+sK_1}} = \lim_{s\to 0} \frac{0.5s+1+K_1}{s(0.5s+1+K_1)+K_A} = \frac{1+K_1}{K_A}$$

Thus,

$$\frac{K_1+1}{K_A} = 0.2 \qquad \text{[determined for part (a)]} \tag{ii}$$

Dividing Eq. (i) by Eq. (ii), we get

$$\sqrt{K_A} = \frac{0.7}{(\sqrt{0.5})(0.2)}$$

$$\Rightarrow \qquad K_A = 24.5 \tag{iii}$$

From Eqs. (ii) and (iii), we get

$$K_1 = 0.2K_A - 1 = 3.9$$

REVIEW QUESTIONS

7.1 Choose the correct answer

(i) Excellent transient and steady state response are provided by
   (A) a proportional action system
   (B) a proportional + differential action system
   (C) a proportional + integral + differential action system
   (D) none of the above

(ii) Derivative controllers are never used alone because of
   (A) their large input impedance
   (B) they amplify noise signal and cause saturation of the actuator
   (C) they worsen the transient response of the system
   (D) none of the above

(iii) Which of the following control actions is to be employed for eliminating the offset?
   (A) Proportional control
   (B) Proportional plus derivative control
   (C) Proportional plus integral control
   (D) None of these

(iv) By the use of PD control to a second order system, the rise time
   (A) decreases (B) increases
   (C) remains the same (D) has no effect

(v) Derivative feedback control
   (A) increases rise time
   (B) increases overshoot
   (C) decreases the steady-state error
   (D) does not affect the steady-state error

7.2 (a) Why is derivative controller never used alone?
   (b) What is the effect of derivative controller on peak overshoot of a second order underdamped system?

7.3 (a) Derive the expression for PI, PD and PID controller.
   (b) Write down the advantages and disadvantages of PI, PD and PID control action.

7.4 Discuss with block diagram PI, PD and PID controllers on transient response of a second order system.

7.5 (a) Explain the function of a PID controller enumerating the benefits of using a PID controller.
   (b) Draw the circuit diagram of a PID controller using op-amp.

7.6 Describe the procedure adopted in the design of a PID controller consisting of a PI portion connected in cascade with a PD portion.

7.7 Consider the system with

$$G(s) = \frac{10^{-4}}{s^2 - 1.21} \quad \text{and} \quad H(s) = 1$$

Design a proportional plus derivative controller $G_C(s) = K_p(1 + K_d s)$ such that the damping ratio $\zeta = 0.7$ and undamped natural frequency $\omega_n = 0.5$ rad/s for the closed loop system.

7.8 Figure 7.16 shows a feedback control system employing a proportional plus error derivative controller.

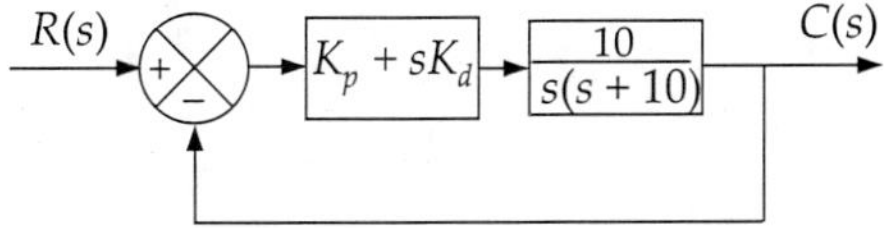

Figure 7.16: Feedback control system (Question 7.8).

   (a) State the type of the plant. Determine the closed-loop transfer function and state the order of the system.

(b) Find out the values of K_p and K_d so that the following time domain specifications are met: e_{ss} to a unit ramp ≤ 0.02 and % peak overshoot ≤ 15.

(c) How does the zero in the closed-loop transfer function (introduced by the error derivative action) improve the speed of response of the system?

Answers to Selected Questions

7.1. (i) (C) (ii) (B) (iii) (C) (iv) (B) (v) (A) 7.7. $K_p = 1.46 \times 10^4 \quad K_d = 0.48$

7.8. (a) Type 1, CLTF $= \dfrac{10(K_p + sK_d)}{s^2 + 10s(K_d + 1) + 10K_p}$, Order 2 (b) $K_p = 50, \; K_d = 1.73$

8
Stability Concepts

When a system receives an input, its output passes through an initial transient stage. After the transients die down, it should reach a steady state. The question that a designer is concerned with is whether the steady state is stable or not. In fact, the designer should be interested to test the stability of the system, not only before activating it, but even before it is constructed.

The question is: What is the precise definition of stability?

8.1 DIFFERENT KINDS OF STABILITY

In connection with the stability of a linear time-invariant system, a few terms are generally used. These are defined in Table 8.1.

Table 8.1: Definition of different kinds of stability

Kind of stability	*Definition*
Bounded Input Bounded Output (BIBO) stability	Output of the system should be bounded i.e. should remain confined within a limited range when a bounded input is applied to it.
Asymptotic (or zero input) stability	In the absence of any input, there should be no output from the system.
Absolute stability	If a bounded input is applied, the system remains stable for all values of system parameters.
Relative stability	It indicates how close the system is to instability. The values of *gain margin* and *phase margin* are measures of relative stability.
Conditional stability	Stability of a system for a particular range of any/some parameter(s).

8.2 STABILITY CONDITION FOR A BIBO SYSTEM

Let us consider a transfer function defined as

$$G(s) = \frac{C(s)}{R(s)} = \frac{b_0 s^m + b_1 s^{m-1} + b_2 s^{m-2} + \cdots + b_{m-1} s^1 + b_m}{a_0 s^n + a_1 s^{n-1} + a_2 s^{n-2} + \cdots + a_{n-1} s + a_n} \tag{8.1}$$

We know, for the impulse input $R(s) = 1$, and, therefore, $C(s) = G(s)$. Taking inverse Laplace transform, $c(t) = g(t)$. This is true only for the impulse input. But it can be shown that for any input $r(t)$, the output $c(t)$ can be written in the form of convolution

$$c(t) = \int_0^\infty g(\tau) r(t - \tau)\, d\tau \tag{8.2}$$

where $g(\tau)$ is the impulse response of the system. If $r(t)$ remains bounded (i.e., finite) and $c(t)$ has to remain bounded for all t, we find from Eq. (8.2) that it is necessary

$$\int_0^\infty g(\tau)\, d\tau = \text{finite} \tag{8.3}$$

We wrote Eq. (8.3) intuitively. Mathematically, it can be derived as follows. We take absolute value on both sides of Eq. (8.2) to get

$$|c(t)| = \left| \int_0^{\infty} g(\tau) r(t-\tau)\, d\tau \right| \tag{8.4}$$

A theorem states that the absolute value of an integral is no greater than the integral of the absolute value of the integrand. Therefore, it follows from Eq. (8.4) that

$$\begin{aligned} |c(t)| &\leq \int_0^{\infty} |g(\tau) r(t-\tau)|\, d\tau \\ &\leq \int_0^{\infty} |g(\tau)||r(t-\tau)|\, d\tau \end{aligned} \tag{8.5}$$

If $r(t)$ is bounded, this means that

$$|r(t)| \leq M < \infty \quad \text{for} \quad t \geq t_0 \tag{8.6}$$

Substitution of Eq. (8.6) in Eq. (8.5) implies

$$\begin{aligned} |c(t)| &\leq \int_0^{\infty} M |g(\tau)\, d\tau \\ &\leq M \int_0^{\infty} |g(\tau)\, d\tau \end{aligned} \tag{8.7}$$

If now $c(t)$ has to remain bounded, it means

$$|c(t)| \leq N < \infty \quad \text{for} \quad t \geq t_0 \tag{8.8}$$

Combining Eqs. (8.7) and (8.8), we get

$$\boxed{\int_0^{\infty} |g(\tau)\, d\tau \leq P < \infty} \tag{8.9}$$

Equation (8.9) is the same as that we arrived at Eq. (8.3) from intuitive reasoning.

So, this result indicates that stability of a system can be studied by considering the impulse response of the system.

8.3 EFFECT OF LOCATION OF POLES ON STABILITY

We know that the overall transfer function of a system is defined as

$$T(s) = \frac{C(s)}{R(s)} \tag{8.10}$$

where terms have their usual significance. Poles are the values of s for which $T(s) \to \infty$. That happens when $R(s)$[1] $= 0$. So, by solving the characteristic equation $R(s) = 0$, we can get the poles of the transfer function of the system. We consider the stability of systems by placing poles at different locations in the s-plane and mark our observations as shown in Table 8.2.

[1]Must not be confused with the reference input. Here, $T(s)$ is the closed-loop transfer function. Its denominator equated to zero is nothing but the characteristic equation of the system.

Table 8.2: Stability vis-a-vis Location of Poles

	Pole plot	*Response*	*Stability*
Single Pole at the Origin $G(s) = \frac{1}{s}$ $c(t) = 1$			Stable
A Pair of Poles at the Origin $G(s) = \frac{1}{s^2}$ $c(t) = t$	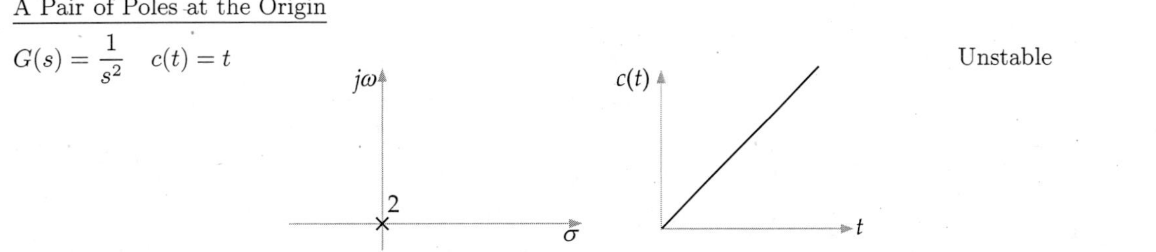		Unstable
Single Real Pole at the Right Hand Plane (RHP) $G(s) = \frac{1}{s-1}$ $c(t) = e^{t}$	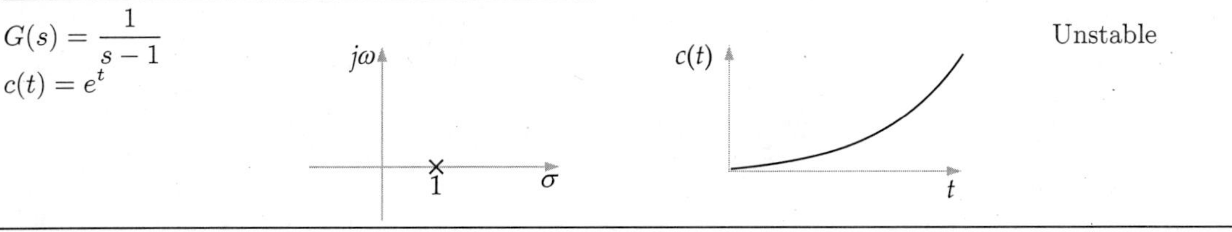		Unstable
Single Real Pole at the Left Hand Plane (LHP) $G(s) = \frac{1}{s+1}$ $c(t) = e^{-t}$	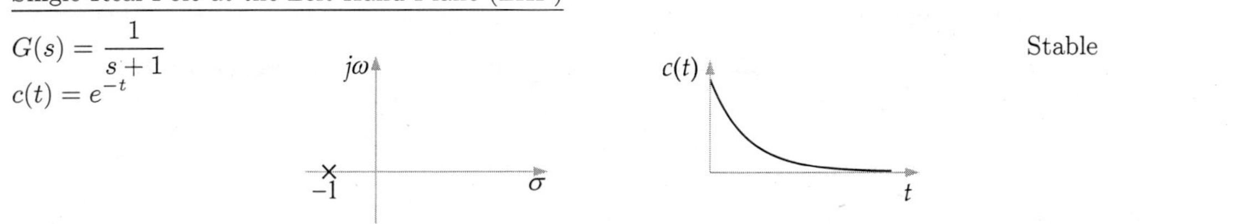		Stable
A Pair of Poles on the Imaginary Axis $G(s) = \frac{1}{s^2+1}$ $c(t) = \sin t$	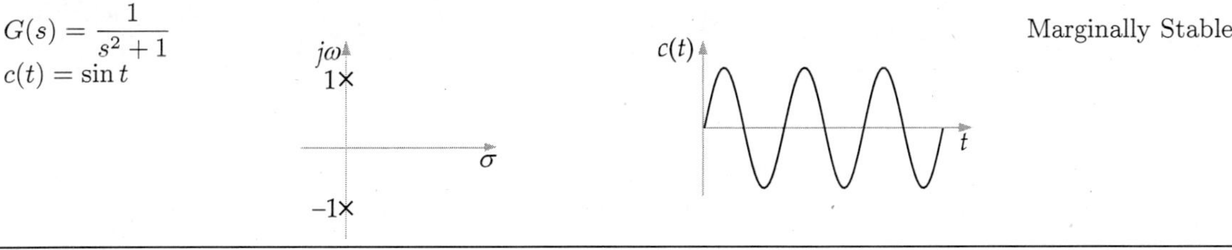		Marginally Stable

Continued on next page

Table 8.2 – *Continued from previous page*

	Pole plot	*Response*	*Stability*
A Pair of Complex Poles at the LHP $G(s) = \dfrac{1}{(s-1)^2 - 1}$ $c(t) = e^{-t}\sin t$			Stable
Two Pairs of Repeated Poles on the Imaginary Axis $G(s) = \dfrac{1}{(s^2+1)^2}$ $c(t) = t\sin t$	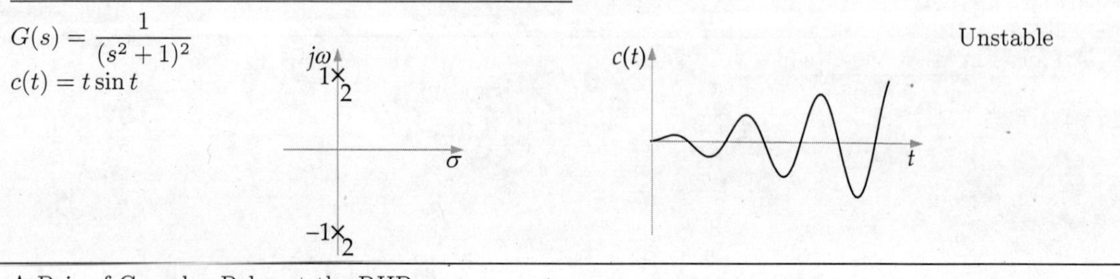		Unstable
A Pair of Complex Poles at the RHP $G(s) = \dfrac{1}{(s-1)^2 + 1}$ $c(t) = e^{t}\sin t$	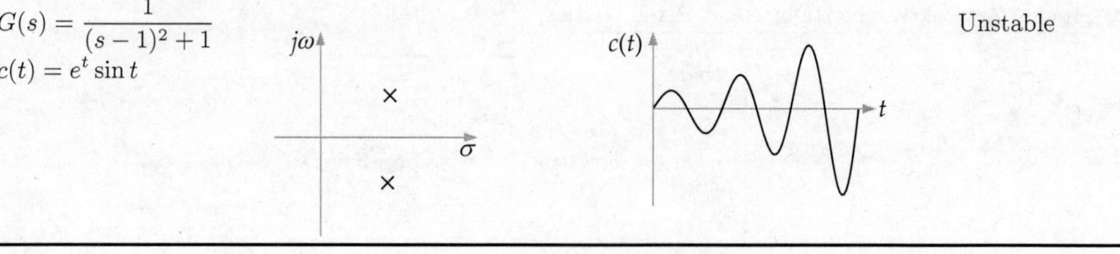		Unstable

Note: The number 2 near a pole mark indicates that there are as many poles at that location.

Table 8.3 summarizes what we observed from the impulse response plots of transfer functions having poles at different locations of the s-plane vis-a-vis the stability of the systems.

Location of poles	*Stability*
Left half of the s-plane, i.e. LHP	Stable
Right half of the s-plane, i.e. RHP	Unstable
Conjugate pair of poles on the imaginary axis	Marginally stable
Repeated poles on the imaginary axis	Unstable
Repeated poles at the origin	Unstable

Table 8.3: Summary of observations

Now let us examine a few stability problems in the light of these observations. And then in the following chapters we will see how the stability of systems can be studied more elaborately by using different methods.

EXAMPLE 8.1. Comment on the stability of the systems the open-loop transfer functions of which are given by

(a) $G(s) = \dfrac{10(s+2)}{(s+1)(s+4)(s+5)}$

(b) $G(s) = \dfrac{5(s-3)}{(s+1)(s^2+4)}$

(c) $G(s) = \dfrac{Ks}{(s^2+1)^2(s+5)}$

(d) $G(s) = \dfrac{20}{(s-2)(s+3)}$

(e) $G(s) = \dfrac{30}{s^4+30s^3+s^2+10s}$

Solution

(a) Poles are at $s = -1,\ -4,\ -5$ and zero at $s = -2$. Stable.
(b) Poles are at $s = -1,\ \pm\jmath 2$ and zero at $s = 3$. Marginally stable.
(c) Poles are at $s = -5, \pm\jmath$ (double). Unstable.
(d) Poles are at $s = 2, -3$. Unstable.
(e) One pole is at $s = 0$. May be stable.

EXAMPLE 8.2. Match the roots in the s-plane with the impulse responses shown in the adjacent figure.

(A) Two imaginary roots
(B) Two complex roots in the LHP
(C) A single root on the negative real axis
(D) A single root at the origin

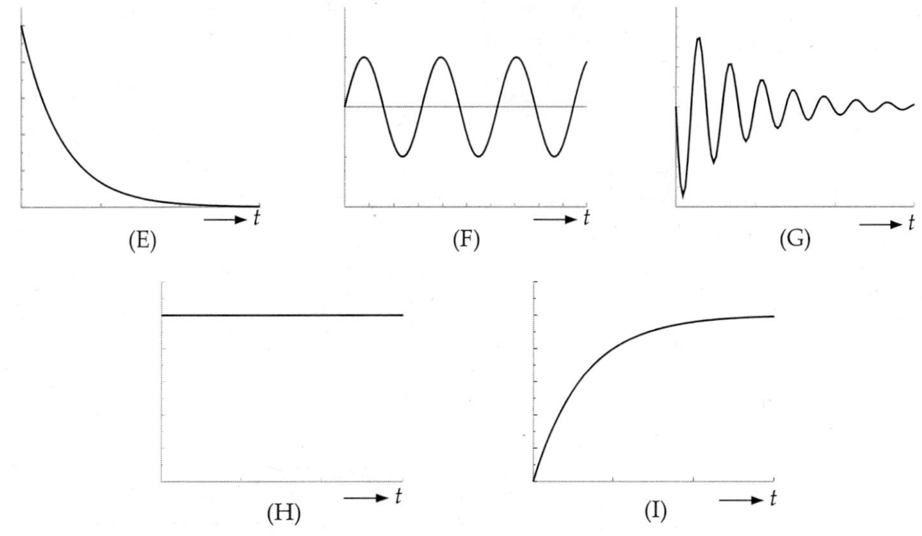

Solution $(A) \to (F)$, $(B) \to (G)$, $(C) \to (E)$, $(D) \to (H)$.

EXAMPLE 8.3. The system shown in the following figure is

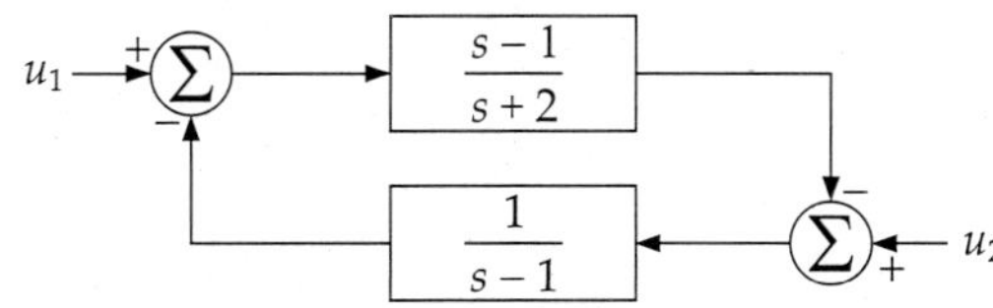

(A) stable
(B) unstable

(C) conditionally stable
(D) stable for input u_1, but unstable for input u_2.

Solution While figuring out the transfer function for the input u_1, we ignore the input u_2. The block diagram then looks like Figure 8.1a.

The corresponding transfer function is

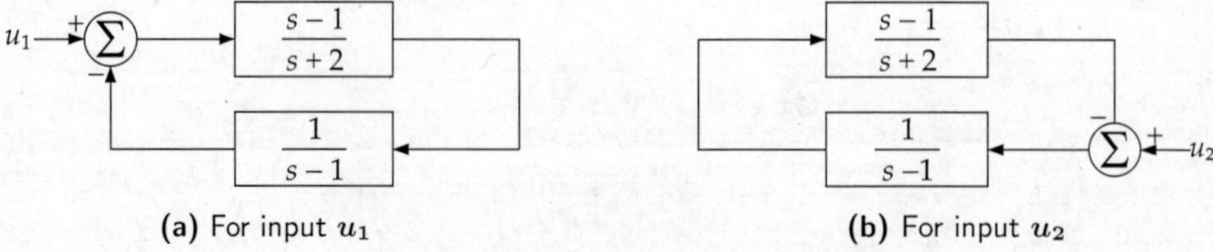

Figure 8.1: Block diagrams for two inputs.

$$T_1(s) = \frac{\dfrac{s-1}{s+2}}{1+\left(\dfrac{s-1}{s+2}\right)\cdot\left(\dfrac{1}{s-1}\right)} = \frac{s-1}{s+3}$$

Pole of the system lies at $s=-3$, i.e. at the LHP of the s-plane. So, it is stable.

For the input u_2, the system ($u_1 = 0$) looks like Figure 8.1b. The corresponding transfer function is

$$T_2(s) = \frac{\dfrac{1}{s-1}}{1+\left(\dfrac{1}{s-1}\right)\cdot\left(\dfrac{s-1}{s+2}\right)} = \frac{s+2}{(s-1)(s+3)}$$

One pole of the system is lying in the right half of the s-plane. So, the system is unstable.
Ans. (D)

EXAMPLE 8.4. A continuous-time LTI system with system function $H(j\omega)$ has the pole-zero plot shown in Figure 8.2a. For this system, which of the alternatives is TRUE?

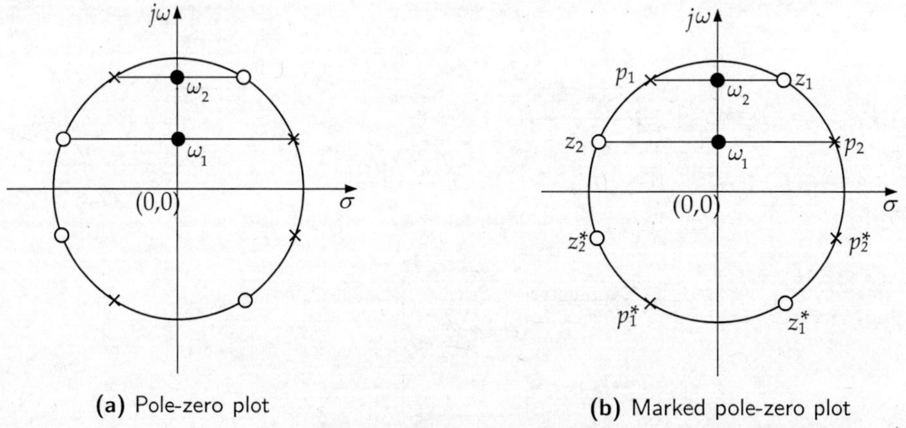

Figure 8.2: Example 8.4.

(A) $|H(0)| > |H(\omega)|;\ |\omega| > 0$
(B) $|H(\omega)|$ has multiple maxima, at ω_1 and ω_2
(C) $|H(0)| < |H(\omega)|;\ |\omega| > 0$
(D) $|H(\omega)| = \text{constant};\ -\infty < \omega < \infty$

Solution We observe from the marked pole-zero plot [Figure 8.2b] that the transfer function can be written as

$$H(s) = K\frac{(s - z_1)(s - z_1^*)(s - z_2)(s - z_2^*)}{(s - p_1)(s - p_1^*)(s - p_2)(s - p_2^*)}$$

or,
$$|H(\jmath\omega)| = K\frac{\sqrt{\omega^2 + z_1^2}\sqrt{\omega^2 + z_1^2}\sqrt{\omega^2 + z_2^2}\sqrt{\omega^2 + z_2^2}}{\sqrt{\omega^2 + p_1^2}\sqrt{\omega^2 + p_1^2}\sqrt{\omega^2 + p_2^2}\sqrt{\omega^2 + p_2^2}}$$

Since $|z_1| = |p_2|$ [from Figure 8.2b]

and $|z_2| = |p_1|$ [ditto]

We have $|H(\jmath\omega)| = K$ [constant]

Ans. (D)

8.4 METHODS OF STUDY OF STABILITY

Next, we will consider the following methods that are generally used to study the stability of an LTI system. While two of them deal with the study of the problem in the s-plane, the others do it in the $\jmath\omega$-plane as shown in Table 8.4.

Table 8.4: Methods of study of the stability

s-plane (time-domain)	*$\jmath\omega$-plane (frequency-domain)*
1. Routh's rule	1. Bode-plot
2. Root-locus plot	2. Polar- and Nyquist-plot

REVIEW QUESTIONS

8.1 Choose the correct answer:

(i) A system which has some roots with real parts equal to zero, but none with positive real parts is

(A) relatively stable (B) marginally stable
(C) absolutely stable (D)absolutely unstable

(ii) The system, represented by its transfer function, has some poles lying on imaginary axis of s-plane. The system is

(A) unconditionally stable (B) conditionally stable
(C) stable (D) marginally stable

(iii) A system is stable if

(A) bounded inputs produce bounded outputs
(B) unbounded inputs produce bounded outputs
(C) bounded inputs produce unbounded outputs
(D) all bounded inputs produce bounded outputs

(iv) If the poles corresponding to the transfer function of a system lie on the left half of the s-plane, the system

(A) is unstable (B) is marginally stable
(C) achieves stability with time (D) behaviour is unpredictable

(v) If a conjugate pair of poles corresponding to the transfer function of a system lie on the imaginary axis of the s-plane, the system

(A) is unstable (B) is marginally stable
(C) achieves stability with time (D) behaviour is unpredictable

(vi) If repeated poles corresponding to the transfer function of a system lie on the imaginary axis of the s-plane, the system

(A) is unstable (B) is marginally stable
(C) achieves stability with time (D) behaviour is unpredictable

(vii) A system is always stable if in the s-plane

(A) all poles are in the left half and zeros are in the right half
(B) all poles and zeros are in the left half
(C) all poles and zeros are in the right half
(D) all poles in the right half and zeros are in the left half

(viii) The system whose characteristic equation has the following roots is marginally stable

(A) $-\jmath, \jmath, -1$ (B) $-3, -2, 0$
(C) $-2 + \jmath 3, -2 - \jmath 3, -2$ (D) $-3, -2, -1$

(ix) An example of bounded input is

(A) e^{-4t} (B) e^{2t} (C) t (D) $e^{t} \sin t$

8.2 Define the following:

(a) Absolutely stable system
(b) Conditionally stable system
(c) Asymptotically stable system

8.3 (a) What is the necessary condition of stability?
(b) Define stable, unstable and limitedly stable systems.
(c) What do you mean by bounded-input bounded-output (BIBO) stability?

8.4 Show that for a bounded input bounded output stable system $\int_0^{\infty} |g(\tau)| d\tau$ is finite, where $g(\tau)$ is the impulse response of the system.

Answers to Selected Questions

8.1. (i) (B) (ii) (B) (iii) (A) (iv) (C) (v) (B) (vi) (A) (vii) (B) (viii) (A) (ix) (A)

9

Hurwitz-Routh Stability Criteria

We know that the closed-loop transfer function of a linear time-invariant system can be written as

$$\frac{C(s)}{R(s)} = \frac{b_0 s^m + b_1 s^{m-1} + \cdots + b_{m-1}s + b_m}{a_0 s^n + a_1 s^{m-1} + \cdots + a_{n-1}s + a_n}$$

which generates the characteristic equation

$$a_0 s^n + a_1 s^{m-1} + \cdots + a_{n-1}s + a_n = 0 \tag{9.1}$$

The roots of this equation represent closed-loop poles, on the location of which the stability of a system depends. Conditions, though not sufficient, for having no such poles in the RHP are as follows:

1. All coefficients, $a_0,\ a_1, \ldots a_n$ must have the same sign.
2. All powers of s must be present in descending order.

Since these conditions are *not sufficient*, one has to solve the characteristic equation. Hurwitz and Routh developed criteria which allow one to determine the stability without solving the characteristic equation.

9.1 HURWITZ'S CRITERION

Hurwitz[1] defined certain determinants out of coefficients of the characteristic equation. The sufficient condition is that these determinants must be positive for having all roots of the characteristic equation in the LHP.

The Hurwitz determinants are formed as follows. The main determinant, D_n, is given by

$$D_n = \begin{vmatrix} a_1 & a_3 & a_5 & \cdots & a_{2n-1} \\ a_0 & a_2 & a_4 & \cdots & a_{2n-2} \\ 0 & a_1 & a_3 & \cdots & a_{2n-3} \\ 0 & a_0 & a_2 & \cdots & a_{2n-4} \\ \vdots & \vdots & \vdots & \vdots & \vdots \\ 0 & 0 & 0 & \cdots & a_n \end{vmatrix}$$

Thus,

$$D_1 = |a_1| \qquad D_2 = \begin{vmatrix} a_1 & a_3 \\ a_0 & a_2 \end{vmatrix}$$

$$D_3 = \begin{vmatrix} a_1 & a_3 & a_5 \\ a_0 & a_2 & a_4 \\ 0 & a_1 & a_3 \end{vmatrix} \qquad D_4 = \begin{vmatrix} a_1 & a_3 & a_5 & a_7 \\ a_0 & a_2 & a_4 & a_6 \\ 0 & a_1 & a_3 & a_5 \\ 0 & a_0 & a_2 & a_4 \end{vmatrix}$$

[1] Adolf Hurwitz (1859–1919): German mathematician.

The condition for stability is,

$$D_1 > 0,\ D_2 > 0,\ D_3 > 0 \ldots D_k > 0$$

Hurwitz's criterion is rather not so useful in comparison to other methods because of the following shortcomings:

1. It is time-consuming and cumbersome to solve large determinants for higher order systems.
2. It does not yield exact number of poles located in the RHP.
3. Marginal stability prediction, by this method, becomes tough.

Nevertheless, we consider a few examples to illustrate the method.

EXAMPLE 9.1. Using Hurwitz's criterion, study the stability of unity feedback (a) first and (b) second order systems.

Solution (a) For the first order system

$$G(s) = \frac{K}{\tau s + 1} \qquad \Rightarrow T(s) = \frac{G(s)}{1 + G(s)} = \frac{K}{\tau s + K + 1}$$

The characteristic equation is $\tau s + K + 1 = 0$. This gives

$$a_0 = \tau \qquad a_1 = K + 1$$

$$\therefore \quad D_1 = |K + 1| = K + 1$$

$D_1 > 0$ for $K \geq 0$. Hence the system is stable.

(b) For the second order system

$$G(s) = \frac{K\omega_n^2}{s^2 + 2\zeta\omega_n s + \omega_n^2} \qquad \Rightarrow T(s) = \frac{G(s)}{1 + G(s)} = \frac{K\omega_n^2}{s^2 + 2\zeta\omega_n s + (K+1)\omega_n^2}$$

The characteristic equation is $s^2 + 2\zeta\omega_n s + (K+1)\omega_n^2 = 0$. This gives,

$$a_0 = 1 \qquad a_1 = 2\zeta\omega_n \qquad a_2 = (K+1)\omega_n^2$$

The Hurwitz determinants are

$$D_1 = |a_1| = 2\zeta\omega_n \qquad D_2 = \begin{vmatrix} a_1 & a_3 \\ a_0 & a_2 \end{vmatrix} = \begin{vmatrix} 2\zeta\omega_n & 0 \\ 1 & (K+1)\omega_n^2 \end{vmatrix} = 2\zeta\omega_n^2(K+1)$$

Now

$$\begin{aligned} D_1 &> 0 \quad \text{for} \quad \zeta > 0 \\ D_2 &> 0 \quad \text{for} \quad \zeta > 0 \quad \text{and} \quad K \geq 0 \end{aligned}$$

The conditions are normally true. Hence the system is stable.

EXAMPLE 9.2. Find, using Hurwitz's criterion, whether a system having characteristic equation

$$2s^4 + s^3 + 3s^2 + 4s + 10 = 0$$

is stable or not.

Solution Here,

$$a_0 = 2, \qquad a_1 = 1, \qquad a_2 = 3, \qquad a_3 = 4, \qquad a_4 = 10$$

Therefore,

$$D_1 = |a_1| = 1$$

$$D_2 = \begin{vmatrix} a_1 & a_3 \\ a_0 & a_2 \end{vmatrix} = \begin{vmatrix} 1 & 4 \\ 2 & 3 \end{vmatrix} = -5$$

We need not work out D_3 and D_4, because already $D_2 < 0$. The system is unstable.

9.2 ROUTH'S RULE

Routh[2] gave an alternative but much simpler sufficient condition for finding the stability of a system. In this method, the coefficients of the characteristic equation Eq. (9.1) are arranged in an array, called the *Routh array*, as given in the following representation.

$$\begin{array}{c|ccccc} s^n & a_0 & a_2 & a_4 & a_6 & \cdots \\ s^{n-1} & a_1 & a_3 & a_5 & a_7 & \cdots \\ s^{n-2} & b_1 & b_2 & b_3 & b_4 & \cdots \\ s^{n-3} & c_1 & c_2 & c_3 & c_4 & \cdots \\ \vdots & \vdots & & & & \\ s^0 & a_n & & & & \end{array}$$

where,

$$b_1 = \frac{a_1a_2 - a_0a_3}{a_1} \qquad b_2 = \frac{a_1a_4 - a_0a_5}{a_1} \qquad b_3 = \frac{a_1a_6 - a_0a_7}{a_1} \qquad b_4 = \frac{a_1a_8 - a_0a_9}{a_1}$$

$$c_1 = \frac{b_1a_3 - a_1b_2}{b_1} \qquad c_2 = \frac{b_1a_5 - a_1b_3}{b_1} \qquad \text{and so on}$$

To obtain this representation, the following rules have to be observed.

- The characteristic equation has to be written in *descending* power of s.
- In case there is a missing power, its coefficient has to be taken as zero. For example, if the characteristic equation is $s^4 + 3s^2 + 5s + 1 = 0$, we note that s^3 term is missing. So we have to write the characteristic equation as $s^4 + 0s^3 + 3s^2 + 5s + 1 = 0$.
- In the first row of the Routh array, the zeroth and even numbered coefficients of the characteristic equation are written.
- In the second row, odd numbered coefficients are written.
- The elements of the subsequent rows are formed by using the above formulae.

If the connectivities shown in the figure are followed, the pattern of formation of the formulae will be clear.

The stability of a system can be predicted from the array according to the following rules:

1. For the system to be *stable*, *all the elements in the first column* of the Routh array *must have the same sign.*
2. In case there is a *change of sign*, the system is *unstable.*
3. The *number of changes of sign* indicates the *number of roots lying in the RHP* of the CLTF.

In fact, Routh's derivation of these criteria follows from Hurwitz's formulation for the problem. This is why, these criteria are often referred to as *Hurwitz-Routh criteria.*

Let us work out a few examples here to show how Hurwitz-Routh criteria are utilized to study stability problems.

EXAMPLE 9.3. Find, using Routh's rule, whether a system, having the characteristic equation

$$2s^4 + s^3 + 3s^2 + 4s + 10 = 0$$

is stable or not.

[2] Edward John Routh (1831–1907): English mathematician.

Solution The Routh array is

$$
\begin{array}{c|ccc}
s^4 & 2 & 3 & 10 \\
s^3 & 1 & 4 & 0 \\
s^2 & -5(b_1) & 10(b_2) & 0 \\
s^1 & 6(c_1) & 0(c_2) & \\
s^0 & 10(d_1) & &
\end{array}
$$

The calculation of elements from the third row onwards is as follows:

$$b_1 = \frac{(1)(3) - (2)(4)}{1} = -5 \qquad b_2 = \frac{(1)(10) - (2)(0)}{1} = 10 \qquad c_1 = \frac{(-5)(4) - (1)(10)}{-5} = 6$$

$$c_2 = \frac{(-5)(0) - (1)(0)}{-5} = 0 \qquad d_1 = \frac{(6)(10) - (-5)(0)}{6} = 10$$

There is change of sign in the 1st column of the Routh array. So, the system is unstable. In fact, there are two changes of sign, from + to − and then from − to +. So, two roots lie in the RHP.

EXAMPLE 9.4. The characteristic equation of a closed-loop system is $s^2 + Ks + 4K - 1 = 0$. For K greater than what value is the system stable?

Solution The Routh array is given by

$$
\begin{array}{c|cc}
s^2 & 1 & 4K-1 \\
s^1 & K & 0 \\
s^0 & 4K-1 &
\end{array}
$$

For stability of the system

$$4K - 1 > 0$$

or,

$$K > \frac{1}{4}$$

EXAMPLE 9.5. The condition for stability of a closed-loop system with a characteristic equation $s^3 + bs^2 + cs + 1 = 0$, with positive coefficients is

(A) $b + c > 1$ (B) $bc > 1$ (C) $b = c$ (D) $b > c$

Solution The Routh array for the characteristic equation is

$$
\begin{array}{c|cc}
s^3 & 1 & c \\
s^2 & b & 1 \\
s^1 & c - \dfrac{1}{b} & 0 \\
s^0 & 1 &
\end{array}
$$

For stability of the system

$$c > \frac{1}{b}$$

or,

$$bc > 1$$

Ans. (B)

EXAMPLE 9.6. Consider the unity negative feedback system shown in Figure 9.1, where

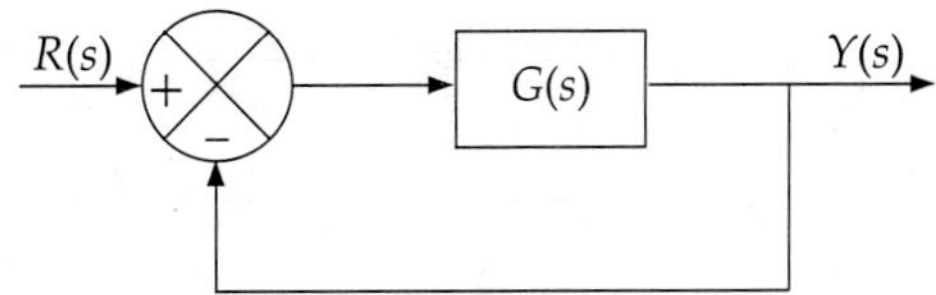

Figure 9.1: System (Example 9.6).

$$G(s) = \frac{K}{(1+0.5s)(1+0.5s+0.25s^2)}$$

Using Routh's criterion, find the value of K for which the system is stable.

Solution Here,

$$T(s) = \frac{Y(s)}{R(s)} = \frac{G(s)}{1+G(s)} = \frac{K}{(1+0.5s)(1+0.5s+0.25s^2)+K}$$

The characteristic equation is

$$(1+0.5s)(1+0.5s+0.25s^2)+K = 0$$
$$0.125s^3 + 0.5s^2 + s + (K+1) = 0$$

The Routh array is

$$\begin{array}{c|ccc} s^3 & 0.125 & 1.0 & 0 \\ s^2 & 0.5 & K+1 & 0 \\ s^1 & 1.0-0.25(K+1) & 0 & \\ s^0 & K+1 & & \end{array}$$

The calculation of elements from the third row onwards is shown as follows:

$$b_1 = \frac{(0.5)(1.0)-(0.125)(K+1)}{0.5} = 1.0 - 0.25(K+1)$$

$$b_2 = \frac{(0.5)(0)-(0.125)(0)}{0.5} = 0$$

$$c_1 = \frac{[1.0-0.25(K+1)](K+1)-(0.5)(0)}{1.0-0.25(K+1)} = K+1$$

For the system to be stable, Routh's criterion necessitates that the third row first column factor should be positive. Thus,

$$1.0 - 0.25(K+1) > 0$$
$$0.25(K+1) < 1.0$$
$$K+1 < 4$$
$$K < 3$$

and the fourth row first column factor should be positive. This means,

$$K+1 > 0 \qquad \Rightarrow \qquad K > -1$$

Therefore, for the stability of the system

$$3 > K > -1$$

EXAMPLE 9.7. For the feedback control system shown in Figure 9.2, using Routh's tabulation, determine the range of scalar variable K for which the system is stable.

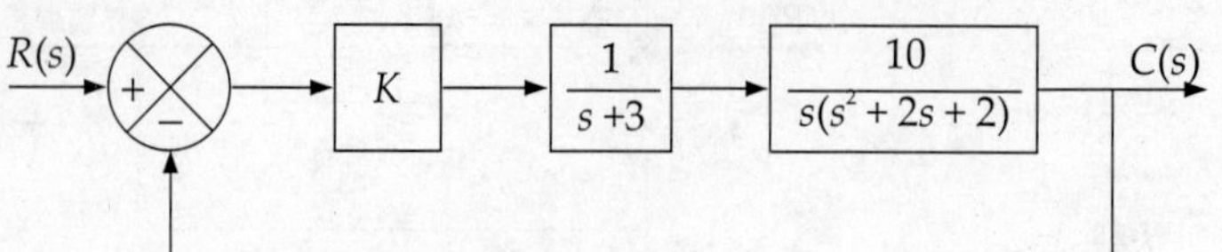

Figure 9.2: System (Example 9.7).

Solution The closed-loop transfer function is given by

$$\frac{10K}{(s+3)[s(s^2+2s+2)]+10K}$$

The characteristic equation is

$$s^4+5s^3+8s^2+6s+10K=0$$

Therefore, the Routh table is as follows:

$$\begin{array}{c|ccc} s^4 & 1 & 8 & 10K \\ s^3 & 5 & 6 & \\ s^2 & 6.8 & 10K & \\ s^1 & 6-7.35K & 0 & \\ s^0 & 10K & & \end{array}$$

For stability of the system, we have

$$6-7.35K>0 \qquad \text{or,} \quad K<\frac{6}{7.35}<0.816$$

and

$$10K>0 \qquad \text{or,} \quad K>0$$

Thus, the required range is $0<K<0.816$.

EXAMPLE 9.8. The closed-loop control system, shown in Figure 9.3, has $\tau_1>0$. The system will remain stable for all τ_1 in the range of

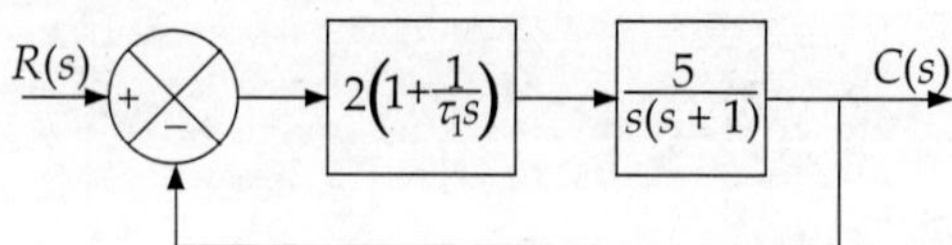

Figure 9.3: System (Example 9.8).

(A) $\tau_1<0.5$ (B) $0.5<\tau_1<1.0$ (C) $\tau_1<1.0$ (D) $\tau_1>1.0$

Solution The closed-loop transfer function is given by

$$\frac{C(s)}{R(s)}=\frac{10\dfrac{[1+(1/\tau_1 s)]}{s(s+1)}}{1+10\dfrac{[1+(1/\tau_1 s)]}{s(s+1)}}=\frac{10\left(1+\dfrac{1}{\tau_1 s}\right)}{s(s+1)+10\left(1+\dfrac{1}{\tau_1 s}\right)}$$

The characteristic equation is

$$s(s+1)+10\left(1+\frac{1}{\tau_1 s}\right)=0$$

or,

$$\tau_1 s^3+\tau_1 s^2+10\tau_1 s+10=0$$

The Routh array is

s^3	τ_1	$10\tau_1$
s^2	τ_1	10
s^1	$10\tau_1 - 10$	0
s^0	10	

The stability condition is

$$10\tau_1 - 10 > 0 \qquad \text{or} \quad \tau_1 > 1$$

Ans. (D)

9.3 SPECIAL CASES

Row of Zeros

When all the elements in any row of the Routh array are zero, it indicates that there are symmetrically located roots in the s-plane. That means there exists

- either a pair of real roots with opposite signs;
- and/or a pair of conjugate roots on the imaginary axis;
- and/or complex conjugate roots forming quadrates in the s-plane.

These are indeed indications of a marginally stable system which crosses the imaginary axis temporarily to attain stability but reverts to unstable state again.

In such cases, an auxiliary equation is to be formed by the elements of the row just above the row of zeros of the Routh array. The first derivative of the auxiliary polynomial gives the elements of the row containing zeros. We consider a few examples to show how to handle this kind of problem.

EXAMPLE 9.9. The characteristic equation of a system is given by

$$s^6 + 2s^5 + 8s^4 + 12s^3 + 18s^2 + 12s + 12 = 0$$

Find the stability of the system.

Solution The Routh array is

s^6	1	8	18	12
s^5	2	12	12	
s^4	2	12	12	
s^3	0	0	0	

We find that all the elements in the s^3-row are zeros. Therefore, we need to form the auxiliary polynomial from the s^4-row with the elements as coefficients. The auxiliary polynomial thus formed is

$$\begin{aligned} A(s) &= 2s^4 + 12s^2 + 12 \\ &= s^4 + 6s^2 + 6 \end{aligned}$$

Mark that we have divided all the coefficients of the polynomial by 2. This is allowed since dividing or multiplying coefficients by the same number does not affect the roots of the corresponding characteristic equation.

Now,

$$\frac{dA(s)}{ds} = 4s^3 + 12s$$

So, 4 and 12 are the elements for the s^3-row. The Routh array now becomes

$$\begin{array}{c|cccc} s^6 & 1 & 8 & 18 & 12 \\ s^5 & 1 & 6 & 6 & \\ s^4 & 1 & 6 & 6 & \\ s^3 & 4 & 12 & 0 & \\ s^2 & 3 & 6 & & \\ s^1 & 4 & & & \\ s^0 & 6 & & & \end{array}$$

The auxiliary equation is

$$s^4 + 6s^2 + 6 = 0$$

This yields

$$s^2 = \frac{-6 \pm \sqrt{(6)^2 - 4(1)(6)}}{2} = -3 \pm \sqrt{3}$$
$$= -4.732 \quad \text{and} \quad -1.268$$

Therefore,

$$s = \pm\, \jmath\, 2.175 \quad \text{and} \quad \pm\, \jmath\, 1.126$$

These are also among the roots of the characteristic equation. And there is no sign change in the elements of the first column of the Routh array formed with the help of the auxiliary polynomial. Because there are multiple, non-repeated roots on the imaginary-axis, we conclude that the system is marginally stable.

EXAMPLE 9.10. The partial Routh array of the characteristic equation of a system is given by

$$\begin{array}{c|ccc} s^4 & 1 & a & 8 \\ s^3 & 3 & 12 & \end{array}$$

The system oscillates with a frequency of 2 rad/s. The value of the parameter a of the system is

(A) 6 (B) 2 (C) 8 (D) 12

Solution The next row of the Routh array is

$$s^2 \mid a - 4 \quad 8$$

Therefore, the auxiliary equation is

$$(a - 4)s^2 + 8 = 0$$

or, $$a - 4 = \frac{8}{\omega^2} = \frac{8}{4} = 2 \qquad [\because s^2 = -\omega^2 \text{ and } \omega = 2 \text{ (given)}]$$

or, $$a = 6$$

Ans. (A)

EXAMPLE 9.11. The Routh-Hurwitz array for a third order characteristic equation is given as

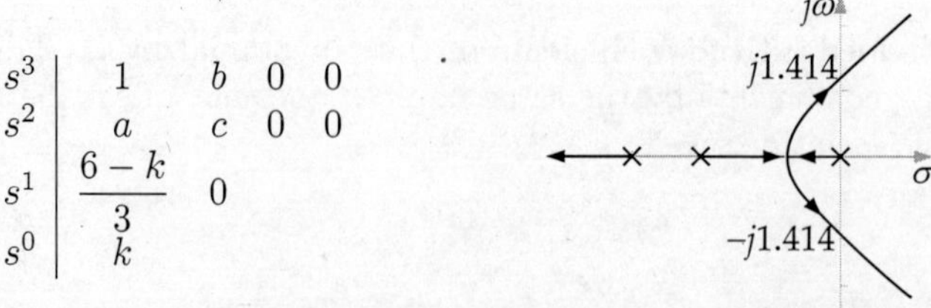

The coefficients $a,\ b,\ c$ and k are such that $a = 3,\ b > 0,\ c > 0,\ k > 0$ and c is a function of k. The root locus[3] for the corresponding characteristic equation is as shown in the given figure. The values of k and c for critical stability are

(A) 6 and 6 (B) 6 and 2 (C) 3 and 2 (D) 3 and 3

Solution From the Routh-Hurwitz array, the auxiliary equation is

$$as^2 + c = 0 \tag{i}$$

The root locus intersects the $\jmath\omega$-axis at $s = \pm\jmath 1.414 = \pm\jmath\sqrt{2}$ from which we get

$$s = \pm\jmath\sqrt{2} \qquad \text{or,} \quad s^2 + 2 = 0 \tag{ii}$$

Comparing Eqs. (i) and (ii), we obtain

$$\frac{c}{a} = 2 \qquad \text{or,} \quad c = 6 \qquad [\because a = 3 \text{ (given)}]$$

For the critical stability of the system, we get from the Routh-Hurwitz array

$$\frac{6-k}{3} = 0 \qquad \text{or,} \quad k = 6$$

Ans. (A)

EXAMPLE 9.12. The first two rows of Routh's tabulation of a third order equation are as follows:

$$\begin{array}{c|cc} s^3 & 2 & 2 \\ s^2 & 4 & 4 \end{array}$$

This means that there are:

(A) two roots at $s = \pm\jmath$ and one root in right half s-plane
(B) two roots at $s = \pm\jmath 2$ and one root in left half s-plane
(C) two roots at $s = \pm\jmath 2$ and one root in right half s-plane
(D) two roots at $s = \pm\jmath$ and one root in left half s-plane

Solution From Routh's tabulation we construct the auxiliary equation and find the roots on the $\jmath\omega$-axis as follows:

$$4s^2 + 4 = 0 \qquad \text{or,} \quad s = \pm\jmath \tag{i}$$

Also, from the two rows of Routh's tabulation we construct the characteristic equation and extract other roots therefrom as follows:

$$2s^3 + 4s^2 + 2s + 4 = 0 \quad \Rightarrow s^3 + 2s^2 + s + 2 = 0 \quad \Rightarrow (s^2+1)(s+2) = 0 \tag{ii}$$

From Eqs. (i) and (ii) we note that the choice given in (D) is true.

EXAMPLE 9.13. The open-loop transfer function of a unity feedback system is given by

$$G(s) = \frac{K}{(s+2)(s+4)(s^2+6s+25)}$$

By applying the Hurwitz-Routh criterion, discuss the stability of the closed-loop system as a function of K. Determine the values of K which will cause sustained oscillation in the closed-loop system. What are the corresponding frequencies?

[3]See the next chapter for the meaning of the term.

Solution Here,

$$T(s) = \frac{C(s)}{R(s)} = \frac{G(s)}{1+G(s)} = \frac{K}{(s+2)(s+4)(s^2+6s+25)+K}$$

The characteristic equation is

$$(s+2)(s+4)(s^2+6s+25)+K=0$$

or,

$$s^4+12s^3+69s^2+198s+(200+K)=0$$

The Routh array is

$$\begin{array}{c|ccc} s^4 & 1 & 69 & 200+K \\ s^3 & 12 & 198 & 0 \\ s^2 & 52.5 & 200+K & 0 \\ s^1 & 152.3-0.23K & 0 & \\ s^0 & 200+K & & \end{array}$$

$$b_1 = \frac{(12)(69)-(1)(198)}{12} = 52.5$$

$$b_2 = \frac{(12)(200+K)-(1)(0)}{12} = 200+K$$

$$c_1 = \frac{(52.5)(198)-(12)(200+K)}{52.5} = 152.3-0.23K$$

$$d_1 = \frac{(152.3-0.23K)(200+K)-(52.5)(0)}{152.3-0.23K} = 200+K$$

For stability:

$$200+K>0 \quad \Rightarrow \quad K>-200$$

and

$$152.3-0.23K>0 \quad \Rightarrow \quad K<662.17$$

Therefore,

$$-200<K<662.17$$

For marginal stability, $K = 662.17$ so that the last two rows of the Routh array become zeros. Then,

$$\begin{array}{c|ccc} s^4 & 1 & 69 & 200+K \\ s^3 & 12 & 198 & 0 \\ s^2 & 52.5 & 200+K & 0 \\ s^1 & 0 & 0 & 0 \\ s^0 & 0 & & \end{array}$$

The auxiliary equation is formed as

$$A(s) = 52.5s^2+(200+K)=0$$

or,

$$52.5s^2 = -862.17 \quad [\because K = 662.17]$$

Therefore,

$$s = \pm \jmath\, 4.05 \equiv \jmath\,\omega$$

Hence, the frequency of oscillation ω is 4.05 rad/s or 0.6446 Hz.

Zero in the First Term of a Row

When the first term in any row of the Routh array is zero while the rest of the row has at least one non-zero term, the problem can be solved by one of three methods. Let us work out one example to make ourselves familiar with different methods.

EXAMPLE 9.14. Examine the stability of the system represented by the characteristic equation

$$s^5 + s^4 + 2s^3 + 2s^2 + 3s + 5 = 0$$

Solution The Routh array is

$$\begin{array}{c|ccc} s^5 & 1 & 2 & 3 \\ s^4 & 1 & 2 & 5 \\ s^3 & 0 & -2 & 0 \end{array}$$

We cannot proceed further because there is a '0' in the first column. This problem can be solved by any of the following three methods.

Method 1. We replace '0' with a small positive number ε. Then we examine the sign change by taking $\varepsilon \to 0$. Then the Routh array is

$$\begin{array}{c|ccc} s^5 & 1 & 2 & 3 \\ s^4 & 1 & 2 & 5 \\ s^3 & 0 = \varepsilon & -2 & 0 \\ s^2 & \lim\limits_{\varepsilon \to 0} \dfrac{2\varepsilon + 2}{\varepsilon} = \infty & 5 & \\ s^1 & \lim\limits_{\varepsilon \to 0} \dfrac{-4\varepsilon - 4 - 5\varepsilon^2}{2\varepsilon + 2} = -2 & 0 & \\ s^0 & 5 & & \end{array}$$

There is a change of sign in the first column. The system is, therefore, unstable.

Method 2. We substitute $s = 1/z$. Next we transform the characteristic equation, construct Routh array and conclude therefrom. With transformation, the characteristic equation becomes

$$\frac{1}{z^5} + \frac{1}{z^4} + \frac{2}{z^3} + \frac{2}{z^2} + \frac{3}{z} + 5 = 0$$

$$\Rightarrow \quad 5z^5 + 3z^4 + 2z^3 + 2z^2 + z + 1 = 0$$

The corresponding Routh array is

$$\begin{array}{c|ccc} z^5 & 5 & 2 & 1 \\ z^4 & 3 & 2 & 1 \\ z^3 & -4/3 & -2/3 & \\ z^2 & 1/2 & 1 & \\ z^1 & 2 & 0 & \\ z^0 & 1 & & \end{array}$$

There is a change of sign in the first column. Thus, the system is unstable.

Method 3. We multiply the characteristic equation by a factor $(s + \alpha)$ where $\alpha > 0$. Normally, the factor $(s + 1)$ is used. Then we check the stability from the Routh array of the resulting equation. In the present example, the equation, on multiplication by $(s + 1)$, becomes

$$s^6 + 2s^5 + 3s^4 + 4s^3 + 5s^2 + 8s + 5 = 0$$

Its Routh array is

$$
\begin{array}{c|cccc}
s^6 & 1 & 3 & 5 & 5 \\
s^5 & 2 & 4 & 8 & 0 \\
s^4 & 1 & 1 & 5 & \\
s^3 & 2 & -2 & & \\
s^2 & 2 & 5 & & \\
s^1 & -7 & & & \\
s^0 & 5 & & &
\end{array}
$$

There is a change of sign in the first column. The system is, therefore, unstable.

Systems with Transportation Lag

We derived the transfer function for transportation lag in Section 4.4 at page 99. Here, we will consider how to study stability of systems with transportation lag through the Routh-Hurwitz criteria. To do that, we will have to have resort to the following approximations:

1. For small values of T, we can write from the expansion of the exponential function as

$$e^{-sT} \simeq 1 - sT \tag{9.2}$$

2. Alternatively, we can use the following relation:

$$e^{-sT} = e^{-\frac{sT}{2}} \cdot e^{-\frac{sT}{2}} = \frac{e^{-\frac{sT}{2}}}{e^{\frac{sT}{2}}} = \frac{1 - \frac{sT}{2} + \frac{1}{2} \cdot \left(\frac{sT}{2}\right)^2 - \ldots}{1 + \frac{sT}{2} + \frac{1}{2} \cdot \left(\frac{sT}{2}\right)^2 - \ldots}$$

If T is small

$$\cong \frac{1 - (sT/2)}{1 + (sT/2)} = \frac{2 - sT}{2 + sT} \tag{9.3}$$

In fact, Eq. (9.3) is nothing but the first order Padé[4] approximant.

Equation (9.2) adds a zero in the RHP of the system transfer function while Eq. (9.3) adds a zero in the RHP and pole in the LHP. Though Eq.(9.2) is recommended by some authors, it is indeed a poor approximation because it gives a completely wrong step response of the system. Other authors on classical control system theory usually provide a relation with equal degree for the numerator and the denominator as given in Eq. (9.3). This approximation is better though it generates a puzzling jump at $t = 0$. The results will be evident from Figure 9.4 where we have plotted the step response of the following open-loop transfer function for both the approximations.

$$G(s) = \frac{10e^{-s}}{s+1} \tag{9.4}$$

We work out one example using both the approximations.

[4]Henri Eugéne Padé (1863–1953) was a French mathematician who developed approximation techniques for functions using rational functions.

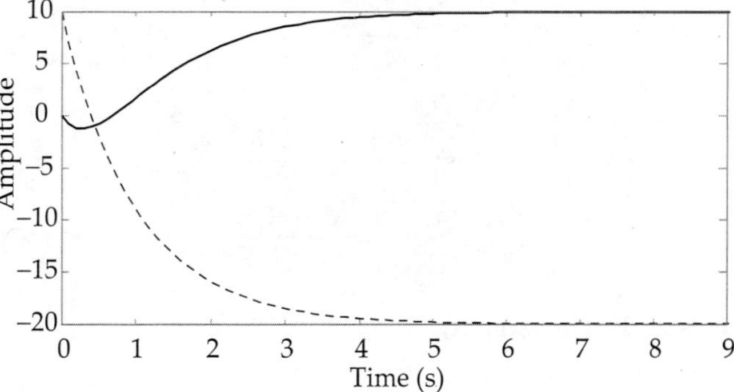

Figure 9.4: Step response - - - -, using Eq. (9.2) for the transportation lag and ——, using Eq. (9.3).

EXAMPLE 9.15. A unity feedback system has an open-loop transfer function

$$G(s)H(s) = \frac{Ke^{-s}}{s(s^2+5s+9)}$$

Determine, by using the Hurwitz-Routh stability criteria, the range of K for which the closed-loop system will be stable.

Solution Here,

$$T(s) = \frac{C(s)}{R(s)} = \frac{G(s)}{1+G(s)} = \frac{Ke^{-s}}{s(s^2+5s+9)+Ke^{-s}}$$

The characteristic equation is

$$s(s^2+5s+9)+Ke^{-s} = 0 \tag{i}$$

Using Eq. (9.2): Equation (i) can be written as

$$s(s^2+5s+9)+K(1-s) = 0 \qquad \Rightarrow \qquad s^3+5s^2+(9-K)s+K = 0$$

The Routh array is

$$\begin{array}{c|ccc} s^3 & 1 & 9-K & 0 \\ s^2 & 5 & K & 0 \\ s^1 & \dfrac{45-6K}{5} & 0 & \\ s^0 & K & & \end{array}$$

$$b_1 = \frac{5(9-K)-K}{5} = \frac{45-6K}{5}$$

The stability criteria are

$$K > 0$$

and

$$\frac{45-6K}{5} > 0 \qquad \Rightarrow \qquad 45-6K > 0 \qquad \Rightarrow \qquad K < 7.5$$

Therefore,

$$0 < K < 7.5$$

Using Eq. (9.3): Again, Eq. (i) can be rewritten as

$$s(s^2+5s+9)+\frac{K(2-s)}{2+s} = 0 \qquad \Rightarrow \qquad s^4+7s^3+19s^2+(18-K)s+2K = 0$$

The Routh array is

$$\begin{array}{c|ccc} s^4 & 1 & 19 & 2K \\ s^3 & 7 & 18-K & 0 \\ s^2 & \dfrac{115+K}{7} & 2K & \\ s^1 & \dfrac{2070-11K-K^2}{7} & & \\ s^0 & 2K & & \end{array}$$

The stability criteria are

$$K > 0$$

and $\quad 2070 - 11K - K^2 > 0 \quad \Rightarrow \quad -(K-40.3)(K+51.3) \quad > 0 \quad \Rightarrow \quad 40.3 > K > -51.3$

Therefore,

$$40.3 > K > 0$$

Let us now consider a few additional assorted examples and then see how relative stability can be studied by Hurwitz-Routh criteria.

EXAMPLE 9.16. The system shown in Figure 9.5 has oscillation of 2.5 rad/s. Determine the values of K and p. There are no poles in the RHP.

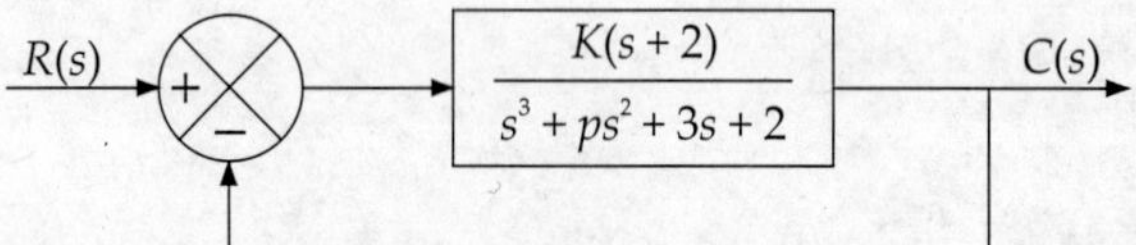

Figure 9.5: System (Example 9.16).

Solution The characteristic equation, $1 + G(s) = 0$, is

$$s^3 + ps^2 + 3s + 2 + K(s+2) = 0 \qquad \Rightarrow \qquad s^3 + ps^2 + +(3+K)s + (2K+2) = 0$$

The Routh array is

$$\begin{array}{c|cc} s^3 & 1 & 3+K \\ s^2 & p & 2K+2 \\ s^1 & (3+K) - \dfrac{2K+2}{p} & 0 \\ s^0 & 2K+2 & \end{array}$$

Since there is no pole in the RHP and the system oscillates (i.e. it is marginally stable),

$$(3+K) - \frac{2K+2}{p} = 0 \qquad \Rightarrow \qquad 3 + K = \frac{2K+2}{p} \tag{i}$$

The auxiliary equation is

$$ps^2 + 2K + 2 = 0 \qquad \Rightarrow \quad s^2 = -\frac{2K+2}{p} \qquad \Rightarrow \qquad s = \pm \jmath\sqrt{\frac{2K+2}{p}} \equiv \pm \jmath\,\omega \tag{ii}$$

From Eq. (ii) we get that the frequency of oscillation is

$$\omega = \sqrt{\frac{2K+2}{p}} = 2.5 \quad \text{(given)} \qquad \Rightarrow \qquad \frac{2K+2}{p} = 6.25 \tag{iii}$$

Solving Eqs. (i) and (iii), we get $K = 3.25$ and $p = 1.36$.

EXAMPLE 9.17. For a system having the characteristic equation $2s^4 + 4s^2 + 1 = 0$, find the following:

(a) The number of roots in the LHP
(b) The number of roots in the RHP
(c) The number of roots on the imaginary axis

Use the Routh-Hurwitz criterion.

Solution The characteristic equation is

$$2s^4 + 0s^3 + 4s^2 + 0s + 1 = 0$$

Therefore, the Routh array is

$$\begin{array}{c|ccc} s^4 & 2 & 4 & 1 \\ s^3 & (0)\ 1 & (0)\ 1 & 0 \\ s^2 & 2 & 1 & \\ s^1 & \frac{1}{2} & 0 & \\ s^0 & 1 & & \end{array}$$

Here, the elements of the s^3-row are zero (indicated in parentheses). So, we check the auxiliary polynomial

$$2s^4 + 4s^2 + 1$$

and get its derivative

$$8s^3 + 8s \equiv s^3 + s$$

The coefficients of the derivative are the revised elements of the row, which are 1, 1.

Now, the roots of the auxiliary equation

$$2s^4 + 4s^2 + 1 = 0$$

can be found as[5]

$$\pm\, \jmath\, 1.3065 \quad \text{and} \quad \pm\, \jmath\, 0.541$$

We observe that:

(a) No root lies in the LHP.
(b) No root lies in the RHP.
(c) All four roots lie on the imaginary axis, but they are not repetitive.

Since the Routh array shows no change of sign in the first column, we conclude that the system is marginally stable.

Note: In this case, the auxiliary equation is the same as the characteristic equation. Therefore, the results could be obtained without having recourse to the Routh-Hurwitz criterion.

EXAMPLE 9.18. Find the stability of a system having the characteristic equation

$$s^6 + 2s^5 + 7s^4 + 10s^3 + 14s^2 + 8s + 8 = 0$$

[5]The equation can be solved by putting $s^2 = x$.

Solution The Routh array is

s^6	1	7	14	8
s^5	2	10	8	0
s^4	2	10	8	
s^3	(0) 2	(0) 5	0	
s^2	5	8		
s^1	1.8	0		
s^0	8			

Since there is no change of sign in the 1st column of the Routh array, the system is not unstable. Here, the row of s^3 contains zero elements. So, we check the auxiliary polynomial which is

$$2s^4 + 10s^2 + 8$$

and its derivative is

$$8s^3 + 20s \equiv 2s^3 + 5s$$

The revised elements of the s^3-row are therefore, 2 and 5. The roots of the auxiliary equation

$$2s^4 + 10s^2 + 8 = 0$$

are

$$\pm\, j1 \quad \text{and} \quad \pm\, j2$$

However, the roots on the imaginary axis are not repetitive, which indicates that the system is marginally stable.

9.4 RELATIVE STABILITY STUDY VIA ROUTH'S RULE

So far we have studied absolute stability of systems using the Hurwitz-Routh criteria. A qualitative analysis of stability that determines the location of roots of the characteristic equation i.e. poles of the closed-loop transfer function in the s-plane, is termed *absolute stability*.

Relative stability, on the other hand, means a comparative analysis of a system stability, which determines the parameters like (i) maximum overshoot to a step input, (ii) damping ratio, (iii) gain margin and (iv) phase margin of a system.

Routh-Hurwitz criteria can be utilized to study the relative stability of systems to a limited extent. This is done by shifting the origin of the s-plane to another location s' by substituting

$$s = z + s'$$

and checking by the application of Routh's rule how many poles lie on what side of the shifted origin. Figure 9.6 shows what the substitution means for $s' = -\alpha$.

The following examples will make the procedure clear.

EXAMPLE 9.19. Determine how many roots of the characteristic equation

$$s^3 + 10s^2 + 37s + 52 = 0$$

are more negative than $s = -1$.

Solution Substituting $s = z - 1$ in the equation we get

$$(z-1)^3 + 10(z-1)^2 + 37(z-1) + 52 = 0$$

or,

$$z^3 + 7z^2 + 20z + 24 = 0$$

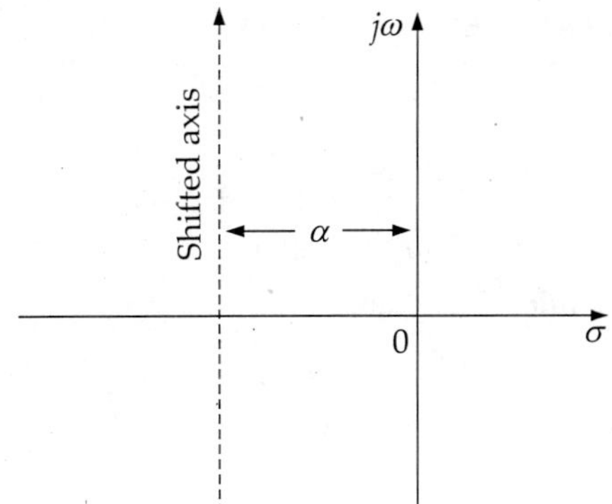

Figure 9.6: Shifting of origin of the s-plane.

Its Routh array is

$$\begin{array}{c|cc} z^3 & 1 & 20 \\ z^2 & 7 & 24 \\ z^1 & 16.57 & \\ z^0 & 24 & \end{array}$$

There is no sign change in the first column of the Routh array. It indicates that no root lies to the right of the shifted axis of $s = -1$. Nor do they lie on the line of $s = -1$, because in that case there would have been a marginally stable condition. So, all the roots are more negative than $s = -1$.

EXAMPLE 9.20. Determine the range of values of K ($K > 0$) such that the characteristic equation

$$s^3 + 3(K+1)s^2 + (7K+5)s + (4K+7) = 0$$

has roots more negative than $s = -1$. Apply the Routh-Hurwitz criterion.

Solution Substituting $s = z - 1$ in the equation we get

$$(z-1)^3 + 3(K+1)(z-1)^2 + (7K+5)(z-1) + (4K+7) = 0$$

or,

$$z^3 + 3Kz^2 + (K+2)z + 4 = 0$$

Its Routh array is given by

$$\begin{array}{c|cc} z^3 & 1 & K+2 \\ z^2 & 3K & 4 \\ z^1 & \dfrac{3K(K+2)-4}{3K} & \\ z^0 & 4 & \end{array}$$

For stability of the system,

$$3K > 0 \qquad \text{(i)}$$

$$3K(K+2) - 4 = 0$$

The second condition is a quadratic equation in K which yields

$$K = -2.63 \quad \text{and} \quad 0.53$$

Considering Eq. (i), i.e. $K > 0$, we observe that the acceptable value for K is 0.53. Therefore, the required range is $K > 0.53$.

EXAMPLE 9.21. The open-loop transfer function of a unity feedback system is given by

$$G(s) = \frac{K}{(s+1)(s+3)(s+5)}$$

Find the limiting value of K such that all the closed-loop poles are on the left of the straight line $s = -1$ in the s-plane.

Solution The charactéristic equation of the system is

$$(s+1)(s+3)(s+5) + K = 0 \qquad \text{(i)}$$

Substituting $s = z - 1$ in Eq. (i), we get

$$z(z+2)(z+4) + K = 0$$

or,

$$z^3 + 6z^2 + 8z + K = 0 \qquad \text{(ii)}$$

The Routh array constructed from Eq. (ii) is

$$\begin{array}{c|cc} z^3 & 1 & 8 \\ z^2 & 6 & K \\ z^1 & 8 - \dfrac{K}{6} & 0 \\ z^0 & K & \end{array}$$

The stability conditions yield

$$K > 0$$

and

$$8 - \frac{K}{6} > 0 \quad \Rightarrow \quad 8 > \frac{K}{6} \quad \Rightarrow \quad 48 > K$$

Therefore, the required limiting values are $0 < K < 48$.

EXAMPLE 9.22. A unity feedback control system is characterized by the open-loop transfer function

$$G(s) = \frac{K(s+13)}{s(s+3)(s+7)}$$

(a) Calculate the range of values of K for the stability of the system. Use the Hurwitz-Routh criteria.

(b) For $K = 1$, check if all the roots of the characteristic equation have damping ratio greater than 0.5.

Solution Since this is a unity feedback system, the characteristic equation is given by

$$1 + G(s) = 0$$

$$\Rightarrow \qquad 1 + \frac{K(s+13)}{s(s+3)(s+7)} = 0$$

$$\Rightarrow \qquad s(s+3)(s+7) + K(s+13) = 0$$

or,

$$s^3 + 10s^2 + (21+K)s + 13K = 0$$

The Routh array is, therefore,

$$\begin{array}{c|cc} s^3 & 1 & 21+K \\ s^2 & 10 & 13K \\ s^1 & \dfrac{210-3K}{10} & \\ s^0 & 13K & \end{array}$$

(a) For stability of the system

$$13K > 0 \quad \Rightarrow \quad K > 0$$

and
$$\frac{210 - 3K}{10} > 0 \quad \Rightarrow \quad K < 70$$

Therefore,
$$0 < K < 70$$

(b) At $K = 1$, the characteristic equation turns out to be

$$s^3 + 10s^2 + 22s + 13 = 0 \tag{i}$$

We need to find out the roots of Eq. (i) to check the damping ratio. It being a cubic equation, we examine a trial solution $s = -1$. We find that the left hand polynomial becomes zero for this s value. So, this is one root. To find out the other two roots, we divide the left hand polynomial by $s + 1$ to get the quadratic equation

$$s^2 + 9s + 13 = 0$$

whose solutions are

$$\frac{-9 \pm \sqrt{9^2 - 4(13)}}{2} = -7.1926, \ -1.8074$$

We observe that all the three roots are lying on the negative side of the real axis. Therefore, all of them subtend an angle of $\theta = 180°$. Now, the damping ratio is

$$\left| \frac{\sqrt{1 - \zeta^2}}{\zeta} \right| = \tan\theta = \tan 180° = 0$$

$\Rightarrow$
$$\zeta = 1$$

which is > 0.5 for all the roots.

REVIEW QUESTIONS

9.1 Choose the correct answer:

(i) If the Routh array for a system is given by

s^4	1	2	5
s^3	1	3	0
s^2	−1	5	
s^1	−8		
s^0	5		

the system is

(A) stable (B) marginally stable
(C) unstable (D) conditionally stable

(ii) The Routh array for a system is given by

s^4	1	3	3
s^3	1	2	0
s^2	1	3	
s^1	−1		
s^0	3		

The number of poles lying in the RHP is

(A) 0 (B) 1 (C) 2 (D) 3

(iii) The number of sign changes in the first column of the Routh array indicates
(A) the number of open-loop zeros in the RHP
(B) the number of closed-loop zeros in the RHP
(C) the number of open-loop poles in the RHP
(D) the number of closed-loop poles in the RHP

(iv) The Routh array for a system is

s^3	0.25	1
s^2	1	4
s^1	0	0
s^0	0	

The auxiliary equation for this array is
(A) $4s^2 + 1 = 0$ (B) $0.25s^3 + s = 0$
(C) $s^2 + 4 = 0$ (D) $s^3 + 0.25s = 0$

(v) If the changes of signs of the elements of the first column of Routh array are two for the fifth order characteristic equation, the location of roots in the s-plane are as
(A) 2 in the left half and 3 in the right half
(B) 2 in the right half and 3 in the left half
(C) all 5 in the left half
(D) all 5 in the right half

(vi) Which of the systems represented by their characteristic equations, is stable? a, b, c, d are positive constants.
(A) $as^2 + bs + c = 0$ (B) $as^4 + bs^2 + cs + d = 0$
(C) $-as^2 + bs - c = 0$ (D) $as^2 - bs - c = 0$

(vii) The Routh-Hurwitz criterion gives
(A) relative stability (B) absolute stability
(C) gain margin (D) phase margin

9.2 Find the possible range of K given in Figure 9.7 for the closed-loop system to be stable.

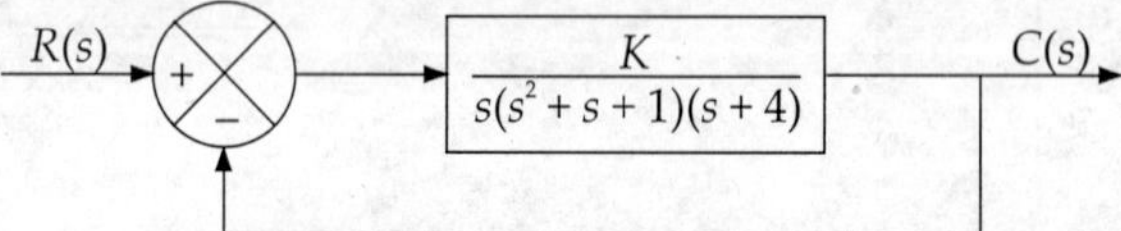

Figure 9.7: Question 9.2.

9.3 (a) With the help of Routh-Hurwitz stability criteria, determine the range of values of $K(K > 0)$ such that the characteristic equation $s^3 + 3(K+1)s^2 + (7K+5)s + (4K+7) = 0$ has roots more negative than $s = -1$.

(b) Using Routh-Hurwitz criteria, check whether the system represented by the following characteristic equation is stable or not.

$$s^4 + 2s^3 + 6s^2 + 8s + 8 = 0$$

Comment on the location of roots.

9.4 Check the stability of the system given in Figure 9.8 using Routh-Hurwitz stability criterion.

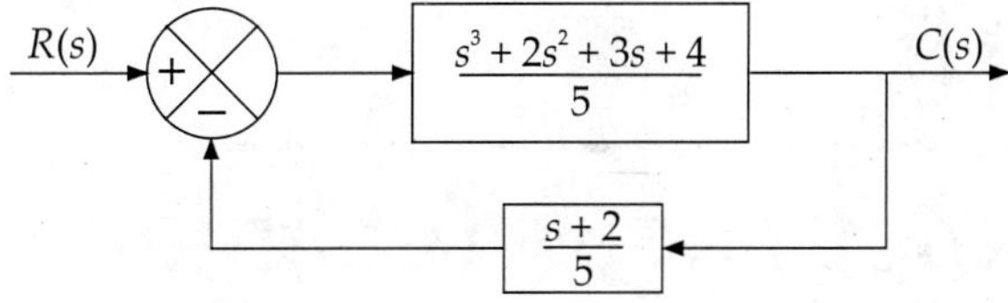

Figure 9.8: Question 9.4.

9.5 Determine the range of K such that the feedback system having characteristic equation

$$s(s^2 + s + 1)(s + 4) + K = 0$$

will be stable.

9.6 (a) State and explain Routh's stability criterion.
(b) Consider the characteristic equation

$$s^4 + Ks^3 + s^2 + s + 1 = 0$$

Using Routh's stability criterion determine the range of K for stability.

9.7 A unity feedback system is characterized by the open-loop transfer function

$$G(s) = \frac{K(s + 13)}{s(s + 3)(s + 7)}$$

Check whether for $K = 1$, all roots of the characteristic equation are more negative than $s = -2$.

9.8 The open-loop transfer function of a control system is given by

$$G(s)H(s) = \frac{Ke^{-s}}{s(s^2 + 2s + 1)}$$

Determine the maximum value of K for the closed-loop stability.

Answers to Selected Questions

9.1. (i) (C) (ii) (C) (iii) (D) (iv) (C) (v) (B) (vi) (A) (vii) (A) 9.2. $0 < K < 3.36$
9.3. (a) $K > 0.53$ (b) Marginally stable. Roots are at $s = 0 \pm \jmath 2,\ -1 \pm \jmath 1$.
9.4. Stable. 9.5. $0 < K < 0.96$ 9.6. (b) $K > 3$ 9.7. All roots. 9.8. $K = \frac{2}{3}$, assuming $e^{-s} \cong (1 - s)$

10

Root Locus Technique

10.1 INTRODUCTION

We have seen that the location of poles in the s-plane determines the stability as well as transient response of a closed-loop system. In the root locus technique, introduced by WR Evans[1] in 1948, the loci of the roots of the characteristic equation, i.e. poles, are plotted by varying a system parameter, generally the system gain-factor K, from 0 to ∞. The resulting graph, called the *root locus*, helps us study the stability of the system. Root locus is indeed a powerful graphical method for analysis and design of control systems. Given a feedback control system, the root loci illustrate how the poles of the closed-loop system vary with the closed-loop gain. This chapter describes the root locus, its properties, and a step-by-step procedure for constructing a root locus. Examples are included to illustrate how to use the method to study the stability of a control system.

10.2 DEPENDENCE OF ROOT LOCUS ON K VARIATION

Consider the Single Input Single Output (SISO)system represented by Figure 10.1.

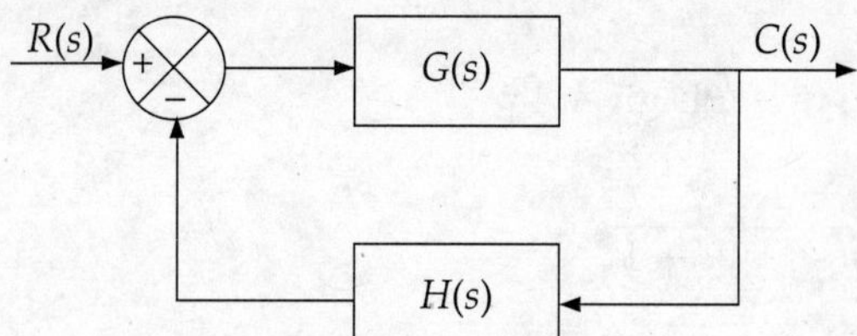

Figure 10.1: Block diagram of a SISO.

The overall transfer function of the system is

$$T(s) = \frac{C(s)}{R(s)} = \frac{G(s)}{1 + G(s)H(s)}$$

So, the characteristic equation is

$$1 + G(s)H(s) = 0 \tag{10.1}$$

or,

$$G(s)H(s) = -1 \tag{10.2}$$

Since s is generally a complex quantity, we can write Eq. (10.2) as

$$|G(s)H(s)|\angle G(s)H(s) = |1|\angle \pm (2q+1)180^\circ \qquad q = 0,\ 1,\ 2\ldots$$

Thus,

$$|G(s)H(s)| = |1| \tag{10.3}$$

and

$$\angle G(s)H(s) = \pm(2q+1)180^\circ \qquad q = 0,\ 1,\ 2\ldots \tag{10.4}$$

[1]Walter R Evans (1920–1999) was an American engineer.

Equations (10.3) and (10.4) are called the *magnitude and angle condition* of Evans. We know that the transfer function can be expressed as

$$G(s)H(s) = \frac{K(s-b_1)(s-b_2)\cdots(s-b_m)}{(s-a_1)(s-a_2)\cdots(s-a_n)} \tag{10.5}$$

where K is the gain-factor, m is the number of zeros and n is the number of poles. Now, from Eqs. (10.1) and (10.5)

$$1+\frac{K(s-b_1)(s-b_2)\cdots(s-b_m)}{(s-a_1)(s-a_2)\cdots(s-a_n)} = 0 \tag{10.6}$$

or,
$$(s-a_1)(s-a_2)\cdots(s-a_n)+K(s-b_1)(s-b_2)\cdots(s-b_m) = 0 \tag{10.7}$$

Normally, $n > m$. Then the order of the characteristic equation is n and it will possess n roots. It is also apparent from Eq. (10.7) that the value of roots will depend on the value of K. As the value of K changes, roots alter positions in the s-plane. So, the variation of K from 0 to ∞ will generate loci of roots in the s-plane.

We note from Eq. (10.7) that

- When $K = 0$: The second term on the left hand side (LHS) of Eq. (10.7) does not exist and hence the roots of the resulting characteristic equation are simply the *poles of the open-loop transfer function* $G(s)H(s)$.
- When $K \to \infty$: The first term on the LHS of Eq. (10.7) can be neglected and the roots of the resulting characteristic equation are simply the *zeros of the open-loop transfer function* $G(s)H(s)$.

> Therefore, we observe that the loci of the roots of the characteristic equation *begin at the poles* and *end at the zeros* of open-loop transfer function $G(s)H(s)$ as K goes from 0 to ∞.

Root locus construction. As regards the root locus construction procedure, we may note that the exact loci can be calculated by varying K from 0 to ∞ and finding the roots of the characteristic equation. These positions of roots, when plotted on the s-plane, give the exact root locus plot of the system. That is, precisely, what is done by software like Matlab. However, this procedure is cumbersome for manual calculation and consequent plotting of the root locus.

For manual drawing of the loci, a few rules are followed to find information like:

- **Part of the Re-axis** belonging to the root locus (A—A in Figure 10.2);
- **Angles of asymptotes** that the loci will follow when K values are high (B, B in Figure 10.2);
- **Centroid** from where the asymptotes will diverge (C in Figure 10.2);
- **Breakaway point** for loci originating from poles facing each other (D in Figure 10.2);
- **Angle of departure** for loci originating from complex poles (E, E in Figure 10.2);
- **Point of intersection** on the Im-axis for complex loci (F, F in Figure 10.2).

Once these pieces of information are collected, approximate root loci are drawn manually. One such drawing is shown in Figure 10.2 to show the cardinal points as mentioned above.

10.3 ROOT LOCUS CONSTRUCTION PROCEDURE

To plot a root locus, we must have the *open-loop transfer function* $G(s)H(s)$ *with* K *as the variation parameter.* Then we follow the steps as laid down here to construct the root locus.

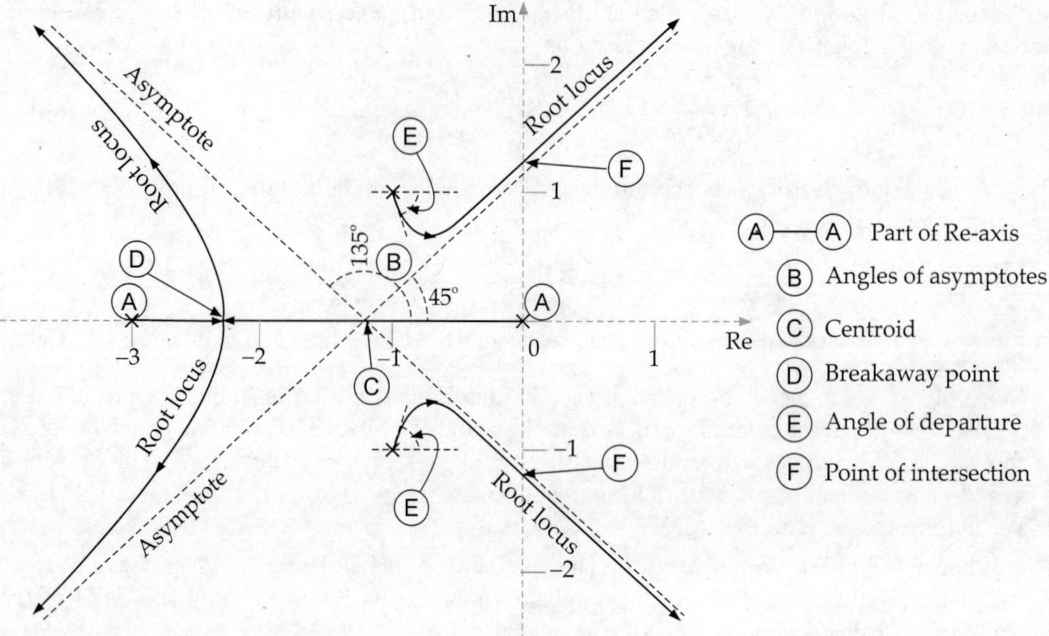

Figure 10.2: A root locus diagram showing cardinal points and asymptotes which help us draw it. Root loci are shown with black thick lines.

Step 1. Find all poles and zeros of the system and plot them on the *s*-plane. Then find which part(s) of the Re-axis will belong to the root locus.

The following rules are to be followed while figuring out the part(s) of the Re-axis constituting part(s) of the root locus:

Rule 1: A point on the real axis will lie on the root locus if and only if the sum of the number of poles and zeros *to its right* is *odd* when $K > 0$ and *even* when $K < 0$.

Rule 2: The root locus is always symmetrical about the real axis. That is obvious because complex roots occur in conjugate pairs.

Rule 3: All root loci start from poles and terminate at zeros. If there are t more poles than zeros then t loci will terminate at ∞ as $K \to \infty$.

Rule 4: Generally, the transfer function is proper and, therefore, $n > m$, where $n =$ number of poles and $m =$ number of zeros. Then, the number of root loci is n.

Note: The **Rule 1** owes its origin to Eq. (10.4) which indicates that the sum of angles contributed by poles and zeros should be odd multiples of 180°. The sum of angles contributed by poles and zeros to the left of the point is zero because phasors originating from the poles and zeros on the left of the point concerned are directed towards the Re-axis (therefore, angle $= 0°$) while those originating from poles and zeros on the right are directed towards the opposite of the Re-axis (therefore, angle $= 180°$). We observe that in Figure 10.3a, the phasors from poles p_1, p_2 and zero z_1 directed towards the pole p_3 subtend a total angle of $3 \times 180°$. Similarly the angle subtended by the phasor from p_1 to p_2 is 180°. Therefore, portions of the Re-axis between (p_2 and p_1) and (p_3 and z_1) constitute parts of the root locus.

Thus, the angle contributed by poles and zeros to the right of the point has to be odd multiple of 180° to satisfy Eq. (10.4). That is possible only when their number is odd when $K > 0$ and even when $K < 0$.

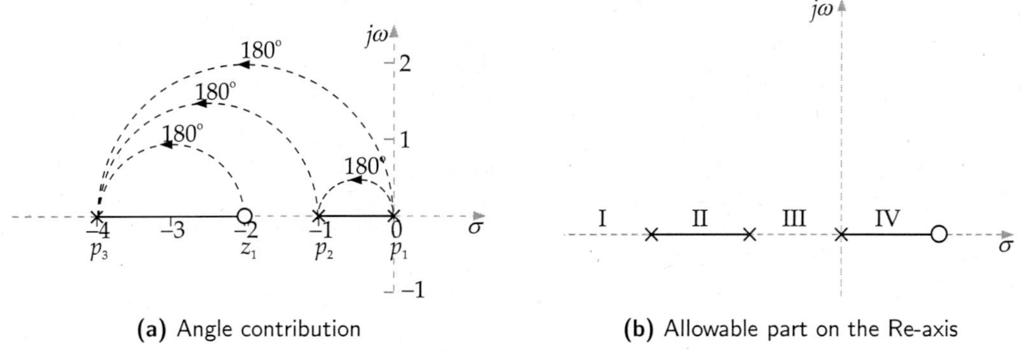

(a) Angle contribution

(b) Allowable part on the Re-axis

Figure 10.3: Angle contribution of poles and zeros and allowable part of the Re-axis in the root locus. Root loci are shown with black thick lines.

In the plot of Figure 10.3b, segments I and III are not part of the root locus because $(m+n)$ for them are 4 and 2, respectively, while segments II and IV, for which $(m+n)$ are 3 and 1, respectively, are parts of the root locus.

Step 2. Find the number of asymptotes[2] their angles and centroids (or, centres of gravity[3]). Draw results.

The centroid always lies on the real axis.d

Rule 5: The number of asymptotes is $n - m$.
Rule 6: Angles of asymptotes are given by

$$\boxed{\theta = \begin{cases} \dfrac{2q+1}{n-m} \times 180^\circ & \text{for } K > 0 \\ \dfrac{2q}{n-m} \times 180^\circ & \text{for } K < 0 \end{cases}}$$

where $q = 0,\ 1,\ 2,\ \ldots,\ (n-m)$.

Rule 7: The centroid (or, centre of gravity) of asymptotes will lie on the Re-axis and its coordinate χ_c is calculated from the relation

$$\boxed{\chi_c = \frac{\Sigma\text{Re (poles)} - \Sigma\text{Re (zeros)}}{n-m}}$$

where Σ Re (poles) means sum of real parts of coordinates (i.e., s-values) of poles.

EXAMPLE 10.1. Calculate the angles of asymptotes and their centroid for the system having

$$G(s)H(s) = \frac{K(s+2)}{s(s+1)(s+4)} \quad K > 0$$

Solution Here, poles are at $s = 0,\ -1$ and -4, $n = 3$. Zero is at $s = -2$, $m = 1$

[2] *Asymptote* is a line which touches the root locus at ∞.

[3] *Centroid* (or, *centre of gravity*) is the point on the σ-axis (i.e., real axis) from where all asymptotes start (i.e. where they are centred).

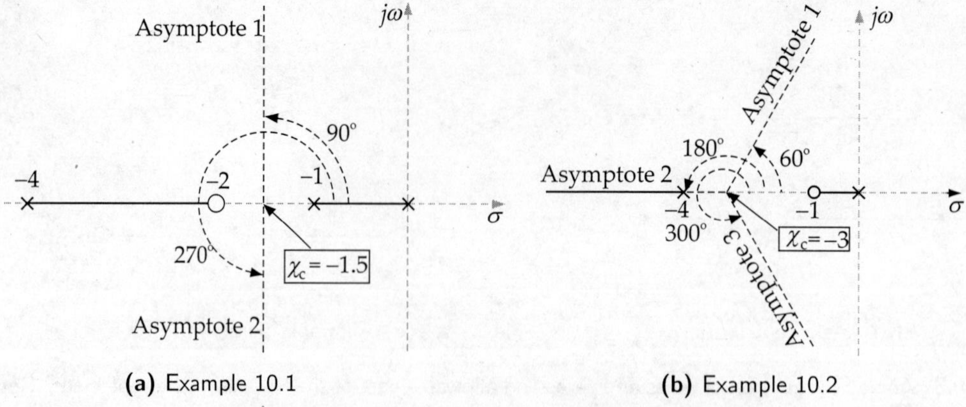

Figure 10.4: Root locus plots along with asymptotes.

The number of asymptotes $= n - m = 3 - 1 = 2$. Their angles are

$$\theta = \frac{2(0)+1}{2} \times 180^\circ = 90^\circ \quad \text{for} \quad q = 0$$

$$= \frac{2(1)+1}{2} \times 180^\circ = 270^\circ \quad \text{for} \quad q = 1$$

Centroid:
$$\chi_c = \frac{(-1-4)-(-2)}{2} = -1.5$$

The angles of asymptotes and the location of the centroid are shown in Figure 10.4a.

EXAMPLE 10.2. Draw the asymptotes for the system whose open-loop transfer function is given as

$$G(s)H(s) = \frac{s+1}{s(s+4)(s^2+6s+10)}$$

Solution The transfer function can be written as

$$G(s)H(s) = \frac{s+1}{s(s+4)(s+3-\jmath 1)(s+3+\jmath 1)}$$

Therefore, poles are at $s = 0,\ -4,\ -3+\jmath 1$ and $-3-\jmath 1$. $n = 4$. Zero is at $s = -1$. $m = 1$. The number of asymptotes $= 4 - 1 = 3$. Their angles are

$$\theta = \frac{2q+1}{n-m} \times 180^\circ \quad \text{for} \quad q = 0,\ 1,\ 2 \quad (\because\ K = 1)$$

$$= 60^\circ,\ 180^\circ,\ 300^\circ$$

Centroid:
$$\chi_c = \frac{(-4-3-3)-(-1)}{3} = -3$$

The relevant plot is given in Figure 10.4b.

EXAMPLE 10.3. The open-loop transfer function $G(s)$ of a unity feedback control system is given as

$$G(s) = \frac{K\left[s+(2/3)\right]}{s^2(s+2)}$$

From the root locus, it can be inferred that when K tends to positive infinity,

(A) three roots with nearly equal real parts exist on the left half of the s-plane
(B) one real root is found on the right half of the s-plane
(C) the root loci cross the $j\omega$-axis for a finite value of K; $K \neq 0$
(D) three real roots are found on the right half of the s-plane

Solution The poles are at $s = 0$ (two) and $s = -2$. Zero exists at $s = -(2/3)$. Therefore, the coordinate of the centroid is

$$\chi_c = \frac{-2 - \left[-(2/3)\right]}{3-1} = \frac{-6+2}{6} = -\frac{2}{3}$$

The number of asymptotes is $(n - m) = (3 - 1) = 2$. Their angles are

$$\theta_1 = \frac{2(0)+1}{2} \times 180^\circ = 90^\circ \qquad \text{for} \qquad q = 0$$

$$\theta_2 = \frac{2(1)+1}{2} \times 180^\circ = 270^\circ \qquad \text{for} \qquad q = 1$$

These indicate that 2 loci will start from the poles at the origin and will follow the asymptote at 90° while another locus will start from $s = -2$ and will go to $-\infty$ along the asymptote at 270°.
Ans. (A)

EXAMPLE 10.4. A unity feedback system has an open-loop transfer function

$$G(s) = \frac{K}{s^2}$$

The root locus plot (the 2 number near the pole location indicates the presence of as many poles there) is

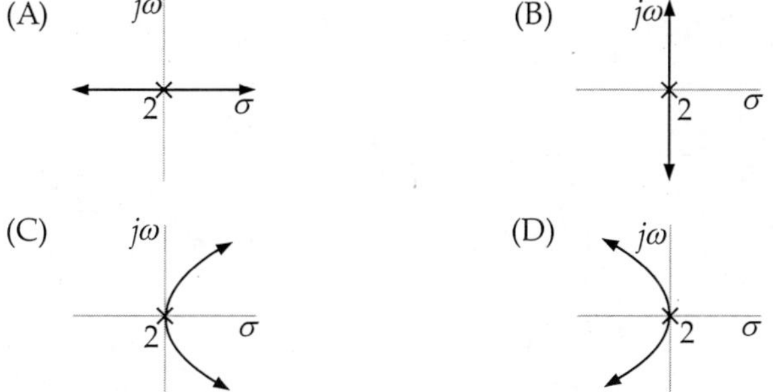

Solution From the given open-loop transfer function, we get 2 poles at $s = 0$ and no zero. So, there are 2 asymptotes for $q = 0$. Their angles are

$$\theta = \frac{\pm[(2 \times 0) + 1]}{2-0} \times 180^\circ = \pm 90^\circ$$

Option (B) satisfies this criterion.
Ans. (B)

Step 3. Determine the Breakaway/Break-in point, if present, and mark the point on the graph.

Breakaway point: It is the point at which the root locus breaks away from (i.e., comes out of) the real axis. The root locus leaves the real axis where there is a multiplicity of poles, typically two poles. It can be shown that K is an extremum at the breakaway point. So, the roots of the equation

$$\boxed{\frac{dK}{ds} = 0}$$

are the breakaway points.

Proof. Let us write Eq. (10.6) as

$$1 + K\frac{Z(s)}{P(s)} = 0 \tag{10.8}$$

Then Eq. (10.5) can be written as

$$G(s)H(s) = \frac{KZ(s)}{P(s)} \tag{10.9}$$

At the breakaway point, it is necessary that

$$\frac{d}{ds}[G(s)H(s)] = 0 \tag{10.10}$$

Differentiating Eq. (10.9), we get

$$\frac{d}{ds}[G(s)H(s)] = K\left[\frac{P(s)Z'(s) - P'(s)Z(s)}{[P(s)]^2}\right] = 0$$

or,

$$P(s)Z'(s) - P'(s)Z(s) = 0 \tag{10.11}$$

Again, from Eq. (10.8)

$$K\frac{Z(s)}{P(s)} = -1$$

or,

$$K = -\frac{P(s)}{Z(s)}$$

Therefore,

$$\frac{dK}{ds} = \frac{P(s)Z'(s) - P'(s)Z(s)}{[Z(s)]^2} \tag{10.12}$$

Substituting the result of Eq. (10.11) in Eq. (10.12), we get

$$\frac{dK}{ds} = 0$$

It is necessary that K is *real* at the breakaway points. If a is the number of branches that approach the breakaway point, the locus leaves the point at an angle of $\pm\, 180°/a$.

Break-in point: It is the point at which the root locus enters (breaks in) the real axis. Table 10.1 indicates where to expect breakaway or break-in points.

EXAMPLE 10.5. For

$$G(s)H(s) = \frac{K}{s(s+2)(s+3)}$$

determine the coordinates of valid breakaway/break-in points.

Table 10.1: Diagnostics of breakaway/break-in points

Indication	*Inference*	*Illustration*
Two adjacent poles on the real axis and the segment of the real axis is a part of the root locus.	At least one *breakaway point* exists between the poles.	$j\omega$ Breakaway point σ
Two adjacent zeros on the real axis and the segment of the real axis is a part of the root locus.	At least one *break-in point* exists between the zeros.	$j\omega$ Break-in point σ
No poles and zeros to the left of a zero on the real axis and the segment of the real axis is a part of the root locus.	At least one *break-in point* exists there.	$j\omega$ Break-in point Breakaway point σ

Solution The characteristic equation is

$$1 + G(s)H(s) = 0$$

$\Rightarrow$
$$s(s+2)(s+3) + K = 0$$

or,
$$K = -s^3 - 5s^2 - 6s \qquad \text{(i)}$$

Therefore,

$$\frac{dK}{ds} = -3s^2 - 10s - 6 = 0$$

The roots are

$$s = \frac{-10 \pm \sqrt{100-72}}{6} = -0.7847, \; -2.5485$$

The corresponding values of K, namely 2.113 and -0.631 [from Eq. (i)], are real. Poles are located at $s = 0, -2$ and -3. There is no zero (see Figure 10.5a). Segment I, where one breakaway point ($s = -2.55$) could be located, is not a part of the root locus, because an even number of poles exists on its right. Hence, the only breakaway point is at $s = -0.785$. Number of branches approaching the breakaway point is 2. Therefore, the loci leave the real axis at angles $\pm\, 180°/2 = \pm\, 90°$.

EXAMPLE 10.6. Find the breakaway point for a system represented by

$$G(s)H(s) = \frac{K(s+1)}{s(s+2)(s+3)}$$

Solution The poles of the system are at $s = 0, -2$ and -3. There is a zero at $s = -1$. These are plotted in Figure 10.5b. The plot shows that the segment between poles at $s = -2$ and -3

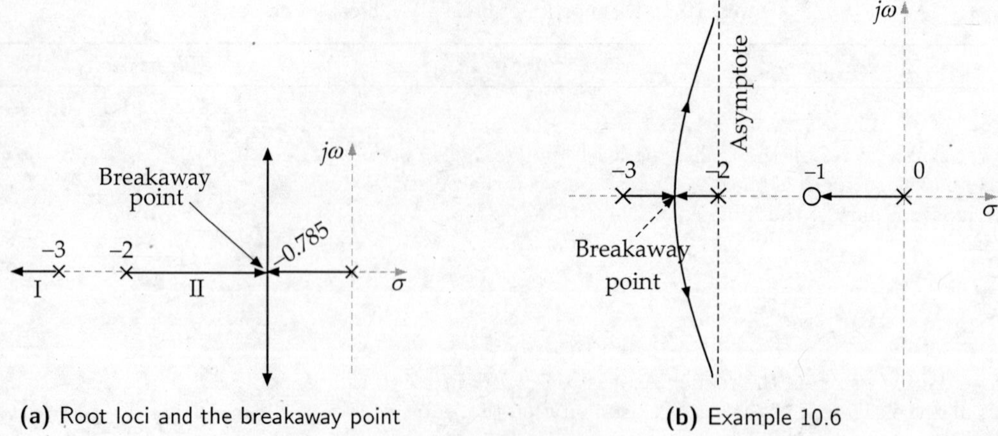

Figure 10.5: Root loci and breakaway point.

is a part of the root locus. So, the loci originating from these two poles have to breakaway at a point in between.

The characteristic equation is given by

$$s(s+2)(s+3)+K(s+1)=0$$

or,
$$K=\frac{-s(s+2)(s+3)}{(s+1)} \tag{i}$$

$$\frac{dK}{ds}=\frac{d}{ds}\left[\frac{-s(s+2)(s+3)}{(s+1)}\right]=\frac{(s^3+5s^2+6s)-(s+1)(3s^2+10s+6)}{(s+1)^2}=0$$

$\Rightarrow$
$$2s^3+8s^2+10s+6=0 \tag{ii}$$

Equation (ii) can be solved with the help of a calculator given here[4]. The values obtained are: -2.46557, $-0.76721+\jmath 0.79255$, $-0.76721-\jmath 0.79255$. Since we know that the breakaway point lies between $s=-2$ and $s=-3$, we accept the value $s\approx -2.47$.

EXAMPLE 10.7. The gain at the breakaway point of the root-locus of a unity feedback system with open-loop transfer function

$$G(s)=\frac{Ks}{(s-1)(s-4)}$$

is

(A) 1 (B) 2 (C) 5 (D) 9

Solution The given open-loop transfer function gives us the following characteristic equation

$$1+G(s)H(s)=0$$

or,
$$1+\frac{Ks}{(s-1)(s-4)}=0$$

or,
$$Ks+(s^2-5s+4)=0$$

or,
$$K=-\frac{s^2-5s+4}{s}=-\left[s-5+\frac{4}{s}\right] \tag{i}$$

[4] https://www.calculatorsoup.com/calculators/algebra/cubicequation.php. Scientific pocket calculators also provide this facility.

For the breakaway point

$$\frac{dK}{ds} = 0$$

or, $$= 1 - 0 - \frac{4}{s^2} \qquad \text{[from Eq. (i)]}$$

or, $$s = \pm 2$$

The solution $s = -2$ is ruled out because the part of the Re-axis belonging to the locus lies between $s = 1$ and $s = 4$. So, $s = 2$.

Now, the gain at $s = 2$ is

$$K = -\left[s - 5 + \frac{4}{s}\right]_{s=2} = 1$$

Alternatively

$$K = \frac{\Pi \, \text{distance of all poles from } s = 2}{\Pi \, \text{distance of all zeros from } s = 2}$$
$$= \frac{|4-2||1-2|}{|2-0|} = 1$$

Ans. (A)

EXAMPLE 10.8. A unity feedback system has the transfer function

$$\frac{K(s+b)}{s^2(s+20)}$$

The value of b for which the loci of all the three roots of the closed-loop characteristic equation meet at a single point is

(A) $\frac{10}{9}$ (B) $\frac{20}{9}$ (C) $\frac{30}{9}$ (D) $\frac{40}{9}$

Solution Since $H(s) = 1$, the characteristic equation is given by

$$1 + \frac{K(s+b)}{s^2(s+20)} = 0$$

or, $$s^2(s+20) + K(s+b) = 0$$

Therefore, $$K = -\frac{s^2(s+20)}{s+b}$$

If all loci meet at a point, then it must be a breakaway point. So, we have

$$\frac{dK}{ds} = \frac{s^2(s+20) - (3s^2+40s)(s+b)}{(s+b)^2} = 0$$

or, $$s^2(s+20) - s(3s+40)(s+b) = 0$$

or, $$2s^2 + (3b+20)s + 40b = 0 \qquad \text{(i)}$$

Equation (i) yields

$$s = \frac{-(3b+20) \pm \sqrt{(3b+20)^2 - 320b}}{4} \qquad \text{(ii)}$$

The condition for s to be a single point is

$$(3b+20)^2 = 320b$$

or,
$$9b^2 - 200b + 400 = 0 \tag{iii}$$

Equation (iii) yields two values for b, namely

$$b = 20 \text{ and } \frac{20}{9}$$

For $b = 20$, the breakaway point turns out to be at $s = -20$ from Eq. (ii). This value is discarded because from the pole-zero position of the open-loop transfer function we expect a breakaway point between 0 and -20. So, the second value, i.e. $b = 20/9$ which yields a breakaway point at $s = -6.67$ is accepted.
Ans. (B)

Step 4. Find the angle of departure (θ_d) from a pole and angle of arrival (θ_a) at a zero for complex poles and zero.

This can be done either *analytically* or *graphically.* We describe both the methods below with a typical example.

EXAMPLE 10.9. Find the angle of departure for

$$G(s)H(s) = \frac{K(s+3)}{(s+2+\jmath1)(s+2-\jmath1)}$$

where $K > 0$.

Solution Analytical method: The relevant equations are:

$$\boxed{\theta_d = \begin{cases} 180° + \arg|G(s)H(s)| & \text{for poles} \\ 180° - \arg|G(s)H(s)| & \text{for zeros} \end{cases}}$$

where, $\arg|G(s)H(s)|$ is the angle of $G(s)H(s)$ excluding the *pole* or *zero*, where the angle is to be calculated.

For the pole at $s = -2+\jmath1$,

$$\arg|G(s)H(s)|_{s=-2+\jmath1} = \arg\left|\frac{K(s+3)}{(s+2+\jmath1)}\right|_{s=-2+\jmath1} = \arg\left|\frac{K(1+\jmath1)}{2\jmath1}\right| = \frac{\angle 45°}{\angle 90°} = \angle -45°$$

Therefore,
$$\theta_d = 180° - 45° = 135°$$

For the other pole, it will obviously be $-135°$.

Graphical method: The relevant formula is

$$\boxed{\begin{aligned} \theta_d = 180° - \{&\Sigma \text{ angles to a complex pole from other poles} \\ &- \Sigma \text{ angles to a complex pole from other zeros}\} \end{aligned}} \tag{i}$$

Here, poles are at $s = -2+\jmath1$ and $s = -2-\jmath1$. Zero is at $s = -3$. To calculate θ_d at the pole at $s = -2+\jmath1$, angles subtended by the pole at $s = -2-\jmath1$ and the zero at $s = -3$ are plotted in Figure 10.6. Therefore, according to Eq. (i)

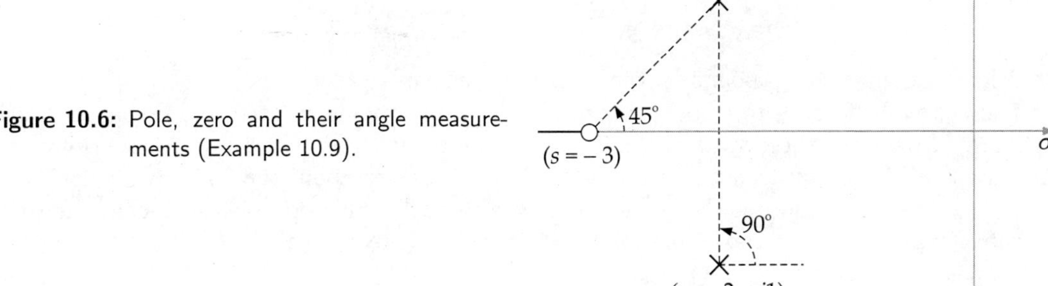

Figure 10.6: Pole, zero and their angle measurements (Example 10.9).

$$\theta_d = 180^\circ - (90^\circ - 45^\circ) = 135^\circ$$

So, it is clear that the angle of departure can be calculated either by the analytical or the graphical method. We will be using either of these two methods while solving problems later in this chapter.

Step 5. Find the intersection of the root loci with the imaginary axis.

Since the roots lying on the imaginary axis will make the system marginally stable, these roots can be found with the help of the Routh array as already discussed. To recapitulate, the steps are as follows:

1. Form the Routh array.
2. Determine $K = K_{\text{marginal}}$ which creates a row of zeros in the array.
3. Solve the auxiliary equation to get the required s-values.

EXAMPLE 10.10. Find the points of intersection of the root locus with the imaginary axis for the system

$$G(s)H(s) = \frac{K}{s(s+4)(s^2+4s+13)}$$

Solution The characteristic equation is

$$1 + G(s)H(s) = 0$$

$$\Rightarrow \quad s(s+4)(s^2+4s+13) + K = 0$$

$$\Rightarrow \quad s^4 + 8s^3 + 29s^2 + 52s + +K = 0$$

The Routh array is

$$\begin{array}{c|ccc}
s^4 & 1 & 29 & K \\
s^3 & 8 & 52 & \\
s^2 & 22.5 & K & \\
s^1 & 52 - 0.35556K & 0 & \\
s^0 & K & &
\end{array}$$

To determine K_{marginal}, we put

$$52 - 0.35556\, K\big|_{K=K_{\text{marginal}}} = 0$$

Therefore,

$$K_{\text{marginal}} = \frac{52}{0.35556} = 146.25$$

The auxiliary equation is

$$22.5s^2 + 146.25 = 0 \qquad \Rightarrow \quad s = \pm\, \jmath\, 2.55$$

Therefore, the intersection points on the imaginary axis are $\pm\, \jmath\, 2.55$.

These are all the steps to follow for finding out cardinal points related to drawing root loci manually. Obviously, these steps *will not give all the points to draw graphs of loci.* However, these steps will give enough *hints for making educated guesses* about the forms of the loci.

With this background knowledge, let us solve a few full-fledged problems to learn the root locus technique. In the first example, we indicate the steps sequentially.

EXAMPLE 10.11. Sketch the root locus of

$$G(s)H(s) = \frac{K}{s(s+1+\jmath 1)(s+1-\jmath 1)} \qquad K > 0$$

Solution Here, the poles are at $s = 0,\ (-1-\jmath 1),\ (-1+\jmath 1)$. Therefore, $n = 3$. No zero. Therefore, $m = 0$.

Step 1: The real axis locus is present for $-\infty < s < 0$.

Step 2: *Asymptotes*: $n - m = 3$. Therefore, the number of asymptotes = 3 and $q = 0,\ 1,\ 2$. The angles of asymptotes are:

$$\theta_0 = \frac{1}{3} \times 180^\circ = 60^\circ \qquad \theta_1 = \frac{3}{3} \times 180^\circ = 180^\circ \qquad \theta_2 = \frac{5}{3} \times 180^\circ = 300^\circ$$

The centroid coordinate is: $\chi_c = \dfrac{-1-1}{3} = -\dfrac{2}{3} = -0.67$

Step 3: The *breakaway point* calculation is not necessary since the loci do not break away from the real axis.

Step 4: *Angle of departure at* $s = -1+\jmath 1$*:*

<u>Analytical calculation:</u>

$$\arg\left|G(s)H(s)\right|_{s=-1+\jmath 1} = \arg\left|\frac{K}{s(s+1+\jmath 1)}\right|_{s=-1+\jmath 1} = \arg\left|\frac{K}{(-1+\jmath 1)(2\jmath 1)}\right|$$

$$= \frac{1}{\angle(\pi - 45^\circ) + (\angle 90^\circ)} = \angle - 225^\circ$$

Therefore, $\qquad \theta_d = 180^\circ - 225^\circ = -45^\circ$

Step 5: *Intersection at the* $\jmath\omega$*-axis*:

$$1 + G(s)H(s) = 0$$

$$\Rightarrow \qquad s(s+1+\jmath 1)(s+1-\jmath 1) + K = 0$$

$$\Rightarrow \qquad s^3 + 2s^2 + 2s + K = 0$$

The Routh array is

$$\begin{array}{c|cc} s^3 & 1 & 2 \\ s^2 & 2 & K \\ s^1 & 2 - \dfrac{K}{2} & 0 \\ s^0 & K & \end{array}$$

For marginal stability

$$2 - \frac{K_{\text{mar}}}{2} = 0$$

$$\Rightarrow \qquad K_{\text{mar}} = 4$$

The auxiliary equation $\qquad 2s^2 + 4 = 0$

$$\Rightarrow \qquad s = \pm\, \jmath\, 1.414$$

With the data that we obtained in steps 1 to 5, the root locus is drawn and shown in Figure 10.7.

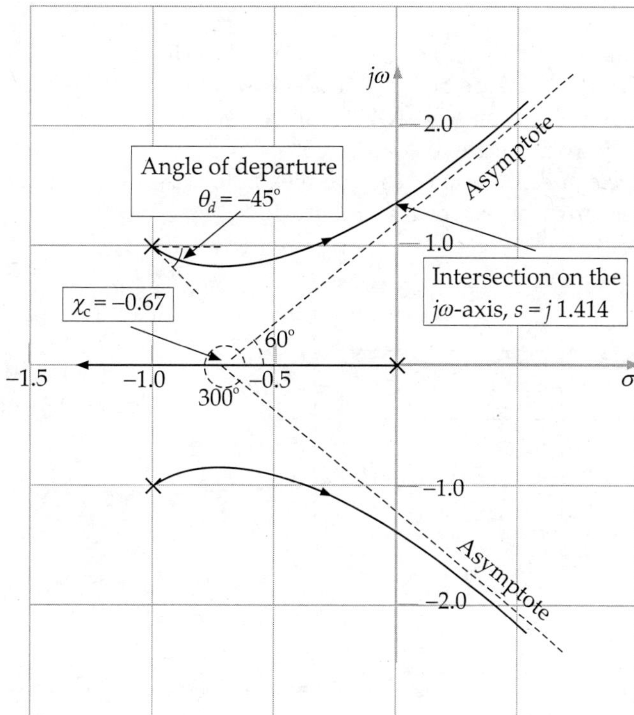

Figure 10.7: Root locus (Example 10.11).

EXAMPLE 10.12. Draw the root locus of the system having the open-loop transfer function as

$$G(s)H(s) = \frac{K}{s(s+3)(s^2+2s+2)}$$

Solution The open-loop transfer function can be written as

$$G(s)H(s) = \frac{K}{s(s+3)(s+1-\jmath 1)(s+1+\jmath 1)}$$

Poles are at $s = 0,\ -3,\ -1+\jmath 1,\ -1-\jmath 1$. Therefore, $n = 4$. No zero. Thus, $m = 0$.
Part of the Re-axis: The real axis locus is present for $0 < s < -3$.
Asymptotes: The number of asymptotes $= (4-0) = 4$. Therefore, $q = 0,\ 1,\ 2,\ 3$.
Angles of asymptotes are:

$$\theta_0 = \frac{1}{4} \times 180^\circ = 45^\circ \qquad \theta_1 = \frac{3}{4} \times 180^\circ = 135^\circ$$

$$\theta_2 = \frac{5}{4} \times 180^\circ = 225^\circ \qquad \theta_3 = \frac{7}{4} \times 180^\circ = 315^\circ$$

Centroid: $\chi_c = \dfrac{-3-1-1}{4} = -1.25$

Breakaway point: Since there are two adjacent poles at $s = 0$ and -3, and this segment of the real axis is a part of the root locus, a breakaway point is expected in this segment. To calculate the point, we proceed from the characteristic equation as follows:

$$s(s+3)(s^2+2s+2)+K=0$$

$$\Rightarrow \qquad K = -s^4 - 5s^3 - 8s^2 - 6s$$

Therefore, $$\frac{dK}{ds} = 4s^3 + 15s^2 + 16s + 6 = 0 \qquad \text{(i)}$$

Equation (i) is a cubic equation which has the roots: -2.28858, $-0.73071+\jmath 0.34856$, $-0.73071-\jmath 0.34856$[5]. We do not expect breakaway points in the complex plane. Therefore, the only breakaway point is at $s = -2.289$. Since two poles are involved, the angle at which their loci break away from the Re-axis is $\pm\, 180° \div 2 = \pm\, 90°$.

Angle of departure for the complex pole at $s = -1+\jmath 1$: For the graphical calculation, the angle measurement is shown in Figure 10.8.

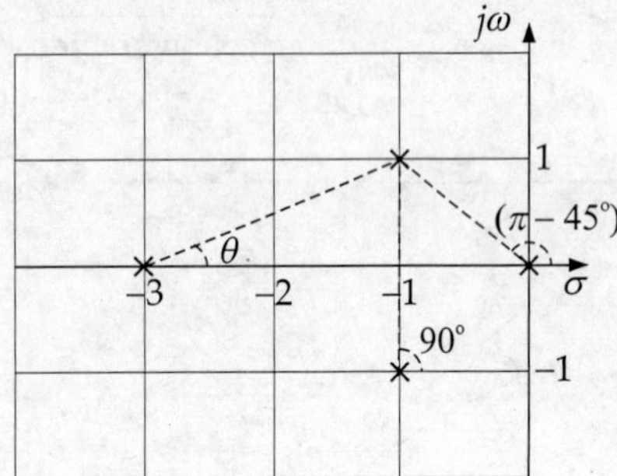

Figure 10.8: Angle measurement (Example 10.12).

$$\tan\theta = \frac{1}{2} \qquad \text{[see Figure 10.8]}$$

Therefore, $$\theta = \tan^{-1}\frac{1}{2} = 26.565°$$

Thus, $$\theta_d = 180° - [\angle(\pi - 45°) + \angle 26.56° + \angle 90°] = -71.56°$$

Intersection on the $\jmath\omega$-axis: The characteristic equation is

$$s^4 + 5s^3 + 8s^2 + 6s + K = 0$$

Therefore, the Routh array is

$$\begin{array}{c|ccc} s^4 & 1 & 8 & K \\ s^3 & 5 & 6 & \\ s^2 & 6.8 & K & \\ s^1 & 6-0.735K & 0 & \\ s^0 & K & & \end{array}$$

Thus, for marginal stability $$K = \frac{6}{0.735} = 8.16$$

[5]Using the calculator at https://www.calculatorsoup.com/calculators/algebra/cubicequation.php.

Substituting this value in the auxiliary equation, we get

$$6.8s^2 + 8.16 = 0$$

$$\Rightarrow \qquad s = \pm\, \jmath 1.1$$

Having obtained these data, we plot the root locus as shown in Figure 10.9.

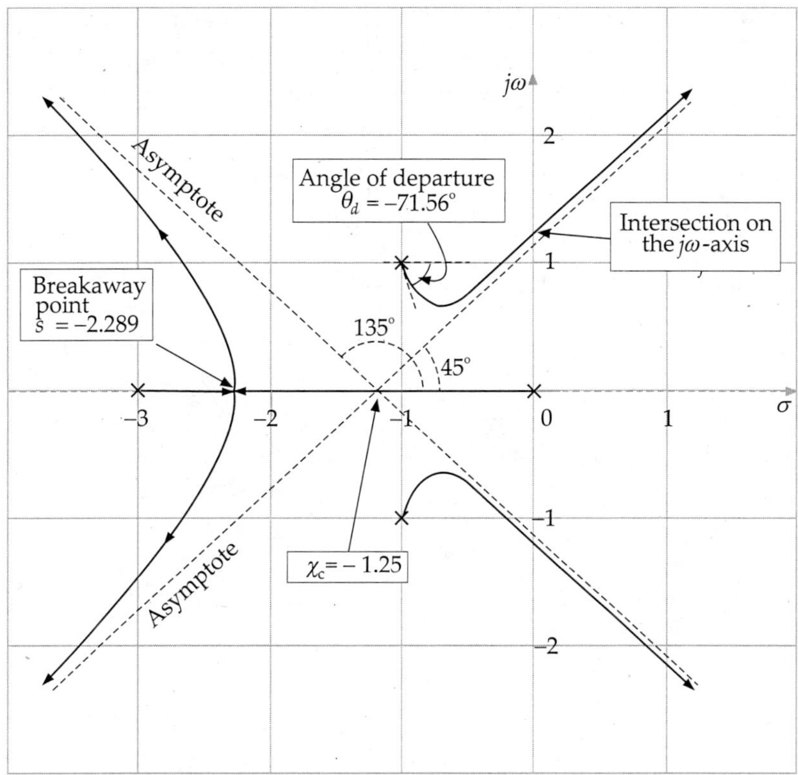

Figure 10.9: Root locus plot (Example 10.12).

EXAMPLE 10.13. The system is given by Figure 10.10. Draw the root locus and find the gain factor K at the breakaway point.

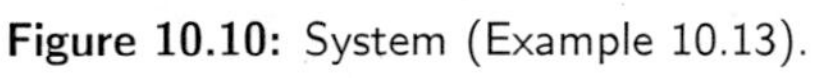

Figure 10.10: System (Example 10.13).

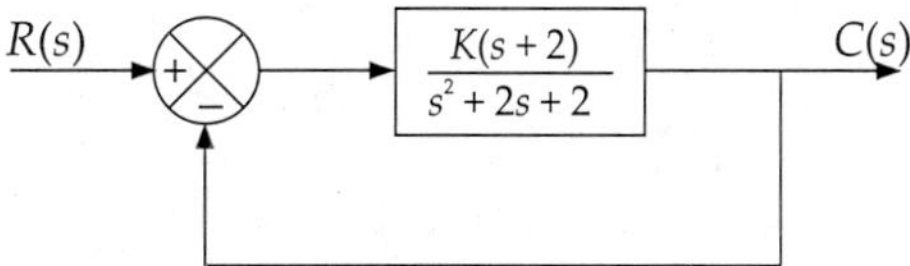

Solution The open-loop transfer function can be written as

$$G(s)H(s) = \frac{K(s+2)}{(s+1+\jmath 1)(s+1-\jmath 1)} \qquad \text{(i)}$$

Poles are at $s = -1 + \jmath 1,\ -1 - \jmath 1$, Therefore, $n = 2$. Zero is at $s = -2$. Thus, $m = 1$.

Two root loci will start from two poles. One will terminate at zero (i.e. $s = -2$) and the other at $s \to -\infty$. So, the real axis locus is present for $-\infty > s > -2$.

Asymptotes: The number of asymptotes is $n - m = 1$. Hence, its angle is 180° $(q = 0)$.
Breakaway point: The characteristic equation is

$$s^2 + 2s + 2 + K(s+2) = 0$$

$$\Rightarrow \quad K = -\frac{s^2 + 2s + 2}{s+2} \tag{ii}$$

$$\frac{dK}{ds} = \frac{(s^2 + 2s + 2) - (2s+2)(s+2)}{(s+2)^2} = 0$$

The roots of the equation are obtained as follows:

$$(s^2 + 2s + 2) - (2s+2)(s+2) = 0$$

$$\Rightarrow \quad s^2 + 4s + 2 = 0$$

Therefore, $s = -0.586$ and -3.414

The former point does not fall on the root locus on the real axis. So, the latter is the *break-in point.*

Note: This is a *break-in* rather than *breakaway* point.

Angle of departure at $s = -1 + \jmath 1$:

$$\arg\left|\frac{K(s+2)}{s+1+\jmath 1}\right|_{s=-1+\jmath 1} = \arg\left|\frac{K(1+\jmath 1)}{2\jmath 1}\right| = \frac{\angle 45°}{\angle 90°} = \angle -45°$$

Therefore, $\theta_d = 180° - 45° = 135°$

To figure out the locus in the complex plane, let us choose a complex point $s = \sigma + \jmath\omega$ on the root locus and invoke the angle condition as follows:

$$\angle\{G(\sigma + \jmath\omega)H(\sigma + \jmath\omega)\} = \arg\left|\frac{\sigma + \jmath\omega + 2}{(\sigma + \jmath\omega + 1 + \jmath 1)(\sigma + \jmath\omega + 1 - \jmath 1)}\right| = \pm\pi$$

$$\Rightarrow \quad \tan^{-1}\frac{\omega}{\sigma+2} - \tan^{-1}\frac{\omega+1}{\sigma+1} - \tan^{-1}\frac{\omega-1}{\sigma+1} = -\tan^{-1}(0)$$

$$\Rightarrow \quad \tan^{-1}(0) + \tan^{-1}\frac{\omega}{\sigma+2} = \tan^{-1}\frac{\omega+1}{\sigma+1} + \tan^{-1}\frac{\omega-1}{\sigma+1}$$

$$\Rightarrow \quad \tan^{-1}\frac{0 + \dfrac{\omega}{\sigma+2}}{1 - (0)\dfrac{\omega}{\sigma+2}} = \tan^{-1}\frac{\dfrac{2\omega}{\sigma+1}}{1 - \dfrac{\omega^2}{(\sigma+1)^2}}$$

Taking tangent of both sides, we get, by diagonal multiplication, the relation

$$\frac{2\cancel{\omega}}{\sigma+1} = \frac{\cancel{\omega}}{\sigma+2}\left[1 - \frac{\omega^2}{(\sigma+1)^2}\right]$$

$$\Rightarrow \quad 2(\sigma+1)(\sigma+2) = (\sigma+1)^2 - \omega^2 + 1$$

$$\Rightarrow \quad \sigma^2 + 4\sigma + 2 + \omega^2 = 0$$

$$\Rightarrow \quad (\sigma+2)^2 + (\omega+0)^2 = 2 \tag{iii}$$

Equation (iii) represents a circle of radius $\sqrt{2}$, having its centre at $(-2, 0)$. Having obtained these data, we draw the root locus as given in Figure 10.11.

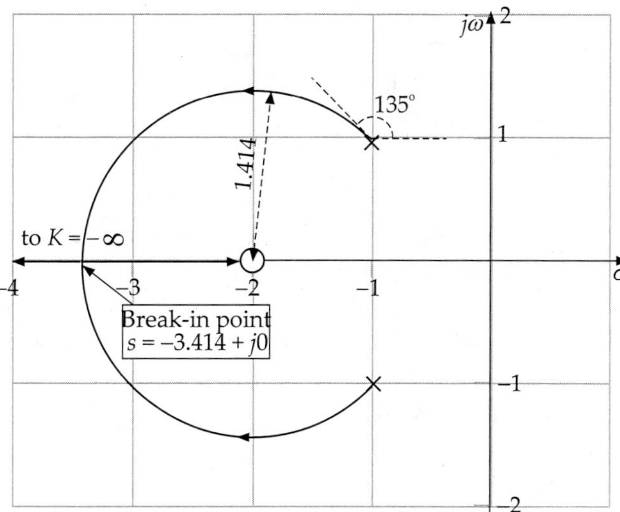

Figure 10.11: Root locus (Example 10.13).

Value of K at the break-in point: Analytical calculation From Eq. (ii)

$$K = -\frac{(-3.414)^2 + 2(-3.414) + 2}{-3.414 + 2} = 4.828$$

Graphical calculation

$$K = \frac{\prod \text{ vector lengths from poles to break-in point}}{\prod \text{ vector lengths from zeros to break-in point}}$$

$$= \frac{\left(\sqrt{2.414^2 + 1^2}\right)\left(\sqrt{2.414^2 + 1^2}\right)}{1.414}$$

$$= 4.828$$

EXAMPLE 10.14. Plot the root locus for

$$G(s)H(s) = \frac{K}{s(s^2 + 6s + 8)}$$

Then find the value of K:

(a) For marginal stability
(b) At breakaway point
(c) For damping ratio $\zeta = 0.5$

Solution The open-loop transfer function can be written as

$$G(s)H(s) = \frac{K}{s(s+2)(s+3)}$$

Therefore, the poles are at $s = 0,\ -2,\ -4$ and $n = 3$. No zero. Therefore, $m = 0$. The root locus exists on the $j\omega$-axis between $s = 0$ and -2, and $s = -4$ and $-\infty$. The root loci start from poles and end at ∞, since there is no zero.

Asymptotes: Number $= 3 - 0 = 3$. Therefore, $q = 0,\ 1,\ 2$. The angles of asymptotes are

$$\theta_0 = \frac{1}{3} \times 180^\circ = 60^\circ \qquad \theta_1 = \frac{3}{3} \times 180^\circ = 180^\circ \qquad \theta_3 = \frac{5}{3} \times 180^\circ = 300^\circ$$

and the centroid coordinate is $\chi_c = \dfrac{-2-4}{3} = -2$

Breakaway point: It should exist between $s = 0$ and -2. The characteristic equation is

$$s^3 + 6s^2 + 8s + K = 0$$

$$\Rightarrow \quad K = -s^3 - 6s^2 - 8s \qquad \text{(i)}$$

$$\frac{dK}{ds} = 3s^2 + 12s + 8 = 0$$

$$\Rightarrow \quad s = -0.845, \ -3.154$$

The point-3.154 is ruled out because it does not lie on the root locus. Therefore, the breakaway point is at $s = -0.845$.

Intersection at the$\jmath\omega$-axis: The Routh array is

$$\begin{array}{c|cc} s^3 & 1 & 8 \\ s^2 & 6 & K \\ s^1 & 8 - \dfrac{K}{6} & \\ s^0 & K & \end{array}$$

For marginal stability,

$$8 - \frac{K_{\text{mar}}}{6} = 0$$

$$\Rightarrow \quad K_{\text{mar}} = 48$$

The auxiliary equation is

$$6s^2 + 48 = 0$$

$$\Rightarrow \quad s = \pm \jmath 2.83$$

With the help of these available data, the root locus, as shown in Figure 10.12, is drawn. Thus,

(a) The value of K for marginal stability is 48.

(b) The value of K at the breakaway point can be calculated as follows:
Analytically: From Eq. (i)

$$K|_{s=-0.845} = -(-0.845)^3 - 6(-0.845)^2 - 8(-0.845) = 3.079$$

(c) For damping ratio $\zeta = 0.5$, the phase angle

$$\phi = \tan^{-1}\frac{\sqrt{1-\zeta^2}}{\zeta} = \tan^{-1}\frac{\sqrt{1-(1/2)^2}}{1/2} = \tan^{-1}\sqrt{3} = 60^\circ$$

We draw a line in the second quadrant at an angle of 60° with the negative part of the real axis from the origin. It intercepts the locus at $-0.6 + \jmath 1.3$. The K value at this point is

$$\begin{aligned} K &= \frac{\prod \text{vector lengths from poles to } (-0.6 + \jmath 1.3)}{\prod \text{vector lengths from zeros to } (-0.6 + \jmath 1.3)} \\ &= \left[\sqrt{(0+0.6)^2 + 1.3^2}\right]\left[\sqrt{(-2+0.6)^2 + 1.3^2}\right]\left[\sqrt{(-4+0.6)^2 + 1.3^2}\right] \\ &= 9.957 \end{aligned}$$

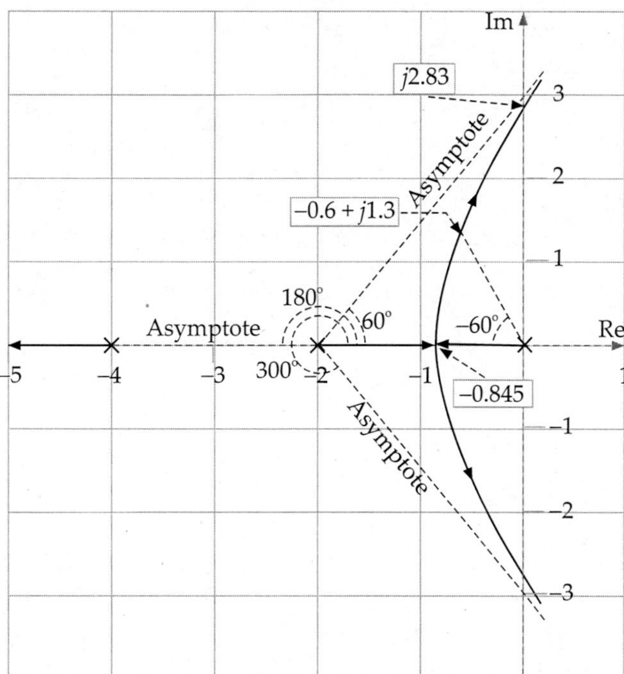

Figure 10.12: Root locus (Example 10.14).

EXAMPLE 10.15. A unity feedback system has an open loop transfer function of

$$G(s) = \frac{K}{s(s+4)(s^2+4s+20)}$$

Sketch the root locus of the system by determining the following:

(a) Number and angles of asymptotes, centroid
(b) Angle of departure of root locus from poles
(c) Breakaway point
(d) The value of K and the frequency at which the root locus crosses the $j\omega$-axis
(e) Hence find the value of K so that the system has a damping factor of 0.707

Solution The transfer function can be written as

$$G(s)H(s) = \frac{K}{s(s+4)(s+2+j4)(s+2-j4)}$$

So, the number of poles, $n = 4$. No zero, $m = 0$.

(a) *Number of asymptotes*: $n - m = 4$. *Angles of asymptotes*:

$$\theta_0 = \frac{1}{4} \times 180° = 45° \qquad \theta_1 = \frac{3}{4} \times 180° = 135°$$

$$\theta_2 = \frac{5}{4} \times 180° = 225° \qquad \theta_3 = \frac{7}{4} \times 180° = 315°$$

$$\text{Centroid: } \chi_c = \frac{-4-2-2}{4} = -2$$

(b) *Angle of departure*: Analytically,

$$\arg[G(s)H(s)]_{s=-2+\jmath 4} = \arg\left|\frac{K}{s(s+4)(s+2+\jmath 4)}\right|_{s=-2+\jmath 4} = \arg\left|\frac{K}{(-2+\jmath 4)(2+\jmath 4)(2\jmath 4)}\right|$$

$$= \frac{1}{(\pi - \tan^{-1}2) + \tan^{-1}2 + 90°} = \angle -270°$$

Therefore, $\theta_d = 180° - 270° = -90°$

(c) *Breakaway points*: The characteristic equation is

$$s(s+4)(s^2+4s+20) + K = 0$$

$$\Rightarrow \quad s^4 + 8s^3 + 36s^2 + 80s + K = 0 \qquad \text{(i)}$$

Therefore,

$$K = -s^4 - 8s^3 - 36s^2 - 80s$$

$$\Rightarrow \quad \frac{dK}{ds} = -4s^3 - 24s^2 - 72s - 80 = 0$$

$$\Rightarrow \quad s^3 + 6s^2 + 18s + 20 = 0$$

$$\Rightarrow \quad (s+2)(s^2+4s+10) = 0 \qquad \text{(ii)}$$

From Eq, (ii), we get the following coordinates of the breakaway points:

$$s = -2, \ -2 \pm \jmath\, 2.45$$

(d) *Intersection on the $\jmath\omega$ axis:* The Routh array is constructed from Eq. (i) as follows:

s^4	1	36	K
s^3	8	80	0
s^2	26	K	
s^1	$80 - \frac{8}{26}K$		
s^0	K		

For marginal stability,

$$80 - \frac{8}{26}K = 0$$

$$\Rightarrow \quad K = 80 \times \frac{26}{8} = 260$$

Auxiliary equation

$$26s^2 + 260 = 0$$

$$\Rightarrow \quad s = \pm\jmath\sqrt{10} = \pm\jmath\, 3.16$$

With these data, we construct the root locus diagram as shown in Figure 10.13.

Note: There are breakaway points with complex values of s where the two loci, breaking away from the Re-axis again break away from two loci coming from $s = -2 \pm \jmath 4$. Their angles of departure are $-90°$ and $+90°$, respectively.

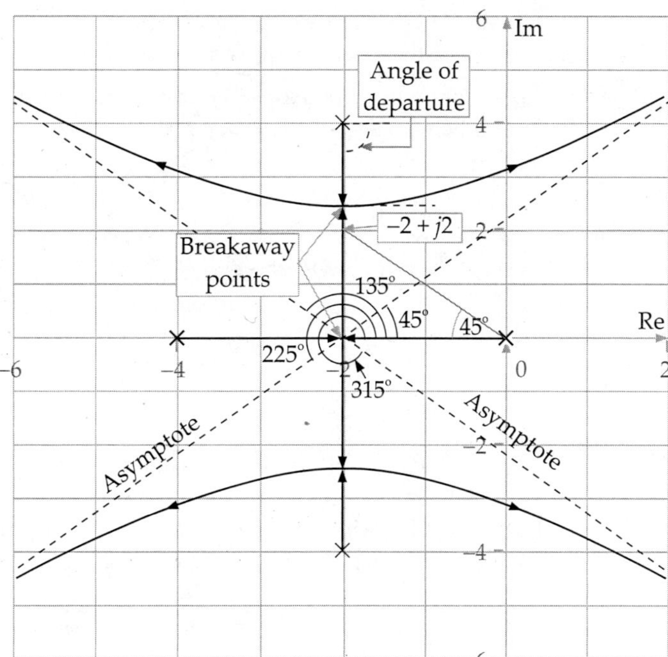

Figure 10.13: Root locus (Example 10.15).

To find the value of K at $\zeta = 0.707 = 1/\sqrt{2}$, we note that the corresponding angle ϕ on the negative side of the Re-axis is

$$\phi = \tan^{-1}\frac{\sqrt{1-\zeta^2}}{\zeta} = \tan^{-1}\frac{1/\sqrt{2}}{1/\sqrt{2}} = 45^\circ$$

This line intersects the root locus at $s = -2 + \jmath 2$. The K value there is calculated from Eq. (i) as follows:

$$\begin{aligned} K &= -(-2+\jmath 2)(-2+\jmath 2+4)[(-2+\jmath 2)^2 + 4(-2+\jmath 2) + 20] \\ &= -(-4-4)(4-4-4\jmath 2 - 8 + 4\jmath 2 + 20) = 96 \end{aligned}$$

EXAMPLE 10.16. Sketch the root locus of the unity feedback control system whose open loop transfer function is given by

$$G(s)H(s) = \frac{K}{s(s+2)(s^2+4s+13)}$$

Find

(a) the number, angle and centroid of asymptotes
(b) the angle of departure
(c) the breakaway point
(d) the condition for marginal stability
(e) the value of K so that the system has a damping factor of 0.5

Solution The open-loop transfer function can be written as

$$G(s)H(s) = \frac{K}{s(s+2)(s+2+\jmath 3)(s+2-\jmath 3)}$$

The pole-zero map shows two poles on the Re-axis at $s = 0$ and $s = -2$. So part of the Re-axis between them is part of the root locus.

(a) *Asymptotes*: Number of poles $n = 4$. No zero, $m = 0$. Therefore, number of asymptotes: $n - m = 4$. Their angles are:

$$\theta_0 = \frac{1}{4} \times 180^\circ = 45^\circ \qquad \theta_1 = \frac{3}{4} \times 180^\circ = 135^\circ$$

$$\theta_2 = \frac{5}{4} \times 180^\circ = 225^\circ \qquad \theta_3 = \frac{7}{4} \times 180^\circ = 315^\circ$$

Centroid: $\chi_c = \dfrac{-2-2-2}{4} = -1.5$

(b) *Angle of departure*: Graphically,

$$\theta_d = 180^\circ - \left[90^\circ + 90^\circ + \left(\pi - \tan^{-1}\frac{3}{2}\right)\right] = -123.7^\circ$$

(c) *Breakaway point*: Starting from the characteristic equation

$$s(s+2)(s^2 + 4s + 13) + K = 0 \qquad \text{(i)}$$

$$\Rightarrow \quad K = -s^4 - 6s^3 - 21s^2 - 26s$$

$$\frac{dK}{ds} = 4s^3 + 18s^2 + 42s + 26 = 0$$

$$\Rightarrow \quad s = -0.89, \ -1.8 \pm \jmath 2^6$$

Of these three breakaway points, only $s = -0.89$ is valid because it falls on the Re-axis. There is no question of having complex breakaway points here.

(d) *Intersection on $\jmath\omega$-axis*: From Eq. (i) we get

$$s^4 + 6s^3 + 21s^2 + 26s + K = 0$$

So, the Routh array is constructed as follows:

$$\begin{array}{c|ccc} s^4 & 1 & 21 & K \\ s^3 & 6 & 26 & \\ s^2 & 16.67 & K & \\ s^1 & 26 - \dfrac{6}{16.67}K & & \\ s^0 & K & & \end{array}$$

For marginal stability,

$$26 - \frac{6}{16.67}K = 0 \qquad \therefore \quad K = 26 \times \frac{16.67}{6} = 72.24$$

From the auxiliary equation

$$16.67s^2 + 26 \times \frac{16.67}{6} = 0 \qquad \Rightarrow \quad s = \sqrt{-\frac{26}{6}} = \pm\jmath 2.08$$

With these available data, the root locus, as shown in Figure 10.14, is drawn.

[6] Obtained using the calculator at https://www.calculatorsoup.com/calculators/algebra/cubicequation.php.

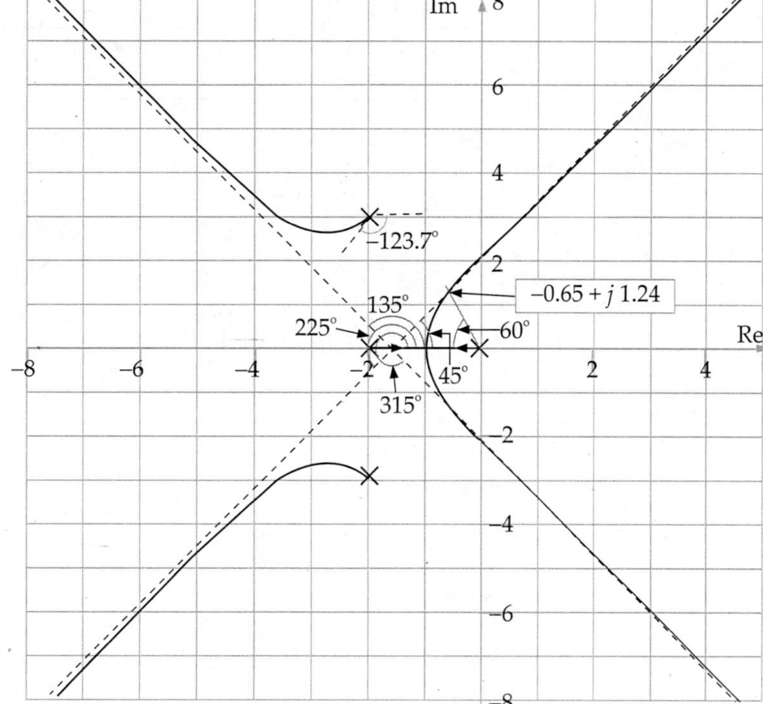

Figure 10.14: Root locus (Example 10.16).

To determine the value of K where $\zeta = 0.5$, we draw a line at an angle $\tan^{-1}\frac{\sqrt{1-0.5^2}}{0.5} = \tan^{-1}\sqrt{3} = 60°$ on the negative Re-axis. The line cuts the root locus at $s = -0.65 + \jmath 1.24$. The value of K at this point is calculated from Eq. (i) as follows:

$$\begin{aligned} K &= |-0.65+\jmath 1.24||-0.65+\jmath 1.24+2||-0.65+\jmath 1.24+2+\jmath 3||-0.65+\jmath 1.24+2-\jmath 3| \\ &= \sqrt{0.65^2+1.24^2}\times\sqrt{1.35^2+1.24^2}\times\sqrt{1.35^2+4.24^2}\times\sqrt{1.35^2+1.76^2} \\ &= (1.4)(1.833)(4.4497)(2.2181) = 25.3 \end{aligned}$$

EXAMPLE 10.17. Draw the root locus for the unity feedback system whose open-loop transfer function is

$$G(s) = \frac{K(s+1)}{(s-1)(s+2)(s+4)}$$

Find the range of K for which the system is stable. Show all relevant steps of calculation.

Solution Parts of the Re-axis which belong to the root locus are between $s = -4$ and -2, and $s = -1$ and 1. The number of poles: $n = 3$; number of zeros: $m = 1$. Therefore, number of asymptotes: $n - m = 2$.

Asymptotes:

$$\theta_0 = \frac{180°}{3-1} = 90° \qquad \theta_1 = \frac{180° \times 3}{3-1} = 270°$$

Centroid: $\chi_c = \dfrac{1-2-4-1}{3-1} = -2$

Breakaway point: We start from the characteristic equation

$$(s-1)(s+2)(s+4) + K(s+1) = 0 \tag{i}$$

$$K = -\frac{(s-1)(s+2)(s+4)}{s+1} = -\frac{s^3+5s^2+2s-8}{s+1}$$

$$\frac{dK}{ds} = -\left[\frac{(s+1)(3s^2+10s+2)-(s^3+5s^2+2s-8)}{(s+1)^2}\right] = 0$$

$$\Rightarrow \qquad 2s^3+8s^2++10s+10=0$$

$$\Rightarrow \qquad s^3+4s^2+5s+5=0$$

$$\Rightarrow \qquad s=-2.864, \ -0.568 \pm \jmath 1.193$$

Of these, only $s = -2.864$ is acceptable because from the pole-zero plot, we do not expect any complex breakaway point.

Intersection on the $\jmath\omega$-axis: From the characteristic equation given by Eq. (i), we get

$$s^3+5s^2+(2+K)s+(-8+K)=0 \tag{ii}$$

We construct the Routh array from Eq. (ii) as follows:

$$\begin{array}{c|cc} s^3 & 1 & 2+k \\ s^2 & 5 & -8+K \\ s^1 & \dfrac{2+6K}{5} & \\ s^0 & -8+K & \end{array}$$

For marginal stability,

$$-8+K=0 \qquad \Rightarrow \qquad K=8$$

So, the stability condition is $K > 8$. The auxiliary equation is

$$5s^2-8+8=0 \qquad \Rightarrow \qquad s=0$$

This means that the root loci do not intersect the $\jmath\omega$-axis.

With these available data, the root locus plot, as shown in Figure 10.15, is drawn.

EXAMPLE 10.18. The open-loop transfer function of a feedback control system is given by

$$G(s)H(s) = \frac{K(s+6)}{s(s+4)}$$

(a) Sketch the root locus with K as a variable parameter and show that the loci of complex roots are part of a circle.
(b) Determine the breakaway/break-in points, if any.
(c) Determine the range of K for which the system is underdamped.
(d) Determine the value of K for critical damping.
(e) Determine the minimum value of the damping ratio.

Solution (a) The number of poles $n = 2$, number of zero $m = 1$. The pole-zero map (see Figure 10.16) shows that the part of the Re-axis between $s = 0$ and $s = -4$ belongs to the root locus. Since $n - m = 1$, there exists only one asymptote at 180°.

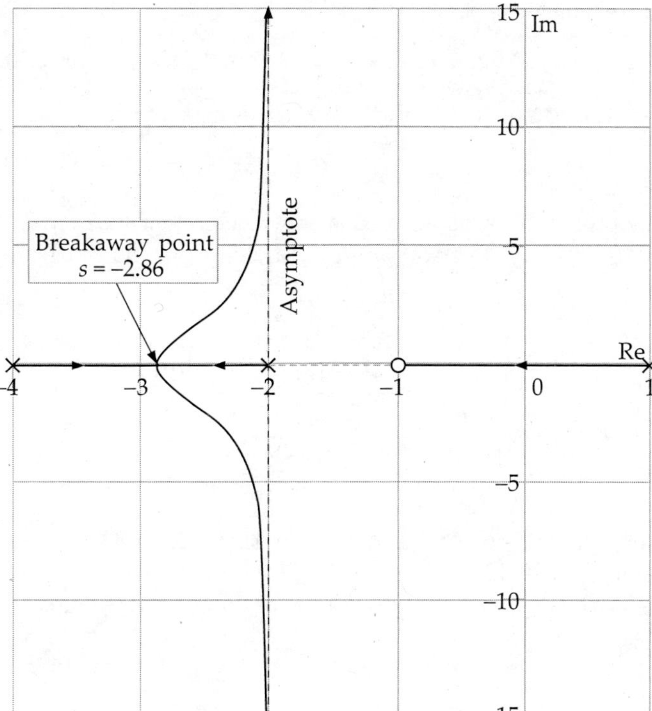

Figure 10.15: Root locus (Example 10.17).

To show that the loci of complex roots are part of a circle, let $s = \sigma + \jmath\omega$ be a complex point on the locus. The angle criterion of Evans must be satisfied by this point since it is on the root locus. This means that

$$\angle G(\sigma + \jmath\omega) = \arg\left|\frac{K(\sigma + \jmath\omega + 6)}{(\sigma + \jmath\omega)(\sigma + \jmath\omega + 4)}\right| = \pm\pi$$

$$\tan^{-1}\frac{\omega}{\sigma + 6} - \tan^{-1}\frac{\omega}{\sigma} - \tan^{-1}\frac{\omega}{\sigma + 4} = -\pi$$

or, $$\pi + \tan^{-1}\frac{\omega}{\sigma + 6} = \tan^{-1}\frac{\omega}{\sigma} + \tan^{-1}\frac{\omega}{\sigma + 4}$$

or, $$\tan^{-1}0 + \tan^{-1}\frac{\omega}{\omega + 6} = \tan^{-1}\frac{\omega}{\sigma} + \tan^{-1}\frac{\omega}{\sigma + 4} \qquad (\because \pi = \tan^{-1}0)$$

or, $$\tan^{-1}\frac{0 + \dfrac{\omega}{\sigma + 6}}{1 - (0)\dfrac{\omega}{\sigma + 6}} = \tan^{-1}\frac{\dfrac{\omega}{\sigma} + \dfrac{\omega}{\sigma + 4}}{1 - \dfrac{\omega^2}{\sigma(\sigma + 4)}} \qquad \text{(i)}$$

Taking tangent of both sides and multiplying the sides diagonally, we get from Eq. (i) the relation

$$\frac{\omega}{\sigma} + \frac{\omega}{\sigma + 4} = \frac{\omega}{\sigma + 6}\left[1 - \frac{\omega^2}{\sigma(\sigma + 4)}\right]$$

$\Rightarrow$ $$\frac{2\sigma + 4}{\sigma(\sigma + 4)} = \frac{\sigma^2 + 4\sigma - \omega^2}{\sigma(\sigma + 4)(\sigma + 6)}$$

$\Rightarrow$ $$(2\sigma + 4)(\sigma + 6) = \sigma^2 + 4\sigma - \omega^2 \qquad \text{(ii)}$$

On rearrangement, Eq. (ii) yields

$$\sigma^2 + 12\sigma + 24 + \omega^2 = 0 \tag{iii}$$

Converting Eq. (iii) to the following form

$$(\sigma + 6)^2 + (\omega + 0)^2 = 36 - 24 = 12$$

we observe that this is an equation of a circle with a radius of $\sqrt{12}$ having its centre at $(-6, 0)$.
(b) *Breakaway/break-in points:* Starting from the characteristic equation, we get

$$1 + \frac{K(s+6)}{s(s+4)} = 0 \qquad \Rightarrow \quad K = -\frac{s(s+4)}{s+6} \tag{iv}$$

$$\frac{dK}{ds} = \frac{(s^2+4s) - (s+6)(2s+4)}{(s+6)^2} = 0 \qquad \Rightarrow \quad s^2 + 12s + 24 = 0$$

Therefore, $$s = \frac{-12 \pm \sqrt{48}}{2} = -2.54, \; -9.46$$

From the pole-zero plot, it is clear that the breakaway point is $s = -2.54$ and the break in point, $s = -9.46$.

With these data, we draw the root locus as presented in Figure 10.16.

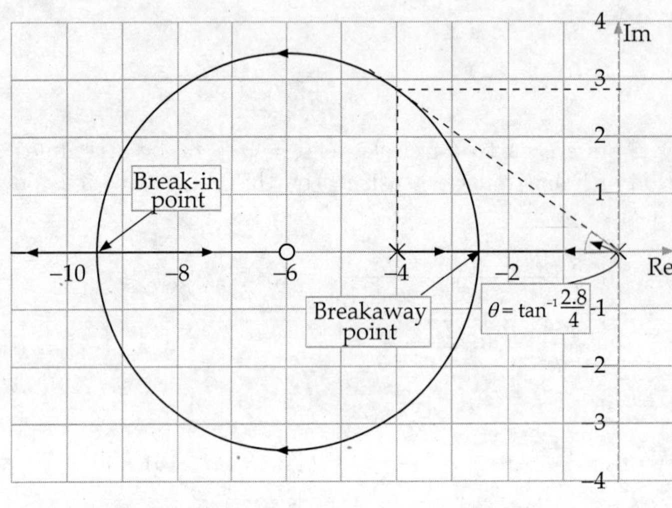

Figure 10.16: Root locus (Example 10.18).

(c) *Value of K for which the system is underdamped*: We know from Eq. (6.16) that if the system is underdapmed, i.e. $\zeta < 1$, s assumes complex values. We observe from the root locus that s assumes complex values between the breakaway and break-in points. K values at these points can be calculated as follows.

Analytical calculation: From Eq. (iv), we get

$$K_{\text{breakaway}} = -\left.\frac{s(s+4)}{s+6}\right|_{s=-2.54} = -\frac{-2.54(-2.54+4)}{-2.54+6} = 1.072$$

$$K_{\text{break-in}} = -\left.\frac{s(s+4)}{s+6}\right|_{s=-9.46} = -\frac{-9.46(-9.46+4)}{-9.46+6} = 14.928$$

Graphical calculation:

$$K_{\text{breakaway}} = \frac{\prod \text{distance of poles from the breakaway point}}{\prod \text{distance of zeros from the breakaway point}}$$
$$= \frac{(2.54-0)(4-2.54)}{6-2.54} = 1.072$$
$$K_{\text{break-in}} = \frac{\prod \text{distance of poles from the break-in point}}{\prod \text{distance of zeros from the break-in point}}$$
$$= \frac{(9.46-0)(9.46-4)}{9.46-6} = 14.928$$

Therefore, the system remains underdamped in the interval when $14.928 > K > 1.072$.

(d) *Value of K for the critical damping*: For critical damping, $\theta = \tan^{-1}\frac{\sqrt{1-1^2}}{1} = 0°$. The root locus except its detour from the Re-axis between the breakaway and break-in points satisfies this criterion. So for critical damping $K < 1.072$ and $K > 14.928$.

(e) *Minimum value of the damping ratio*: Since $\frac{\sqrt{1-\zeta^2}}{\zeta} = \tan\theta$, the minimum value of the damping ratio occurs when θ is the largest. From Figure 10.16 we observe that

$$\theta_{\max} = \tan^{-1}\frac{2.8}{4} = 34.992°$$

Hence,
$$\frac{\sqrt{1-\zeta_{\min}^2}}{\zeta_{\min}} = \tan 34.992° = 0.7$$

$\Rightarrow$
$$1.49\,\zeta_{\min}^2 = 1$$

$\Rightarrow$
$$\zeta_{\min} = 0.82$$

EXAMPLE 10.19. The open-loop transfer function of a unity feedback system is given by

$$G(s) = \frac{K(s+0.5)}{s^2(s+4.5)}$$

Sketch the root locus.

Solution Poles are at $s = 0$ (two) and $s = -4.5$. Therefore $n = 3$. Zero at $s = -0.5$. Therefore $m = 1$. The pole-zero map (see Figure 10.17) shows that the part of the Re-axis between $s = -4.5$ and $s = -0.5$ belongs to the root locus.

Asymptotes: Number $= n - m = 2$.

$$\theta_0 = \frac{180°}{3-1} = 90° \qquad \theta_1 = \frac{180° \times 3}{3-1} = 270°$$

Centroid: $\chi_c = \dfrac{0.5-4.5}{2} = -2$

Breakaway/break-in points: The characteristic equation is

$$1 + \frac{K(s+0.5)}{s^2(s+4.5)} = 0 \tag{i}$$

$\Rightarrow$
$$s^3 + 45s^2 + Ks + 0.5K = 0 \tag{ii}$$

From Eq. (i), we get

$$K = -\frac{s^2(s+4.5)}{s+0.5}$$

$$\frac{dK}{ds} = -\left[\frac{(s+0.5)(3s^2+9s)-(s^3+4.5s^2)}{(s+0.5)^2}\right] = -\frac{2s\left(s^2+3s+\frac{9}{4}\right)}{(s+0.5)^2} = -\frac{2s\left(s+\frac{3}{2}\right)^2}{(s+0.5)^2} = 0$$

$$\therefore \qquad s = 0,\ -\frac{3}{2},\ -\frac{3}{2}$$

Here,

(i) $s = 0$, where two poles exist, is a breakaway point,

(ii) $s = -\frac{3}{2}$ has to be a break-in as well as breakaway point in order that the loci follow the asymptotes.

(iii) The two poles at the origin break away at an angle of $\frac{180°}{2} = 90°$.

The plotted root locus is shown in Figure 10.17.

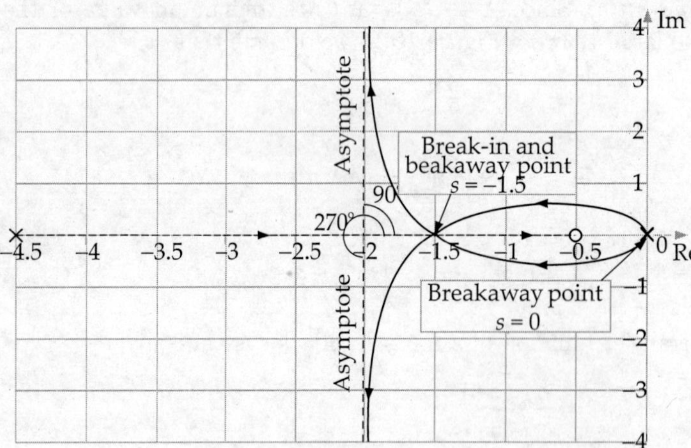

Figure 10.17: Root locus (Example 10.19).

The root locus does not cross the Im-axis though it touches it at $s = 0$ where, we find from Eq. (ii), $K = 0$. So, the system is marginally stable at $K = 0$, and stable for all other values of K.

Having considered construction of complete root locus plots for a few typical problems, we will now work out a few root locus based problems.

EXAMPLE 10.20. The root locus plot of a unity feedback system is shown in Figure 10.18a. The system transfer function is of the form

(A) $\dfrac{K}{s(s+4)(s+8)}$ (B) $\dfrac{K(s+8)}{s(s+4)}$ (C) $\dfrac{K(s+4)}{s(s+8)}$ (D) $\dfrac{Ks}{(s+4)(s+8)}$

Solution From the pole-zero map we observe that the transfer function has poles at $s =$ 0 and -4 and a zero at $s = -8$. Of the given choices, (B) conforms to this observation.
Ans. (B)

EXAMPLE 10.21. The characteristic equation of a closed-loop system is $s(s+1)(s+3)+k(s+2) = 0,\ k > 0$. Which of the following statements is true?

(A) Its roots are always real.

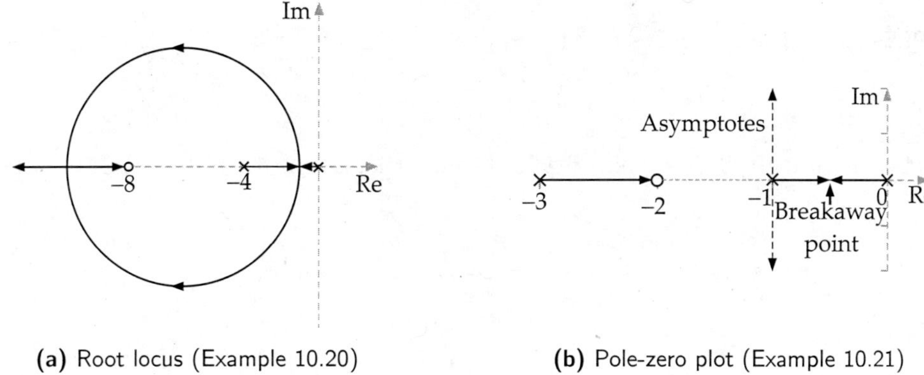

(a) Root locus (Example 10.20)   **(b)** Pole-zero plot (Example 10.21)

Figure 10.18: Root locus and pole-zero plot.

(B) It cannot have a breakaway point in the range $-1 < \text{Re}[s] < 0$.
(C) Two of its roots tend to infinity along the asymptotes $\text{Re}[s] = -1$.
(D) It may have complex roots in the right half plane.

Solution The open-loop transfer function corresponding to the given characteristic equation is

$$G(s)H(s) = \frac{k(s+2)}{s(s+1)(s+3)}$$

It consists of 3 poles at $s = 0,\ -1,\ -3$ and a zero at $s = -2$. Therefore, there are 2 asymptotes at angles

$$\theta_0 = \frac{2 \times 0 + 1}{2} \cdot 180° = 90° \qquad \theta_1 = \frac{2 \times 1 + 1}{2} \cdot 180° = 270°$$

The centroid is at $s = \dfrac{-1-3+2}{2} = -1$. The pole-zero plot along with asymptotes is shown in Figure 10.18b. It is obvious from the plot that the 2 adjacent poles at $s = 0$ and -1 break away and tend to go to ∞ along asymptotes at $s = -1$. So, the statement (C) is true.
Ans. (C)

EXAMPLE 10.22. The breakaway point of the root locus from the real axis for a closed-loop system with a loop gain

$$G(s)H(s) = \frac{K(s+10)}{(s+2)(s+5)}$$

lies

(A) between -2 and origin   (B) between -2 and -5
(C) between -10 and $-\infty$ (D) at $-\infty$

Solution The pole-zero map, shown in Figure 10.19, indicates that the part of the real axis between $s = -2$ and $s = -5$ belongs to the root locus.

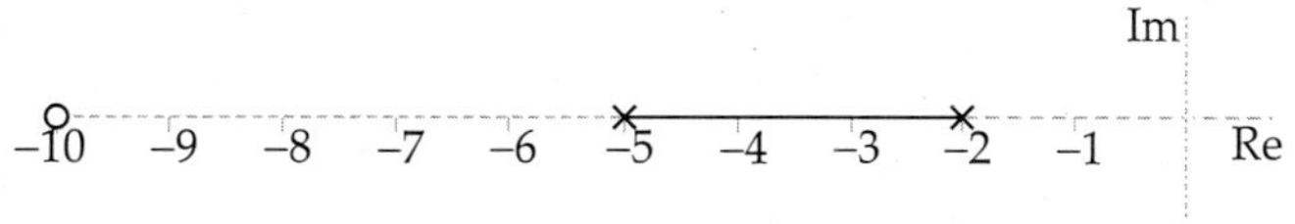

Figure 10.19: Pole-zero map (Example 10.22).

Im
−10 −9 −8 −7 −6 −5 −4 −3 −2 −1 Re

Obviously, the breakaway point must exist between $s = -2$ and $s = -5$.
Ans. (B)

EXAMPLE 10.23. The root locus plot for a system with transfer function

$$\frac{2}{s(s+1)(s+2)}$$

is shown in Figure 10.20a. A unity feedback proportional control system is built using this system.

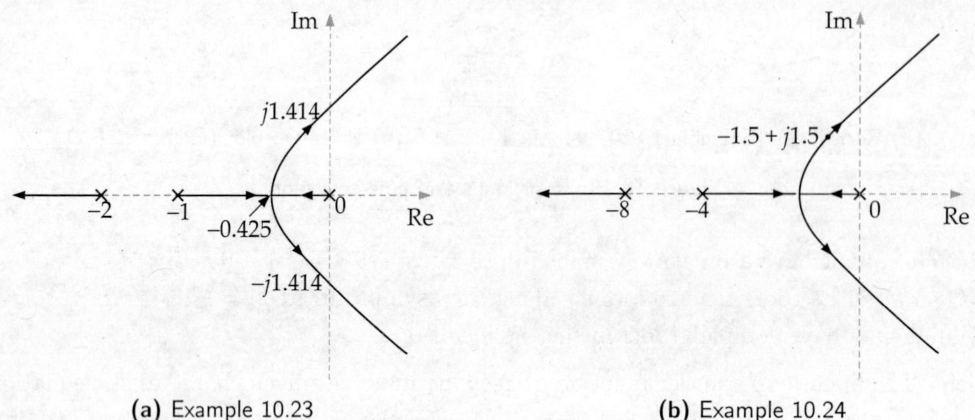

(a) Example 10.23 (b) Example 10.24

Figure 10.20: Root locus plots.

(a) The maximum possible controller gain, for which the unity feedback system is stable, is approximately

(A) 6.0 (B) 3.0 (C) 0.4 (D) 0.2

(b) The maximum possible controller gain, for which the unity feedback system exhibits a non-oscillatory response to a unit step is

(A) 6.0 (B) 3.0 (C) 0.4 (D) 0.2

Solution (a) The closed-loop transfer function is given by

$$\frac{C(s)}{R(s)} = \frac{K}{1+\dfrac{2K}{s(s+1)(s+2)}} = \frac{K}{s(s+1)(s+2)+2K}$$

From the characteristic equation

$$s(s+1)(s+2)+2K = 0$$

we get

$$|K| = \frac{s(s+1)(s+2)}{2}$$

Therefore,

$$|K|_{s=j\sqrt{2}} = \frac{\sqrt{2}\sqrt{1+2}\sqrt{2^2+2}}{2} = 3$$

Ans. (B)

(b) The root locus breaks away from the Re-axis at the point $s = -0.425$ and enters the complex space when its response becomes oscillatory. So the maximum gain when the system exhibits a non-oscillatory response occurs at $s = -0.425$. Therefore,

$$|K|_{s=0.425} = \left|\frac{-0.425(-0.425+1)(-0.425+2)}{2}\right| \approx 0.2$$

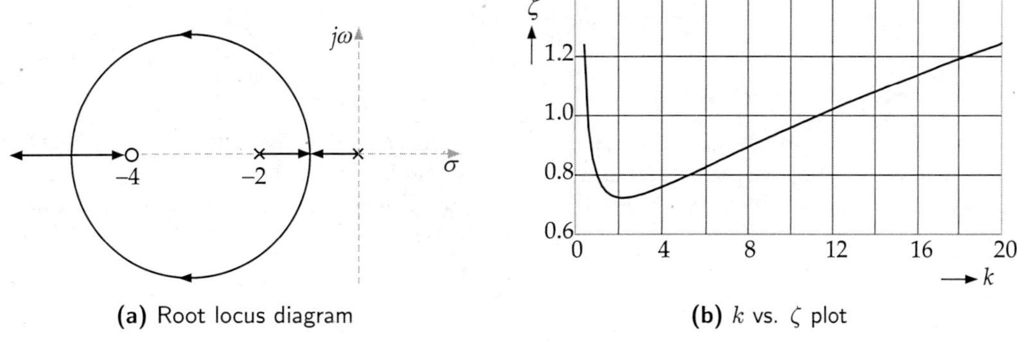

(a) Root locus diagram

(b) k vs. ζ plot

Figure 10.21: Example 10.25.

Ans. (D)

EXAMPLE 10.24. The root locus of a plant is given in Figure 10.20b. The root locus crosses imaginary axis at $\omega = 4\sqrt{2}$ rad/s with gain $K = 384$. It is observed that the point $s = -1.5 + \jmath 1.5$ lies on the root locus. The gain K at $s = -1.5 + \jmath 1.5$ is computed as

(A) 11.3 (B) 21.2 (C) 41.3 (D) 61.2

Solution From the pole-zero map, we can write the open-loop transfer function as

$$G(s)H(s) = \frac{K'}{s(s+4)(s+8)}$$

The gain at $s = \jmath\omega = \jmath 4\sqrt{2}$ is given to be 384. We know, this gain is calculated as

$$K = 384 = K' \frac{\prod \text{distance of all poles from } s = 0 + \jmath 4\sqrt{2}}{\prod \text{distance of all zeros from } s = 0 + \jmath 4\sqrt{2}}$$

$$= K'(4\sqrt{2})\left(\sqrt{4^2 + (4\sqrt{2})^2}\right)\left(\sqrt{8^2 + (4\sqrt{2})^2}\right)$$

$$= K'(384)$$

Therefore, $K' = 1$

Thus, the gain at $s = -1.5 + \jmath 1.5$ is

$$K\Big|_{s=-1.5+\jmath 1.5} = \left(\sqrt{1.5^2 + 1.5^2}\right)\left(\sqrt{(4-1.5)^2 + 1.5^2}\right)\left(\sqrt{(8-1.5)^2 + 1.5^2}\right)$$

$$= 41.2568$$

Ans. (C)

EXAMPLE 10.25. Consider the root locus diagram of a system [Figure 10.21a] and the following statements:

1. The open-loop system is a second order system
2. The system is overdamped for $k > 1$
3. The system is absolutely stable for all values of k

Of the statements:

(A) 1, 2 and 3 are correct (B) 1 and 3 are correct

(C) 1 and 2 are correct (D) 2 and 3 are correct

Solution From the pole-zero plot, the open-loop transfer function is written as

$$G(s)H(s) = \frac{k(s+4)}{s(s+2)}$$

It is obvious that the open-loop system is of second order. Secondly, since the root locus lies on the LHP of the Im-axis, the system is absolutely stable. Now, the question is whether the system is overdamped for $k > 1$. To determine that, we observe that the characteristic equation is given by

$$1 + G(s)H(s) = 0$$

or,

$$s(s+2) + k(s+4) = 0$$

or,

$$s^2 + (2+k)s + 4k = 0 \qquad \text{(i)}$$

Comparing Eq. (i) with the characteristic equation of a standard 2nd order system, viz. $s^2 + 2\zeta\omega_n s + \omega_n^2 = 0$, we get

$$\omega_n^2 = 4k \qquad \Rightarrow \qquad \omega_n = 2\sqrt{k}$$

$$2\zeta\omega_n = 2 + k \qquad \Rightarrow \qquad \zeta = \frac{2+k}{2\omega_n} = \frac{2+k}{4\sqrt{k}} \qquad \text{(ii)}$$

The k vs. ζ plot, as obtained from Eq. (ii), is shown in Figure 10.21b.

It is clear from the plot that the system is critically damped for $k \approx 1$ and 12. In fact, these are k-values at the breakaway ($s = -1.172$) and break-in ($s = -6.828$) points. For $k < 1$ and $k > 12$ the system is overdamped. So, assuming $k > 1$ but $k < 12$, the system is not overdamped. This means that the second statement is wrong.
Ans. (B)

EXAMPLE 10.26. Figure 10.22 shows the root locus plot (location of poles not given) of a third order system whose open-loop transfer function is

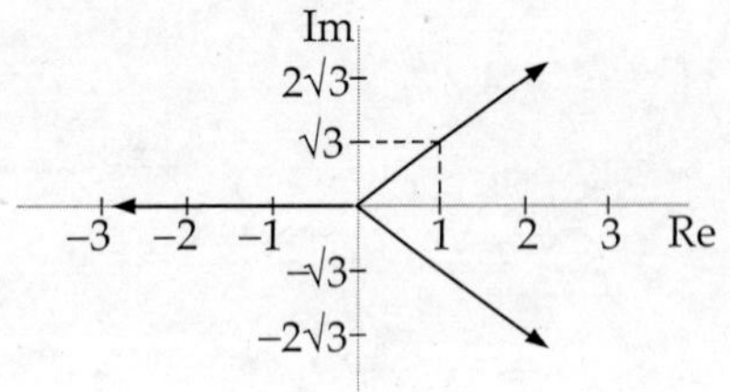

Figure 10.22: Root locus plot (Example 10.26).

(A) $\dfrac{K}{s^3}$ (B) $\dfrac{K}{s^2(s+1)}$ (C) $\dfrac{K}{s(s^2+1)}$ (D) $\dfrac{K}{s(s^2-1)}$

Solution From the figure it is clear that the origin i.e. $s = 0$ is the breakaway point of the three loci. Now if we try the option (A), we find

$$G(s)H(s) = \frac{K}{s^3}$$

$\therefore$ the characteristic equation is

$$1 + G(s)H(s) = 0 = 1 + \frac{K}{s^3}$$

or,

$$s^3 + K = 0$$

So,

$$\frac{dK}{ds} = 3s^2 = 0$$

This gives

$$s = 0, 0$$

No other option will yield the breakaway point at the origin.
Ans. (A)

Systems with Transportation Lag

We considered stability of systems with transportation lag through Routh-Hurwitz formalism in Section 9.3 at page 208. In the following example, we study the same through the root locus technique.

EXAMPLE 10.27. The open-loop transfer function of a unity feedback system is given by

$$G(s) = \frac{Ke^{-1.5s}}{6s+1}$$

Sketch its root locus.

Solution Using Eq. (9.3), the given open-loop transfer function can be written as

$$G(s)H(s) = \frac{Ke^{-1.5s}}{6s+1} = \frac{K(2-1.5s)}{(2+1.5s)(6s+1)} = \frac{-K(1.5s-2)}{(1.5s+2)(6s+1)} \quad \text{(i)}$$

The poles are at $s = -1.33$ and $s = -0.17$ Hence, $n = 2$. Zero is at $s = 1.33$. Therefore, $m = 1$. We note that the gain of the open-loop transfer function is negative, i.e. $K < 0$. Therefore, part of the Re-axis lying between $s = -1.33$ and $s = -0.17$ belongs to the root locus because number of poles and zeros to the right of $s = -1.33$ is *even*.

Breakaway point: The characteristic equation is given by

$$1 + G(s)H(s) = 1 - \frac{K(1.5s-2)}{(1.5s+2)(6s+1)} = 0 \quad \text{(ii)}$$

$$\Rightarrow \qquad 9s^2 + (13.5 - 1.5K)s + (2 + 2K) = 0 \quad \text{(iii)}$$

Therefore,

$$K = \frac{(1.5s+2)(6s+1)}{1.5s-2} = \frac{9s^2+13.5s+2}{1.5s-2} \quad \text{[from Eq. (ii)]}$$

$$\frac{dK}{ds} = \frac{18s+13.5}{1.5s-2} - \frac{(9s^2+13.5s+2)1.5}{(1.5s-2)^2} = 0$$

$$\Rightarrow \qquad \frac{27s^2 - 15.8s - 27 - 13.5s^2 - 20.25s - 3}{(1.5s-2)^2} = 0$$

$$13.5s^2 - 36s - 30 = 0$$

or,

$$s = 3.33, \ -0.67$$

Of the two points, $s = -0.67$ is the breakaway point and $s = 3.33$ is both the break-in and breakaway point.

Intersection on the $j\omega$-axis: From Eq. (iii), we construct the Routh array as follows:

$$\begin{array}{c|cc} s^2 & 9 & 2+2K \\ s^1 & 13.5-1.5K & 0 \\ s^0 & 2+2K & \end{array}$$

For marginal stability,

$$2 + 2K = 0 \quad \Rightarrow \quad K = -1$$

$$13.5 - 1.5K = 0 \quad \Rightarrow \quad K = 9$$

When $K = 9$, we get from the auxiliary equation, the relation

$$9s^2 + 20 = 0 \qquad \Rightarrow s^2 = -\frac{20}{9} \qquad \therefore s = \pm j1.49$$

Shape of the loci in the complex plane: For this purpose we choose a point $s = \sigma + j\omega$ and apply the angle condition for $K < 0$ to the open-loop transfer function as follows:

$$\angle G(s)H(s) = 2q(180^\circ) = \arg\left|\frac{-K[1.5(\sigma + j\omega) - 2]}{[1.5(\sigma + j\omega) + 2][6(\sigma + j\omega) + 1]}\right| \qquad \text{[from Eq. (i)]}$$

$$= \arg\left|\frac{-K[(1.5\sigma - 2) + j1.5\omega]}{[1.5\sigma + 2) + j1.5\omega][(6\sigma + 1) + j\,6\omega]}\right|$$

$$= \tan^{-1}\frac{1.5\omega}{1.5\sigma - 2} - \left[\tan^{-1}\frac{1.5\omega}{1.5\sigma + 2} + \tan^{-1}\frac{6\omega}{6\sigma + 1}\right]$$

$$0^\circ_{\text{when } q=0} = \tan^{-1}\frac{1.5\omega}{1.5\sigma - 2} - \left[\tan^{-1}\frac{\dfrac{1.5\omega}{1.5\sigma + 2} + \dfrac{6\omega}{6\sigma + 1}}{1 - \dfrac{1.5\omega}{1.5\sigma + 2}\cdot\dfrac{6\omega}{6\sigma + 1}}\right]$$

or, $$\tan^{-1}\frac{1.5\omega}{1.5\sigma - 2} = \tan^{-1}\frac{(18\sigma + 13.5)\omega}{9\sigma^2 + 13.5\sigma + 2 - 9\omega^2}$$

Taking tangent of both sides and multiplying the quantities diagonally, we get

$$27\sigma^2 + 20.25\sigma - 36\sigma - 27 = 13.5\sigma^2 + 20.25\sigma + 3 - 13.5\omega^2$$

$$\Rightarrow \qquad 13.5\sigma^2 - 36\sigma + 13.5\omega^2 - 30 = 0$$

$$\Rightarrow \qquad \sigma^2 - 2.67\sigma + (1.33)^2 - (1.33)^2 + \omega^2 - 2.22 = 0$$

$$\Rightarrow \qquad (\sigma - 1.33)^2 + (\omega - 0)^2 = 2^2 \qquad \text{(iv)}$$

Equation (iv) represents a circle of radius 2, having its centre at (1.33, 0).

With these data, the root locus is drawn and presented in Figure 10.23.

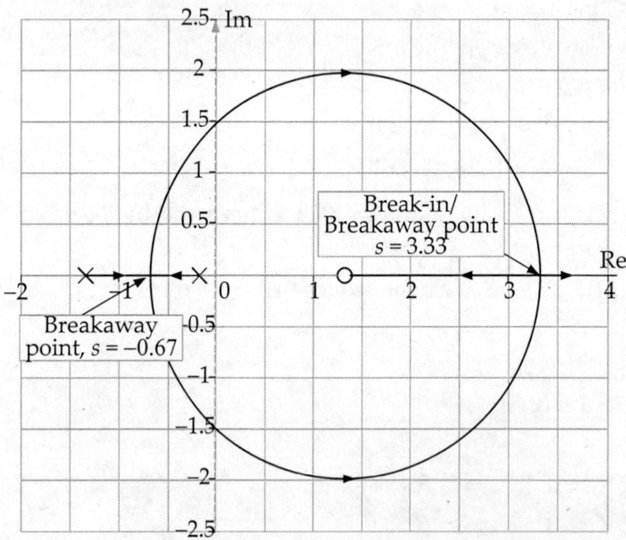

Figure 10.23: Root locus for system with transportation lag (Example 10.27).

10.4 POLE ZERO CANCELLATION RULE

The pole zero cancellation rule is:

> Poles of $G(s)$ cannot be cancelled by zeros of $H(s)$ though poles of $H(s)$ can be cancelled by zeros of $G(s)$.

We explain the reason behind the rule through examples. Suppose we have

$$G(s) = \frac{1}{(s+a)(s+b)}$$
$$H(s) = \frac{s(s+a)}{(s+c)}$$

If we allow pole zero cancellation, the characteristic equation becomes

$$1 + G(s)H(s) = 0$$

$$\Rightarrow \qquad 1 + \frac{s\cancel{(s+a)}}{\cancel{(s+a)}(s+b)(s+c)} = (s+b)(s+c) + s = 0 \qquad (10.13)$$

whereas the closed-loop transfer function is

$$\frac{C(s)}{R(s)} = \frac{G(s)}{1+G(s)H(s)} = \frac{\dfrac{1}{(s+a)(s+b)}}{1 + \dfrac{s(s+a)}{(s+b)(s+c)(s+a)}}$$
$$= \frac{(s+c)}{(s+a)[s+(s+b)(s+c)]}$$

And the characteristic equation from the closed-loop transfer function is given by

$$(s+a)[s+(s+b)(s+c)] = 0 \qquad (10.14)$$

Comparing Eq. (10.14) with Eq. (10.13), we find that the closed-loop pole at $s = -a$ is missing in the latter because of pole zero cancellation. Therefore, this cancellation is not permitted.

Next, let us consider

$$G(s) = \frac{(s+c)}{(s+a)(s+b)}$$
$$H(s) = \frac{s}{(s+c)}$$

Allowing zero pole cancellation, the characteristic equation becomes

$$1 + G(s)H(s) = 0$$

$$\Rightarrow \qquad 1 + \frac{s\cancel{(s+c)}}{(s+a)(s+b)\cancel{(s+c)}} = (s+a)(s+b) + s = 0 \qquad (10.15)$$

Here, the closed-loop transfer function is

$$\frac{C(s)}{R(s)} = \frac{G(s)}{1+G(s)H(s)} = \frac{\dfrac{(s+c)}{(s+a)(s+b)}}{1 + \dfrac{s(s+c)}{(s+a)(s+b)(s+c)}}$$
$$= \frac{(s+c)}{(s+c)[s+(s+a)(s+b)]} = \frac{1}{s+(s+a)(s+b)}$$

Consequently, the characteristic equation is

$$s + (s + a)(s + b) = 0 \tag{10.16}$$

We find Eq. (10.16) perfectly tallies with Eq. (10.15) which was derived allowing zero pole cancellation.

EXAMPLE 10.28. Sketch the root locus for a system having

$$G(s) = \frac{K}{s+1} \qquad H(s) = \frac{s+1}{s^2 + 4s + 5}$$

Solution We note that if we simply multiply $G(s)$ by $H(s)$, one pole of $G(s)$, namely that at $s = -1$, gets cancelled by a zero of $H(s)$. This is not permitted. So, to keep the characteristic equation intact, we restore the pole and write the open-loop transfer function as

$$G(s)H(s) = \frac{K}{(s+1)(s^2 + 4s + 5)} = \frac{K}{(s+1)(s+2-\jmath 1)(s+2+\jmath 1)}$$

Poles are at $s = -1$, $-2 + \jmath 1$ and $-2 - \jmath 1$. Therefore, $n = 3$. No zero. Thus, $m = 0$. Part of the real axis on the root locus: $-\infty < s < -1$.

Asymptotes: Number = 3, angles = 60°, 180°, 300°. Centroid: $\chi_c = \dfrac{-1-2-2}{3} = -1.67$
Angle of departure for pole at $s = -2 + \jmath 1$: Graphically,

$$\theta_d = 180° - [90° + (\pi - 45°)] = -45°$$

Intersection at the $\jmath\omega$*-axis*: The characteristic equation is

$$s^3 + 5s^2 + 9s + 5 + K = 0$$

The Routh array is

$$\begin{array}{c|cc} s^3 & 1 & 9 \\ s^2 & 5 & 5+K \\ s & 8-0.2K & 0 \\ s^0 & K & \end{array}$$

For marginal stability,

$$8 - 0.2K_{\text{mar}} = 0 \quad \Rightarrow \quad K_{\text{mar}} = 40$$

From the auxiliary equation

$$5s^2 + 45 = 0 \quad \Rightarrow \quad s = \pm \jmath 3$$

With the help of these data, we plot the root locus of Figure 10.24.

10.5 STABILITY MARGINS THROUGH ROOT LOCUS

The relative stability of a control system is pretty often determined from a study of two parameters, called *gain margin* and *phase margin*.

Gain Margin

The maximum allowable loop gain in decibels before the system becomes unstable is called the *gain margin*.

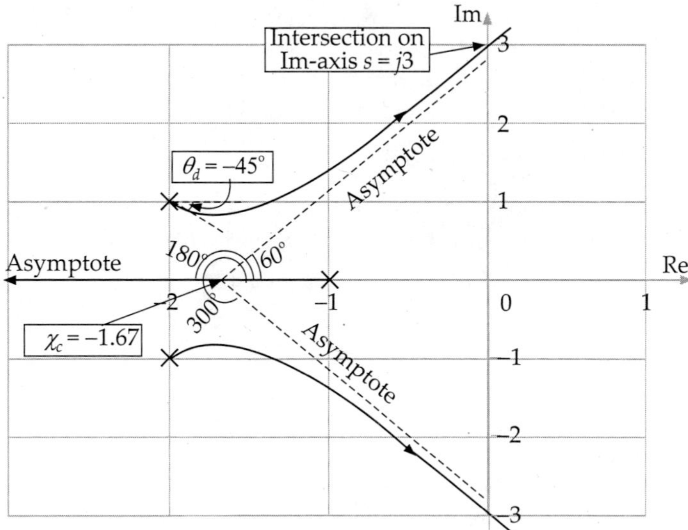

Figure 10.24: Root locus (Example 10.28).

We know that a CLTF pole placed on the RHP of the $\jmath\omega$-axis makes the system unstable. In a root locus study, the locus of a root is plotted by increasing the value of the gain factor K. While doing so, once the locus touches the $\jmath\omega$-axis for a certain value of K, the system becomes marginally stable. Beyond that value of K, the system crosses the Im-axis and becomes unstable.

So, the gain margin can be calculated from the root locus by finding out the ratio of K value for marginal stability (i.e. at the crossover of the imaginary axis) to that required by the design. Thus,

$$\text{Gain margin (Gm)} = 20 \log \frac{K_{\text{mar}}}{K_{\text{required}}} \text{ dB} \tag{10.17}$$

In case the root locus does not cross the $\jmath\omega$-axis, an infinite value is assigned to K.

Phase Margin

The maximum allowable increment of phase angle of $G(s)H(s)$ in degrees up to which the system remains stable is called the *phase margin.*

A $-180°$ phase angle makes a system unstable. So, to determine the phase margin, we have to calculate by what angle the phase of a given system falls short of $-180°$. Thus, the phase margin of system can be calculated from the K value by following the procedure given below:

- For the given K value, we calculate ω_g for which $|G(\jmath\omega_g)H(\jmath\omega_g)| = 1$.
- Then we calculate corresponding $\arg|G(\jmath\omega_g)H(\jmath\omega_g)|$.

The phase margin is then calculated from the relation

$$\text{Phase margin (Pm)} = 180° + \arg|G(\jmath\omega_g)H(\jmath\omega_g)|$$

However, it may not be easy to follow this procedure for all systems. It is rather easy to calculate phase margin from Bode plots which we will discuss later.

EXAMPLE 10.29. Sketch the root locus plot for the unity feedback system having open-loop transfer function

$$G(s) = \frac{K}{s(s+1)(s+3)}$$

and find

(a) K for critical damping
(b) K for zero damping
(c) Gain margin for gain = 5

Solution The poles are at $s = 0,\ -1,\ -3$. Therefore, $n = 3$. No zero. Thus, $m = 0$. Part of σ-axis (i.e., Re-axis) in the root locus: between $s = 0$ and -1, and $s = -3$ and $-\infty$.

Asymptotes: Number = 3, therefore, $q = 0,\ 1,\ 2$. Their angles and the centroid are as follows:

$$\theta_0 = \frac{1}{3} \times 180° = 60° \qquad \theta_1 = \frac{3}{3} \times 180° = 180°$$

$$\theta_2 = \frac{5}{3} \times 180° = 300° \qquad \chi_c = \frac{-1-3}{3} = -1.33$$

Breakaway point: It is expected between $s = 0$ and -1 and calculated from the characteristic equation as follows:

$$s(s+1)(s+3) + K = 0$$

$$s^3 + 4s^2 + 3s + K = 0 \tag{i}$$

$$\Rightarrow \quad K = -s^3 - 4s^2 - 3s \tag{ii}$$

$$\frac{dK}{ds} = 3s^2 + 8s + 3 = 0 \tag{10.18}$$

Therefore, $$s = -0.45,\ -2.22$$

Its location at $s = -2.22$ is ruled out because the point does not lie on the root locus. Therefore, the breakaway point is at $s = -0.45$.

Intersection at the $j\omega$-axis: We construct the Routh array from the characteristic equation [Eq. (i)] as follows:

$$\begin{array}{c|cc} s^3 & 1 & 3 \\ s^2 & 4 & K \\ s & 3-0.25K & 0 \\ s^0 & K & \end{array}$$

For marginal stability,

$$3 - 0.25K_{\text{mar}} = 0 \quad \Rightarrow \quad K_{\text{mar}} = 12$$

The auxiliary equation is

$$4s^2 + 12 = 0 \quad \Rightarrow \quad s = \pm j1.73$$

With the help these data, we plot the root locus as shown in Figure 10.25.

(a) For critical damping, $\zeta = 1$. Therefore, $\phi = \tan^{-1}\dfrac{\sqrt{1-\zeta^2}}{\zeta} = 0°$. The root locus intersection at 0° from the origin is at $s = -0.45$. From Eq. (ii)

$$K\Big|_{s=-0.45} = -(-0.45)^3 - 4(-0.45)^2 - 3(-0.45) = 0.63$$

(b) For $\zeta = 0$, $K = K_{\text{mar}} = 12$.

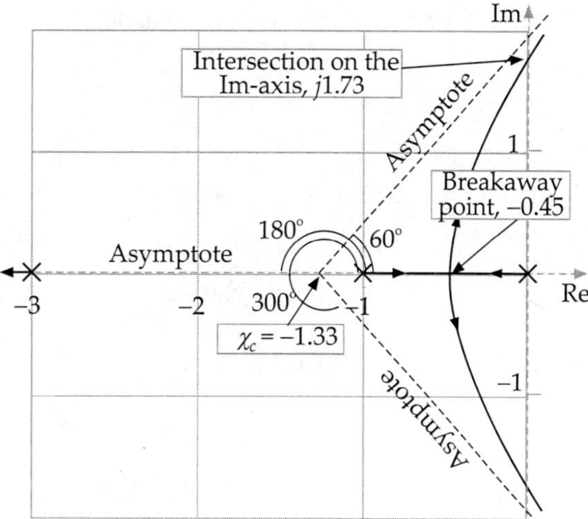

Figure 10.25: Root locus plot (Example 10.29).

(c) From Eq. (10.17)

$$\text{Gm} = 20 \log \frac{K_{\text{mar}}}{K_{\text{required}}}$$

We have already found that $K_{\text{mar}} = 12$. Therefore, the required gain margin is

$$20 \log \left(\frac{12}{5}\right) = 7.6 \text{ dB}$$

EXAMPLE 10.30. The open-loop transfer function for a system is given by

$$G(s)H(s) = \frac{K}{s(s+4)^2}$$

Calculate the phase margin for the system for $K_{\text{design}} = 25$.

Solution Substituting $j\omega$ for s, we get

$$|G(j\omega)H(j\omega)| = \left|\frac{25}{j\omega(j\omega+4)^2}\right| = 1 \qquad \Rightarrow \quad \left|j\omega(j\omega+4)^2\right| = 25$$

$$\Rightarrow \qquad \omega(\omega^2+16) = 25 \qquad \Rightarrow \quad \omega^3 + 16\omega - 25 = 0$$

The real root of this cubic equation[7] is given by

$$|\omega| = 1.39341 \equiv \omega_g$$

The corresponding argument of $G(j\omega)H(j\omega)$ is

$$\arg\left|G(j\omega_g)H(j\omega_g)\right| = \frac{\angle 0^\circ}{\angle 90^\circ + 2\left(\tan^{-1}\dfrac{1.3934}{4}\right)} = -[90^\circ + 2(19.21^\circ)] = -128.41^\circ$$

Therefore, the phase margin is

$$180^\circ - 128.41^\circ = 51.59^\circ$$

[7]Using the calculator at https://www.calculatorsoup.com/calculators/algebra/cubicequation.php.

EXAMPLE 10.31. Draw the root locus for the open-loop transfer function

$$G(s)H(s) = \frac{K}{s(s+2)(s+4)}$$

and determine the gain margin and phase margin for $K = 12$.

Solution The root locus has been worked out in Example 10.14. From there we get $K_{\text{mar}} = 48$. Therefore,

$$\text{Gain margin} = 20 \log \frac{K_{\text{mar}}}{K_{\text{required}}} = 20 \log \frac{48}{12} = 20 \log 4 = 12.04 \text{ dB}$$

For phase margin calculation, we put

$$|G(\jmath\omega)H(\jmath\omega)| = \left| \frac{12}{\jmath\omega(\jmath\omega+2)(\jmath\omega+4)} \right| = 1 \quad \Rightarrow \quad \frac{12}{\omega\sqrt{\omega^2+4}\sqrt{\omega^2+16}} = 1$$

$$\Rightarrow \qquad \omega\sqrt{\omega^4 + 20\omega^3 + 64} = 12 \quad \Rightarrow \quad \omega^6 + 20\omega^4 + 64\omega^2 - 144 = 0 \qquad \text{(i)}$$

Substituting $x = \omega^2$ in Eq. (i), we get the cubic equation

$$x^3 + 20x^2 + 64x - 144 = 0$$

Its three roots are

$$x_1 = 1.49713 \qquad x_2 = -15.14716 \qquad x_3 = -6.34997$$

Evidently, the first root yields real values for ω. Since frequency cannot be negative, $\omega = 1.2236$ rad/s is the gain crossover frequency. The phase margin is, therefore,

$$\text{Pm} = 180° + \frac{\angle 0°}{\angle 90° + \tan^{-1}(1.2236/2) + \tan^{-1}(1.2236/4)}$$
$$= 180° - (90° + 31.46° + 17°) = 41.54°$$

10.6 EFFECTS OF ADDITION OF POLES AND ZEROS ON ROOT LOCUS

We will study the effects by considering one example. Suppose our open-loop transfer function is given by

$$G(s)H(s) = \frac{K}{s(s+3)} \qquad (10.19)$$

The root locus plot of Eq. (10.19) is given in Figure 10.26a.

Addition of Poles

Now we add one pole at $s = -5$ so that

$$G(s)H(s) = \frac{K}{s(s+3)(s+5)} \qquad (10.20)$$

The root locus plot of Eq. (10.20) is given in Figure 10.26b.

It is clear from Figures 10.26a and 10.26b that before the addition of pole, the system was perfectly stable. After the addition of the pole:

1. The complex part of the root loci bends toward the RHP.
2. The angles of asymptotes change from $\pm 90°$ to $\pm 60°$.
3. The breakaway point moves to the right.
4. The system becomes unstable for values of K when the loci cross the Im-axis.

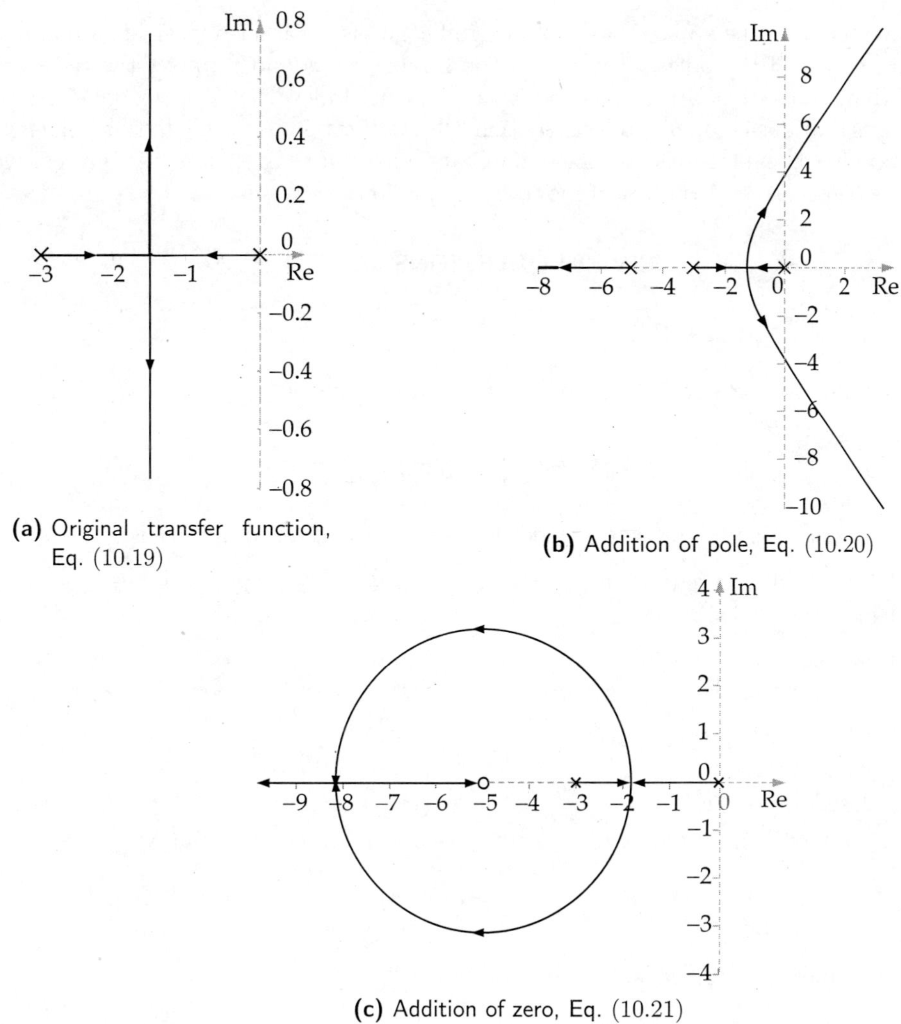

(a) Original transfer function, Eq. (10.19)

(b) Addition of pole, Eq. (10.20)

(c) Addition of zero, Eq. (10.21)

Figure 10.26: Root locus plots to show effect of addition of pole and zero.

Addition of Zeros

Now let us consider the case when we add a zero in the LHP of Eq. (10.19) so that

$$G(s)H(s) = \frac{K(s+5)}{s(s+3)} \tag{10.21}$$

The root locus plot of Eq. (10.21) is given in Figure 10.26c.

Comparing Figure 10.26c with Figure 10.26a we find that the addition of a zero bends the root loci towards the left to form a circle. As a result, the relative stability of the system improves. So, we may infer that the addition of zeros moves root loci more towards the left half of the s-plane.

Although inferred from a particular case, the results are quite general.

Dominant poles and zeros. In this context, we would like to mention that the transfer function poles that have maximum effect on the transient response of the system, are called *dominant poles.* The same definition applies for dominant zeros as well.

The location of poles in the s-plane has a lot of significance as far as the transient response of the system is concerned. LHP poles located near the Im-axis are found to affect the transient response more than those that are located far away from it. Indeed, these poles give rise to transient responses that decay much more slowly than the far-away ones. Owing to this influence, poles in the proximity of the Im-axis are called dominant poles. Of course, closeness is a relative term and much depends on the structure of the transfer function.

REVIEW QUESTIONS

10.1 Choose the correct answer:

(i) The open-loop transfer function of a unity feedback system is given by

$$G(s) = \frac{Ks}{(s+1)(s+2)(s+3)}$$

The number of asymptotes of the root-locus plot that tend to infinity is

(A) 1 (B) 2 (C) 0 (D) 3

(ii) Root loci start from

(A) open-loop poles (B) open-loop zeros
(C) closed-loop poles (D) none of the above

(iii) Root loci terminate at

(A) open-loop poles (B) open-loop zeros
(C) closed-loop poles (D) none of the above

(iv) The open-loop transfer function of a unity feedback system is given by

$$G(s) = \frac{K(s+4)}{(s+1)(s+2)(s+3)}$$

The centroid of asymptotes is located at $s =$

(A) -1 (B) -2 (C) -3 (D) -4

(v) The open-loop transfer function of a unity feedback system is given by

$$G(s) = \frac{Ks}{(s+1)(s+2)(s+3)} \quad (K > 0)$$

The angle(s) of asymptote(s) is/are

(A) 60°, 180°, 270° (B) 90°, 180° (C) 180° (D) 0°

(vi) If there are two adjacent poles on the real axis and the segment of the real axis is a part of the root locus, then

(A) at least one break-in point exists between the poles
(B) at least one breakaway point exists between the poles
(C) at least two break-in points exist between the poles
(D) at least two breakaway points exist between the poles

(vii) At the breakaway point, the gain factor K should be

(A) real (B) imaginary (C) complex (D) none of these

(viii) The root locus diagram

(A) is always symmetric about the real axis

(B) is always symmetric about the imaginary axis

(C) is never symmetric about the real axis

(D) is always symmetric about both the real and imaginary axes

(ix) The root locus diagram can be used to determine

(A) absolute stability (B) relative stability

(C) conditional stability (D) none of the above

(x) The root loci of a system have three asymptotes. The system can have

(A) five poles and two zeros (B) three poles and one zero

(C) four poles and two zeros (D) six poles and four zeros

(xi) The open-loop transfer function for a unity feedback system is

$$\frac{K}{(s+1)(s+2)(s+3)}$$

When K varies from 0 to ∞, the breakaway point in the root locus plot will be between

(A) 0 and -1 (B) -2 and -3 (C) -1 and-2 (D) beyond -3

(xii) A system having 3 zeros and 4 poles has root loci equal to

(A) 3 (B) 4 (C) 1 (D) 7

(xiii) Root locus is plot of

(A) closed-loop poles (B) closed-loop zeros

(C) open-loop zeros (D) open-loop poles

10.2 A point on the real axis lies on the root locus if the number of open-loop poles plus zeros on the real axis to the right of the point is odd—justify the statement.

10.3 "The breakaway points of the root locus are the solutions of $(dK/ds) = 0$". Justify.

10.4 Consider the open-loop transfer function

$$G(s)H(s) = \frac{K}{s(0.2s+1)(0.05s+1)}$$

Draw the root locus and determine for what value of gain K the phase margin will be 40°.

10.5 The open-loop transfer function of a system is given by

$$G(s)H(s) = \frac{5(s+\alpha)}{s(s+2)}$$

Draw the root locus and find the value of α for which the damping ratio is $1/\sqrt{2}$.

10.6 Obtain a root locus of the unity feedback system whose forward transfer function is

$$\frac{K(s+2)(s+3)}{s(s+1)}$$

Determine the break-in and breakaway points and the values of K at those points. State whether the system is underdamped or overdamped for the following values of K: (i) 0.05, (ii) 5 and (iii) 15. Determine the range of K for which the system is stable.

10.7 For the system with the open-loop transfer function

$$G(s)H(s) = \frac{K}{s(s+1)(s+2)} \qquad 0 \le K \le \infty$$

determine if the following points are on the root locus: (i) $s = -0.5$, (ii) $s = 0.5$, (iii) $s = \jmath 1.732$, and (iv) $s = -1 + \jmath$.

10.8 Plot the root locus of a unity feedback control system whose transfer function is given by

$$G(s) = \frac{K}{s(s+2)(s+3)}$$

10.9 (a) What is a root locus?

(b) Explain why generally each branch of a root locus originates from an open-loop pole and terminates to an open-loop zero as K increases from zero to infinity.

(c) Consider a feedback system with the characteristic equation given by

$$1 + \frac{K}{s(s^2+2s+4)} = 0$$

Roughly draw the root locus of the system.

10.10 A unity feedback control system has

$$G(s) = \frac{K}{s(s+2)(s+5)}.$$

Sketch the root locus and show on it a line for $\zeta = 0.5$ and the value of K for this damping.

10.11 The open-loop transfer function of a unity feedback control system is given by

$$G(s) = \frac{K(s+3)}{s(s^2+2s+2)(s+5)(s+6)}$$

(a) Draw the root locus as a function of K.

(b) Determine the value of K which makes the relative damping ratio of the closed-loop complex pole equal to 0.4.

10.12 The open-loop transfer function of a unity feedback control system is given by

$$G(s) = \frac{K}{s(s^2+8s+32)}$$

Sketch the root-locus plot for the system by determining

(a) the number, angle and the centroid of the asymptotes

(b) angle of departure

(c) the breakaway point

(d) the intersection of imaginary axis by the root locus

10.13 (a) State the rules of construction of root loci.

(b) Draw the root loci for the open-loop transfer function

$$G(s)H(s) = \frac{K}{s(s+1)(s+2)}$$

(c) Find the critical value of K for which the system becomes unstable and the critical frequency thereof.

10.14 A unity feedback control system has an open-loop transfer function

$$G(s) = \frac{K}{s(s^2 + 4s + 13)}$$

Sketch the root locus plot of the system by determining the following:

(a) The centroid, number and angle of asymptotes
(b) The angle of departure of root loci from the poles
(c) The breakaway point, if any
(d) The value of K and frequency at which the root loci cross the $\jmath\omega$-axis.

10.15 A control system is characterized by an open-loop transfer function of

$$G(s)H(s) = \frac{K}{s(s+2)(s^2 + 4s + 8)}$$

(a) Draw the root locus of the characteristic equation.
(b) Find the value of K for which the closed-loop system has a pair of poles on the imaginary axis.
(c) Find out the closed-loop poles for the value of K thus obtained.

10.16 The open-loop transfer function of a feedback control system is given by

$$G(s)H(s) = \frac{K}{(s+4)(s^2 + 4s + 8)}$$

(a) Draw the root locus plot of the system by determining the centroid, number and angle of asymptotes, angle of departure of root locus branches from the open loop poles, breakaway points, if any, and the intersection points of the root locus branches with the imaginary axis.
(b) Find out the value of K for which the dominant pole pair will have a damping ratio of 0.5.
(c) Determine the closed-loop poles for the value of K as computed in part (B).

Answers to Selected Questions

10.1. (i) (B) (ii) (A) (iii) (B) (iv) (A) (v) (B) (vi) (B) (vii) (A) (viii) (A) (ix) (B) (x) (A) (xi) (C) (xii) (B) (xiii) (D)

10.4. 5.22 10.5. 4.9

10.6. Break-in at $s = -2.37$, breakaway at $s = -0.63$ (i) critically damped,
(ii) underdamped (iii) critically damped $-5 < K < 0$

10.7. (i) Yes (ii) No (iii) No (iv) No 10.10. 11.4 10.11 (b) 11.8

10.12. (a) Number = 3, angles 60°, 180°, 240°, centroid at $s = -2.67$ (b) 135°
(c) No breakaway point (d) $s = \pm\jmath 5.66$ 10.13. (c) $K = 6$, $\omega = 1.41$ rad/s

10.14. (a) Centroid at $s = -1.33$, number = 3/ angles $60°$, $180°$, $240°$ (b) $146.3°$
(c) No breakaway point (d) $K = 52$, $\omega = 3.6$ rad/s

10.15. (b) 35.56 (c) $s_{1,2} = -3 \pm \jmath 2.08$, $s_{3,4} = 0 \pm \jmath 1.63$

10.16 (a) Centroid at $s = -2.67$, number = 3, angles $60°$, $180°$, $240°$, angle of departure $= 65°$, no breakaway point, intersection points at the Im-axis at $s = \pm \jmath 4.9$
(b) 12.9 (c) $s_1 = -4.99$, $s_{2,3} = -1.5 \pm \jmath 2.59$

11

Frequency Domain Analysis

11.1 WHY FREQUENCY DOMAIN?

While talking about the methods of analysis of stability of a control system in Chapter 8 (Section 8.4) we mentioned that three of the widely used methods are based on analysis in the frequency domain. In this chapter we will discuss what the frequency domain analysis means. What its advantages and disadvantages are given in Appendix *Frequency Domain Analysis.*

On previous occasions, while testing the response of a system, we used four standard inputs, namely (i) impulse, (ii) step, (iii) ramp, and (iv) parabolic inputs. Now, suppose we give a sinusoidal input like

$$x(t) = A \sin \omega t \tag{11.1}$$

to a control system. What happens can of course be studied by the previously laid down procedure in Chapter 6. But a general analysis shows that we need not go into such an elaborate analysis for a sinusoidal input, because for a sinusoidal input, as shown schematically in Figure 11.1, after the transients die out:

- The output remains sinusoidal with the same frequency.
- Only the amplitude and phase of the input change after passing through any LTI system.

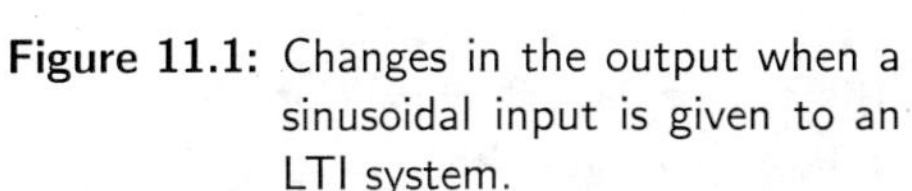

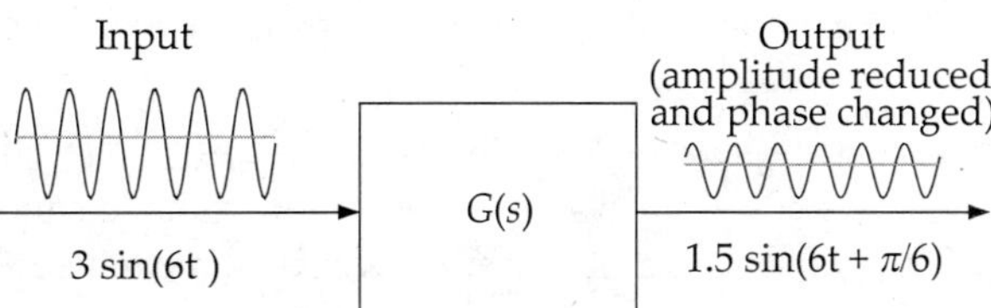

Figure 11.1: Changes in the output when a sinusoidal input is given to an LTI system.

A proof is given in Section C.1 in Appendix *Frequency Domain Analysis.*

HW Bode[1] and H Nyquist[2] developed procedures for studying stability of feedback systems in the frequency domain. In later chapters we will discuss those procedures.

Before that, let us consider frequency response specifications and frequency domain relations.

11.2 FREQUENCY RESPONSE SPECIFICATIONS

The frequency response of a control system is generally characterized in terms eight parameters, namely

[1]Hendrik Wade Bode (1905–1982) developed his technique in 1938. A pioneer of modern control theory and electronic instrumentation, he was an American engineer of Dutch ancestry.

[2]Harry Nyquist (1889–1976), an American (born Swedish) physicist, developed his method in 1934.

1. Resonant peak, M_r
2. Resonant frequency, ω_r
3. Bandwidth, Bw
4. Gain crossover frequency, ω_g
5. Phase crossover frequency, ω_p
6. Cut-off frequency, ω_c
7. Gain margin, Gm
8. Phase margin, Pm

We consider them in more detail now.

1. Resonant Peak, M_r. The closed-loop transfer function in the frequency domain is written as

$$\frac{G(\jmath\omega)}{1+G(\jmath\omega)H(\jmath\omega)} \equiv |M(\jmath\omega)|\angle\phi$$

where, $M(\jmath\omega)$ is the amplitude and ϕ is the phase (angle) of the complex quantity. The maximum value of $|M(\jmath\omega)|$, as ω is varied, is called the *resonant peak* and is denoted as M_r.

M_r value gives an indication about the relative stability of a system. A large M_r indicates a large maximum overshoot, M_p, in step response of the system. Normally M_r lies between 1.1 and 1.5.

2. Resonant Frequency, ω_r. The resonant frequency is the frequency at which the resonant peak occurs.

3. Bandwidth, Bw. As ω is varied, the value of $M(\jmath\omega)$ also varies. Ultimately, its level may drop down by 3 dB below[3] its value at zero frequency which is normally 0 dB level. The range of frequency between 0 dB level and −3 dB level of $M(\jmath\omega)$ is called the *bandwidth*.

While a large Bw may enhance the system response, it makes the system susceptible to noise as well. On the other hand, a low Bw indicates a sluggish system response.

4. Gain Crossover Frequency, ω_g. This is the frequency at which the gain of the system is unity.

5. Phase Crossover Frequency, ω_p. This is the frequency at which the phase angle ϕ of $M(\jmath\omega)$ is 180°.

6. Cut-off Frequency, ω_c. This is the frequency at which the power level falls to half of its value. Since power is proportional to the square of the amplitude, the magnitude of $M(\jmath\omega)$ equals $1/\sqrt{2}$ at ω_c.

7. Gain Margin, Gm. We have already defined[4] that the factor by which the gain can be increased before the system becomes unstable is called its *gain margin* (Gm).

The gain margin of a system can be determined from its open-loop frequency response. It is the reciprocal of the open-loop gain at the phase crossover frequency.

8. Phase Margin, Pm. The gain margin alone is not sufficient to express the relative stability of a system. A system may have a very good gain margin, but it may require a small additional phase lag to render the system unstable. Hence, a further criterion, termed *phase margin* (Pm), is necessary to assess the stability of systems of this type. It is defined as the additional phase lag that will make the system marginally stable[5].

The phase margin is determined from the open-loop frequency response. It is the clockwise angle through which the unit vector must be rotated before it lies on the negative real axis[6]. Hence, phase margin is the difference of the phase shift of the system at the gain crossover frequency plus 180°.

[3]It corresponds to $1/\sqrt{2}$ (= 0.707) times the value of M at zero frequency.

[4]See Section 10.3 at page 254.

[5]See also Section 10.3 at page 254.

[6]The reason why it is so defined will be clear when we will discuss the Nyquist plot.

Utility of gain and phase margins. Control system designers frequently use gain and phase margins to be sure about the frequency response of a system. Of course, *these margins relate to the open-loop systems* only.

A good gain margin is desirable from the point of view of robust stability of the system. However, one should not select an arbitrarily large gain margin because the dynamic performance characteristics of the system are related to it and the response may become unacceptably overdamped and sluggish as a result.

To provide a good overall stability to a system, the rule of thumb is that the gain and phase margins should be around 12 dB and 45° to 60°, respectively. In most of the cases, a good Gm automatically ensures a good Pm and vicé versa. But there are exceptions to this rule as well.

11.3 FREQUENCY DOMAIN RELATIONS

We study these relations for a second order system where they are relevant. For a unity feedback second order system

$$G(s) = \frac{C(s)}{R(s)} = \frac{\omega_n^2}{s^2 + 2\zeta\omega_n s + \omega_n^2} \qquad H(s) = 1$$

Therefore,

$$G(\jmath\omega) = \frac{C(\jmath\omega)}{R(\jmath\omega)} = \frac{\omega_n^2}{-\omega^2 + \jmath 2\zeta\omega_n\omega + \omega_n^2} \qquad [s \equiv \jmath\omega]$$

$$\equiv \frac{1}{1 + \jmath 2\zeta u - u^2} \qquad [\text{where, } u = (\omega/\omega_n)]$$

If we express $G(\jmath\omega)$ as

$$G(\jmath\omega) = |G(\jmath\omega)|\angle G(\jmath\omega) \equiv |M|\angle\phi$$

then,

$$|M| = \frac{1}{\sqrt{(1-u^2)^2 + (2\zeta u)^2}} \tag{11.2}$$

$$\phi = -\tan^{-1}\frac{2\zeta u}{1-u^2} \tag{11.3}$$

The Table 11.1 gives us values of magnitude and phase at certain values of u. The entire plots of M vs. u and ϕ vs. u, as obtained from Eqs. (11.2) and (11.3), are shown in Figure 11.2a and Figure 11.2b, respectively.

Table 11.1: Magnitude and phase values

u	M	ϕ
0	1	0
1	$1/(2\zeta)$	$\pi/2$
∞	0	$-\pi$

Resonant Frequency, ω_r. For resonant frequency

$$\left.\frac{dM(\omega)}{d\omega}\right|_{\omega_r} = \left.\frac{dM(u)}{du}\right|_{u_r} = 0$$

$$\Rightarrow \qquad -\frac{1}{2}\cdot\frac{2(1-u_r^2)(-2u_r) + 8\zeta^2 u_r}{[(1-u_r^2)^2 + 4\zeta^2 u_r^2]^{3/2}} = 0$$

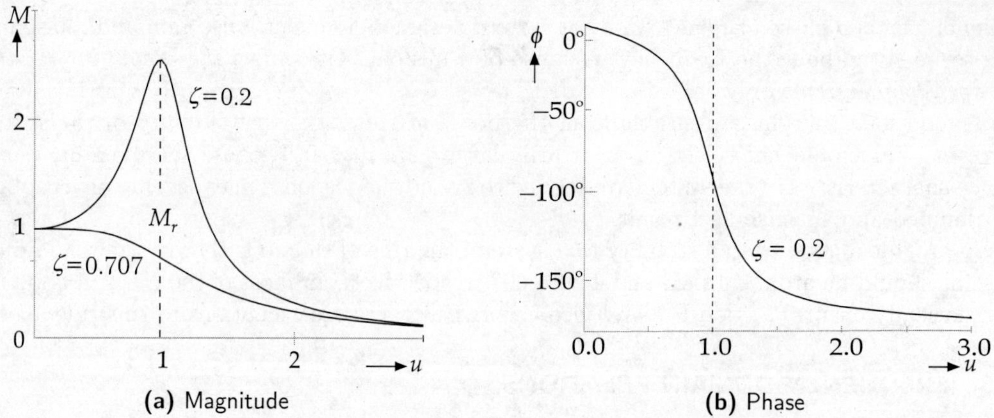

Figure 11.2: Second order system parameter plots.

$$\Rightarrow \qquad u_r^2 = 1 - 2\zeta^2$$

$$\Rightarrow \qquad u_r = \sqrt{1 - 2\zeta^2} \tag{11.4}$$

Therefore,

$$\boxed{\omega_r = \omega_n\sqrt{1 - 2\zeta^2}}$$

Resonant Peak, M_r. From Eqs. (11.4), (11.2) and (11.3) we get

$$\boxed{M_r = \frac{1}{2\zeta\sqrt{1 - \zeta^2}}} \tag{11.5}$$

$$\boxed{\phi_r = -\tan^{-1}\frac{\sqrt{1 - 2\zeta^2}}{\zeta}} \tag{11.6}$$

It is interesting to note from Eqs. (11.5) and (11.6) that

- As $\zeta \to 0$, $M_r \to \infty$ and $\omega_r \to \omega_n$. This is the resonance condition.
- For $\zeta = 1/\sqrt{2} = 0.707$, $\omega_r = 0$ and $M_r = 1$. So, for $0 < \zeta < 0.707$, $\omega_r < \omega_n$ and $M_r > 1$.
- No resonant peak occurs for $\zeta > 0.707$.

Bandwidth, Bw. To calculate the Bw, we find the u_c-value corresponding to $M = 1/\sqrt{2}$. Then,

$$\frac{1}{\sqrt{(1 - u_c^2)^2 + 4\zeta^2 u_c^2}} = \frac{1}{\sqrt{2}}$$

$$\Rightarrow \qquad u_c^2 = 1 - 2\zeta^2 + \sqrt{(2\zeta^2 - 1)^2 + 1}$$

$$\Rightarrow \qquad u_c = \sqrt{1 - 2\zeta^2 + \sqrt{(2\zeta^2 - 1)^2 + 1}}$$

Therefore,

$$\boxed{\omega_c = \omega_n\sqrt{1 - 2\zeta^2 + \sqrt{(2\zeta^2 - 1)^2 + 1}}} \tag{11.7}$$

This cut-off frequency is the measure of the bandwidth. The graph in Figure 11.3a gives an idea about the bandwidth. If we define normalized bandwidth as ω_c/ω_n, then the graph given in Figure 11.3b indicates what it looks like against ζ values. The graph as well as Eq. (11.7) shows that for $\zeta = 1/\sqrt{2}$, the normalized bandwidth equals 1.

From Eq. (11.7) we note the following:

- Bw $= \omega_c$
- Bw decreases as ζ increases [see Figure 11.3b].

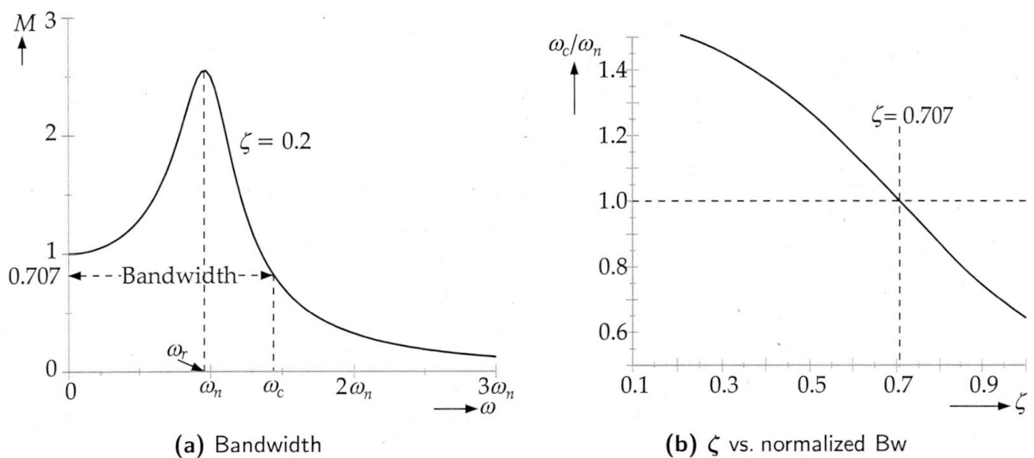

(a) Bandwidth (b) ζ vs. normalized Bw

Figure 11.3: Bandwidth and its variation with ζ.

Comparison between Time domain and Frequency Domain Parameters

For the sake of comparison, we better choose a second order system. In the frequency domain, a system is mainly characterized by its resonant frequency, ω_r and resonant peak, M_r, while in the time domain, the damped frequency, ω_d and the peak overshoot, M_p in response to a step input are major characteristic parameters.

The resonant frequency of the frequency domain and the damped frequency of the time domain are both related to the natural frequency of oscillation, ω_n and the damping ratio, ζ as follows:

$$\omega_r = \omega_n\sqrt{1-2\zeta^2}$$

$$\omega_d = \omega_n\sqrt{1-\zeta^2} \qquad 0 \le \zeta \le 1$$

Therefore,

$$\frac{\omega_r}{\omega_d} = \sqrt{\frac{1-2\zeta^2}{1-\zeta^2}}$$

The interrelation between ω_r and ω_d can be visualized from Figure 11.4a.

Similarly, we observe that the resonant peak, M_r in the frequency domain and the peak overshoot, M_p in the time domain are also ζ-dependent as can be seen from the following

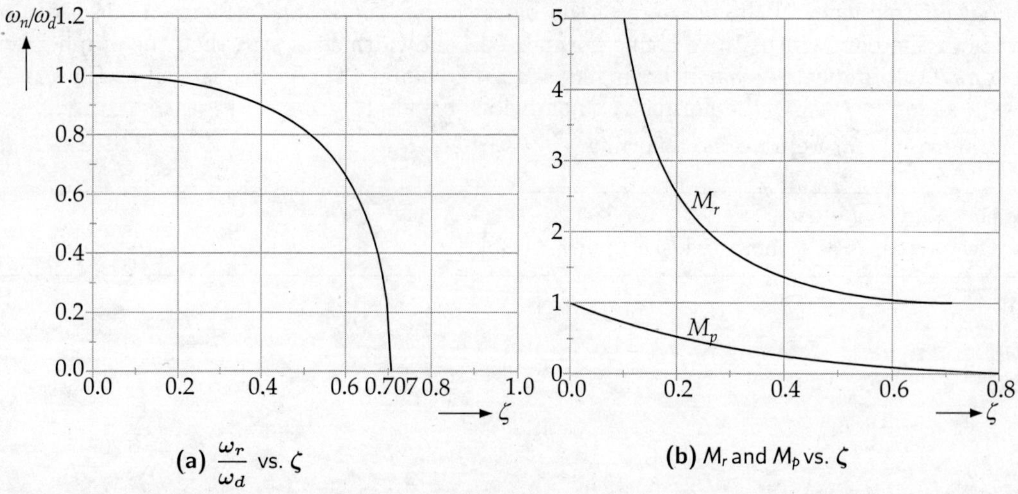

(a) $\frac{\omega_r}{\omega_d}$ vs. ζ (b) M_r and M_p vs. ζ

Figure 11.4: Comparison between frequency- and time-domain parameters.

formulae:

$$M_r = \frac{1}{2\zeta\sqrt{1-\zeta^2}}$$

$$M_p = \exp\left(-\frac{\pi\zeta}{\sqrt{1-\zeta^2}}\right)$$

The visual picture of their interrelation is presented in Figure 11.4b. Of course, this relation exists up to $\zeta = 0.707$ beyond which the resonant peak does not exist.

EXAMPLE 11.1. The transfer function of a system is given by

$$\frac{V_0(s)}{V_1(s)} = \frac{1-s}{1+s}$$

Let the output of the system be $v_o(t) = V_m \sin(\omega t + \phi)$ for the input $v_i(t) = V_m \sin(\omega t)$. Then the minimum and maximum values of ϕ (in radians) are respectively

(A) $-\frac{\pi}{2}$ and $\frac{\pi}{2}$ (B) $-\frac{\pi}{2}$ and 0 (C) 0 and $\frac{\pi}{2}$ (D) $-\pi$ and 0

Solution In frequency domain the transfer function can be written as

$$G(\jmath\omega) = \frac{1-\jmath\omega}{1+\jmath\omega}$$

Therefore, $\angle G(\jmath\omega) = -2\tan^{-1}\omega$

At $\omega = 0$ $\angle G(\jmath\omega) = 0$

At $\omega \to \infty$ $\angle G(\jmath\omega) = -\pi$

Thus, $\phi|_{\min} = -\pi$ and $\phi|_{\max} = 0$.
Ans. (D)

EXAMPLE 11.2. A system has the transfer function

$$\frac{Y(s)}{X(s)} = \frac{s-\pi}{s+\pi}$$

Let $u(t)$ be the unit step function. The input $x(t)$ that results in a steady-state output $y(t) = \sin \pi t$ is

(A) $x(t) = \sin(\pi t)u(t)$ (B) $x(t) \sin\left(\pi t + \dfrac{\pi}{2}\right)u(t)$

(C) $x(t) = \sin\left(\pi t - \dfrac{\pi}{2}\right)u(t)$ (D) $x(t) = \cos\left(\pi t + \dfrac{\pi}{2}\right)u(t)$

Solution We know

$$x(t) \xrightarrow{A \sin(\pi t + \theta)} \boxed{H(s)} \xrightarrow{A' \sin(\pi t + \phi)} y(t)$$

where,

$$A' = A \cdot |H(\jmath\omega)|_{\omega=\omega_0} \quad \text{(i)}$$

$$\phi = \theta + \angle H(\jmath\omega)|_{\omega=\omega_0} \quad \text{(ii)}$$

Given, $\omega_0 = \pi,\ A' = 1,\ \phi = 0.$

Now,

$$|H(\jmath\omega)|_{\omega=\pi} = \frac{\sqrt{\omega^2+\pi^2}}{\sqrt{\omega^2+\pi^2}} = 1 \quad \text{(iii)}$$

$$\angle H(\jmath\omega)|_{\omega=\pi} = \left[180^\circ - \tan^{-1}\left(\frac{\omega}{\pi}\right)\right] - \tan^{-1}\left(\frac{\omega}{\pi}\right)$$

$$= 180^\circ - 45^\circ - 45^\circ = 90^\circ \quad \text{(iv)}$$

Therefore, $A = 1$ [from Eqs. (i) and (iii)]

and $\theta = -90^\circ$ [from Eqs. (ii) and (iv)]

Thus, $x(t) = \sin(\pi t - 90^\circ) = \sin[\pi t - (\pi/2)]$.
Ans. (C)

EXAMPLE 11.3. The transfer function of a system is

$$\frac{Y(s)}{R(s)} = \frac{s}{s+2}$$

The steady-state output $y(t)$ is $M\cos(2t+\phi)$ for the input $\cos(2t)$. The values of M and ϕ are, respectively

(A) $\dfrac{1}{\sqrt{2}},\ -45^\circ$ (B) $\dfrac{1}{\sqrt{2}},\ +45^\circ$ (C) $\sqrt{2},\ -45^\circ$ (D) $\sqrt{2},\ +45^\circ$

Solution In frequency domain, we get from the given transfer function

$$G(\jmath\omega) = \frac{\jmath\omega}{\jmath\omega + 2}$$

Therefore,

$$\left|G(\jmath\omega)\right|_{\omega=2} = \frac{2}{\sqrt{2^2+2^2}} = \frac{1}{\sqrt{2}} \equiv M$$

and

$$\angle G(\jmath\omega)\Big|_{\omega=2} = 90^\circ - \tan^{-1}\frac{2}{2} = 90^\circ - 45^\circ = 45^\circ \equiv \phi$$

Ans. (B)

EXAMPLE 11.4. An open-loop control system results in a response of $e^{-2t}(\sin 5t + \cos 5t)$ for a unit impulse input. The dc gain of the control system is ———.

Solution We note that the impulse response

$$c(t) = e^{-2t}(\sin 5t + \cos 5t)$$

which implies
$$g(t) = e^{-2t}(\sin 5t + \cos 5t)$$

Therefore,
$$\mathscr{L}\{g(t)\} \equiv G(s) = \frac{5}{(s+2)^2+5^2} + \frac{s+2}{(s+2)^2+5^2} \qquad \text{[see Appendix A]}$$

For dc gain
$$|G(s)|_{s=0} = \frac{5}{2^2+5^2} + \frac{2}{2^2+5^2} = \frac{7}{29}$$

Ans. $\frac{7}{29}$

EXAMPLE 11.5. Let a signal $a_1 \sin(\omega_1 t + \phi)$ be applied to a stable linear time-invariant system. Let the corresponding steady-state output be represented as $a_2 F(\omega_2 t + \phi_2)$. Then which of the following statements is true?

(A) F is not necessarily a 'sine' or 'cosine' function but must be periodic with $\omega_1 = \omega_2$.
(B) F must be a 'sine' or 'cosine' function with $a_1 = a_2$.
(C) F must be a 'sine' function with $\omega_1 = \omega_2$ and $\phi_1 = \phi_2$.
(D) F must be a 'sine' or 'cosine' function with $\omega_1 = \omega_2$.

Solution We know that for a linear time-invariant system the input and output wave will be the same with the same frequency though the amplitude and phase may change. So, F must be a 'sine' or 'cosine' function with $\omega_1 = \omega_2$.
Ans. (D)

EXAMPLE 11.6. In the system shown in Figure 11.5, the input is $x(t) = \sin t$. In the steady state, the response $y(t)$ will be

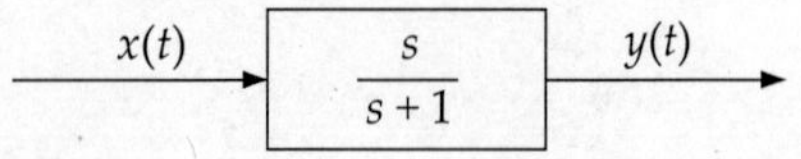

$x(t) \rightarrow \boxed{\dfrac{s}{s+1}} \rightarrow y(t)$

Figure 11.5: System (Example 11.6).

(A) $\frac{1}{\sqrt{2}}\sin(t - 45°)$ (B) $\frac{1}{\sqrt{2}}\sin(t + 45°)$ (C) $\sin(t - 45°)$ (D) $\sin(t + 45°)$

Solution Given, input $x(t) = \sin t$, which implies that frequency $\omega = 1$. From the given transfer function we get

$$G(j\omega) = \frac{j\omega}{j\omega + 1}$$

Given
$$M_{\text{input}} = 1$$

So,
$$M_{\text{output}} = \frac{\omega}{\sqrt{1+\omega^2}} = \frac{1}{\sqrt{2}}$$

$$\phi_{\text{input}} = 0°$$

$$\phi_{\text{output}} = 90° - \tan^{-1}\omega = 90° - 45° = 45°$$

Therefore,
$$y(t) = \frac{1}{\sqrt{2}}\sin(t + 45°)$$

Ans. (B)

EXAMPLE 11.7. A system with zero initial conditions has the closed-loop transfer function

$$T(s) = \frac{s^2+4}{(s+1)(s+4)}$$

The system output is zero at the frequency

(A) 0.5 rad/s (B) 1 rad/s (C) 2 rad/s (D) 4 rad/s

Solution In frequency domain, we get from the given transfer function

$$|T(j\omega)| = \left|\frac{(j\omega)^2+4}{(j\omega+1)(j\omega+4)}\right| = 0$$

Therefore, $-\omega^2 + 4 = 0$

or, $\omega = 2\,\text{rad/s}$

Ans. (C)

EXAMPLE 11.8. Open-loop transfer function of a unity feedback system is

$$G(s) = G_1(s)\cdot e^{-s\tau_D} = \frac{e^{-s\tau_D}}{s(s+1)(s+2)}$$

Given: $|G_1(j\omega)| \approx 1$ when $\omega = 0.466$

(a) Determine the phase margin when $\tau_D = 0$.

(b) Comment in one sentence on the effect of dead-time on the stability of the system.

(c) Determine the maximum value of dead-time τ_D for the closed-loop system to be stable.

Solution (a) The transfer function in frequency domain is

$$G(j\omega) = \frac{e^{-j\omega\tau_D}}{j\omega(j\omega+1)(j\omega+2)}$$

Also, from the given condition we know that the phase cross-over frequency $\omega_c = 0.466$. When $\tau_D = 0$, the phase margin, ϕ is given by

$$\begin{aligned}\phi &= 180^\circ - \left\{90^\circ + \tan^{-1}\omega_c + \tan^{-1}\frac{\omega_c}{2}\right\}\\ &= 180^\circ - \{90^\circ + \tan^{-1}(0.466) + \tan^{-1}(0.233)\}\\ &= 51.9^\circ\end{aligned}$$

(b) The system becomes less stable as the dead-time increases, because with the increase of dead-time gain margin and phase margin both decrease.

(c) Now

$$e^{-j\omega\tau_D} = \cos\omega\tau_D - j\sin\omega\tau_D$$

Corresponding phase

$$\begin{aligned}\phi_D &= -\tan^{-1}\frac{\sin\omega\tau_D}{\cos\omega\tau_D} = -\omega\tau_D\\ &= -\omega\tau_D \times \frac{180}{\pi}\end{aligned}$$

The maximum contribution of this phase can be when the phase margin becomes zero. Thus,

$$51.9 - \omega\tau_D \times \frac{180}{\pi} = 0$$

or,
$$\omega\tau_D = 51.9 \times \frac{\pi}{180^\circ} \approx 0.90$$

or,
$$\tau_D = \frac{0.90}{0.466} \approx 1.94\,\text{s}$$

EXAMPLE 11.9. Consider the following systems:

$$\text{System 1: } G(s) = \frac{1}{2s+1} \qquad \text{System 2: } G(s) = \frac{1}{5s+1}$$

The true statement regarding the systems is:

(A) Bandwidth of System 1 > bandwidth of System 2
(B) Bandwidth of System 1 < bandwidth of System 2
(C) Bandwidths of both the systems are the same
(D) Bandwidths of both the systems are infinite

Solution In frequency domain, the amplitudes of the systems are

$$M_1(j\omega) = \frac{1}{j2\omega_1 + 1} = \frac{1}{\sqrt{4\omega_1^2 + 1}} \qquad M_2 = \frac{1}{j5\omega_2 + 1} = \frac{1}{\sqrt{25\omega_2^2 + 1}}$$

To determine bandwidth, we put $M = \frac{1}{\sqrt{2}}$. So cut-off frequencies are determined as follows:

$$\frac{1}{\sqrt{4\omega_{c1}^2 + 1}} = \frac{1}{\sqrt{2}} \quad \Rightarrow \omega_{c1} = \frac{1}{2} \qquad \frac{1}{\sqrt{25\omega_{c2}^2 + 1}} = \frac{1}{\sqrt{2}} \quad \Rightarrow \omega_{c2} = \frac{1}{5}$$

Obviously, the bandwidth of the System 1 > bandwidth of System 2.
Ans. (A)

EXAMPLE 11.10. A linear time-invariant causal[7] system has a frequency response given in polar form as

$$\frac{1}{\sqrt{1+\omega^2}} \angle -\tan^{-1}\omega$$

For input $x(t) = \sin t$, the output is

(A) $\frac{1}{\sqrt{2}}\cos t$ (B) $\frac{1}{\sqrt{2}}\cos\left(t - \frac{\pi}{4}\right)$ (C) $\frac{1}{\sqrt{2}}\sin t$ (D) $\frac{1}{\sqrt{2}}\sin\left(t - \frac{\pi}{4}\right)$

Solution From $x(t) = \sin t = \sin(1)t$, we know that the input frequency $\omega = 1$. Hence, from the given response, we find the following magnitude M and phase ϕ of the output

$$M = \left.\frac{1}{\sqrt{1+\omega^2}}\right|_{\omega=1} = \frac{1}{\sqrt{2}} \qquad \phi = -\tan^{-1} 1 = -45^\circ$$

Therefore, the output will be

$$y(t) = \frac{1}{\sqrt{2}}\sin\left(t - \frac{\pi}{4}\right)$$

Ans. (D)

---

[7]For definition, see Section 2.3 at page 22.

EXAMPLE 11.11. The open-loop transfer function of a feedback control system is $1/(s+1)^3$. The gain margin of the system is

(A) 16 (B) 8 (C) 4 (D) 2

Solution At the outset, we note that the Gm value is not in dB. Now, the open-loop transfer function in frequency domain yields

$$G(\jmath\omega)H(\jmath\omega) = \frac{1}{(\jmath\omega+1)^3} \equiv \frac{1}{\left(\sqrt{\omega^2+1}\right)^3}\angle -3\tan^{-1}\omega$$

To calculate the phase crossover frequency ω_p, the gain at ω_p and the gain margin, we have

$$-3\tan^{-1}\omega_p = -180^\circ \quad \Rightarrow \tan^{-1}\omega_p = 60^\circ \quad \Rightarrow \omega_p = \tan 60^\circ = \sqrt{3}\ \text{rad/s}$$

$$\text{Gain}\Big|_{\omega_p} = \frac{1}{\left(\sqrt{\left(\sqrt{3}\right)^2+1}\right)^3} = 0.125$$

Therefore,

$$\text{Gm} = \frac{1}{0.125} = 8$$

Ans. (B)

EXAMPLE 11.12. The loop transfer function of a system is given by

$$G(s)H(s) = \frac{10e^{-Ls}}{s}$$

The phase crossover frequency is 5 rad/s. The value of the dead time L is

(A) $\pi/20$ (B) $\pi/10$ (C) $-\pi/20$ (D) zero

Solution From the given transfer function, we get in the frequency domain

$$G(\jmath\omega)H(\jmath\omega) = \frac{10e^{-\jmath L\omega}}{\jmath\omega} = \frac{10(\cos L\omega - \jmath\sin L\omega)}{\jmath\omega} \equiv \frac{10|\cos^2 L\omega + \sin^2 L\omega|\angle -\tan^{-1}\tan L\omega}{|\omega|\angle\tan^{-1}\dfrac{\omega}{0}}$$

$$= \frac{10|1|\angle -L\omega}{\omega\angle 90^\circ} = \frac{10}{\omega}\angle -L\omega - 90^\circ$$

For phase crossover frequency, ω_p we have the angle condition as

$$-L\omega_p - 90^\circ = -180^\circ \quad \Rightarrow L\omega_p = 90^\circ = \frac{\pi}{2}$$

$$\Rightarrow L = \frac{\pi}{2\omega_p} = \frac{\pi}{2\times 5} = \frac{\pi}{10} \qquad [\text{given, } \omega_p = 5\,\text{rad/s}]$$

Ans. (B)

EXAMPLE 11.13. Determine the frequency domain specifications for the unity feedback system given by

$$G(s) = \frac{256}{s(s+8)}$$

Solution The closed-loop transfer function of the system is

$$T(s) = \frac{C(s)}{R(s)} = \frac{G(s)}{1+G(s)H(s)} = \frac{256}{s^2+8s+256} \tag{i}$$

The standard transfer function of a second order system is given by

$$\frac{C(s)}{R(s)} = \frac{K\omega_n^2}{s^2+2\zeta\omega_n s+\omega_n^2} \tag{ii}$$

Comparing the denominators of Eq. (i) and (ii), we get

$$\omega_n^2 = 256 \quad \text{or,} \quad \omega_n = \sqrt{256} = 16 \text{ rad/s}$$

$$2\zeta\omega_n = 8 \quad \text{or,} \quad \zeta = \frac{8}{2(16)} = 0.25$$

Therefore,

$$M_r = \frac{1}{2\zeta\sqrt{1-\zeta^2}} = \frac{1}{2(0.25)\sqrt{1-(0.25)^2}} = 2.066$$

$$\omega_r = \omega_n\sqrt{1-2\zeta^2} = (16)\sqrt{1-2(0.25)^2} = 14.97 \text{ rad/s}$$

$$\text{Bw} = \omega_c = \omega_n\sqrt{1-2\zeta^2+\sqrt{(2\zeta^2-1)^2+1}}$$

$$= (16)\sqrt{1-2(0.25)^2+\sqrt{[2(0.25)^2-1]^2+1}} = 35.26 \text{ rad/s}$$

EXAMPLE 11.14. Determine the frequency domain specifications for the system whose closed-loop transfer function is given by

$$\frac{C(s)}{R(s)} = \frac{50}{s^2+10s+100}$$

Solution Since this is a closed-loop transfer function, the characteristic equation is given by

$$s^2+10s+100 = 0$$

Comparing it with that of a standard second order system, we get

$$\omega_n^2 = 100 \quad \text{or,} \quad \omega_n = \sqrt{100} = 10 \text{ rad/s}$$

$$2\zeta\omega_n = 10 \quad \text{or,} \quad \zeta = \frac{10}{2(10)} = 0.5$$

Therefore,

$$M_r = \frac{1}{2\zeta\sqrt{1-\zeta^2}} = \frac{1}{2(0.5)\sqrt{1-(0.5)^2}} = 1.154$$

$$\omega_r = \omega_n\sqrt{1-2\zeta^2} = (10)\sqrt{1-2(0.5)^2} = 7.07 \text{ rad/s}$$

$$\text{Bw} = \omega_c = \omega_n\sqrt{1-2\zeta^2+\sqrt{(2\zeta^2-1)^2+1}}$$

$$= (10)\sqrt{1-2(0.5)^2+\sqrt{[2(0.5)^2-1]^2+1}} = 12.72 \text{ rad/s}$$

EXAMPLE 11.15. A control system is represented by Figure 11.6. Determine K and α values so that $M_r = 2.0$ and $\omega_r = 10$. Find the corresponding bandwidth and settling time.

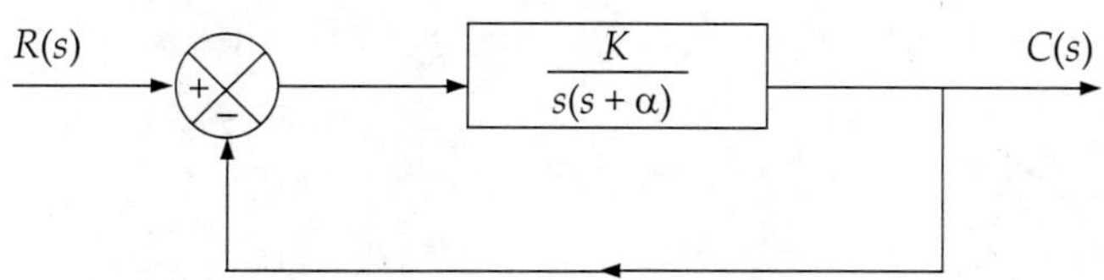

Figure 11.6: System (Example 11.15).

Solution The closed-loop transfer function of the system is

$$\frac{C(s)}{R(s)} = \frac{G(s)}{1+G(s)} = \frac{K}{s(s+\alpha)+K} = \frac{K}{s^2+\alpha s+K} \quad \text{(i)}$$

Comparing the denominator of Eq. (i) with that of a standard second order transfer function, we get

$$\omega_n^2 = K \quad \text{or,} \quad \omega_n = \sqrt{K} \quad \text{(ii)}$$

$$2\zeta\omega_n = \alpha \quad \text{or,} \quad \zeta = \frac{\alpha}{2\sqrt{K}} \quad \text{(iii)}$$

Now,

$$M_r = \frac{1}{2\zeta\sqrt{1-\zeta^2}} = \frac{1}{\dfrac{\alpha}{\sqrt{K}}\sqrt{1-\dfrac{\alpha^2}{4K}}} = 2 \text{ (given)}$$

This yields

$$\frac{2K}{\alpha\sqrt{4K-\alpha^2}} = 2 \quad \text{or,} \quad \alpha\sqrt{4K-\alpha^2} = K \quad \text{or,} \quad K^2 - 4\alpha^2 K + \alpha^4 = 0$$

Thus,

$$K = \frac{4\alpha^2 \pm \sqrt{16\alpha^4 - 4\alpha^4}}{2} = 3.732\alpha^2,\ 0.268\alpha^2 \quad \text{(iv)}$$

Secondly,

$$\omega_r = \omega_n\sqrt{1-2\zeta^2} = (\sqrt{K})\sqrt{1-2\left(\frac{\alpha}{2\sqrt{K}}\right)^2} = 10 \text{ (given)}$$

This yields

$$\sqrt{\frac{2K-\alpha^2}{2}} = 10 \quad \Rightarrow 2K - \alpha^2 = 200 \quad \text{(v)}$$

Of the two values of K, as obtained in Eq. (iv), only $3.732\alpha^2$ yields a real value for α when calculated from Eq. (v) as shown below

$$2(3.732\alpha^2) - \alpha^2 = 200 \quad \Rightarrow \alpha = \sqrt{\frac{200}{2(3.732)-1}} = 5.5624$$

A real value for α is necessary to make K positive. With this value of α, from Eq. (iv) K turns out to be

$$K = 3.732(5.5624)^2 = 115.47$$

Substituting these values of α and K in Eqs. (ii) and (iii), we get

$$\omega_n = \sqrt{115.47} = 10.75 \text{ rad/s}$$

$$\zeta = \frac{5.5624}{2\sqrt{115.47}} = 0.26$$

Therefore, the settling time is

$$t_s = \frac{4}{\zeta\omega_n} = \frac{4}{(0.26)(10.75)} = 1.43 \text{ s}$$

and the bandwidth is

$$\text{Bw} = \omega_c = \omega_n\sqrt{1 - 2\zeta^2 + \sqrt{(2\zeta^2 - 1)^2 + 1}}$$

$$= (10.75)\sqrt{1 - 2(0.26)^2 + \sqrt{[2(0.26)^2 - 1]^2 + 1}} = 15.9 \text{ rad/s}$$

EXAMPLE 11.16. A unity feedback system has an open-loop transfer function of

$$G(s) = \frac{e^{-0.1s}}{1 + 0.5s}$$

(a) Show that the phase crossover frequency of the system is approximately 16.9 rad/s.
(b) What is the gain margin of the system?

Solution Since $H(s) = 1$, the open-loop transfer function in frequency domain is

$$G(\jmath\omega)H(\jmath\omega) = \frac{e^{-\jmath 0.1\omega}}{1 + \jmath 0.5\omega} \equiv \frac{|1|\angle - 0.1\omega}{\sqrt{1 + 0.25\omega^2}\angle \tan^{-1} 0.5\omega} = \frac{1}{\sqrt{1 + 0.25\omega^2}}\angle(-0.1\omega - \tan^{-1} 0.5\omega)$$

(a) Since at the phase crossover frequency ω_p the phase angle equals $-180°$, we have

$$-0.1\omega - \tan^{-1} 0.5\omega = -\pi$$

$$\Rightarrow \quad 0.1\omega_p + \left(\frac{\pi}{2} - \frac{1}{0.5\omega_p}\right) = \pi \qquad \left[\because \tan^{-1} x \cong \frac{\pi}{2} - \frac{1}{x} \text{ for } x > 1\right]$$

$$\Rightarrow \quad 0.1\omega_p - \frac{1}{0.5\omega_p} = \frac{\pi}{2}$$

$$\Rightarrow \quad \frac{0.05\omega_p^2 - 1}{0.5\omega_p} = \frac{\pi}{2} = 1.571$$

or, $$0.05\omega_p^2 - 0.785\omega_p - 1 = 0 \qquad \text{(i)}$$

Equation (i) yields two values of ω_p, namely, 16.9 rad/s and -1.18 rad/s. The latter value is rejected because the frequency cannot be negative.

(b) The gain margin is calculated as follows:

$$\text{Gm} = -20\log\frac{1}{\sqrt{1 + 0.25\omega_p^2}} = -20\log\frac{1}{\sqrt{1 + 0.25 \times 16.9^2}} = 18.6 \text{ dB}$$

EXAMPLE 11.17. Consider a closed-loop system as shown in Figure 11.7. $G_p(s) = 14.4/[s(1 + 0.1s)]$ is the plant transfer function and $G_c(s) = 1$ is the compensator. For a unit step input, the output has damped oscillations. The damped natural frequency is ——— rad/s. (Round off to 2 decimal places).

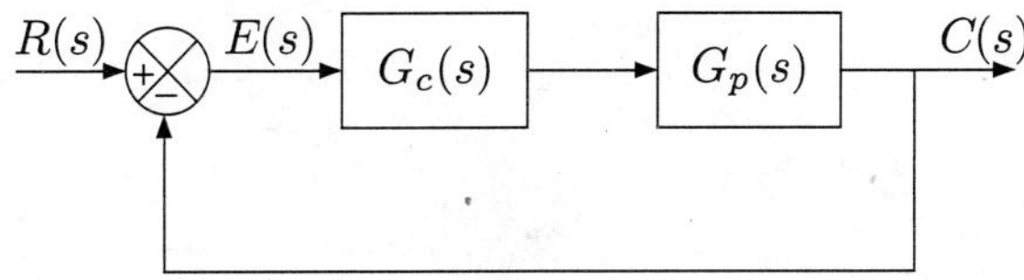

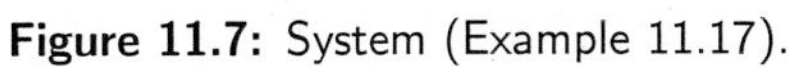
Figure 11.7: System (Example 11.17).

Solution The closed-loop transfer function for the given system is

$$\frac{C(s)}{R(s)} = \frac{G_c(s)G_p(s)}{1+G_c(s)G_p(s)} = \frac{\dfrac{14.4}{s(1+0.1s)}}{1+\dfrac{14.4}{s(1+0.1s)}} = \frac{14.4}{s(1+0.1s)+14.4}$$

$$= \frac{14.4}{0.1s^2+s+14.4} = \frac{144}{s^2+10s+144} \tag{i}$$

Comparing the denominator of Eq. (i) with that of a standard 2nd order system, namely, $s^2+2\zeta\omega_n s+\omega_n^2$, we get

$$\omega_n^2 = 144 \quad \text{or,} \quad \omega_n = 12$$

$$2\zeta\omega_n = 10 \quad \text{or,} \quad \zeta = \frac{10}{2\times 12} = 0.416$$

Therefore, the damped natural frequency is

$$\omega_d = \omega_n\sqrt{1-\zeta^2} = 12\sqrt{1-(0.416)^2} = 10.90\,\text{rad/s}$$

EXAMPLE 11.18. A unity feedback control loop with an open-loop transfer function of the form $K/[s(s+a)]$ has a gain crossover frequency of 1 rad/s and a phase margin of 60°. If an element having a transfer function $(s-\sqrt{3})/(s+\sqrt{3})$ is inserted into the loop, what will be the phase margin?

Solution In frequency domain, the given transfer function can be written as

$$\frac{K}{\omega\sqrt{\omega^2+a^2}}\angle\left(-90^\circ - \tan^{-1}\frac{\omega}{a}\right)$$

The given value of phase margin is 60° and the gain crossover frequency is $\omega_g = 1$ rad/s. So, we have

$$180^\circ - 90^\circ - \tan^{-1}\frac{1}{a} = 60^\circ \qquad [\because \omega_g = 1, \text{ given}]$$

$$\Rightarrow \qquad \tan^{-1}\frac{1}{a} = 30^\circ \qquad \Rightarrow \frac{1}{a} = \tan 30^\circ = \frac{1}{\sqrt{3}} \qquad \Rightarrow a = \sqrt{3}$$

Now, at the gain crossover frequency [given, 1 rad/s], the gain is 1 which in dB is 0. Therefore,

$$\left.\frac{K}{\omega\sqrt{\omega^2+a^2}}\right|_{\omega=1} = 1 \qquad \Rightarrow \frac{K}{\sqrt{1+a^2}} = 1 \qquad \Rightarrow K = \sqrt{a^2+1} = \sqrt{3+1} = 2$$

So, the original open-loop transfer function is given by

$$G(s)H(s) = \frac{2}{s(s+\sqrt{3})}$$

With the insertion of a new element of transfer function $\dfrac{s-\sqrt{3}}{s+\sqrt{3}}$, the new open-loop transfer function becomes

$$G'(s)H(s) = \frac{2}{s(s+\sqrt{3})} \cdot \frac{s-\sqrt{3}}{s+\sqrt{3}} = \frac{2(s-\sqrt{3})}{s(s+\sqrt{3})^2}$$

To determine the Pm of this open-loop transfer function, we first find the gain crossover frequency as follows:

$$\frac{2\sqrt{\omega_g^2+3}}{\omega_g(\omega_g^2+3)} = 1 \quad \Rightarrow \frac{2}{\omega_g\sqrt{\omega_g^2+3}} = 1 \quad \Rightarrow \frac{4}{\omega_g^2+3} = \omega_g^2 \quad \Rightarrow \omega_g^4 + 3\omega_g^2 - 4 = 0 \quad \text{(i)}$$

Substituting $\omega_g^2 = x$, the two values of x that we get from Eq. (i) are 1 and -4. The latter is rejected because a frequency cannot be imaginary. So, $\omega_g = 1$. That ω_g remains unchanged is expected because a zero and a pole have been added to the transfer function on both sides of the $\Im$-axis. Anyway, the phase margin is now calculated as follows:

$$\text{Pm} = 180^\circ + \left(\pi - \tan^{-1}\frac{\omega_g}{\sqrt{3}}\right) - 90^\circ - 2\tan^{-1}\frac{\omega_g}{\sqrt{3}} = 270^\circ - 3\tan^{-1}\frac{1}{\sqrt{3}} = 180^\circ$$

EXAMPLE 11.19. Consider a unity feedback system with open-loop transfer function

$$G(s) = \frac{1+6s}{s^2(1+s)(1+2s)}$$

(i) The phase crossover frequency of the system in radians per second is

(A) 0.125 (B) 0.25 (C) 0.5 (D) 1

(ii) The gain margin of the system is

(A) 0.125 (B) 0.25 (C) 0.5 (D) 1

Solution In the frequency domain, the open-loop transfer function can be written as

$$G(\jmath\omega)H(\jmath\omega) = \frac{1+\jmath 6\omega}{(\jmath\omega)^2(1+\jmath\omega)(1+\jmath 2\omega)}$$
$$\equiv \frac{\sqrt{1+36\omega^2}}{\omega^2\sqrt{1+\omega^2}\sqrt{1+4\omega^2}}\angle\left(\tan^{-1}6\omega - 180^\circ - \tan^{-1}\omega - \tan^{-1}2\omega\right)$$

(i) To find out the phase crossover frequency ω_p, we have

$$\tan^{-1}6\omega_p - 180^\circ - \tan^{-1}\omega_p - \tan^{-1}2\omega_p = -180^\circ$$
$$\Rightarrow \qquad \tan^{-1}6\omega_p = \tan^{-1}\omega_p + \tan^{-1}2\omega_p = \tan^{-1}\frac{3\omega_p}{1-2\omega_p^2}$$

Taking tangents of both sides, we get

$$1-2\omega_p^2 = \frac{3\omega_p}{6\omega_p} = \frac{1}{2} \quad \text{or,} \quad 2\omega_p^2 = \frac{1}{2} \quad \text{or,} \quad \omega_p = \frac{1}{2} = 0.5$$

Ans. (C)

(ii) To calculate the Gm, we observe that

$$\text{Gain}\Big|_{\omega_p} = \left.\frac{\sqrt{1+36\omega^2}}{\omega^2\sqrt{1+\omega^2}\sqrt{1+4\omega^2}}\right|_{\omega=0.5} = \frac{\sqrt{1+\dfrac{36}{4}}}{\dfrac{1}{4}\sqrt{\left(1+\dfrac{1}{4}\right)\left(1+\dfrac{4}{4}\right)}} = \frac{\sqrt{10}}{\dfrac{1}{4}\sqrt{\dfrac{10}{4}}}$$

$$= 8$$

Therefore,

$$\text{Gm} = \frac{1}{8} = 0.125$$

Ans. (A)

With this background on the frequency domain, we move to the study of stability of systems in this domain via Bode, polar and Nyquist plots.

REVIEW QUESTIONS

11.1 Choose the correct answer:

(i) The phase crossover frequency is the frequency at which the phase is

(A) $0°$ (B) $90°$ (C) $-180°$ (D) $180°$

(ii) The phase cross-over frequency of the transfer function $G(s) = 100/(s+1)^3$ in rad/s is

(A) $\sqrt{3}$ (B) $1/\sqrt{3}$ (C) 3 (D) $3\sqrt{3}$

(iii) The gain crossover frequency is at which the gain is

(A) 0 (B) 1 (C) 10 (D) 100

(iv) At the cut-off frequency the magnitude $|M(j\omega)|$ equals

(A) $\frac{1}{2}$ (B) $\frac{1}{\sqrt{2}}$ (C) 1 (D) 0

(v) The bandwidth is equal to

(A) gain crossover frequency (B) phase crossover frequency
(C) cut-off frequency (D) resonant frequency

(vi) If the damping ratio ζ increases, the bandwidth

(A) decreases (B) increases
(C) remains the same (D) none of the above is correct

(vii) No resonant peak occurs for

(A) $\zeta < 0.707$ (B) $\zeta > 0.707$ (C) $\zeta < 0.5$ (D) $\zeta > 0.5$

(viii) At resonance

(A) $M_r \to 0$ (B) $M_r \to 1$ (C) $M_r \to \frac{1}{\sqrt{2}}$ (D) $M_r \to \infty$

(ix) A closed-loop system is stable for

(A) Gm = 15 dB and Pm = $70°$ (B) Gm = -15 dB and Pm = $70°$
(C) Gm = 15 dB and Pm = $-70°$ (D) Gm =-15 dB and Pm = $-70°$

(x) The loop transfer function of a system is given by $G(s) = \dfrac{e^{-0.1s}}{s}$. The phase crossover frequency is given by

(A) $\frac{\pi}{2}$ (B) $\frac{\pi}{10}$ (C) $\frac{\pi}{0.2}$ (D) $\frac{\pi}{4}$

(xi) The open-loop transfer function of a unity feedback system is given by $G(s) = \frac{e^{-0.1s}}{s}$. The gain margin of this system is

(A) 11.95 dB (B) 17.67 dB (C) 21.33 dB (D) 23.9 dB

(xii) A system has a gain margin of -4. The system is

(A) stable (B) unstable
(C) marginally stable (D) asymptotically stable

(xiii) Gain margin is a measure of

(A) relative stability (B) controllability
(C) observability (D) absolute stability

(xiv) Consider an LTI system with transfer function $H(s) = 1/[s(s+4)]$. If the input to the system is $\cos(3t)$ and the steady-state output is $A\sin(3t+\alpha)$, then the value of A is

(A) 1/30 (B) 1/15 (C) 3/4 (D) 4/3

(xv) The open-loop transfer function of a unity feedback system is given by $G(s) = (e^{-0.1s})/s$. The gain margin of the system is

(A) 11.95 dB (B) 17.67 dB (C) 21.33 dB (D) 23.9 dB

11.2 Define the following

(a) Gain margin
(b) Phase margin
(c) Phase crossover frequency
(d) Gain crossover frequency

11.3 Determine frequency domain specifications for a unity feedback control system having

$$G(s) = \frac{144}{s(s+1)}$$

11.4 A control system is represented by Figure 11.8. Determine the values of K and α so that $M_r = 1.05$ and $\omega_r = 12$ rad/s. Find out the corresponding bandwidth.

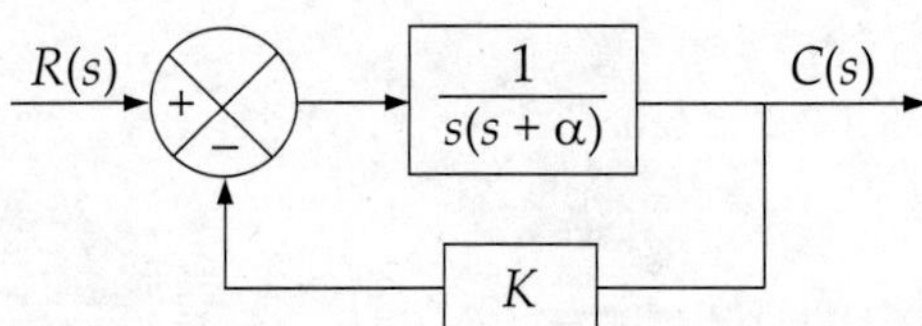

Figure 11.8: System (Question 11.4).

11.5 Consider a linear time-invariant with transfer function

$$H(s) = \frac{1}{s+1}$$

If the input is $\cos(t)$ and the steady-state output is $A\cos(t+\alpha)$, then the value of A is ______.

11.6 A unity feedback system has an open-loop transfer function of

$$G(s) = \frac{10000}{s(s+1)^2}$$

(a) Determine the magnitude of $G(\jmath\omega)$ in dB at an angular frequency of $\omega = 20\,\text{rad/s}$.
(b) Determine the phase margin in degrees.
(c) Determine the gain margin in dB.
(d) Is the system stable or unstable?

Answers to Selected Questions

11.1 (i) (C) (ii) (A) (iii) (B) (iv) (B) (v) (C) (vi) (A) (vii) (B) (viii) (D) (ix) (A) (x)(C) (xi) (D) (xii) (B) (xiii) (A) (xiv) (B) (xv) (D)

11.3. $\omega_r = 11.98$ rad/s $M_r = 12.51$ Bw $= 18.62$ rad/s

11.4. $K = 472.7$ $\alpha = 25.64$ Bw $= 25.25$ rad/s

11.5. 0.707

11.6. (a) 0 dB (b) $-36.88°$ (c) -13.97 dB (d) Unstable

12

Bode Plot

12.1 INTRODUCTION

In Chapter 11, we showed that the study of sinusoidal inputs is important. It is also known how phasor representations of the input and the system transfer function can be used to easily determine system output [see Appendix C]. This is referred to as the *frequency domain behaviour of a system.*

Bode[1] plots are a very useful way of representing the gain and phase of a system as a function of frequency.

What are Bode plots?

A transfer function in frequency domain is a complex quantity, comprising real and imaginary components. We know that a complex quantity can be represented by a magnitude and phase angle. For example,

$$a + \jmath b \equiv |\sqrt{a^2 + b^2}| \angle \tan^{-1} \frac{b}{a}$$

Bode plots are nothing but a way of representing the frequency domain transfer function with two plots—(i) magnitude vs. frequency plot and (ii) phase vs. frequency plot. Plots are logarithmic on the frequency axis, and the magnitude is plotted in dBs (deciBels[2]).

The question arises: Why is logarithmic scale used? Drawing the phase is fairly simple. We can draw each phase term separately, and then simply add them. The magnitude term is not so straightforward because of the fact that the magnitude terms are *multiplied.*

Consider, for example, a simple open-loop transfer function

$$G(s)H(s) = K\frac{s+b}{(s+a_1)(s+a_2)} \equiv K\frac{\jmath\omega + b}{(\jmath\omega + a_1)(\jmath\omega + a_2)}$$

In magnitude and phase form it can be written as

$$G(\jmath\omega)H(\jmath\omega) = K\left|\frac{\sqrt{\omega^2 + b^2}}{\sqrt{\omega^2 + a_1^2}\sqrt{\omega^2 + a_2^2}}\right| \angle\left(\tan^{-1}\frac{\omega}{b} - \tan^{-1}\frac{\omega}{a_1} - \tan^{-1}\frac{\omega}{a_2}\right)$$

Here, we observe that the magnitude terms are multiplied[3]. It would be much easier if they were added. Then we could draw each term on a graph and just add them.

One way to transform multiplication into addition is by using the logarithm. Instead of using a simple logarithm, Bode used decibel or dB. The relationship between a quantity, Q, and its dB representation, X, is given by

$$X = 20 \log_{10} Q \tag{12.1}$$

Table 12.1 gives an idea of how X values vary with Q. So, Bode plots are nothing but logarithmic

[1]Pronounced Boh-dee or Boh-day.

[2]Named after Alexander Graham Bell (1847–1922), the American inventor of the telephone.

[3]Division is also multiplication. It is multiplication of the inverse of the quantity.

Table 12.1: dB values for common quantities

Quantity (Q)	1/100	1/10	1/2	$1/\sqrt{2}$	1	$\sqrt{2}$	2	3.16	5	10	100
dB value (X)	−40	−20	−6	−3	0	3	6	10	14	20	40

magnitude and phase plots of the open-loop transfer function versus the frequency (also in logarithmic scale). Bode derived stability criteria and other things of transfer functions from such plots.

In those days, computers or even electronic calculators were not available. Engineering calculations were to be done with the help of Log Tables or Slide Rules which also utilized logarithmic values of numbers. So, he simplified the procedure by inventing asymptotic plots of poles and zeros of transfer functions and ways of adding them to obtain magnitude and phase of the open-loop transfer function.

12.2 EQUIVALENCE OF SCALES IN LOGARITHMIC PLOT

As told earlier, that the frequency domain transfer function is a complex quantity and hence it can be represented in magnitude and angle form, as follows:

$$G(\jmath\omega) = |G(\jmath\omega)|\angle G(\jmath\omega) = M(\omega)\angle\phi$$

If the gain M is expressed in dB then[4]

$$\text{Gain in dB} = 20\log M(\omega) \tag{12.2}$$

Now, suppose

$$x = \log\omega$$

Therefore,

$$x_2 - x_1 = \log\omega_2 - \log\omega_1 = \log\frac{\omega_2}{\omega_1}$$

$(x_2 - x_1)$ gives an interval between two frequencies in the logarithmic scale. If the interval is an *octave*, i.e. if $\omega_2 = 2\omega_1$, using Eq. (12.2) we find

$$\text{Gain in dB} = 20\log 2 = 6$$

So, it is a 6 dB gain. But if the interval is a decade, i.e. if $\omega_2 = 10\omega_1$, it is a 20 dB gain. Now, since

$$\frac{6}{\log 2} = 20 = \frac{20}{\log 10}$$

we have

$$\boxed{\frac{6\text{ dB}}{\text{octave}} = \frac{20\text{ dB}}{\text{decade}}}$$

when the frequency is plotted in a logarithmic scale.

12.3 BODE PLOT REQUIREMENTS

The open-loop transfer function of a system of type N and gain-factor K can be written as

$$G(s)H(s) = \frac{K(1+sT_1)(1+sT_2)\cdots\omega_n^2}{s^N(1+sT_a)\cdots(s^2+2\zeta\omega_n s+\omega_n^2)}$$

[4]We will write 'log' to indicate $\log_{10}$ and 'ln' to indicate $\log_e$.

Obviously, it contains (i) first order terms in the numerator and (ii) (first + second) order terms in the denominator. The magnitude and phase of its frequency domain representation can be rewritten in terms of dB as

$$
\begin{aligned}
|M| = 20\log|G(\jmath\omega)H(\jmath\omega)| &= 20\log K + 20\log|1+\jmath\omega T_1| + \cdots \\
&\quad - 20N\log|\jmath\omega| - 20\log|1+\jmath\omega T_a| - \cdots \\
&\quad - 20\log\left|\frac{\omega_n^2-\omega^2}{\omega_n^2} + \jmath\frac{2\zeta\omega_n\omega}{\omega_n^2}\right| \qquad (12.3)
\end{aligned}
$$

$$
\begin{aligned}
\phi = \angle G(\jmath\omega)H(\jmath\omega) &= \angle 0^\circ + \tan^{-1}\omega T_1 + \tan^{-1}\omega T_2 + \cdots \\
&\quad - N\angle 90^\circ - \tan^{-1}\omega T_a - \cdots \\
&\quad - \tan^{-1}\left[\frac{2\zeta\omega_n\omega}{\omega_n^2-\omega^2}\right] \qquad (12.4)
\end{aligned}
$$

If we want to draw magnitude and phase plots by continuous curves by calculating them at different frequencies, as is done by software like Matlab, we may do so without going through the following discussion. But if we want to get approximate plots, called *asymptotic plots*, as was suggested by Bode, we have to note the following.

A scrutiny of Eqs. (12.3) and (12.4) reveals that the expressions for magnitude and phase consist of the following factors:

1. Constant: K
2. Zeros or poles at the origin: $(\jmath\omega)^{\pm 1}$
3. Simple zeros or poles: $(1+\jmath\omega T)^{\pm 1}$
4. Quadratic zeros or poles: $(1+\jmath 2\zeta u - u^2)^{\pm 1}$ where $u = \omega/\omega_n$

Let us examine the characteristics of Bode, or asymptotic, plots for each factor individually.

Plot for Constant *K*

The magnitude plot is a horizontal line through $20\log|K|$ and the phase plot is a horizontal line through 0° (see Figures 12.1a and 12.1b).

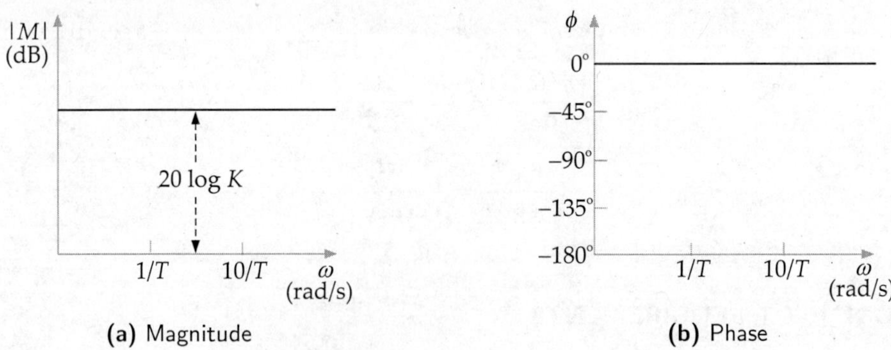

Figure 12.1: Plots for constant.

For a constant

1. the magnitude plot is horizontal line through $20\log|K|$,
2. the phase plot is a horizontal line through $0°$.

Plots for Zero and Pole at Origin

(a) Zero at origin: $\jmath\omega^N$. Here, $|M| = 20N\log\omega$ and $\phi = (+N)\times 90°$. For $\omega = 10$ and $N = 1$, $|M| = 20$ dB.

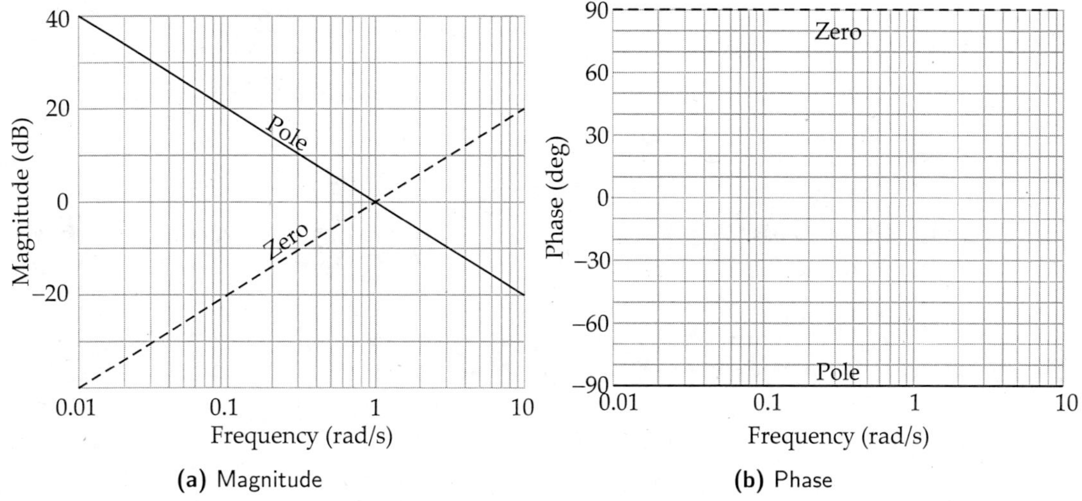

(a) Magnitude **(b)** Phase

Figure 12.2: Bode asymptotic plots for a zero and a pole at the origin.

For a single zero at the origin,

(a) the magnitude plot is a straight line of slope 20 dB/decade passing through 0 dB at $\omega = 1$ [see Figure 12.2a], and

(b) the phase plot is a horizontal line through $90°$ [see Figure 12.2b].

(b) Pole at origin: $\jmath\omega^{-N}$

For obvious reasons, plots for the pole at the origin are mirror images of those for zero. They are shown in Figure 12.2.

Plots for Simple Zero and Simple Pole

(a) Simple zero: $(1 + \jmath\omega T)$

Magnitude. Here,

$$|M| = 20\log\sqrt{1+\omega^2T^2} \tag{12.5}$$

At low frequencies, $\omega T \ll 1$

Therefore, $|M| = 0$ dB

At high frequencies, $\omega T \gg 1$

Therefore, $|M| = 20\log\omega T$ dB (12.6)

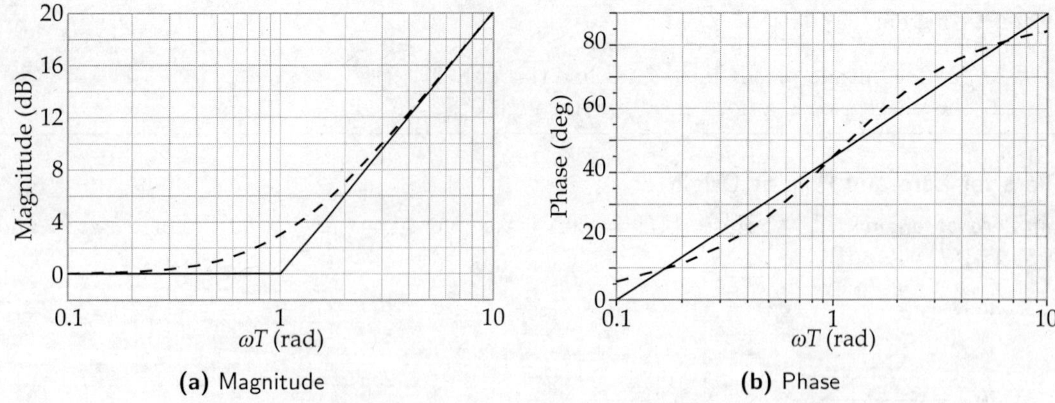

Figure 12.3: Bode plots for a simple zero. — asymptotic, - - - actual.

For $\omega T = 10$, $|M| = 20$ dB. Thus, a simple zero is represented by a horizontal line through 0 dB and then a straight line of slope 20 dB/decade through the *corner* (or *break*) *frequency.*

Corner (or Break) frequency is the intersection point of low and high frequency asymptotes. So, if ω_c is the corner frequency[5], we can equate the $|M|$-value at $\omega_c T$ to that at $\omega T = 1$. Thus,

$$20 \log \omega_c T = 20 \log 1 \qquad \Rightarrow \omega_c T = \frac{1}{T} \tag{12.7}$$

The plot drawn for simple zero, using Eqs. (12.6) and (12.7), is called *asymptotic.* The actual curve can be obtained by using Eq. (12.5) [see broken curves in Figures 12.3a and 12.3b]. The error in the asymptotic value at $\omega_c T = 1$ is

$$\text{Actual} - \text{asymptotic} = 20 \log \sqrt{1+1} - 20 \log 1 = 3.01 \text{ dB}$$

Table 12.2 lists the error for other frequencies. It can be seen that the error is maximum at the corner frequency.

Frequency	$0.5/T$	$1/T$	$2/T$
Approx. error	1 dB	3 dB	1 dB

Table 12.2: Error at frequencies near the corner frequency

Phase. The exact relation is

$$\phi = \tan^{-1} \omega T$$

Approximately, it is 45°/decade because $\phi|_{\omega T = 0.1} = 0°$ and $\phi|_{\omega T=1} = 45°$. Table 12.3 as well as Figure 12.3b shows that the maximum error in approximate plot is about 6°.

ω	$0.1/t$	$1/T$	$2/T$	$10/T$
$\phi_{\text{asymptotic}}$	0°	45°	60°	90°
ϕ_{exact}	5.7°	45°	64°	84.3°

Table 12.3: Exact and asymptotic phase values for simple zero

[5]It should not be confused with the *cut-off frequency* which is also written as ω_c.

(b) Simple pole: $(1+\jmath\omega T)^{-1}$. For obvious reasons, plots for a simple pole are mirror images of those for a zero.

1. For a simple zero
   (a) the magnitude is represented by a horizontal line through 0 dB and then a straight line of slope 20 dB/decade through the corner frequency, and
   (b) the phase is represented by 45°/decade line through the corner frequency which starts at $(0.1\omega_c,\ 0°)$ and ends at $(10\omega_c,\ 90°)$.
2. Plots for a simple pole are mirror images of those for a zero.

Plots for Quadratic Zero and Pole

(a) Quadratic zero

Magnitude. Here,

$$|M| = 20\log\left|\left(1-\frac{\omega^2}{\omega_n^2}\right)+\jmath 2\zeta\frac{\omega}{\omega_n}\right| = 20\log\sqrt{\left(1-\frac{\omega^2}{\omega_n^2}\right)+4\zeta^2\frac{\omega^2}{\omega_n^2}}\ \text{dB} \tag{12.8}$$

For low frequencies, i.e., at $\omega \ll \omega_n$,

$$|M| = 20\log 1 = 0\ \text{dB} \tag{12.9}$$

This means that the $\log\omega$ asymptote is a horizontal line through 0 dB at low frequencies. For high frequencies, i.e., at $\omega \gg \omega_n$,

$$|M| = 20\log\sqrt{\left(\frac{\omega^2}{\omega_n^2}\right)^2} = 40\log\frac{\omega}{\omega_n} \tag{12.10}$$

For a decade, i.e. $\omega/\omega_n = 10$, $|M| = 40$ dB from Eq. (12.10). This implies that the $\log\omega$ asymptote is a straight line of 40 dB/decade (12 dB/octave) at frequencies above the corner frequency. The corner frequency, as can be seen by equating Eqs. (12.9) and (12.10), is $\omega_c = \omega_n$. Therefore, the approximate value of $|M|_{\text{approx}}$ is 0. For $\omega = \omega_n$, the exact value of $|M|$, as determined from Eq. (12.8), is

$$|M|_{\text{exact}} = 20\log\sqrt{4\zeta^2} = 20\log 2\zeta \tag{12.11}$$

$$\text{Error} = |M|_{\text{exact}} - |M|_{\text{approx}} = 20\log 2\zeta \tag{12.12}$$

Equation (12.12) shows that the error at the corner frequency is a function of ζ. Table 12.4 gives an idea of the error at different damping ratios.

Table 12.4: Error vs. damping ratio at the corner frequency

ζ	0.1	0.3	0.5	0.7	0.9	1.0
Error (dB)	−13.98	−4.44	0	2.92	5.10	6.02

The complete asymptotic plot, along with the exact plot, of the magnitude of a quadratic zero is given in Figure 12.4a.

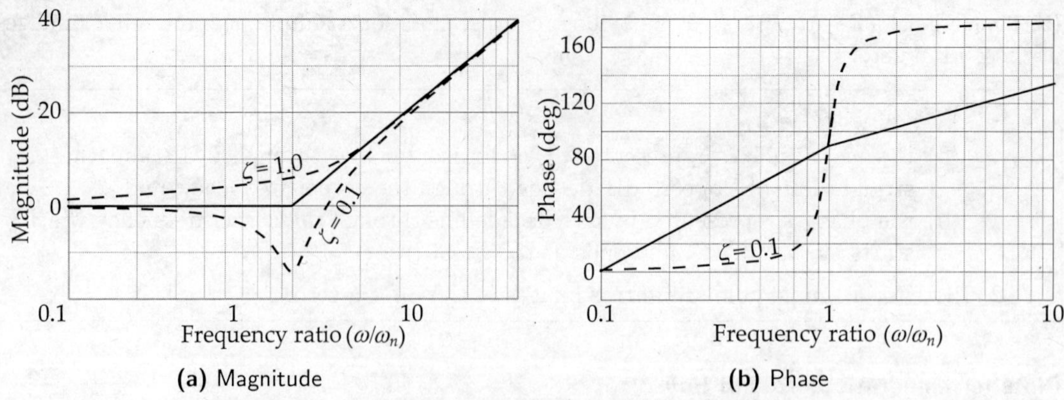

(a) Magnitude **(b)** Phase

Figure 12.4: Plots for quadratic zero. — asymptotic, - - - actual.

Phase. For a quadratic zero the phase is given by

$$\phi = \tan^{-1}\left[\left(2\zeta \cdot \frac{\omega}{\omega_n}\right) \div \left(1 - \frac{\omega^2}{\omega_n^2}\right)\right] \tag{12.13}$$

From Eq. (12.13) we can quickly obtain the following values of frequency and phase:

ω	0	ω_n	∞
ϕ	0°	90°	180°

However, Eq. (12.13) indicates negative values for ϕ for $\omega > \omega_n$. To correct it, we need to add 180° to such values. The graph shows, although asymptotic, Bode plots for the magnitude [Figure 12.4a] at higher ζ are not too bad. However, the asymptotic plot for the phase [Figure 12.4b] of quadratic zero is far from acceptable.

(b) Quadratic pole Plots for a quadratic pole are mirror images of those for the zero.

1. For a quadratic zero
   (a) the magnitude is represented by +40 dB/decade line from $\omega = \omega_n$; if $\zeta < 0.7$, a peak is drawn at frequency $\omega_r = \omega_n\sqrt{1-2\zeta^2}$ with amplitude $20\log\left(2\zeta\sqrt{1-\zeta^2}\right)$, and
   (b) the phase curve is drawn from calculating the phase value from the phase relation.
2. For a quadratic pole, plots are mirror images of those for the zero.

12.4 DRAWING BODE PLOTS

To draw a Bode diagram *for an open-loop transfer function*, we have to take the following steps:

1. We rewrite the transfer function in a form such that first order factors look like $1 + s/a$ and the second order factors look like $1 + bs + s^2/c$, where a, b and c are constants. Then we convert the function to frequency domain form by substituting $s = \jmath\omega$. We consider

an example to show how to do it.

$$G(s)H(s) = 30 \cdot \frac{s+12}{(s+2)(s+60)}$$

$$= 30 \cdot \frac{12\left(1+\frac{s}{12}\right)}{2\left(1+\frac{s}{2}\right)60\left(1+\frac{s}{60}\right)} = 3 \cdot \frac{1+\frac{s}{12}}{\left(1+\frac{s}{2}\right)\left(1+\frac{s}{60}\right)}$$

$$\equiv 3 \cdot \frac{1+\frac{\jmath\omega}{12}}{\left(1+\frac{\jmath\omega}{2}\right)\left(1+\frac{\jmath\omega}{60}\right)}$$

2. We find corner frequencies for different factors of the open-loop transfer function. In the above example, corner frequencies are 2, 12 and 60 rad/s. We have to remember that when we plot the graph, the frequency in the semi-log graph paper[6] must start from a lower value than the lowest of corner frequencies.
3. For the phase plot, we need to estimate the total variation of angle using Eq. (12.4) and accordingly choose the axis in the graph paper. In the above example, the *frequency should start from 0.2 rad/s which is 1/10th of the lowest corner frequency 2 and it should extend to 600 rad/s, which is ten times the highest corner frequency 60.*
4. We plot magnitude and phase for each factor according to prescriptions given in boxes in Sections 12.3 through 12.3.
5. We add results obtained from step 4 to draw the overall Bode diagram.

Here we need to keep it in mind that following these prescriptions we are drawing *asymptotic Bode plots* and *not exact ones.*

Drawing exact Bode plots entails evaluation of magnitude and phase values for the entire open-loop transfer function at many frequencies. Matlab[7] and Python (a free programming language having control tools) precisely do that job.

Nevertheless, asymptotic plots are useful for approximate calculations and, therefore, they find wide use in the design of control systems.

Note: 1. For drawing asymptotic plots for magnitude of quadratic poles and zeros, peaks exist for $0 < \zeta < 0.707$ $(= 1/\sqrt{2})$ and peak frequency is typically very near the corner frequency.

2. For drawing phase plots of quadratic poles and zeros, it is better to calculate phases at a few frequencies and draw exact plots. In fact, even for other factors, similar exact phase plots are better drawn, because summing the different contributions poses problem as we will see later.

3. For nth order pole or zero, we need to make asymptotes, peaks and slopes n times higher (i.e., second order asymptote is -40 dB/decade and phase goes from $0°$ to $-180°$).

EXAMPLE 12.1. Draw the asymptotic Bode diagram for the unity feedback system having

$$G(s) = \frac{10(s+20)}{s^2 + 210s + 2000}$$

[6]The semi-log graph paper can be downloaded from http://www.intmath.com/downloads/Graph-paper.php.

[7]A software produced by *The Mathworks Inc.*, Natick, Massachusetts, USA.

Solution

1. Rewrite the transfer function to a suitable form as discussed earlier. Thus,

$$G(s)H(s) = \frac{10^3(s+20)}{s^2+210s+2000} = \frac{10^3(s+20)}{(s+10)(s+200)}$$

$$= 10^3 \frac{20\left(1+\frac{s}{20}\right)}{10\left(1+\frac{s}{10}\right)200\left(1+\frac{s}{200}\right)} = \frac{10\left(1+\frac{s}{20}\right)}{\left(1+\frac{s}{10}\right)\left(1+\frac{s}{200}\right)}.$$

Therefore, $$G(\jmath\omega)H(\jmath\omega) = \frac{10\left(1+\frac{\jmath\omega}{20}\right)}{\left(1+\frac{\jmath\omega}{10}\right)\left(1+\frac{\jmath\omega}{200}\right)}$$

2. The open-loop transfer function consists of the following components:
   (a) A constant gain $K = 10$.
   (b) A simple zero at $\omega_c = 20$.
   (c) A pair of simple poles at $\omega_c = 10$ and $\omega_c = 200$.

3. Before starting the plot, we need to determine the scales, especially the frequency scale which should be common for both the magnitude and the phase plots. Here, the *phase is the determining factor.* We observe that the lowest and highest corner frequencies are 10 and 200, respectively. So, for the phase plot we need a span of 1 (0.1×10) to 2000 (10×200). Therefore, we choose a span of 1 to 10000 for the frequency.

   For the magnitude we choose a span of +20 to − 20 dB to cover both the constant and poles. For phase, we need to select a span of +90° to − 90°.

   Magnitude plot
   (a) For $K = 10$, $|M| = 20\log 10 = 20$ dB. So, a horizontal line is drawn at 20 dB.
   (b) For the simple zero, a 20 dB/decade line is drawn at $\omega_c = 20$.
   (c) For the two simple poles, two −20 dB/decade lines are drawn at $\omega_c = 10$ and $\omega_c = 200$.

   Phase plot
   (a) A horizontal line at 0° is drawn for the constant K.
   (b) A line joining 0° at $\omega = 2$ [0.1× corner frequency of zero] and 90° at $\omega = 200$ [10× corner frequency of zero] is drawn for the zero at $\omega_c = 20$.
   (c) A line joining 0° at $\omega = 1$ [0.1× corner frequency of zero] and −90° at $\omega = 100$ [10× corner frequency of zero] is drawn for the pole at $\omega_c = 10$.
   (d) A line joining 0° at $\omega = 20$ [0.1× corner frequency of zero] and −90° at $\omega = 2000$ [10× corner frequency of zero] is drawn for the pole at $\omega_c = 200$.

4. Individual contributions are added to obtain the resultant asymptotic plots for magnitude and phase [see Figures 12.5a and 12.5b]. The question is, how to add the individual components?

   For the magnitude plot, we note that the only component that exists till $\omega = 10$ is the 20 dB contribution from the constant. Then the −20 dB/decade contribution from the simple pole at $\omega_c = 10$ starts. So the the resultant curve assumes the same slope till $\omega = 20$ when it encounters a +20 dB/decade contribution from the simple zero. The contributions from the zero and pole cancel each other till the resultant curve reaches $\omega = 200$ when the contribution from another simple pole sets in. So, from this point onwards the resultant curve continues with a slope of −20 dB/decade.

   Contributions from the phase components are added in the same manner.

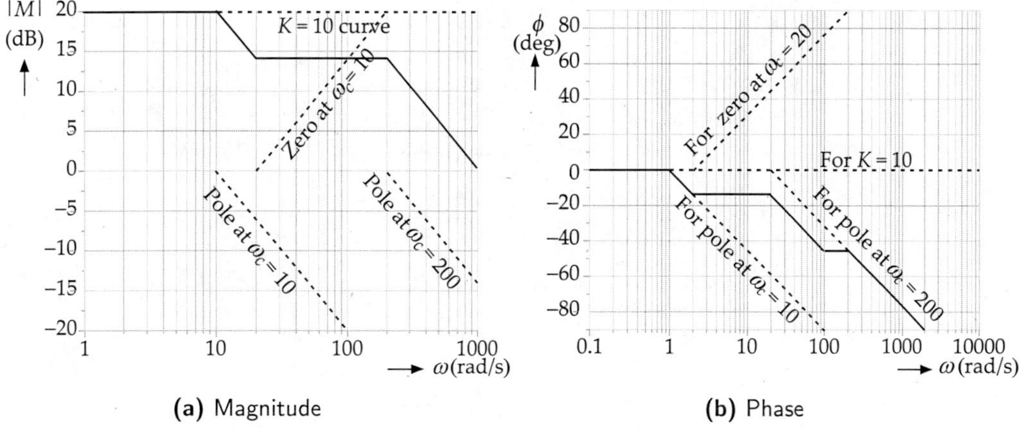

(a) Magnitude **(b)** Phase

Figure 12.5: Asymptotic Bode diagram (Example 12.1): - - -, plots of components, —, resultant asymptotic plots.

EXAMPLE 12.2. Draw Bode diagram for the system having the open-loop transfer function as

$$G(s)H(s) = \frac{2500(s+10)}{s(s+2)(s^2+30s+2500)}$$

Solution Comparing the second order term in the denominator with its standard form, we get

$$\omega_n^2 = 2500 \qquad \text{or,} \qquad \omega_n = 50$$

$$2\zeta\omega_n = 30 \qquad \text{or,} \qquad \zeta = \frac{30}{2\times 50} = 0.3$$

The transfer function can be written in the Bode form as

$$G(s)H(s) = \frac{\cancel{2500}\cdot 10\left(1+\frac{s}{10}\right)}{\cancel{2500}\cdot s\cdot 2\left(1+\frac{s}{2}\right)\left(\frac{s^2}{50^2}+\frac{2\times 15}{50}\cdot\frac{s}{50}+1\right)}$$

Therefore,

$$G(j\omega)H(j\omega) = \frac{5\left(1+\frac{j\omega}{10}\right)}{j\omega\left(1+\frac{j\omega}{2}\right)\left[1+0.6\left(\frac{j\omega}{50}\right)+\left(\frac{j\omega}{50}\right)^2\right]}$$

The transfer function comprises the following factors:

1. A constant gain $K = 5$.
2. A zero at $\omega_c = 10$.
3. Poles at $\omega_c = 0$ and 2.
4. A pair of quadratic (or, complex) poles at $\omega_n = 50$.

Therefore, the frequency span chosen is from 0.1 to 100.

Magnitude plot

1. The constant gain is a horizontal line at $20\log 5 = 14$ dB.
2. The asymptote for zero at $\omega_c = 10$ is drawn with a slope of 20 dB/decade.

3. The asymptote for the pole at the origin extends from 0.1 to ∞. Passing through $\omega = 1$ at 0 dB, it has a slope of -20 dB/decade.
4. The asymptote for pole at $\omega_c = 2$ is drawn with a slope of -20 dB/decade.
5. The asymptote for the quadratic pole at $\omega_n = 50$ has a slope of -40 dB/decade. Since $\zeta = 0.3$, corrections are necessary to determine peak frequency and peak magnitude which are given as

$$\begin{aligned} \omega_r &= \omega_n\sqrt{1-\zeta^2} = 45.3 \text{ rad/s} \\ |G(\jmath\omega_r)H(\jmath\omega_r)| &= -20\log(2\zeta\sqrt{1-\zeta^2}) = 4.85 \text{ dB} \end{aligned}$$

6. Individual contributions are added to draw the final plot.

Phase plot

Since the open-loop transfer function contains a quadratic pole, it is better to calculate the actual phase variation between $\omega = 0.1$ and 100 with the help of Eq. (12.13)[8]. Our calculations from the equation

$$\phi = \tan^{-1}\frac{\omega}{10} - 90^\circ - \tan^{-1}\frac{\omega}{2} - \tan^{-1}\frac{(0.6\omega/10)}{1-(\omega^2/2500)} \tag{i}$$

are given in the following table.

ω	0.1	0.2	1	2	10	20	100
ϕ (deg)[a]	−92.6333	−95.2523	114.2881	−130.5502	−155.6958	−165.8625	31.1299

[a] Calculated from Eq. (i)

The calculated plot is shown with points marked by circles. We have also included a function plot (shown by a continuous line) generated from Eq. (i). The plots are given in Figure 12.6. It may be seen that the plots, though show some difference between $20 < \omega < 100$ rad/s where we did not do manual calculations, the calculated plot is acceptable.

EXAMPLE 12.3. For the transfer function

$$G(s) = \frac{5(s+4)}{s(s+0.25)(s^2+4s+25)}$$

The values of the constant gain term and the highest corner frequency of the Bode plot are, respectively

(A) 3.2, 5.0 (B) 16.0, 4.0 (C) 3.2, 4.0 (D) 16.0, 5.0

Solution The transfer function can be written in the Bode form as

$$\begin{aligned} G(s) &= \frac{5(4)\left(1+\dfrac{s}{4}\right)}{s(0.25)\left(1+\dfrac{s}{0.25}\right)(25)\left(\dfrac{s^2}{5^2}+\dfrac{2\times 2}{5}\cdot\dfrac{s}{5}+1\right)} \\ &= \frac{3.2\left(1+\dfrac{s}{4}\right)}{s\left(1+\dfrac{s}{0.25}\right)\left(\dfrac{s^2}{5^2}+\dfrac{2\times 2}{5}\cdot\dfrac{s}{5}+1\right)} \end{aligned} \tag{i}$$

[8] With a negative sign.

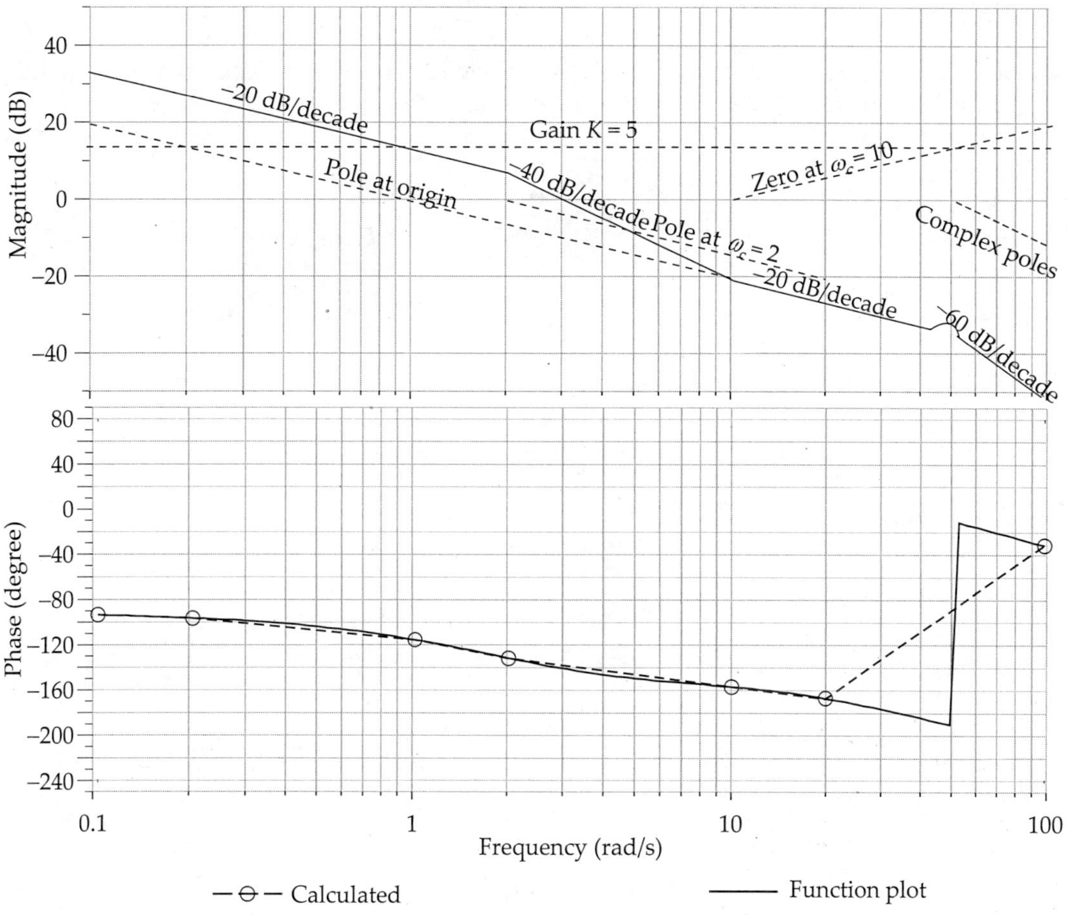

— ⊖ — Calculated ——— Function plot

Figure 12.6: Asymptotic Bode diagram (Example 12.2): (top) Magnitude plot and (bottom) Phase plot. Two phase plots— calculated and function plot—have been given. Legends are at the bottom of the plot.

From Eq. (i) we know that the constant gain is 3.2 and corner frequencies are 0, 0.25, 4 and 5. Ans. (A)

EXAMPLE 12.4. The Bode plot of a transfer function $G(s)$ is shown in Figure 12.7. The gain—

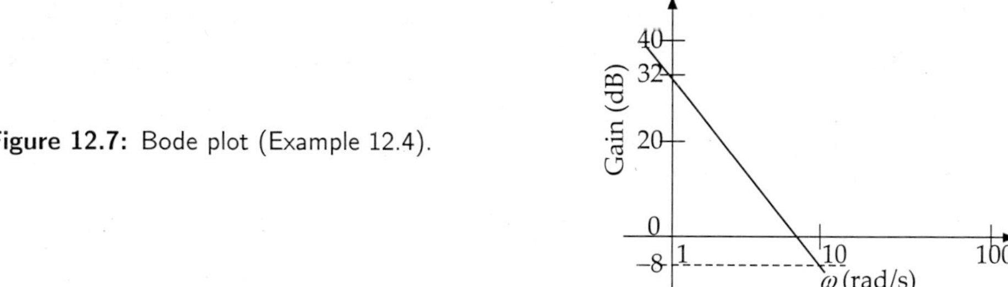

Figure 12.7: Bode plot (Example 12.4).

$20 \log |G(s)|$—is 32 dB and −8 dB at 1 rad/s and 10 rad/s, respectively. The phase is negative for all ω. Then $G(s)$ is

(A) $\dfrac{39.8}{s}$ (B) $\dfrac{39.8}{s^2}$ (C) $\dfrac{32}{s}$ (D) $\dfrac{32}{s^2}$

Solution Since the magnitude starts from 32 dB, obviously there is a constant term, K, in the transfer function given by

$$20 \log K = 32$$

Therefore,

$$K = 10^{\frac{32}{20}} = 39.8$$

The slope of the plot is $[(-8) - 32 = -40]$ dB/decade. So, there are two poles at the origin, i.e.

$$G(s) = \frac{39.8}{s^2}$$

Ans. (B)

EXAMPLE 12.5. The asymptotic Bode plots of the transfer function

$$\frac{K}{1 + \dfrac{s}{a}}$$

are given in Figure 12.8.

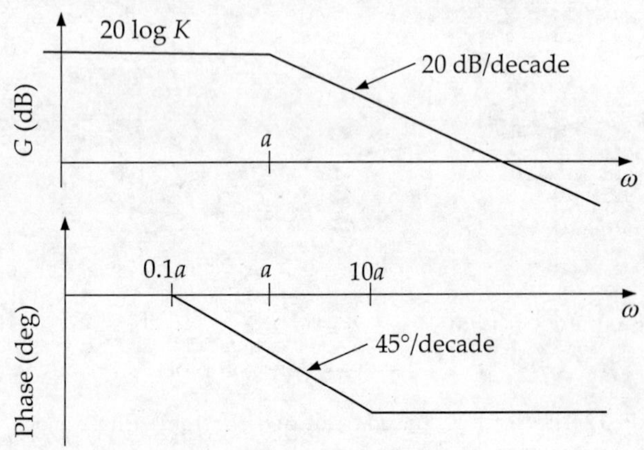

Figure 12.8: Bode plot (Example 12.5).

The error in phase angle and dB gain at a frequency of $\omega = 0.5$ are, respectively

(A) 4.1°, 0.97 dB (B) 5.7°, 3 dB (C) 4.1°, 3 dB (D) 5.7°, 0.97 dB

Solution For the given system, the phase angle is

$$\phi = -\tan^{-1}\frac{\omega}{a}$$

Therefore, $\phi\big|_{\omega=0.5a} = -\tan^{-1} 0.5 = -26.56°$

From the given plot $\phi\big|_{\omega=0.5a} = -22.5°$ [half of the slope 45°/decade]

Therefore, error $= -22.5 - (-26.56) = 4.06°$

Exact gain in dB at $\omega = 0.5a$ is

$$\text{Gain (dB)} = 20 \log K - 20 \log \sqrt{1 + \frac{\omega^2}{a^2}}$$

$$= 20 \log K - 10 \log \left[1 + \frac{(0.5a)^2}{a^2}\right]$$

$$= 20 \log K - 0.97$$

Gain in dB obtained from the asymptotic plot at $\omega = 0.5a = 20\log K$.
Hence, the error in gain $=[20\log K - (20\log K - 0.97)]$ dB=0.97 dB.
Ans. (A)

EXAMPLE 12.6. The Bode magnitude plot of a first order stable system shown in Figure 12.9 is constant with frequency. The asymptotic value of the high frequency phase, for the system, is $-180°$.

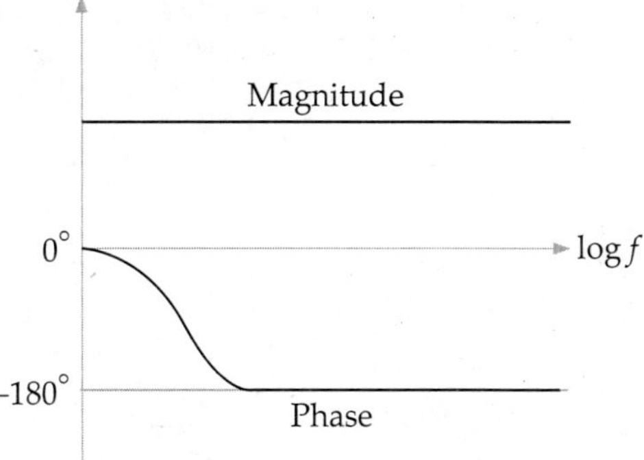

Figure 12.9: Bode plot (Example 12.6).

This system has

(A) one LHP pole and one RHP zero at the same frequency.
(B) one LHP pole and one LHP zero at the same frequency.
(C) two LHP poles and one RHP zero.
(D) two RHP poles and one LHP zero.

Solution The constant magnitude plot for all frequency shows that it is an all pass system, having poles and zeros placed symmetrically about the $\jmath\omega$-axis. The possible transfer functions for all pass system are

$$T_1(s) = \frac{s-1}{s+1} T_2(s) = \frac{1-s}{1+s}$$

Now, the phase plot shows that for $\omega \to \infty \quad \phi = -180°$.

For $T_1(s)$:
$$T_1(\jmath\omega) = \frac{\jmath\omega - 1}{\jmath\omega + 1}$$
$$\angle T_1(\jmath\omega) = \pi - \tan^{-1}\omega - \tan^{-1}\omega = \pi - 2\tan^{-1}\omega$$
At $\omega \to \infty$
$$\angle T_1(\jmath\omega) = 0°$$

For $T_2(s)$:
$$T_2(\jmath\omega) = \frac{1-\jmath\omega}{1+\jmath\omega}$$
$$\angle T_2(\jmath\omega) = -\tan^{-1}\omega - \tan^{-1}\omega = -2\tan^{-1}\omega$$
At $\omega \to \infty$
$$\angle T_2(\jmath\omega) = -180°$$

Hence, the transfer function of the given all pass filter is

$$T_2(s) = \frac{1-s}{1+s}$$

which implies one LHP pole and one RHP zero at the same frequency.
Ans. (A)

EXAMPLE 12.7. The magnitude and phase plots of an LTI system are shown in Figure 12.10. The transfer function of the system is

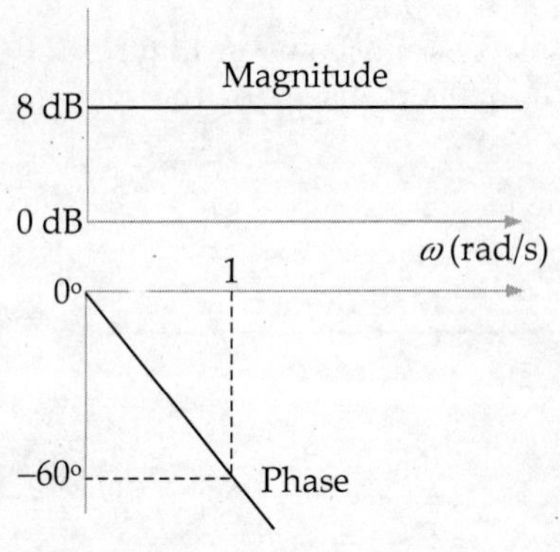

Figure 12.10: System (Example 12.7).

(A) $2.515e^{-0.032s}$ (B) $\dfrac{e^{-2.514s}}{s+1}$ (C) $1.04e^{-2.514s}$ (D) $2.51e^{-1.047s}$

Solution Let the transfer function be $G(s) = Ke^{-sT_d}$. Given,

Magnitude, $M = 8\,\text{dB} = 20 \log K$

Therefore, $K = 2.511$

Angle at $\omega = 1\,\text{rad/s}$ $\phi = -60°$

Again, angle $\phi = -\omega T_d \times \dfrac{180°}{\pi}$

Therefore, $60° = -1 \times T_d \times \dfrac{180°}{\pi}$

or, $T_d = 1.047$

So, the required transfer function is

$$G(s) = 2.511e^{-1.047s}$$

Ans. (D)

12.5 STABILITY STUDY FROM BODE PLOT

The parameters, which give an idea about the stability of a system, are

1. Gain crossover frequency, ω_g
2. Phase crossover frequency, ω_p
3. Gain margin, Gm
4. Phase margin, Pm

We have already defined these parameters in Section 11.2 at page 265. Here, we discuss how these parameters can be found out from Bode plots. In fact, Bode plots are eminently suitable for finding out these parameters.

Gain crossover frequency, ω_g. It is the frequency at which the Bode magnitude plot crosses the 0 dB axis (see Figure 12.11).

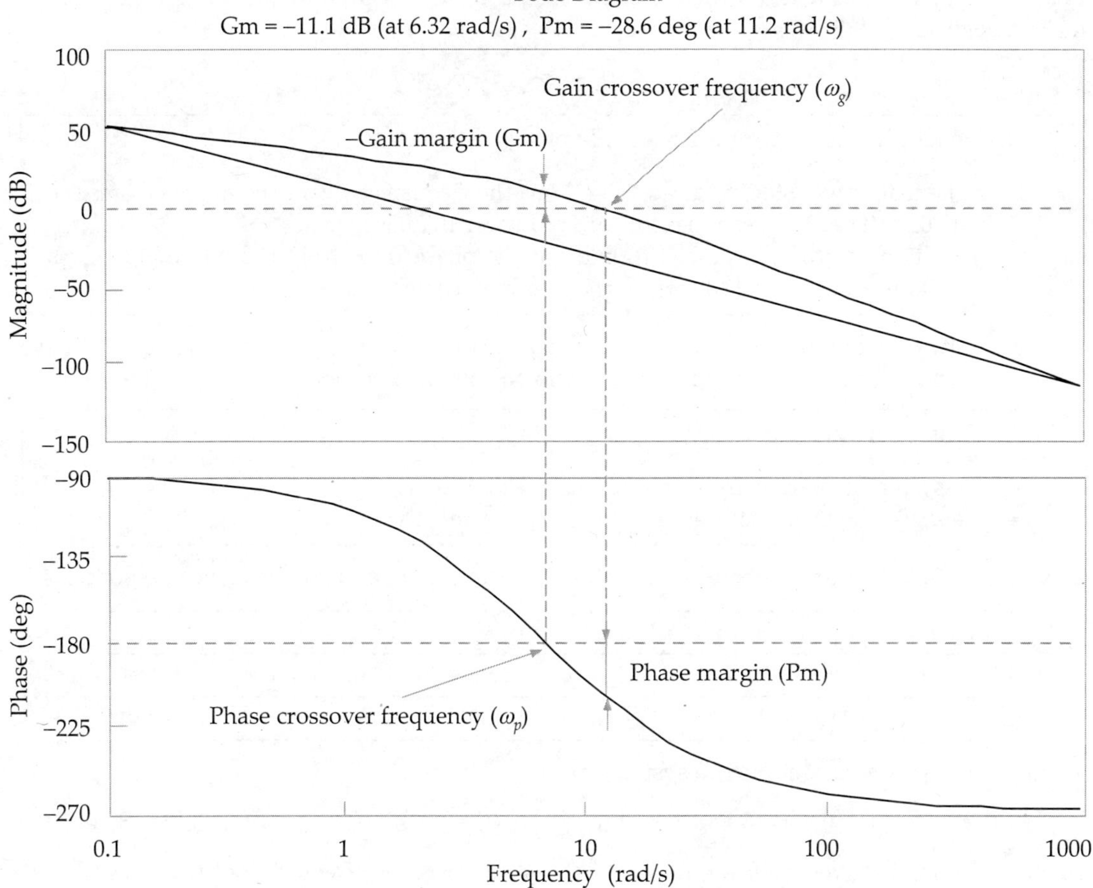

Figure 12.11: Gain crossover frequency, phase crossover frequency, gain margin and phase margin.

Phase crossover frequency, ω_p. It is the frequency at which the Bode phase plot crosses the $-180°$ axis (see Figure 12.11).

Gain margin, Gm. If ω_p is the phase crossover frequency, then the reciprocal of gain at this frequency is the gain margin. Quantitatively, it is defined as

$$\text{Gm} = |M|_{\omega=\omega_p}^{-1} = -20\log\left|G(\jmath\,\omega_p)H(\jmath\,\omega_p\right| \text{ dB}$$

To determine gain margin from Bode plots:

1. We need to determine the frequency where the phase plot crosses the $-180°$ line. This frequency is the phase crossover frequency ω_p.
2. Corresponding to ω_p, we need to find out the value of the gain from the magnitude plot. This value of the gain, *with its sign reversed*, is the gain margin. This means that, if the value of the gain is found negative, the gain margin is positive and vicé versa.

Phase margin, Pm. If ϕ is the phase angle at the gain crossover frequency ω_g, then phase margin is given by $180° + \phi$. Symbolically,

$$\text{Pm} = 180° + \phi|_{\omega=\omega_g} \text{ degree}$$

To determine the phase margin from Bode plots,

1. In the magnitude plot, we need to find out the frequency where the curve crosses the 0 dB line. That frequency is the gain crossover frequency, ω_g.
2. Corresponding to ω_g, we need to find out the phase value from the phase plot. This phase value, added to 180°, gives the phase margin.

Obviously, to determine the stability margins, both the magnitude and the phase plot have to be made in the same frequency range and one on top of the other.

The stability conditions are:

- The system is marginally stable if Gm = Pm = 0.
- The system is unstable if either Gm or Pm is negative.

EXAMPLE 12.8. A unity feedback system has the following open-loop frequency response:

ω (rad/s)	2	3	4	5	6	8	10
$\lvert G(j\omega)\rvert$	7.5	4.8	3.15	2.25	1.70	1.00	0.64
$\angle G(j\omega)$	−118°	−130°	−140°	−150°	−157°	−170°	−180°

The gain and phase margins of the system are:

(A) 0 dB, −180° (B) 3.88 dB, −170° (C) 0 dB, 10° (D) 3.88 dB, 10°

Solution From the given table we find that the phase crossover frequency of the system is 10 rad/s, where the phase angle is −180°. At that frequency the gain is 0.64. Hence,

$$\text{Gm} = -20 \log 0.64 = 3.88 \text{ dB}$$

Also, we find from the table that the gain crossover frequency, where the gain is 1.00 (so that gain = 0 dB), is 8 rad/s. At that frequency, the phase reads −170°. Therefore,

$$\text{Pm} = 180° - 170° = 10°$$

Ans. (D)

EXAMPLE 12.9. Sketch the Bode plot for the system having its open-loop transfer function as

$$G(s)H(s) = \frac{20}{s(s+1)(s+4)}$$

Find the gain margin and phase margin, and comment on the stability of the system.

Solution The transfer function can be written in Bode form as

$$G(s)H(s) = \frac{5}{s\left(1+\frac{s}{1}\right)\left(1+\frac{s}{4}\right)} \quad \Rightarrow G(j\omega)H(j\omega) = \frac{5}{j\omega\left(1+\frac{j\omega}{1}\right)\left(1+\frac{j\omega}{4}\right)}$$

The transfer function comprises the following components:

1. A constant $K = 5$; 2. A pole at the origin; 3. Two simple poles at $\omega_c = 1$ and 4.

Magnitude plot

1. For the constant, a horizontal line is drawn at 20 log 5 = 14 dB.
2. For the pole at the origin, a −20 dB/decade line that passes through $\omega = 1$ at 0 dB is drawn.
3. For two simple poles, two −20 decade/dB lines are drawn at $\omega_c = 1$ and 4.
4. Individual contributions are added to get the resultant curve.

Note: On the left hand corner of the magnitude plot, we have drawn three lines (broken) with marks −20 dB/decade, −40 dB/decade and −60 dB/decade. These lines are not part of the magnitude plot. They are there only as reference lines for help in drawing magnitude lines of those values.

Phase plot

1. For the constant, the phase is 0°.
2. For the pole at the origin, the phase is a horizontal line at −90°.
3. For the pole at $\omega_c = 1$, a line between 0° at $\omega = 0.1$ and −90° at $\omega = 10$ is drawn. For the other pole, a parallel line is drawn between $\omega = 0.4$ and $\omega = 40$.
4. Individual contributions are added to get the final curve.

Note: To find Gm, Pm etc. we have to draw magnitude and phase in the same graph, keeping frequency scale the same.

ω_p, ω_g, *Gm and Pm*

From the asymptotic phase plot (Figure 12.12) we find that the phase curve cuts the −180° line at $\omega = 2.0$ rad/s. This is the phase crossover frequency ω_p.

From the magnitude plot, we observe that the magnitude curve cuts the 0 dB line at $\omega = 2.2$ rad/s. This is the gain crossover frequency ω_g.

Following the procedure laid down in Section 12.5, the $\omega = 2.0$ rad/s line when extended to the magnitude plot, intersects it at 2 dB. Therefore, gain margin = −2 dB.

Similarly, the $\omega = 2.2$ rad/s line is extended to the phase plot. It intersects the phase plot at −182°. Therefore, phase margin = 180° − 182° = −2°.

Since both Gm and Pm are negative, the system is unstable.

Note: The actual plots, done through the Matlab, show $\omega_p = \omega_g = 2.0$ rad/s and Gm = Pm = 0, which indicate a marginally stable system. This variation in conclusion occurs owing to the approximate nature of asymptotic plots.

EXAMPLE 12.10. Draw the Bode plot of the system whose open-loop transfer function is given by

$$G(s) = \frac{64(s+2)}{s(s+0.5)}$$

Find the gain margin and phase margin from the plot.

Solution The transfer function can be written in the Bode form as

$$G(s) = 64 \cdot \frac{2}{0.5} \cdot \frac{1+\frac{s}{2}}{s\left(1+\frac{s}{0.5}\right)} = \frac{256\left(1+\frac{s}{2}\right)}{s\left(1+\frac{s}{0.5}\right)} \qquad \Rightarrow G(j\omega) = \frac{256\left(1+\frac{j\omega}{2}\right)}{j\omega\left(1+\frac{j\omega}{0.5}\right)}$$

The transfer function comprises the following components:

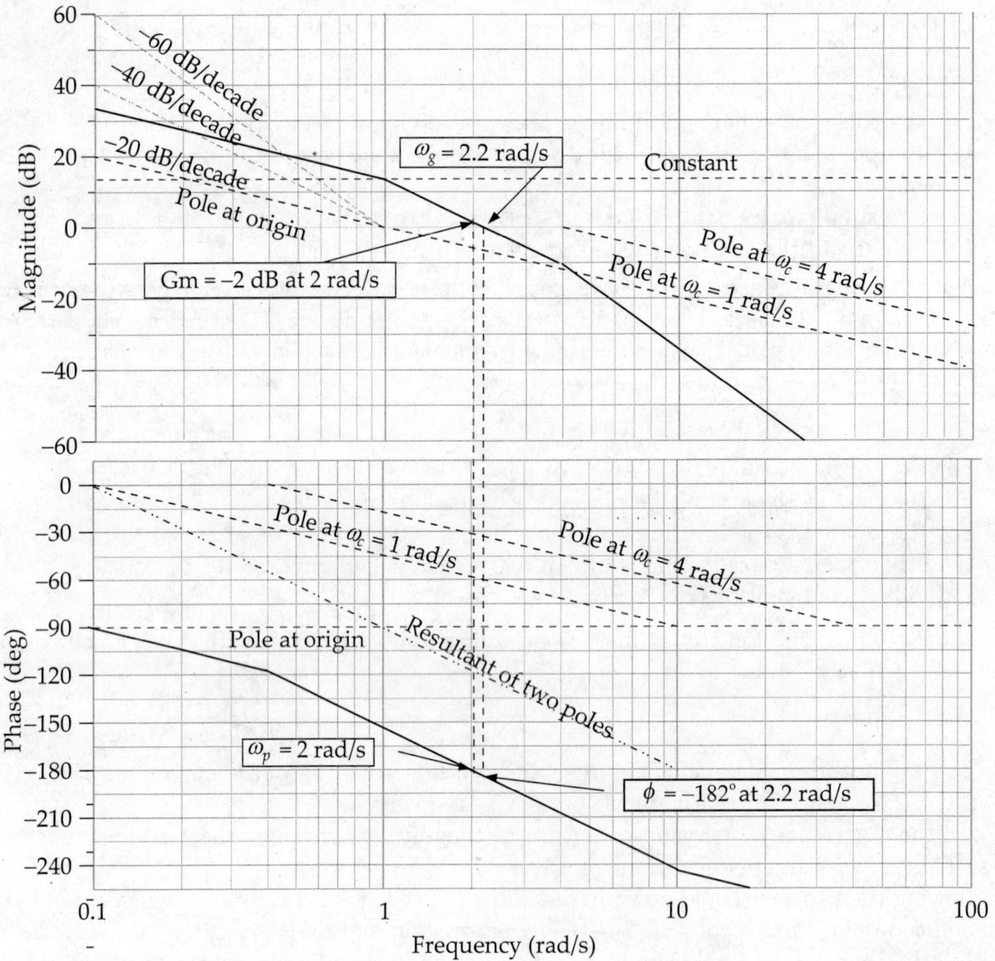

Figure 12.12: Bode plots (Example 12.9): Magnitude plot (top) and Phase plot (bottom).

1. A constant $K = 256$;
2. A pole at the origin;
3. A simple zero at $\omega_c = 2$;
4. A simple pole at $\omega_c = 0.5$.

Magnitude plot

1. For the constant, a horizontal line is drawn at 20 log 256 = 48.2 dB.
2. For the pole at the origin, a -20 dB/decade line that passes through $\omega = 1$ at 0 dB is drawn.
3. For the simple zero, a 20 dB/decade line is drawn at $\omega_c = 2$.
4. For the simple pole, a -20 dB/decade line is drawn at $\omega_c = 0.5$.
5. Individual contributions are added to get the final curve.

Phase plot

1. For the constant, the phase is 0°.
2. For the pole at the origin, the phase is a horizontal line at $-90°$.
3. For the zero at $\omega_c = 2$, a line between 0° at $\omega = 0.2$ and 90° at $\omega = 20$ is drawn.
4. For the pole at $\omega_c = 0.5$, a line between 0° at $\omega = 0.05$ and $-90°$ at $\omega = 5$ is drawn.

5. Individual contributions are added to get the final curve.

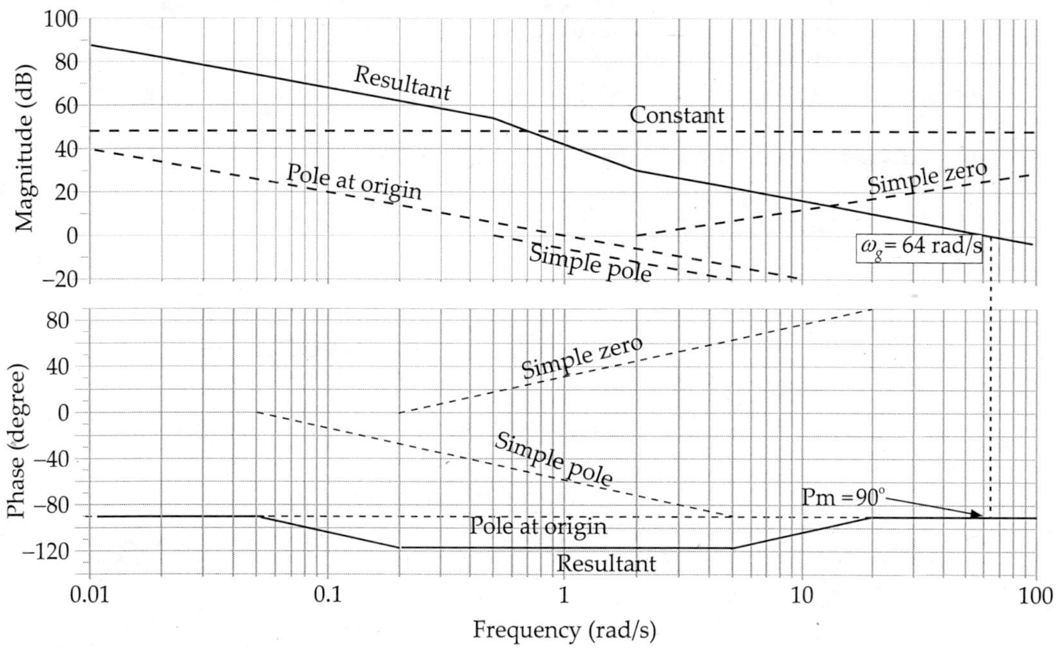

Figure 12.13: Bode plots (Example 12.10): (top) Magnitude plot and (bottom) Phase plot.

From the plots (Figure 12.13) we find that $\omega_p = \infty$, $\omega_g = 64$ and phase at $\omega_g = -90°$. So, gain margin $= \infty$, phase margin $= 180° - 90° = 90°$. Therefore, the system is stable.

EXAMPLE 12.11. The open-loop transfer function of a control system is given by

$$G(s)H(s) = \frac{K}{s(s+4)(s+10)}$$

From its Bode plot find the value of K when the system is marginally stable.

Solution The open-loop transfer function can be written in the Bode form as

$$G(s)H(s) = \frac{K}{4\times 10}\cdot\frac{1}{s\left(1+\frac{s}{4}\right)\left(1+\frac{s}{10}\right)} = \frac{K'}{s\left(1+\frac{s}{4}\right)\left(1+\frac{s}{10}\right)} \qquad \text{where } K' = \frac{K}{40}$$

Therefore,
$$G(\jmath\omega) = \frac{K'}{\jmath\omega\left(1+\frac{\jmath\omega}{4}\right)\left(1+\frac{\jmath\omega}{10}\right)}$$

The transfer function comprises the following:

1. A constant K'; 2. A pole at the origin; 3. Two simple poles at $\omega_c = 4$ and 10.

Magnitude plot

1. The constant is an unknown quantity. So, we ignore it for the time being.
2. For the pole at the origin, a -20 dB/decade line that passes $\omega = 1$ at 0 dB is drawn.
3. For the simple poles, two -20 dB/decade lines are drawn $\omega_c = 4$ and 10.
4. Individual contributions are added to get the final curve.

Phase plot

1. For the constant, the phase is 0°.
2. For the pole at the origin, the phase is a horizontal line at −90°.
3. For the pole at $\omega_c = 4$, a line between 0° at $\omega = 0.4$ and −90° at $\omega = 40$ is drawn. For the second pole at $\omega_c = 10$, another line between 0° at $\omega = 1$ and −90° at $\omega = 100$ is drawn.
4. Individual contributions are added to get the final curve.

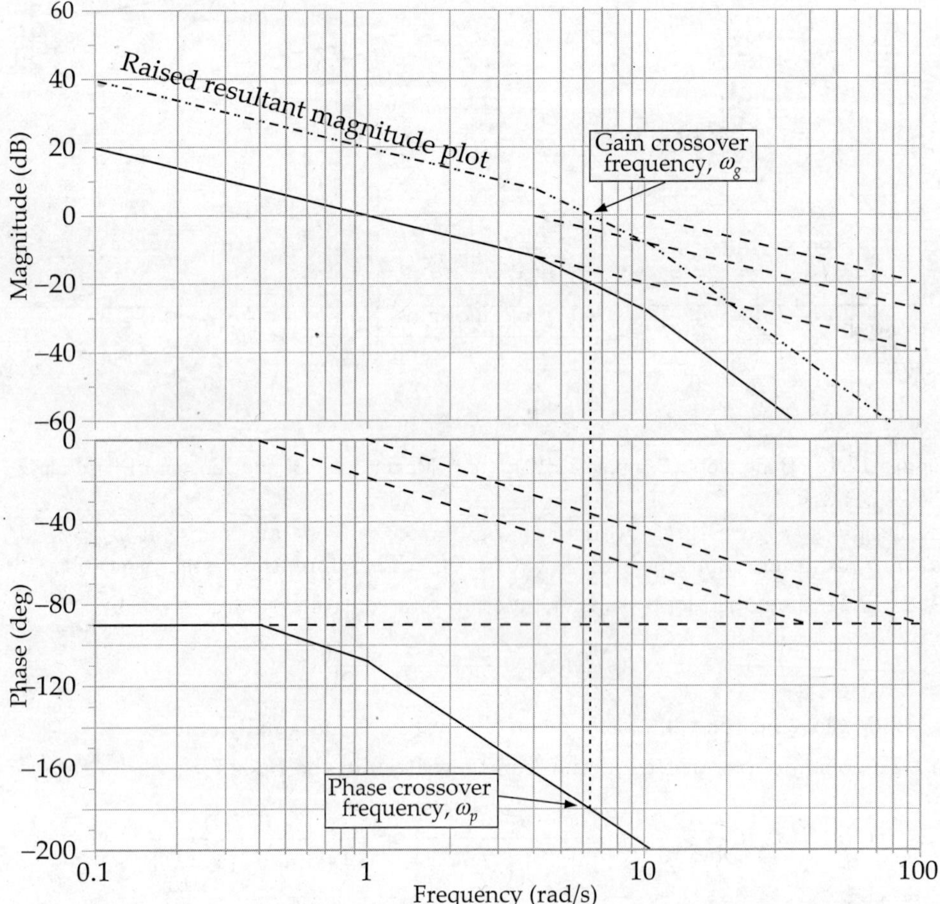

Figure 12.14: Bode plots (Example 12.11): (top) Magnitude plot and (bottom) Phase plot.

Bode magnitude and phase plots are given in Figure 12.14. It may be seen from the phase plot, which is not affected by the unknown value of the constant (because the constant contributes nothing to the phase), that the phase crossover frequency is 7.5 rad/s. Therefore, to make the system marginally stable, the gain crossover frequency has to be made equal to 7.5 rad/s. The magnitude plot shows that at ω_p the gain is −20 dB. So, if we raise the magnitude curve by 20 dB, the gain becomes 0 dB (i.e. unity) to make $\omega_g = \omega_p$. Now,

$$20 \log K' = 20 \text{ dB} \qquad \Rightarrow K' = 10 = \frac{K}{40} \qquad \Rightarrow K = 400$$

EXAMPLE 12.12. Sketch the Bode plot (both gain and phase) for the transfer function given below and determine the gain crossover frequency.

$$\frac{100(1+0.5s)}{s(s^2+16s+100)}$$

Solution Comparing the second order term in the denominator to its standard form, we get

$$\omega_n^2 = 100 \qquad \text{or,} \qquad \omega_n = 10$$

$$2\zeta\omega_n = 16 \qquad \text{or,} \qquad \zeta = \frac{16}{2\times 10} = 0.8$$

The transfer function can be written in the Bode form as

$$G(s)H(s) = \frac{\cancel{100}\left(1+\dfrac{s}{2}\right)}{s(\cancel{100})\left(1+\dfrac{2\times 8}{10}\cdot\dfrac{s}{10}+\dfrac{s^2}{10^2}\right)}$$

$$G(\jmath\omega)H(\jmath\omega) = \frac{\left(1+\dfrac{\jmath\omega}{2}\right)}{\jmath\omega\left[1+2\times 0.8\left(\dfrac{\jmath\omega}{10}\right)+\left(\dfrac{\jmath\omega}{10}\right)^2\right]} \equiv \frac{\left(1+\dfrac{\jmath\omega}{2}\right)}{\jmath\omega\left[1-\dfrac{\omega^2}{100}+0.16\jmath\omega\right]}$$

The transfer function comprises the following factors:

1. A zero at $\omega_c = 2$; 2. A pole at $\omega_c = 0$; 3. A quadratic pole at $\omega_n = 10$.

Magnitude plot

1. The asymptote for zero at $\omega_c = 2$ is drawn with a slope of 20 dB/decade.
2. The asymptote for the pole at the origin extends from 0.1 to ∞. Passing through $\omega = 1$ at 0 dB, it has a slope of -20 dB/decade.
3. The asymptote for the quadratic pole at $\omega_n = 10$ has a slope of -40 dB/decade. Since $\zeta = 0.8$ which is > 0.707, it has no peak. Therefore, it is not necessary to determine peak frequency and peak magnitude and make corrections to the curve.
4. Individual contributions are added to draw the final plot.

Phase plot

Since the open-loop transfer function contains a quadratic pole, we calculate the phase from the following equation

$$\phi = \tan^{-1}\frac{\omega}{2} - 90^\circ - \tan^{-1}\frac{0.16\omega}{1-(\omega^2/100)} \tag{i}$$

ω (rad/s)	0.1	0.2	1.0	2.0	10	20	100
ϕ (deg)[a]	−88.0552	−86.1296	−72.6154	−63.4349	−101.3099	41.137	8.0347

[a] Calculated from Eq. (i)

The plots are given in Figure 12.15. From the magnitude plot, we find that the frequency corresponding to 0 dB gain is 1. Thus, the gain crossover frequency is 1 rad/s.

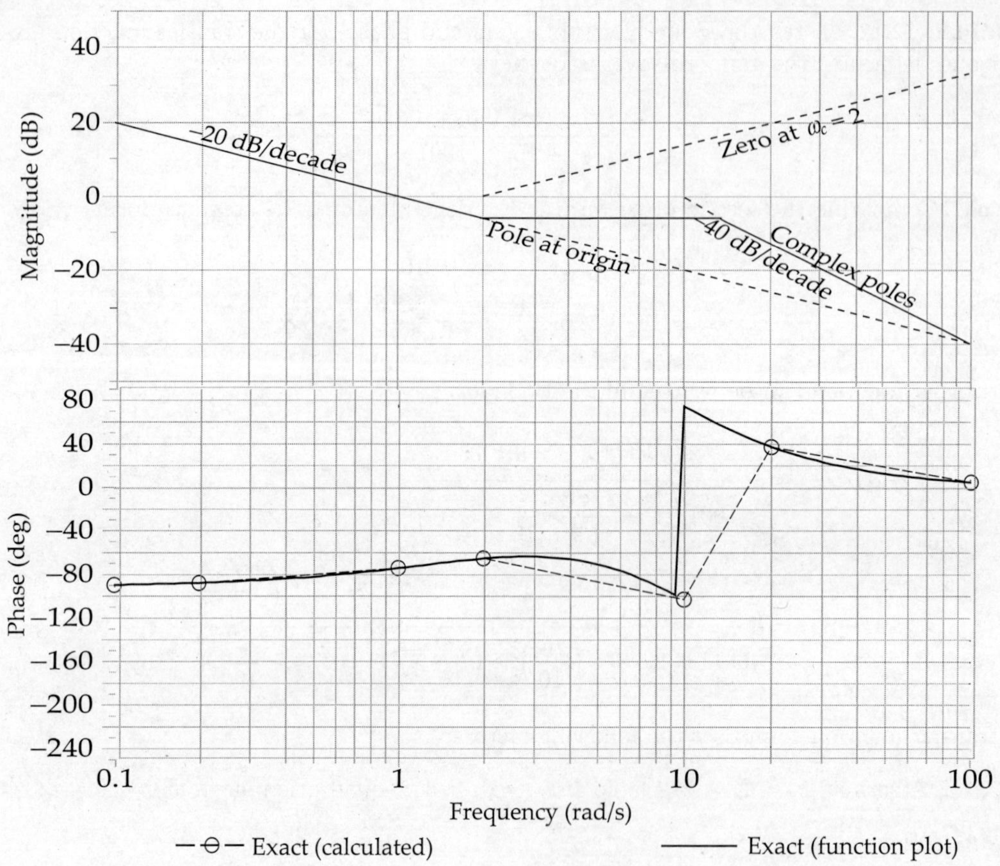

Figure 12.15: Bode plots (Example 12.12): (top) Magnitude plot and (bottom) Phase plot. Legends for different plots are given at the bottom.

EXAMPLE 12.13. Sketch the asymptotic Bode plot for the open loop transfer function

$$G(s)H(s) = \frac{2(s+0.25)}{s^2(s+1)(s+0.5)}$$

Is the system stable?

Solution The transfer function can be written in Bode form as

$$G(s)H(s) = \frac{1+\dfrac{s}{0.25}}{s^2\left(1+\dfrac{s}{1}\right)\left(1+\dfrac{s}{0.5}\right)} \qquad \Rightarrow G(\jmath\omega)H(\jmath\omega) = \frac{1+\dfrac{\jmath\omega}{0.25}}{(\jmath\omega)^2\left(1+\dfrac{\jmath\omega}{1}\right)\left(1+\dfrac{\jmath\omega}{0.5}\right)}$$

The open-loop transfer function comprises the following components:

1. A zero at $\omega_c = 0.25$; 2. Two poles at the origin; 3. Two simple poles at $\omega_c = 0.5$ and 1.

Magnitude plot

1. For the zero, a 20 decade/dB line is drawn at $\omega_c = 0.25$.
2. For two poles at the origin, a -40 dB/decade line that passes through $\omega = 1$ is drawn.
3. For two simple poles, two -20 decade/dB lines are drawn at $\omega_c = 0.5$ and 1.
4. Individual contributions are added to get the final curve.

Phase plot

1. For the zero at $\omega_c = 0.25$, a line between $0°$ at $\omega = 0.025$ and $90°$ at $\omega = 2.5$ is drawn.
2. For two poles at the origin, the phase is a horizontal line at $-180°$.
3. For the pole at $\omega_c = 0.5$, a line between $0°$ at $\omega = 0.05$ and $-90°$ at $\omega = 5$ is drawn. For the other pole a parallel line is drawn between $\omega = 0.1$ and $\omega = 10$.
4. Individual contributions are added to get the final curve.

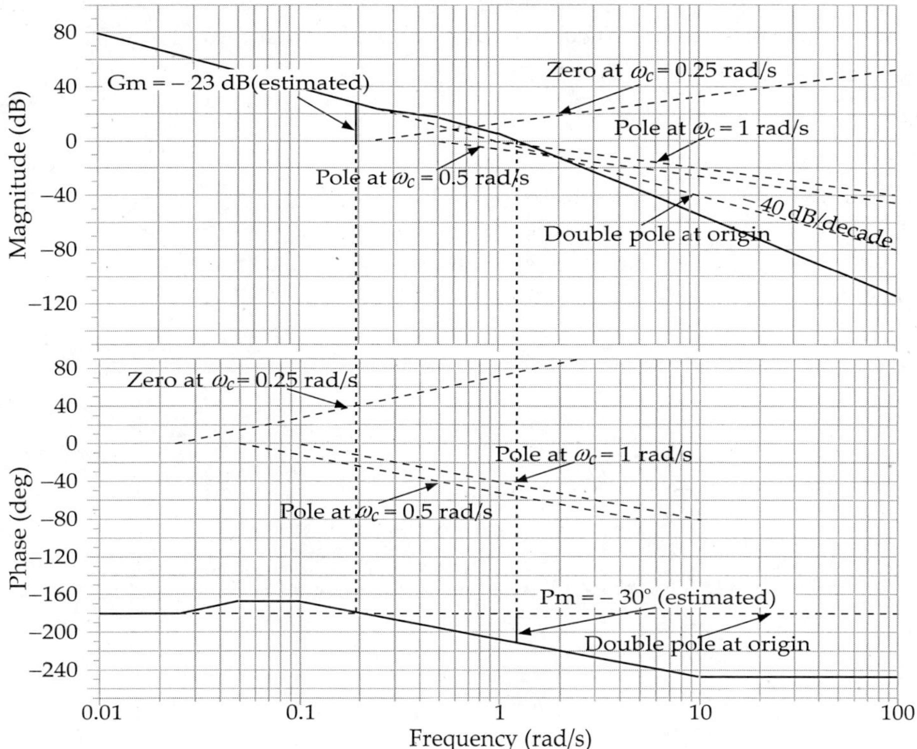

Figure 12.16: Bode plots (Example 12.13): (top) Magnitude plot and (bottom) Phase plot.

From the asymptotic Bode plots (Figure 12.16), we find that $\omega_p = 0.2$ rad/s while $\omega_g = 1.3$ rad/s. Here, gain margin $= -23$ dB and phase margin $= 180° - 210° = -30°$. Therefore, the system is unstable.

Note: However, actual plots, done by the Matlab, show $\omega_g = 1.12$ rad/s, $\omega_p = 0.354$ rad/s, Gm $=$ -20.6 dB and Pm $= -36.7°$.

EXAMPLE 12.14. Sketch the Bode plot for a unity feedback system whose open-loop transfer function is given by

$$G(s)H(s) = \frac{500}{s(s+1)(s+50)}$$

From the Bode plot determine the following: (a) gain and phase cross-over frequencies, (b) gain and phase margin and (c) stability of the closed-loop system.

Solution The transfer function can be written in the Bode form as

$$G(s)H(s) = \frac{10}{s\left(1+\frac{s}{1}\right)\left(1+\frac{s}{50}\right)} \qquad \Rightarrow G(j\omega)H(j\omega) = \frac{10}{j\omega\left(1+\frac{j\omega}{1}\right)\left(1+\frac{j\omega}{50}\right)}$$

The transfer function comprises the following components:

1. A constant $K = 10$; 2. A pole at the origin; 3. Two simple poles at $\omega_c = 1$ and 50.

Magnitude plot

1. For the constant, a horizontal line is drawn at 20 log 10 = 20 dB.
2. For the pole at the origin, a -20 dB/decade line that passes through $\omega = 1$ at 0 dB is drawn.
3. For two simple poles, two -20 decade/dB lines are drawn at $\omega_c = 1$ and 50.
4. Individual contributions are added to get the final curve.

Phase plot

1. For the constant, the phase is 0°.
2. For the pole at the origin, the phase is a horizontal line at $-90°$.
3. For the pole at $\omega_c = 1$, a line between 0° at $\omega = 0.1$ and $-90°$ at $\omega = 10$ is drawn. For the other pole, a parallel line is drawn between $\omega = 5$ and $\omega = 500$.
4. Individual contributions are added to get the final curve.

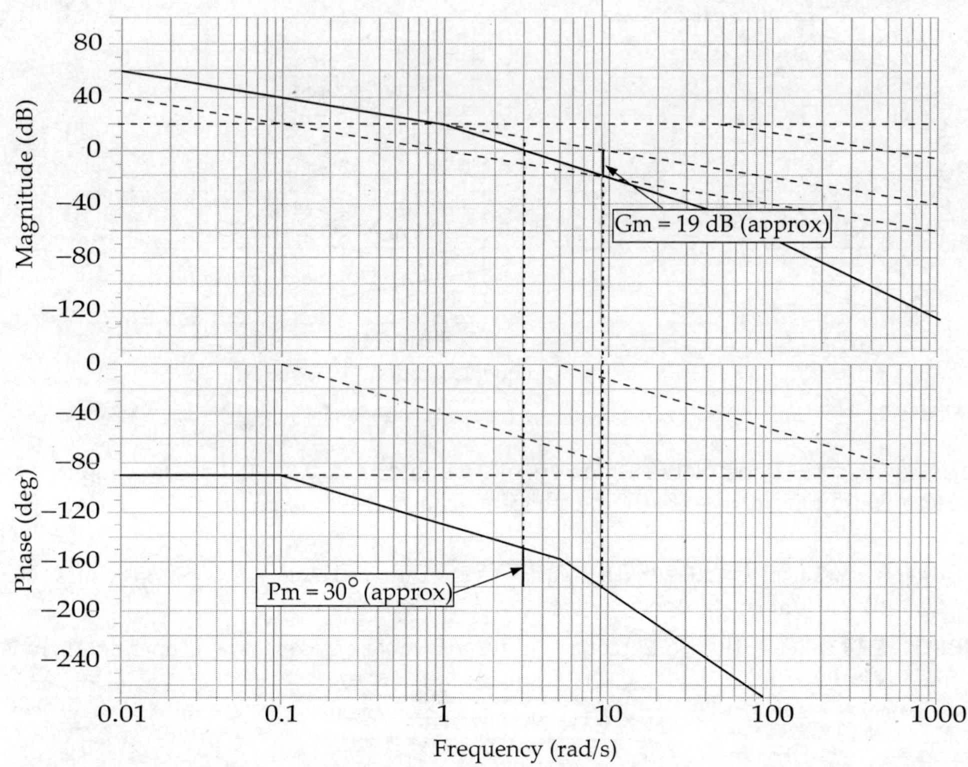

Figure 12.17: Bode plot (Example 12.14): (top) Magnitude plot and (bottom) Phase plot.

(a) From the asymptotic Bode plots (Fig 12.17), we find that $\omega_p = 9.0$ rad/s while $\omega_g = 3.1$ rad/s.

(b) Here, gain margin = 19 dB and phase margin = 180° – 150° = 30°.
(c) Therefore, the system is stable.

EXAMPLE 12.15. Draw Bode plot of the open-loop transfer function

$$G(s)H(s) = \frac{50}{s(1+0.25s)(1+0.1s)}$$

From the graph determine: (a) gain crossover frequency, (b) phase crossover frequency, (c) gain and phase margins, and (d) stability of the system.

Solution The given open-loop transfer function can be written in the Bode form as:

$$G(s) = \frac{50}{s\left(1+\dfrac{s}{4}\right)\left(1+\dfrac{s}{10}\right)} \qquad \Rightarrow G(\jmath\omega) = \frac{50}{\jmath\omega\left(1+\dfrac{\jmath\omega}{4}\right)\left(1+\dfrac{\jmath\omega}{10}\right)}$$

The transfer function consists of the following components:

1. A constant $K = 50$; 2. A pole at the origin; 3. Two simple poles at $\omega_c = 4$ and 10.

Magnitude plot

1. For the constant, a horizontal line is drawn at $20\log 50 = 34$ dB.
2. For the pole at the origin, a -20 dB/decade line that passes through $\omega = 1$ at 0 dB is drawn.
3. For the two simple poles, two -20 dB/decade lines are drawn from $\omega_c = 4$ and 10.
4. Individual contributions are added to get the final magnitude plot.

Phase plot

1. For the pole at the origin, the phase is a horizontal line at $-90°$.
2. For the pole at $\omega_c = 4$ rad/s, a line is drawn connecting 0° at $\omega = 0.4$ rad/s and $-90°$ at $\omega = 40$ rad/s.
3. For the pole at $\omega_c = 10$ rad/s, a line is drawn connecting 0° at $\omega = 1$ rad/s and $-90°$ at $\omega = 100$ rad/s.
4. Individual contributions are added to get the final phase plot.

The Bode plot is presented in Figure 12.18. From the plot, we get the following data:

(a) $\omega_g = 12$ rad/s (b) $\omega_p = 7.5$ rad/s (c) Gm = -10 dB Pm = $-20°$

(d) Since Gm and Pm values are negative, the system is unstable.

EXAMPLE 12.16. For the open-loop transfer function of a unity feedback system is given by

$$G(s) = \frac{K(20+s)}{(s+1)(s+2)(s+10)}$$

Construct the Bode plot for $K = 10$. Determine: (a) gain crossover frequency, (b) phase crossover frequency, (c) gain margin, and (d) phase margin.

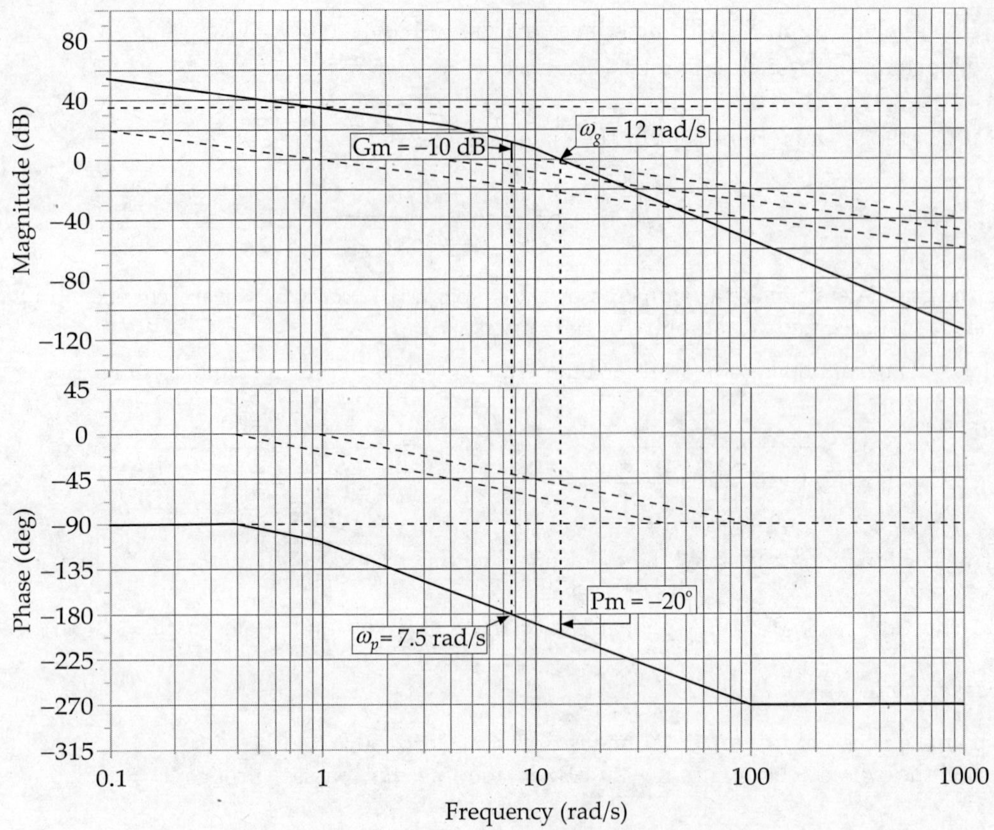

Figure 12.18: Bode plot (Example 12.15).

Solution The given open-loop transfer function can be written in the Bode form as:

$$G(s) = \frac{10(20)\left(1+\dfrac{s}{20}\right)}{(1+s)(2)\left(1+\dfrac{s}{2}\right)(10)\left(1+\dfrac{s}{10}\right)} = \frac{10\left(1+\dfrac{s}{20}\right)}{(1+s)\left(1+\dfrac{s}{2}\right)\left(1+\dfrac{s}{10}\right)}$$

$$G(\jmath\omega) = \frac{10\left(1+\dfrac{\jmath\omega}{20}\right)}{(1+\jmath\omega)\left(1+\dfrac{\jmath\omega}{2}\right)\left(1+\dfrac{\jmath\omega}{10}\right)}$$

The transfer function consists of the following components:

1. A constant $K = 10$; 2. A zero at $\omega_c = 20$; 3. Three simple poles at $\omega_c = 1,\ 2$ and 10.

Magnitude plot

1. For the constant, a horizontal line is drawn at $20\log 10 = 20$ dB.
2. For the three simple poles, three -20 dB/decade lines are drawn from $\omega_c = 1, 2$ and 10.
3. Individual contributions are added to get the final magnitude plot.

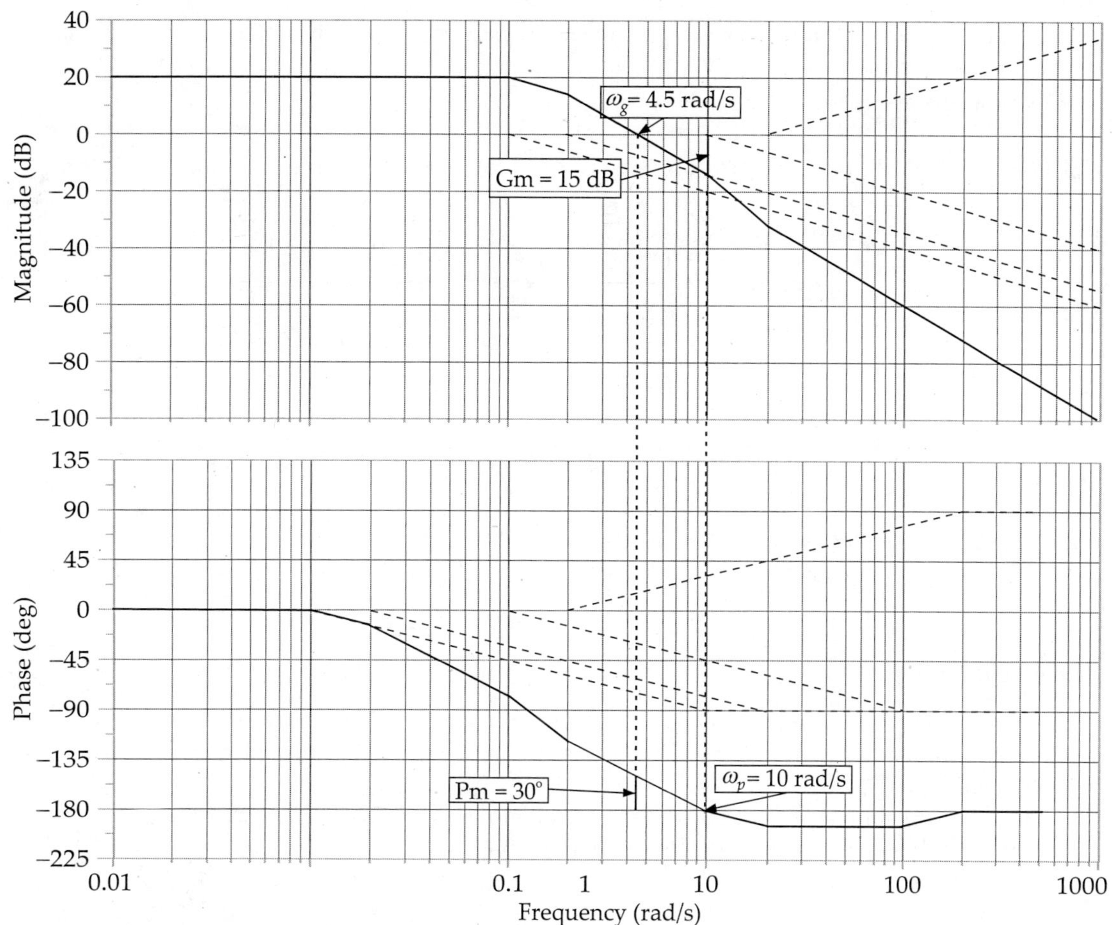

Figure 12.19: Bode plot (Example 12.16).

Phase plot

1. For the zero at $\omega_c = 20$ rad/s, a line is drawn connecting 0° at $\omega = 2$ rad/s and 90° at $\omega = 200$ rad/s.
2. For the pole at $\omega_c = 1$ rad/s, a line is drawn connecting 0° at $\omega = 0.1$ rad/s and −90° at $\omega = 10$ rad/s.
3. For the pole at $\omega_c = 2$ rad/s, a line is drawn connecting 0° at $\omega = 0.2$ rad/s and −90° at $\omega = 20$ rad/s.
4. For the pole at $\omega_c = 10$ rad/s, a line is drawn connecting 0° at $\omega = 1$ rad/s and −90° at $\omega = 100$ rad/s.
5. Individual contributions are added to get the final phase plot.

The Bode plot is presented in Figure 12.19. From the plot, we get the following data:

(a) $\omega_g = 4.5$ rad/s (b) $\omega_p = 10$ rad/s (c) Gm = 15 dB (d) Pm = 30°

Since both Gm and Pm values are positive, the system is stable.

EXAMPLE 12.17. Construct the Bode plot for the unity feedback control system having

$$G(s) = \frac{10(s+10)}{s(s+2)(s+5)}$$

(a) From the plot, obtain gain margin, phase margin, gain crossover frequency, and phase crossover frequency.
(b) Comment on the stability of the system.

Solution The given open-loop transfer function can be written in the Bode form as:

$$G(s) = \frac{10(10)\left(1+\dfrac{s}{10}\right)}{s(2)\left(1+\dfrac{s}{2}\right)(5)\left(1+\dfrac{s}{5}\right)} = \frac{10\left(1+\dfrac{s}{10}\right)}{s\left(1+\dfrac{s}{2}\right)\left(1+\dfrac{s}{5}\right)}$$

$$\Rightarrow \quad G(j\omega) = \frac{10\left(1+\dfrac{j\omega}{10}\right)}{j\omega\left(1+\dfrac{j\omega}{2}\right)\left(1+\dfrac{j\omega}{5}\right)}$$

The transfer function consists of the following components:

1. A constant $K = 10$; 2. A zero at $\omega_c = 10$;
3. A pole at the origin; 4. Two simple poles at $\omega_c = 2$ and 5.

Magnitude plot

1. For the constant, a horizontal line is drawn at $20 \log 10 = 20$ dB.
2. For the pole at the origin, a -20 dB/decade line that passes through $\omega = 1$ at 0 dB is drawn.
3. For the two simple poles, two -20 dB/decade lines are drawn from $\omega_c = 2$ and 5.
4. Individual contributions are added to get the final magnitude plot.

Phase plot

1. For the zero at $\omega_c = 10$ rad/s, a line is drawn connecting 0° at $\omega = 1$ rad/s and 90° at $\omega = 100$ rad/s.
2. For the pole at the origin, the phase is a horizontal line at $-90°$.
3. For the pole at $\omega_c = 2$ rad/s, a line is drawn connecting 0° at $\omega = 0.2$ rad/s and $-90°$ at $\omega = 20$ rad/s.
4. For the pole at $\omega_c = 5$ rad/s, a line is drawn connecting 0° at $\omega = 0.5$ rad/s and $-90°$ at $\omega = 50$ rad/s.
5. Individual contributions are added to get the final phase plot.

The Bode plot is presented in Figure 12.20. From the plot, we get the following data:
(a) Gm = 10 dB; Pm = 10°; $\omega_g = 4.5$ rad/s; $\omega_p = 7$ rad/s.
(b) Since both Gm and Pm values are positive, the system is stable.

EXAMPLE 12.18. Sketch the plot showing the magnitude in dBs and phase angle in degrees as a function of frequency in logarithm scale for the transfer function given by

$$G(s) = \frac{10}{s(1+0.5s)(1+0.1s)}$$

Hence determine the gain margin and phase margin of the system. Comment on the stability of the system.

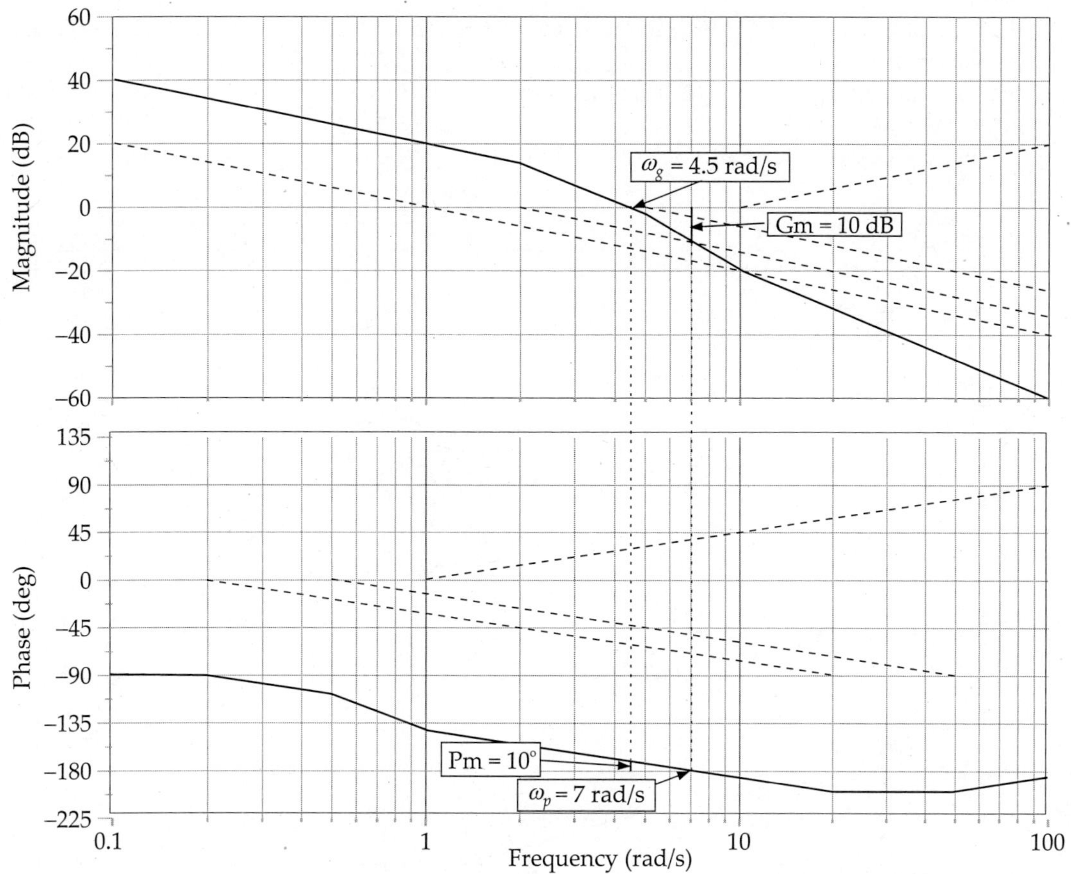

Figure 12.20: Bode plot (Example 12.17).

Solution The given open-loop transfer function can be written in the Bode form as:

$$G(s) = \frac{10}{s\left(1+\dfrac{s}{2}\right)\left(1+\dfrac{s}{10}\right)} \qquad \Rightarrow G(\jmath\omega) = \frac{10}{\jmath\omega\left(1+\dfrac{\jmath\omega}{2}\right)\left(1+\dfrac{\jmath\omega}{10}\right)}$$

The transfer function consists of the following components:

1. A constant $K = 10$; 2. A pole at the origin; 3. Two simple poles at $\omega_c = 2$ and 10.

Magnitude plot

1. For the constant, a horizontal line is drawn at $20 \log 10 = 20$ dB.
2. For the pole at the origin, a -20 dB/decade line that passes through $\omega = 1$ at 0 dB is drawn.
3. For the two simple poles, two -20 dB/decade lines are drawn from $\omega_c = 2$ and 10.
4. Individual contributions are added to get the final magnitude plot.

Phase plot

1. For the pole at the origin, the phase is a horizontal line at $-90°$.

2. For the pole at $\omega_c = 2$ rad/s, a line is drawn connecting 0° at $\omega = 0.2$ rad/s and −90° at $\omega = 20$ rad/s.
3. For the pole at $\omega_c = 10$ rad/s, a line is drawn connecting 0° at $\omega = 1$ rad/s and −90° at $\omega = 100$ rad/s.
4. Individual contributions are added to get the final phase plot.

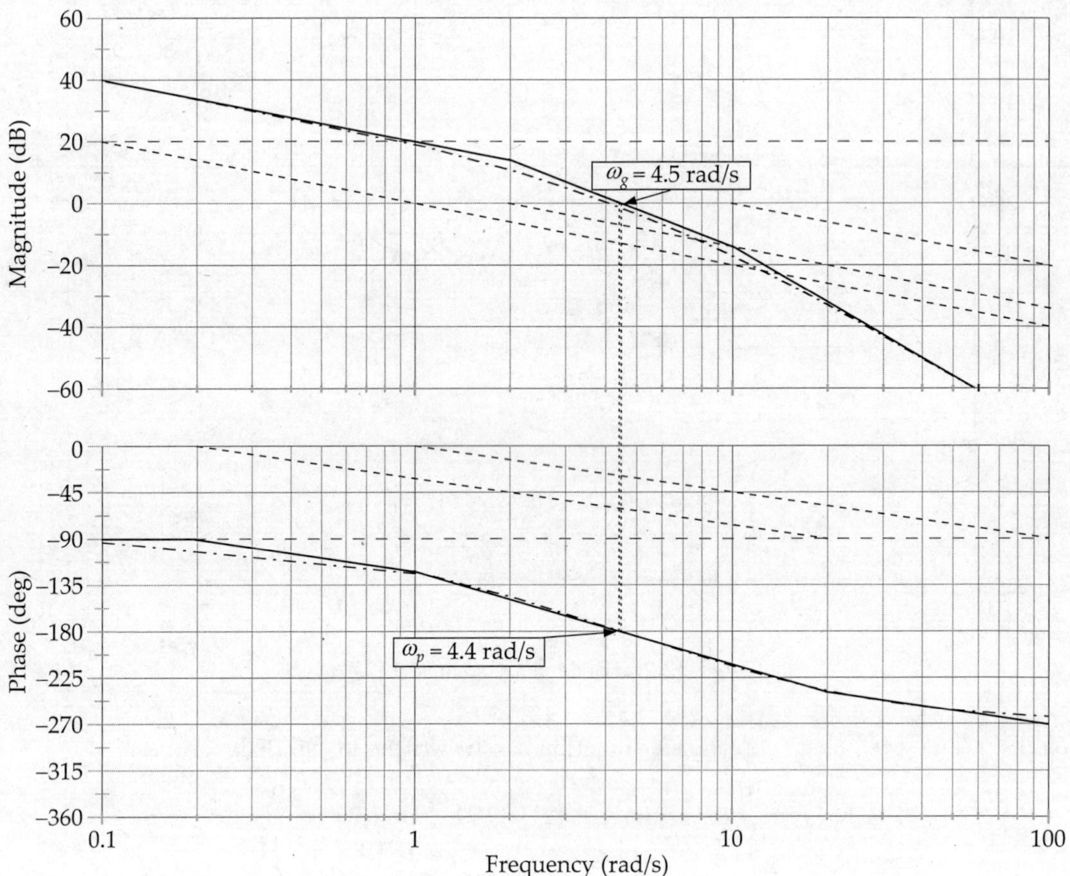

Figure 12.21: Bode plot (Example 12.18). Dash-dot curves indicate exact plots.

The Bode plot is presented in Figure 12.21. From the plot, we get the following data:

$$\text{Gm} = \text{Pm} \approx 0 \qquad [\because \omega_g \approx \omega_p]$$

Since both Gm and Pm values are zeros, the system is marginally stable.

Note: The Matlab shows a stable system with Gm = 1.58 dB at $\omega_p = 4.47$ rad/s and Pm = 10.7° at $\omega_g = 4.08$ rad/s. Our crossover frequencies are pretty close, but owing to the approximate plots, we have arrived at a wrong conclusion about the stability of the system.

Normally, the asymptotic plots have large errors near corner frequencies which may invert the conclusion. To investigate that we calculated the actual magnitude and phase for a few frequencies from the relations

$$M = 20\log\left(\frac{10}{\omega\sqrt{1+0.25\omega^2}\sqrt{1+0.01\omega^2}}\right)$$
$$\phi = -90^\circ - \tan^{-1}0.5\omega - \tan^{-1}0.1\omega \tag{12.14}$$

The results obtained are given in Table 12.5.

Table 12.5: Magnitude and phase values as obtained from Eq. (12.14)

ω (rad/s)	M (dB)	ϕ (deg)	ω (rad/s)	M (dB)	ϕ (deg)
0.1	20	−93	20	−33.15	−238
0.3	30.36	−100	30	−43.1	−248
0.5	25.75	−107	40	−50.45	−253
1.0	18.98	−122	50	−55.9	−256
1.5	5.27	−135	60	−60.8	−259
2	10.8	−146	70	−64.8	−260
5	−3.6	−185	80	−68.2	−261
10	−17.1	−214	100	−74.0	−263

The exact curves obtained from these values are shown by dash-dot curves in Figure 12.21. It can be seen from the figure that a small change in the vital area of the magnitude plot has made $\omega_g > \omega_p$ in the asymptotic plots whereas the situation is opposite in the exact plots.

EXAMPLE 12.19. The asymptotic Bode magnitude plot of a minimum phase transfer function $G(s)$ is shown in Figure 12.22.

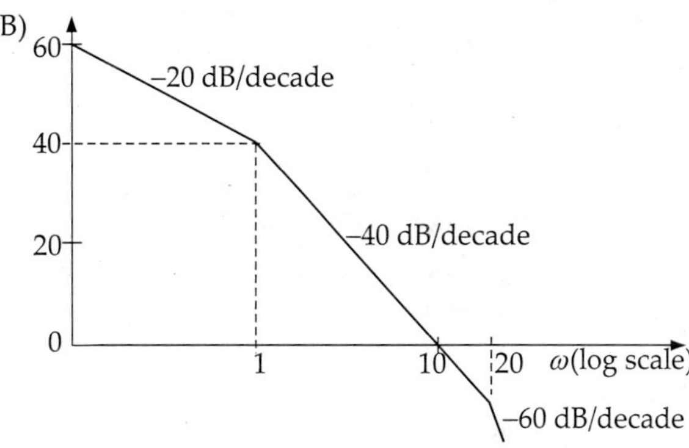

Figure 12.22: Magnitude plot (Example 12.19).

Consider the following two statements:

Statement I: Transfer function $G(s)$ has three poles and one zero.

Statement II: At very high frequency ($\omega \to \infty$), the phase angle $\angle G(j\omega) = -\dfrac{3\pi}{2}$.

Which one of the following options is correct?

(A) Statement I is false and statement II is true.

(B) Both the statements are true.

(C) Both the statements are false.

(D) Statement I is true and statement II is false.

Solution From the given Bode plot we observe that the magnitude plot changes slope by -20 dB/decade at the origin and at $\omega = 1$ and 20. So, there are 3 poles and no zero. Hence,

the statement I is false. Next, we note that the three poles contribute a total phase of $-270°$ at $\omega \to \infty$. So, the statement II is true.
Ans. (A)

EXAMPLE 12.20. The closed-loop transfer function of a position control system is given by

$$T(s) = \frac{9}{s^2 + 1.2s + 9}$$

(a) Draw the Bode plot of the system.
(b) Find out the resonant frequency ω_r and resonant peak M_r from the plot. Also find out the phase lag at ω_r.
(c) From the plot find out the 3 dB bandwidth of the system.

Solution

(a) We note that this is a closed-loop transfer function, though Bode plots are made for open-loop transfer functions. But here, instead of finding the stability of the system, we are asked to find some frequency domain parameters for the system through the Bode-type plots. So, we go for exact magnitude and phase plots in the Bode scheme and do not look for asymptotic plots because the denominator consists of complex poles. To do that, we have to write the transfer function in magnitude and phase form as

$$T(s) = \frac{9}{-\omega^2 + 1.2\jmath\omega + 9} = \left|\frac{9}{\sqrt{(9-\omega^2)^2 + (1.2\omega)^2}}\right| \angle - \tan^{-1}\frac{1.2\omega}{9-\omega^2}$$

or,
$$M \equiv 20\log\left|\frac{9}{\sqrt{(9-\omega^2)^2 + (1.2\omega)^2}}\right| \text{ dB}$$

and
$$\phi \equiv -\left(\frac{180}{\pi}\cdot\tan^{-1}\frac{1.2\omega}{9-\omega^2}\right) \text{ degree}$$

Now, we have to calculate magnitude in dB and phase in degrees. By comparison with the standard second order transfer function, we note that the undamped natural frequency of the system is 3 rad/s. So, the resonant frequency will be close to it and therefore, while tabulating data, we have to find magnitude and phase at close intervals near 3 rad/s. The calculated magnitude and phase are given in Table 12.6.

ω (rad/s)	M (dB)	ϕ (degree)	ω (rad/s)	M (dB)	ϕ (degree)
1.0	0.9268	−8.5308	3.0	7.9588	−90.0
2.0	4.2048	−25.641	3.1	7.5589	−99.3124
2.5	6.8937	−47.4896	3.2	7.2459	−107.896
2.6	7.3962	−54.3236	3.3	6.2393	−115.514
2.7	7.8067	−62.1759	3.5	4.5819	−127.733
2.8	8.0689	−70.7534	4.0	0.5094	−145.561
2.85	8.1274	−75.6095	5.0	−5.5687	−159.444
2.9	8.1301	−80.3776			

Table 12.6: Magnitude and phase values (Example 12.20)

These values are presented in Bode plot form in Figure 12.23.

(b) From the magnitude and phase plot we get: resonant frequency = 2.9 rad/s, resonant peak = 8.1 dB, and phase lag at $\omega_r = -80°$

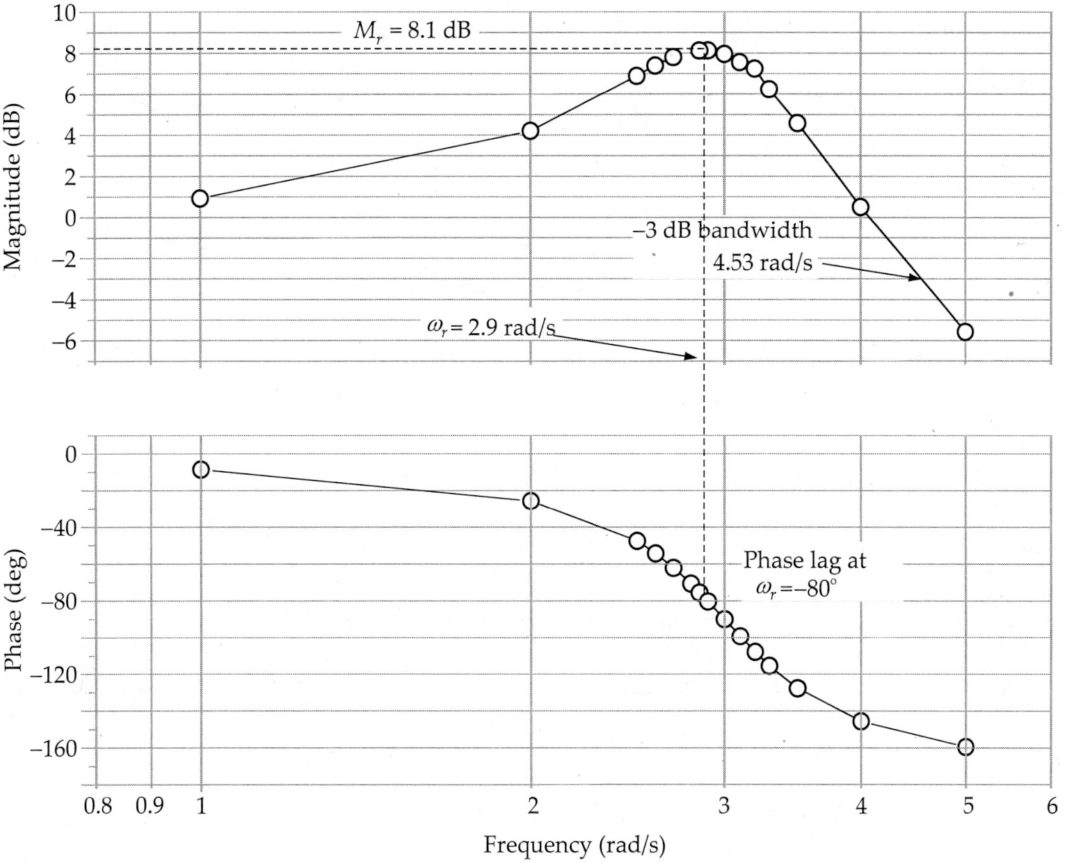

Figure 12.23: Bode plot of (top) magnitude and (bottom) phase (Example 12.20).

(c) From the magnitude plot we find that the −3 dB bandwidth is 4.53 rad/s.

EXAMPLE 12.21. The Bode magnitude plot [Figure 12.24b] of the transfer function $V_o(s)/V_i(s)$ of the circuit [Figure 12.24a] is as shown. The value of R is ———Ω (round off to 2 decimal places).

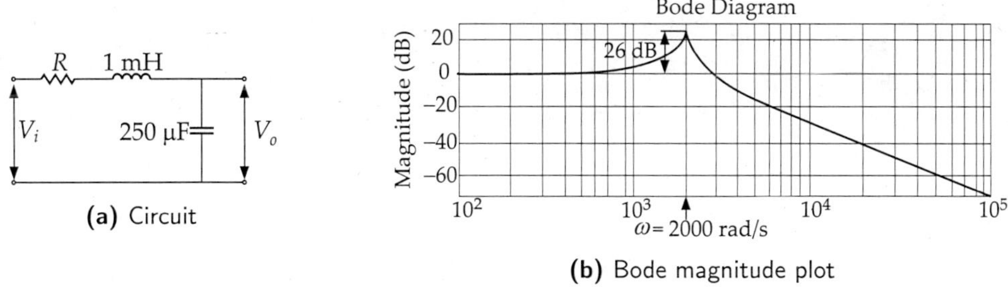

(a) Circuit

(b) Bode magnitude plot

Figure 12.24: Example 12.21.

Solution For the given RLC circuit, the gain $V_o(s)/V_i(s)$ is given 26 dB. We have to figure out the value of R. At the operating frequency $\omega = 2000\,\text{rad/s}$,

$$Z_L = \jmath\omega L = \jmath \times 2000 \times 1 \times 10^{-3} = \jmath 2$$

$$Z_c = -\frac{\jmath}{\omega C} = -\frac{\jmath}{200 \times 250 \times 10^{-6}} = -\jmath 2$$

$$V_o(\jmath\omega) = \frac{Z_c}{R + Z_L + Z_c} V_i(\jmath\omega)$$

or, $$\left|\frac{V_o(\jmath\omega)}{V_i(\jmath\omega)}\right| = \frac{-\jmath 2}{R + \jmath 2 - \jmath 2} = \frac{2}{R} \tag{i}$$

Given, $$20\log\left|\frac{V_o(\jmath\omega)}{V_i(\jmath\omega)}\right| = 26$$ [see Figure 12.24b]

or, $$\left|\frac{V_o(\jmath\omega)}{V_i(\jmath\omega)}\right| = \text{antilog}\,\frac{26}{20} = 19.95 \tag{ii}$$

From Eqs. (i) and (ii) $$\frac{2}{R} = 19.95$$

Therefore, $$R = \frac{2}{19.95} \approx 0.10\,\Omega$$

12.6 BODE PLOT FOR SYSTEMS WITH TRANSPORTATION LAG

We have seen in Eq. (4.64) that the transfer function for transportation lag can be written as

$$G(s) = e^{-sT}$$

where T is the time delay in seconds. In the frequency domain, this transfer function becomes

$$G(\jmath\omega) = e^{-\jmath\omega T} \tag{12.15}$$

Putting Eq. (12.15) in magnitude and phase form, we obtain

$$\begin{aligned} G(\jmath\omega) &= \cos\omega T - \jmath\,\sin\omega T \\ &= \left|\sqrt{\cos^2\omega T + \sin^2\omega T}\right| \angle -\tan^{-1}\frac{\sin\omega T}{\cos\omega T} \\ &= |1|\,\angle -\omega T \end{aligned}$$

In dB, the magnitude turns out to be 0 while the phase angle, which varies linearly with the frequency ω for a given T, is in degrees

$$\phi = -\frac{180}{\pi}\omega T \approx -57.3\omega T \text{ degree}$$

EXAMPLE 12.22. Draw the Bode plot for

$$G(s)H(s) = \frac{e^{-0.4s}}{s(s+2)}$$

and find gain and phase margins.

Solution Written in the Bode form, the open-loop transfer function is

$$G(\jmath\omega)H(\jmath\omega) = \frac{0.5e^{-\jmath 0.4\omega}}{\jmath\omega\left(1 + \frac{\jmath\omega}{2}\right)}$$

We observe that the transfer function comprises the following components:

1. A constant = 0.5;
2. A transportation lag of 0.4 s;
3. A pole at the origin;
4. A pole at $\omega_c = 2$.

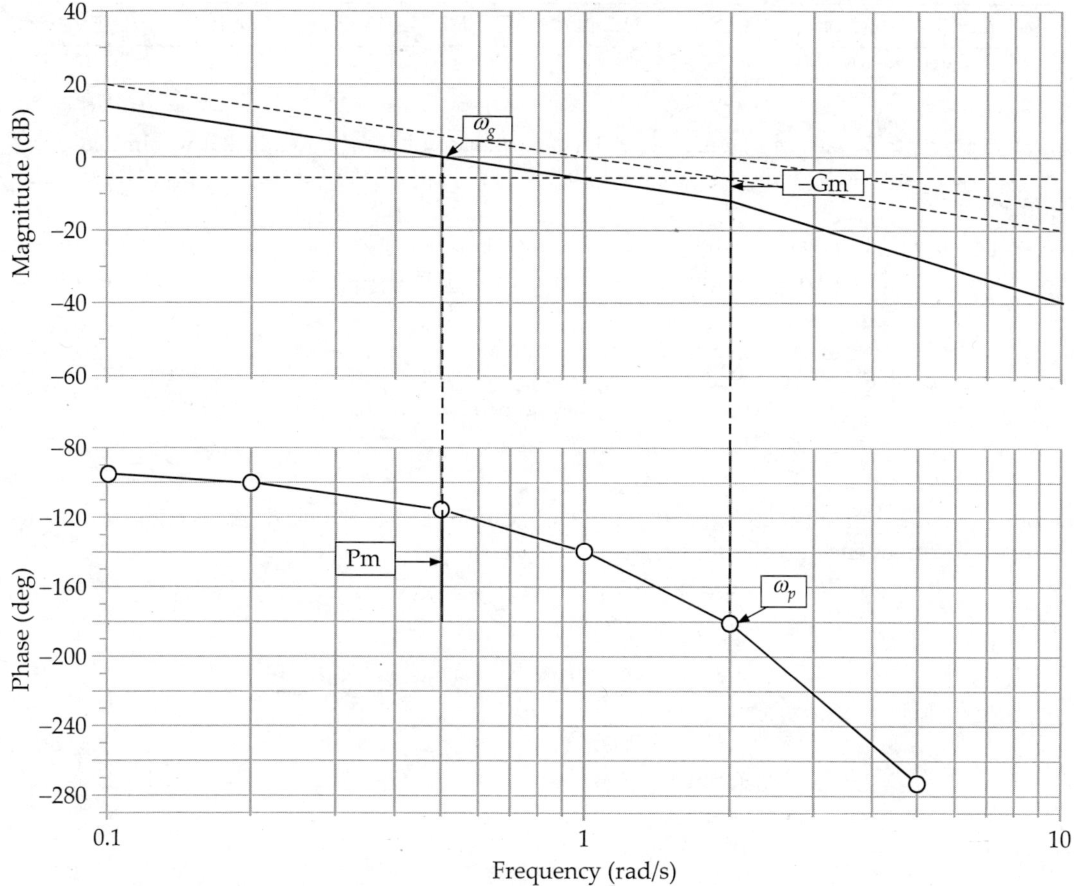

Figure 12.25: Bode plot (Example 12.22): (top) asymptotic magnitude plot and (bottom) actual phase plot.

Magnitude plot

1. For the constant, a horizontal line is drawn at 20 log 0.5 = −6 dB.
2. The transportation lag has no contribution to magnitude.
3. For the pole at the origin, a −20 dB/decade line that passes through $\omega = 1$ at 0 dB is drawn.
4. For the simple pole, −20 dB/decade line is drawn at $\omega_c = 2$.
5. Individual contributions are added to get the final curve.

Phase plot

The transportation lag adds a phase lag of $\angle e^{-j0.4\omega} = -57.3(0.4\omega) = -22.92\omega$. Therefore, the phase angle is given by the relation

$$\phi = \angle G(j\omega)H(j\omega) = -90^\circ - \tan^{-1}\frac{\omega}{2} - 22.92\omega \tag{i}$$

The actual Bode phase diagram is drawn from the values given in Table 12.7.

Table 12.7: Data for actual phase plot (Example 12.22)

ω (rad/s)	0.1	0.2	0.5	1.0	2.0	5.0	10.0
ϕ (deg)[a]	−95.154	−100.295	−115.496	−139.485	−180.84	−272.799	−397.89

[a] Calculated from Eq. (i)

From the asymptotic plot of magnitude and actual plot of phase [Figure 12.25] we find that $\omega_g = 5$ rad/s, $\omega_p = 2$ rad/s, Gm = 14 dB and Pm = $(180° - 118°) = 62°$. Therefore, the system is stable.

EXAMPLE 12.23. Consider the unity feedback control system shown in Figure 12.26. The value of K that results in a phase margin of the system to be 30° is ———. (Give the answer up to two decimal places.)

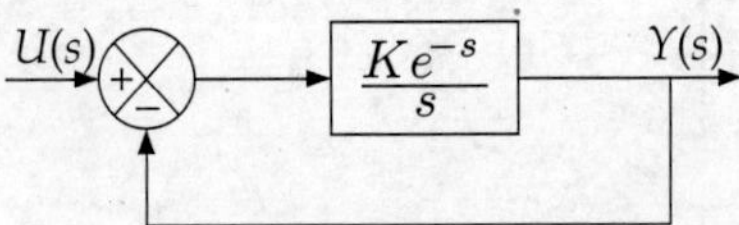

Figure 12.26: System (Example 12.23).

Solution Required phase margin = 30°. Which implies that the required phase angle, ϕ at the gain-crossover frequency ω_g should be

$$\text{Phase margin} = 180° + \phi$$

or,

$$30° = 180° + \phi$$

or,

$$\phi = -(180 - 30)° = -150° = \angle G(\jmath\omega)\big|_{\omega=\omega_g} \tag{i}$$

Again,

$$e^{-\jmath\omega} = \cos(\omega) - \jmath\sin(\omega) = |\sqrt{cos^{\omega} + \sin^2\omega}|\angle - \tan^{-1}\frac{\sin\omega}{\cos\omega} = |1|\angle - \omega$$

Therefore,

$$\angle G(\jmath\omega)\big|_{\omega_g} = -90° - \frac{180°}{\pi}\omega_g = -90° - 57.3\omega_g^{\circ} \tag{ii}$$

We know that at ω_g, the magnitude of the transfer function equals 1. Therefore,

$$|G(\jmath\omega)|_{\omega_g} = 1 = \left|\frac{K \times 1}{\omega_g}\right|$$

or,

$$K = \omega_g$$

or,

$$-90° - 57.3K° = -150° \qquad \text{[from Eqs. (i) and (ii)]}$$

Therefore,

$$K = \frac{60}{57.3} = 1.047$$

12.7 TRANSFER FUNCTION FROM BODE PLOT

If the system is of minimum phase type[9], the transfer function of a system can be derived from the Bode magnitude plot by following the inverse procedure. We consider a few examples to see how it is done.

EXAMPLE 12.24. The Bode asymptotic magnitude plot of a transfer function is shown in Figure 12.27a. In the frequency range shown, the transfer function has

[9]See Section 2.1 at page 13.

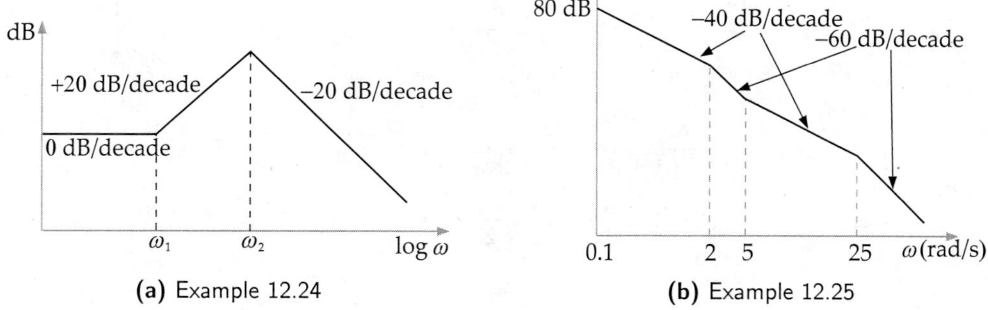

(a) Example 12.24

(b) Example 12.25

Figure 12.27: Bode plots.

(A) 3 poles and 1 zero
(B) 1 pole and 2 zeros
(C) 2 poles and 1 zero
(D) 2 poles and 2 zeros

Solution Since the origin of the graph is not indicated and that the initial portion of the plot shows a horizontal straight line, it may be the resultant of a simple pole and a simple zero, or caused by a constant gain. The second part is created by a simple zero. The third part, which reverses the slope, must have been caused by the existence of two poles or a quadratic pole. So, the nearest answer is (A), assuming that the initial part is caused by a constant and that there is no quadratic pole in the third part.
Ans. (A)

EXAMPLE 12.25. The asymptotic approximation of the log-magnitude vs. frequency plot of a system containing only real poles and zeros is shown in Figure 12.27b. Its transfer function is

(A) $\dfrac{10(s+5)}{s(s+2)(s+25)}$
(B) $\dfrac{1000(s+5)}{s^2(s+2)(s+25)}$
(C) $\dfrac{100(s+5)}{s(s+2)(s+25)}$
(D) $\dfrac{80(s+5)}{s^2(s+2)(s+25)}$

Solution The initial slope of -40 dB/decade suggests the existence of two poles at the origin. The slope gets enhanced to -60 dB/decade at the corner frequency 2 rad/s, which suggests the existence of a simple pole there. The slope improves to -40 dB/decade at the corner frequency 5 rad/s owing obviously to the existence of a zero there. Finally, at the corner frequency 25 rad/s, the slope again becomes -40 dB/decade as a consequence of the existence of a simple pole there. So, let our open-loop transfer function be

$$G(s)H(s) = \frac{K(s+5)}{s^2(s+2)(s+25)} \equiv \frac{K(\jmath\omega+5)}{(\jmath\omega)^2(\jmath\omega+2)(\jmath\omega+25)}$$
$$\equiv \frac{K\sqrt{\omega^2+25}}{\omega^2\sqrt{\omega^2+4}\sqrt{\omega^2+625}}\angle\left(\tan^{-1}\frac{\omega}{5} - 180^\circ - \tan^{-1}\frac{\omega}{2} - \tan^{-1}\frac{\omega}{25}\right)$$

From the given figure, we observe that the magnitude is 80 dB at $\omega = 0.1$ rad/s. Therefore, we have

$$20\log \frac{K\sqrt{\omega^2+25}}{\omega^2\sqrt{\omega^2+4}\sqrt{\omega^2+625}}\bigg|_{\omega=0.1} = 80$$

$$\Rightarrow \quad \log \frac{K\sqrt{0.01+25}}{0.01\sqrt{0.01+4}\sqrt{0.01+625}} = 4$$

$$\Rightarrow \quad \log \frac{5K}{0.01\times 2\times 25} = 4 \qquad [\because 4,\ 25,\ 625 \gg 0.01]$$

$$\Rightarrow \quad \frac{K}{0.1} = \text{antilog } 4 = 10000$$

$$\Rightarrow \quad K = 1000$$

Ans. (B)

EXAMPLE 12.26. The asymptotic Bode plot for the gain magnitude of a minimum phase system $G(s)$ is shown in Figure 12.28a.

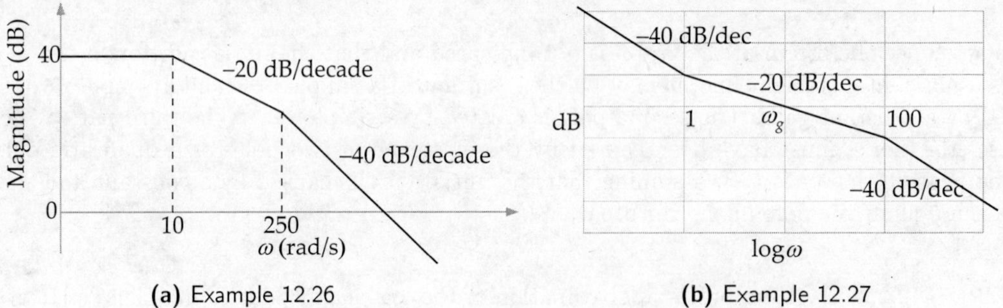

(a) Example 12.26 **(b)** Example 12.27

Figure 12.28: Bode plots.

The transfer function $G(s)$ is given by

(A) $\dfrac{100}{[1+(s/10)][1+(s/250)]}$ (B) $\dfrac{40}{s(s+250)}$

(C) $\dfrac{100}{(s+10)(s+250)}$ (D) $\dfrac{100s}{(s+10)(s+250)}$

Solution From the given magnitude plot we get the following data: There are two simple poles at corner frequencies 10 and 250 rad/s, and there is a gain factor K which makes the initial part of the plot a straight line parallel to the ω-axis. To find out the value of K, we have

$$20\log K = 40 \quad \Rightarrow \quad \log K = 2 \quad \Rightarrow \quad K = 100$$

So,

$$G(s) = \frac{100}{\left(1+\dfrac{s}{10}\right)\left(1+\dfrac{s}{250}\right)}$$

Ans. (A)

EXAMPLE 12.27. The asymptotic Bode magnitude plot of an open-loop system $G(S)$ with $K > 0$ and all poles and zeros on the left hand side of the s-plane is shown Figure 12.28b. It is completely symmetric about ω_g.

The maximum absolute phase angle contribution by $G(s)$ is given by

(A) $-78.6°$ (B) $-90°$ (C) $-101.4°$ (D) $-180°$

Solution We note that the initial slope of the magnitude plot is -40 dB/decade. This indicates the existence of a double pole at the origin. Then, at the corner frequency 1, the slope becomes -20 dB/decade indicating the presence of a zero there. Thereafter, at the corner frequency 100, the slope again becomes -40 dB/decade which suggests the presence of a pole at that frequency. Since, we need to determine the phase, we need not bother about the constant factor because, a constant contributes nothing to the phase. Our transfer function is, therefore

$$G(s) = \frac{K(1+s)}{s^2\left(1+\dfrac{s}{100}\right)}$$

In frequency domain, its phase factor ϕ is given by

$$\phi = \tan^{-1}\omega - 180° - \tan^{-1}\frac{\omega}{100}$$

The phase of a transfer function, where a zero is symmetrically sandwiched between two poles, will obviously have a bell shape in the middle, i.e. at ω_g. The phase will have a maximum there. Since $\omega_g = 10$ (as can be seen from the plot), the maximum value of phase is

$$\phi_{\max} = \tan^{-1}10 - 180° - \tan^{-1}0.1 = -101.4°$$

Ans. (C)

EXAMPLE 12.28. The approximate Bode magnitude plot of a lead network with its pole and zero on the left half of the s-plane is shown in Figure 12.29a.

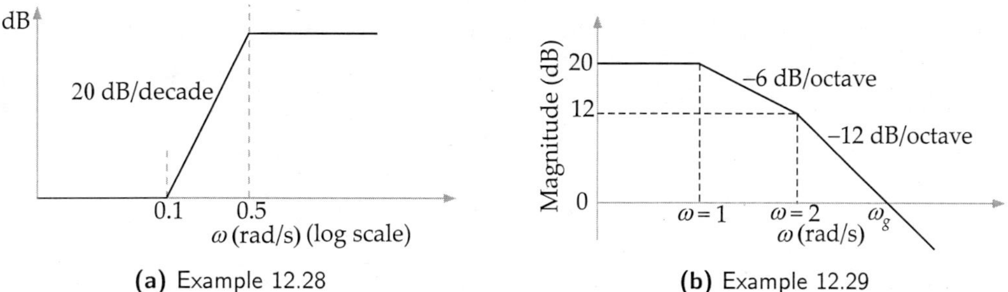

(a) Example 12.28 (b) Example 12.29

Figure 12.29: Bode plots.

The frequency at which the phase angle of the network is maximum (in rad/s) is

(A) $\dfrac{3}{\sqrt{10}}$ (B) $\dfrac{1}{\sqrt{20}}$ (C) $\dfrac{1}{20}$ (D) $\dfrac{1}{30}$

Solution The shape of the magnitude curve suggests the existence of a zero at the corner frequency 0.1 rad/s and a pole at the corner frequency 0.5 rad/s. So, let the transfer function be given by

$$G(s)H(s) = \frac{K\left(1+\dfrac{s}{0.1}\right)}{1+\dfrac{s}{0.5}} = \frac{K(0.1+s)}{0.5+s} \equiv \frac{K(0.1+\jmath\omega)}{0.5+\jmath\omega}$$

The phase factor of this function in the frequency domain is given by

$$\phi = \tan^{-1}\frac{\omega}{0.1} - \tan^{-1}\frac{\omega}{0.5} = \tan^{-1}10\omega - \tan^{-1}2\omega$$

To maximize phase, it is necessary that

$$\frac{d\phi}{d\omega} = \frac{d}{d\omega}(\tan^{-1} 10\omega) - \frac{d}{d\omega}(\tan^{-1} 2\omega) = 0$$

$$\Rightarrow \qquad \frac{10}{1+100\omega^2} - \frac{2}{1+4\omega^2} = 0$$

$$\Rightarrow \qquad 1+100\omega^2 = 5+20\omega^2$$

$$\Rightarrow \qquad \omega = \frac{1}{\sqrt{20}}$$

Ans. (B)

EXAMPLE 12.29. The asymptotic approximation to the log-magnitude versus frequency plot (Bode plot) of a unity feedback control system is shown in Figure 12.29b. This is a minimum phase system.

Determine: (a) Gain crossover frequency, in rad/s, (b) Gain margin, in dB, and (c) Phase margin, in degrees.

Solution Since the plot begins with a constant magnitude of 20 dB, there exists a dc gain of antilog $(20 \div 20) = 10$ in the transfer function of the system. -6 dB/octave[10] at $\omega = 1$ and -12 dB/octave at $\omega = 2$ indicate the existence of two simple poles at corner frequencies $\omega = 1$ and $\omega = 2$. Therefore, the open-loop transfer function is given by

$$G(s)H(s) = \frac{10}{(1+s)[1+(s/2)]} \equiv \frac{10}{(1+\jmath\omega)[1+(\jmath\omega/2)]} \equiv \frac{10}{\sqrt{1+\omega^2}\sqrt{1+\dfrac{\omega^2}{4}}}$$

(a) To find the gain crossover frequency ω_g, we equate the magnitude part to 0 dB. Thus,

$$20\log \frac{10}{\sqrt{1+\omega_g^2}\sqrt{1+\dfrac{\omega_g^2}{4}}} = 20\log 1$$

$$\Rightarrow \qquad \sqrt{1+\omega_g^2}\sqrt{1+\frac{\omega_g^2}{4}} = 10$$

$$\Rightarrow \qquad 4+5\omega_g^2+\omega_g^4 = 400 \qquad \text{(i)}$$

Equation (i) yields two values for ω_g^2, namely, 17.56 and -22.56. The latter value is discarded because the square of a frequency cannot be negative. So, $\omega_g = \sqrt{17.56} = 4.19$ rad/s.

(b) To determine the gain margin, we need to know the phase crossover frequency ω_p. We do that by equating the phase factor to $-180°$. Thus,

$$-\tan^{-1}\omega_p - \tan^{-1}\frac{\omega_p}{2} = -180°$$

$$\Rightarrow \qquad \tan^{-1}\frac{\omega_p + (\omega_p/2)}{1-\omega_p\cdot(\omega_p/2)} = 180°$$

$$\Rightarrow \qquad \frac{(3/2)\omega_p}{1-(\omega_p^2/2)} = \tan 180° = 0 \qquad \text{(ii)}$$

The RHS of Eq. (ii) becomes 0 if $\omega_p = 0$ or $\omega_p = \infty$. The former value, being impractical, is rejected. So, $\omega_p = \infty$ is accepted. This frequency is substituted in the magnitude factor

[10]$\equiv -20$ dB/decade.

to find the gain and then we need to convert its inverse to dB to find the gain margin. Now,

$$|M| = \left.\frac{10}{\sqrt{1+\omega_p^2}\sqrt{1+(\omega_p^2/4)}}\right|_{\omega_p=\infty} = \frac{1}{\infty} = 0 \qquad \Rightarrow \frac{1}{|M|} = \infty$$

Thus, $$\text{Gm} = 20\log\infty = \infty$$

(c) The phase margin is calculated as:

$$\phi\Big|_{\omega_g=4.19} = -\tan^{-1}4.19 - \tan^{-1}\frac{4.19}{2} = -141.06^\circ$$

Therefore, $$\text{Pm} = 180^\circ - 141.06^\circ = 38.9^\circ$$

EXAMPLE 12.30. Figure 12.30a shows the Bode magnitude and phase plots of a system (not to scale). Determine from the plot the transfer function of the system.

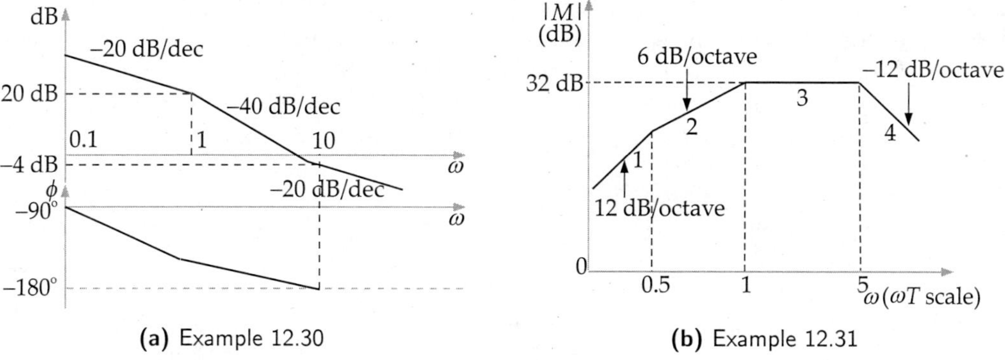

(a) Example 12.30

(b) Example 12.31

Figure 12.30: Bode plots.

Solution The magnitude plot starts with a −20 dB/decade slope. The slope increases to −40 dB/decade at the corner frequency 1 and then again decreases to −20 dB/decade at the unknown corner frequency x. Obviously, a zero is present at the corner frequency x. So, let us write the open-loop transfer function as

$$G(s)H(s) = \frac{K\left(1+\dfrac{s}{x}\right)}{s\left(1+\dfrac{s}{1}\right)}$$

Since the value of the magnitude is 20 dB at $\omega = 1$ rad/s, we have

$$20\log K = 20 \qquad \Rightarrow \log K = 1 \qquad \Rightarrow K = 10$$

The plot indicates that the phase crossover frequency line intersects the magnitude plot at –4 dB.

So, Gm = 4 dB at $\omega_p = 10$ rad/s. Therefore,

$$\text{Gm} = 4 \text{ dB} = -20 \log |G(\jmath\omega)H(\jmath\omega)|\Big|_{\omega=10} = -20 \log \frac{10\sqrt{1+\omega^2/x^2}}{\omega\sqrt{1+\omega^2}}\Bigg|_{\omega=10}$$

$$\Rightarrow \qquad \text{antilog}\left(-\frac{4}{20}\right) = \frac{10\sqrt{1+100/x^2}}{10\sqrt{1+100}}$$

This yields
$$\frac{100}{x^2} = 39.2182$$

$$x = \sqrt{\frac{100}{39.2182}} = 1.5968$$

The required transfer function is, therefore, given by

$$G(s)H(s) = \frac{10[1+(s/1.5968)]}{s(1+s)} = \frac{10(1+0.6262s)}{s(1+s)}$$

EXAMPLE 12.31. Obtain the transfer function of a system whose Bode magnitude plot is given in Figure 12.30b.

Solution The curve contains four segments which we have numbered as 1, 2, 3 and 4. We will designate transfer functions corresponding to these segments as G_1, $G_{12}\ldots$ etc. meaning thereby the transfer function up to that segment of the curve.

At $\omega T = 0$, the slope is 12 dB/octave, i.e. 40 dB/decade. The slope is positive which indicates there are two zeros there, because we know, each zero contributes a slope of 20 dB/decade. Therefore,

$$G_1 = Ks^2$$

where K is the gain.

At $\omega T = 0.5$, the slope is 6 dB/octave, i.e. 20 dB/decade. So, the change in slope is -20 dB/decade, indicating the existence of a simple pole at that corner frequency. Therefore,

$$G_{12} = \frac{Ks^2}{1+\dfrac{s}{0.5}}$$

At $\omega T = 1$, the slope becomes zero which indicates the existence of another simple pole that nullifies the existing 6 dB/octave slope. Therefore,

$$G_{123} = \frac{Ks^2}{\left(1+\dfrac{s}{0.5}\right)(1+s)}$$

Now we observe that at $\omega T = 1$

(i) The resultant magnitude plot is horizontal.

(ii) Its magnitude is 32 dB.

This indicates that the magnitude 32 dB comes from the contribution of the constant—poles and zeros having annulled their contributions here. Thus,

$$20 \log K = 32 \qquad \Rightarrow K = 39.8$$

At $\omega T = 5$, the slope becomes -12 dB/octave, i.e. -40 dB/decade. This indicates the existence of a double pole there. Combining all these, we get

$$G_{1234} = \frac{39.8s^2}{\left(1+\frac{s}{0.5}\right)(1+s)\left(1+\frac{s}{5}\right)^2}$$
$$= (0.5 \times 25) \cdot \frac{39.8s^2}{(s+0.5)(s+1)(s+5)^2}$$

Therefore, the required transfer function is $\dfrac{497.6s^2}{(s+0.5)(s+1)(s+5)^2}$.

EXAMPLE 12.32. The magnitude and phase plots shown in Figure 12.31 match with the transfer function of

(A) $\dfrac{10000}{s^2+2s+10000}$ (B) $\dfrac{10000e^{-0.05s}}{s^2+2s+10000}$

(C) $\dfrac{10000e^{-0.05\times 10^2 s}}{s^2+2s+10000}$ (D) $\dfrac{100}{s^2+2s+100}$

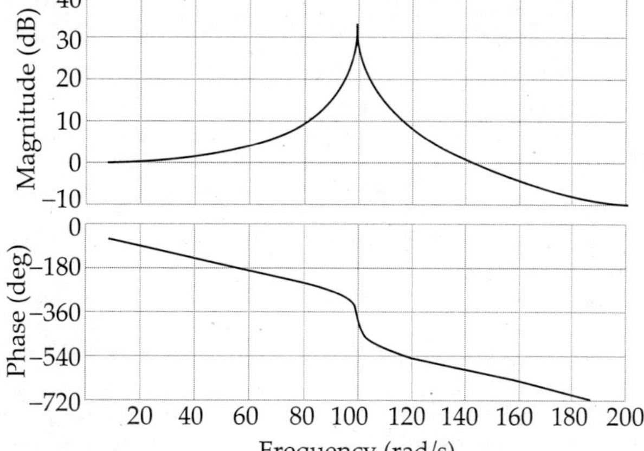

Figure 12.31: Magnitude and phase plot (Example 12.32).

Solution From the phase plot, it is clear that there is a transportation lag. Hence, options (B) or (C) could be the answer. Also, from the magnitude plot, we get the resonance frequency $\omega_n = 100$ rad/s where the phase is $\approx -360°$.

Except the transportation lag factor, from the rest of the transfer function of (B) and (C), we get at $\omega_n = 100$

$$G(\jmath\omega)H(\jmath\omega) = \frac{100^2}{-\omega^2 + 2\jmath\omega + 100^2}$$

Therefore,
$$\phi = -\tan^{-1}\frac{2\omega_n}{100^2 - \omega_n^2} = \infty = -90°$$

Now, considering the (C) option, the transportation lag element is

$$e^{-0.05\times 10^2 s} = e^{-5\jmath\omega}$$

Hence, its contribution to phase is $= -5$ rad $= -\dfrac{180°}{\pi} \times 5 = -286.5°$.
Therefore, the total phase at $\omega = \omega_n$ is

$$\phi = -90° - 286.5° = -376.5°$$

which tallies with the value obtained from the graph.
Ans. (C)

EXAMPLE 12.33. From the asymptotic magnitude (in dB) versus frequency (log scale) plot of Figure 12.32, find the associated transfer function. Assume that no right half plane poles or zeros are present.

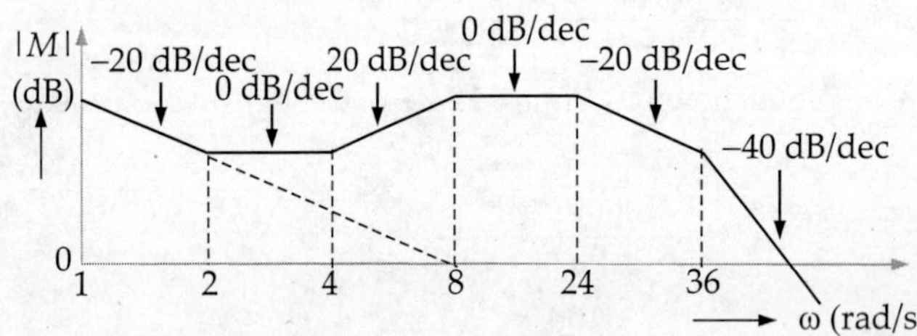

Figure 12.32: Asymptotic magnitude plot (Example 12.33).

Solution At $\omega_c = 1$, the slope is -20 dB/decade. This indicates that there is a pole at the origin. Therefore,

$$G_1 = \frac{K}{s}$$

At $\omega_c = 2$, the slope changes to 0 dB/decade, indicating the existence of a simple zero at that corner frequency. Therefore,

$$G_{12} = \frac{K\left(1+\frac{s}{2}\right)}{s}$$

At $\omega_c = 4$, the slope changes to + 20 dB/decade, which indicates the existence of another simple zero there. Therefore,

$$G_{123} = \frac{K\left(1+\frac{s}{2}\right)\left(1+\frac{s}{4}\right)}{s}$$

Now, we observe that at $\omega_c = 8$ the resultant magnitude plot is horizontal. This indicates the existence of a pole at $\omega_c = 8$, which neutralizes the + 20 dB/decade slope. Therefore,

$$G_{1234} = \frac{K\left(1+\frac{s}{2}\right)\left(1+\frac{s}{4}\right)}{s\left(1+\frac{s}{8}\right)}$$

At $\omega_c = 24$, the slope becomes -20 dB/decade. This indicates the existence of a pole there. Therefore,

$$G_{12345} = \frac{K\left(1+\frac{s}{2}\right)\left(1+\frac{s}{4}\right)}{s\left(1+\frac{s}{8}\right)\left(1+\frac{s}{24}\right)}$$

Finally, at $\omega_c = 36$, the slope becomes -40 dB/decade, indicating the existence of another pole there. Therefore,

$$G_{123456} = \frac{K\left(1+\frac{s}{2}\right)\left(1+\frac{s}{4}\right)}{s\left(1+\frac{s}{8}\right)\left(1+\frac{s}{24}\right)\left(1+\frac{s}{36}\right)} \qquad \text{(i)}$$

Evaluation of K: We observe that for the pole at origin, the -20 dB/decade line should have passed through 0 dB at $\omega_c = 1$. Instead, it does so at $\omega_c = 8$. That could happen because of the contribution of 20 log K dB contribution of the constant. The -20 dB/decade line has a drop of 20 log 8 dB from $\omega_c = 1$ to $\omega_c = 8$. Therefore,

$$20 \log K = 20 \log 8 \text{ dB} \qquad \Rightarrow K = 8$$

Substituting this value of K in Eq. (i) yields, on simplification, the transfer function

$$G(s) = \frac{6912(2+s)(4+s)}{s(8+s)(24+s)(36+s)}$$

EXAMPLE 12.34. From the asymptotic magnitude (in dB) versus frequency (log scale) plot of Figure 12.33, find the associated transfer function. Assume that the system is of minimum phase type.

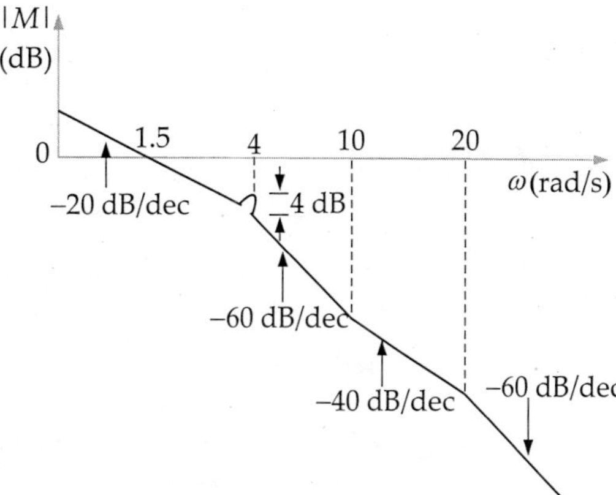

Figure 12.33: Asymptotic Bode magnitude plot (Example 12.34).

Solution Since the system is of minimum phase type, there is no pole or zero in the RHP.

The starting line on the left possesses a slope of -20 dB/decade, which indicates the existence of a pole at the origin. But this line, instead of passing through 0 dB at $\omega_c = 1$, passes through $\omega_c = 1.5$. That indicates a contribution from the gain K. So, our initial transfer function is

$$G_1 = \frac{K}{s}$$

At $\omega_c = 1.5$, the value of the -20 dB/decade line is 20 log 1.5. This should equal the contribution from the constant to make the resultant 0 dB. Therefore,

$$20 \log K = 20 \log 1.5 \qquad \Rightarrow K = 1.5$$

Now, we observe that:

(i) There exists a peak at $\omega_c = 4$ with a peak height of 4 dB.

(ii) The slope of the magnitude plot has changed by -40 dB/decade.

These indicate the existence of a quadratic pole at that corner frequency. We know from Eq. (12.11) that for a quadratic pole[11], $\omega_c = \omega_n$ and

$$|M| = 20 \log \left(\frac{1}{2\zeta}\right) = 4 \text{ dB (given)} \qquad \therefore \zeta = 0.315$$

[11]Note that Eq. (12.11)pertains to a quadratic zero. For a quadratic pole, inverse of the term under the radical sign in Eq. (12.8) has to be considered.

So, the transfer function now becomes

$$G_{12} = \frac{K\omega_n^2}{s(s^2 + 2\zeta\omega_n s + \omega_n^2)} = \frac{(1.5)(4)^2}{s[s^2 + 2(0.315)(4)s + 4^2]} = \frac{24}{s(s^2 + 2.52s + 16)}$$

Next, we observe that the slope changes from −60 dB/decade to −40 dB/decade, indicating the existence of a zero at the corner frequency of 10 rad/s. So, the transfer function now becomes

$$G_{123} = \frac{24\left(1 + \dfrac{s}{10}\right)}{s(s^2 + 2.52s + 16)}$$

Finally, the slope again changes to −60 dB/decade at 20 rad/s, indicating the existence of another pole there. Hence, the transfer function is

$$G_{1234} = \frac{24\left(1 + \dfrac{s}{10}\right)}{s\left(1 + \dfrac{s}{20}\right)(s^2 + 2.52s + 16)}$$

On simplification, the required transfer function turns out to be

$$G(s) = \frac{48(s + 10)}{s(s + 20)(s^2 + 2.52s + 16)}$$

12.8 STATIC ERROR CONSTANTS FROM BODE PLOT

The static error constants K_p, K_v and K_a can be determined from the Bode plot. We will show how to do that with simple examples.

Let an open-loop transfer function be defined as

$$G(s)H(s) = \frac{1}{s^N}$$

In frequency domain, its magnitude in dB is written as

$$|M| = 20\log|G(\jmath\omega)H(\jmath\omega)| = 20\log\left|\frac{K}{(\jmath\omega)^N}\right| = 20\log K - 20N\log\omega \tag{12.16}$$

Static Position Error Constant

Let us consider a type 0 system with $N = 0$. Substituting this value in Eq. (12.16), we find

$$|M| = 20\log K$$

Now we recall that the static position error constant is defined as

$$K_p = \lim_{s\to 0} G(s)H(s)$$

which in the frequency domain becomes

$$K_p = \lim_{\omega\to 0} G(\jmath\omega)H(\jmath\omega)$$

Thus the initial magnitude of the Bode plot, corresponding to the magnitude of the constant gain K equals the static position error constant.

Static Velocity Error Constant

Let us consider a type 1 system having $N = 1$. Putting this value in Eq. 12.16 we get at the intersection of the Bode magnitude plot with the 0 dB line

$$0 = 20\log K - 20\log\omega \qquad \Rightarrow K = \omega$$

The static velocity error constant is defined as

$$K_v = \lim_{s\to 0} sG(s)H(s)$$

For a type 1 system, the s-factor in the numerator and denominator cancel out yielding just the constant K at $\omega \to 0$. In other words,

$$K_v = K = \omega$$

Thus, for a type 1 system, the value of ω at the 0 dB intersection of the initial −20 dB/decade line gives the value of the static velocity error constant. A type 1 system has an initial slope of −20 dB/decade. We need to extend that line to 0 dB and find the corresponding ω which gives the value of K_v [see Figure 12.34a].

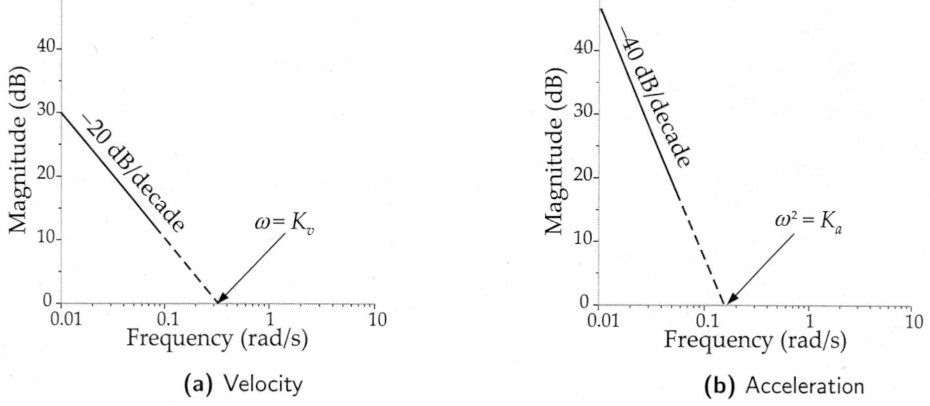

(a) Velocity (b) Acceleration

Figure 12.34: Static error constant determination from Bode magnitude plot.

Static Acceleration Error Constant

Consider a type 2 system having $N = 2$. As before, putting this value in Eq. (12.16) we get at the intersection of the Bode magnitude plot with the 0 dB line

$$0 = 20\log K - 20\log\omega^2 \qquad \Rightarrow K = \omega^2$$

So, following the same argument that we put forward for the velocity error constant, we can find out the static acceleration constant by extending the initial −40 dB/decade line (a type 2 system has that slope initially) to the 0 dB line [see Figure 12.34b] and figuring out the square of the corresponding frequency value.

12.9 MINIMUM AND NON-MINIMUM PHASE SYSTEMS

We know systems having no poles or zeros in the right-half of the s-plane (RHP) of their transfer functions are called *minimum phase systems*, while those having poles and/or zeros in the RHP are called *non-minimum phase systems*. The modulus of the phase response for a non-minimum-phase system is then always larger than that for a system with minimum-phase behaviour though both may have the same amplitude response.

Let us consider, for example, a minimum phase system having the transfer function as

$$G_{\text{min}}(s) = \frac{1+sT}{1+sT_1} \tag{12.17}$$

and a non-minimum phase system having a transfer function of

$$G_{\text{non-min}}(s) = \frac{1-sT}{1+sT_1} \tag{12.18}$$

where $T_1 > T > 0$. The pole-zero plots of the systems are shown in Figures 12.35a and 12.35b.

(a) Minimum **(b)** Non-minimum

Figure 12.35: Pole-zero plot of minimum and non-minimum phase systems.

The frequency domain transfer functions of Eqs. (12.17) and (12.18) yield the same amplitude of

$$|M| = \frac{\sqrt{1+\omega^2T^2}}{\sqrt{1+\omega^2T_1^2}}$$

while the phases, which are different, are given by

$$\phi_{\text{min}} = \tan^{-1}\omega T - \tan^{-1}\omega T_1 = -\tan^{-1}\frac{\omega(T_1-T)}{1+\omega^2T_1T}$$

$$\phi_{\text{non-min}} = -\tan^{-1}\omega T - \tan^{-1}\omega T_1 = -\tan^{-1}\frac{\omega(T_1+T)}{1-\omega^2T_1T}$$

respectively.

Their plots, shown in Figures 12.35a and 12.35b, depict a huge phase sweep for the non-minimum phase system. In fact, given a particular magnitude of response, the minimum phase system offers less phase shift than all other systems. A non-minimum phase system transfer function, such as that given by Eq. (12.18), can always be represented as a cascaded minimum phase and a pure phase shift element (also called *all-pass* system) as follows.

$$G(s) = \frac{1+sT}{1+sT_1}\cdot\frac{1-sT}{1+sT}$$

This phase-shift element,

$$\frac{1-sT}{1+sT}$$

which consists of reflected pole-zero in the s-plane, has a magnitude of unity, but a phase shift of

$$\phi = -2\tan^{-1}\omega T$$

Thus, this all-pass element which introduces a phase shift of 0° to −180°, increases the phase sweep of non-minimum phase systems.

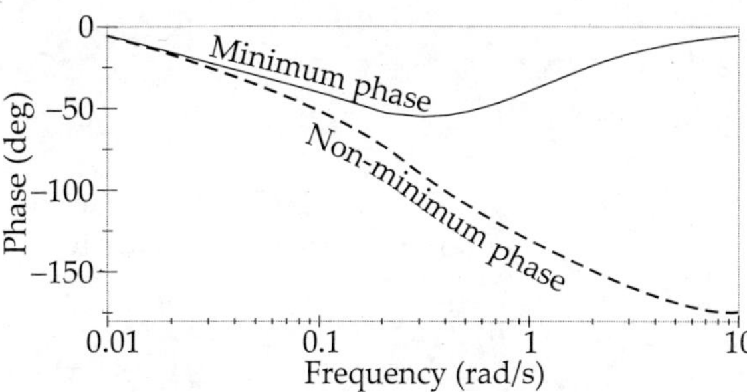

Figure 12.36: Phase shifts of minimum and non-minimum phase systems.

Because of the high phase shift, non-minimum phase systems are slow in response and so they are not desirable in a control system.

So far, we have centred our discussion on first order systems. In general, if m and n denote degrees of polynomials of the numerator and denominator of a transfer function, for minimum phase systems,

$$\phi|_{\omega\to\infty} = -90°(n-m)$$

where ϕ is the phase angle. This criterion does not hold good for non-minimum phase systems, though for both,

$$M|_{\omega\to\infty} = -20 \text{ dB/decade}$$

where M is the magnitude. So, from the behaviour of the phase at $\omega \to \infty$, a non-minimum-phase system can be recognized.

REVIEW QUESTIONS

12.1 Choose the correct answer

(i) 20 dB/decade corresponds to

(A) 3 dB/octave (B) 6 dB/octave (C) 9 dB/octave (D) none of these

(ii) A system has 4 poles and 1 zero. Its high frequency asymptote in magnitude plot has a slope of

(A) −100 dB/decade (B) 100 dB/decade
(C) −60 dB/decade (D) 60 dB/decade

(iii) Slope of asymptotes in Bode plot for a second order system is

(A) 18 dB/octave (B) −12 dB/octave (C) 6 dB/octave (D) −3 dB/octave

(iv) The frequency 2 decade above a frequency of 2 rad/s is

(A) 22 rad/s (B) 102 rad/s (C) 200 rad/s (D) 32 rad/s

(v) The gain of a system is 10. In terms of dB it is

(A) 0 dB (B) 1 dB (C) 20 dB (D) 100 dB

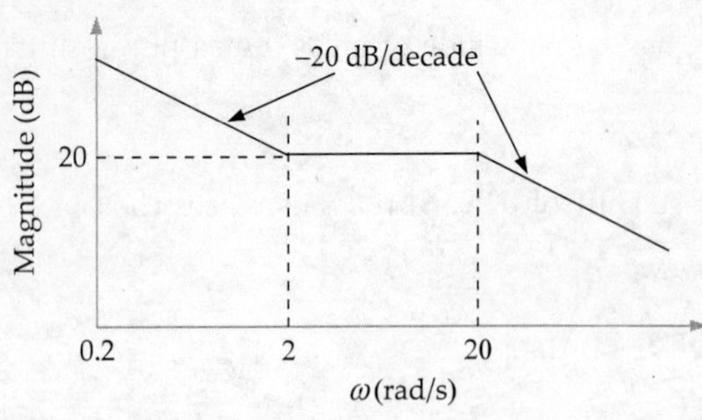

Figure 12.37: Bode magnitude plot [Question 12.1(vi)].

(vi) Figure 12.37 shows the Bode magnitude plot of a system. The minimum phase transfer function of the system is given by

(A) $\dfrac{80\left(\dfrac{s}{2}+1\right)}{s\left(\dfrac{s}{20}+1\right)}$ (B) $\dfrac{20\left(\dfrac{s}{2}+1\right)}{s\left(\dfrac{s}{20}+1\right)}$

(C) $\dfrac{8(s+2)}{s(s+20)}$ (D) $\dfrac{80(s+2)}{s(s+20)}$

(vii) The Bode plot of the transfer function $G(s) = s$ is

(A) zero magnitude and zero phase shift
(B) 20 dB/decade and phase shift π
(C) 20 dB/decade and phase shift $\pi/2$
(D) -20 dB/decade and phase shift $-\pi/2$

(viii) The initial slope of Bode plot for a transfer function having simple pole at origin is

(A) 20 dB/decade (B) -40 dB/decade (C) 40 dB/decade (D) -20 dB/decade

(ix) The error at the corner frequency of an asymptotic Bode plot for the term $(1+\jmath\omega T)^N$ is

(A) $\pm 5N$ dB (B) ± 3 dB (C) ± 6 dB (D) $\pm 3N$ dB

(x) Consider the asymptotic Bode magnitude plot (ω is in rad/s) of Figure 12.38. Which

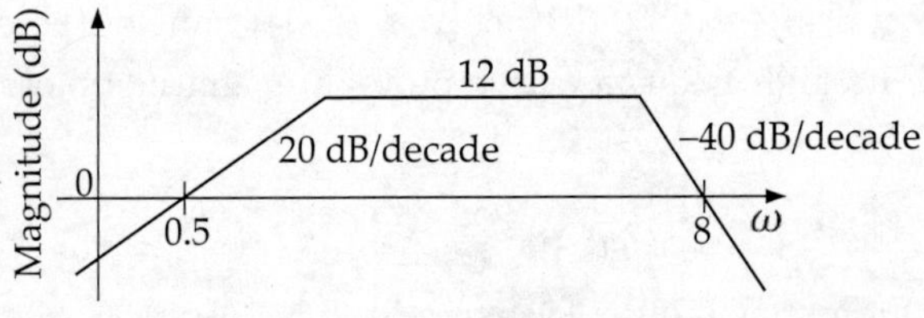

Figure 12.38: Bode magnitude plot [Question 12.1(x)].

of the following transfer functions is best represented by the above Bode magnitude plot?

(A) $\dfrac{2s}{(1+0.5s)(1+0.25s)^2}$ (B) $\dfrac{4(1+0.5s)}{s(1+0.25s)}$

(C) $\dfrac{2s}{(1+2s)(1+4s)}$ (D) $\dfrac{4s}{(1+2s)(1+4s)^2}$

(xi) The Bode magnitude plot of

$$H(\jmath\omega) = \frac{10^4(1+\jmath\omega)}{(10+\jmath\omega)(100+\jmath\omega)}$$

is

(A)

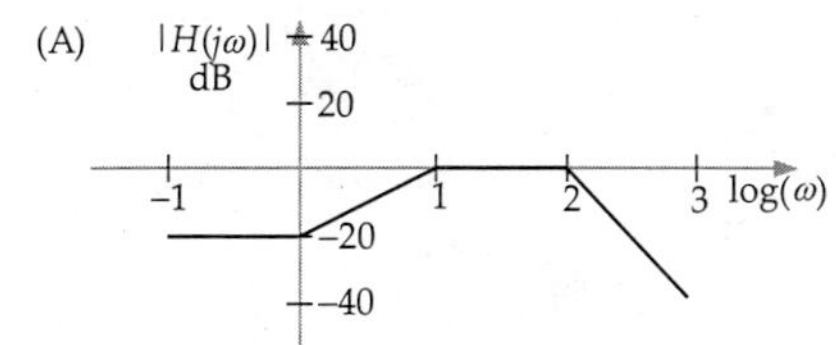

(B) |H(jω)| dB 40 20 –1 1 2 3 log(ω) –20 –40

(C)

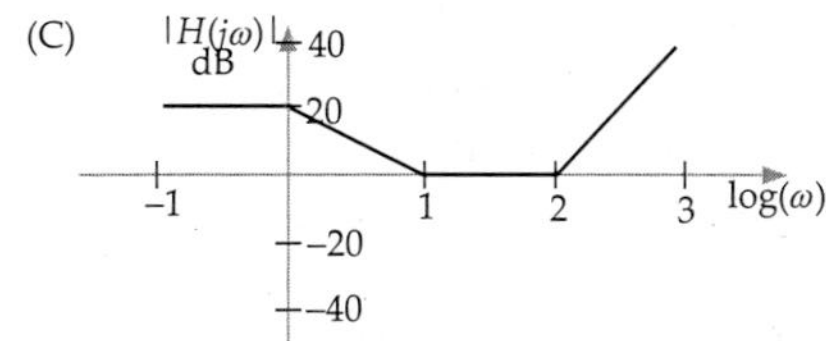

(D)

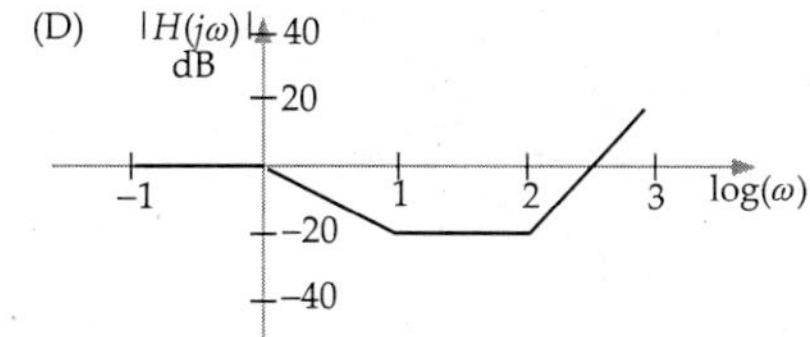

12.2 (a) Draw the Bode plot for a system having

$$G(s)H(s) = \frac{100}{s(s+1)(s+2)}$$

(b) For the above problem, find graphically gain margin, phase margin, gain crossover frequency and phase crossover frequency.

(c) Comment on the stability of the system.

12.3 (a) Define gain margin and phase margin. How are they used to determine the amount of relative stability of a closed-loop system?

(b) How can you evaluate the Gm and Pm of a control system from its Bode plot?

12.4 (a) Sketch the asymptotic Bode plot for the following open-loop transfer function with unity feedback

$$G(s) = \frac{10}{s(s+1)(s+5)}$$

(b) Calculate the gain margin and phase margin from the Bode plot. Determine the gain cross-over frequency.

12.5 Sketch the Bode plots showing the magnitude in decibels and phase angle in degrees as function of log frequency for the transfer function

$$G(s)H(s) = \frac{10(s+1)}{s(s+2)(s+10)}$$

12.6 Draw Bode plots for a control system whose loop transfer function is

$$G(s)H(s) = \frac{80}{s(1+0.02s)(1+0.05s)}$$

Determine the gain margin and phase margin and thus check the stability.

12.7 The transfer function of a control system is

$$G(s) = \frac{1000(s+3)}{s(s+12)(s+50)}$$

Draw the Bode plot of amplitude versus log ω and phase versus log ω.

Answers to Selected Questions

12.1. (i) (B) (ii) (C) (iii) (B) (iv) (A) (v) (C) (vi) (D) (vii) (C) (viii) (D) (ix) (D) (x) (A) (xi) (A)

12.2. (b) Gm = −24 dB, Pm = −53°, $\omega_g = 4.5$ rad/s, $\omega_p = 1.4$ rad/s (c) Unstable

12.4. (b) Gm = 9.5 dB, Pm = 25°, $\omega_g = 1.2$ rad/s 12.6. Gm = −1.2 dB, Pm = −3.4°. Unstable

13

Polar Plots

We know that the frequency domain transfer function is a complex quantity and like any other complex quantity, the complex transfer function can be expressed in magnitude and phase (angle) form. It is well-known that a magnitude and angle can be plotted in polar coordinates. The polar plot is, therefore, a plot of the magnitude of $G(\jmath\omega)$ vs. angle of $G(\jmath\omega)$ in *polar coordinates* as ω is varied from 0 to ∞.

> In polar plots
>
> - Angles are positive if they are measured from the positive real axis in the *counter-clockwise* direction and vicè versa.
> - The positive imaginary axis is the line at $+90°$, and the negative imaginary axis, at $-90°$.
> - The positive real axis is the line at $0°$ (or $360°$) and the negative real axis, at $+180°$ (or -180).

To familiarize ourselves with this kind of plots, let us draw plots of a few standard components of frequency domain transfer functions.

13.1 POLAR PLOTS OF STANDARD FUNCTIONS

Integral and Derivative Factors

The magnitude and angle form of an integral factor, $G(s) = 1/s$, is given by

$$G(\jmath\omega) = \frac{1}{\jmath\omega} \equiv \left|\frac{1}{\omega}\right| \angle - \tan^{-1}\frac{1}{0} = \left|\frac{1}{\omega}\right| \angle - 90° \tag{13.1}$$

The magnitude of Eq. (13.1) varies with ω, as shown in Table 13.1, but the angle remains constant at $-90°$. So, the polar plot [Figure 13.1a] is the negative part of the imaginary axis[1].

Table 13.1: Magnitude variation of derivative and integral factors

$\lvert\omega\rvert$	0.01	0.1	1	∞
$\lvert 1/\omega\rvert$	100	10	1	0

The magnitude and angle form of the derivative factor, $G(s) = s$, is as follows:

$$G(\jmath\omega) = \jmath\omega \equiv |\omega|\angle 90°$$

Obviously, the positive part of the imaginary axis represents the plot [Figure 13.1b].

[1] Real and Imaginary axes are written as Re and Im. They are also symbolically written as $\Re$ and $\Im$.

(a) Integral (b) Derivative

Figure 13.1: Polar plots of integral and derivative factors.

Simple Pole and Simple Zero

A simple pole, which is represented in the s-plane by the transfer function $G(s) = 1/(1 + sT)$, is given by its magnitude and angle form in the frequency domain as

$$G(\jmath\omega) = \frac{1}{1 + \jmath\omega T} \equiv \left|\frac{1}{\sqrt{1 + \omega^2 T^2}}\right| \angle - \tan^{-1}(\omega T)$$

In this case, both the magnitude and angle of $G(\jmath\omega)$ vary with ω. Their values at some points are given in Table 13.2. The polar plot, [Figure 13.2a], is a semi-circle of radius 0.5 having its centre at (0.5, 0).

ω	*Pole*		*Zero*	
	$\lvert G(\jmath\omega)\rvert$	$\angle G(\jmath\omega)$	$\lvert G(\jmath\omega)\rvert$	$\angle G(\jmath\omega)$
0	1	0°	1	0°
$1/T$	$1/\sqrt{2}$	−45°	$\sqrt{2}$	45°
∞	0	−90°	∞	−90°

Table 13.2: Magnitude and angle of simple pole and simple zero

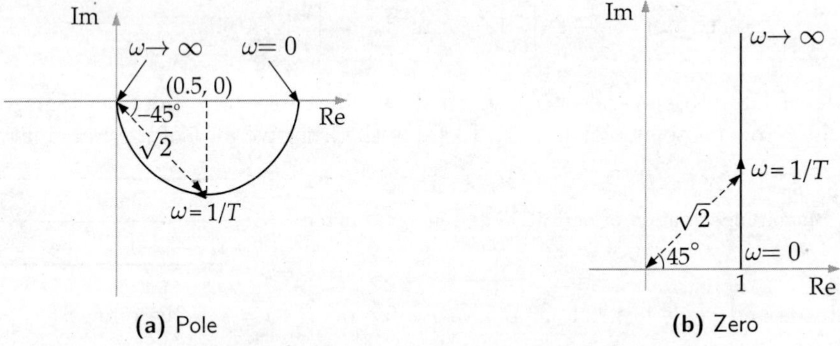

(a) Pole (b) Zero

Figure 13.2: Simple pole and simple zero plots.

A simple zero, which is given by the transfer function $G(s) = (1 + sT)$ is given by its magnitude and angle form in the frequency domain as

$$G(\jmath\omega) = 1 + \jmath\omega T \equiv |\sqrt{1 + \omega^2 T^2}| \angle \tan^{-1}(\omega T)$$

The values of magnitude and angle for a few ω values are given in Table 13.2. The polar plot, [Figure 13.2b], is a straight line parallel to the positive part of the imaginary axis, originating from (1, 0).

Quadratic Pole and Quadratic Zero

The magnitude and angle form of the standard transfer function of a quadratic pole, given by $G(s) = \omega_n^2/(s^2 + 2\zeta\omega_n s + \omega_n^2) \quad \zeta > 0$, can be derived from its complex form as follows:

$$G(\jmath\omega) = \frac{1}{1 + \jmath 2\zeta(\omega/\omega_n) - (\omega/\omega_n)^2)}$$
$$\equiv \left| \frac{1}{\sqrt{\left[1 - (\omega/\omega_n)\right]^2 + 4\zeta^2(\omega/\omega_n)^2}} \right| \angle - \tan^{-1} \frac{2\zeta(\omega/\omega_n)}{1 - (\omega/\omega_n)^2}$$

The values of magnitude and angle for three cardinal ω values are given in Table 13.3.

Table 13.3: Magnitude and angle of quadratic pole and quadratic zero

ω	*Pole*		*Zero*	
	$\lvert G(\jmath\omega)\rvert$	$\angle G(\jmath\omega)$	$\lvert G(\jmath\omega)\rvert$	$\angle G(\jmath\omega)$
0	1	0°	1	0°
ω_n	$1/2\zeta$	−90°	2ζ	90°
∞	0	−180°	∞	180°

The polar plot is shown in Figure 13.3a.

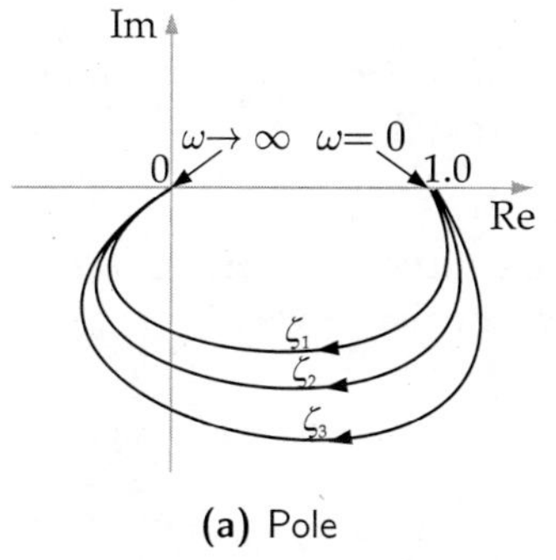

(a) Pole

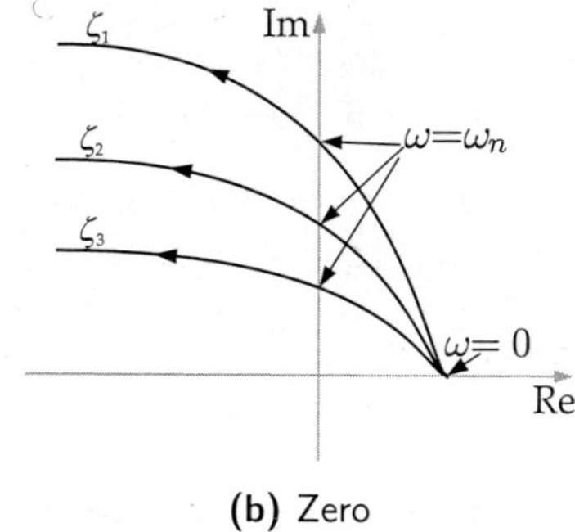

(b) Zero

Figure 13.3: Quadratic pole and quadratic zero plots for different ζ ($\zeta_1 > \zeta_2 > \zeta_3$).

Similarly, the polar form of a quadratic zero, which in the s-plane is given by $G(s) = (s^2 + 2\zeta\omega_n s + \omega_n^2)/\omega_n^2 \quad \zeta > 0$, can be obtained as follows:

$$G(\jmath\omega) = 1 + \jmath 2\zeta(\omega/\omega_n) - (\omega/\omega_n)^2$$
$$\equiv \left| \sqrt{\left(1 - \frac{\omega}{\omega_n}\right)^2 + 4\zeta^2\left(\frac{\omega}{\omega_n}\right)^2} \right| \angle \tan^{-1} \frac{2\zeta(\omega/\omega_n)}{1 - (\omega/\omega_n)^2}$$

The values of magnitude and angle for three cardinal ω values are given in Table 13.3. The polar plot is shown in Figure 13.3b.

Transportation Lag

We have seen in Section 4.4 at page 99 that the transfer function for the transportation lag is given by

$$G(s) = e^{-sT}$$

Its polar representation is obtained as follows:

$$G(\jmath\omega) = e^{-\jmath\omega T} = \cos\omega T - \jmath\sin\omega T \equiv 1\angle -\omega T$$

Obviously, its polar plot (Figure 13.4) is a circle, because its magnitude is constant for all angles.

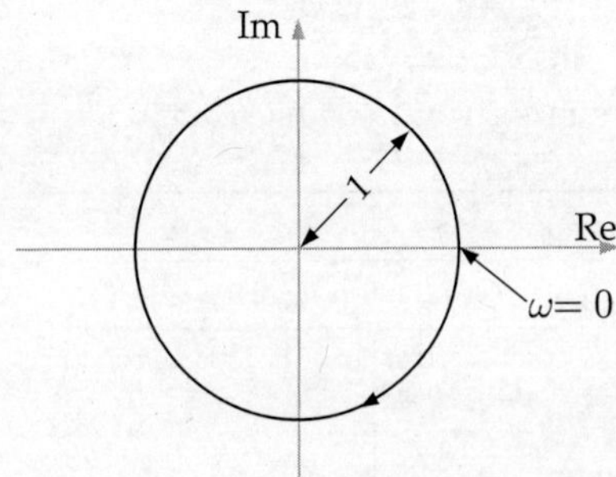

Figure 13.4: Polar plot of transportation lag.

13.2 DRAWING POLAR PLOTS

Based on what we have done so far to draw polar plots of a few standard simple transfer functions, we now delineate the steps for drawing them in general.

1. Convert the transfer function to frequency domain form by replacing s with $\jmath\omega$.
2. Convert the frequency domain transfer function to magnitude and angle form.
3. Magnitudes $|G(\jmath\omega)|$ are found at $\omega = 0$ and $\omega = \infty$. In the same way phase angles are calculated at $\omega = 0$ and $\omega = \infty$.
4. Break the complex transfer function $G(\jmath\omega)$ into real and imaginary parts through rationalization.
5. Equate the real part to zero and find out the corresponding frequency. Evaluate the imaginary part of $|G(\jmath\omega)|$ at this value of the frequency. This gives the intercept of the plot on the imaginary axis.
6. Equate the imaginary part, obtained in *Step 4*, to zero and find out the corresponding frequency. Evaluate the real part of $|G(\jmath\omega)|$ at this value of the frequency. This gives the intercept of the plot on the real axis.
7. From the data obtained so far, draw the polar plot.

It is true that we have to apply an educated guess to draw the entire plot because not many points will be obtained in the foregoing steps. But with a little bit of practice, it will not be a difficult task to do.

We will draw a few polar plots before we discuss how we can utilize them in our stability studies.

Type 0 System

Let the transfer function for a type 0 system be given by

$$G(s) = \frac{1}{(1+sT_1)(1+sT_2)}$$

In the frequency domain the transfer function is as follows:

$$G(\jmath\omega) = \frac{1}{(1+\jmath\omega T_1)(1+\jmath\omega T_2)}$$

Its magnitude and angle form is given by

$$G(\jmath\omega) = \frac{1}{\sqrt{1+(\omega T_1)^2}\sqrt{1+(\omega T_2)^2}}\angle(-\tan^{-1}\omega T_1 - \tan^{-1}\omega T_2) \tag{13.2}$$

The limiting values of the magnitude and phase, calculated from Eq. (13.2), are given in Table 13.4.

Table 13.4: Limiting values of magnitude and phase (type 0)

ω	$\lvert G(\jmath\omega)\rvert$	$\angle G(\jmath\omega)$
0	1	$0°$
∞	0	$-180°$
$\frac{1}{\sqrt{T_1T_2}}$	$\left\lvert\frac{\sqrt{T_1T_2}}{T_1+T_2}\right\rvert$	$-90°$

Now we split the complex transfer function into real and imaginary parts as follows:

$$\begin{aligned}
G(\jmath\omega) &= \frac{1}{(1+\jmath\omega T_1)(1+\jmath\omega T_2)}\cdot\frac{(1-\jmath\omega T_1)(1-\jmath\omega T_2)}{(1-\jmath\omega T_1)(1-\jmath\omega T_2)}\\
&= \frac{1-\omega^2T_1T_2-\jmath\omega(T_1+T_2)}{(1+\omega^2T_1^2)(1+\omega^2T_2^2)}\\
&= \frac{1-\omega^2T_1T_2}{(1+\omega^2T_1^2)(1+\omega^2T_2^2)} - \jmath\frac{\omega(T_1+T_2)}{(1+\omega^2T_1^2)(1+\omega^2T_2^2)}
\end{aligned} \tag{13.3}$$

Equating the real part to zero, we obtain the corresponding frequency as follows:

$$\begin{aligned}
\Re\{G(\jmath\omega)\} &= \frac{1-\omega^2T_1T_2}{(1+\omega^2T_1^2)(1+\omega^2T_2^2)} = 0\\
\Rightarrow\quad 1-\omega^2T_1T_2 &= 0\\
\Rightarrow\quad \omega &= \frac{1}{\sqrt{T_1T_2}}
\end{aligned} \tag{13.4}$$

Substituting ω from Eq. (13.4) in the imaginary part of Eq. (13.3), we get

$$\begin{aligned}
\Im\{G(\jmath\omega)\} &= -\jmath\frac{\frac{1}{\sqrt{T_1T_2}}(T_1+T_2)}{\left(1+\frac{T_1^2}{T_1T_2}\right)\left(1+\frac{T_2^2}{T_1T_2}\right)} = -\jmath\frac{\frac{1}{\sqrt{T_1T_2}}(T_1+T_2)}{\left(\frac{T_1+T_2}{T_2}\right)\left(\frac{T_1+T_2}{T_1}\right)}\\
&= -\jmath\frac{\sqrt{T_1T_2}}{T_1+T_2} \equiv \frac{\sqrt{T_1T_2}}{T_1+T_2}\angle-90°
\end{aligned}$$

We now equate the imaginary part, obtained from Eq. (13.3), to 0 and find out the corresponding frequency as follows:

$$\Im\{G(\jmath\omega)\} = \frac{\omega(T_1+T_2)}{(1+\omega^2T_1^2)(1+\omega^2T_2^2)} = 0$$

$$\Rightarrow \qquad \omega = 0,\ \infty$$

We already know the magnitude and angle values at these frequencies from Table 13.4. The polar plot is presented in Figure 13.5.

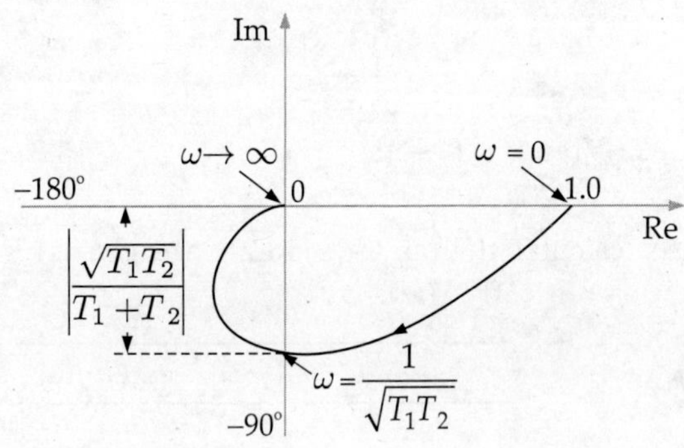

Figure 13.5: Polar plot for the type 0 system.

Type 1 System

Let the transfer function for the type 1 system be

$$G(s) = \frac{1}{s(1+sT_1)(1+sT_2)}$$

We now follow the steps, as shown before, to make its polar plot.

$$\begin{aligned} G(\jmath\omega) &= \frac{1}{\jmath\omega(1+\jmath\omega T_1)(1+\jmath\omega T_2)} \\ &= \frac{1}{\omega\sqrt{1+(\omega T_1)^2}\sqrt{1+(\omega T_2)^2}}\angle(-90^\circ - \tan^{-1}\omega T_1 - \tan^{-1}\omega T_2) \qquad (13.5) \end{aligned}$$

The data given in Table 13.5 can be obtained from Eq. (13.5).

ω	$\lvert G(\jmath\omega)\rvert$	$\angle G(\jmath\omega)$
0	∞	90°
∞	0	−270°
$\frac{1}{\sqrt{T_1T_2}}$	$\frac{\sqrt{T_1T_2}}{T_1+T_2}$	−0°

Table 13.5: Limiting magnitude and angle obtained from Eq. (13.5) (type 1)

Splitting Eq. (13.4) into real and imaginary parts:

$$\begin{aligned} G(\jmath\omega) &= \frac{1}{\jmath\omega(1+\jmath\omega T_1)(1+\jmath\omega T_2)} \cdot \frac{-\jmath(1-\jmath\omega T_1)(1-\jmath\omega T_2)}{-\jmath(1-\jmath\omega T_1)(1-\jmath\omega T_2)} \\ &= \frac{-\jmath[1-\omega^2T_1T_2 - \jmath\omega(T_1+T_2)]}{\omega(1-\omega^2T_1^2)(1-\omega^2T_2^2)} \\ &= \frac{-\omega(T_1+T_2)}{\omega(1-\omega^2T_1^2)(1-\omega^2T_2^2)} - \jmath\frac{1-\omega^2T_1T_2}{\omega(1-\omega^2T_1^2)(1-\omega^2T_2^2)} \qquad (13.6) \end{aligned}$$

$$\Re\{G(\jmath\omega)\} = \frac{-\omega(T_1+T_2)}{\omega(1-\omega^2T_1^2)(1-\omega^2T_2^2)} = 0$$

$$\Rightarrow \qquad \omega = \infty$$

This means that the polar plot does not intercept the imaginary axis. The angle at $\omega = \infty$ is known to us from Eq. (13.5) and given in Table 13.5.

$$\Im\{G(\jmath\omega)\} = \frac{1-\omega^2T_1T_2}{\omega(1-\omega^2T_1^2)(1-\omega^2T_2^2)} = 0$$

$$\Rightarrow \qquad 1-\omega^2T_1^2 = 0$$

$$\Rightarrow \qquad \omega = \frac{1}{\sqrt{T_1T_2}} \tag{13.7}$$

Substituting ω from Eq. (13.7) in the real part of Eq. (13.6), we get

$$\Re\{G(\jmath\omega)\} = -\frac{T_1+T_2}{\left(1+\frac{T_1^2}{T_1T_2}\right)\left(1+\frac{T_2^2}{T_1T_2}\right)} = -\frac{T_1+T_2}{\left(\frac{T_1+T_2}{T_2}\right)\left(\frac{T_1+T_2}{T_1}\right)}$$

$$= -\frac{T_1T_2}{T_1+T_2} \equiv \frac{T_1T_2}{T_1+T_2}\angle 0^\circ$$

Based on these data, the polar plot, as shown in Figure 13.6a, is drawn.

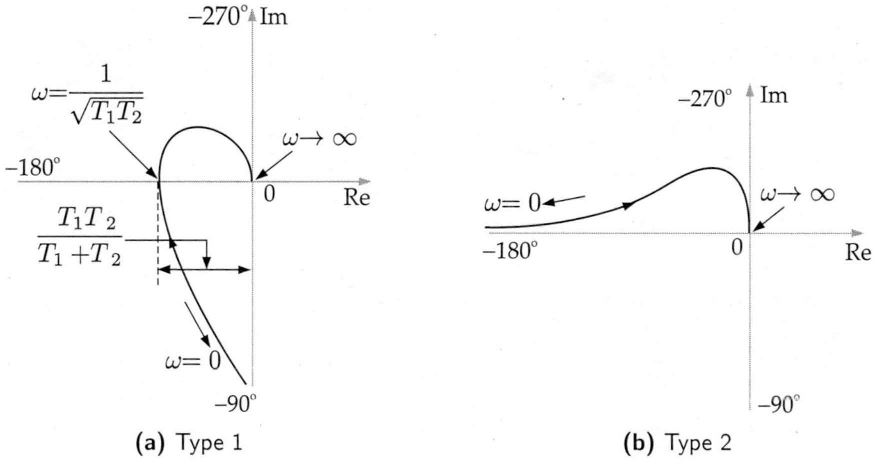

Figure 13.6: Polar plots of type 1 and type 2 systems.

Type 2 System

Let the transfer function of a type 2 system be given by

$$G(s) = \frac{1}{s^2(1+sT)}$$

We follow the steps to obtain the indicative data points for drawing the polar plot.

$$\begin{aligned} G(\jmath\omega) &= \frac{1}{(\jmath\omega)^2(1+\jmath\omega T)} \\ &= \frac{1}{\omega^2\sqrt{1+(\omega T)^2}}\angle(-180^\circ - \tan^{-1}\omega T) \end{aligned} \tag{13.8}$$

From Eq. (13.8), we can generate the data given in Table 13.6.

ω	$\lvert G(\jmath\omega)\rvert$	$\angle G(\jmath\omega)$
0	∞	-180°
∞	0	-270°

Table 13.6: Limiting magnitude and angle values obtained from Eq. (13.8) (type 2)

Equation (13.8) can be split into real and imaginary parts as follows:

$$\begin{aligned} G(\jmath\omega) &= \frac{1}{-\omega^2(1+\jmath\omega T)}\cdot\frac{(1-\jmath\omega T)}{(1-\jmath\omega T)} = \frac{(1-\jmath\omega T)}{-\omega^2(1+\jmath\omega T)} \\ &= -\frac{1}{\omega^2(1+\omega^2T^2)} - \jmath\frac{\omega T}{\omega^2(1+\omega^2T^2)} \end{aligned}$$

$$\Re\{G(\jmath\omega)\} = -\frac{1}{\omega^2(1+\omega^2T^2)} = 0$$

$$\Rightarrow \qquad \omega = \infty$$

This means that the polar plot does not intercept the imaginary axis. The angle at $\omega = \infty$ is known to us from Eq. (13.8).

$$\Im\{G(\jmath\omega)\} = \frac{\omega T}{\omega^2(1+\omega^2T^2)} = 0$$

$$\Rightarrow \qquad \omega = \infty$$

This again means that the polar plot intercepts the real axis only at the origin. The angle at $\omega = \infty$ is known to us from Eq. (13.8) and given in Table 13.6. Based on these data, the polar plot, as shown in Figure 13.6b, is drawn.

Note: Polar plots are important especially for making Nyquist plots of Chapter 14. So, we work out a few examples to familiarize ourselves with polar plots.

EXAMPLE 13.1. Figure 13.7a shows the polar plot of a system. The transfer function of the system is

(A) $5(1+0.1s)$ (B) $1+0.5s$ (C) $5(1+10s)$ (D) $5(1+s)$

Solution Let the transfer function be given by

$$G(s) = K(1+sT)$$

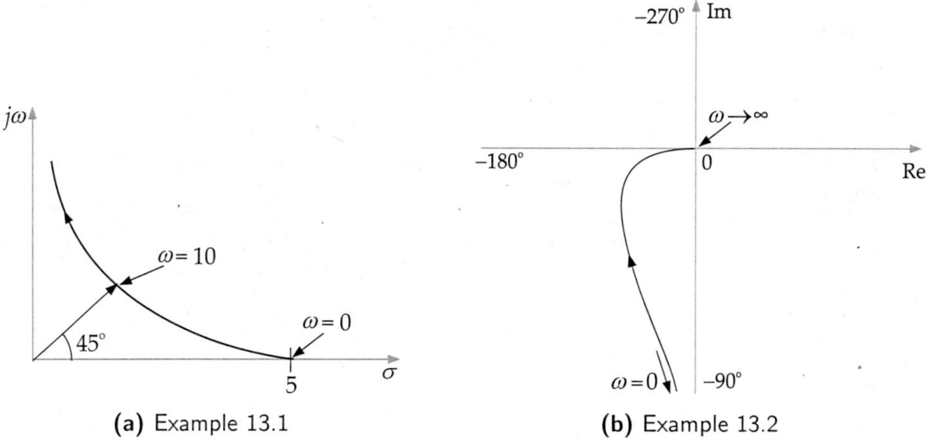

(a) Example 13.1 **(b)** Example 13.2

Figure 13.7: Figures as mentioned.

Then, in the frequency domain

$$G(\jmath\omega) = K(1 + \jmath\omega T) \equiv K\sqrt{1+\omega^2T^2}\angle\tan^{-1}\omega T$$

$$\phi = \tan^{-1}10T = 45^\circ \qquad \text{[from the given diagram]}$$

$$\Rightarrow \quad 10T = \tan 45^\circ = 1$$

$$\Rightarrow \quad T = 0.1$$

Now, $K\sqrt{1+0} = 5$ [from the given diagram]

$$\Rightarrow \quad K = 5$$

Therefore, $G(s) = 5(1 + 0.1s)$

Ans. (A)

EXAMPLE 13.2. Sketch the polar plot for the following transfer function:

$$G(s) = \frac{20}{s(s+1)}$$

Solution We substitute $s \equiv \jmath\omega$ and convert the transfer function to magnitude and angle form as follows:

$$G(\jmath\omega) = \frac{20}{\jmath\omega(\jmath\omega+1)} \equiv \frac{20}{\omega\sqrt{\omega^2+1}}\angle - (90^\circ + \tan^{-1}\omega)$$

Evaluating the magnitude and angle at $\omega = 0$ and ∞, we get the data given in Table 13.7.

Table 13.7: Magnitude and angle (Example 13.2)

ω	$\lvert G(\jmath\omega)\rvert$	$\angle G(\jmath\omega)$
0	∞	-90°
∞	0	-180°

Now, we check if there are intersections on the real and/or imaginary axes. To do that, we have to break the transfer function into real and imaginary parts as follows:

$$G(\jmath\omega) = \frac{20}{\jmath\omega(\jmath\omega+1)} \cdot \frac{-\jmath\omega(\jmath\omega-1)}{-\jmath\omega(\jmath\omega-1)} = \frac{20\omega^2}{\omega^2(\omega^2+1)} + \jmath\frac{20\omega}{\omega^2(\omega^2+1)}$$

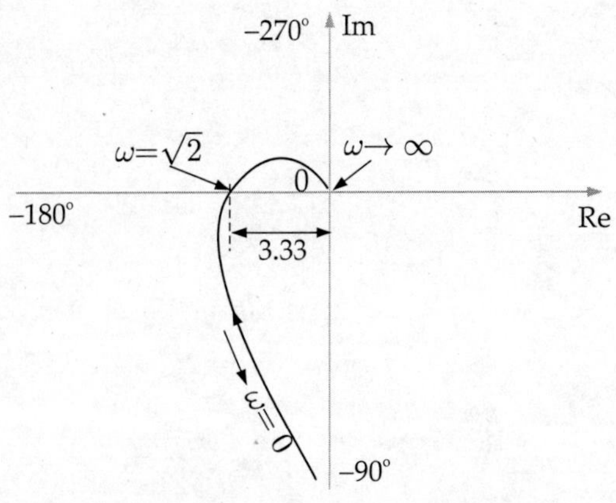

Figure 13.8: Polar plot (Example 13.3).

Table 13.8: M and ϕ (Example 13.3)

ω	$\lvert M\rvert$	ϕ
0	∞	$-90°$
∞	0	$-180°$

Equating the real part to zero, we get

$$\Re\{G(\jmath\omega)\} = \frac{20\omega^2}{\omega^2(\omega^2+1)} = 0$$

$$\Rightarrow \qquad \omega = \infty$$

This indicates that the polar plot does not intersect the imaginary axis. Next we equate the imaginary part to zero to find out the intersection at the real axis.

$$\Im\{G(\jmath\omega)\} = \frac{20\omega}{\omega^2(\omega^2+1)} = 0$$

$$\Rightarrow \qquad \omega = \infty$$

The result indicates that the plot intersects the real axis only at the origin. From these data we construct the polar plot as given in Figure 13.7b.

EXAMPLE 13.3. Sketch the polar plot for the following transfer function:

$$G(s) = \frac{20}{s(s+1)(s+2)}$$

Solution We substitute $s \equiv \jmath\omega$ and convert the transfer function to the magnitude and angle form as follows:

$$G(\jmath\omega) = \frac{20}{\jmath\omega(\jmath\omega+1)(\jmath\omega+2)} = \frac{20}{\omega\sqrt{\omega^2+1}\sqrt{\omega^4+4}}\angle\left(-90° - \tan^{-1}\omega - \tan^{-1}\frac{\omega}{2}\right) \qquad (13.9)$$

Evaluating the magnitude and angle at $\omega = 0$ and ∞ from Eq. (i), we get the values given in Table 13.8.

Now, we check if there are intersections on the real and/or imaginary axes. To do that, we have to break the transfer function into real and imaginary parts as follows:

$$\begin{aligned} G(\jmath\omega) &= \frac{20}{\jmath\omega(\jmath\omega+1)(\jmath\omega+2)} \cdot \frac{-\jmath\omega(\jmath\omega-1)(\jmath\omega-2)}{\jmath\omega(\jmath\omega-1)(\jmath\omega-2)} \\ &= \frac{-60\omega^2}{\omega^2(\omega^2+1)(\omega^2+4)} + \jmath\frac{20\omega(\omega^2-2)}{\omega^2(\omega^2+1)(\omega^2+4)} \end{aligned} \qquad \text{(i)}$$

Equating the real part to zero, we get

$$\Re\{G(\jmath\omega)\} = \frac{-60\omega^2}{\omega^2(\omega^2+1)(\omega^2+4)} = 0$$

$$\Rightarrow \qquad \omega = \infty$$

This indicates that the polar plot does not intersect the imaginary axis. Next we equate the imaginary part to zero to find out the intersection at the real axis.

$$\Im\{G(\jmath\omega)\} = \frac{20\omega(\omega^2-2)}{\omega^2(\omega^2+1)(\omega^2+4)} = 0$$

$$\Rightarrow \qquad \omega = \pm\sqrt{2}$$

The result indicates that the plot intersects the real axis at $\omega = \sqrt{2}$ rad/s, the negative frequency value being meaningless. Substituting this value in Eq. (i), we get

$$G(\jmath\sqrt{2}) = \frac{-60}{(2+1)(2+4)} = -3.33$$

From these data we construct the polar plot as given in Figure 13.8.

13.3 ALTERNATIVE METHOD FOR POLAR PLOT

This method is based on plotting the tip of the vector $M(\omega)e^{\jmath\omega\phi}$ in the G-plane by varying ω from 0 to ∞. The following example will make the procedure clear.

EXAMPLE 13.4. Sketch the polar plot for

$$G(s) = \frac{10(s+1)}{s+10}$$

Solution We substitute $s \equiv \jmath\omega$ and convert the transfer function to the magnitude and angle form as follows.

$$G(\jmath\omega) = \frac{10(\jmath\omega+1)}{\jmath\omega+10} \equiv \left|\frac{10\sqrt{\omega^2+1}}{\sqrt{\omega^2+100}}\right| \angle\left(\tan^{-1}\omega - \tan^{-1}\frac{\omega}{10}\right) \qquad \text{(i)}$$

$$\equiv |M|\angle\phi$$

We generate the data of Table 13.9 from Eq. (i).

Then we plot $M\cos\phi$ and $M\sin\phi$ along the real and imaginary axes of Figure 13.9.

13.4 STABILITY STUDY FROM POLAR PLOT

A polar plot of the open-loop transfer function can be utilized to assess the stability of a system by studying the gain and phase margins.

Gain and Phase Crossover Frequencies

In the polar plot shown in Figure 13.10a, the plot crosses the negative Re-axis at A and the unit circle at B. A and B denote phase crossover and gain crossover points respectively, because at A the plot crosses −180° phase and at B, the unit gain.

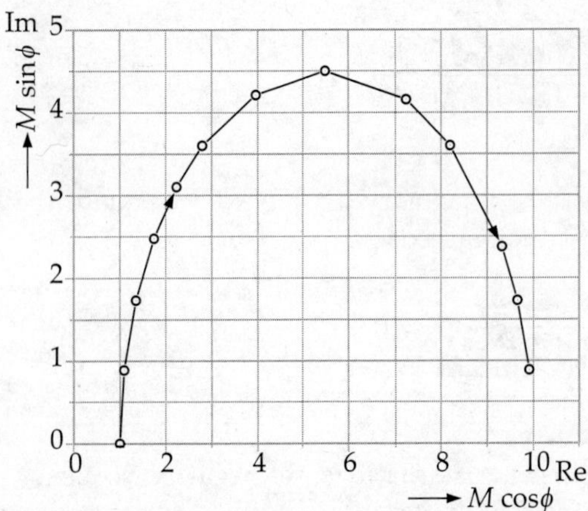

Figure 13.9: Polar plot (Example 13.4.).

Table 13.9: Data for polar plot of Example 13.4

ω (rad/s)	$\|M\|$	$\angle\phi$ (deg)	$M\cos\phi + \jmath M\sin\phi$
0	1.0	0.0	$1 + \jmath 0$
1	1.407	39.289	$1.089 + \jmath 0.891$
2	2.193	52.125	$1.346 + \jmath 1.731$
3	3.029	54.866	$1.743 + \jmath 2.477$
4	3.828	54.162	$2.241 + \jmath 3.103$
5	4.561	52.125	$2.800 + \jmath 3.600$
7	5.793	46.878	$3.960 + \jmath 4.210$
10	7.106	39.289	$5.500 + \jmath 4.500$
15	8.339	29.876	$7.231 + \jmath 4.154$
20	8.955	23.703	$8.200 + \jmath 3.600$
35	9.619	14.309	$9.321 + \jmath 2.377$
50	9.808	10.164	$9.654 + \jmath 1.729$
100	9.951	5.138	$9.911 + \jmath 0.891$
200	9.988	2.576	$9.978 + \jmath 0.449$

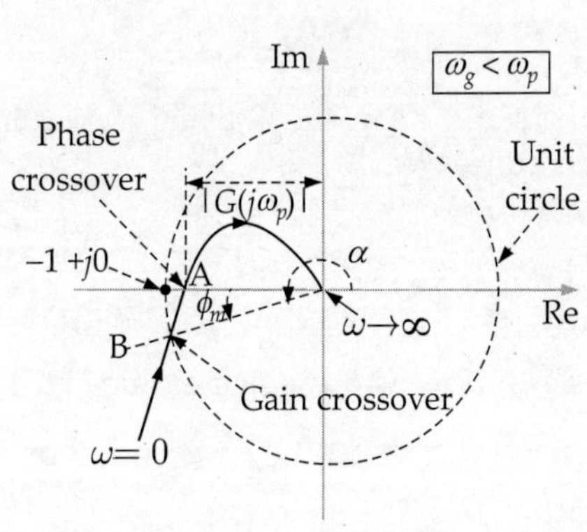

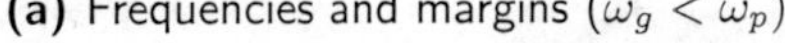

(a) Frequencies and margins ($\omega_g < \omega_p$)

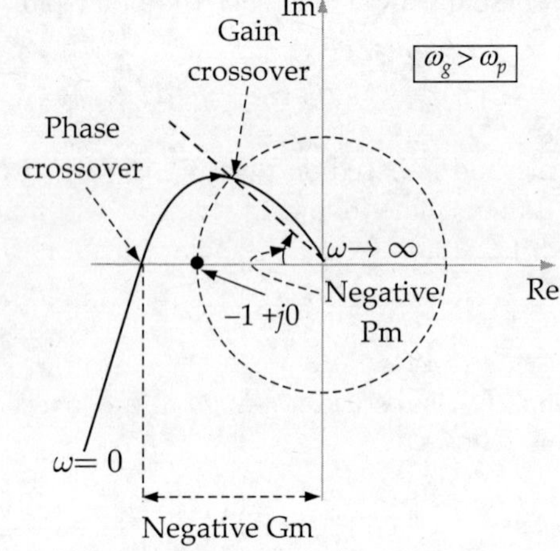

(b) Unstable condition ($\omega_g > \omega_p$)

Figure 13.10: Gain and phase crossover frequencies, and gain and phase margins in a polar plot.

Gain Margin

If $|G(\jmath\omega)|$ denotes the gain at the phase crossover frequency ω_p, then according to the definition

$$\boxed{\text{Gm} = 20\log\frac{1}{|G(\jmath\omega)|} = -20\log|G(\jmath\omega)| \text{ dB}} \tag{13.10}$$

Four cases may emerge from polar plots of systems. The plot characteristic, gain margin and the stability of a system corresponding to the cases are given in Table 13.10.

Phase Margin

By definition, the phase margin is

$$\text{Pm} = 180^\circ + \alpha \tag{13.11}$$

Table 13.10: Polar plots vis-a-vis gain margins and stabilities

Case	*Plot characteristic*	*Gain margin*	*Stability of the system*
1	The plot may not cross the negative Re-axis at all except at $\omega = \infty$ where $\lvert G(\jmath\omega)\rvert = 0$.	Gm $= \infty$	Stable
2	The plot may cross the negative Re-axis between the origin and $-1 + \jmath 0$. Figure 13.10a is such a case. Here, since $\lvert G(\jmath\omega)\rvert$ is a fraction, its logarithm will be negative.	Gm ≥ 0	Stable
3	The plot may cross the negative Re-axis beyond $-1 + \jmath 0$ [see Figure 13.10b]. Here the point $-1 + \jmath 0$ lies to the right of the curve as we move with ω from 0 to ∞ *in a clockwise direction.* It is said the plot *encloses* $-1 + \jmath 0$ [see Figure 13.11a to know the region of enclosure]. Obviously, here $\lvert G(\jmath\omega)\rvert > 1$	Gm < 0	Unstable
4	The plot may pass through $-1 + \jmath 0$. So, a slight increase in $\lvert G(\jmath\omega)\rvert$ will make it positive.	Gm $= 0$	Marginally stable

However (see Figure 13.10a)

$$\alpha = 180^\circ + \phi_m \tag{13.12}$$

Combining Eqs. (13.11) and (13.12),

$$\boxed{\text{Pm} = 180^\circ + 180^\circ + \phi_m = \phi_m} \tag{13.13}$$

So, the *angle measured in the counter-clockwise direction* from the negative Re-axis (i.e. 180°) to the gain crossover point is the measure of the phase margin.

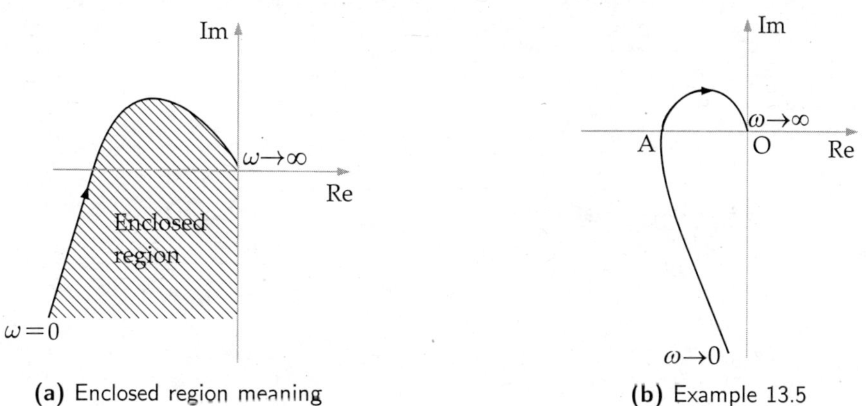

(a) Enclosed region meaning (b) Example 13.5

Figure 13.11: Polar plots.

We consider a couple of examples to see how stability conditions can be worked out from polar plots.

EXAMPLE 13.5. Sketch the polar plot for the unity feedback system given by

$$G(s) = \frac{K}{s(s+2)(s+5)}$$

and determine the range of K for the stability of the system.

Solution Since it is a unity feedback system, the open-loop transfer function in the frequency domain is

$$G(\jmath\omega)H(\jmath\omega) = \frac{K}{\jmath\omega(\jmath\omega+2)(\jmath\omega+5)} \equiv \frac{K}{\omega\sqrt{\omega^2+4}\sqrt{\omega^2+25}} \angle -\left(90^\circ + \tan^{-1}\frac{\omega}{2} + \tan^{-1}\frac{\omega}{5}\right) \tag{i}$$

From Eq. (i), we get the data given in Table 13.11.

ω	$\lvert G(\jmath\omega)H(\jmath\omega)\rvert$	$\angle G(\jmath\omega)H(\jmath\omega)$
0	∞	-90°
∞	0	-270°

Table 13.11: Data for polar plot (Example 13.5)

The sketch of the polar plot is given in Figure 13.11b. To find the gain margin, we have to calculate the phase crossover frequency ω_p at A.

At point A, where $\omega = \omega_p$, we have

$$\phi = -180^\circ = -90^\circ - \tan^{-1}\frac{\omega_p}{2} - \tan^{-1}\frac{\omega_p}{5} \tag{13.14}$$

which gives
$$90^\circ = \tan^{-1}\frac{\omega_p}{2} + \tan^{-1}\frac{\omega_p}{5} = \tan^{-1}\frac{\frac{\omega_p}{2}+\frac{\omega_p}{5}}{1-\frac{\omega_p}{2}\cdot\frac{\omega_p}{5}}$$

$$\Rightarrow \quad \tan 90^\circ = \tan\left(\tan^{-1}\frac{7\omega_p}{10-\omega_p^2}\right)$$

$$\Rightarrow \quad \infty = \frac{7\omega_p}{10-\omega_p^2} \tag{ii}$$

Equation (ii) indicates

$$10 - \omega_p^2 = 0$$

$$\Rightarrow \quad \omega_p = 3.162 \text{ rad/s}$$

Therefore,
$$\text{OA} = |G(\jmath\omega)H(\jmath\omega)|_{\omega=\omega_p} = \frac{K}{(3.162)\sqrt{(3.162)^2+4}\sqrt{(3.162)^2+25}} = \frac{K}{70}$$

For the system to be marginally stable, OA = 1. Hence,

$$1 = \frac{K}{70}$$

$$\Rightarrow \quad K = 70 \tag{iii}$$

From Eq. (iii) we get the following stability conditions:

$$\text{For } K \begin{cases} < 70 & \text{the system is stable} \\ = 70 & \text{the system is marginally stable} \\ > 70 & \text{the system is unstable} \end{cases}$$

EXAMPLE 13.6. Sketch the polar plot of the unity feedback system given by

$$G(s) = \frac{1}{(s+1)^2}$$

and find the phase margin from there.

Solution The open-loop transfer function in the frequency domain is

$$G(\jmath\omega)H(\jmath\omega) = \frac{1}{(\jmath\omega+1)^2} \equiv \frac{1}{\omega^2+1}\angle -2\tan^{-1}\omega \tag{i}$$

From Eq. (i), we get the data given in Table 13.12 to draw the polar plot.

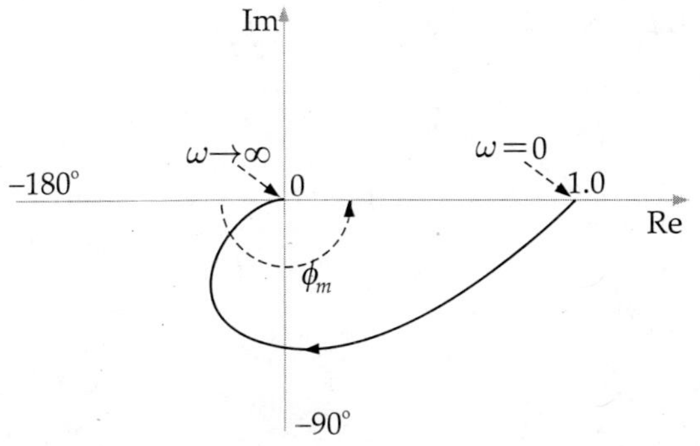

Figure 13.12: Polar plot (Example 13.6).

Table 13.12: Data for polar plot (Example 13.6)

ω	$\lvert G(\jmath\omega)H(\jmath\omega)\rvert$	$\angle G(\jmath\omega)H(\jmath\omega)$
0	1	$-0°$
∞	0	$-180°$

The polar plot is presented in Figure 13.12. To figure out the phase margin from the plot, we need to draw unit circle and find at what angle from the 180°-line (i.e. negative real axis) the polar plot intercepts it. It is clear from the above table as well as Figure 13.12 that the unit gain occurs at $\omega = 0$, where the phase angle is 0° and consequently the phase margin $\phi_m = 180°$.

EXAMPLE 13.7. The frequency response of

$$G(s) = \frac{1}{s(s+1)(s+2)}$$

plotted in the complex plane $G(\jmath\omega)$-plane (for $0 < \omega < \infty$ is

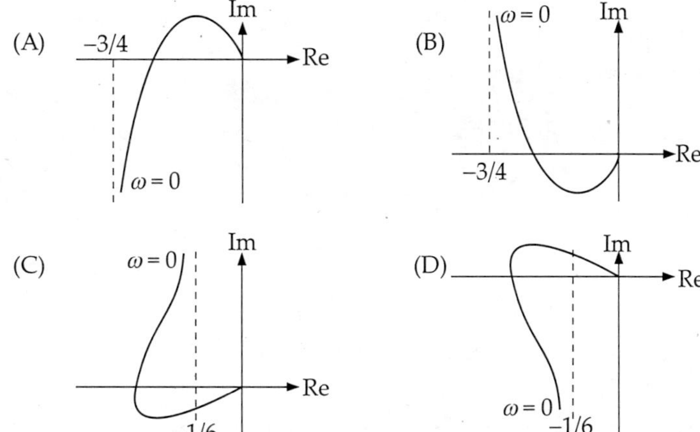

Solution In frequency domain, the given transfer function is

$$G(\jmath\omega) = \frac{1}{\jmath\omega(\jmath\omega + 1)(\jmath\omega + 2)}$$

Therefore,
$$|M| = \frac{1}{\omega\sqrt{\omega^2+1}\sqrt{\omega^2+2}} \tag{i}$$

and
$$\angle\phi = -90^\circ - \tan^{-1}\omega - \tan^{-1}\frac{\omega}{2} \tag{ii}$$

From Eqs. (i) and (ii), we get the data given in Table 13.13 to draw the polar plot.

ω	$\lvert M\rvert$	$\angle\phi$
0	∞	-90°
∞	0	-270°

Table 13.13: Data for polar plot (Example 13.7)

To determine where the curve intercepts the Re-axis, we note that here the phase angle $\angle\phi = -180^\circ$. This gives from Eq. (ii)

$$-90^\circ - \tan^{-1}\omega - \tan^{-1}\frac{\omega}{2} = -180^\circ$$

or,
$$-\tan^{-1}\frac{\omega + \dfrac{\omega}{2}}{1 - \dfrac{\omega^2}{2}} = 270^\circ$$

or,
$$\frac{\omega + \dfrac{\omega}{2}}{1 - \dfrac{\omega^2}{2}} = \tan 270^\circ = \infty$$

Therefore,
$$1 - \frac{\omega^2}{2} = 0$$

or,
$$\omega = \sqrt{2}$$

Now, from Eq. (i),

$$|M|_{\omega=\sqrt{2}} = \frac{1}{\sqrt{2}\sqrt{3}\sqrt{2+4}} = \frac{1}{\sqrt{36}} = \frac{1}{6} \text{ which is } < \left(\frac{3}{4}\right)$$

Ans. (A)

With this background in polar plots, we now move to Nyquist plots.

REVIEW QUESTIONS

13.1 Choose the correct answer:

A polar plot is a plot of

(A) magnitude and phase of frequency domain transfer function in the G-plane when ω varies from 0 to ∞

(B) pole and zero plot in the s-plane

(C) path traversed by poles when gain varies from 0 to ∞

(D) magnitude (in dB) and phase of frequency domain open-loop transfer function

13.2 Consider the transfer function

$$G(\jmath\omega) = \frac{1}{\jmath\omega(1+\jmath\omega T)}$$

Draw the polar plot of the function. Graphically find the gain crossover frequency and phase margin of the transfer function.

13.3 A feedback control system is characterised by the open-loop transfer function

$$G(s)H(s) = \frac{100}{s(s+4)(s+20)}$$

(a) Draw the polar plot of the open-loop system. Calculate the magnitude and phase at 10 equispaced frequency points (including the point where the plot intersects the imaginary axis and use them to construct the plot.

(b) From the plot determine the gain margin, phase margin, gain crossover frequency and phase crossover frequency, and comment on the system stability.

Answers to Selected Questions

13.1. (A) 13.3. (b) Gm = 25.7 dB, Pm = 69.9°, $\omega_g = 1.2$ rad/s, $\omega_p = 8.9$ rad/s. Stable

14

Nyquist Plot

14.1 INTRODUCTION

A Nyquist plot combines the two types of Bode plot—magnitude and phase—on a single graph, with frequency as a parameter along the curve. It is useful for assessing the *stability of a closed-loop system.*

The Nyquist plot shows the amplification/attenuation and phase-shift of the transfer function in the complex plane. The phase-shift of frequency-domain transfer function is represented by the argument and the magnitude is represented by the length of a vector from the origin in the direction described by the argument.

The method is essentially based on the following procedures and principle:

- Drawing a Nyquist *path* (or *contour*) for the open-loop transfer function
- Mapping of the path to $G(s)H(s)$-plane following the polar plot scheme
- Applying the *principle of argument.*

We have had some idea of polar plots from Chapter 13. Now we consider the other two aspects of Nyquist plots before going to construct the plots themselves.

14.2 MAPPING IN THE $G(s)H(s)$-PLANE

Consider, for example, an open-loop transfer function

$$G(s)H(s) = s^2 + 2$$

Suppose we consider a point $S_1 = 2 + \jmath 3$ in the s-plane [see Figure 14.1a]. The corresponding value of $G(s)H(s)$ is

$$G(s)H(s) = (2 + \jmath 3)^2 + 2 = -3 + \jmath 12 \tag{14.1}$$

So, if we consider a $G(s)H(s)$-plane, the point G_1 [see Figure 14.1b] corresponds to Eq. (14.1). It is said that the point S_1 of the s-plane maps to the point G_1 of the $G(s)H(s)$-plane.

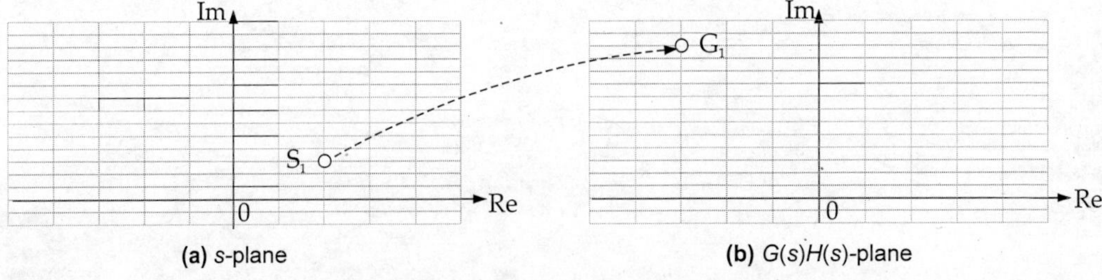

Figure 14.1: Mapping of a point in s-plane to $G(s)H(s)$-plane.

14.3 PRINCIPLE OF ARGUMENT

Let us consider an analytic[1] and single-valued[2] function $F(s)$ which can be written as

$$F(s) = K\frac{(s-z_1)(s-z_2)\cdots(s-z_m)}{(s-p_1)(s-p_2)\cdots(s-p_n)} \qquad m < n \tag{14.2}$$

Let us draw the plot of zeros z_1, z_2 and poles p_1, p_2 of $F(s)$ in the s-plane. Now, we draw in this plane a closed arbitrary contour which does not enclose any of the poles or zeros [Figure 14.2a]. As **s** moves in the clockwise (CW) direction along the contour, phasors of zeros and poles, namely $(\mathbf{s}-\mathbf{z_1})$, $(\mathbf{s}-\mathbf{z_2})$, $(\mathbf{s}-\mathbf{p_1})$, $(\mathbf{s}-\mathbf{p_2})$, describe separate paths, phasor-heads always remaining on the contour, but none of the phasors describes a net angle. So, the total contribution of angle to $F(s)$ is zero.

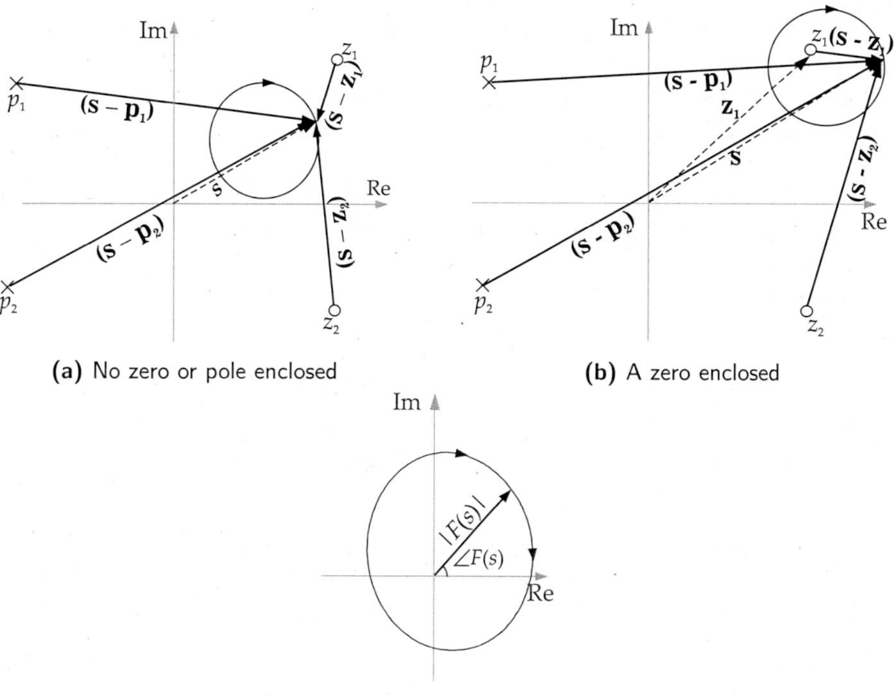

(a) No zero or pole enclosed

(b) A zero enclosed

(c) Mapping in the $1 + G(s)H(s)$ plane

Figure 14.2: Principle of argument illustration.

Next we consider the situation when the s-plane contour encloses a zero, namely, z_1 [see Figure 14.2b]. Then as the contour completes its path in the CW-direction, the phasor $(\mathbf{s}-\mathbf{z_1})$ describes a net angle of -2π radians[3], while other zeros and poles of $F(s)$, as in the previous case, do not contribute any net angle to $F(s)$.

Now, these poles and zeros are related to $F(s)$ through Eq. (14.2) from which the point

[1] A function, whose value and the value of its derivative are defined in the prescribed domain, is called an *analytic function* within the domain.

[2] The function $F(s) = \sqrt{2}$ is not single-valued while $F(s) = s$ is.

[3] The angle is negative because we are moving in the clockwise direction.

$F(s)$ can be written as

$$|F(s)| = K\frac{|s-z_1||s-z_2|\ldots|s-z_m|}{|s-p_1||s-p_2|\ldots|s-p_n|}$$

$$\angle F(s) = \angle(s-z_1) + \angle(s-z_2) + \cdots + \angle(s-z_m) \\ - \angle(s-p_1) - \angle(s-p_2) - \cdots - \angle(s-p_n)$$

So, if along with the rotation of the **s** vector in the s-plane we simultaneously rotate the **s** vector in the $F(s)$-plane, an angle of -2π radians will be described *around the origin there*, s having started from the origin.

Thus, this enclosure of a zero by the s-plane contour will generate a CW-encirclement of the origin in the $F(s)$-plane [see Figure 14.2c]. Let us call the contour in the s-plane Γ_s and that in the $F(s)$-plane, $\Gamma_{F(s)}$. Then we can say that the CW Γ_s contour enclosing a zero of $F(s)$ maps to a CW encirclement of the origin by the $\Gamma_{F(s)}$ contour. If two zeros are enclosed by the Γ_s contour, there will result two encirclements of the origin by the $\Gamma_{F(s)}$ contour, because then the angles described by phasors $(\mathbf{s}-\mathbf{z_1})$ and $(\mathbf{s}-\mathbf{z_2})$ will add up to -4π radians.

What happens if, instead of a zero, a pole is enclosed by the Γ_s contour? This time, too, the $(\mathbf{s}-\mathbf{p_1})$ phasor will describe a net angle of -2π radians. But the corresponding angle contribution to $F(s)$ will be $+2\pi$ radians because $(\mathbf{s}-\mathbf{p_1})$ lies in the denominator of $F(s)$.

In other words, the CW enclosure of a pole by the Γ_s contour maps to a counter-clockwise (CCW) encirclement of the origin by the $\Gamma_{F(s)}$ contour. As in the case of zeros, enclosure of two poles will generate two such encirclements, and so on.

Thus, if

N = number of clockwise encirclements of the origin of the $F(s)$-plane
Z = number of zeros of $F(s)$ enclosed by the clockwise Γ_s contour
P = number of poles enclosed by the clockwise Γ_s contour

then

$$N = Z - P$$

This relation between the enclosure of zeros and poles by contours in the s-plane and encirclements of the origin in the $F(s)$-plane by the $\Gamma_{F(s)}$ contour is commonly known as the *principle of argument* derived from a theorem of Cauchy[4]. The Nyquist plot utilizes this principle to determine the stability of control systems. But before we proceed to consider such plots, let us have a look at the relation between the principle of argument and the stability of control systems. To do that, in turn, we have to know what are

1. Nyquist criterion
2. Nyquist path

14.4 NYQUIST CRITERION

We know that the *open-loop transfer function* of a control system can be written as

$$G(s)H(s) = K\frac{(s-z_1)(s-z_2)\cdots(s-z_m)}{(s-p_1)(s-p_2)\cdots(s-p_n)} \qquad m < n \tag{14.3}$$

[4]Augustin Louis Cauchy (1789–1857) was a French mathematician.

The corresponding closed-loop transfer function is given by

$$\frac{C(s)}{R(s)} = \frac{G(s)}{1+G(s)H(s)} \tag{14.4}$$

and the characteristic equation is given by

$$1+G(s)H(s)=0 \tag{14.5}$$

We know that the roots of the characteristic equation are the poles of the closed-loop transfer function.

We have seen before that for the stability of a control system, the poles of the CLTF, i.e. the roots of the characteristic equation, must not lie in the RHP of the s-plane.

Now, combining Eqs. (14.3) and (14.5), we get

$$\begin{aligned} F(s) \equiv 1+G(s)H(s) &= 1+K\frac{(s-z_1)(s-z_2)\cdots(s-z_m)}{(s-p_1)(s-p_2)\cdots(s-p_n)} \qquad m<n \\ &= \frac{(s-p_1)(s-p_2)\cdots(s-p_n)+K(s-z_1)(s-z_2)\cdots(s-z_m)}{(s-p_1)(s-p_2)\cdots(s-p_n)} \\ &\equiv \frac{(s-z_1')(s-z_2')\cdots(s-z_m')}{(s-p_1)(s-p_2)\cdots(s-p_n)} \end{aligned} \tag{14.6}$$

Comparing Eqs. (14.3) and (14.6) we observe that

- The poles of $F(s)$ are the same as poles of the open-loop transfer function
- But *zeros of $F(s)$ are indeed the poles of the closed-loop transfer function* if we calculate it by substituting results of Eq. (14.6) in Eq. (14.4).

The stability condition dictates that *though a pole of the open-loop transfer function in the RHP of the s-plane may not destabilize the system per se, a pole of the closed-loop transfer function, and hence, a zero of $F(s)$, must not lie in the RHP for stability.* This probably indicates that an open-loop unstable system may turn out to be stable through feedback.

Therefore, to check the stability of a control system, we need to map the $F(s)$-plane by an s-plane contour which encompasses the entire RHP and check for encirclements of the origin by the $F(s)$-plane contour. While drawing the s-plane contour—we call it Γ_s contour—we need to make small detours if poles or zeros of $F(s)$ lie on the imaginary-axis. Such a contour is called the *Nyquist path.*

14.5 NYQUIST PATH

Nyquist proposed a contour in the s-plane that goes in the following manner:

- A path travelling up the $j\omega$-axis, from $(0, -j\infty)$ to $(0, +j\infty)$.
- A semicircular arc, with radius $R \to \infty$, that starts at $(0, +j\infty)$ and travels clockwise to $(0, -j\infty)$.

The contour as specified above is called the *Nyquist path.* The path encloses all poles and zeros (i.e. poles of CLTF) of the $1+G(s)H(s)$ that are located in the RHP. The path should not pass through any pole [see Figure 14.3a] but it may contain them [as shown in Figure 14.3b]. Mathematical representations corresponding to the plots are given in Table 14.1.

Applying Cauchy's theorem to control problem,

- If a closed path in a *specified region* is mapped to the $1+G(s)H(s)$-plane,

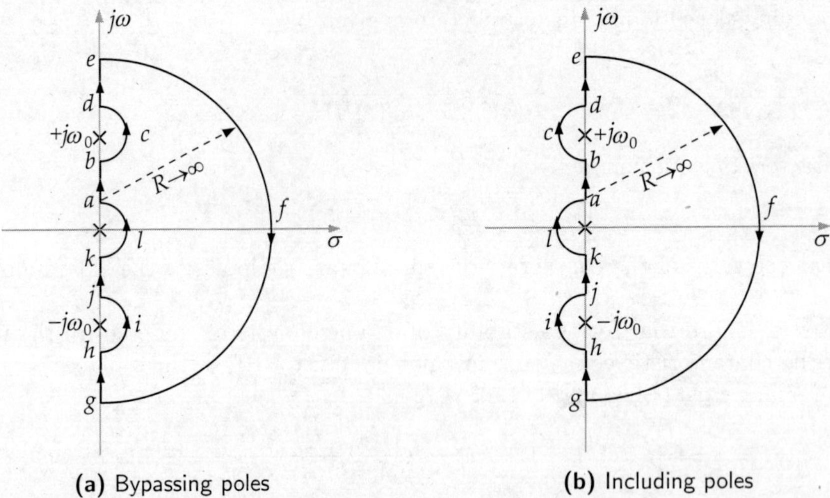

Figure 14.3: General Nyquist paths.

Table 14.1: Mathematical representation of segments of Nyquist paths

Segment	*s-value*	*Limits*
ab	$j\omega$	$\omega = 0$ to ω_0
bcd	$\lim_{\rho\to 0}(\rho e^{j\theta}) + j\omega_0$	$\theta = -90°$ to $+90°$
de	$j\omega$	$\omega = \omega_0$ to ∞
efg	$\lim_{R\to\infty} Re^{j}\theta$	$\theta = +90°$ to $-90°$
gh	$j\omega$	$\omega = -\infty$ to $-\omega_0$
hij	$\lim_{\rho\to 0}(\rho e^{j\theta}) - j\omega_0$	$\theta = -90°$ to $+90°$
jk	$j\omega$	$\omega = -\omega_0$ to 0
kla	$\lim_{\rho\to 0}(\rho e^{j\theta})$	$\theta = -90°$ to $+90°$

- If the number of zeros (i.e. poles of CLTF) is Z and the number of poles enclosed in the RHP of the s-plane is P, then the change in the argument (i.e. phase) is given by

$$\Delta \arg = -2\pi(Z - P) \text{ rad}$$

EXAMPLE 14.1. In the Nyquist plot of the open-loop transfer function

$$G(s)H(s) = \frac{3s+5}{s-1}$$

corresponding to the feedback loop shown in Figure 14.4a, the infinite semi-circular arc of the Nyquist contour in the s-plane is mapped into a point at

(A) $G(s)H(s) = \infty$ (B) $G(s)H(s) = 0$ (C) $G(s)H(s) = 3$ (D) $G(s)H(s) = -5$

Solution The Nyquist contour is shown in Figure 14.4b. We need to find the value of the $G(s)H(s)$ that maps with $R \to \infty$. Thus,

$$\begin{aligned} G(s)H(s) &= \lim_{R\to\infty} \frac{3Re^{j\theta}+5}{Re^{j\theta}-1} \\ &= 3 \end{aligned}$$

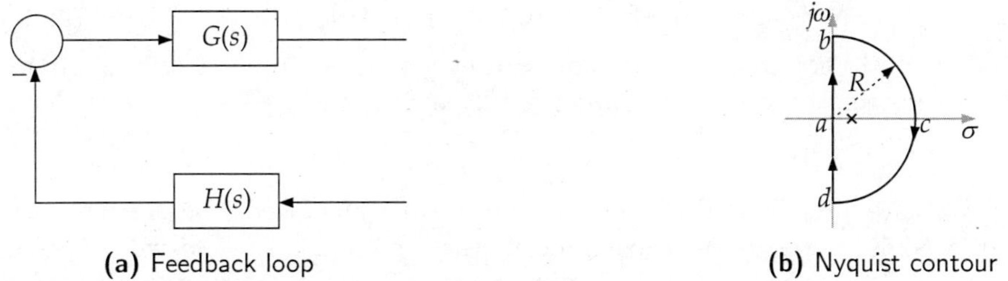

(a) Feedback loop (b) Nyquist contour

Figure 14.4: Example 14.1.

Ans. (C)

Three terms, namely (i) specified region, (ii) enclosure, and (iii) encirclement, require clarification in this context.

Specified Region. This is the right-hand side of the s-plane (i.e. RHP), where the existence of roots makes the system unstable.

Enclosure. The term *enclosed in the clockwise direction* means if a path is followed in this direction and a point is found to lie *to the right* of the path, the point is enclosed by the path. The point A in Figure 14.5a is enclosed by the path BO moving in the clockwise direction.

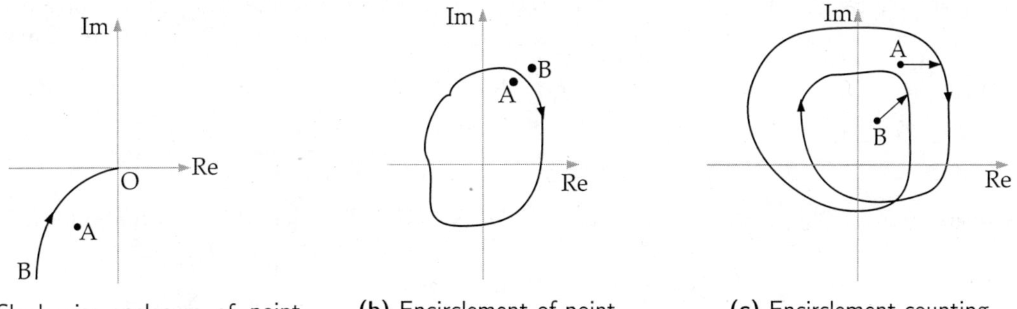

(a) Clock-wise enclosure of point (b) Encirclement of point (c) Encirclement counting

Figure 14.5: Enclosure, encirclement and counting of encirclements illustration.

EXAMPLE 14.2. Consider the following Nyquist plots of loop transfer function over $\omega = 0$ to $\omega = \infty$. Which of these plots represents a stable closed-loop system?

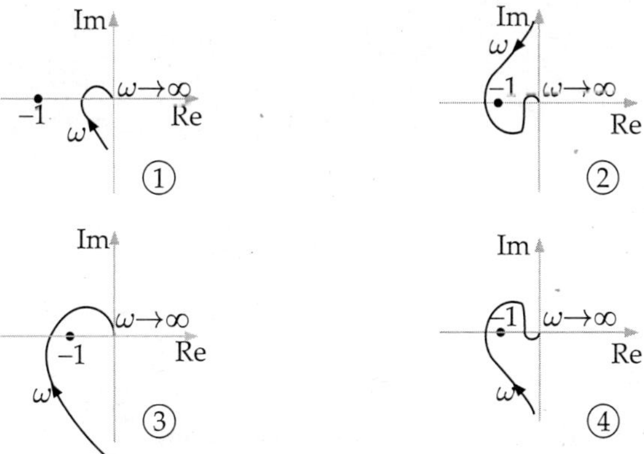

(A) ① only (B) all, except ① (C) all, except ③ (D) ① and ② only

Solution ① is obviously not enclosed. ② looks not enclosed, because the critical point is on the left-hand side, but since the path is counter-clockwise it is enclosed.
Ans. (A)

Encirclement. We have already used the term *encirclement* which needs elaboration now. The encirclement of a point by a prescribed path means that the point is found within the closed path. In Figure 14.5b, the point A is encircled by the closed path, but the point B is not.

Positive and negative encirclements. We will consider it a positive encirclement if the path moves in a clockwise direction and negative encirclement if it is a counter-clockwise path.

Counting encirclements. To determine the number of encirclements, we can draw a vector from the point to the path and observe how many revolutions the vector suffers while moving along the path until it returns to its starting point. Following this method, we find the point A in Figure 14.5c is encircled once while the point B, twice.

14.6 STABILITY ANALYSIS

If we map the Nyquist path to the $G(s)H(s)$-plane, it will include roots of $G(s)H(s) = 0$ and *not* of the characteristic equation $1 + G(s)H(s) = 0$. So, such a mapping can determine the stability of the open-loop transfer function and *not* the closed-loop transfer function. However, we note that

$$\begin{aligned} 1 + G(s)H(s) &= 0 \\ G(s)H(s) &= -1 \end{aligned} \tag{14.7}$$

Critical Point

Equations (14.7) suggest that an encirclement around (0, 0) in the $G(s)H(s)$-mapping for the OLTF is equivalent to an encirclement around $(-1 + \jmath 0)$ in the $G(s)H(s)$-mapping for the CLTF. The point $(-1 + \jmath 0)$ is often called the *critical point*[5].

Stability

Given a Nyquist contour Γ_s in the s-plane, the mapped contour $\Gamma_{F(s)}$ in the $F(s)$-plane shall, for a stable feedback system, encircle the point $(-1 + \jmath 0)$ in CW-direction a number of times N such that

$$\boxed{N = Z - P} \tag{14.8}$$

where, P is the number of poles of $F(s)$ (and therefore, poles of OLTF) enclosed by Γ_s, and Z is the number of zeros of $F(s)$ (i.e. poles of CLTF) enclosed by Γ_s.

Stated in clear terms, Nyquist stability criterion[6] states that

> A closed-loop control system will be stable *iff* $Z = 0$ as calculated from Eq. (14.8).

[5]Sometimes this is written as 'Q-point'.

[6]Some texts claim that any encirclement of the point $(-1 + \jmath 0)$ causes the system to become unstable. This is not strictly accurate. Given the equation $N = Z - P$, only Z must be zero to ensure that the system is stable. An OLTF may have a pole in the RHP, without compromising stability.

EXAMPLE 14.3. The loop transfer function of a system has a pole in the right half of the s-plane. The Nyquist plot makes one clockwise encirclement of the $(-1,\ 0)$ point. The closed-loop system is

(A) unstable (B) stable
(C) marginally stable (D) asymptotically stable

Solution Since the open-loop transfer function has a pole in the RHP, $P = 1$. Now the Nyquist plot makes one CW-encirclement of the critical point. So, $N = 1$. Therefore, from Eq. (14.8) we find

$$1 = Z - 1 \qquad \Rightarrow Z = 2$$

This means that there are 2 poles in the RHP of the closed-loop transfer function and therefore, the system is unstable.
Ans. (A)

EXAMPLE 14.4. The open-loop part of a unity feedback control system $G(s)$ is unstable and has two poles on the right hand side of the s-plane. However, the closed-loop system is stable. The number of encirclements made by the Nyquist plot of the $(-1,\ 0)$ point in the $G(s)$-plane is

(A) 2 (B) 0 (C) -3 (D) -1

Solution From the given data, $P = 2$. However, the encirclement queries do not mention if they are CW or CCW. Assuming, CCW encirclement, for the stability of the system

$$-N = Z - P = 0 - 2$$

Ans. (A)

EXAMPLE 14.5. Consider a negative unity feedback system with forward path transfer function

$$G(s) = \frac{K}{(s+a)(s-b)(s+c)}$$

where K, a, b, c are positive real numbers. For a Nyquist path enclosing the entire imaginary axis and right half of the s-plane in the clockwise direction, the Nyquist plot of $[1 + G(s)]$, encircles the origin of $[1+G(s)]$-plane once in the clockwise direction and never passes through the origin for a certain value of K. Then, the number of poles of $\{G(s)/[1+G(s)]\}$ lying in the open right half of the s-plane is ———.

Solution From the given data, we find

Number of open-loop poles	$P = 1$
Number of clockwise encirclements	$N = 1$
Therefore, number of closed-loop poles on the RHP	$Z = N + P = 1 + 1 = 2$ [see Eq. (14.8)]

Ans. 2

EXAMPLE 14.6. The number of times the Nyquist plot of

$$G(s)H(s) = \frac{1(s-1)(s-2)}{2(s+1)(s+2)}$$

encircles the critical point is ———.

Solution From the given OLTF, we find that

(i) The number of poles in the RHP, $P = 0$.
(ii) The number of zeros in the RHP, $Z = 2$.

Therefore, from Eq. (14.8), we get

$$N = Z - P = 2 - 0 = 2$$

Ans. 2 CW encirclements

14.7 NYQUIST PLOT DRAWING AND INFERENCE

A step-by-step procedure for drawing Nyquist plots is as follows:

1. Find from the OLTF the number of poles at the origin or on the $\jmath\omega$-axis.
2. Draw the Nyquist path in the s-plane.
3. Utilizing Table 14.1, map the path from a to e onto the $G(s)H(s)$-plane.
4. Draw a mirror image of the map obtained in #3 and connect the two.
5. Count the number of CW encirclements about $(-1 + \jmath 0)$.
6. Determine stability using the criterion stated in Eq. (14.8).

We consider two examples to familiarize ourselves with the procedure.

EXAMPLE 14.7. Draw the Nyquist plot for

$$G(s)H(s) = \frac{1}{(s+2)}$$

and determine the stability of the closed-loop system.

Solution

1. There is only one pole at $s = -2$. No pole at the origin or on the $\jmath\omega$-axis.
2. The Nyquist path is, therefore, as given in Figure 14.6a.
3. The $G(s)H(s)$-plane mapping can be done as follows:
   For ab-segment: $s = \jmath\omega$ and $0 < \omega < \infty$

$$G(\jmath\omega)H(\jmath\omega) = \frac{1}{\jmath\omega + 2} \equiv \frac{1}{\sqrt{4+\omega^2}} \angle -\tan^{-1}\left(\frac{\omega}{2}\right)$$

   which gives the data of Table 14.2. With the help of these data, we plot the lower segment in the Nyquist diagram [Figure 14.6b].

ω	$\lvert G(\jmath\omega)H(\jmath\omega)\rvert$	$\angle G(\jmath\omega)H(\jmath\omega)$
$+0$	0.5	0°
$+\infty$	0	−90°

Table 14.2: Data for Example 14.7

4. A mirror image (upper segment) is drawn and the two segments are connected.
5. Since there is no encirclement about the critical point, $N = 0$.
6. There is no pole in the RHP of the open-loop transfer function. Therefore, $P = 0$. Thus, from the Nyquist criterion $Z = 0$ and the system is stable.

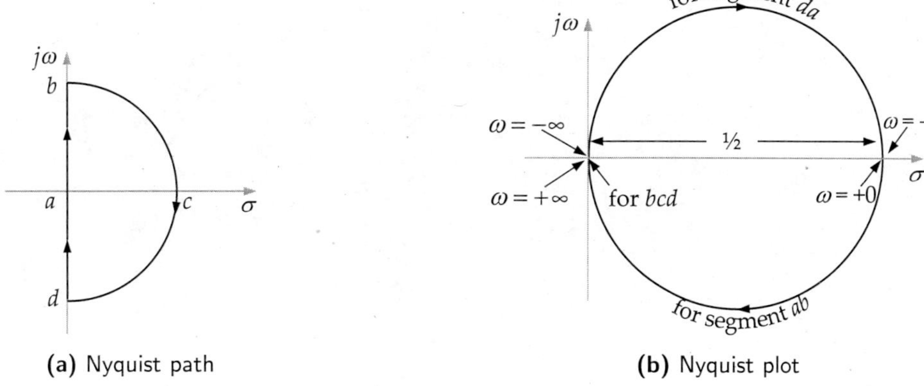

(a) Nyquist path

(b) Nyquist plot

Figure 14.6: Example 14.7.

EXAMPLE 14.8. Draw Nyquist plot for

$$G(s)H(s) = \frac{1}{s(s+2)}$$

and determine stability.

Solution

1. Two poles exist at $s = 0, -2$. So, there exists one pole at the origin and none on the Im-axis.
2. The Nyquist path is given in Figure 14.7a.
3. The mapping is as follows:

   For ab-segment: $s \equiv \jmath\omega \quad 0 < \omega < \infty$

$$G(\jmath\omega)H(\jmath\omega) = \frac{1}{\jmath\omega(\jmath\omega+2)} \equiv \left|\frac{1}{\omega\sqrt{\omega^2+4}}\right| \angle\left(-90^\circ - \tan^{-1}\frac{\omega}{2}\right)$$

   This yields the Table 14.3:

Table 14.3: Data for Example 14.8

ω	$\lvert G(\jmath\omega)H(\jmath\omega)\rvert$	$\angle G(\jmath\omega)H(\jmath\omega)$
0	∞	-90°
∞	0	-180°

   For de-segment: A mirror image of the plot for the *ab*-segment.

   For bcd-segment: $s \equiv \lim_{R\to\infty} Re^{\jmath\theta} \quad +90^\circ < \theta < -90^\circ$

$$|G(s)H(s)| = \lim_{R\to\infty} \frac{1}{Re^{\jmath\theta}\left(Re^{\jmath\theta}+2\right)} = 0$$

   Therefore, it maps to a point at the origin.

   For efa-segment: $s \equiv \lim_{\rho\to 0} \rho e^{\jmath\theta} \quad -90^\circ < \theta < +90^\circ$

$$|G(s)H(s)| = \lim_{\rho\to 0}\left\{\frac{1}{\rho e^{\jmath\theta}\left(\rho e^{\jmath\theta}+2\right)}\right\} = \infty$$

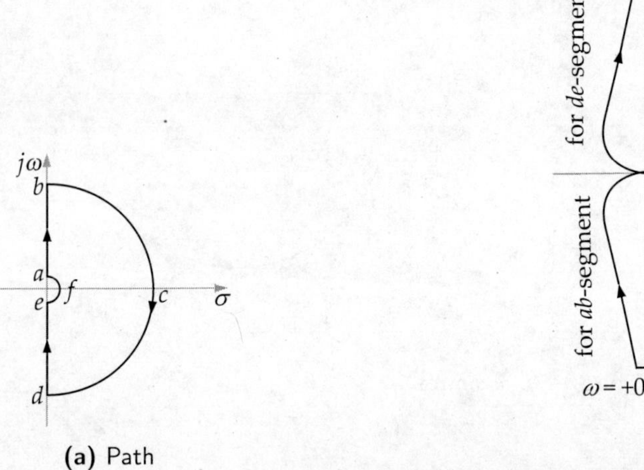

(a) Path **(b)** Plot

Figure 14.7: Nyquist diagrams (Example 14.8).

To determine $\angle G(s)H(s)$ we observe that

$$\angle \lim_{\rho\to 0} \rho e^{\jmath\theta} = \lim_{\rho\to 0}\left(\tan^{-1}\frac{\rho\sin\theta}{\rho\cos\theta}\right) = \tan^{-1}\tan\theta = \theta$$

$$\angle \lim_{\rho\to 0}\left(\rho e^{\jmath\theta}+2\right) = \lim_{\rho\to 0}\frac{\rho\sin\theta}{2+\rho\cos\theta} = 0$$

Therefore, the total angle in the denominator $= \theta + 0 = \theta$. When brought to the numerator it is $-\theta$. Since θ varies from $-90°$ to $+90°$, the phase of the magnitude varies from $+90°$ to $-90°$. The complete Nyquist plot is given in Figure 14.7b.

4. No encirclement about $(-1+\jmath 0)$. Therefore, $N = 0$.
5. No OLTF poles in the RHP. Therefore, $P = 0$.
6. Since, $N = 0,\ P = 0,\ Z = 0$. Therefore, the system is stable.

Closing Nyquist Plots

For $s = -\jmath 0$ to $s = +\jmath 0$, ω varies from $\omega = -0$ to $\omega = +0$, since s equals $\jmath\omega$ in the frequency domain. This region is marked by the indented portion [efa in Figure 14.7a] of the Nyquist path. We showed in Example 14.8 how to calculate directly the magnitude and phase of its mapping in the $G(s)H(s)$-plane so as to complete or close plots generated through mapping of ab and de segments of the Nyquist path.

However, without doing this exercise every time, a scheme may be developed as to how to close the plots. It is observed from the last two examples that the closure or completion of the plots depends on the *type of the open-loop transfer function*, i.e. number of poles at the origin. Suppose the type of the open-loop transfer function is n and the transfer function is as follows:

$$G(s)H(s) = \frac{K(sT_a+1)(sT_b+1)\cdots}{s^n(sT_p+1)(sT_q+1)\cdots}$$

For $s = \lim_{\rho\to 0} \rho e^{j\theta}$, the factor $sT_a + 1$ becomes

$$
\begin{aligned}
sT_a + 1 &= \lim_{\rho\to 0} \rho e^{j\theta} T_a + 1 = \lim_{\rho\to 0} \rho(\cos\theta T_a + j \sin\theta T_a) + 1 \\
&\equiv \lim_{\rho\to 0} \left|\sqrt{(\rho\cos\theta T_a + 1)^2 + (\rho\sin\theta T_a)^2}\right| \angle \tan^{-1}\frac{\rho\sin\theta T_a}{\rho\cos\theta T_a + 1} \\
&= |1|\angle 0^\circ
\end{aligned}
$$

Therefore, for $s = \lim_{\rho\to 0} \rho e^{j\theta}$ all the terms $(sT_a + 1)$, $(sT_b + 1), \cdots (sT_p + 1)$, $(sT_q + 1) \cdots$ contribute magnitudes of 1 and phases of 0°.

However, for the term s^n, we find

$$
\begin{aligned}
s^n &= \lim_{\rho\to 0} \rho^n e^{jn\theta} = \lim_{\rho\to 0} (\rho^n \cos n\theta + j\rho^n \sin n\theta) \\
&= \lim_{\rho\to 0} \left|\sqrt{\rho^{2n}\left(\cos^2 n\theta + \sin^2 n\theta\right)}\right| \angle \tan^{-1}\frac{\rho^n \sin n\theta}{\rho^n \cos n\theta} \\
&= |0|\angle \tan^{-1}\tan n\theta = |0|\angle n\theta
\end{aligned}
$$

if θ variation is positive, i.e. in the counter-clockwise direction as shown in Figure 14.3a.

However, if θ variation is negative, as shown in Figure 14.3b, the angle will be $-n\theta$.

Now, if the variation of s from $-j0$ to $+j0$ takes place in the indented portion of the Nyquist path in the positive (i.e. CCW) direction and it is represented by the rotation of a vector $\rho \to 0$ along a semicircular arc that describes angles from $-\pi/2$ to $+\pi/2$, the corresponding phase variation in the $G(s)H(s)$-plane will be

$$n\theta = \left(-\frac{n\pi}{2}\right) - \left(+\frac{n\pi}{2}\right) = -n\pi$$

The negative sign indicates that the phase variation of $G(s)H(s)$ takes place in the *clockwise* direction.

If the Nyquist path included the pole at the origin and, therefore, it were indented to the left, s would have varied in the negative (i.e. CW) direction and would be represented by the rotation of a vector $\rho \to 0$ along a semicircular arc that describes angles from $-\pi/2$ to $+\pi/2$ in the clockwise direction, the corresponding phase variation in the $G(s)H(s)$-plane will be

$$-n\theta = -\left[\left(-\frac{n\pi}{2}\right) - \left(+\frac{n\pi}{2}\right)\right] = +n\pi$$

The positive sign indicates that the phase variation of $G(s)H(s)$ would take place in the *counter-clockwise* direction.

Therefore, we conclude that

1. For a type n system, the Nyquist plots will close from $\omega = -0$ to $\omega = +0$
   (a) in the *clockwise* direction for the Nyquist path *excluding poles* on the path
   (b) in the *counter-clockwise* direction for the Nyquist path *including poles* on the path

   with an angle of $n\pi$ and an arc of infinite radius.
2. For a type 0 system, the points $\omega = -0$ and $\omega = +0$ will always coincide.

Henceforward, we will follow this rule to close Nyquist plots generated by mapping of *ab* and *de* segments [see Figure 14.7a] of the Nyquist path.

14.8 STABILITY MARGINS

In addition to determining the absolute stability of a system, the Nyquist diagrams provide qualitative information as to the degree of its stability. The $(-1+\jmath 0)$ point plays the same role in the Nyquist diagram as the imaginary axis does in the root locus diagram.

We know, the quantitative measures of relative stability are provided by gain and phase margins. These are found in the same way as for the polar plots. We recapitulate them here for convenience.

Gain Margin

By definition, the gain margin is the reciprocal of the open-loop gain at the frequency where phase is $-180°$ (known as phase crossover frequency, ω_p). In a typical Nyquist diagram (Figure 14.8), A is the point on the plot where the phase is $-180°$. The corresponding open-loop gain is OA. Therefore, Gm = 1/OA. Expressed in dB the gain margin is given by

$$\mathrm{Gm} = -20\log(\mathrm{OA})\ \mathrm{dB}$$

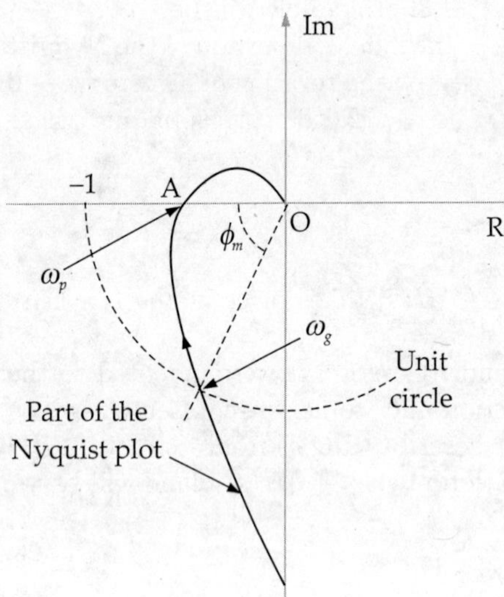

Figure 14.8: Gain and phase margins from Nyquist plot.

Phase Margin

By definition, it is the difference of the phase shift of the system at the gain crossover frequency (i.e. where the gain is unity) plus 180°. So, to determine the phase margin, a circle of radius unity is drawn. The phase margin is given by the subtended angle ϕ_m in Figure 14.8 [see Eq. (13.13)], where the unit circle cuts the Nyquist plot because there the gain is unity.

The following examples will make the procedure clear.

EXAMPLE 14.9. The Nyquist plot of loop transfer function $G(s)H(s)$ of a closed-loop control system passes through the point $(-1, \jmath 0)$ in the $G(s)H(s)$ plane. The phase margin of the system is

(A) 0° (B) 45° (c) 90° (D) 180°

Solution Since, at the point $(-1, \jmath 0)$ the phase angle $\angle G(\jmath\omega)H(\jmath\omega)|_{\omega=\omega_g} = -180°$, we have

$$\text{Pm} = \angle G(\jmath\omega)H(\jmath\omega)|_{\omega=\omega_g} + 180° = -180° + 180° = 0°$$

Ans. (A)

EXAMPLE 14.10. Draw a Nyquist plot, decide stability and calculate gain margin for the system characterized by it open-loop transfer function

$$G(s)H(s) = \frac{12}{s(s+1)(s+2)}$$

Solution The OLTF shows one pole at the origin (Type 1 system) and none on the $\jmath\omega$-axis. Therefore, the Nyquist path (or, contour) is as given in Figure 14.9a.

Mapping: (i) *For ab-segment:* $s \equiv \jmath\omega \quad 0 < \omega < \infty$

$$G(\jmath\omega)H(\jmath\omega) = \frac{12}{\jmath\omega(\jmath\omega+1)(\jmath\omega+2)} \equiv \left|\frac{12}{\omega\sqrt{\omega^2+1}\sqrt{\omega^2+4}}\right| \angle\left(-90° - \tan^{-1}\omega - \tan^{-1}\frac{\omega}{2}\right)$$

This equation generates values of magnitude and angle of the frequency-domain open-loop transfer function, which are listed in Table 14.4.

Table 14.4: Magnitude and angle (Example 14.10)

ω	$\lvert G(\jmath\omega)H(\jmath\omega)\rvert$	$\angle G(\jmath\omega)H(\jmath\omega)$
$+0$	∞	$-90°$
$+\infty$	0	$-270°$

The mapping of the segment is shown in Figure 14.9b with an appropriate label.

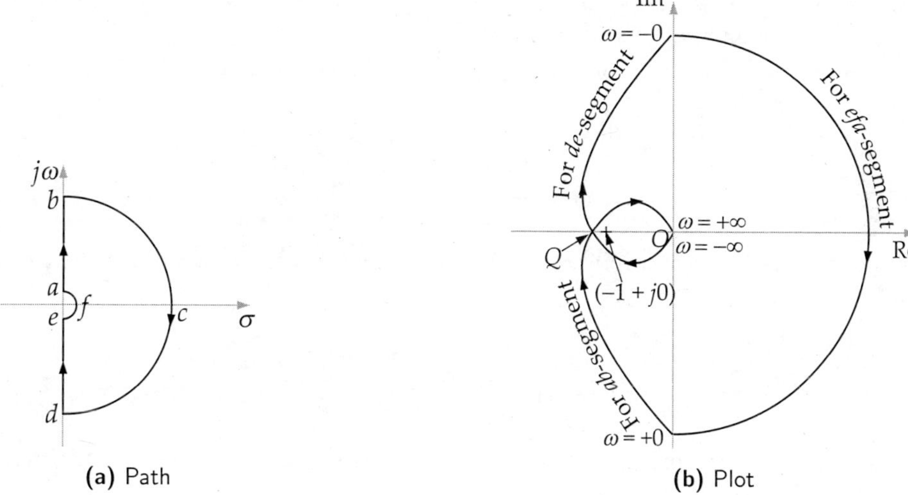

(a) Path

(b) Plot

Figure 14.9: Nyquist plots (Example 14.10).

(ii) *For bcd-segment:* $s \equiv \lim_{R\to\infty} Re^{\jmath\theta} \quad +90° < \theta < -90°$

$$G(s)H(s) = \lim_{R\to\infty}\left\{\frac{12}{Re^{\jmath\theta}(Re^{\jmath\theta}+1)(Re^{\jmath\theta}+2)}\right\} = 0$$

Therefore, this segment maps onto a point at the origin.
(iii) *For de-segment:* The mirror image of the mapping of the *ab*-segment.
(iv) *For efa-segment:* $s \equiv \lim_{\rho \to 0} \rho e^{j\theta} \quad -90^\circ < \theta < +90^\circ$

As this is a type 1 system, the Nyquist plots will close from $\omega = -0$ to $\omega = +0$ sweeping an angle of π, i.e. 180° in the clockwise direction (see *Closing Nyquist Plots* at page 364). Therefore, the semicircular curve of infinite radius from $\omega = -0$ to $\omega = +0$ maps the segment.
Q-point: Here the curve crosses the Re-axis. Therefore, $\Im|G(j\omega)H(j\omega)| = 0$. Now,

$$G(j\omega)H(j\omega) = \frac{12}{j\omega(j\omega+1)(j\omega+2)} = \frac{12}{-3\omega^2 + j(2\omega - \omega^3)} = \frac{12\left\{-3\omega^2 - j(2\omega - \omega^3)\right\}}{(-3\omega)^2 + (2\omega - \omega^3)^2}$$

Therefore,
$$\Im\{G(j\omega)H(j\omega)\}_{\omega_p} = \left.\frac{-12\,(2\omega - \omega^3)}{(-3\omega)^2 + (2\omega - \omega^3)^2}\right|_{\omega_p} = 0$$

$$\Rightarrow \qquad 2 - \omega_p^2 = 0$$

$$\Rightarrow \qquad \omega_p = \sqrt{2}$$

At this frequency the magnitude is

$$|G(j\omega)H(j\omega)|_{\omega=\sqrt{2}} = \left|\frac{12}{\omega\sqrt{\omega^2+1}\sqrt{\omega^2+4}}\right|_{\omega=\sqrt{2}} = \frac{12}{\sqrt{2}\sqrt{3}\sqrt{6}} = 2$$

Therefore,
$$Q = (-2 + j0)$$

The entire mapping is given in Figure 14.9b. We find that there is CW encirclement of $(-1+j0)$ point. Now, since $P = 0$, we get from Eq. (14.8) $Z > 0$, which indicates the existence of CLTF pole in the RHP. Therefore, the system is unstable.

$$\text{Gm} = -20\log(OQ) = -20\log(2) = -6.2\,\text{dB}$$

EXAMPLE 14.11. For a unity feedback system having open-loop transfer function

$$G(s) = \frac{K}{s(s+2)(s+6)}$$

determine, using the Nyquist stability criterion, the range of gain K for which the closed-loop system will be stable.

Solution Given, $H(s) = 1$. Therefore,

$$G(s)H(s) = \frac{K}{s(s+2)(s+6)}$$

So, there is one pole at the origin and none on the $j\omega$-axis. Therefore, the Nyquist path is the same as given in Figure 14.9a.
Mapping: (i) *For ab-segment:* $s \equiv j\omega \quad 0 < \omega < \infty$

$$G(j\omega)H(j\omega) = \frac{K}{j\omega(j\omega+2)(j\omega+6)} \equiv \frac{K}{\omega\sqrt{\omega^2+4}\sqrt{\omega^2+36}} \angle\left(-90^\circ - \tan^{-1}\frac{\omega}{2} - \tan^{-1}\frac{\omega}{6}\right)$$

The above equation generates the values of magnitude and angle of Table 14.5. The mapping of the segment is shown in Figure 14.10a with an appropriate label.

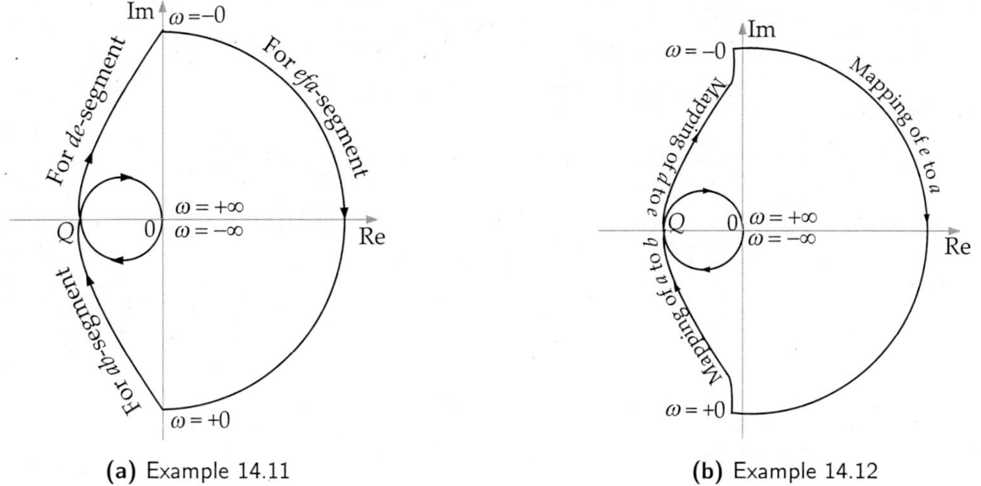

(a) Example 14.11

(b) Example 14.12

Figure 14.10: Nyquist plots.

Table 14.5: Magnitude and angle (Example 14.11)

ω	$\lvert G(\jmath\omega)H(\jmath\omega)\rvert$	$\angle G(\jmath\omega)H(\jmath\omega)$
$+0$	∞	$-90°$
$+\infty$	0	$-270°$

(ii) *For de-segment:* The mirror image of *ab*-segment mapping.
(iii) *For bcd-segment:* $s \equiv \lim_{R\to\infty} Re^{\jmath\theta} \quad +90° < \theta < -90°$

$$G(s)H(s) = \lim_{R\to\infty}\left\{\frac{K}{Re^{\jmath\theta}(Re^{\jmath\theta}+2)(Re^{\jmath\theta}+6)}\right\} = 0$$

Therefore, this segment maps onto a point at the origin.
(iv) *For efa-segment:* $s \equiv \lim_{\rho\to 0}\rho e^{\jmath\theta} \quad -90° < \theta < +90°$
As this is a type 1 system, the Nyquist plots will close from $\omega = -0$ to $\omega = +0$ sweeping an angle of π, i.e. 180° in the clockwise direction (see *Closing Nyquist Plots* at page 364). Therefore, the semicircular curve of infinite radius from $\omega = -0$ to $\omega = +0$ maps the segment [see Figure 14.10a].
Q-point: Here the curve crosses the Re-axis. Therefore, $\Im\{G(\jmath\omega)H(\jmath\omega)\}_{\omega_p} = 0$. Now,

$$G(\jmath\omega)H(\jmath\omega) = \frac{K}{\jmath\omega(\jmath\omega+2)(\jmath\omega+6)} = \frac{K}{-8\omega^2+\jmath(12\omega-\omega^3)} = \frac{K\{-8\omega^2-\jmath(12\omega-\omega^3)\}}{(-8\omega^2)^2+(12\omega-\omega^3)^2}$$

Therefore,
$$\Im\{G(\jmath\omega)H(\jmath\omega)\}_{\omega_p} = \frac{-K\left(12\omega_p-\omega_p^3\right)}{\left(-8\omega_p^2\right)^2+\left(12\omega_p-\omega_p^3\right)^2} = 0$$

or,
$$12-\omega_p^2 = 0$$

or,
$$\omega_p = 2\sqrt{3}$$

Alternatively

Here the phase of the open-loop transfer function becomes $-180°$. Therefore,

$$-180° = -90° - \tan^{-1}\frac{\omega_p}{2} - \tan^{-1}\frac{\omega_p}{6}$$

or,
$$90° = \tan^{-1}\frac{\omega_p}{2} + \tan^{-1}\frac{\omega_p}{6} = \tan^{-1}\frac{\frac{4\omega_p}{6}}{1-\frac{\omega_p^2}{12}}$$

or,
$$\frac{\frac{4\omega_p}{6}}{1-\frac{\omega_p^2}{12}} = \tan 90° = \infty$$

$\Rightarrow$
$$1 - \frac{\omega_p^2}{12} = 0$$

Therefore,
$$\omega_p = 2\sqrt{3}$$

At this frequency the magnitude is

$$|G(\jmath\omega)H(\jmath\omega)|_{\omega_p} = \left.\frac{K}{\omega\sqrt{\omega^2+4}\sqrt{\omega^2+36}}\right|_{\omega_p} = \frac{K}{(2\sqrt{3})(\sqrt{16})(\sqrt{48})} = \frac{K}{96}$$

For stability, this magnitude should be < 1. Therefore, the required condition is $K < 96$.

EXAMPLE 14.12. Given

$$G(s)H(s) = \frac{K}{s(1+sT_1)(1+sT_2)}$$

Find the condition of stability for the system from the Nyquist plot.

Solution The system has one pole at the origin and none on the $\jmath\omega$-axis. Therefore, the Nyquist path is the same as given in Figure 14.9a.

Mapping: (i) *For ab-segment:* $s \equiv \jmath\omega \quad 0 < \omega < \infty$

$$G(\jmath\omega)H(\jmath\omega) = \frac{K}{\jmath\omega(1+\jmath\omega T_1)(1+\jmath\omega T_2)}$$
$$\equiv \frac{K}{\omega\sqrt{1+\omega^2T_1^2}\sqrt{1+\omega^2T_2^2}}\angle\left(-90° - \tan^1\omega T_1 - \tan^1\omega T_2\right)$$

The above equation helps generate the data given in Table 14.6.

ω	$\lvert G(\jmath\omega)H(\jmath\omega)\rvert$	$\angle G(\jmath\omega)H(\jmath\omega)$
$+0$	∞	$-90°$
$+\infty$	0	$-270°$

Table 14.6: Magnitude and angle data (Example 14.12)

The corresponding mapping is shown in Figure 14.10b with an appropriate label.
(ii) *For de-segment:* The mirror image of *ab*-segment mapping.
(iii) *For bcd-segment:* $s \equiv \lim_{R\to\infty} Re^{\jmath\theta} \quad +90° < \theta < -90°$

$$G(s)H(s) = \lim_{R\to\infty}\frac{K}{Re^{\jmath\theta}(1+Re^{\jmath\theta}T_1)(1+Re^{\jmath\theta}T_2)} = 0$$

(iv) *For efa-segment:* $s \equiv \lim_{\rho\to 0}\rho e^{\jmath\theta} \quad -90° < \theta < +90°$

As this is a type 1 system, the Nyquist plots will close from $\omega = -0$ to $\omega = +0$ sweeping an angle of π i.e. $180°$ in the clockwise direction (see *Closing Nyquist Plots* at page 364).

Therefore, the semicircular curve of infinite radius from $\omega = -0$ to $\omega = +0$ maps the segment [see Figure 14.10b].

Q-point: Here the curve crosses the Re-axis. Therefore, $\Im\{G(\jmath\omega)H(\jmath\omega)\}_{\omega_p} = 0$. Now,

$$G(\jmath\omega)H(\jmath\omega) = \frac{K}{\jmath\omega(1+\jmath\omega T_1)(1+\jmath\omega T_2)} = \frac{K}{-(T_1+T_2)\omega^2 + \jmath(\omega - T_1T_2\omega^3)}$$
$$= \frac{K\{-(T_1+T_2)\omega^2 - \jmath(\omega - T_1T_2\omega^3)\}}{(T_1+T_2)^2\omega^4 + (\omega - T_1T_2\omega^3)^2}$$

Therefore,
$$\Im\{G(\jmath\omega)H(\jmath\omega)\}_{\omega_p} = \left.\frac{(\omega - T_1T_2\omega^3)}{(T_1+T_2)^2\omega^4 + (\omega - T_1T_2\omega^3)^2}\right|_{\omega_p} = 0$$

or,
$$(1 - T_1T_2\omega_p^2) = 0 \qquad \text{or,} \quad \omega_p = \frac{1}{\sqrt{T_1T_2}}$$

Alternatively

The phase of the OLTF is $-180°$ when it crosses the Re-axis. The corresponding frequency is the phase crossover frequency ω_p. Therefore,

$$-180° = -90° - \tan^{-1}\omega_pT_1 - \tan^{-1}\omega_pT_2$$

or,
$$90° = \tan^{-1}\frac{\omega_p(T_1+T_2)}{1-\omega_p^2T_1T_2}$$

or,
$$\frac{\omega_p(T_1+T_2)}{1-\omega_p^2T_1T_2} = \tan 90° = \infty$$

or,
$$1 - \omega_p^2T_1T_2 = 0$$

Therefore,
$$\omega_p = \frac{1}{\sqrt{T_1T_2}}$$

At this frequency the magnitude is

$$\left|G(\jmath\omega)H(\jmath\omega)\right|_{\omega=\frac{1}{\sqrt{T_1T_2}}} = \left.\frac{K}{\omega\sqrt{1+\omega^2T_1^2}\sqrt{1+\omega^2T_2^2}}\right|_{\omega=\frac{1}{\sqrt{T_1T_2}}}$$
$$= \frac{K}{\left(\frac{1}{\sqrt{T_1T_2}}\right)\left(\sqrt{1+\frac{T_1}{T_2}}\right)\left(\sqrt{1+\frac{T_2}{T_1}}\right)}$$
$$= \frac{KT_1T_2}{T_1+T_2}$$

For stability, this magnitude should be less than 1. Therefore, the required condition is

$$K < \frac{T_1+T_2}{T_1T_2}$$

EXAMPLE 14.13. Determine, through Nyquist's criterion, the stability of the system having open-loop transfer function as

$$G(s)H(s) = \frac{1}{s^2(s+2)}$$

Solution The system has two poles at the origin and none on the $\jmath\omega$-axis. Therefore, the Nyquist path is the same as given in Figure 14.9a.

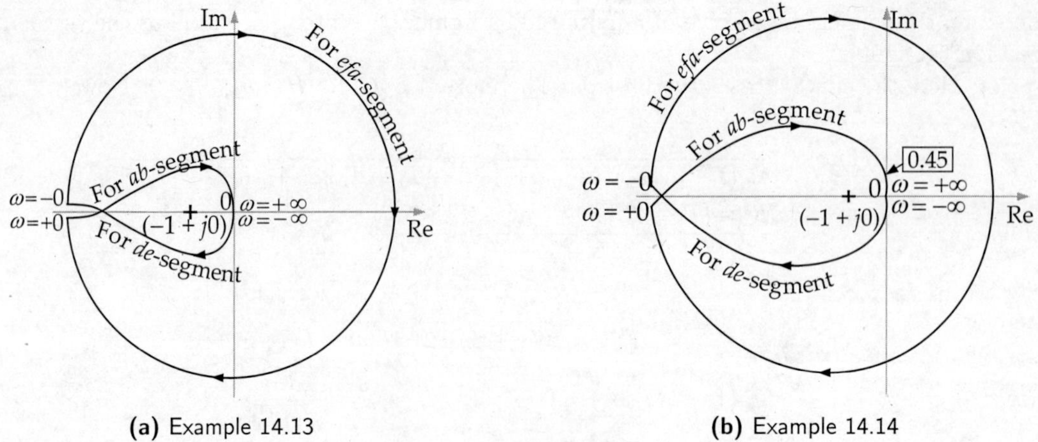

(a) Example 14.13 **(b)** Example 14.14

Figure 14.11: Nyquist plots.

Mapping: (i) *For ab-segment:* $s \equiv \jmath\omega \quad 0 < \omega < \infty$

$$G(\jmath\omega)H(\jmath\omega) = \frac{1}{(\jmath\omega)^2(\jmath\omega + 2)} \equiv \frac{1}{\omega^2\sqrt{\omega^2+4}}\angle\left(-180^\circ - \tan^{-1}\frac{\omega}{2}\right)$$

The above equation helps generate the data given in Table 14.7.

The corresponding mapping is shown in Figure 14.11a with an appropriate label.

ω	$\lvert G(\jmath\omega)H(\jmath\omega)\rvert$	$\angle G(\jmath\omega)H(\jmath\omega)$
$+0$	∞	-180°
$+\infty$	0	-270°

Table 14.7: Magnitude and angle data (Example 14.13)

(ii) *For de-segment:* The mirror image of *ab*-segment mapping.
(iii) *For bcd-segment:* $s \equiv \lim\limits_{R\to\infty} Re^{\jmath\theta} \quad +90^\circ < \theta < -90^\circ$

$$G(s)H(s) = \lim_{R\to\infty}\left\{\frac{1}{R^2e^{\jmath\,2\theta}(Re^{\jmath\theta}+2)}\right\} = 0$$

Therefore, this segment maps onto a point at the origin.
(iv) *For efa-segment:* $s \equiv \lim\limits_{\rho\to 0}\rho e^{\jmath\theta} \quad -90^\circ < \theta < +90^\circ$

As this is a type 2 system, the Nyquist plots will close from $\omega = -0$ to $\omega = +0$ sweeping an angle of 2π i.e. 360° in the clockwise direction (see *Closing Nyquist Plots* at page 364). Therefore, the circular curve of infinite radius from $\omega = -0$ to $\omega = +0$ maps the segment [see Figure 14.11a].

The Nyquist plot shows encirclement of the $(-1+\jmath 0)$ point twice while the number of RHP poles P in the OLTF is zero. Therefore, from Eq. (14.8), we get

$$2 = Z - 0$$

which indicates that the system, having two poles in the RHP of the CLTF, is unstable.

Note: The points at $\omega + 0$ and $\omega = -0$ are separate in the plot and therefore, the two encirclements are not apparent. Actually, the two points are one and the same and apart from the inner encirclement, there is another by the outer circle.

EXAMPLE 14.14. Check the stability of the closed-loop system by the Nyquist criterion when open-loop transfer function of the unity feedback system is given by

$$G(s) = \frac{10}{s^2(1+0.2s)(1+0.5s)}$$

Solution The open-loop transfer function possesses no poles in the RHP and two poles at the origin. Therefore, $P = 0$ and the Nyquist path is as shown in Figure 14.9a.

Mapping: (i) *For ab-segment:* $s \equiv \jmath\omega \quad 0 < \omega < \infty$. The open-loop transfer function can then be written as

$$\begin{aligned} G(\jmath\omega)H(\jmath\omega) &= \frac{10}{(\jmath\omega)^2(1+\jmath 0.2\omega)(1+\jmath 0.5\omega)} \\ &= \frac{10}{\omega^2\sqrt{1+0.04\omega^2}\sqrt{1+0.25\omega^2}}\angle\left(-180^\circ - \tan^{-1}0.2\omega - \tan^{-1}0.5\omega\right) \qquad \text{(i)} \end{aligned}$$

From Eq. (i), we can construct Table 14.8.

Table 14.8: Magnitude and phase data (Example 14.14)

ω	$\lvert G(\jmath\omega)H(\jmath\omega)\rvert$	$\angle G(\jmath\omega)H(\jmath\omega)$
$+0$	∞	-180°
$+\infty$	0	-360°

From the two angles it is apparent that if a polar plot has to traverse from -180° at magnitude ∞ to -360° at magnitude 0, it has to intersect the Im-axis somewhere. So, we calculate the intercept by first rationalizing the frequency-domain open-loop transfer function and then equating the real part of the numerator to zero to find the corresponding frequency. This frequency, for which the system is marginally stable is, in fact, its natural frequency of oscillation, ω_n because here the damping ratio $\zeta = 0$. From this frequency, we calculate the magnitude. Thus,

$$\begin{aligned} G(\jmath\omega)H(\jmath\omega) &= \frac{10}{(\jmath\omega)^2(1+\jmath 0.2\omega)(1+\jmath 0.5\omega)} \\ &= \frac{10(1-\jmath 0.2\omega)(1-\jmath 0.5\omega)}{(\jmath\omega)^2(1+\jmath 0.2\omega)(1+\jmath 0.5\omega)(1-\jmath 0.2\omega)(1-\jmath 0.5\omega)} \\ &= \frac{10[(1-0.1\omega^2)-\jmath 0.7\omega]}{(\jmath\omega)^2(1+\jmath 0.2\omega)(1+\jmath 0.5\omega)(1-\jmath 0.2\omega)(1-\jmath 0.5\omega)} \end{aligned}$$

So, $\quad 1 - 0.1\omega_n^2 = 0 \qquad \Rightarrow \omega_n^2 = \dfrac{1}{0.1} \qquad \Rightarrow \omega_n = \sqrt{10}$

Alternatively

At the point where the polar plot intersects the Im-axis, its phase is -270°. So, we can calculate ω_n from the relation

$$-270^\circ = -180^\circ - \tan^{-1}0.2\omega_n - \tan^{-1}0.5\omega_n$$

leading to the same result. The corresponding magnitude is

$$\lvert G(\jmath\omega)H(\jmath\omega)\rvert_{\omega_n} = \left.\frac{10}{\omega^2\sqrt{1+0.04\omega^2}\sqrt{1+0.25\omega^2}}\right|_{\omega_n} = \frac{10}{10\sqrt{1.4}\sqrt{3.5}} = 0.45$$

With these available data, we draw the polar plot corresponding to *ab*-segment of the Nyquist path. It is shown with an appropriate label in Figure 14.11b.
(ii) *For de-segment:* The mapping for this segment is the mirror image of that for the *ab*-segment.

These two plots complete the encirclement of the critical point once. We are not calculating the contribution of the *bcd*-segment which will be a point at the origin.
(iii) *For efa-segment:* $s \equiv \lim_{\rho \to 0} \rho e^{j\theta} \quad -90° > \theta > +90°$
Since this is a type 2 system, the Nyquist plots will close from $\omega = -0$ to $\omega = +0$ sweeping an angle of 2π, i.e. 360° in the clockwise direction (see *Closing Nyquist Plots* at page 364). Therefore, the circular curve of infinite radius from $\omega = -0$ to $\omega = +0$ maps the segment.

The complete Nyquist plot is shown in Figure 14.11b. From the plot we observe that the curve is CW throughout and that there are two encirclements of the critical point, i.e. $N = 2$. So, from the Nyquist criterion $N = Z - P$, we get

$$2 = Z - 0$$

$$\Rightarrow \qquad Z = 2 \qquad \text{(ii)}$$

Equation (ii) indicates the presence of two poles in the RHP of the CLTF. Therefore, the closed-loop system is unstable.

EXAMPLE 14.15. A unity feedback system has open-loop transfer function

$$G(s) = \frac{1}{s(1+2s)(1+s)}$$

Sketch the Nyquist plot for the system and obtain gain and phase margins.

Solution The system has one pole at the origin and none on the $j\omega$-axis. Therefore, the Nyquist path is the same as given in Figure 14.9a.

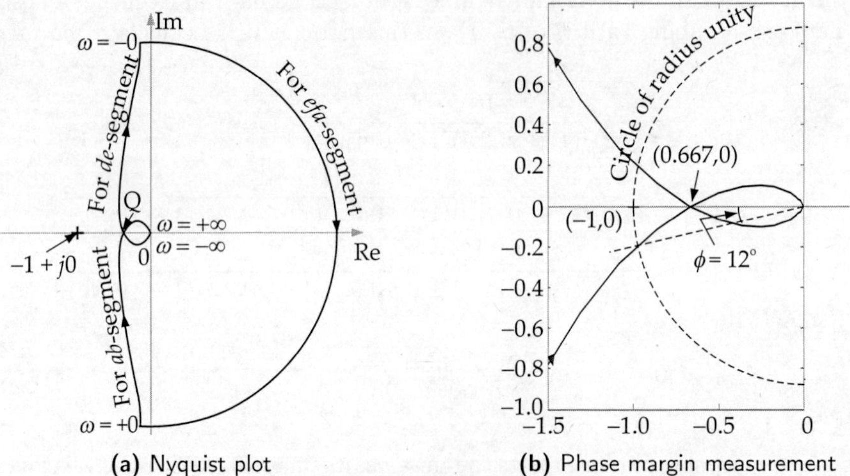

(a) Nyquist plot **(b)** Phase margin measurement

Figure 14.12: Nyquist plot and details of phase margin measurement (Example 14.15).

Mapping: (i) *For ab-segment:* $s \equiv j\omega \quad 0 < \omega < \infty$

$$G(j\omega)H(j\omega) = \frac{1}{j\omega(1+j2\omega)(1+j\omega)}$$

$$\equiv \frac{1}{\omega\sqrt{1+4\omega^2}\sqrt{1+\omega^2}} \angle\left(-90° - \tan^1 2\omega - \tan^1 \omega\right)$$

The above equation helps generate the data given in Table 14.9.

Table 14.9: Magnitude and angle data (Example 14.15)

ω	$\lvert G(\jmath\omega)H(\jmath\omega)\rvert$	$\angle G(\jmath\omega)H(\jmath\omega)$
$+0$	∞	-90°
$+\infty$	0	-270°

The corresponding mapping is shown in Figure 14.12a with an appropriate label.
(ii) *For de-segment:* The mirror image of *ab*-segment mapping.
(iii) *For bcd-segment:* $s \equiv \lim_{R\to\infty} Re^{\jmath\theta} \quad +90^\circ < \theta < -90^\circ$

$$G(s)H(s) = \lim_{R\to\infty}\left\{\frac{1}{Re^{\jmath\theta}(1+2Re^{\jmath\theta})(1+Re^{\jmath\theta})}\right\} = 0$$

(iv) *For efa-segment:* $s \equiv \lim_{\rho\to 0} \rho e^{\jmath\theta} \quad -90^\circ < \theta < +90^\circ$
As this is a type 1 system, the Nyquist plots will close from $\omega = -0$ to $\omega = +0$ sweeping an angle of π i.e. 180° in the clockwise direction. Therefore, the semicircular curve of infinite radius from $\omega = -0$ to $\omega = +0$ maps the segment [see Figure 14.12a].
Q-point: Here the curve crosses the Re-axis. Therefore, $\Im\{G(\jmath\omega)H(\jmath\omega)\}_{\omega_p} = 0$. Now,

$$G(\jmath\omega)H(\jmath\omega) = \frac{1}{\jmath\omega(1+\jmath 2\omega)(1+\jmath\omega)} = \frac{1}{-3\omega^2+\jmath(\omega-2\omega^3)} = \frac{-3\omega^2-\jmath(\omega-2\omega^3)}{(-3\omega^2)^2+(\omega-2\omega^3)^2}$$

Therefore,
$$\text{Im}\{G(\jmath\omega)H(\jmath\omega)\}_{\omega_p} = \frac{(\omega_p-2\omega_p^3)}{(-3\omega_p^2)^2+(\omega_p-2\omega_p^3)^2} = 0$$

or,
$$1-2\omega_p^2 = 0$$

or,
$$\omega_p = \frac{1}{\sqrt{2}}$$

Alternatively

The phase of the OLTF is -180° when it crosses the Re-axis. The corresponding frequency is the phase crossover frequency ω_p. Therefore,

$$-180^\circ = -90^\circ - \tan^{-1}2\omega_p - \tan^{-1}\omega_p$$

or,
$$90^\circ = \tan^{-1}\frac{3\omega_p}{1-2\omega_p^2}$$

or,
$$\frac{3\omega_p}{1-2\omega_p^2} = \tan 90^\circ = \infty$$

or,
$$1-2\omega_p^2 = 0$$

Thus,
$$\omega_p = \frac{1}{\sqrt{2}}$$

At this frequency the magnitude is

$$\left|G(\jmath\omega)H(\jmath\omega)\right|_{\omega_p} = \left.\frac{1}{\omega\sqrt{1+4\omega^2}\sqrt{1+\omega^2}}\right|_{\omega_p}$$

$$= \frac{1}{\left(\frac{1}{\sqrt{2}}\right)(\sqrt{1+2})\left(\sqrt{1+\frac{1}{2}}\right)} = \frac{2}{3}$$

Therefore,
$$\text{Gm} = -20\log\left(\frac{2}{3}\right)\text{ dB} = 3.521\text{ dB}$$

For measuring the phase margin, a unit circle is drawn with its centre at the origin. It intersects the Nyquist plot at 12° beyond 180°. Hence, the phase margin is

$$\text{Pm} = \phi = 12^\circ$$

Note: To measure phase margin from the Nyquist diagram, an accurate plot has to be made with many points on it. It is indeed a time consuming and laborious process. We have shown the relevant area of the plot in Figure 14.12b.

EXAMPLE 14.16. Sketch the Nyquist plot and determine stability of the unity feedback control system given by

$$G(s) = \frac{K}{(1+sT_1)(1+sT_2)}$$

Solution The system has no poles at the origin (type 0 system) and none on the $\jmath\omega$-axis. Therefore, the Nyquist path is as given in Figure 14.13a.

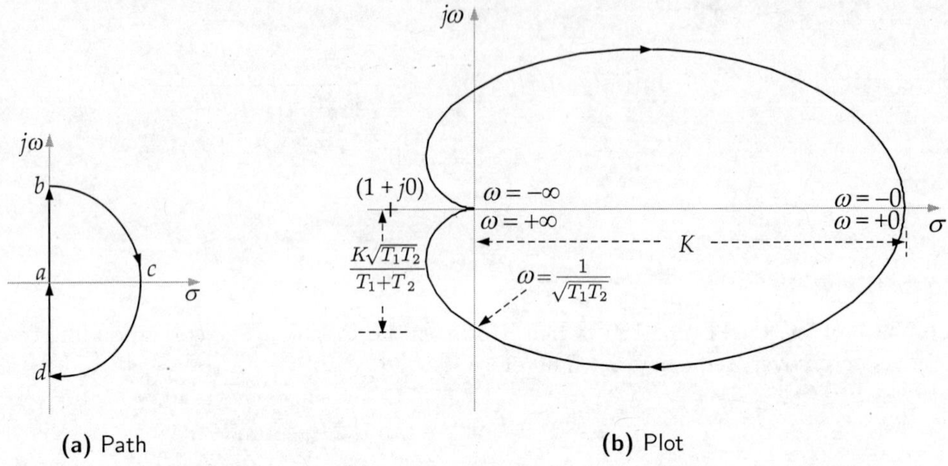

Figure 14.13: Nyquist diagrams (Example 14.16).

Mapping: (i) *For ab-segment:* $s \equiv \jmath\omega \quad 0 < \omega < \infty$

$$G(\jmath\omega)H(\jmath\omega) = \frac{K}{(1+\jmath\omega T_1)(1+\jmath\omega T_2)}$$

$$\equiv \frac{K}{\sqrt{1+\omega^2T_1^2}\sqrt{1+\omega^2T_2^2}}\angle\left(-\tan^{-1}\omega T_1 - \tan^{-1}\omega T_2\right) \qquad \text{(i)}$$

Table 14.10: Magnitude and angle (Example 14.16)

ω	$\lvert G(\jmath\omega)H(\jmath\omega)\rvert$	$\angle G(\jmath\omega)H(\jmath\omega)$
$+0$	K	0°
$+\infty$	0	-180°

Equation (i) helps generate the data of Table 14.10. These data indicate that the curve will cross the Im-axis. To find the intercept on the Im-axis, we can isolate the real and imaginary parts of the frequency domain transfer function, equate the real part to 0, and solve the resulting equation to find out the frequency. The magnitude of the frequency domain transfer function at that frequency gives the intercept. Thus,

$$G(\jmath\omega)H(\jmath\omega) = \frac{K}{(1+\jmath\omega T_1)(1+\jmath\omega T_2)} = \frac{K(1-\jmath\omega T_1)(1-\jmath\omega T_2)}{(1+\jmath\omega T_1)(1+\jmath\omega T_2)(1-\jmath\omega T_1)(1-\jmath\omega T_2)}$$
$$= \frac{K(1-\omega^2 T_1T_2)}{(1+\omega^2T_1^2)(1+\omega^2T_2^2)} - \jmath\left\{\frac{K\omega(T_1+T_2)}{(1+\omega^2T_1^2)(1+\omega^2T_2^2)}\right\}$$

Equating the real part to zero to obtain the frequency, ω_n, for marginal stability of the system, we get

$$\left.\frac{K(1-\omega^2T_1T_2)}{(1+\omega^2T_1^2)(1+\omega^2T_2^2)}\right|_{\omega_n} = 0 \qquad \Rightarrow 1-\omega_n^2T_1T_2 = 0 \qquad \Rightarrow \omega_n = \frac{1}{\sqrt{T_1T_2}}$$

Alternatively

We can find ω_n from the consideration that the phase where the OLTF crosses the Im-axis is -270°. Therefore,

$$-270^\circ = -\tan^{-1}\omega_nT_1 - \tan^{-1}\omega_nT_2 = -\tan^{-1}\frac{\omega_n(T_1+T_2)}{1-\omega_n^2T_1T_2}$$

$$\Rightarrow \frac{\omega_n(T_1+T_2)}{1-\omega_n^2T_1T_2} = \tan 270^\circ = \infty \qquad \Rightarrow 1-\omega_n^2T_1T_2 = 0 \qquad \Rightarrow \omega_n = \frac{1}{\sqrt{T_1T_2}}$$

The magnitude at this frequency is

$$\left.\frac{K}{\sqrt{1+\omega^2T_1^2}\sqrt{1+\omega^2T_2^2}}\right|_{\omega_n=\frac{1}{\sqrt{T_1T_2}}} = \frac{K\sqrt{T_1T_2}}{T_1+T_2}$$

(ii) *For bcd-segment:* $\quad s \equiv \lim_{R\to\infty} Re^{\jmath\theta} \quad +90^\circ < \theta < -90^\circ$

$$G(s)H(s) = \lim_{R\to\infty}\frac{K}{(1+T_1Re^{\jmath\theta})(1+T_2Re^{\jmath\theta})} = \lim_{R\to\infty}\frac{K}{R\left(\frac{1}{R}+T_1e^{\jmath\theta}\right)R\left(\frac{1}{R}+T_2e^{\jmath\theta}\right)} = 0$$

(iii) *For de-segment:* The mirror image of ab-segment mapping.
Since this is a type 0 system, the points $\omega = +0$ and $\omega = -0$ are coincident (see *Closing Nyquist Plots* at page 364). Taking these factors into consideration, the Nyquist plot, as shown in Figure 14.13b, is constructed.

The plot shows no encirclement of the $(1+\jmath 0)$ point, i.e. $N = 0$, while $P = 0$. Therefore, $Z = 0$ and hence, the system is stable.

EXAMPLE 14.17. Draw the Nyquist plot for the unity feedback control system with the open-loop transfer function of

$$G(s) = \frac{10(s+2)}{s(s-1)}$$

Using the Nyquist stability criterion, determine the stability of the closed-loop system.

Solution The open-loop transfer function has a pole at the origin and another pole in the RHP. So, $P = 1$ and the Nyquist path is as shown in Figure 14.14a.

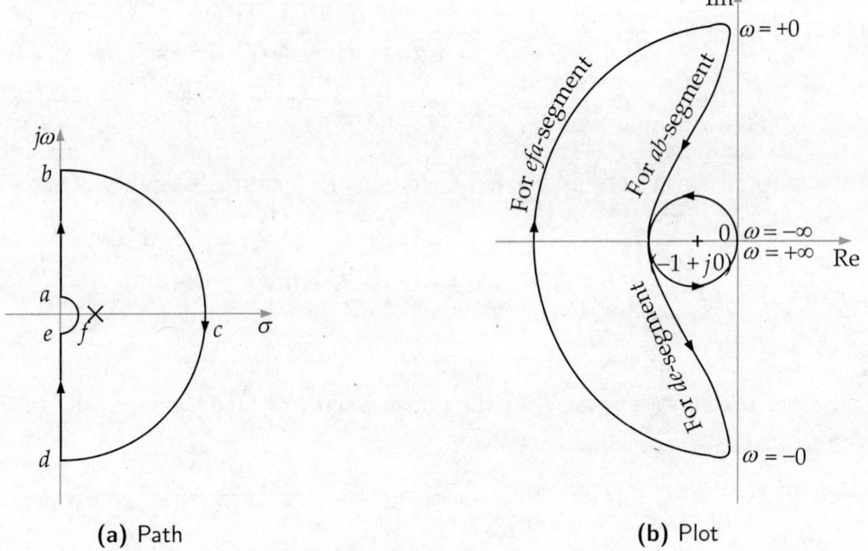

(a) Path **(b)** Plot

Figure 14.14: Nyquist diagrams (Example 14.17).

Mapping: (i) *For ab-segment:* $s \equiv \jmath\omega \quad 0 < \omega < \infty$

$$G(\jmath\omega)H(\jmath\omega) = \frac{10(\jmath\omega + 3)}{\jmath\omega(\jmath\omega - 1)} \tag{i}$$

$$= \frac{10\sqrt{\omega^2+9}}{\omega\sqrt{\omega^2+1}} \angle\left[\tan^{-1}\frac{\omega}{3} - 90^\circ - (\pi - \tan^{-1}\omega)\right] \tag{ii}$$

From Eq. (ii), we can construct Table 14.11.

ω	$\lvert G(\jmath\omega)H(\jmath\omega)\rvert$	$\angle G(\jmath\omega)H(\jmath\omega)$
+0	∞	−270°
+∞	0	−90°

Table 14.11: Magnitude and angle (Example 14.17)

The data help us draw the plot for the segment as labelled in Figure 14.14b. It is obvious from the plot that the curve intersects the Re-axis. To find the intercept on the Re-axis, we rationalize the RHS of Eq. (i), pick out the imaginary part of the numerator, and equate it to zero to find the phase cross-over frequency ω_p as follows:

$$G(j\omega)H(j\omega) = \frac{10(j\omega+3)}{j\omega(j\omega-1)} = \frac{10(j\omega+3)(-j)(-j\omega-1)}{\omega(j\omega-1)(-j\omega-1)} = \frac{10[-4\omega+j(3-\omega^2)]}{\omega(j\omega-1)(-j\omega-1)}$$

or,
$$3-\omega_p^2 = 0$$

or,
$$\omega_p = \sqrt{3}$$

Alternatively

The phase of the OLTF is $-180°$ when it crosses the Re-axis. The corresponding frequency is the phase crossover frequency ω_p. Therefore,

$$-180° = -90° - \pi + \tan^{-1}\frac{\omega_p}{3} + \tan^{-1}\omega_p$$

or,
$$90° = \tan^{-1}\frac{(4/3)\omega_p}{1-(\omega_p^2/3)}$$

or,
$$\frac{(4/3)\omega_p}{1-(\omega_p^2/3)} = \tan 90° = \infty$$

or,
$$1-\frac{\omega_p^2}{3} = 0$$

Therefore,
$$\omega_p = \sqrt{3}$$

The magnitude at ω_p is

$$\left.\frac{10\sqrt{\omega^2+9}}{\omega\sqrt{\omega^2+1}}\right|_{\omega_p} = \frac{10\sqrt{12}}{\sqrt{3}\sqrt{4}} = 10$$

(ii) *For bcd-segment:* $s \equiv \lim_{R\to\infty} Re^{j\theta} \quad +90° < \theta < -90°$

$$G(s)H(s) = \lim_{R\to\infty}\frac{10\left(Re^{j\theta}+3\right)}{Re^{j\theta}\left(Re^{j\theta}-1\right)} = \lim_{R\to\infty}\frac{10R\left(e^{j\theta}+\dfrac{3}{R}\right)}{R^2e^{j\theta}\left(e^{j\theta}-\dfrac{1}{R}\right)} = 0$$

Therefore, the segment maps onto a point at the origin.
(iii) *For the de-segment:* A mirror image of the curve obtained for the *ab*-segment is drawn.
(iv) *For the efa-segment:* $s \equiv \lim_{\rho\to 0}\rho e^{j\theta} \quad -90° > \theta > +90°$

Since this is a type 1 system, the Nyquist plots will close from $\omega = -0$ to $\omega = +0$ sweeping an angle of π, i.e. 180° in the clockwise direction (see *Closing Nyquist Plots* at page 364). Therefore, the semicircular curve of infinite radius from $\omega = -0$ to $\omega = +0$ maps the segment [see Figure 14.14b].

We observe from the figure that there is *CCW encirclement* of the critical point by the Nyquist plot. So, $N = -1$. From the Nyquist criterion $N = Z - P$, we have

$$-1 = Z - 1$$

$\Rightarrow$
$$Z = 0$$

Thus, the system is stable.

Note: The example shows that, despite the presence of a pole in the RHP in the open-loop transfer function, the closed-loop system can be stable. We will consider more such examples.

EXAMPLE 14.18. The open-loop transfer function of a unity feedback system is given by

$$G(s) = \frac{8s}{(s-1)(s-2)}$$

Find the closed-loop stability of the system by the Nyquist method.

Solution The open-loop transfer function possesses two poles in the RHP. Therefore $P = 2$. There is a zero at the origin. So the Nyquist path is as shown in Figure 14.15a.

Mapping: (i) *For ab-segment:* $s \equiv \jmath\omega \quad 0 < \omega < \infty$

$$G(\jmath\omega)H(\jmath\omega) = \frac{\jmath 8\omega}{(\jmath\omega - 1)(\jmath\omega - 2)} \tag{i}$$

$$= \frac{8\omega}{\sqrt{\omega^2+1}\sqrt{\omega^2+4}} \angle\left[90^\circ - \left(\pi - \tan^{-1}\omega\right) - \left(\pi - \tan^{-1}\frac{\omega}{2}\right)\right]$$

$$= \frac{8\omega}{\sqrt{\omega^2+1}\sqrt{\omega^2+4}} t\angle\left(-270^\circ + \tan^{-1}\omega + \tan^{-1}\frac{\omega}{2}\right) \tag{ii}$$

Observe that we have written -270° in lieu of 90° (they are the same) for convenience in the angle-magnitude calculation. From Eq. (ii), we can construct Table 14.12.

ω	$\lvert G(\jmath\omega)H(\jmath\omega)\rvert$	$\angle G(\jmath\omega)H(\jmath\omega)$
$+0$	0	-270°
$+\infty$	0	-90°

Table 14.12: Magnitude and phase (Example 14.18)

We note that the magnitude is 0 both when $\omega \to 0$ and $\omega \to \infty$ while the angle varies from -270° to -90°. Therefore, it is evident that the polar plot has to cross the Re-axis at a point. *Intersection in the Re-axis:* To find the intercept, as before, we rationalize the complex open-loop transfer function and equate the imaginary part of the numerator to zero. The phase crossover frequency ω_p found thereby is used to calculate the magnitude of the intercept. Thus,

$$G(\jmath\omega)H(\jmath\omega) = \frac{\jmath 8\omega}{(\jmath\omega - 1)(\jmath\omega - 2)} = \frac{8(\jmath\omega)(-\jmath\omega - 1)(-\jmath\omega - 2)}{(\jmath\omega - 1)(\jmath\omega - 2)(-\jmath\omega - 1)(-\jmath\omega - 2)}$$

$$= \frac{8(\jmath\omega)(-\omega^2 + 2 + \jmath 3\omega)}{(\jmath\omega - 1)(\jmath\omega - 2)(-\jmath\omega - 1)(-\jmath\omega - 2)} = \frac{8\left[-3\omega^2 + \jmath\omega(-\omega^2 + 2)\right]}{(\jmath\omega - 1)(\jmath\omega - 2)(-\jmath\omega - 1)(-\jmath\omega - 2)}$$

Equating the imaginary part of the numerator to zero, we get

$$\omega_p^2 = 2 \qquad \text{or,} \qquad \omega_p = \sqrt{2}$$

Alternatively

The phase of the OLTF is -180° when it crosses the Re-axis. The corresponding frequency is the phase crossover frequency ω_p. Therefore ω_p can be calculated from the following equation as well

$$-180^\circ = -270^\circ + \tan^{-1}\omega_p + \tan^{-1}\frac{\omega_p}{2}$$

leading to the same result. The magnitude at ω_p is

$$\frac{8\sqrt{2}}{\sqrt{2+1}\sqrt{2+4}} = \frac{8}{3} = 2.67$$

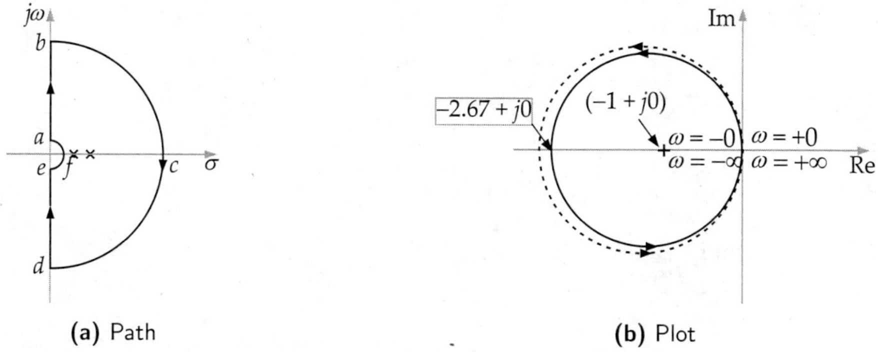

(a) Path (b) Plot

Figure 14.15: Nyquist diagrams (Example 14.18). The dashed curve, which maps to the *de*-segment of the Nyquist path should actually coincide with the full curve that maps to the *ab*-segment. To show its existence, it has been displaced to some extent.

So, the ab-segment of the Nyquist path maps to a complete curve (because it starts from 0 and ends at 0) intersecting the Re-axis at $(-2.67 + \jmath 0)$, as shown in Figure 14.15b. Thereby it encircles the critical point in a CCW direction.

The de-segment, which maps to form a mirror image of that of the ab-segment, also encircles the critical point in a CCW direction. Since this is a type 0 system, the points $\omega = -0$ and $\omega = +0$ are coincident (see *Closing Nyquist Plots* at page 364).

Thus, $N = -2$, and according to the Nyquist criterion $N = Z - P$

$$-2 = Z - 2 \qquad \text{or,} \qquad Z = 0$$

The result indicates that the closed-loop system is stable.

EXAMPLE 14.19. The open-loop transfer function of a unity feedback system is given by

$$G(s) = \frac{s+2}{(s+1)(s-1)}$$

Examine the stability of the closed-loop system by Nyquist criteria.

Solution The open-loop transfer function consists of a pole in the RHP and none at the origin or on the Im-axis. Therefore, $P = 1$ and the Nyquist path is as shown in Figure 14.16a.

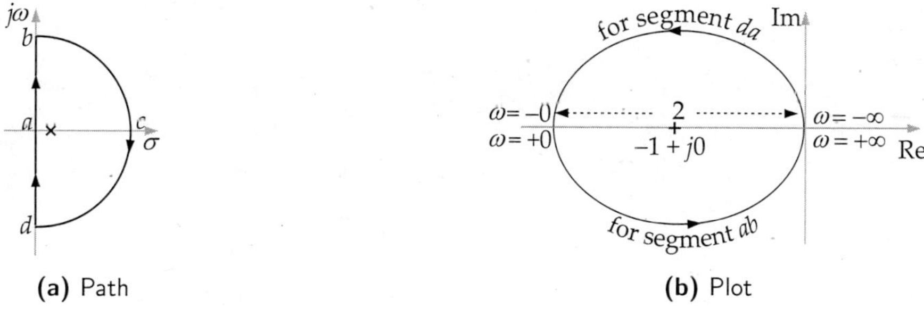

(a) Path (b) Plot

Figure 14.16: Nyquist diagrams (Example 14.19).

Mapping: (i) *For ab-segment:* $s \equiv \jmath\omega \quad 0 < \omega < \infty$. The frequency-domain open-loop transfer function can be written as follows:

$$G(\jmath\omega)H(\jmath\omega) = \frac{\sqrt{\omega^2+4}}{\sqrt{\omega^2+1}\sqrt{\omega^2+1}}\angle\left[\tan^{-1}\frac{\omega}{2} - \tan^{-1}\omega - (\pi - \tan^{-1}\omega)\right]$$

$$= \frac{\sqrt{\omega^2+4}}{\sqrt{\omega^2+1}\sqrt{\omega^2+1}}\angle\left[\tan^{-1}\frac{\omega}{2} - \pi\right] \qquad \text{(i)}$$

From Eq. (i) we can build Table 14.13.

ω	$\lvert G(\jmath\omega)H(\jmath\omega)\rvert$	$\angle G(\jmath\omega)H(\jmath\omega)$
$+0$	2	$-180°$
$+\infty$	0	$-90°$

Table 14.13: Magnitude and phase (Example 14.19)

These data help us construct the polar plot for the *ab*-segment [see 'for segment *ab*'-labelled part of Figure 14.16b].

(ii) *For da-segment:* The *da*-segment of the Nyquist path maps to form a mirror image of that of the *ab*-segment. Since this is a type 0 system, the points $\omega = -0$ and $\omega = +0$ are coincident (see *Closing Nyquist Plots* at page 364).

Thus, the Nyquist plot shows a *CCW encirclement* of the critical point so that $N = -1$. From the Nyquist criterion $N = Z - P$, we get

$$-1 = Z - 1 \qquad \text{or,} \qquad Z = 0$$

Therefore, the system is stable.

EXAMPLE 14.20. A unity feedback system has the open-loop transfer function of

$$G(s) = \frac{2}{s(1-2s)}$$

Draw the Nyquist plot and determine the closed-loop stability.

Solution The open-loop transfer function consists of one pole in the RHP and another at the origin. Therefore, $P = 1$ and the Nyquist path is as shown in Figure 14.14a.

Mapping: (i) *For ab-segment:* $s \equiv \jmath\omega \quad 0 < \omega < \infty$. The open-loop transfer function can then be written as

$$G(\jmath\omega)H(\jmath\omega) = \frac{2}{\jmath\omega(1-\jmath 2\omega)} = \frac{2}{\omega\sqrt{1+4\omega^2}}\angle[-90° + \tan^{-1}2\omega] \qquad \text{(i)}$$

From Eq. (i), we can construct Table 14.14.

ω	$\lvert G(\jmath\omega)H(\jmath\omega)\rvert$	$\angle G(\jmath\omega)H(\jmath\omega)$
$+0$	∞	$-90°$
$+\infty$	0	$0°$

Table 14.14: Magnitude and phase (Example 14.20)

The corresponding polar plot is shown with an appropriate label in Figure 14.17a.

(ii) *For bcd-segment:* $s \equiv \lim_{R\to\infty} Re^{\jmath\theta} \quad +90° > \theta > -90°$

$$G(s)H(s) \equiv \lim_{R\to\infty} \frac{2}{Re^{\jmath\theta}(1-2Re^{\jmath\theta})} = 0$$

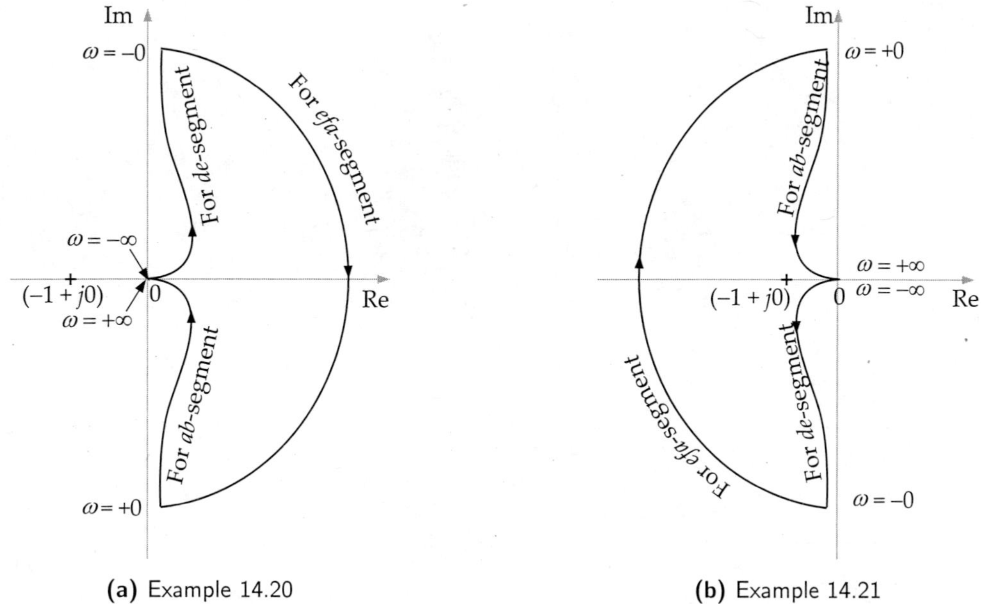

(a) Example 14.20 **(b)** Example 14.21

Figure 14.17: Nyquist plots.

Thus, this segment of the Nyquist path maps to a point at the origin in the Nyquist plot.
(iii) *For de-segment:* This segment maps to a mirror image of that of the ab-segment.
(iv) *For efa-segment:* $s \equiv \lim_{\rho\to 0} \rho e^{j\theta} \quad -90^\circ > \theta > +90^\circ$

Since this is a type 1 system, the Nyquist plots will close from $\omega = -0$ to $\omega = +0$ sweeping an angle of π, i.e. 180° in the clockwise direction (see *Closing Nyquist Plots* at page 364). Therefore, the semicircular curve of infinite radius from $\omega = -0$ to $\omega = +0$ maps the segment [see Figure 14.17a].

From the plot we observe that the curve is CW throughout and that there is no encirclement of the critical point, i.e. $N = 0$. So, from the Nyquist criterion $N = Z - P$, we get

$$0 = Z - 1$$

or,

$$Z = 1 \tag{ii}$$

Equation (ii) indicates the presence of a pole in the RHP of the CLTF. Therefore, the closed-loop system is unstable.

Note: This example demonstrates that a system can be unstable even if there is no encirclement of the critical point by the Nyquist plot.

EXAMPLE 14.21. The open-loop transfer function of a feedback control system is given by

$$G(s)H(s) = \frac{-1}{2s(1-20s)}$$

Determine the stability of the closed-loop system by Nyquist criteria.

Solution The open-loop transfer function consists of a pole in the RHP and another at the origin. So, $P = 1$ and the Nyquist path is as shown in Figure 14.14a.

Mapping: (i) *For ab-segment:* $s \equiv j\omega \quad 0 < \omega < \infty$. The open-loop transfer function can then be written as

$$G(\jmath\omega)H(\jmath\omega) = \frac{-1}{\jmath 2\omega(1-\jmath 20\omega)} \tag{14.9}$$

$$= \frac{1}{2\omega\sqrt{1+400\omega^2}} \angle\left[\tan^{-1}\frac{0}{-1} - 90^\circ + \tan^{-1} 20\omega\right] \tag{14.10}$$

$$= \frac{1}{2\omega\sqrt{1+400\omega^2}} \angle\left[(\pi - 0^\circ) - 90^\circ + \tan^{-1} 20\omega\right] \tag{14.11}$$

$$= \frac{1}{2\omega\sqrt{1+400\omega^2}} \angle\left(180^\circ - 90^\circ + \tan^{-1} 20\omega\right) \tag{i}$$

From Eq. (i), we can construct Table 14.15.

ω	$\lvert G(\jmath\omega)H(\jmath\omega)\rvert$	$\angle G(\jmath\omega)H(\jmath\omega)$
$+0$	∞	90°
$+\infty$	0	180°

Table 14.15: Magnitude and phase (Example 14.21)

With these available data, we draw the polar plot corresponding to *ab*-segment of the Nyquist path. It is shown with an appropriate label in Figure 14.17b.

(ii) *For bcd-segment:* $s \equiv \lim_{R\to\infty} Re^{\jmath\theta} \quad +90^\circ > \theta > -90^\circ$

$$G(s)H(s) \equiv \lim_{R\to\infty} \frac{-1}{2Re^{\jmath\theta}(1-20Re^{\jmath\theta})} = 0$$

Thus, this segment of the Nyquist path maps onto a point at the origin in the Nyquist plot.

(iii) *For de-segment:* This segment maps to a mirror image of that of the *ab*-segment.

(iv) *For efa-segment:* $s \equiv \lim_{\rho\to 0} \rho e^{\jmath\theta} \quad -90^\circ > \theta > +90^\circ$

Since this is a type 1 system, the Nyquist plots will close from $\omega = -0$ to $\omega = +0$ sweeping an angle of π i.e. 180° in the clockwise direction (see *Closing Nyquist Plots* at page 364). Therefore, the semicircular curve of infinite radius from $\omega = -0$ to $\omega = +0$ maps the segment.

The complete Nyquist plot is shown in Figure 14.17b. From the plot we observe that the CW curve encircles the critical point once, i.e. $N = 1$. So, from the Nyquist criterion $N = Z - P$, we get

$$1 = Z - 1$$

$$\Rightarrow \quad Z = 2 \tag{ii}$$

Equation (ii) indicates the presence of two poles in the RHP of the CLTF. Therefore, the closed-loop system is unstable.

EXAMPLE 14.22. The open-loop transfer function of a system is given by

$$G(s)H(s) = \frac{K(1+s)}{1-s}$$

Determine through the Nyquist formalism the value of K that will make the closed-loop system stable.

Solution The open-loop transfer function possesses one pole in the RHP and no pole at the origin. Therefore, $P = 1$ and the Nyquist path is as shown in Figure 14.14a.

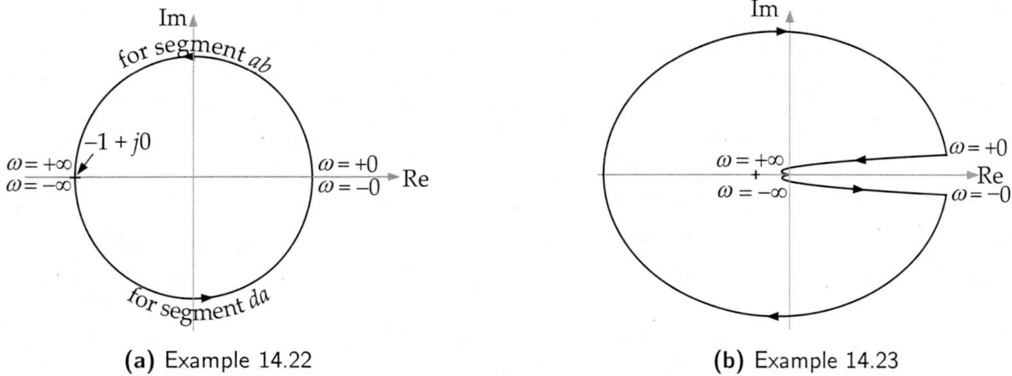

(a) Example 14.22

(b) Example 14.23

Figure 14.18: Nyquist Plots. The critical point is shown with a + sign on the negative Re-axis.

Mapping: *For ab-segment:* $s \equiv \jmath\omega \quad 0 < \omega < \infty$. The open-loop transfer function can then be written as

$$G(\jmath\omega)H(\jmath\omega) = \frac{K(1+\jmath\omega)}{1-\jmath\omega} = \frac{K\sqrt{1+\omega^2}}{\sqrt{1+\omega^2}}\angle 2\tan^{-1}\omega \tag{i}$$

We can construct Table 14.16 from Eq. (i).

Table 14.16: Magnitude and phase (Example 14.22)

ω	$\lvert G(\jmath\omega)H(\jmath\omega)\rvert$	$\angle G(\jmath\omega)H(\jmath\omega)$
+0	K	0°
1	K	90°
$+\infty$	K	180°

The polar plot for the above data is a semicircular curve of radius K.

The *da*-segment will map to form a mirror image of the this curve. Since this is a type 0 system, the points $\omega = -0$ and $\omega = +0$ are coincident (see *Closing Nyquist Plots* at page 364).

So, the complete Nyquist plot is given by Figure 14.18a. It shows a CCW encircling. For $K = 1$, it passes through the critical point making the system marginally stable. The system is stable if $K > 1$ when the Nyquist plot encircles the critical point and therefore $N = -1$. Then from the Nyquist criterion $N = Z - P$, we get

$$-1 = Z - 1$$

$$\Rightarrow \qquad Z = 0$$

This means that for $K > 1$ there is no RHP pole in the CLTF and therefore, then the system will be stable.

EXAMPLE 14.23. Sketch the Nyquist plot for the open-loop transfer function

$$G(s)H(s) = \frac{s+1}{s^2(s-2)}$$

and determine the closed-loop stability of the system.

Solution This is a type 2 system having a pole in the RHP. Therefore, $P = 1$. The frequency domain transfer function can be written as

$$G(\jmath\omega)H(\jmath\omega) = \frac{\jmath\omega+1}{-\omega^2(\jmath\omega-2)} \equiv \frac{\omega^2+1}{\omega^2\sqrt{\omega^2+4}}\angle\left[\tan^{-1}\omega - 180° - \left(\pi - \tan^{-1}\frac{\omega}{2}\right)\right] \tag{i}$$

The Nyquist path for the system is given Figure 14.14a.

Mapping: (i) *For ab-segment:* From Eq. (i) we get the data of Table 14.17.

ω	M	ϕ
$+0$	∞	$-360°$
$+\infty$	0	$-180°$

Table 14.17: Magnitude and phase (Example 14.23)

The polar plot of these data is shown between $\omega = +0$ and $\omega = +\infty$ in Figure 14.18b.

(ii) *For bcd-segment:* In the usual way it can be seen that this segment maps onto a point at the origin.

(iii) *For de-segment:* This segment maps to the plot between $\omega = -\infty$ and $\omega = -0$ in Figure 14.18b, which is a mirror image of the mapping of the *ab*-segment.

(iv) *For efa-segment* $s \equiv \lim_{\rho \to 0} \rho e^{j\theta} \quad -90° > \theta > +90°$

Since this is a type 2 system, the Nyquist plots will close from $\omega = -0$ to $\omega = +0$ sweeping an angle of 2π, i.e. 360° in the clockwise direction (see *Closing Nyquist Plots* at page 364). Therefore, the circular curve of infinite radius from $\omega = -0$ to $\omega = +0$ maps the segment.

The complete Nyquist plot is shown in Figure 14.18b. From the plot we observe that the curve is CW throughout and that there is one encirclement of the critical point, i.e. $N = 1$. So, from the Nyquist criterion $N = Z - P$, we get

$$1 = Z - 1$$

$$\Rightarrow \qquad Z = 2 \qquad \text{(i)}$$

Equation (i) indicates the presence of two poles in the RHP of the CLTF. Therefore, the closed-loop system is unstable.

EXAMPLE 14.24. Construct the complete Nyquist plot for a unity feedback control system whose open-loop transfer function is

$$G(s)H(s) = \frac{K}{s(s^2 + 2s + 2)}$$

Find the maximum value of K for which the system is stable.

Solution This is a type 1 system with no OLTF pole in the RHP. The quadratic pole factorizes to $(s+1+\jmath 1)$ and $(s+1-\jmath 1)$. Therefore, there exists no pole on the Im-axis. So, the Nyquist path is as shown in Figure 14.9a.

Mapping: (i) *For ab-segment:* The frequency domain representation of the open-loop transfer function is

$$G(\jmath\omega)H(\jmath\omega) = \frac{K}{\jmath\omega(-\omega^2 + \jmath 2\omega + 2} \equiv \frac{K}{\omega\sqrt{(2-\omega^2)^2 + 4\omega^2}} \angle\left(-90° - \tan^{-1}\frac{2\omega}{2-\omega^2}\right) \qquad \text{(i)}$$

From Eq. (i) we can construct Table 14.18. The data are plotted in Figure 14.19 with proper labelling.

ω	M	ϕ
$+0$	∞	$-90°$
$+\infty$	0	$-270°$[7]

Table 14.18: Magnitude and phase (Example 14.24)

[7]The calculation is as follows:

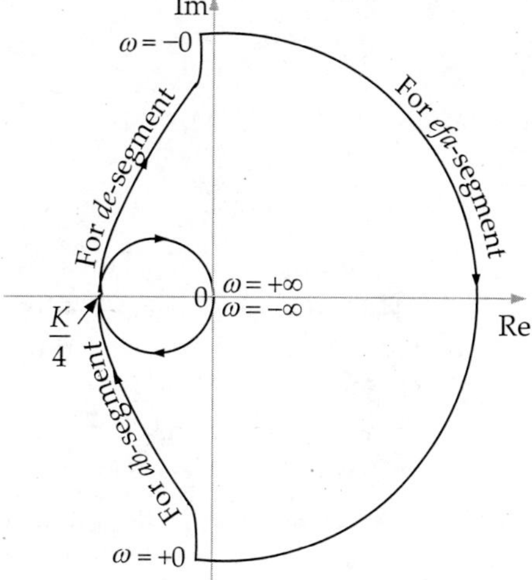

Figure 14.19: Nyquist plot (Example 14.24).

(ii) *For bcd-segment:* $s = \lim_{R\to\infty} Re^{\jmath\theta} \quad +90^\circ > \theta > -90^\circ$

Since this is a strictly proper transfer function, this segment maps onto a point at the origin as usual.

(iii) *For de-segment:* The mapping of this segment is a mirror image of that of the *ab*-segment. In Figure 14.19, it is the curve marked with $\omega = -\infty$ to $\omega = -0$.

(iv) *For efa-segment:* $s = \lim_{\rho\to 0} \rho e^{\jmath\theta} \quad -90^\circ > \theta > +90^\circ$

Since this is a type 1 system, the Nyquist plots will close from $\omega = -0$ to $\omega = +0$ sweeping an angle of π, i.e. 180° in the clockwise direction (see *Closing Nyquist Plots* at page 364). Therefore, the semicircular curve of infinite radius from $\omega = -0$ to $\omega = +0$ maps the segment.

To find where the curve intersects the Re-axis, we find the phase crossover frequency ω_p and calculate the magnitude M at that frequency as follows:

$$-90^\circ - \tan^{-1}\frac{2\omega_p}{2-\omega_p^2} = -180^\circ$$

$$\Rightarrow \qquad \frac{2\omega_p}{2-\omega_p^2} = \tan 90^\circ = \infty$$

$$\Rightarrow \qquad -\omega_p^2 = 0$$

$$\Rightarrow \qquad \omega_p = \sqrt{2}$$

Thus, $$M\Big|_{\omega_p=\sqrt{2}} = \frac{K}{\sqrt{2}\sqrt{8}} = \frac{K}{4} \qquad \text{[from Eq. (i)]}$$

The complete Nyquist plot is shown in Figure 14.19. For the stability of the system, this

$$-90^\circ - \tan^{-1}\frac{2}{\frac{2}{\omega}-\omega}\Big|_{\omega=\infty} = -90^\circ - \tan^{-1}\frac{2}{-\omega}\Big|_{\omega=\infty}$$
$$= -90^\circ - \pi + \tan^{-1}\frac{2}{\omega}\Big|_{\omega=\infty}$$
$$= -90^\circ - \pi = -270^\circ$$

intercept magnitude should be less than the magnitude of the critical point, i.e.

$$\frac{K}{4} < 1 \qquad \Rightarrow K < 4$$

EXAMPLE 14.25. (a) Draw the Nyquist plot for the system

$$G(s) = \frac{s+1}{s(s-1)}$$

by plotting G along the closed contour shown in Figure 14.20a. Note the indentation of this contour at the origin is to the left.

(b) What is the encirclement criterion for stability of this system with the type of contour chosen?

(c) Using the criterion, determine the range of the gain K of the proportional controller over which the closed-loop system will be stable.

Solution (a) This is a type 1 system with an OLTF pole in the RHP. Therefore, $P = 1$. In frequency domain, the open-loop transfer function can be written as

$$G(\jmath\omega)H(\jmath\omega) = \frac{\jmath\omega+1}{\jmath\omega(\jmath\omega-1)} = \frac{\sqrt{\omega^2+1}}{\omega\sqrt{\omega^2+1}}\angle[\tan^{-1}\omega - 90^\circ - (\pi - \tan^{-1}\omega)] \qquad \text{(i)}$$

Mapping: (i) *For ab-segment:* $s = +\jmath 0$ to $+\jmath\infty$, i.e. $\omega = +0$ to $+\infty$. We get the data of Table 14.19 from Eq. (i).

ω	M	ϕ
$+0$	∞	-270°
$+\infty$	0	-90°

Table 14.19: Magnitude and phase (Example 14.24)

The corresponding curve is shown in Figure 14.20b with proper labelling.

(ii) *For bcd-segment:* $s = \lim_{R\to\infty} Re^{\jmath\theta} \quad +90^\circ > \theta > -90^\circ$

Since this is a strictly proper transfer function , this segment maps onto a point at the origin as usual.

(iii) *For de-segment:* The mapping of this segment is a mirror image of that of the *ab*-segment. In Figure 14.20b, it is the curve marked with $\omega = -\infty$ to $\omega = -0$.

(iv) *For efa-segment:* $s = \lim_{\rho\to 0} \rho e^{\jmath\theta} \quad -90^\circ > \theta > +90^\circ$

Since this is a type 1 system, and that the Nyquist path includes the pole at the origin, the Nyquist plots will close from $\omega = -0$ to $\omega = +0$ sweeping an angle of π, i.e. 180° in the *counter-clockwise* direction (see *Closing Nyquist Plots* at page 364). Therefore, the semicircular curve of infinite radius from $\omega = -0$ to $\omega = +0$ maps the segment.

To find where the curve intersects the Re-axis, we find the phase crossover frequency ω_p and calculate the magnitude M at that frequency as follows:

$$2\tan^{-1}\omega_p - 270^\circ = -180^\circ$$

$$\Rightarrow \qquad \tan^{-1}\omega_p = 45^\circ$$

Therefore,

$$\omega_p = 1$$

So,

$$M\Big|_{\omega_p=\sqrt{2}} = \frac{1}{1} = 1 \qquad \text{[from Eq. (i)]}$$

The complete Nyquist plot is shown in Figure 14.20(b).

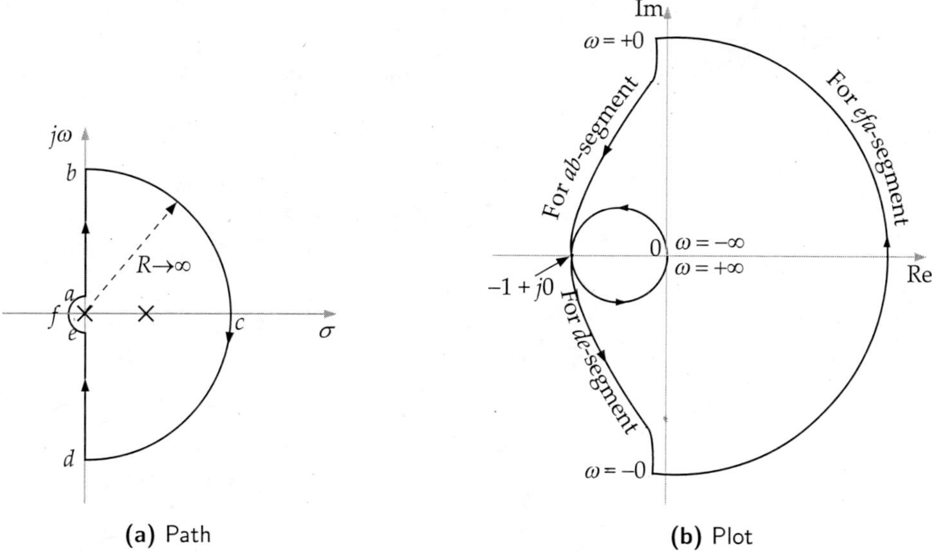

(a) Path (b) Plot

Figure 14.20: Nyquist diagrams (Example 14.25).

Note: The opposite indentation in the Nyquist path makes no difference while constructing the Nyquist plot.

(b) The Nyquist criterion for stability of this system with this type of contour chosen is the same as given by Eq. (14.8) stated in page 360.

(c) Since there are two CCW encirclements, $N = -2$ and that one open-loop transfer function pole is included on the Nyquist path, $P = 2$. So, from Eq. (14.8) we get

$$-2 = Z - 2 \qquad \text{or,} \qquad Z = 0$$

As can be seen from Figure 14.20b, the plot passes through the critical point. Therefore, the system as such is marginally stable. If a proportional controller of gain K were included, the condition for stability would be $K < 1$.

EXAMPLE 14.26. The open-loop transfer function of a unity feedback system is given by

$$G(s) = \frac{\pi e^{-0.25s}}{s}$$

In the $G(s)$-plane, the Nyquist plot of $G(s)$ passes through the real axis at the point

(A) $(-1.5, \jmath 0)$ (B) $(-0.5, \jmath 0)$ (C) $(-0.75, \jmath 0)$ (D) $(-1.25, \jmath 0)$

Solution

$$G(\jmath\omega) = \frac{\pi e^{-0.25\jmath\omega}}{\jmath\omega} = \frac{\pi(\cos 0.25\omega - \jmath \sin 0.25\omega)}{\jmath\omega} = \left|\frac{\pi}{\omega}\right| \angle(-90^\circ - \tan^{-1}\tan 0.25\omega)$$

$$= \left|\frac{\pi}{\omega}\right| \angle(-90^\circ - 0.25\omega)$$

Since when the Nyquist plot passes through the negative Re-axis, its phase becomes -180°, we have

$$-180^\circ = -90^\circ - 0.25\omega \qquad \Rightarrow \omega = 2\pi$$

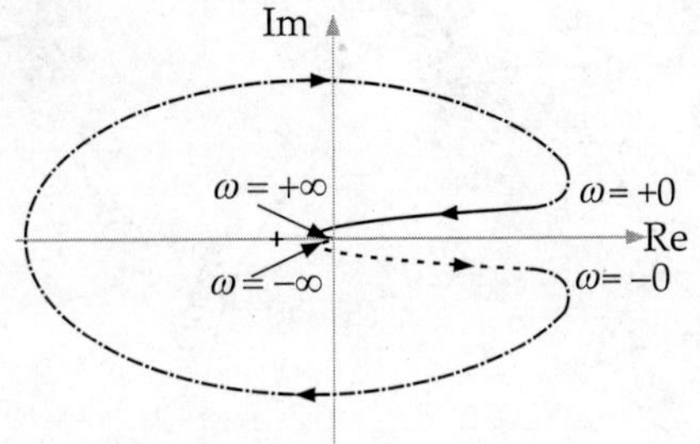

Figure 14.21: Nyquist plot (Example 14.27).

Table 14.20: M and ϕ (Example 14.27)

ω	$\|M\|$	ϕ
0	∞	0°
∞	0	180°
3	1/9	90°

So, the magnitude at this frequency is

$$\left|\frac{\pi}{\omega}\right| = \frac{\pi}{2\pi} = \frac{1}{2}$$

Therefore, the Nyquist plot passes through the negative Re-axis at $(-0.5, \jmath 0)$.
Ans. (B)

EXAMPLE 14.27. Loop transfer function of a feedback system is

$$G(s)H(s) = \frac{s+3}{s^2(s-3)}$$

Take the Nyquist contour in the clockwise direction. Then, the Nyquist plot of $G(s)$ encircles $(-1+\jmath 0)$

(A) Once in the clockwise direction
(B) Twice in the clockwise direction
(C) Once in the counter-clockwise direction
(D) Twice in the counter-clockwise direction

Solution This is a type 2 system having a pole at the RHP. From the given transfer function, we get in the frequency domain

$$|GH| = M = \frac{\sqrt{\omega^2+9}}{\omega^2\sqrt{\omega^2+9}} = \frac{1}{\omega^2} \tag{i}$$

$$\angle GH = \phi = \left[\tan^{-1}\frac{\omega}{3}\right] - \left[180^\circ + \left(\pi - \tan^{-1}\frac{\omega}{3}\right)\right] = 2\tan^{-1}\frac{\omega}{3} \tag{ii}$$

From Eqs. (i) and (ii), we construct Table 14.20. Having obtained these points, we draw the Nyquist plot. First we make the polar plot (continuous line) from $\omega = +0$ to $\omega = +\infty$. Then we draw its mirror image (dotted curve). Then, following the rules for closing Nyquist plots, we connect $\omega = -0$ with $\omega = +0$ (dash-dot curve). The complete plot is shown in Figure 14.21.

We observe that the plot encircles the critical point (shown by +) once in the clockwise direction.
Ans. (A)

EXAMPLE 14.28. Nyquist plot of the functions $G_1(s)$ and $G_2(s)$ are shown in the figure below.

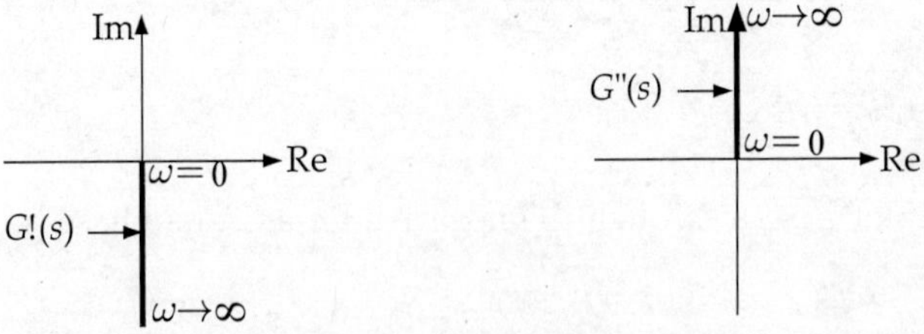

Nyquist plot of the product of $G_1(s)$ and $G_2(s)$ is

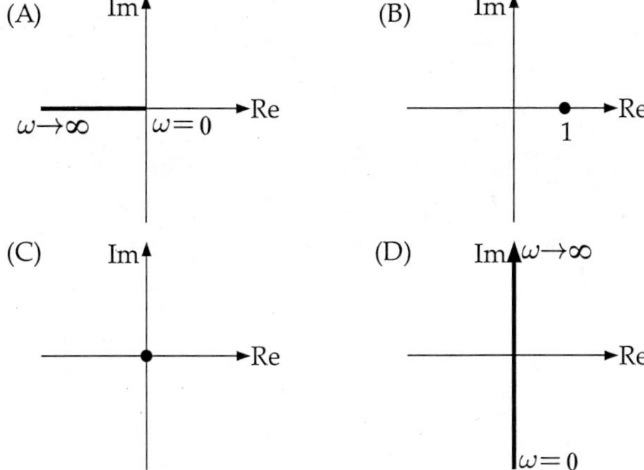

Solution The given diagrams indicate

$$G_1(s) = \frac{1}{s} \qquad G_2(s) = s$$

Therefore, $$G_1(s) \cdot G_2(s) = 1$$

Ans. (B)

Drawing Nyquist Plot from Polar Plot

A Nyquist plot can be drawn from a given polar plot and the stability of the system can be predicted. Example 14.29 shows how to do it.

EXAMPLE 14.29. The polar plot of an open-loop stable system is shown in Figure 14.22a. The closed-loop system is

(A) always stable
(B) marginally stable
(C) unstable with one pole on the RH s-plane
(D) unstable with two poles on the RH s-plane

Solution Since it is a polar plot, it is not possible to predict how many poles are in the RH s-plane. So, we have to convert it to a Nyquist plot. We do that by plotting a mirror image of the given plot (shown in broken lines) and then connecting $\omega = -0$ point to the $\omega = +0$ point as shown in Figure 14.22b.

Here, $P = 0$ [since the system is stated to be stable], $N = 2$. Therefore,

$$Z = N + P = 2 + 0 = 2$$

which means, in the closed-loop transfer function there are two poles in the RH of s-plane. Hence the system is unstable.

Ans. (D)

Nyquist Plot for System with Transportation Lag

The following example will make the procedure clear.

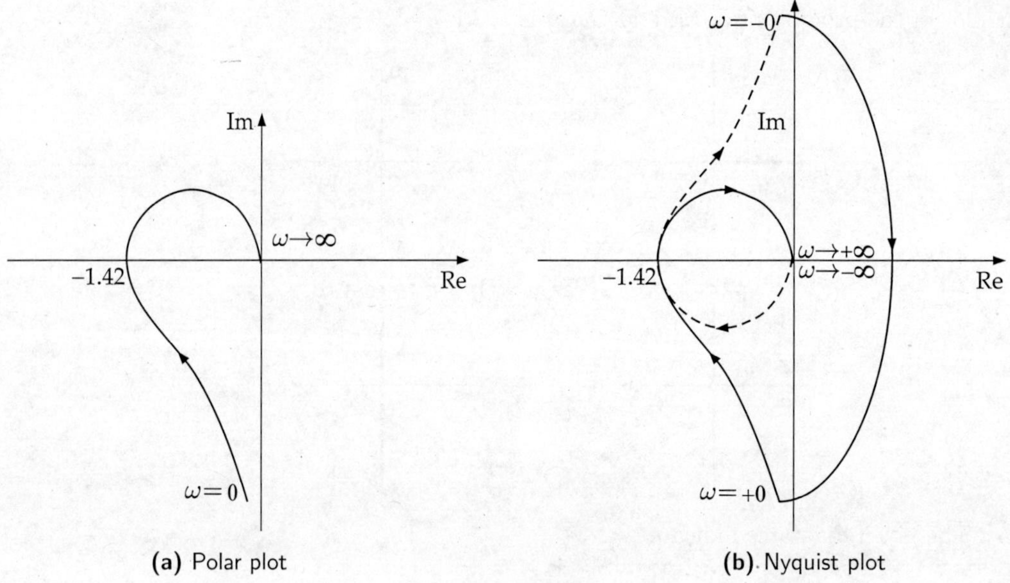

(a) Polar plot

(b) Nyquist plot

Figure 14.22: Example 14.29.

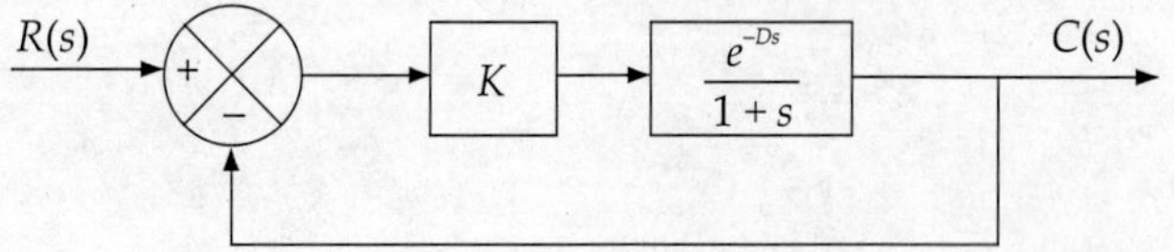

Figure 14.23: Feedback system (Example 14.30).

EXAMPLE 14.30. For the unity feedback system shown Figure 14.23, the phase crossover occurs at a frequency of $\omega = 2.448$ rad/s.

(a) Determine the transport delay D.

(b) Find the limiting value of the proportional controller gain K for the stable operation and sketch the corresponding Nyquist plot.

Solution (a) The open-loop transfer function is given by

$$G(s)H(s) = \frac{Ke^{-Ds}}{1+s}$$

$$\Rightarrow \quad G(j\omega)H(j\omega) = \frac{Ke^{-jD\omega}}{1+j\omega} \equiv \frac{K}{\sqrt{1+\omega^2}}\angle(-D\omega - \tan^{-1}\omega) \qquad \text{(i)}$$

$$\left[\because\ e^{-jD\omega} = \cos D\omega - j\sin D\omega \equiv |\sqrt{\sin^2 D\omega + \cos^2 D\omega}|\angle \tan^{-1}\frac{-\sin D\omega}{\cos D\omega} = |1|\angle - D\omega\right]$$

Corresponding to the phase crossover frequency ω_p, we get from Eq. (i) the relation

$$-D\omega_p - \tan^{-1}\omega_p = -180° = -\pi$$

$$\Rightarrow \qquad 2.448D + \tan^{-1} 2.448 = \pi \qquad [\text{Given, } \omega_p = 2.448\,\text{rad/s}]$$

Therefore,
$$D = \frac{\pi - 1.183}{2.448} = 0.8 \text{ s}$$

(b) The open-loop transfer function in the frequency domain is given by

$$G(\jmath\omega)H(\jmath\omega) = \frac{K}{\sqrt{1+\omega^2}}\angle(-0.8\omega - \tan^{-1}\omega) \tag{ii}$$

From Eq. (ii), we get the data of Table 14.21.

Table 14.21: Magnitude and phase (Example 14.30)

ω	M	ϕ
$+0$	K	$0°$
$+\infty$	0	$-\infty$

From Eq. (ii), we also observe that as ω varies between 0 and $+\infty$, the phase angle ϕ varies continuously between 0° and $-\infty$. This variation indicates that the Nyquist plot will assume the form of a spiral as shown in Figure 14.24a. The points of intersection of the plot at the

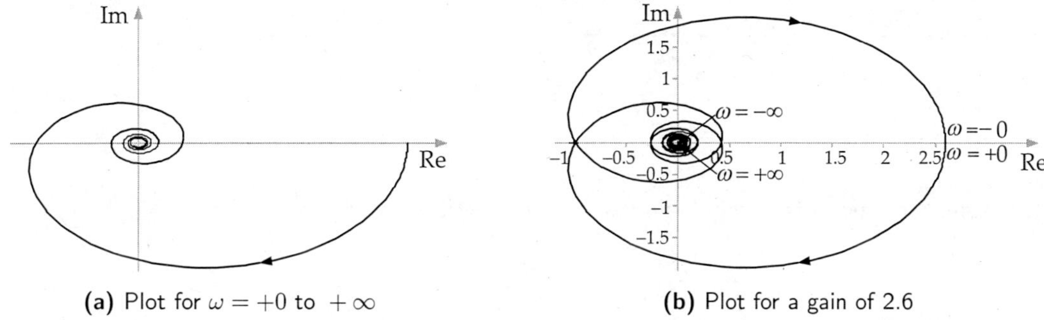

(a) Plot for $\omega = +0$ to $+\infty$ **(b)** Plot for a gain of 2.6

Figure 14.24: Nyquist plots (Example 14.30).

negative Re-axis are determined by the relation

$$-0.8\omega - \tan^{-1}\omega = -(2n+1)\pi \qquad n = 0,\ 1,\ 2, \cdots$$

The first intercept for $\omega > 0$ will take place when $n = 0$. This condition yields

$$-0.8\omega - \tan^{-1}\omega = -\pi \tag{iii}$$

To solve this equation, we use the approximate relation

$$\tan^{-1}x \cong \frac{\pi}{2} - \frac{1}{x} \qquad \text{[valid for } x > 1\text{]}$$

Thus, from Eq. (iii),

$$\frac{\pi}{2} - \frac{1}{\omega} + 0.8\omega = \pi$$

$$\Rightarrow \qquad 0.8\omega^2 - \frac{\pi}{2}\omega - 1 = 0$$

Therefore, $$\omega = 2.4696,\ -0.5061$$

The negative frequency being unrealistic, we get, from the magnitude condition for stability

$$\left.\frac{K}{\sqrt{1+\omega^2}}\right|_{\omega=2.4696} < 1$$

or, $$\frac{K}{2.6644} < 1$$

or, $$K < 2.6644$$

Therefore, the limiting value of the proportional controller gain K is approximately 2.66. The Nyquist plot for a gain of 2.6 is displayed in Figure 14.24b.

14.9 CLOSED-LOOP FREQUENCY RESPONSE FROM OPEN-LOOP TRANSFER FUNCTION

Consider a closed-loop transfer function of a unity feedback system in the frequency domain, given by

$$\frac{C(\jmath\omega)}{R(\jmath\omega)} = \frac{G(\jmath\omega)}{1+G(\jmath\omega)} \tag{14.12}$$

$G(\jmath\omega)$ being a complex quantity, it can be written in the form of

$$G(\jmath\omega) \equiv x + \jmath y$$

Then, Eq. (14.12) can be written as

$$\frac{C(\jmath\omega)}{R(\jmath\omega)} \equiv \frac{x+\jmath y}{1+x+\jmath y} \tag{14.13}$$

M-Circles

The magnitude part of the complex quantity given by Eq. (14.13) is

$$M = \left|\frac{x+\jmath y}{1+x+\jmath y}\right| = \frac{\sqrt{x^2+y^2}}{\sqrt{(1+x)^2+y^2}}$$

or,

$$M\left(\sqrt{(1+x)^2+y^2}\right) = \sqrt{x^2+y^2} \tag{14.14}$$

Squaring both sides of Eq. (14.14) and then regrouping the terms, we get

$$M^2(1+x^2+2x+y^2) = x^2+y^2$$

or,

$$(1-M^2)x^2 + (1-M^2)y^2 - 2M^2x = M^2 \tag{14.15}$$

Dividing both sides of Eq. (14.15) by $(1-M^2)$, we obtain

$$x^2+y^2-\frac{2M^2}{1-M^2}x = \frac{M^2}{1-M^2} \tag{14.16}$$

Adding $\left(\frac{M^2}{1-M^2}\right)^2$ to both sides of Eq. (14.16) and then rearranging, we have

$$x^2+y^2-\frac{2M^2}{1-M^2}x+\left(\frac{M^2}{1-M^2}\right)^2 = \frac{M^2}{1-M^2}+\left(\frac{M^2}{1-M^2}\right)^2$$

or,

$$\left(x-\frac{M^2}{1-M^2}\right)^2+y^2 = \frac{M^2}{(1-M^2)^2} \tag{14.17}$$

Equation (14.17) represents a circle having its centre at

$$\left(\frac{M^2}{1-M^2},\ 0\right)$$

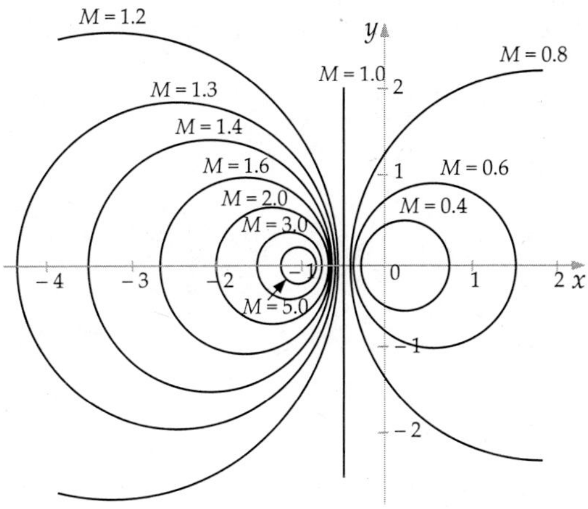

Figure 14.25: M-circles.

Table 14.22: Centres and radii

M	Centre $x = \dfrac{M^2}{1-M^2}$, $y = 0$ (Only x values given)	Radius $r = \left\|\dfrac{M}{1-M^2}\right\|$
0.4	0.190	0.476
0.6	0.563	0.938
0.8	1.778	2.222
1.2	−3.273	2.727
1.3	−2.449	1.884
1.4	−2.042	1.458
1.6	−1.641	1.026
2.0	−1.333	0.067
3.0	−1.125	0.375
5.0	−1.042	0.208

and radius

$$r = \left|\frac{M}{1-M^2}\right|$$

However, Eq. (14.17) is invalid for $M = 1$. But substituting $M = 1$ in Eq. (14.15), we get $x = -\frac{1}{2}$ which represents a straight line parallel to the Im-axis passing through the point $(-\frac{1}{2},\ \jmath 0)$ in the $G(\jmath\omega)$-plane.

For other values of M, we get a family of circles in the $G(\jmath\omega)$-plane. These are called *M-circles* or *M-loci*. To figure out the characteristics of these circles we have calculated the location of centres and values of radii of a few loci and presented them in Table 14.22 and shown the circles in Figure 14.25.

It can be observed from Table 14.22 as well as Figure 14.25 that

If $M > 1$:	The centres of circles lie to the left of $(-1 + \jmath 0)$. The radii of circles become smaller as M grows bigger.
If $M < 1$:	The centres of circles lie to the right of $(-1 + \jmath 0)$. The radii of circles become bigger as M grows bigger.
If $M \to \infty$:	The circle degenerates to a point at $(-1 + \jmath 0)$. This is in agreement with the fact that when the polar plot passes through $(-1 + \jmath 0)$, the system becomes marginally stable (or marginally unstable) and the resonant peak M_r becomes infinite.

The following closed-loop data may be derived from the polar plot of $G(\jmath\omega)$ along with the M-circle grid superposed on it:

Closed-loop data	*How to obtain*
The value of the magnitude M (i.e. $\|G(\jmath\omega)\|$) at the frequency denoted by the $\|G(\jmath\omega)\|$-curve	From the intersections of $G(\jmath\omega)$-plot and constant M-circles
The value of the resonant peak M_r and resonant frequency at the tangent point of the $G(\jmath\omega)$-curve	From the M-circle with the smallest radius which is tangent to the $G(\jmath\omega)$-curve
The value of the bandwidth as determined from the frequency denoted on the $G(\jmath\omega)$-curve	From the intersection of the polar plot with the $M = 0.707$ circle

Figure 14.26a shows polar plot of $G(\jmath\omega)$ for a control system for different values of K. We can determine

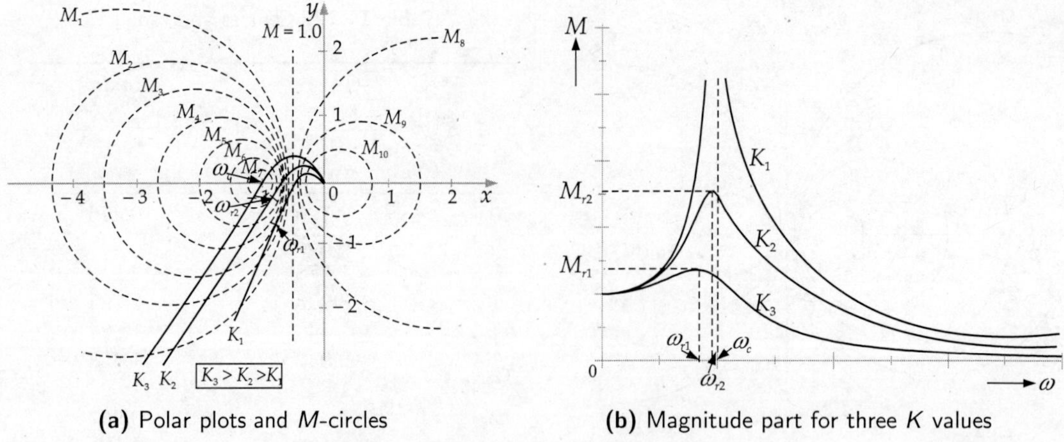

(a) Polar plots and M-circles

(b) Magnitude part for three K values

Figure 14.26: Polar and magnitude plot.

- The corresponding M_r and ω_r values by noting the points where each polar plot touches M-circles.
- The M values at frequencies denoted on the $G(\jmath\omega)$-curves from the intersection points of each polar plot with different M-loci.

In this way we can determine the magnitude part of the closed-loop frequency response for each value of K as shown in Figure 14.26b.

N-Circles

For the phase part of the closed-loop transfer function, we get from Eq. (14.13) the relation

$$\phi = \arg\left|\frac{x+\jmath y}{1+x+\jmath y}\right| = \tan^{-1}\frac{y}{x} - \tan^{-1}\frac{y}{1+x} = \tan^{-1}\frac{\dfrac{y}{x}-\dfrac{y}{1+x}}{1+\left(\dfrac{y}{x}\right)\left(\dfrac{y}{1+x}\right)}$$

$$= \tan^{-1}\frac{y}{x^2+x+y^2} \tag{14.18}$$

Taking tangents of both sides of Eq. (14.18), we get

$$\tan\phi = \frac{y}{x^2+x+y^2} \tag{14.19}$$

Let us put $\tan\phi = N$. Then, from Eq. (14.19)

$$x^2+x+y^2-\frac{y}{N}=0 \tag{14.20}$$

Adding $\dfrac{1}{4}+\dfrac{1}{4N^2}$ to both sides of Eq. (14.20), we obtain

$$x^2+x+y^2-\frac{y}{N}+\frac{1}{4}+\frac{1}{4N^2}=\frac{1}{4}+\frac{1}{4N^2}$$

or,

$$\left(x+\frac{1}{2}\right)^2+\left(y-\frac{1}{2N}\right)^2=\sqrt{\frac{N^2+1}{4N^2}} \tag{14.21}$$

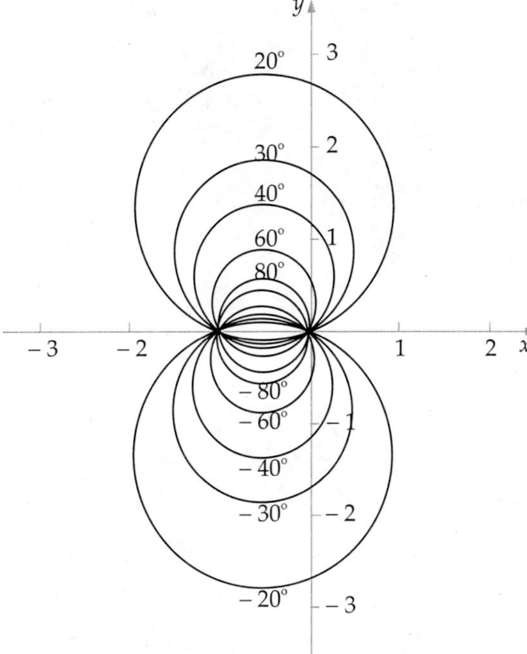

Figure 14.27: Constant N-circles.

Table 14.23: Data for constant N-circles

ϕ (degree)	$N = \tan\phi$	*Centre* (y-values only)	*Radius*
20.	0.364	1.374	1.462
30.	0.577	0.867	1.001
40.	0.839	0.596	0.778
60.	1.732	0.289	0.577
80.	5.671	0.087	0.501
−20.	−0.364	−1.374	1.462
−30.	−0.577	−0.867	1.001
−40.	−0.839	−0.596	0.778
−60.	−1.732	−0.289	0.577
−80.	−5.671	−0.087	0.501

Equation (14.21) represents a circle having its

$$\text{centre at } \left(-\frac{1}{2}, \frac{1}{2N}\right) \text{ and radius } r = \left|\sqrt{\frac{N^2+1}{4N^2}}\right|$$

For different values of N, they produce a family of circles which are called *N-circles* or *N-loci*. Values of centres and radii for different values of ϕ and hence different values of N are presented in Table 14.23. Corresponding N-circles are shown in Figure 14.27.

Equation (14.21) is satisfied by $(x = 0,\ y = 0)$ and $(x = 1,\ y = 0)$ irrespective of the value of N. Hence all circles pass through the origin and the $(-1 + \jmath 0)$ point.

As discussed in the case of M-circles, we can determine the phase part of the closed-loop frequency response from the polar plot of $G(\jmath\omega)$ with constant N-circles. However, M-circles which provide information of M_r, ω_r and Bw, are generally of more use in studies of relative stability of systems.

Nichols Chart

The Nichols[8] Chart is a tool for the designer to read off closed-loop gain and phase directly from a plot of open-loop logarithmic gain and phase, parametrized by frequency. It has proved to be arguably the most useful closed-loop system design tool in the control field.

In polar plots, magnitude and phase are plotted in the G-plane. The M-circle and N-circle grid can be superimposed on it to gather information on closed-loop frequency response of a system. In design problems, it is frequently needed to alter loop gain or add series and feedback compensators[9] to the system. In such cases, the polar plot has to be recalculated. The Bode

[8]Nathaniel B Nichols (1914–97) was an American engineer and researcher.

[9]See Chapter 15 at page 402.

plot, in which the plot is shifted upwards or downwards depending upon the change in gain, has a distinct advantage in this respect. But Bode plots as such do not allow calculation of closed-loop frequency response from plots of open-loop transfer functions. Nichols combined the advantages of the Bode plot and polar plot by devising a plot of magnitude in dB vs. phase in degrees (also known as *gain-phase plot*) and superimposing the circles of M (in dB) and N (in degree) on it.

Suppose we have the open-loop transfer function of a system given by

$$G(s)H(s) = \frac{K}{(s+a)(s+b)(s+c)}$$

Its magnitude vs. phase plot looks like what is shown in Figure 14.28a.

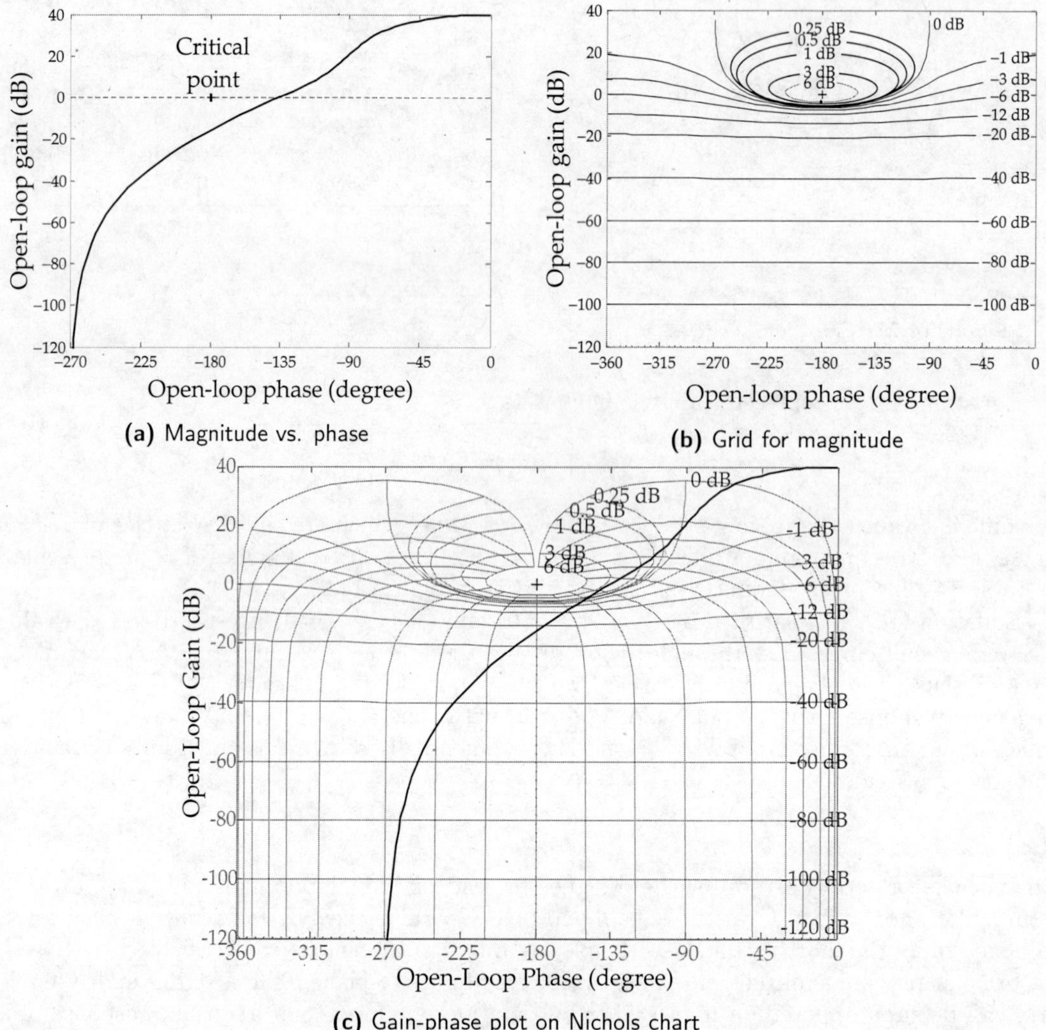

(a) Magnitude vs. phase

(b) Grid for magnitude

(c) Gain-phase plot on Nichols chart

Figure 14.28: Nichols chart components.

The Nichols chart for magnitude is shown in Figure 14.28b. These are M-circles plotted in dBs. A similar chart is prepared for the phase in degrees. This grid, when superimposed on the gain-phase plot (see Figure 14.28c), generates data for the closed-loop frequency response, much in the same way M- and N-circles do for the polar plot.

REVIEW QUESTIONS

14.1 Choose the correct answer:

(i) In the plot shown in Figure Question 1(i), the numbers of encirclement of A and B are
(A) 2, 2 (B) 2, 1 (C) 1, 2 (D) 1, 1

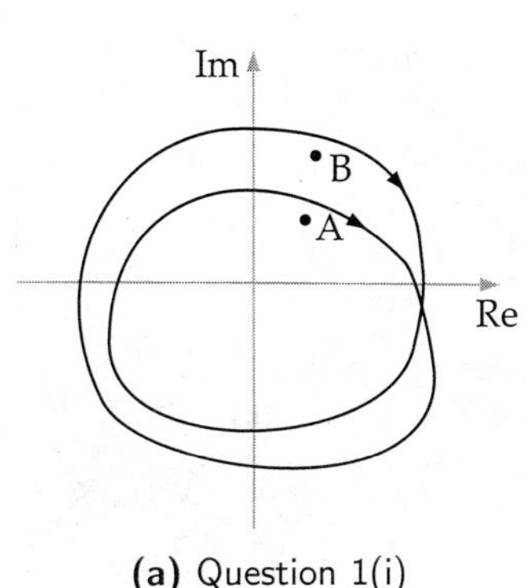

(a) Question 1(i)

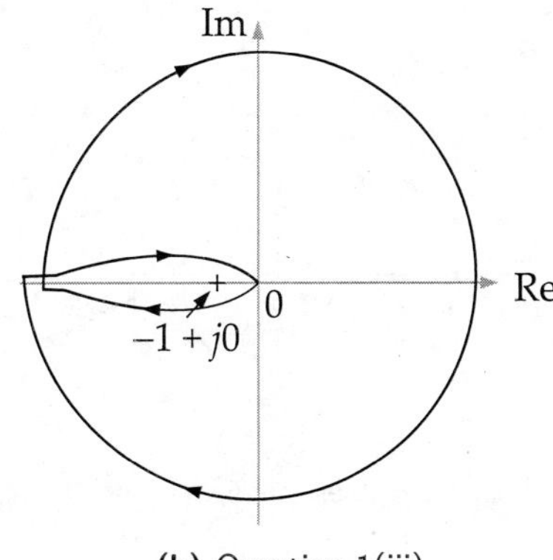

(b) Question 1(iii)

(ii) The Nichols chart helps us determine
(A) open-loop frequency response
(B) closed-loop frequency response
(C) open-loop response in time domain
(D) closed-loop response in time domain

(iii) For the Nyquist plot given in Figure Question 1(iii) the system is
(A) stable (B) unstable
(C) marginally stable (D) none of the above

(iv) With the help of the Nyquist criterion, the stability of the closed-loop system
(A) cannot be determined
(B) can be determined by studying the behaviour of the plot around $(1 + \jmath 0)$
(C) can be determined by studying the behaviour of the plot around $(1 + \jmath 0)$
(D) can be determined by studying whether the plot crosses the Im-axis

(v) The loop transfer function of a negative feedback system is

$$G(s)H(s) = \frac{1}{s(s-2)}$$

The Nyquist plot for the above system
(A) does not encircle $(-1 + \jmath 0)$ point
(B) encircles $(-1 + \jmath 0)$ point once in the clockwise direction
(C) encircles $(-1 + \jmath 0)$ point twice in the counter-clockwise direction
(D) encircles $(-1 + \jmath 0)$ point once in the counter-clockwise direction

(vi) Taking N as positive for clockwise encirclement, otherwise negative, the number of encirclements N of $(-1, \jmath 0)$ in the Nyquist plot of $G(s)H(s) = 3/(s-1)$ is

(A) -1 (B) 1 (C) -2 (D) 2

(vii) The Nyquist contour in the s-plane encloses the entire right half plane and a small neighbourhood in the left half plane as shown in Figure Question 1(vii). The number of encirclements of the point $(-1+\jmath 0)$ by the Nyquist plot of $G(s)H(s)$, corresponding to the Nyquist contour, is denoted by N. Then N equals

(A) 3 (B) 2 (C) 0 (D) 1

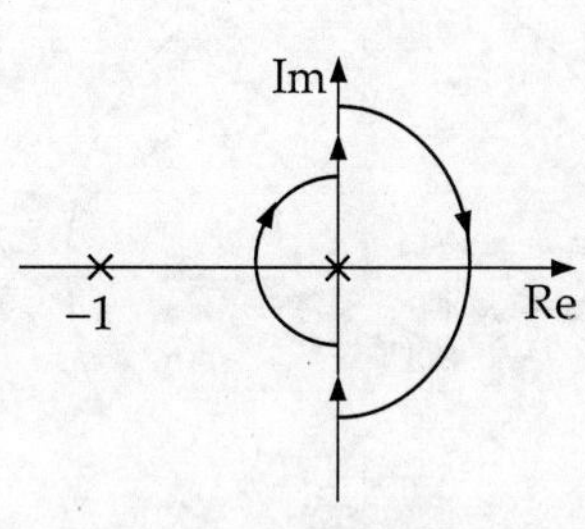

(a) Question 1(vii)

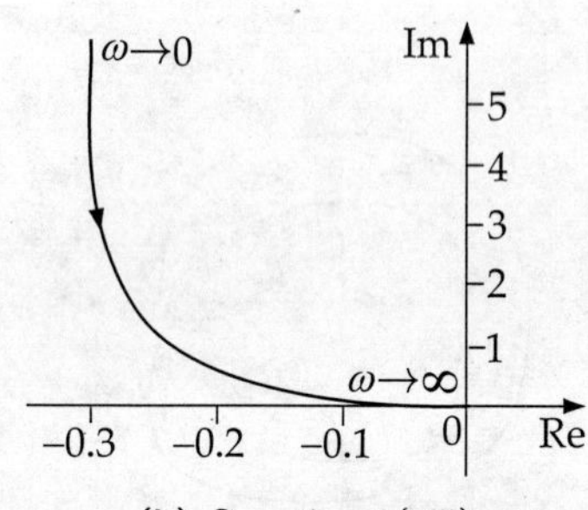

(b) Question 1(viii)

(viii) A transfer function $G(s)$ with the degree of its numerator polynomial zero and the degree of its denominator polynomial two has a Nyquist plot shown in Figure Question 1(viii). The transfer function represents

(A) a stable, type-0 system (B) a stable, type-1 system
(C) an unstable, type-0 system (D) an unstable, type-1 system

14.2 Find the Nyquist plot for

$$G(s)H(s) = \frac{K}{s^2(1+s\tau)}$$

where $K,\ \tau > 0$.

14.3 For the transfer function

$$G(\jmath\omega) = \frac{K}{(1+\jmath 2\omega)(1+\jmath 4\omega)(1+\jmath 6\omega)}$$

draw the Nyquist plot.

14.4 Examine the closed-loop stability of the system whose open-loop transfer function is

$$G(s)H(s) = \frac{50}{s(s+1)(s+2)}$$

using the Nyquist criterion.

14.5 State and explain the principle of argument.

14.6 (a) State the principle of argument and its extension to Nyquist criterion.

(b) Draw the Nyquist plot for the following system

$$G(s)H(s) = \frac{5}{s(1-s)}$$

Discuss the closed-loop stability of the system.

14.7 (a) Define the Nyquist stability criterion. State the principle from which the criterion originated.

(b) The open-loop transfer function of a unity feedback control system is given by

$$G(s) = \frac{K}{(1+s)(1+2s)(1+3s)}$$

(i) Sketch the Nyquist plot.

(ii) From the plot find the value of K above which the system becomes unstable. Also find the critical frequency ω_c.

14.8 (a) State the Nyquist stability criterion.

(b) Consider a system with loop gain

$$G(s)H(s) = \frac{K(s+2)}{s^2(s+1)}$$

Obtain the Nyquist plot and comment on its open-loop stability.

14.9 Consider a system with an open-loop transfer function

$$G(s)H(s) = \frac{2s+1}{s^2(s+1)(s+2)}$$

(a) Draw the Nyquist contour and the corresponding mapping of $G(s)H(s)$.

(b) Comment on the closed-loop stability of the system.

14.10 Determine, through Nyquist's criterion, the stability of the system having open-loop transfer function as

$$G(s)H(s) = \frac{1+4s}{s^2(1+s)(1+2s)}$$

14.11 Examine by the Nyquist method the stability of a unity feedback system having the open-loop transfer function of

$$G(s) = \frac{2(s+3)}{s(s-1)}$$

14.12 Write short notes on:

(a) Nyquist stability criterion
(b) Principle of arguments
(c) Constant M-circles
(d) Nyquist plot
(e) Nichols chart

Answers to Selected Questions

14.1. (i) (B) (ii) (A) (iii) (B) (iv) (C) (v) (A) (vi) (A) (vii) (D) (viii) (D)

14.6. (b) Stable

14.7. (b) $K < 10$, $\omega_g = 1$ rad/s

14.8. (b) Unstable

14.9. (b) Unstable

14.10. Plot similar to Example 14.13. $\omega_p = 1/2\sqrt{2}$ rad/s, Gm $= -38.6$ dB, Unstable.

14.11. Plot similar to Example 14.17. $\omega_p = \sqrt{3}$ rad/s, Stable.

15

Compensators

15.1 WHAT THEY ARE

The performance of a control system is judged from its following characteristics:

1. Speed of response
2. Steady-state error
3. Damping factor
4. Peak overshoot
5. Stability.

In case the performance of a control system falls short of expectations, an additional device may be added to the system to make good, or compensate, for the shortfall of performance. Such devices are called *compensators*.

15.2 TYPES AND EFFECTS

Depending upon the location of the compensating network, compensation is divided into the following three categories:

1. Series or cascade compensation
2. Feedback or parallel compensation
3. Load or series-parallel compensation

Their configurations and functions are elaborated in Table 15.1.

Table 15.1: Comparison of compensations

Type	*Definition*	*Effects*
Series or Cascade Compensation	If the compensating network is included in feed-forward path, it is called *series* or *cascade compensation*. The corresponding block diagram is shown in Figure 15.1a.	The addition of a compensating network in the forward path basically adjusts the gain of a system. As a result, 1. The speed of response increases. 2. Peak overshoot decreases. 3. The stability of the system may worsen.

continued on next page

Continued from previous page

Type	*Definition*	*Effects*
Feedback or Parallel Compensation	When the compensating network is included in the feedback path [Figure 15.1b], it is called *feedback* or *parallel compensation.*	A compensating network in the feedback path may 1. Lower the steady-state error 2. Lower the speed of response 3. Improve the stability of the system
Load or Series-parallel Compensation	Some situations may call for a combination of series and feedback compensation. Such type of compensation is called *load* or *series-parallel compensation.* The corresponding block diagram is shown in Figure 15.1c.	

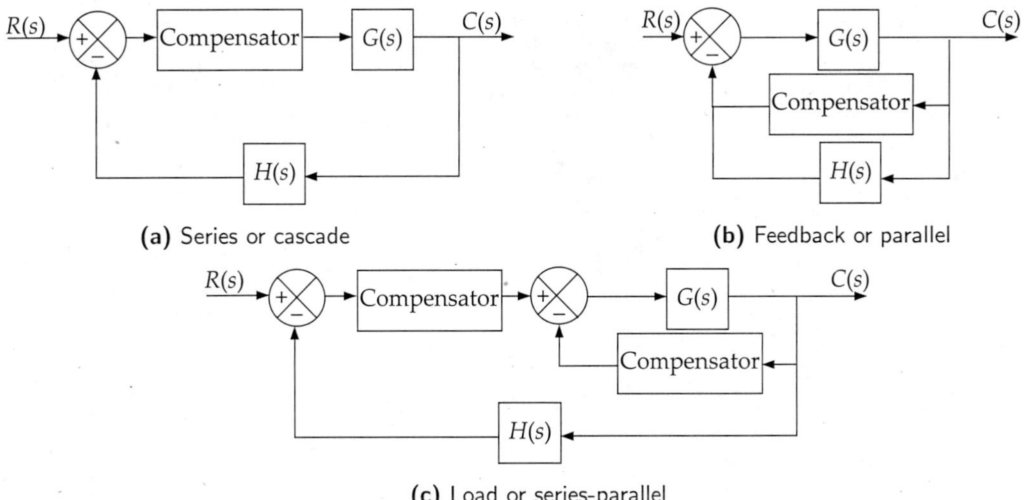

Figure 15.1: Compensating networks.

Compensating networks introduce additional poles and/or zeros in the transfer function of the system. The effects of such additions are shown in the following table.

Addition of poles	*Addition of zeros*
1. Pulls the root locus towards the RHP. As a result, the stability is lowered.	1. Pulls the root locus towards LHP. As a result, the stability improves.
2. Increases settling time.	2. Lowers settling time.
3. Lowers steady-state error, i.e., improves accuracy.	3. Increases steady-state error, i.e., lowers accuracy.

EXAMPLE 15.1. If the compensated system shown in the following figure has a phase margin of 60° at the crossover frequency of 1 rad/s, then the value of the gain K is

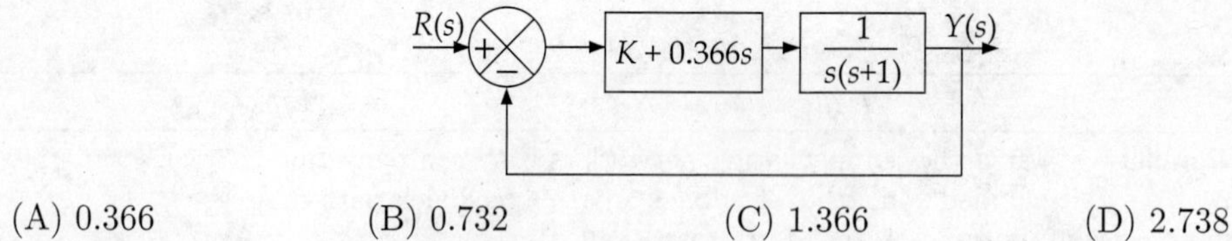

(A) 0.366 (B) 0.732 (C) 1.366 (D) 2.738

Solution Open-loop transfer function of the system is given by

$$G(s)H(s) = \frac{K + 0.366s}{s(s+1)}$$

Therefore,
$$G(\jmath\omega)H(\jmath\omega) = \frac{K + 0.366\jmath\omega}{\jmath\omega(\jmath\omega+1)}$$

$$\angle G(\jmath\omega)H(\jmath\omega) = \tan^{-1}\left(\frac{0.366\omega}{K}\right) - 90^\circ - \tan^{-1}\omega$$

Since,
$$\text{Pm} = 180^\circ + \angle G(\jmath\omega_g)H(\jmath\omega_g)$$

$$60^\circ = 180^\circ + \tan^{-1}\left(\frac{0.366\omega_g}{K}\right) - 90^\circ - \tan^{-1}\omega_g$$

$$= 90^\circ + \tan^{-1}\left(\frac{0.366}{K}\right) - \tan^{-1}(1) \qquad [\because \omega_g = 1\,\text{rad/s}]$$

or,
$$15^\circ = \tan^{-1}\left(\frac{0.366}{K}\right)$$

or,
$$\frac{0.366}{K} = \tan 15^\circ \tag{15.1}$$

or,
$$K = \frac{0.366}{0.267} = 1.366$$

Ans. (C)

15.3 TYPES OF COMPENSATION NETWORKS

With this background on compensation and its effects, let us study the three types of compensating networks (or compensators) that are used in electrical systems. They are as follows:

1. Lead compensation network
2. Lag compensation network
3. Lead-lag compensation network

15.4 LEAD COMPENSATION NETWORK

A simple circuit is shown in Figure 15.2.

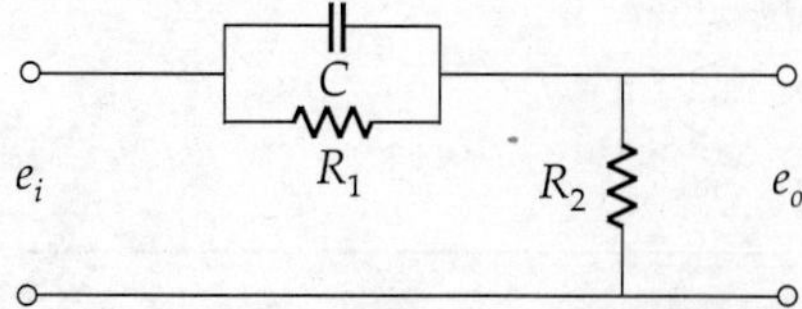

Figure 15.2: Lead compensator.

Considering its Laplace transform equivalent, the transfer function of the circuit can be worked out as follows:

$$I(s) = \frac{E_i(s)}{\left(R_1 \| \dfrac{1}{sC}\right) + R_2} = \frac{E_i(s)}{\dfrac{R_1}{sR_1C+1} + R_2} = \frac{E_i(s)(sR_1C+1)}{sR_1R_2C + R_1 + R_2} \tag{15.2}$$

Also,

$$I(s) = \frac{E_o(s)}{R_2} \tag{15.3}$$

Combining Eqs. (15.2) and (15.3), we get

$$G(s) \equiv \frac{E_o(s)}{E_i(s)} = \frac{R_2(sR_1C+1)}{sR_1R_2C + R_1 + R_2} = \frac{R_1R_2C\left(s + \dfrac{1}{R_1C}\right)}{R_1R_2C\left(s + \dfrac{R_1+R_2}{R_1R_2C}\right)} \equiv \frac{s + \dfrac{1}{T}}{s + \dfrac{1}{\alpha T}} \tag{15.4}$$

where $\quad T = R_1C \quad$ and $\quad \alpha = \dfrac{R_2}{R_1+R_2}$

We may note here that T is the time-constant and α, which is < 1, is an attenuation constant. In frequency domain, the transfer function can be written as

$$G(\jmath\omega) = \frac{\jmath\omega + (1/T)}{\jmath\omega + (1/\alpha T)} = \frac{\alpha(1 + \jmath\omega T)}{1 + \jmath\alpha\omega T} \qquad \alpha < 1 \tag{15.5}$$

From Eqs. (15.4) and (15.5), we observe the following:

- The lead network has a zero at $s = -1/T$ and a pole at $s = -1/\alpha T$. Since $\alpha < 1$, the *zero is dominant*, i.e. located nearer to the imaginary axis [see Figure 15.3a].
- The numerator introduces a phase angle of $\tan^{-1}\omega T$ and the denominator, $-\tan^{-1}\alpha\omega T$. Since $\alpha < 1$, the net phase angle is positive and hence *the output leads the input.*

Other observations are:

1. The polar plot of the network transfer function can be constructed [see Figure 15.3b] from

$$G(\jmath\omega) = \left|\frac{\alpha\sqrt{1+\omega^2T^2}}{\sqrt{1+\alpha^2\omega^2T^2}}\right| \angle\left(\tan^{-1}\omega T - \tan^{-1}\alpha\omega T\right) \tag{15.6}$$

   Equation (15.6) helps us generate the data given in Table 15.2.

Table 15.2: Magnitude and phase data

ω	$\lvert G(\jmath\omega)\rvert$	$\angle G(\jmath\omega)$
0	α	0°
$1/T$ $(\alpha = 0.9)$	0.95	3°
∞	1	0°

2. The phase angle ϕ is given by

$$\phi = \tan^{-1}\omega T - \tan^{-1}\alpha\omega T \tag{15.7}$$

   If ω_m is the frequency for which $\phi = \phi_{\max}$, then

$$\left.\frac{d\phi}{d\omega}\right|_{\omega=\omega_m} = 0$$

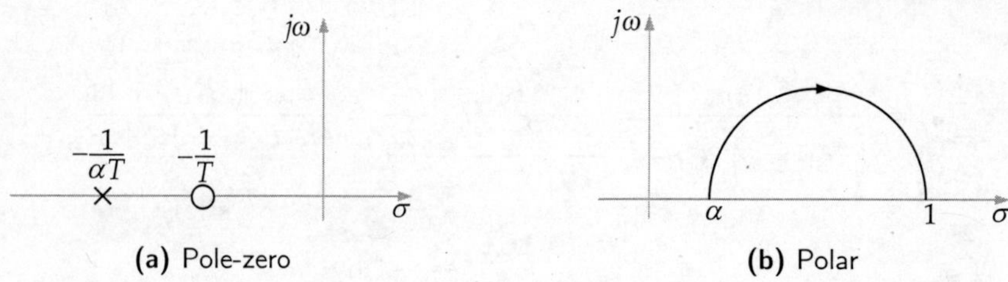

Figure 15.3: Lead network: pole-zero plot and polar plot of the transfer function.

which gives

$$\Rightarrow \frac{T}{1+\omega_m^2 T^2} - \frac{\alpha T}{1+\alpha^2\omega_m^2 T^2} = 0 \qquad \Rightarrow \frac{T}{1+\omega_m^2 T^2} = \frac{\alpha T}{1+\alpha^2\omega_m^2 T^2}$$

$$\Rightarrow \omega_m^2(\alpha T^2 - \alpha^2 T^2) = 1-\alpha \qquad \Rightarrow \omega_m = \frac{1}{T\sqrt{\alpha}} = \left(\frac{1}{\sqrt{T}}\right)\left(\frac{1}{\sqrt{\alpha T}}\right) \tag{15.8}$$

Again, from Eq. (15.7)

$$\phi = \tan^{-1}\frac{\omega T(1-\alpha)}{1+\alpha\omega^2 T^2} \qquad \Rightarrow \tan\phi = \frac{\omega T(1-\alpha)}{1+\alpha\omega^2 T^2} \tag{15.9}$$

Plugging in the value of ω_m in Eq. (15.9), we get

$$\tan\phi_{\max} = \frac{\frac{1}{\sqrt{\alpha}}(1-\alpha)}{1+1} = \frac{1-\alpha}{2\sqrt{\alpha}}$$

Therefore,

$$\sin\phi_{\max} = \frac{1-\alpha}{1+\alpha}$$

$\Rightarrow$

$$\alpha = \frac{1-\sin\phi_{\max}}{1+\sin\phi_{\max}} \tag{15.10}$$

From Eq. (15.8) we note that the frequency ω_m corresponding to the maximum phase angle is the geometric mean of the two corner frequencies. Also from Eqs. (15.6) and (15.8), at ω_m

$$|G(\jmath\omega)|_{\omega=\omega_m} = \frac{\alpha\sqrt{1+\omega_m^2 T^2}}{\sqrt{1+\alpha^2\omega_m^2 T^2}} = \frac{\alpha\sqrt{1+(1/\alpha)}}{\sqrt{1+\alpha}} = \sqrt{\alpha}$$

Therefore,

$$20\log|G(\jmath\omega)|_{\omega=\omega_m} = 20\log(\sqrt{\alpha}) = -10\log\left(\frac{1}{\alpha}\right)\ \text{dB} \tag{15.11}$$

3. From the above discussions, we can draw the Bode plot for the network transfer function as in Figure 15.4.
4. At $\omega = 0$ (dc), the gain is α which is < 1. So, the circuit acts like an attenuator.

Effects of Incorporating Lead Network

1. Bandwidth (hence speed) improves because the network moves the system to a higher gain crossover frequency. This also improves the relative stability of the system because the phase margin increases as a result of increasing ω_g.
2. Damping increases because of addition of a dominant zero and a pole. As a result, peak overshoot and settling time decrease.

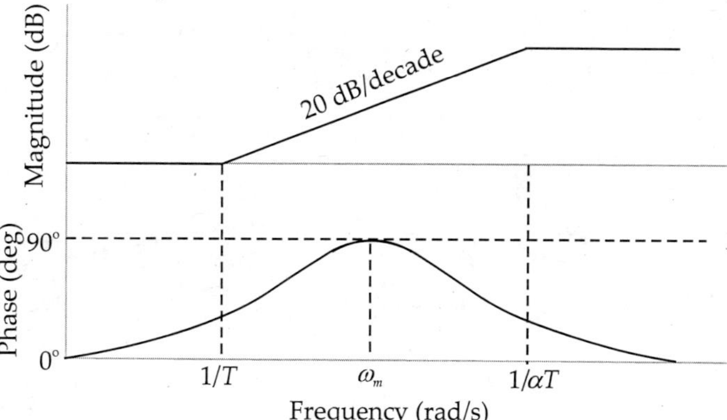

Figure 15.4: Bode plot for lead network transfer function.

Designing Lead Network

1. Draw the Bode plot of the system open-loop transfer function with gain constant K adjusted according to the prescribed steady-state error, if necessary.
2. Determine the gain and phase margins of the uncompensated system from the Bode plot. If ϕ is the Pm of the uncompensated system, ϕ_r is the required Pm and τ is the tolerance introduced, then
$$\phi_m = \phi_r - \phi + \tau$$
Since each lead network normally provides a maximum phase shift of 60°, if $\phi_m > 60°$, two identical networks, each contributing $\phi_m/2$, are required.
3. The value of the α-parameter is calculated using Eq. (15.10).
4. The corresponding required gain is calculated using Eq. (15.11).
5. Once the required gain is figured out, the corresponding ω_m is found from the Bode plot.
6. Now that α and ω_m are known, the corner frequencies for pole and zero namely, $1/\alpha T$ and $1/T$, are found using Eq. (15.8).
7. The transfer function of the lead network is written using corner frequencies for the pole and zero.

The following examples will make the procedure clear.

EXAMPLE 15.2. Design a phase lead compensator for a unity feedback control system having open-loop transfer function
$$G(s) = \frac{50}{s(1+0.4s)}$$
so as to have a phase margin of 30°.

Solution The frequency domain transfer function is
$$G(j\omega) = \frac{50}{j\omega\left(1+\dfrac{j\omega}{2.5}\right)}$$
The corner frequencies are 0 and 2.5 for the pole factors.

The phase angle is
$$\phi = \angle G(j\omega) = -90° - \tan^{-1}(0.4\omega)$$
The above relation gives the following data:

ω	0.1	0.5	1.0	5.0	10.0	11.0	15.0	50.0	100.0
ϕ	−92.3°	−101.3°	−111.8°	−153.4°	−166.0°	−167.0°	−170.5°	−177.1°	−178.5

The Bode plots (Figure 15.5) give $\omega_g = 11$ rad/s and $\omega_p = \infty$. Phase margin = $180° - 167° = 13°$.

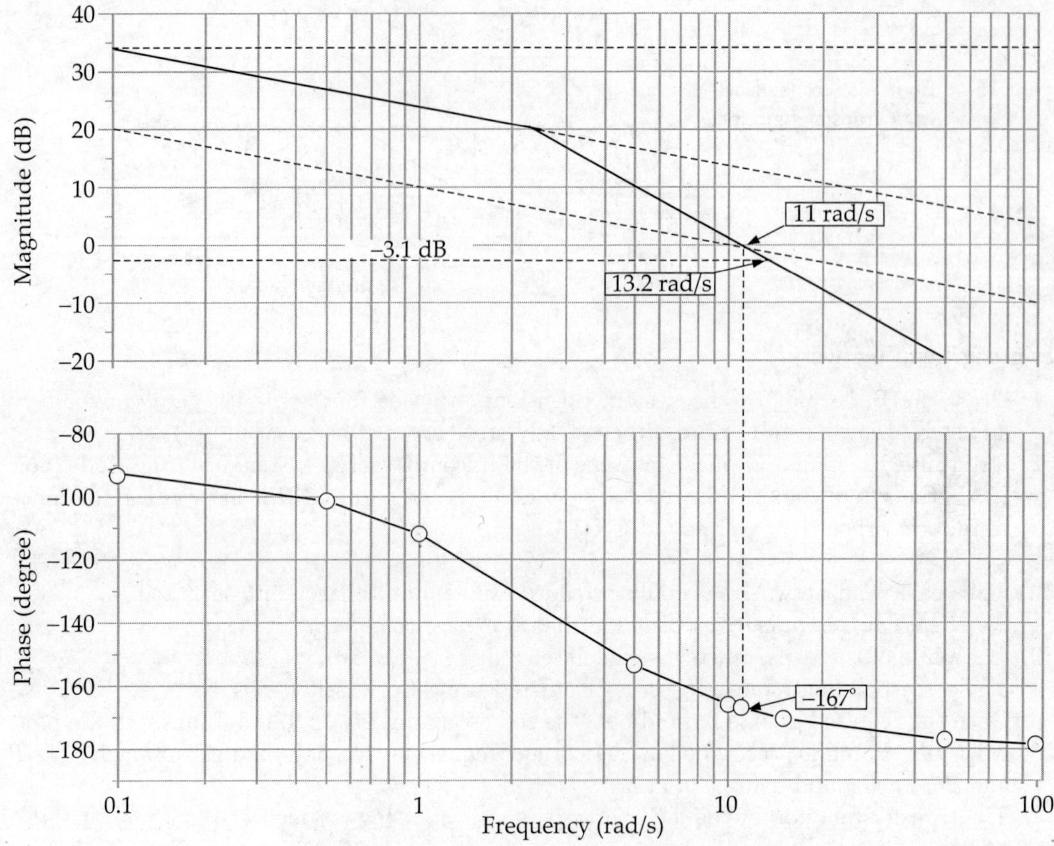

Figure 15.5: Bode plots (Example 15.2).

To increase Pm to 30°, ω_g has to be moved to a higher value. The required phase lead is $\phi_m = (30° - 13°) + 3°$ (tolerance) $= 20°$. The corresponding value of the α-parameter of the lead network is given by [see Eq. (15.10)]

$$\alpha = \frac{1 - \sin\phi_{\max}}{1 + \sin\phi_{\max}} = \frac{1 - \sin 20°}{1 + \sin 20°} = 0.49$$

Now, from Eq. (15.11),

$$20 \log |G(\jmath\omega)|_{\omega=\omega_m} = -10 \log\left(\frac{1}{\alpha}\right) \text{ dB} = -3.1 \text{ dB}$$

From the Bode plot, −3.1 dB gain occurs at $\omega_m = 13.2$ rad/s. Therefore, for the lead network, upper and lower corner frequencies become [see Eq. (15.8)]

$$\frac{1}{\alpha T} = \frac{\omega_m\sqrt{\alpha}}{\alpha} = \frac{\omega_m}{\sqrt{\alpha}} = \frac{13.2}{\sqrt{0.49}} = 18.86 \text{ rad/s}$$

$$\frac{1}{T} = \omega_m\sqrt{\alpha} = 13.2\sqrt{0.49} = 9.24 \text{ rad/s}$$

Therefore, the required open-loop transfer function for the lead network is

$$G'(s) = \frac{s + \dfrac{1}{T}}{s + \dfrac{1}{\alpha T}} = \frac{s + 9.24}{s + 18.86} = 0.49 \times \frac{1 + \dfrac{s}{9.24}}{1 + \dfrac{s}{18.86}}$$

EXAMPLE 15.3. Design a cascade compensation for a unity feedback system whose transfer function is

$$G(s) = \frac{K}{s(1 + 0.1s)(1 + 0.001s)}$$

It should fulfil the following specifications:

(i) Phase margin $\geq 45°$

(ii) Velocity error constant $K_v = 1000\ \text{s}^{-1}$

Solution We know that

$$K_v = \lim_{s \to 0}\{s \cdot G(s)H(s)\} \qquad \text{[see Eq. (6.53) at page 146]}$$

$$= \lim_{s \to 0}\left\{s \cdot \frac{K}{s(1 + 0.1s)(1 + 0.001s)}\right\}$$

Therefore, $\qquad K_v = K$

Since $K_v = 1000$ (given), we have

$$G(s) = \frac{1000}{s(1 + 0.1s)(1 + 0.001s)}$$

The frequency domain transfer function is

$$G(j\omega) = \frac{1000}{j\omega(1 + 0.1j\omega)(1 + 0.001j\omega)}$$

Corner frequencies are 0, 10 and 1000 for the pole factors. The phase angle is

$$\phi = \angle G(j\omega) = -90° - \tan^{-1}(0.1\omega) - \tan^{-1}(0.001\omega)$$

The above relation generates the following data:

ω	1.0	5.0	10.0	50.0	100.0	200.0	300.0	500.0	1000.0
ϕ	−95.77	−116.85	−135.57	−171.55	−180	−188.45	−194.79	−205.42	−224.43

The Bode plots (Figure 15.6) give $\omega_g = 100$ rad/s and $\omega_p = 100$ rad/s. Phase margin $= 180° - 180° = 0°$. To increase Pm to 45°, ω_g has to be moved to a higher value. The required phase lead is

$$\phi_{\max} = 45° + 5°\ (\text{tolerance}) = 50°$$

The corresponding value of the α-parameter of the lead network is given by [see Eq. (15.10)]

$$\alpha = \frac{1 - \sin\phi_{\max}}{1 + \sin\phi_{\max}} = \frac{1 - \sin 50°}{1 + \sin 50°} = 0.132$$

Now, from Eq. (15.11)

$$20\log|G(j\omega)|_{\omega=\omega_m} = -10\log\left(\frac{1}{\alpha}\right)\ \text{dB} = -8.79\ \text{dB}$$

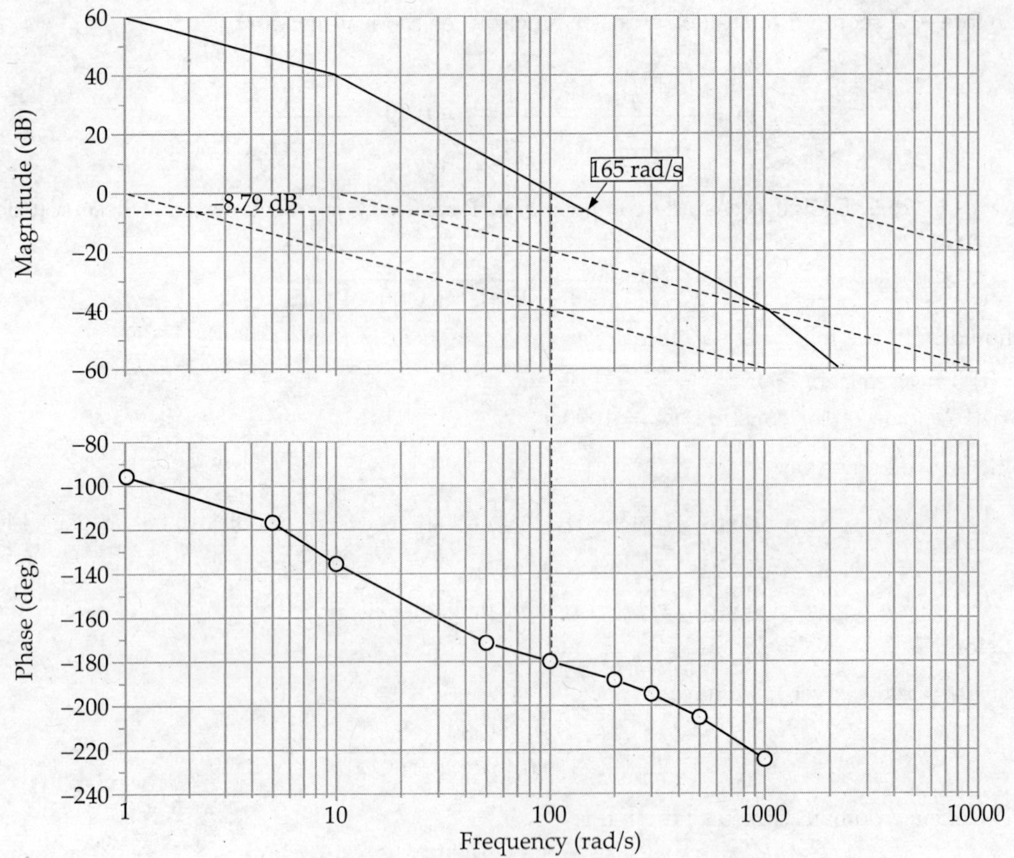

Figure 15.6: Bode plot (Example 15.3).

From the Bode plot, −8.79 dB gain occurs at $\omega_m = 165$ rad/s. Therefore, for the lead network, upper and lower corner frequencies become [see Eq. (15.8)]

$$\frac{1}{\alpha T} = \frac{\omega_m\sqrt{\alpha}}{\alpha} = \frac{\omega_m}{\sqrt{\alpha}} = \frac{165}{\sqrt{0.132}} = 454.147 \text{ rad/s}$$

$$\frac{1}{T} = \omega_m\sqrt{\alpha} = 165(\sqrt{0.132}) = 59.947 \text{ rad/s}$$

Therefore, the required open-loop transfer function for the lead network is

$$G'(s) = \frac{s+\dfrac{1}{T}}{s+\dfrac{1}{\alpha T}} = \frac{s+59.947}{s+454.147} = 0.13 \times \frac{1+\dfrac{s}{59.9}}{1+\dfrac{s}{454.1}}$$

EXAMPLE 15.4. The transfer function of a phase lead compensator is given by

$$D(s) = \frac{3\left[s+(1/3T)\right]}{\left[s+(1/T)\right]}$$

The frequency (in rad/s), at which $\angle D(\jmath\omega)$ is maximum, is

(A) $\sqrt{3T^2}$ (B) $\sqrt{3/T^2}$ (C) $\sqrt{3T}$ (D) $\sqrt{1/3T^2}$

Solution From the given transfer function we get the locations of pole and zero as $s = -(1/T)$ and $s = -(1/3T)$, respectively. Their magnitudes are the corner frequencies of the Bode plot. Now, we know that the frequency ω_m corresponding to the maximum phase angle is the geometric mean of the two corner frequencies [see Eq. (15.8)]. Therefore,

$$\omega_m = \sqrt{\left(\frac{1}{3T}\right)\left(\frac{1}{T}\right)} = \sqrt{\frac{1}{3T^2}}$$

Ans. (D)

EXAMPLE 15.5. The transfer function of a compensator is given by

$$G_c(s) = \frac{s+a}{s+b}$$

(i) $G_c(s)$ is a lead compensator if
(A) $a = 1,\ b = 2$ (B) $a = 3,\ b = 2$ (C) $a = -3,\ b = -1$ (D) $a = 3,\ b = 1$

(ii) The phase of the above lead compensator is maximum at
(A) $\sqrt{2}$ rad/s (B) $\sqrt{3}$ rad/s (C) $\sqrt{6}$ rad/s (D) $1/\sqrt{3}$ rad/s

Solution

(i) For a lead compensator, the zero should be dominant, which implies that $a < b$, and both zero and pole should be in the LHP.
Ans. (A)

(ii) $\omega_{\max} = \sqrt{ab} = \sqrt{2}$ rad/s.
Ans. (A)

EXAMPLE 15.6. A lead compensator used for a closed-loop controller has the following transfer function

$$\frac{K[1+(s/a)]}{[1+(s/b)])}$$

For such a compensator, the condition will be
(A) $a < b$ (B) $b < a$ (C) $a > Kb$ (D) $a < Kb$

Solution The transfer function can be written as

$$\frac{Kb}{a} \cdot \frac{a+s}{b+s} \equiv \frac{Kb}{a} \cdot \frac{a+j\omega}{b+j\omega}$$

Therefore, ϕ the phase of the transfer function is

$$\angle\phi = \tan^{-1}\frac{\omega}{a} - \tan^{-1}\frac{\omega}{b}$$

For a lead compensator, $\angle\phi > 0$. Hence $a < b$.
Ans. (A)

EXAMPLE 15.7. The Bode magnitude plot of a network is shown Figure 15.7. The maximum phase angle ϕ_m and the corresponding gain G_m respectively, are

(A) $-30°$ and 1.73 dB (B) $-30°$ and 4.77 dB
(C) $+30°$ and 4.77 dB (D) $+30°$ and 1.73 dB

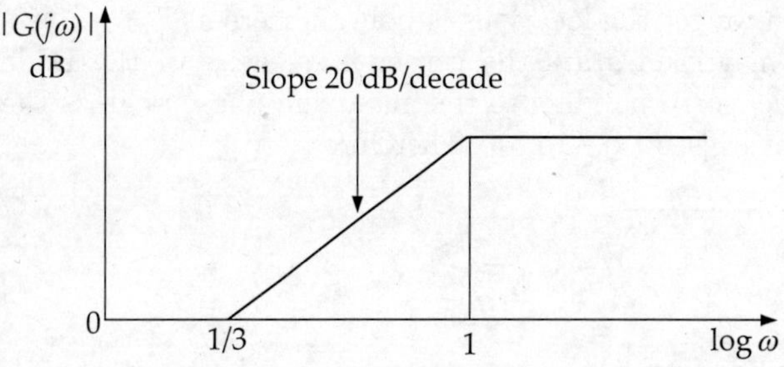

Figure 15.7: Bode magnitude plot (Example 15.7.)

Solution We observe that the slope increases at 1/3 at the rate of 20 dB/decade and decreases at 1 at the same rate. This implies the existence of a zero at 1/3 and a pole at 1. So, the frequency response function can be written as

$$G(\jmath\omega) = \frac{\frac{1}{3} + \jmath\omega}{1 + \jmath\omega} \tag{i}$$

or,

$$= \frac{\frac{1}{3}(1 + 3\jmath\omega)}{1 + \jmath\omega} \tag{ii}$$

For the lead compensator

$$G(\jmath\omega) = \frac{1 + \jmath\omega T}{1 + \alpha\jmath\omega T} \qquad \text{where } \alpha < 1 \tag{iii}$$

Maximum phase angle, ϕ_m is

$$\sin^{-1}\left(\frac{1-\alpha}{1+\alpha}\right) \quad \text{at frequency} \quad \omega_m = \frac{1}{T\sqrt{\alpha}} \text{ rad/s}$$

Comparing Eqs. (ii) and (iii), we get

$$T = 3 \qquad \alpha T = 1 \quad \Rightarrow \alpha = \frac{1}{3}$$

Therefore, maximum phase angle and frequency are

$$\phi_m = \sin^{-1}\left(\frac{1-\frac{1}{3}}{1+\frac{1}{3}}\right) = \sin^{-1}\left(\frac{1}{2}\right) = 30^\circ$$

$$\omega_m = \frac{1}{T\sqrt{\alpha}} = \frac{1}{\sqrt{3}} \tag{iv}$$

From Eq. (i), we get

$$|G(\jmath\omega)| = \left|\frac{\sqrt{(1/3)^2 + \omega^2}}{\sqrt{1^2 + \omega^2}}\right| \tag{v}$$

Substituting the value of ω_m obtained from Eq. (iv) in Eq. (v), we get

$$G_m = 20\log\left(\sqrt{\left(\frac{1}{\sqrt{3}}\right)^2 + \left(\frac{1}{3}\right)^2}\right) - 20\log\sqrt{\left(1^2 + \left(\frac{1}{\sqrt{3}}\right)^2\right)} \approx 4.77\,\text{dB}$$

Ans. (C)

EXAMPLE 15.8. Maximum phase-lead of the compensator

$$D(s) = \frac{0.5s + 1}{0.05s + 1}$$

is

(A) 52° at 4 rad/s (B) 52° at 10 rad/s
(C) 55° at 12 rad/s (D) None of the answers is correct

Solution We know that for the phase-lead compensator

$$D(s) = \frac{s + (1/T)}{s + (1/\alpha T)} \quad \text{(i)}$$

ϕ_m and ω_m are given by

$$\phi_m = \sin^{-1}\frac{1-\alpha}{1+\alpha}$$

$$\omega_m = \frac{1}{T\sqrt{\alpha}}$$

Comparing the given $D(s)$ with Eq. (i) we get, $T = 0.5$ and $\alpha = 0.1$. So,

$$\phi_m = \sin^{-1}\frac{0.9}{1.1} = 54.9° \approx 55°$$

$$\omega_m = \frac{1}{0.5\sqrt{0.1}} = 6.32\,\text{rad/s}$$

Although ϕ_m tallies with that given in (C), ω_m does not.
Ans. (D)

15.5 LAG COMPENSATION NETWORK

A simple lag compensation circuit is shown in Figure 15.8.

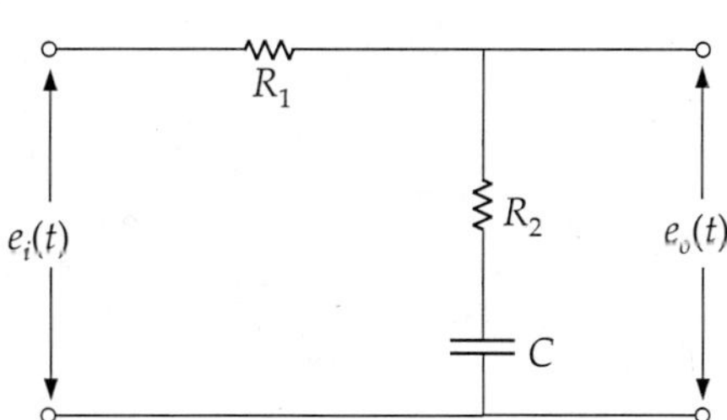

Figure 15.8: Lag compensator circuit.

Applying KVL to the Laplace transformed circuit, we obtain

$$E_i(s) = \left(R_1 + R_2 + \frac{1}{sC}\right) I(s) \quad (15.12)$$

$$E_o(s) = \left(R_2 + \frac{1}{sC}\right) I(s) \quad (15.13)$$

From Eqs. (15.12) and (15.13), we get, for the transfer function of the lag compensation network, the equation

$$G(s) = \frac{E_o(s)}{E_i(s)} = \frac{R_2 + \dfrac{1}{sC}}{R_1 + R_2 + \dfrac{1}{sC}} = \frac{R_1}{R_1 + R_2} \cdot \frac{s + \dfrac{1}{R_2 C}}{s + \dfrac{1}{(R_1 + R_2)C}} \equiv \frac{1}{\beta} \cdot \frac{s + \dfrac{1}{T}}{s + \dfrac{1}{\beta T}} \qquad (15.14)$$

where

$$\beta = \frac{R_1 + R_2}{R_2} \quad \text{which is always } > 1, \qquad T = R_2 C = \text{time constant} \qquad (15.15)$$

The frequency domain transfer function is, therefore,

$$G(\jmath\omega) = \frac{1}{\beta} \cdot \frac{\jmath\omega + \dfrac{1}{T}}{\jmath\omega + \dfrac{1}{\beta T}} = \frac{\jmath\omega T + 1}{\jmath\omega\beta T + 1} \qquad (15.16)$$

From Eqs. (15.13) and (15.15) we observe the following:

- The lag network transfer function has a zero at $s = -1/T$ and a pole at $-1/\beta T$. Since $\beta > 1$, *the pole is dominant* because it is located nearer to the Im-axis [see Figure 15.9a].
- The phase angles introduced by the numerator and denominator are $\tan^{-1}\omega T$ and $\tan^{-1}\beta\omega T$ respectively. Since $\beta > 1$, the latter angle is bigger and therefore, *the net phase angle is negative.* Hence the name *lag network.*

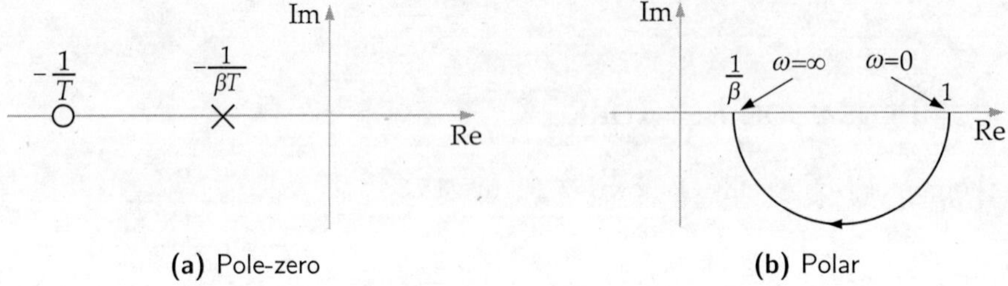

(a) Pole-zero **(b)** Polar

Figure 15.9: Plots of the lag network transfer function.

We note also,

1. The polar plot of the lag network transfer function can be constructed from

$$G(\jmath\omega) = \frac{\jmath\omega T + 1}{\jmath\omega\beta T + 1} \equiv \frac{\sqrt{\omega^2 T^2 + 1}}{\sqrt{\omega^2\beta^2 T^2 + 1}} \angle \left(\tan^{-1}\omega T - \tan^{-1}\omega\beta T\right) \qquad (15.17)$$

   Equation (15.17) yields the data of Table 15.3. The consequent polar plot is shown in Figure 15.9b.

2. The phase angle ϕ is given by

$$\phi = \tan^{-1}\omega T - \tan^{-1}\omega\beta T = \tan^{-1}\frac{\omega T(1 - \beta)}{1 + \beta\omega^2 T^2}$$

Table 15.3: Magnitude and phase

ω	$\lvert G(\jmath\omega)\rvert$	$\angle G(\jmath\omega)$
0	1	0°
$1/T$ ($\beta = 1.1$)	−0.64	−3°
∞	$1/\beta$	0°

If ω_m is the frequency for which $\phi = \phi_{\text{max}}$, then

$$\left.\frac{d\phi}{d\omega}\right|_{\omega=\omega_m} = 0$$

$$\Rightarrow \quad \frac{T}{1+\omega_m^2 T^2} - \frac{\beta T}{1+\omega_m^2\beta^2 T^2} = 0$$

$$\Rightarrow \quad \frac{T}{1+\omega_m^2 T^2} = \frac{\beta T}{1+\omega_m^2\beta^2 T^2}$$

$$\Rightarrow \quad \omega_m^2(\beta T^2 - \beta^2 T^2) = 1-\beta$$

$$\Rightarrow \quad \omega_m = \frac{1}{T\sqrt{\beta}} = \sqrt{\left(\frac{1}{T}\right)\left(\frac{1}{\beta T}\right)} \tag{15.18}$$

We note that Eqs. (15.18) and (15.8) are similar. Therefore, we can write the following equations as well:

$$\tan\phi_{\text{max}} = \frac{1-\beta}{2\sqrt{\beta}} \quad \Rightarrow \sin\phi_{\text{max}} = \frac{1-\beta}{1+\beta} \quad \Rightarrow \beta = \frac{1-\sin\phi_{\text{max}}}{1+\sin\phi_{\text{max}}}$$

From Eq. (15.18) we note that the frequency ω_m, corresponding to the maximum phase angle, is the geometric mean of the two corner frequencies. Also, from Eq. (15.17) at ω_m,

$$\left|G(\jmath\omega)\right|_{\omega=\omega_m} = \frac{\sqrt{\omega^2T^2+1}}{\sqrt{\omega^2\beta^2T^2+1}} = \frac{\sqrt{(1/\beta)+1}}{\beta+1} = \frac{1}{\sqrt{\beta}}$$

Therefore, $$20\log\left|G(\jmath\omega)\right|_{\omega=\omega_m} = 20\log\left(\frac{1}{\sqrt{\beta}}\right) = 10\log\left(\frac{1}{\beta}\right) \text{ dB}$$

3. From the above discussion, Bode plots for the lag network can be drawn as those shown in Figure 15.10.
4. For $\omega = 0$ (dc), the circuit gain is unity, while at high frequency, the gain is $1/\beta$. We know, $\beta > 1$. Its value is generally 10 or higher.

Effects of Incorporating Lag Compensator

The effects of incorporating lag compensators are:

1. The network allows low frequencies but attenuates high frequencies. As a result, it improves the steady-state performance.
2. The gain crossover frequency is lowered. Which is why the bandwidth (i.e. the speed of response) decreases, increasing the peak overshoot and settling time.

Designing Lag Network

To design a lag network, the following steps may be followed:

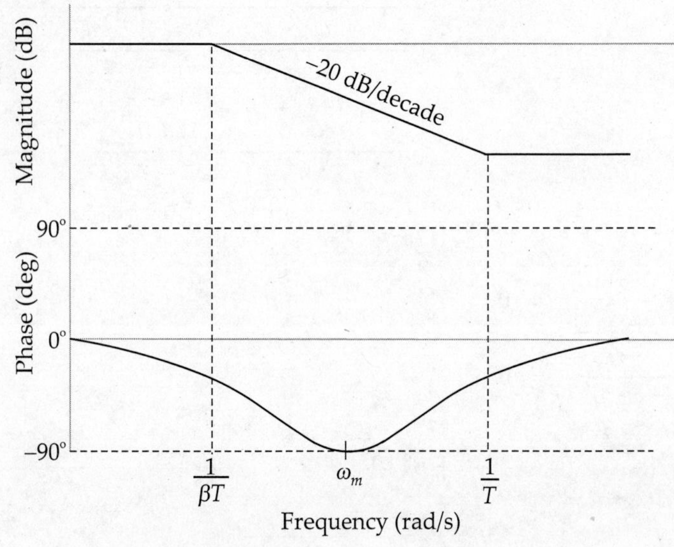

Figure 15.10: Bode plots for the lag network.

1. Draw the Bode plot of the system open-loop transfer function with gain constant K adjusted according to the prescribed steady-state error, if necessary.
2. Determine the phase margin ϕ of the uncompensated system from the Bode plot. If ϕ_r is the required Pm and τ is the tolerance to be introduced, then

$$\phi = \phi_r + \tau$$

3. Determine the frequency from the phase curve where ϕ is the phase margin. This is the new gain crossover frequency ω_g'.
4. Check the gain at ω_g' from the gain curve of the Bode plot. The gain curve has to be adjusted to make it 0 dB at ω_g'. The necessary attenuation has to be provided by the factor $1/\beta$ in the transfer function [see Eq. (15.14)] to keep the shape of the Bode plot unchanged. Therefore,

$$\text{Change necessary in gain} = 20\log(1/\beta) = -20\log\beta$$

Thus, β is obtained from this relation.
5. Usually, the upper corner frequency $1/T$ is placed about a decade below the new gain crossover frequency ω_g'. So, calculate T from the relation

$$\frac{1}{T} = \frac{\omega_g'}{10}$$

6. Once β and T are known, the corner frequencies for pole and zero,viz. $1/\beta T$ and $1/T$, are known, and hence, the transfer function for the network can be written.

EXAMPLE 15.9. Design a phase-lag compensating network for the unity feedback system given by

$$G(s) = \frac{K}{s(1+0.1s)(1+0.2s)}$$

so that

$$K_v = 30\ \text{s}^{-1}, \qquad \text{Pm} \geq 40°$$

Solution We know from Eq. (6.53) that $K_v = \lim_{s\to 0} sG(s)H(s)$. Here, since $H(s) = 1$,

$$K_v = \lim_{s\to 0} sG(s) = \lim_{s\to 0}\left\{s \cdot \frac{K}{s(1+0.1s)(1+0.2s)}\right\} = K$$

Thus, the transfer function becomes

$$G(s) = \frac{30}{s\left(1+\dfrac{s}{10}\right)\left(1+\dfrac{s}{5}\right)}$$

The Bode plots of the function are given in Figure 15.11.

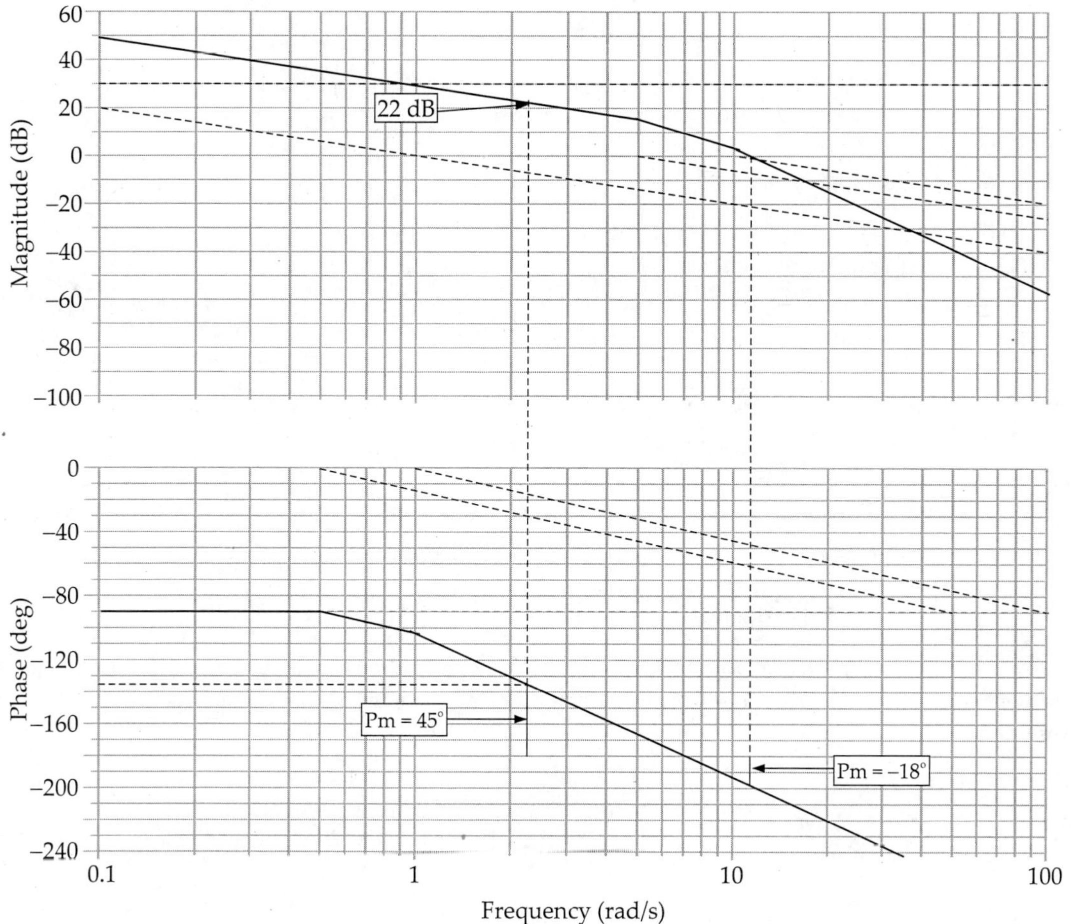

Figure 15.11: Bode plot (Example 15.9).

From the Bode plot we find that the uncompensated Pm is $-18°$. But our required Pm is $40°$, and the tolerance to be introduced is $5°$. So, the PM to be made is $= 40° + 5° = 45°$.

This Pm is available at $\omega = 2.3$ rad/s (see the Bode phase plot). The corresponding Gain is 22 dB. Thus, we need to move the gain plot by -22 dB so that it crosses the 0 dB line there to make 2.3 rad/s the gain crossover frequency and, consequently, the Pm $= 45°$. Therefore,

$$-20\log\beta = -22 \qquad \Rightarrow \beta = 12.59 \tag{i}$$

Also,

$$T = \frac{10}{2.3} = 4.35 \qquad \text{(ii)}$$

With the help of available data from Eqs. (i) and (ii), we can write the transfer function for the lag compensating network as [see Eq. (15.14)]

$$G(s) = \frac{1}{\beta} \cdot \frac{s + \dfrac{1}{T}}{s + \dfrac{1}{\beta T}} = \frac{1}{12.59} \cdot \frac{s + \dfrac{1}{4.35}}{s + \dfrac{1}{(4.35)(12.59)}} = \frac{4.35s + 1}{54.77s + 1}$$

EXAMPLE 15.10. The transfer functions of two compensators are given below

$$C_1(s) = \frac{10(s+1)}{s+10}, \qquad C_2(s) = \frac{s+10}{10(s+1)}$$

Which of the following statements is correct?

(A) C_1 is a lead compensator and C_2 is a lag compensator.
(B) C_1 is a lag compensator and C_2 is a lead compensator.
(C) Both C_1 and C_2 are lead compensators.
(D) Both C_1 and C_2 are lag compensators.

Solution The pole-zero plots for the two compensators are

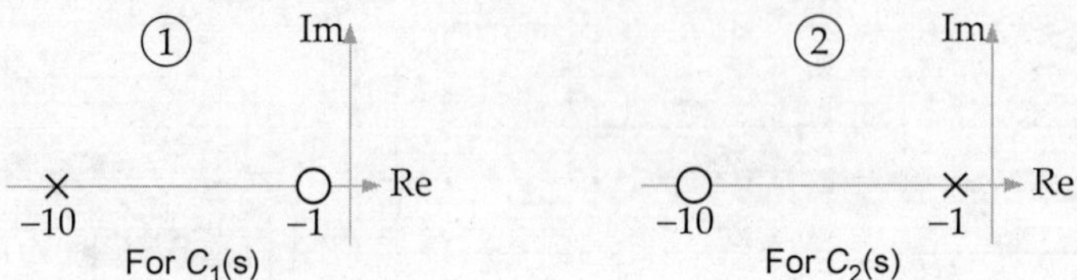

The above figure shows that the zero dominates the pole in ① which, therefore, is a lead compensator and in ②, the pole is dominant indicating that it is a lag compensator.
Ans. (A)

Alternatively

The phases of the two transfer functions are

$$\angle C_1(j\omega) = \tan^{-1}\omega - \tan^{-1}\frac{\omega}{10}$$

$$\angle C_2(j\omega) = \tan^{-1}\frac{\omega}{10} - \tan^{-1}\omega$$

Clearly the phase of C_1 leads while that of C_2 lags.
Ans. (A)

15.6 LEAD-LAG COMPENSATION NETWORK

A simple electrical circuit for the lead-lag compensation network is shown in Figure 15.12.

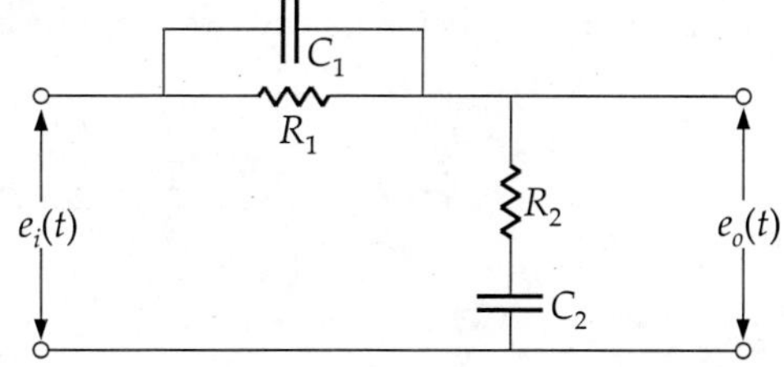

Figure 15.12: Simple lead-lag compensation circuit.

Applying KVL to the Laplace transform equivalent of the circuit, we get

$$E_i(s) = \left[\left(R_1 \| \frac{1}{sC_1}\right) + R_2 + \frac{1}{sC_2}\right] I(s) = \left[\frac{R_1}{sR_1C_1+1} + R_2 + \frac{1}{sC_2}\right] I(s)$$

$$= \frac{sR_1C_2 + sR_2C_2(sR_1C_1+1) + (sR_1C_1+1)}{sC_2(sR_1C_1+1)} I(s) \tag{15.19}$$

$$E_o(s) = \left[R_2 + \frac{1}{sC_2}\right] I(s) = \frac{sR_2C_2+1}{sC_2} I(s) \tag{15.20}$$

With the help of Eqs. (15.19) and (15.20), we get the transfer function of the circuit as

$$G(s) \equiv \frac{E_o(s)}{E_i(s)} = \frac{sR_2C_2+1}{sC_2} I(s) \cdot \frac{sC_2(sR_1C_1+1)}{sR_1C_2 + sR_2C_2(sR_1C_1+1) + (sR_1C_1+1)}$$

$$= \frac{(sR_2C_2+1)(sR_1C_1+1)}{sR_1C_2 + (sR_2C_2+1)(sR_1C_1+1)}$$

$$= \frac{(R_1C_1)(R_2C_2)}{(R_1C_1)(R_2C_2)} \cdot \frac{\left(s + \frac{1}{R_1C_1}\right)\left(s + \frac{1}{R_2C_2}\right)}{\left[\frac{s}{R_2C_1} + \left(s + \frac{1}{R_1C_1}\right)\left(s + \frac{1}{R_2C_2}\right)\right]}$$

$$= \frac{\left(s + \frac{1}{R_1C_1}\right)\left(s + \frac{1}{R_2C_2}\right)}{s^2 + s\left(\frac{1}{R_1C_1} + \frac{1}{R_2C_2} + \frac{1}{R_2C_1} + \frac{1}{R_1R_2C_1C_2}\right)} \tag{15.21}$$

We can write Eq. (15.21) as

$$G(s) \equiv \frac{\left(s + \frac{1}{T_1}\right)\left(s + \frac{1}{T_2}\right)}{\left(s + \frac{1}{\alpha T_1}\right)\left(s + \frac{1}{\beta T_2}\right)} \tag{15.22}$$

where,

$$T_1 = R_1C_1 \tag{15.23}$$

$$T_2 = R_2C_2 \tag{15.24}$$

$$\frac{1}{\alpha T_1} + \frac{1}{\beta T_2} = \frac{1}{R_1C_1} + \frac{1}{R_2C_2} + \frac{1}{R_2C_1} \tag{15.25}$$

Therefore, comparing the denominators of Eq. (15.21) and Eq. (15.25), we get

$$\alpha\beta T_1T_2 = R_1R_2C_1C_2$$

which, when compared with Eqs. (15.23) and (15.24), gives

$$\alpha\beta = 1 \tag{15.26}$$

Equation (15.26) helps us rewrite Eq. (15.22) as

$$G(s) \equiv \frac{\left(s+\dfrac{1}{T_1}\right)\left(s+\dfrac{1}{T_2}\right)}{\left(s+\dfrac{\beta}{T_1}\right)\left(s+\dfrac{1}{\beta T_2}\right)} = \beta \cdot \frac{(sT_1+1)(sT_2+1)}{(sT_1+\beta)(s\beta T_2+1)} \tag{15.27}$$

This is a lead and lag compensator combination, which is displayed by the pole-zero (Figure 15.13a) and polar plots (Figure 15.13b).

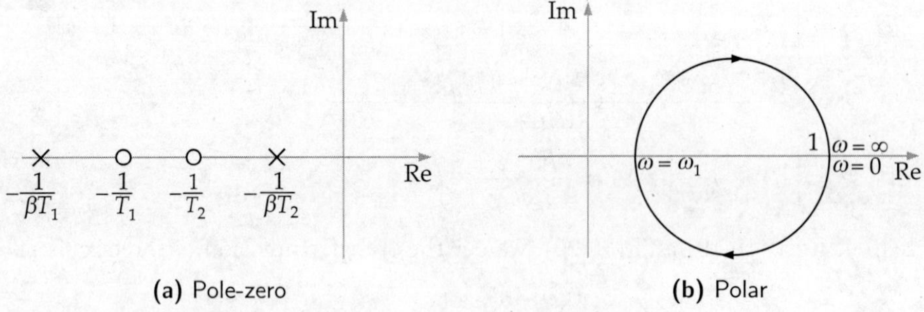

(a) Pole-zero (b) Polar

Figure 15.13: Plots of lead-lag compensator.

The same combination is also discernible in the Bode plot (Figure 15.14).

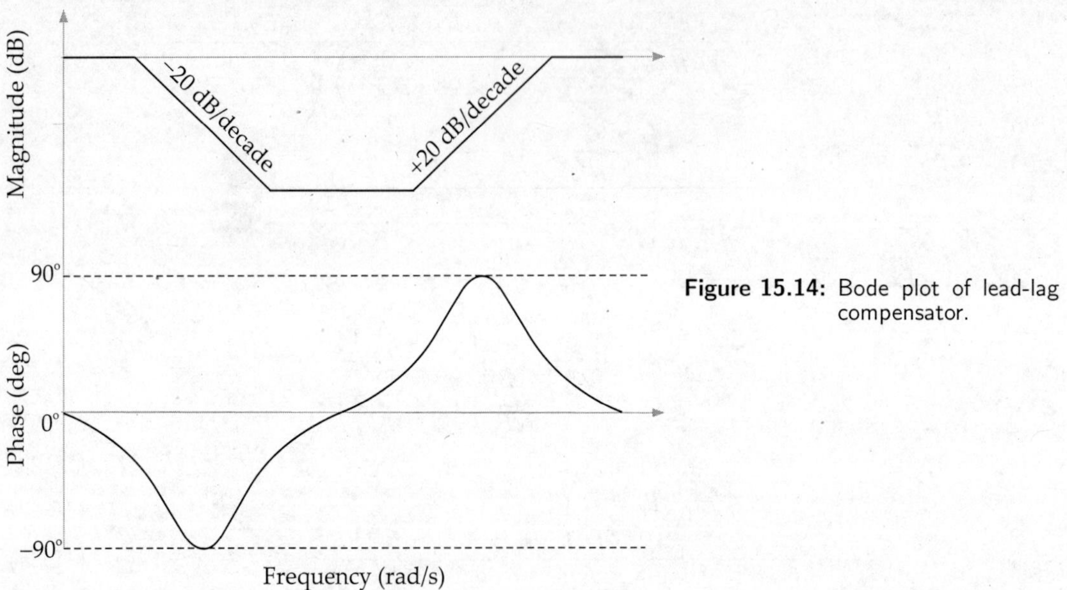

Figure 15.14: Bode plot of lead-lag compensator.

Effects of Lead-Lag Compensation

1. Bandwidth (i.e. speed) improves owing to the existence of the lead network.
2. Steady-state performance (i.e. accuracy) improves owing to the lag network.

EXAMPLE 15.11. A lead-lag compensating network transfer function is given by

$$G(s) = K\beta \cdot \frac{(1+sT_1)(1+sT_2)}{(\beta+sT_2)(1+s\beta T_2)}$$

Show that the phase angle of $G(j\omega)$ becomes zero at the frequency $\omega = \dfrac{1}{\sqrt{T_1T_2}}$.

Solution The frequency domain transfer function is

$$\begin{aligned} G(j\omega) &= K\beta \cdot \frac{(1+j\omega T_1)(1+j\omega T_2)}{(\beta+j\omega T_1)(1+j\omega\beta T_2)} \\ &\equiv K\beta \frac{\sqrt{1+\omega^2T_1^2}\sqrt{1+\omega^2T_2^2}}{\sqrt{\beta^2+\omega^2T_1^2}\sqrt{1+\omega^2\beta^2T_2^2}} \\ &\quad \angle\left[\tan^{-1}(\omega T_1)+\tan^{-1}(\omega T_2)-\tan^{-1}\left(\frac{\omega T_1}{\beta}\right)-\tan^{-1}(\omega\beta T_2)\right] \end{aligned}$$

Thus, the phase angle ϕ is given by

$$\begin{aligned} \phi &= \tan^{-1}(\omega T_1)+\tan^{-1}(\omega T_2)-\tan^{-1}\left(\frac{\omega T_1}{\beta}\right)-\tan^{-1}(\omega\beta T_2) \\ &= \tan^{-1}\frac{\omega(T_1+T_2)}{1-\omega^2T_1T_2} - \tan^{-1}\frac{\omega\left(\dfrac{T_1}{\beta}+\beta T_2\right)}{1-\omega^2T_1T_2} \end{aligned}$$

We observe that the denominators of two inverse functions are the same. At $\omega = \dfrac{1}{\sqrt{T_1T_2}}$ it turns out to be

$$1-\omega^2T_1T_2 = 1-\left(\frac{1}{\sqrt{T_1T_2}}\right)^2 T_1T_2 = 0$$

Therefore,

$$\phi = \tan^{-1}\infty - \tan^{-1}\infty = 90^\circ - 90^\circ = 0^\circ$$

REVIEW QUESTIONS

15.1 Choose the correct answer

(i) The transfer function of a compensator is given by $\dfrac{1+s}{1+0.5s}$. The compensator is a
(A) lag-lead network (B) lead network
(C) lag network (D) none of the above

(ii) The transfer function of a compensator is given by $G(s) = \dfrac{s+z}{s+p}$. It will act as a phase lag compensator if
(A) $z > p$ (B) $z < p$
(C) $z = p$ (D) none of the above

(iii) A lead compensator
(A) increases the bandwidth (B) decreases the bandwidth
(C) has nothing to do with bandwidth (D) makes the bandwidth negative

(iv) $G(s) = \dfrac{s+5}{s+15}$ is the transfer function of a lead compensator. The maximum phase shift that can be obtained from it is

(A) 15° (B) 30° (C) 45° (D) 60°

(v) A lag compensator

(A) allows low frequencies but attenuates high frequencies
(B) allows high frequencies but attenuates low frequencies
(C) allows low and high frequencies equally
(D) attenuates low and high frequencies equally

15.2 A unity feedback system is characterized by an open-loop transfer function of

$$G(s) = \frac{10}{s(1+0.5s)(1+0.1s)}$$

(a) Draw the asymptotic Bode plot of the system and determine the phase and gain margins.
(b) A lead compensator having a transfer function of $(1+0.25s)/(1+0.025s)$ is introduced in series with the plant. Determine the new phase and gain margins.
(c) Comment on the improvement in system response.

15.3 The open-loop transfer function of a unity feedback system is

$$G(s) = \frac{K}{s(1+0.2s)}$$

It is required that the velocity error constant should at least be 20 and the phase margin, 44°. Does the system meet the required specifications? If not, design the compensating network to satisfy the required specifications.

15.4 Write short notes on the following:

(a) Lead compensator
(b) Lag compensator

Answers to Selected Questions

15.1. (i) (B) (ii) (A) (iii) (A) (iv) (B) (v) (A)

15.2. (a) Gm = 1.58 dB, $\omega_p = 4.47$ rad/s, Pm = 3.94°, $\omega_g = 4.08$ rad/s
(b) Gm = 17.5 dB, $\omega_p = 17.4$ rad/s, Pm = 38.6°, $\omega_g = 5.19$ rad/s

15.3. $G'(s) = \dfrac{s+5.36}{s+10.93}$

16

State-space Analysis

16.1 INTRODUCTION

So far, we have considered SISO linear time-invariant systems which could be dealt with by the conventional transfer function approach. However, this method is not suitable for application to multiple input multiple output (MIMO) systems as well as time-varying and nonlinear systems. With n inputs and m outputs, we would have to write down mn Laplace transforms to encode all the information about a MIMO system. The state-space representation provides a convenient and compact way to model and analyze those systems.

In contrast to the conventional theory which is a predominantly frequency domain approach, the state-space analysis is a time domain approach. And, unlike the frequency domain approach, the use of the state-space representation is not limited to systems with linear components and zero initial conditions. "State-space" refers to the space whose axes are the state variables. The state of the system can be represented as a vector within that space.

Before we discuss the approach further, we will familiarize ourselves with a few definitions.

16.2 STATE-SPACE DEFINITIONS

State. If there exists a smallest set of variables—called the *state variables*—whose values and interrelationships at $t = t_0$ along with the input for $t \geq t_0$, completely determine the dynamic behaviour of a system at $t \geq 0$, then this set is known as the *state* of the system.

Suppose we have a set of variables—$x_1(t),\ x_2(t),\ \ldots,\ x_n(t)$—which completely describe the behaviour of a dynamic system such that with a knowledge of the initial values at $t = t_0$ and the input at $t \geq t_0$, the future behaviour of the system is completely determinable. Then these variables are called state variables, and the set, the state.

State vector. The n state variables that completely describe the behaviour of a particular system constitute n components of a vector called the *state vector.*

State-space. One can postulate an n-dimensional space having n-axes representing n state variables and indicate points there to represent a state of the system at a certain time. This space is called the *state-space.*

16.3 STATE-SPACE REPRESENTATION

Let us consider a MIMO system as given in Figure 16.1. Here, $u_1,\ u_2,\ \ldots,\ u_m$ inputs produce $y_1,\ y_2,\ \ldots,\ y_n$ outputs and $x_1,\ x_2,\ \ldots,\ x_p$ state variables describe the system behaviour.

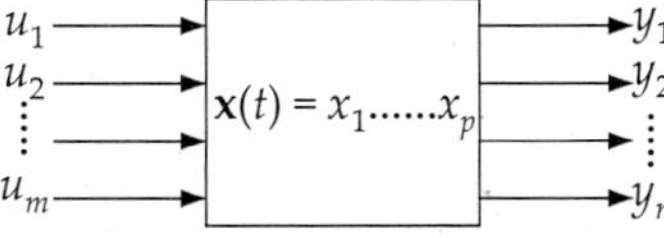

Figure 16.1: MIMO system.

Input, output and state variables are all time-dependent. We define them in vector forms as

$$\mathbf{u}(t) = \begin{bmatrix} u_1(t) \\ u_2(t) \\ \vdots \\ u_m(t) \end{bmatrix} \qquad \mathbf{y}(t) = \begin{bmatrix} y_1(t) \\ y_2(t) \\ \vdots \\ y_n(t) \end{bmatrix} \qquad \mathbf{x}(t) = \begin{bmatrix} cx_1(t) \\ x_2(t) \\ \vdots \\ x_p(t) \end{bmatrix} \tag{16.1}$$

Let these vectors be interrelated as

$$\dot{\mathbf{x}}(t) = \mathbf{A}(t)\mathbf{x}(t) + \mathbf{B}(t)\mathbf{u}(t) \quad \text{[state equation]} \tag{16.2}$$

$$\mathbf{y}(t) = \mathbf{C}(t)\mathbf{x}(t) + \mathbf{D}(t)\mathbf{u}(t) \quad \text{[output equation]} \tag{16.3}$$

Equations (16.2) and (16.3) together are called state-space representation (Figure 16.2 gives a block diagrammatic representation). In these equations,

$\mathbf{u}$ = *input* (or *control*) vector $\quad$ $\mathbf{x}$ = *state* vector $\quad$ $\mathbf{y}$ = *output* vector
$\mathbf{A}$ = *state* (or *system*) matrix $\quad$ $\mathbf{B}$ = *input* (or *control*) matrix $\quad$ $\mathbf{C}$ = *output* matrix
$\mathbf{D}$ = *direct transmission* matrix[1]

D is usually chosen to be the zero matrix in control systems where power amplification is desired. This implies that the system input is chosen not to have direct feedthrough or transmission to the output.

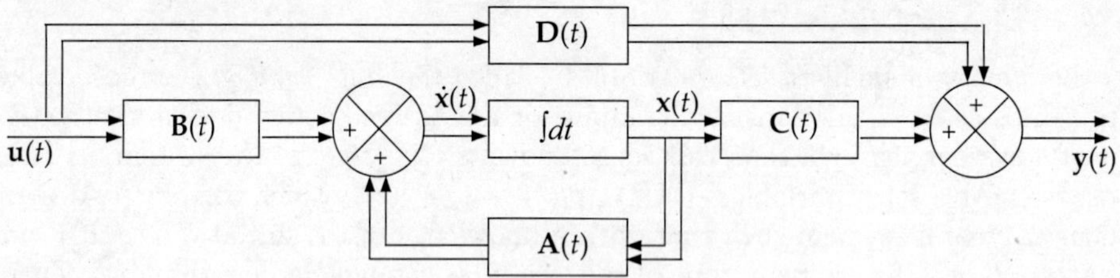

Figure 16.2: Block diagram of state-space representation. Double lines indicate multiple signals.

Salient Features of State-Space Representation

The salient features of the state-space representation are:

1. The system is described in terms of n first order differential equations.
2. The equations may form a vector-matrix differential equation.
3. The use of vector-matrix notation simplifies the mathematical representation of the system. Multiple inputs, outputs and state variables can be dealt with without difficulty or complication.
4. The method is particularly suitable for solution using digital computers.

Apart from the SISO continuous time-invariant systems, which we will use to gain familiarity with the approach, there are other forms of the state-space model (see Table 16.1).

How the equations are framed can be gauged from a consideration of LTI systems. Here, derivatives of state variables at any instant t are linear combinations of system states and

[1] Also called *feedthrough* (or *feedforward*) matrix

Table 16.1: Different state-space models

System type	*State-space model*
Continuous time-invariant	$\dot{\mathbf{x}}(t) = \mathbf{Ax}(t) + \mathbf{Bu}(t)$ $\mathbf{y}(t) = \mathbf{Cx}(t) + \mathbf{Du}(t)$
Continuous time-variant	$\dot{\mathbf{x}}(t) = \mathbf{A}(t)\mathbf{x}(t) + \mathbf{B}(t)\mathbf{u}(t)$ $\mathbf{y}(t) = \mathbf{C}(t)\mathbf{x}(t) + \mathbf{D}(t)\mathbf{u}(t)$
Discrete time-invariant	$\dot{\mathbf{x}}(k+1) = \mathbf{Ax}(k) + \mathbf{Bu}(k)$ $\mathbf{y}(k) = \mathbf{Cx}(k) + \mathbf{Du}(k)$
Discrete time-variant	$\dot{\mathbf{x}}(k+1) = \mathbf{A}(k)\mathbf{x}(k) + \mathbf{B}(k)\mathbf{u}(k)$ $\mathbf{y}(k) = \mathbf{C}(k)\mathbf{x}(k) + \mathbf{D}(k)\mathbf{u}(k)$
Laplace domain of continuous time-invariant	$s\mathbf{X}(s) = \mathbf{AX}(s) + \mathbf{BU}(s)$ $\mathbf{Y}(s) = \mathbf{CX}(s) + \mathbf{DU}(s)$
z-domain of discrete time-invariant	$z\mathbf{X}(z) = \mathbf{AX}(z) + \mathbf{BU}(z)$ $\mathbf{Y}(z) = \mathbf{CX}(z) + \mathbf{DU}(z)$

inputs. So, they can be written as

$$\begin{aligned}
\dot{x}_1(t) &= [a_{11}x_1(t) + a_{12}x_2(t) + \cdots + a_{1n}x_n(t)] + [b_{11}u_1(t) + b_{12}u_2(t) + \cdots + b_{1m}u_m(t)] \\
\dot{x}_2(t) &= [a_{21}x_1(t) + a_{22}x_2(t) + \cdots + a_{2n}x_n(t)] + [b_{21}u_1(t) + b_{22}u_2(t) + \cdots + b_{2m}u_m(t)] \\
&\vdots \\
\dot{x}_n(t) &= [a_{n1}x_1(t) + a_{n2}x_2(t) + \cdots + a_{nn}x_n(t)] + [b_{n1}u_1(t) + b_{n2}u_2(t) + \cdots + b_{nm}u_m(t)]
\end{aligned} \tag{16.4}$$

We can write Eq. (16.4) in matrix form as

$$\dot{\mathbf{x}}(t) = \mathbf{Ax}(t) + \mathbf{Bu}(t) \tag{16.5}$$

where

$$\mathbf{A} = \begin{bmatrix} a_{11} & a_{12} & \cdots & a_{1n} \\ a_{21} & a_{22} & \cdots & a_{2n} \\ \vdots & \vdots & \vdots & \vdots \\ a_{n1} & a_{n2} & \cdots & a_{nn} \end{bmatrix} \quad \text{is an } n \times n \text{ matrix}$$

$$\mathbf{B} = \begin{bmatrix} b_{11} & b_{12} & \cdots & b_{1m} \\ b_{21} & b_{22} & \cdots & b_{2m} \\ \vdots & \vdots & \vdots & \vdots \\ b_{n1} & b_{n2} & \cdots & b_{nm} \end{bmatrix} \quad \text{is an } n \times m \text{ matrix}$$

and $\dot{\mathbf{x}}$, $\mathbf{x}$ and $\mathbf{u}$ are column vectors of dimensions $n \times 1$, $n \times 1$ and $m \times 1$, respectively as given in Eq. (16.1). In a similar way, we can write the output variables as linear combinations of system states and inputs at any instant t as

$$\begin{aligned}
y_1(t) &= [c_{11}x_1(t) + c_{12}x_2(t) + \cdots + c_{1n}x_n(t)] + [d_{11}u_1(t) + d_{12}u_2(t) + \cdots + d_{1m}u_m(t)] \\
y_2(t) &= [c_{21}x_1(t) + c_{22}x_2(t) + \cdots + c_{2n}x_n(t)] + [d_{21}u_1(t) + d_{22}u_2(t) + \cdots + d_{2m}u_m(t)] \\
&\vdots \\
y_p(t) &= [c_{n1}x_1(t) + c_{n2}x_2(t) + \cdots + c_{pn}x_n(t)] + [d_{n1}u_1(t) + d_{n2}u_2(t) + \cdots + d_{pm}u_m(t)]
\end{aligned} \tag{16.6}$$

Equation (16.6) can be written in matrix form as

$$\mathbf{y}(t) = \mathbf{Cx}(t) + \mathbf{Du}(t) \tag{16.7}$$

where

$$\mathbf{C} = \begin{bmatrix} c_{11} & c_{12} & \cdots & c_{1n} \\ c_{21} & c_{22} & \cdots & c_{2n} \\ \vdots & \vdots & \vdots & \vdots \\ c_{n1} & c_{n2} & \cdots & c_{pn} \end{bmatrix} \quad \text{is a } p \times n \text{ matrix}$$

$$\mathbf{D} = \begin{bmatrix} d_{11} & d_{12} & \cdots & d_{1m} \\ d_{21} & d_{22} & \cdots & d_{2m} \\ \vdots & \vdots & \vdots & \vdots \\ d_{p1} & d_{p2} & \cdots & d_{pm} \end{bmatrix} \quad \text{is a } p \times m \text{ matrix}$$

and $\mathbf{y}(t)$ is a column vector of dimension $p \times 1$. Thus, Eqs. (16.5) and (16.7) together constitute state equations for an LTI MIMO system. More often than not, $\mathbf{D}$ is a null matrix and $\mathbf{C}$ is a $1 \times n$ row vector.

We work out a few examples to see how to set up state-space models for systems.

EXAMPLE 16.1. Obtain the state-space model for the system defined by

$$\frac{d^3c(t)}{dt^3} + 5\frac{d^2c(t)}{dt^2} + 8\frac{dc(t)}{dt} + 12c(t) = 5r(t)$$

where $c(t)$ = output and $r(t)$ = input.

Solution Let us choose state variables as

$$x_1 = c(t), \qquad x_2 = \frac{dc(t)}{dt}, \qquad x_3 = \frac{d^2c(t)}{dt^2} \tag{i}$$

Thus, we have the following set of equations

$$\dot{x}_1 = x_2$$
$$\dot{x}_2 = x_3$$
$$\dot{x}_3 = -5x_3 - 8x_2 - 12x_1 + 5r(t)$$

This set of equations can be written in the following matrix form:

$$\begin{bmatrix} \dot{x}_1 \\ \dot{x}_2 \\ \dot{x}_3 \end{bmatrix} = \begin{bmatrix} 0 & 1 & 0 \\ 0 & 0 & 1 \\ -12 & -8 & -5 \end{bmatrix} \begin{bmatrix} x_1 \\ x_2 \\ x_3 \end{bmatrix} \begin{bmatrix} 0 \\ 0 \\ 5 \end{bmatrix} [r(t)]$$

$$c(t) = \begin{bmatrix} 1 & 0 & 0 \end{bmatrix} \begin{bmatrix} x_1 \\ x_2 \\ x_3 \end{bmatrix} \qquad \text{[from Eq. (i)]}$$

These matrices are in the standard form of the state-space

$$\dot{\mathbf{x}}(t) = \mathbf{A}\mathbf{x}(t) + \mathbf{B}\mathbf{u}(t)$$
$$\mathbf{y}(t) = \mathbf{C}\mathbf{x}(t) + \mathbf{D}\mathbf{u}(t)$$

where

$$\mathbf{A} = \begin{bmatrix} 0 & 1 & 0 \\ 0 & 0 & 1 \\ -12 & -8 & -5 \end{bmatrix} \qquad \mathbf{B} = \begin{bmatrix} 0 \\ 0 \\ 5 \end{bmatrix}$$

$$\mathbf{C} = \begin{bmatrix} 1 & 0 & 0 \end{bmatrix} \qquad \mathbf{D} = [0] = \text{null matrix}$$

The signal flow graph of Figure 16.3 also represents the set of equations.

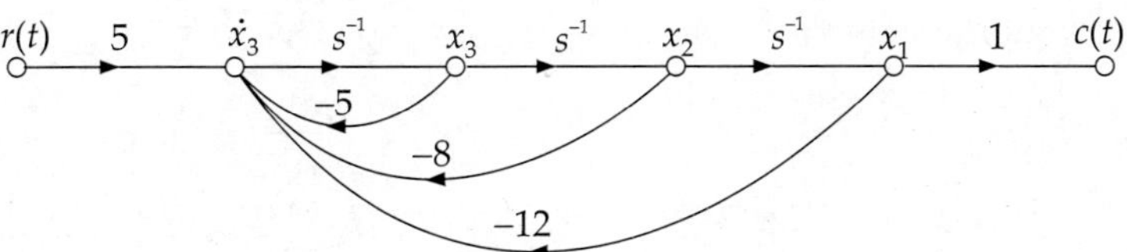

Figure 16.3: Signal flow graph (Example 16.1).

EXAMPLE 16.2. Obtain the state model for the network given in Figure 16.4.

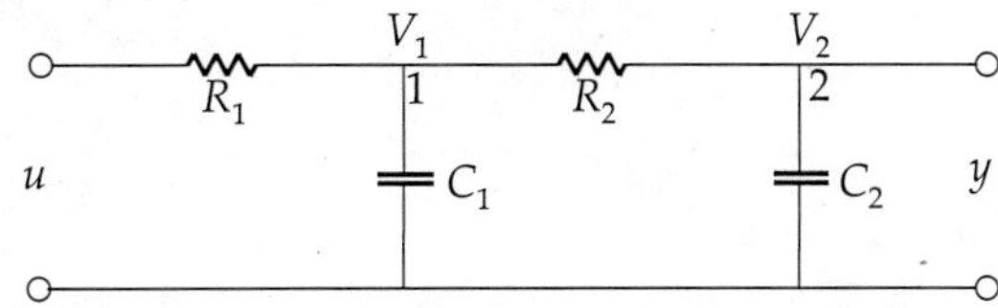

Figure 16.4: Network (Example 16.2).

Solution Considering the circuit, we can choose V_1 and V_2, voltages across two capacitors (energy storage devices) as state vectors. Applying KCL at node 1, we get

$$\frac{V_1 - u}{R_1} + C_1\frac{dV_1}{dt} + \frac{V_1 - V_2}{R_2} = 0$$

$$\Rightarrow \qquad \frac{dV_1}{dt} = -\left(\frac{1}{R_1C_1} + \frac{1}{R_2C_1}\right)V_1 + \frac{1}{R_2C_1}V_2 + \frac{1}{R_1C_1}u \qquad \text{(i)}$$

Applying KCL at node 2, we get

$$\frac{V_2 - V_1}{R_2} + C_2\frac{dV_2}{dt} = 0$$

$$\Rightarrow \qquad \frac{dV_2}{dt} = \frac{1}{R_2C_2}V_1 - \frac{1}{R_2C_2}V_2 \qquad \text{(ii)}$$

and,

$$y = V_2 \qquad \text{(iii)}$$

To obtain the state-space representation, we write Eqs. (i) to (iii) in matrix form as follows:

$$\begin{bmatrix}\dot{V}_1\\ \dot{V}_2\end{bmatrix} = \begin{bmatrix} -\left(\dfrac{1}{R_1C_1} + \dfrac{1}{R_2C_1}\right) & \dfrac{1}{R_2C_1} \\ \dfrac{1}{R_2C_2} & -\dfrac{1}{R_2C_2}\end{bmatrix}\begin{bmatrix}V_1\\ V_2\end{bmatrix} + \begin{bmatrix}\dfrac{1}{R_1C_1}\\ 0\end{bmatrix}[u]$$

$$[y] = \begin{bmatrix}0 & 1\end{bmatrix}\begin{bmatrix}V_1\\ V_2\end{bmatrix} + [0][u]$$

EXAMPLE 16.3. Write state variables in matrix form for the circuit given in Figure 16.5.

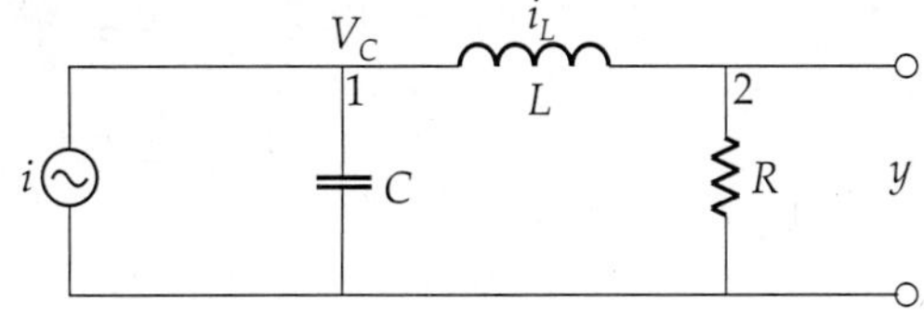

Figure 16.5: Circuit (Example 16.3).

Solution In the given circuit, there are two energy storage devices[2]—the capacitor C and the inductor L. Therefore, we can choose the voltage across the capacitor V_c and the current

[2]Actually, we are looking for derivative factors which are necessary to construct the first equation of the state model. The current passing through a capacitor, $i_C = C(dV_C/dt)$ and the voltage across an inductor, $V_L = L(di_L/dt)$ are suitable in this respect.

through the inductor i_L as state vectors. Now, at node 1, from KCL,

$$C\frac{dV_C}{dt} + i_L + i = 0$$

$$\Rightarrow \qquad \frac{dV_C}{dt} = -\frac{1}{C}i_L - \frac{1}{C}i \qquad \text{(i)}$$

At node 2, from KVL,
$$L\frac{di_L}{dt} + i_L R - V_C = 0$$

$$\Rightarrow \qquad \frac{di_L}{dt} = -\frac{R}{L}i_L + \frac{1}{L}V_C \qquad \text{(ii)}$$

and,
$$y = Ri_L \qquad \text{(iii)}$$

To obtain the state-space representation, we express Eqs. (i) to (iii) in matrix form as follows:

$$\begin{bmatrix} \dot{V}_C \\ \dot{i}_L \end{bmatrix} = \begin{bmatrix} 0 & -\dfrac{1}{C} \\ \dfrac{1}{L} & -\dfrac{R}{L} \end{bmatrix} \begin{bmatrix} V_C \\ i_L \end{bmatrix} + \begin{bmatrix} -\dfrac{1}{C} \\ 0 \end{bmatrix} [i]$$

$$[y] = \begin{bmatrix} 0 & R \end{bmatrix} \begin{bmatrix} V_c \\ i_L \end{bmatrix} + [0][i]$$

EXAMPLE 16.4. Obtain a state variable representation of the system governed by the differential equation:

$$\frac{d^2y}{dt^2} + \frac{dy}{dt} - 2y = u(t)e^{-t}$$

with the choice of state variables as $x_1 = y$, $x_2 = \left(\dfrac{dy}{dt} - y\right)e^t$.

Solution Given,

$$\frac{d^2y}{dt^2} + \frac{dy}{dt} - 2y = u(t)e^{-t} \qquad \text{(i)}$$

$$x_1 = y \qquad \text{(ii)}$$

$$x_2 = \left(\frac{dy}{dt} - y\right)e^t \qquad \text{(iii)}$$

We have to find a state variable representation as

$$\dot{\mathbf{x}} = \mathbf{Ax} + \mathbf{Bu}$$

Now,
$$\dot{x_1} = \frac{dy}{dt} = x_2 e^{-t} + y \qquad \text{[from Eq. (iii)]}$$

$$= x_1 + x_2 e^{-t} \qquad \text{[from Eq. (ii)] (iv)}$$

$$\dot{x_2} = \frac{d}{dt}\left[\left(\frac{dy}{dt} - y\right)e^t\right] = \frac{d^2y}{dt^2}e^t - \cancel{\frac{dy}{dt}e^t} - ye^t + \cancel{e^t\frac{dy}{dt}}$$

$$= \left[u(t)e^{-t} - \frac{dy}{dt} + 2y\right]e^t - ye^t \qquad \text{[using Eq. (i)]}$$

$$= u(t) - \frac{dy}{dt}e^t + ye^t$$

$$= u(t) - x_2 \qquad \text{[using Eq. (iii)]}$$

Thus,
$$\dot{x_2} = u(t) - x_2 \qquad \text{(v)}$$

Hence, we have from Eqs. (iv) and (v)

$$\dot{x_1} = x_1 + x_2 e^{-t}$$
$$\dot{x_2} = -x_2 + u(t)$$

$$\Rightarrow \qquad \begin{bmatrix} \dot{x_1} \\ \dot{x_2} \end{bmatrix} = \begin{bmatrix} 1 & e^{-t} \\ 0 & -1 \end{bmatrix} \begin{bmatrix} x_1 \\ x_2 \end{bmatrix} + \begin{bmatrix} 0 \\ 1 \end{bmatrix} u(t)$$

EXAMPLE 16.5. The following equation defines a separately excited dc motor in the form of a differential equation

$$\frac{d^2\omega}{dt^2} + \frac{B}{J}\frac{d\omega}{dt} + \frac{K^2}{LJ}\omega = \frac{K}{LJ}V_a$$

The above equation may be organized in the state-space form as follows

$$\begin{bmatrix} \dfrac{d^2\omega}{dt^2} \\ \dfrac{d\omega}{dt} \end{bmatrix} = P \begin{bmatrix} \dfrac{d\omega}{dt} \\ \omega \end{bmatrix} + QV_a$$

where the P matrix is given by

(A) $\begin{bmatrix} -\dfrac{B}{J} & -\dfrac{K^2}{LJ} \\ 1 & 0 \end{bmatrix}$ (B) $\begin{bmatrix} -\dfrac{K^2}{LJ} & -\dfrac{B}{J} \\ 0 & 1 \end{bmatrix}$ (C) $\begin{bmatrix} 0 & 1 \\ -\dfrac{K^2}{LJ} & -\dfrac{B}{J} \end{bmatrix}$ (D) $\begin{bmatrix} 1 & 0 \\ -\dfrac{B}{J} & -\dfrac{K^2}{LJ} \end{bmatrix}$

Solution We can write the given differential equation as

$$\frac{d^2\omega}{dt^2} = -\frac{B}{J}\frac{d\omega}{dt} - \frac{K^2}{LJ}\omega + \frac{K}{LJ}V_a \tag{i}$$

and, obviously

$$\frac{d\omega}{dt} = 1.\frac{d\omega}{dt} + 0.\omega \tag{ii}$$

Equations (i) and (ii) can be written as

$$\begin{bmatrix} \dfrac{d^2\omega}{dt^2} \\ \dfrac{d\omega}{dt} \end{bmatrix} = \begin{bmatrix} -\dfrac{B}{J} & -\dfrac{K^2}{LJ} \\ 1 & 0 \end{bmatrix} \begin{bmatrix} \dfrac{d\omega}{dt} \\ \omega \end{bmatrix} + \begin{bmatrix} \dfrac{K}{LJ} \\ 0 \end{bmatrix} V_a$$

Ans. (B)

16.4 NON-UNIQUENESS OF STATE-SPACE MODEL

The representation of any system in the state-space model is not unique. Alternative representations are possible for a given system. Consider a system given by

$$\dot{\mathbf{x}} = \mathbf{Ax} + \mathbf{Bu} \tag{16.8}$$

$$\mathbf{y} = \mathbf{Cx} + \mathbf{Du} \tag{16.9}$$

where $\dot{\mathbf{x}}$, $\mathbf{x}$, $\mathbf{u}$, $\mathbf{y}$ are all functions of time t.

We will see that matrices $\mathbf{A}$, $\mathbf{B}$, $\mathbf{C}$, $\mathbf{D}$ and the state variables $\mathbf{x}$ are not unique for the interrelationship between $\mathbf{u}$ and $\mathbf{y}$.

Consider a transformation

$$\mathbf{x} = \mathbf{Mz}$$

where $\mathbf{M}$ is a non-singular constant matrix. Using this transformation, Eq. (16.8) can be written as

$$\begin{aligned} \mathbf{M}\dot{\mathbf{z}} &= \mathbf{AMz} + \mathbf{Bu} \\ \Rightarrow \qquad \dot{\mathbf{z}} &= \mathbf{M}^{-1}\mathbf{AMz} + \mathbf{M}^{-1}\mathbf{Bu} \\ &\equiv \mathbf{Pz} + \mathbf{Qu} \end{aligned} \tag{16.10}$$

where $\quad \mathbf{P} \equiv \mathbf{M}^{-1}\mathbf{AM} \quad \text{and} \quad \mathbf{Q} \equiv \mathbf{M}^{-1}\mathbf{B}$

Similarly, Eq. (16.9) can be written as

$$\mathbf{y} = \mathbf{CMz} + \mathbf{Du} \equiv \mathbf{Rz} + \mathbf{Du} \tag{16.11}$$

where $\quad \mathbf{R} \equiv \mathbf{CM}$

Thus, from Eqs. (16.10) and (16.11), we see that the state and output equations can be written as

$$\begin{aligned} \dot{\mathbf{z}} &= \mathbf{Pz} + \mathbf{Qu} \\ \mathbf{y} &= \mathbf{Rz} + \mathbf{Du} \end{aligned}$$

where the state variable vector as well as state, input and output matrices are all different.

16.5 TRANSFER FUNCTION AND STATE MODEL CANONICAL REALIZATIONS

The state model can be readily derived from the transfer function of the system through its signal flow graph. However, for the same transfer function there is more than one set of state variables and, therefore, there is more than one possible form of the signal flow graphs. In general, a strictly proper[3] transfer function can be written as

$$G(s) \equiv \frac{Y(s)}{U(s)} = \frac{s^m + b_{m-1}s^{m-1} + \cdots + b_1 s + b_0}{s^n + a_{n-1}s^{n-1} + \cdots + a_1 s + a_0} \tag{16.12}$$

where $n \geq m$ and all the coefficients are real positive numbers. If we multiply the numerator and denominator of Eq. (16.12) by s^{-n}, we get

$$G(s) = \frac{s^{-(n-m)} + b_{m-1}s^{-(n-m+1)} + \cdots + b_1 s^{-(n-1)} + b_0 s^{-n}}{1 + a_{n-1}s^{-1} + \cdots + a_1 s^{-(n-1)} + a_0 s^{-n}} \tag{16.13}$$

Equation (16.13) is in the form of Mason's gain formula[4] which reduces to the following form when all the forward paths touch the feedback loops:

$$G(s) = \frac{\sum_k p_k \Delta_k}{1 - \sum_q L_q} = \frac{\text{Sum of forward (gain} \times \text{path factors)}}{1 - \text{Sum of feedback loop gains}}$$

A number of signal flow graphs can represent a transfer function. Out of them we will consider only two which are of interest to us.

[3]See Section *Kinds of Transfer Functions* at page 11 for definition.

[4]See Chapter 3, Eq. (3.7) at page 43.

Let us consider a transfer function given by

$$G(s) = \frac{b_3 s^3 + b_2 s^2 + b_1 s + b_0}{s^4 + a_3 s^3 + a_2 s^2 + a_1 s + a_0} = \frac{b_3 s^{-1} + b_2 s^{-2} + b_1 s^{-3} + b_0 s^{-4}}{1 + a_3 s^{-1} + a_2 s^{-2} + a_1 s^{-3} + a_0 s^{-4}} \quad (16.14)$$

From Eq. (16.14) we note that

- The system is of 4th order. Therefore, we need four state variables—x_1, x_2, x_3, x_4.
- The signal flow graph must have four integrators because the power of s^{-1}, which indicates integration, has a maximum value of 4.

Now, we will see how $G(s)$ of Eq. 16.14 can be represented by two signal flow graph forms, namely,

1. Controllable canonical form, and
2. Observable canonical form

Let us consider these two forms in detail and work out a few problems thereafter.

Controllable Canonical Form (CCF)

The signal flow graph, shown in Figure 16.6, may represent the transfer function. This is readily verified by noting that the product of forward gains and path factors are $b_3 s^{-1}$, $b_2 s^{-2}$, $b_1 s^{-3}$ and $b_0 s^{-4}$ as required to provide the numerator of the transfer function and the denominator is equal to 1 minus the sum of the loop gains.

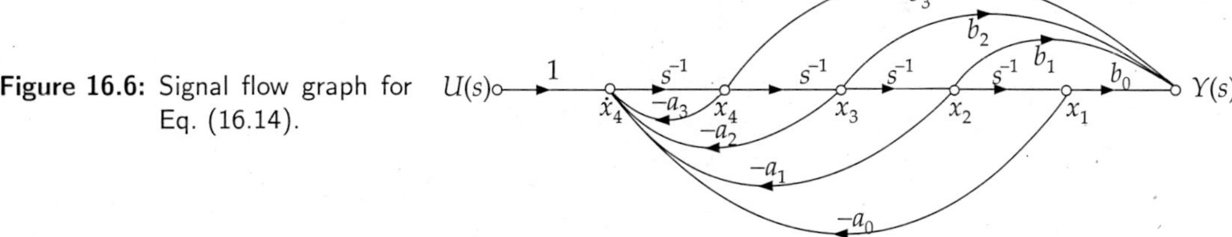

Figure 16.6: Signal flow graph for Eq. (16.14).

The state variables are identified in Figure 16.6 as the output of each energy storage element, i.e. the output of each integrator. In order to obtain the set of first order differential equations representing the state model of Figure 16.6, we have written $x_4 = \dot{x}_3$, $x_3 = \dot{x}_2$, $x_2 = \dot{x}_1$ because they automatically follow from the signal flow graph. Thus, we can obtain the following set of first order differential equations describing the state model

$$\begin{aligned}
\dot{x}_1 &= x_2 \\
\dot{x}_2 &= x_3 \\
\dot{x}_3 &= x_4 \\
x_4 &= -a_0 x_1 - a_1 x_2 - a_2 x_3 - a_3 x_4 + u
\end{aligned} \quad (16.15)$$

Furthermore, the output is simply

$$y(t) = b_0 x_1 + b_1 x_2 + b_2 x_3 + b_3 x_4$$

In matrix form, we have from Eq. (16.15)

$$\dot{\mathbf{x}} = \begin{bmatrix} 0 & 1 & 0 & 0 \\ 0 & 0 & 1 & 0 \\ 0 & 0 & 0 & 1 \\ -a_0 & -a_1 & -a_2 & -a_3 \end{bmatrix} \mathbf{x} + \begin{bmatrix} 0 \\ 0 \\ 0 \\ 1 \end{bmatrix} \mathbf{u}$$

$$\mathbf{y} = \begin{bmatrix} b_0 & b_1 & b_2 & b_3 \end{bmatrix} \mathbf{x} + [0]\mathbf{u}$$

This state-space realization, which incorporates *phase variable* feedback and zeros in the *output* path, is called *controllable canonical form* (CCF) because the resulting model is guaranteed to be controllable[5]. The realization is also known as *phase-variable canonical form.*

Observable Canonical Form (OCF)

The signal flow graph structure of Figure 16.6 is not a unique representation of Eq. (16.14). Another signal flow graph which represents Eq. (16.14) equally well is shown in Figure 16.7. In this case, the forward path factors are obtained by feeding forward the signal $U(s)$. Then the output signal $y(t)$ is equal to the first state variable $x_1(t)$. The flow graph has product of forward gains and path factors as b_0s^{-4}, b_1s^{-3}, b_2s^{-2}, b_3s^{-1}, and all forward paths touch feedback loops. Therefore, the resulting transfer function is indeed equal to Eq. (16.14).

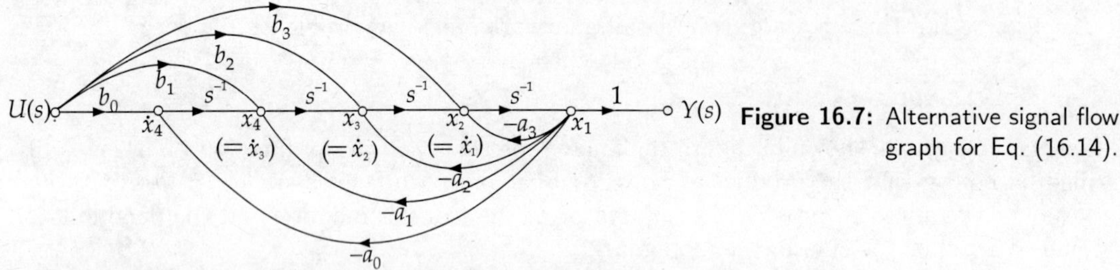

Figure 16.7: Alternative signal flow graph for Eq. (16.14).

Using the signal flow graph of Figure 16.7 we obtain the following set of equations

$$\begin{aligned}\dot{x}_1 &= -a_3x_1 + x_2 + b_3u\\ \dot{x}_2 &= -a_2x_1 + x_3 + b_2u\\ \dot{x}_3 &= -a_1x_1 + x_4 + b_1u\\ \dot{x}_4 &= -a_0x_1 + b_0u\end{aligned} \tag{16.16}$$

$$y(t) = x_1 \tag{16.17}$$

Equations (16.16) and (16.17) can be expressed in matrix forms as follows

$$\dot{\mathbf{x}} = \begin{bmatrix} -a_3 & 1 & 0 & 0\\ -a_2 & 0 & 1 & 0\\ -a_1 & 0 & 0 & 1\\ -a_0 & 0 & 0 & 0\end{bmatrix}\mathbf{x} + \begin{bmatrix} b_3\\ b_2\\ b_1\\ b_0\end{bmatrix}\mathbf{u}$$

$$\mathbf{y} = \begin{bmatrix}1 & 0 & 0 & 0\end{bmatrix}\mathbf{x} + [0]\mathbf{u}$$

This state-space realization, which incorporates *output* feedback and zeros in the *input* path, is called *observable canonical form* (OCF) because the resulting model is guaranteed to be observable.

Note: The signal flow graphs presented in Figures 16.6 and 16.7 represent the same transfer function given by Eq. (16.14). But the state variables of two graphs are not identical because the structures of the two graphs are different.

Let us consider a few examples to gain familiarity with the two forms of state models.

EXAMPLE 16.6. A unity feedback control system is shown in Figure 16.8. Obtain state matrices for its transfer function through CCF and OCF.

---

[5]See Section 16.9 at page 464 for the meaning of *Controllability* and *Observability.*

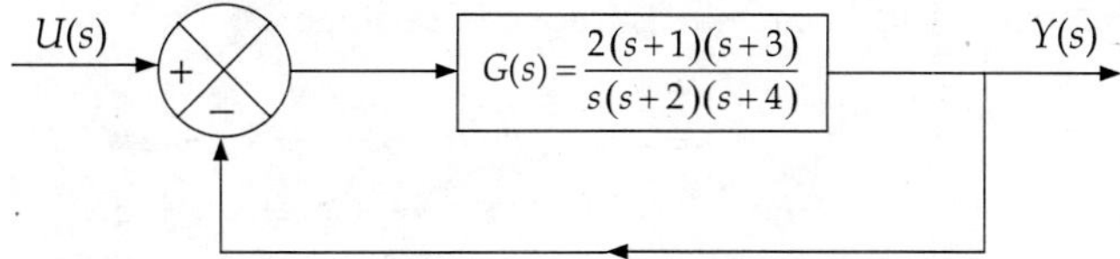

Figure 16.8: Control system (Example 16.6).

Solution The closed-loop transfer function of the system is

$$T(s) \equiv \frac{U(s)}{Y(s)} = \frac{2s^2 + 8s + 6}{s^3 + 8s^2 + 16s + 6} = \frac{2s^{-1} + 8s^{-2} + 6s^{-3}}{1 + 8s^{-1} + 16s^{-2} + 6s^{-3}}$$

The CCF signal flow graph is given in Figure 16.9a. From this graph, we get the following equations and their matrix representations:

$$\begin{aligned} \dot{x}_1 &= x_2 \\ \dot{x}_2 &= x_3 \\ \dot{x}_3 &= -6x_1 - 16x_2 - 8x_3 + u \\ y &= 6x_1 + 8x_2 + 2x_3 \end{aligned}$$

$$\Rightarrow \qquad \begin{bmatrix} \dot{x}_1 \\ \dot{x}_2 \\ \dot{x}_3 \end{bmatrix} = \begin{bmatrix} 0 & 1 & 0 \\ 0 & 0 & 1 \\ -6 & -16 & -8 \end{bmatrix} \begin{bmatrix} x_1 \\ x_2 \\ x_3 \end{bmatrix} + \begin{bmatrix} 0 \\ 0 \\ 1 \end{bmatrix} [u]$$

$$[y] = \begin{bmatrix} 6 & 8 & 2 \end{bmatrix} \begin{bmatrix} x_1 \\ x_2 \\ x_3 \end{bmatrix} + [0][u]$$

The OCF signal flow graph is given in Figure 16.9b. From this graph, we get the following

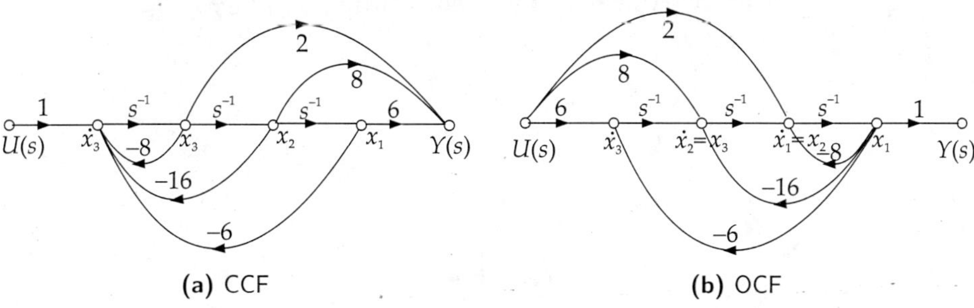

Figure 16.9: Signal flow graphs (Example 16.6).

equations and their matrix representations.

$$\dot{x}_1 = -8x_1 + x_2 + 2u$$
$$\dot{x}_2 = -16x_1 + x_3 + 8u$$
$$\dot{x}_3 = -6x_1 + 6u$$
$$y = x_1$$

$$\Rightarrow \qquad \begin{bmatrix} \dot{x}_1 \\ \dot{x}_2 \\ \dot{x}_3 \end{bmatrix} = \begin{bmatrix} -8 & 1 & 0 \\ -16 & 0 & 1 \\ -6 & 0 & 0 \end{bmatrix} \begin{bmatrix} x_1 \\ x_2 \\ x_3 \end{bmatrix} + \begin{bmatrix} 2 \\ 8 \\ 6 \end{bmatrix} [u]$$

$$[y] = \begin{bmatrix} 1 & 0 & 0 \end{bmatrix} \begin{bmatrix} x_1 \\ x_2 \\ x_3 \end{bmatrix} + [0][u]$$

EXAMPLE 16.7. Express the following closed-loop transfer function in CCF and OCF and obtain corresponding state matrices.

$$\frac{Y(s)}{U(s)} = \frac{1}{s^2 + 2s + 3}$$

Solution The transfer function can be written in the following form:

$$\frac{Y(s)}{U(s)} = \frac{s^{-2}}{1 + 2s^{-1} + 3s^{-2}}$$

The CCF signal flow graph is given in Figure 16.10a.

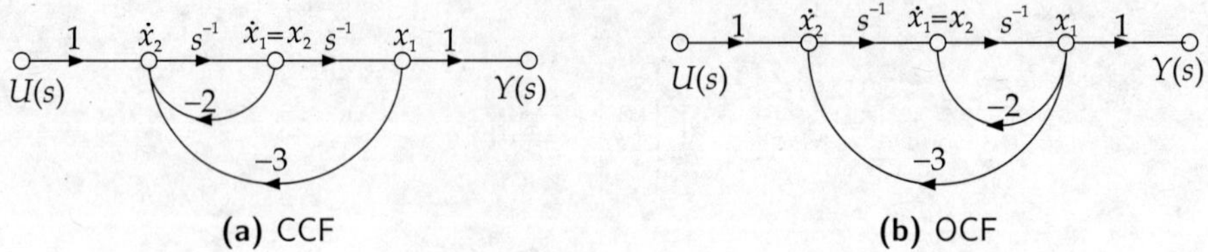

Figure 16.10: Example 16.7.

From this graph, we get the following equations and their matrix representations:

$$\dot{x}_1 = x_2$$
$$\dot{x}_2 = -3x_1 - 2x_2 + u$$
$$y = x_1$$

$$\begin{bmatrix} \dot{x}_1 \\ \dot{x}_2 \end{bmatrix} = \begin{bmatrix} 0 & 1 \\ -3 & -2 \end{bmatrix} \begin{bmatrix} x_1 \\ x_2 \end{bmatrix} + \begin{bmatrix} 0 \\ 1 \end{bmatrix} [u]$$

$$[y] = \begin{bmatrix} 1 & 0 \end{bmatrix} \begin{bmatrix} x_1 \\ x_2 \end{bmatrix} + [0][u]$$

The OCF signal flow graph is given in Figure 16.10b. From this graph, we get the following equations and their matrix representations:

$$\dot{x}_1 = -2x_1 + x_2$$
$$\dot{x}_2 = -3x_1 + u$$
$$y = x_1$$
$$\begin{bmatrix}\dot{x}_1\\ \dot{x}_2\end{bmatrix} = \begin{bmatrix}-2 & 1\\ -3 & 0\end{bmatrix}\begin{bmatrix}x_1\\ x_2\end{bmatrix} + \begin{bmatrix}0\\ 1\end{bmatrix}[u]$$
$$[y] = \begin{bmatrix}1 & 0\end{bmatrix}\begin{bmatrix}x_1\\ x_2\end{bmatrix} + [0][u]$$

16.6 TRANSFER FUNCTION DECOMPOSITION FOR STATE REPRESENTATION

We have seen how we can find state representations from transfer functions through their canonical realizations. Now we will see how we can find them from the decomposition of transfer functions. The decomposition is done in two ways as follows:

1. Cascade decomposition
2. Parallel decomposition

Let us consider the methods in detail.

Cascade Decomposition

Suppose the transfer function is given as the product of first-order terms as follows:

$$\frac{Y(s)}{U(s)} = K\frac{(s+b_1)(s+b_2)}{(s+a_1)(s+a_2)} \tag{16.18}$$

Equation (16.18) can be written as

$$\frac{Y(s)}{U(s)} = K\frac{1+b_1s^{-1}}{1+a_1s^{-1}} \cdot \frac{1+b_2s^{-1}}{1+a_2s^{-1}} \tag{16.19}$$

The signal flow graph for Eq. (16.19) can be drawn as shown in Figure 16.11.

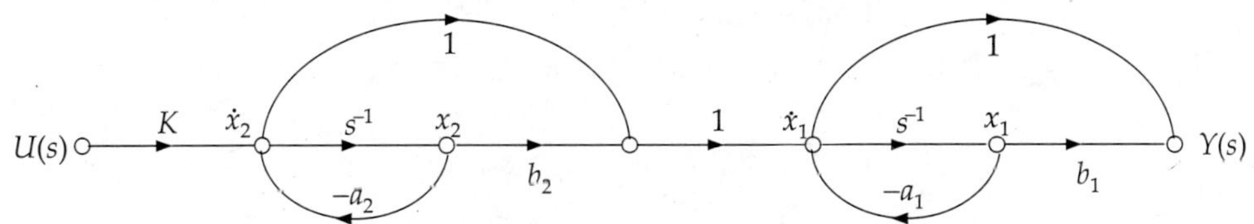

Figure 16.11: Signal flow graph for Eq. (16.19).

From the signal flow graph, $\dot{x}_1$, $\dot{x}_2$ and y can be found.

Let us consider a few examples.

EXAMPLE 16.8. Realize a state model representation by cascade decomposition of the following transfer function:

$$\frac{Y(s)}{U(s)} = \frac{1}{(s+2)(s+3)(s+4)}$$

Solution The transfer function can be written as

$$\frac{Y(s)}{U(s)} = \frac{s^{-1}}{1+2s^{-1}} \cdot \frac{s^{-1}}{1+3s^{-1}} \cdot \frac{s^{-1}}{1+4s^{-1}}$$

The corresponding signal flow graph can be drawn as shown in Figure 16.12.

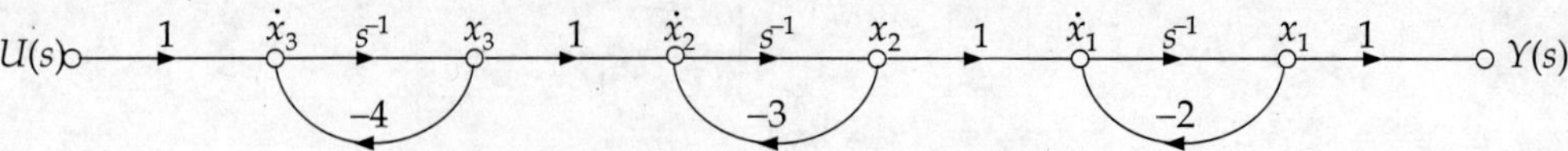

Figure 16.12: Signal flow graph (Example 16.8).

From the above graph we can write the following equations:

$$\begin{aligned} \dot{x}_3 &= -4x_3 + u \\ \dot{x}_2 &= -3x_2 + x_3 \\ \dot{x}_1 &= -2x_1 + x_2 \\ y &= x_1 \end{aligned}$$

Their matrix representations are as follows:

$$\begin{bmatrix} \dot{x}_1 \\ \dot{x}_2 \\ \dot{x}_3 \end{bmatrix} = \begin{bmatrix} -2 & 1 & 0 \\ 0 & -3 & 1 \\ 0 & 0 & 4 \end{bmatrix} \begin{bmatrix} x_1 \\ x_2 \\ x_3 \end{bmatrix} + \begin{bmatrix} 0 \\ 0 \\ 1 \end{bmatrix} [u]$$

$$[y] = \begin{bmatrix} 1 & 0 & 0 \end{bmatrix} \begin{bmatrix} x_1 \\ x_2 \\ x_3 \end{bmatrix}$$

EXAMPLE 16.9. Realize a state model representation by cascade decomposition of the transfer function

$$\frac{Y(s)}{U(s)} = \frac{5(s+1)(s+2)}{(s+4)(s+5)}$$

Solution The transfer function can be written as

$$\frac{Y(s)}{U(s)} = 5 \cdot \frac{1+s^{-1}}{1+4s^{-1}} \cdot \frac{1+2s^{-1}}{1+5s^{-1}}$$

The corresponding signal flow graph is shown in Figure 16.13.

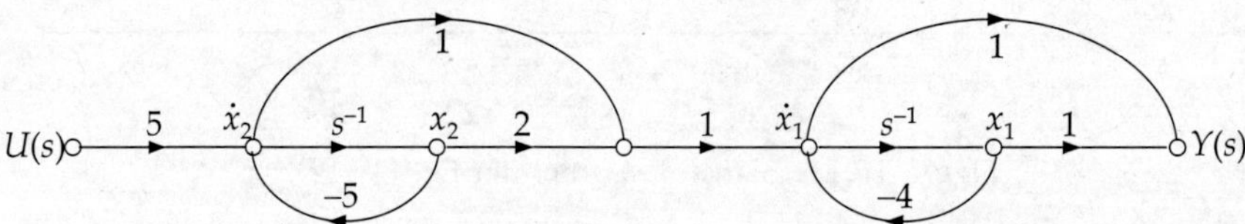

Figure 16.13: Signal flow graph (Example 16.9).

From the graph we can write the following equations:

$$\begin{aligned} \dot{x}_2 &= -5x_2 + 5u \\ \dot{x}_1 &= -4x_1 + 2x_2 \\ y &= x_1 \end{aligned}$$

Therefore, the matrix or state model representation of the transfer function is as follows:

$$\begin{bmatrix} \dot{x}_1 \\ \dot{x}_2 \end{bmatrix} = \begin{bmatrix} -4 & 2 \\ 0 & -5 \end{bmatrix} \begin{bmatrix} x_1 \\ x_2 \end{bmatrix} + \begin{bmatrix} 0 \\ 5 \end{bmatrix} [u]$$

$$[y] = \begin{bmatrix} 1 & 0 \end{bmatrix} \begin{bmatrix} x_1 \\ x_2 \end{bmatrix}$$

Parallel Decomposition

The parallel decomposition method is based on decomposition of the transfer function in partial fractions. We present two examples to explain the procedure.

EXAMPLE 16.10. Realize a state model representation for the following transfer function by its parallel decomposition:

$$\frac{Y(s)}{U(s)} = \frac{1}{(s+2)(s+3)(s+4)}$$

Solution The transfer function can be written as

$$\frac{Y(s)}{U(s)} \equiv \frac{A}{s+2} + \frac{B}{s+3} + \frac{C}{s+4}$$

The constants A, B and C can be found out as follows:

$$A = (s+2)\left.\frac{Y(s)}{U(s)}\right|_{s=-2} = \left.\frac{1}{(s+3)(s+4)}\right|_{s=-2} = \frac{1}{2}$$

$$B = (s+3)\left.\frac{Y(s)}{U(s)}\right|_{s=-3} = \left.\frac{1}{(s+2)(s+4)}\right|_{s=-3} = -1$$

$$C = (s+4)\left.\frac{Y(s)}{U(s)}\right|_{s=-4} = \left.\frac{1}{(s+2)(s+3)}\right|_{s=-4} = \frac{1}{2}$$

So, finally the transfer function can be written as:

$$\frac{Y(s)}{U(s)} = \frac{1}{2} \cdot \frac{s^{-1}}{1+2s^{-1}} - \frac{s^{-1}}{1+3s^{-1}} + \frac{1}{2} \cdot \frac{s^{-1}}{1+4s^{-1}}$$

The corresponding signal flow graph can be drawn as shown in Figure 16.14.

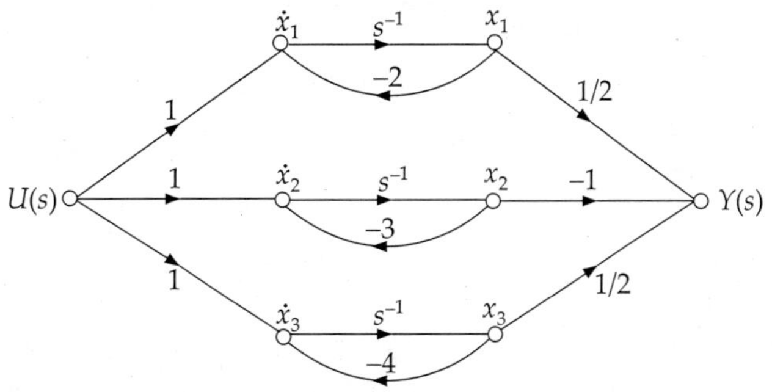

Figure 16.14: Signal flow graph (Example 16.10).

From the graph we can write the following equations:

$$\dot{x}_1 = -2x_1 + u$$
$$\dot{x}_2 = -3x_2 + u$$
$$\dot{x}_3 = -4x_3 + u$$
$$y = \frac{1}{2}x_1 - x_2 + \frac{1}{2}x_3$$

Therefore, the matrix or state model representation of the transfer function is as follows:

$$\begin{bmatrix} \dot{x}_1 \\ \dot{x}_2 \\ \dot{x}_3 \end{bmatrix} = \begin{bmatrix} -2 & 0 & 0 \\ 0 & -3 & 0 \\ 0 & 0 & -4 \end{bmatrix} \begin{bmatrix} x_1 \\ x_2 \\ x_3 \end{bmatrix} + \begin{bmatrix} 1 \\ 1 \\ 1 \end{bmatrix} [u]$$

$$[y] = \begin{bmatrix} \frac{1}{2} & -1 & \frac{1}{2} \end{bmatrix} \begin{bmatrix} x_1 \\ x_2 \\ x_3 \end{bmatrix}$$

EXAMPLE 16.11. Realize a state model representation for the following transfer function by parallel decomposition:

$$\frac{Y(s)}{U(s)} = \frac{(s+1)}{(s+2)(s+3)(s+5)}$$

Solution The transfer function can be written as

$$\frac{Y(s)}{U(s)} \equiv \frac{A}{s+2} + \frac{B}{s+3} + \frac{C}{s+5}$$

The constants A, B and C can be figured out as follows:

$$A = (s+2)\left.\frac{Y(s)}{U(s)}\right|_{s=-2} = \left.\frac{(s+1)}{(s+3)(s+5)}\right|_{s=-2} = -\frac{1}{3}$$
$$B = (s+3)\left.\frac{Y(s)}{U(s)}\right|_{s=-3} = \left.\frac{(s+1)}{(s+2)(s+5)}\right|_{s=-3} = 1$$
$$C = (s+5)\left.\frac{Y(s)}{U(s)}\right|_{s=-5} = \left.\frac{(s+1)}{(s+2)(s+3)}\right|_{s=-5} = -\frac{2}{3}$$

So, finally the transfer function can be written as:

$$\frac{Y(s)}{U(s)} = -\frac{1}{3} \cdot \frac{s^{-1}}{1+2s^{-1}} + \frac{s^{-1}}{1+3s^{-1}} - \frac{2}{3} \cdot \frac{s^{-1}}{1+5s^{-1}}$$

The corresponding signal flow graph can be drawn as shown in Figure 16.15. From the graph we can write the following equations:

$$\dot{x}_1 = -2x_1 + u$$
$$\dot{x}_2 = -3x_2 + u$$
$$\dot{x}_3 = -5x_3 + u$$
$$y = -\frac{1}{3}x_1 + x_2 - \frac{2}{3}x_3$$

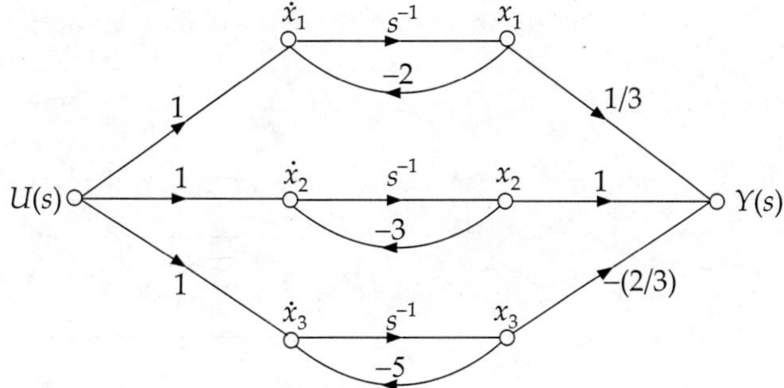

Figure 16.15: Signal flow graph (Example 16.11).

Therefore, the state matrices are

$$\begin{bmatrix} \dot{x}_1 \\ \dot{x}_2 \\ \dot{x}_3 \end{bmatrix} = \begin{bmatrix} -2 & 0 & 0 \\ 0 & -3 & 0 \\ 0 & 0 & -5 \end{bmatrix} \begin{bmatrix} x_1 \\ x_2 \\ x_3 \end{bmatrix} + \begin{bmatrix} 1 \\ 1 \\ 1 \end{bmatrix} [u]$$

$$[y] = \begin{bmatrix} -\dfrac{1}{3} & 1 & -\dfrac{2}{3} \end{bmatrix} \begin{bmatrix} x_1 \\ x_2 \\ x_3 \end{bmatrix}$$

16.7 TRANSFER FUNCTION FROM STATE MODEL

We know that by definition,

$$\text{Transfer function} = \left.\frac{\text{Laplace transform of output}}{\text{Laplace transform of input}}\right|_{\text{initial conditions}=0}$$

Let the state equations be

$$\dot{\mathbf{x}} = \mathbf{Ax} + \mathbf{Bu}$$
$$\mathbf{y} = \mathbf{Cx} + \mathbf{Du}$$

The Laplace transform of the equations are

$$s\mathbf{X}(s) = \mathbf{AX}(s) + \mathbf{BU}(s) \tag{16.20}$$
$$\mathbf{Y}(s) = \mathbf{CX}(s) + \mathbf{DU}(s) \tag{16.21}$$

where we have assumed that $\mathbf{X}(0) = 0$. Now, from Eq. (16.20), we get

$$[s\mathbf{I} - \mathbf{A}]\mathbf{X}(s) = \mathbf{BU}(s)$$

or,

$$\mathbf{X}(s) = [s\mathbf{I} - \mathbf{A}]^{-1}\mathbf{BU}(s) \tag{16.22}$$

where $\mathbf{I}$ is the identity matrix. Substituting Eq. (16.22) in Eq. (16.21), we get

$$\mathbf{Y}(s) = \mathbf{C}[s\mathbf{I} - \mathbf{A}]^{-1}\mathbf{BU}(s) + \mathbf{DU}(s)$$

Therefore, the transfer function

$$\mathbf{G}(s) = \frac{\mathbf{Y}(s)}{\mathbf{U}(s)} = \mathbf{C}[s\mathbf{I} - \mathbf{A}]^{-1}\mathbf{B} + \mathbf{D}$$

Clearly, $\mathbf{G}(s)$ must have $m \times n$ dimensionality, and thus has mn elements. So, for every input there are m transfer functions with one for each output. Which is why the state-space representation can easily be the preferred choice for multiple-input, multiple-output (MIMO) systems.

If $\mathbf{D}$ is a null matrix, which is true when $\mathbf{Y}$ and $\mathbf{U}$ are not directly connected, then

$$\boxed{\mathbf{G}(s) = \frac{\mathbf{Y}(s)}{\mathbf{U}(s)} = \mathbf{C}[s\mathbf{I} - \mathbf{A}]^{-1}\mathbf{B}} \tag{16.23}$$

From the definition of an inverse matrix[6], we get from Eq. (16.23)

$$\mathbf{G}(s) = \frac{\mathbf{Y}(s)}{\mathbf{U}(s)} = \mathbf{C}\frac{\text{Adj}[s\mathbf{I} - \mathbf{A}]}{|s\mathbf{I} - \mathbf{A}|}\mathbf{B}$$

Thus, the characteristic equation of the system is given by

$$\boxed{|s\mathbf{I} - \mathbf{A}| = 0} \tag{16.24}$$

Obviously, the roots of the characteristic equation are the eigenvalues of the matrix $\mathbf{A}$.

We work out a few examples to familiarize ourselves with the procedure. In these examples, we use a simple formula given in D.1 in Appendix D to invert 2×2 matrices.

EXAMPLE 16.12. The state representation of a system is given by

$$\dot{\mathbf{x}} = \begin{bmatrix} 0 & 1 \\ 0 & -3 \end{bmatrix}\mathbf{x} + \begin{bmatrix} 1 \\ 0 \end{bmatrix}\mathbf{u}$$
$$\mathbf{y} = \begin{bmatrix} 1 & 0 \end{bmatrix}\mathbf{x}$$

The transfer function $Y(s)/U(s)$ of the system is

(A) $\dfrac{1}{s}$ (B) $\dfrac{1}{s(s+3)}$ (C) $\dfrac{1}{s+3}$ (D) $\dfrac{1}{s^2}$

Solution From the given state representation, we get $\mathbf{C} = \begin{bmatrix} 1 & 0 \end{bmatrix}$. Also,

$$[s\mathbf{I} - \mathbf{A}] = \begin{bmatrix} s & -1 \\ 0 & s+3 \end{bmatrix} \qquad [s\mathbf{I} - \mathbf{A}]^{-1} = \frac{1}{s(s+3)}\begin{bmatrix} s+3 & 1 \\ 0 & s \end{bmatrix}$$

Therefore, from Eq. (16.23), we get

$$\mathbf{G}(s) = \begin{bmatrix} 1 & 0 \end{bmatrix}\frac{1}{s(s+3)}\begin{bmatrix} s+3 & 1 \\ 0 & s \end{bmatrix}\begin{bmatrix} 1 \\ 0 \end{bmatrix} = \frac{1}{s(s+3)}\begin{bmatrix} 1 & 0 \end{bmatrix}\begin{bmatrix} s+3 \\ 0 \end{bmatrix} = \frac{1}{s(s+3)} \times (s+3) = \frac{1}{s}$$

Ans. (A)

EXAMPLE 16.13. Find the transfer function for the system given by

$$\begin{bmatrix} \dot{x}_1 \\ \dot{x}_2 \end{bmatrix} = \begin{bmatrix} 0 & 1 \\ -3 & -2 \end{bmatrix}\begin{bmatrix} x_1 \\ x_2 \end{bmatrix} + \begin{bmatrix} 0 \\ 1 \end{bmatrix}[u]$$
$$[y] = \begin{bmatrix} 1 & 0 \end{bmatrix}\begin{bmatrix} x_1 \\ x_2 \end{bmatrix}$$

[6]See Appendix *Matrix* at page 530.

Solution Here, $\mathbf{A} = \begin{bmatrix} 0 & 1 \\ -3 & -2 \end{bmatrix}$ $\quad \mathbf{B} = \begin{bmatrix} 0 \\ 1 \end{bmatrix}$ $\quad \mathbf{C} = \begin{bmatrix} 1 & 0 \end{bmatrix}$

Therefore, from Eq. (16.23)

$$\mathbf{G}(s) = \begin{bmatrix} 1 & 0 \end{bmatrix} [s\mathbf{I} - \mathbf{A}]^{-1} \begin{bmatrix} 0 \\ 1 \end{bmatrix}$$

Now, $$[s\mathbf{I} - \mathbf{A}] = \begin{bmatrix} s & 0 \\ 0 & s \end{bmatrix} - \begin{bmatrix} 0 & 1 \\ -3 & -2 \end{bmatrix} = \begin{bmatrix} s & -1 \\ 3 & s+2 \end{bmatrix}$$

So, $$[s\mathbf{I} - \mathbf{A}]^{-1} = \frac{1}{s^2 + 2s + 3} \begin{bmatrix} s+2 & 1 \\ -3 & s \end{bmatrix}$$

Therefore, the transfer function is

$$\mathbf{G}(s) = \begin{bmatrix} 1 & 0 \end{bmatrix} \frac{1}{s^2 + 2s + 3} \begin{bmatrix} s+2 & 1 \\ -3 & s \end{bmatrix} \begin{bmatrix} 0 \\ 1 \end{bmatrix} = \begin{bmatrix} 1 & 0 \end{bmatrix} \frac{1}{s^2 + 2s + 3} \begin{bmatrix} 1 \\ s \end{bmatrix}$$

or, $$G(s) = \frac{1}{s^2 + 2s + 3}$$

EXAMPLE 16.14. A control system is described by the state equations

$$\begin{bmatrix} \dot{x}_1 \\ \dot{x}_2 \end{bmatrix} = \begin{bmatrix} -5 & -1 \\ 3 & -1 \end{bmatrix} \begin{bmatrix} x_1 \\ x_2 \end{bmatrix} + \begin{bmatrix} 2 \\ 5 \end{bmatrix} [u]$$

$$[y] = \begin{bmatrix} 1 & 2 \end{bmatrix} \begin{bmatrix} x_1 \\ x_2 \end{bmatrix}$$

Obtain the transfer function of the system.

Solution Here, $\mathbf{A} = \begin{bmatrix} -5 & -1 \\ 3 & -1 \end{bmatrix}$ $\quad \mathbf{B} = \begin{bmatrix} 2 \\ 5 \end{bmatrix}$ $\quad \mathbf{C} = \begin{bmatrix} 1 & 2 \end{bmatrix}$

Therefore, from Eq. (16.23)

$$\mathbf{G}(s) = \begin{bmatrix} 1 & 2 \end{bmatrix} [s\mathbf{I} - \mathbf{A}]^{-1} \begin{bmatrix} 2 \\ 5 \end{bmatrix}$$

Now, $$[s\mathbf{I} - \mathbf{A}] = \begin{bmatrix} s & 0 \\ 0 & s \end{bmatrix} - \begin{bmatrix} -5 & -1 \\ 3 & -1 \end{bmatrix} = \begin{bmatrix} s+5 & 1 \\ -3 & s+1 \end{bmatrix}$$

So, $$[s\mathbf{I} - \mathbf{A}]^{-1} = \frac{1}{s^2 + 6s + 8} \begin{bmatrix} s+1 & -1 \\ 3 & s+5 \end{bmatrix}$$

Therefore, the transfer function is

$$\mathbf{G}(s) = \begin{bmatrix} 1 & 2 \end{bmatrix} \frac{1}{s^2 + 6s + 8} \begin{bmatrix} s+1 & -1 \\ 3 & s+5 \end{bmatrix} \begin{bmatrix} 2 \\ 5 \end{bmatrix} = \begin{bmatrix} 1 & 2 \end{bmatrix} \frac{1}{s^2 + 6s + 8} \begin{bmatrix} 2(s+1) - 5 \\ 6 + 5(s+5) \end{bmatrix}$$

or, $$G(s) = \frac{2(s+1) - 5 + 12 + 10(s+5)}{s^2 + 6s + 8} = \frac{12s + 59}{s^2 + 6s + 8}$$

EXAMPLE 16.15. For a dynamic system, the matrices for a state space model are given by

$$\mathbf{A} = \begin{bmatrix} 0 & 1 \\ -2 & -3 \end{bmatrix}, \quad \mathbf{B} = \begin{bmatrix} 0 \\ 1 \end{bmatrix}, \quad \mathbf{C} = \begin{bmatrix} 1 & 0 \end{bmatrix}$$

The system poles are located at

(A) −2, −1 (B) −2, −3 (C) −3, −1 (D) −1, 0

Solution We may calculate the transfer function to locate the poles as follows:

$$[s\mathbf{I}-\mathbf{A}] = \begin{bmatrix} s & 0 \\ 0 & s \end{bmatrix} - \begin{bmatrix} 0 & 1 \\ -2 & -3 \end{bmatrix} = \begin{bmatrix} s & -1 \\ 2 & s+3 \end{bmatrix}$$

$$[s\mathbf{I}-\mathbf{A}]^{-1} = \frac{1}{s^2+3s+2}\begin{bmatrix} s+3 & 1 \\ -2 & s \end{bmatrix}$$

$$\mathbf{G}(s) = \frac{1}{s^2+3s+2}\begin{bmatrix} 1 & 0 \end{bmatrix}\begin{bmatrix} s+3 & 1 \\ -2 & s \end{bmatrix}\begin{bmatrix} 0 \\ 1 \end{bmatrix} = \frac{1}{s^2+3s+2}\begin{bmatrix} 1 & 0 \end{bmatrix}\begin{bmatrix} 1 \\ s \end{bmatrix} = \frac{1}{(s+1)(s+2)}$$

Therefore, the poles are at −1, −2.
Ans. (A)

EXAMPLE 16.16. Consider the single-input single-output systems with state and output equations:

System 1:
$$\begin{bmatrix} \dot{x}_1 \\ \dot{x}_2 \end{bmatrix} = \begin{bmatrix} 0 & 1 \\ -2 & -3 \end{bmatrix}\begin{bmatrix} x_1 \\ x_2 \end{bmatrix} + \begin{bmatrix} 0 \\ 2 \end{bmatrix} u_1(t)$$

$$y_1 = \begin{bmatrix} 1 & 2 \end{bmatrix}\begin{bmatrix} x_1 \\ x_2 \end{bmatrix}$$

System 2:
$$\begin{bmatrix} \dot{x}_3 \\ \dot{x}_4 \end{bmatrix} = \begin{bmatrix} -3 & 1 \\ 0 & -3 \end{bmatrix}\begin{bmatrix} x_3 \\ x_4 \end{bmatrix} + \begin{bmatrix} 0 \\ 1 \end{bmatrix} u_2(t)$$

$$y_2 = \begin{bmatrix} 1 & 1 \end{bmatrix}\begin{bmatrix} x_3 \\ x_4 \end{bmatrix}$$

The two systems are connected in cascade by applying the output of system 1 to system 2, that is by making $u_2(t) = y_1(t)$. Determine the transfer function of the combined system.

Solution System 1: $\mathbf{A} = \begin{bmatrix} 0 & 1 \\ -2 & -3 \end{bmatrix}$ $\mathbf{B} = \begin{bmatrix} 0 \\ 2 \end{bmatrix}$ $\mathbf{C} = \begin{bmatrix} 1 & 2 \end{bmatrix}$

$$[s\mathbf{I}-\mathbf{A}] = \begin{bmatrix} s & 0 \\ 0 & s \end{bmatrix} - \begin{bmatrix} 0 & 1 \\ -2 & -3 \end{bmatrix} = \begin{bmatrix} s & -1 \\ 2 & s+3 \end{bmatrix}$$

$$[s\mathbf{I}-\mathbf{A}]^{-1} = \frac{1}{s^2+3s+2}\begin{bmatrix} s+3 & 1 \\ -2 & s \end{bmatrix}$$

Thus,
$$G_1(s) = \begin{bmatrix} 1 & 2 \end{bmatrix}\frac{1}{s^2+3s+2}\begin{bmatrix} s+3 & 1 \\ -2 & s \end{bmatrix}\begin{bmatrix} 0 \\ 2 \end{bmatrix}$$

$$= \frac{1}{s^2+3s+2}\begin{bmatrix} 1 & 2 \end{bmatrix}\begin{bmatrix} 2 \\ 2s \end{bmatrix} = \frac{2(1+2s)}{s^2+3s+2}$$

System 2: $\mathbf{A} = \begin{bmatrix} -3 & 1 \\ 0 & -3 \end{bmatrix}$ $\mathbf{B} = \begin{bmatrix} 0 \\ 1 \end{bmatrix}$ $\mathbf{C} = \begin{bmatrix} 1 & 1 \end{bmatrix}$

$$[s\mathbf{I}-\mathbf{A}] = \begin{bmatrix} s & 0 \\ 0 & s \end{bmatrix} - \begin{bmatrix} -3 & 1 \\ 0 & -3 \end{bmatrix} = \begin{bmatrix} s+3 & -1 \\ 0 & s+3 \end{bmatrix}$$

$$[s\mathbf{I}-\mathbf{A}]^{-1} = \frac{1}{s^2+6s+10}\begin{bmatrix} s+3 & 1 \\ 0 & s+3 \end{bmatrix}$$

Thus,
$$G_2(s) = \begin{bmatrix} 1 & 1 \end{bmatrix}\frac{1}{s^2+6s+10}\begin{bmatrix} s+3 & 1 \\ 0 & s+3 \end{bmatrix}\begin{bmatrix} 0 \\ 1 \end{bmatrix}$$

$$= \frac{1}{s^2+6s+10}\begin{bmatrix} 1 & 1 \end{bmatrix}\begin{bmatrix} 1 \\ s+3 \end{bmatrix} = \frac{s+4}{s^2+6s+10}$$

The transfer function of the cascaded systems is, therefore

$$G(s) = G_1(s)G_2(s) = \frac{2(1+2s)(s+4)}{(s^2+3s+2)(s^2+6s+10)}$$

EXAMPLE 16.17. A state space representation for the transfer function

$$\frac{Y(s)}{U(s)} = \frac{s+6}{s^2+5s+6}$$

is

$$\mathbf{x} = \mathbf{Ax} + \mathbf{Bu}$$

$$\mathbf{y} = \mathbf{Cx}$$

where

$$\mathbf{A} = \begin{bmatrix} 0 & 1 \\ -6 & -5 \end{bmatrix} \quad \mathbf{B} = \begin{bmatrix} 0 \\ 1 \end{bmatrix}$$

The value of $\mathbf{C}$ is

(A) $\begin{bmatrix} 1 \\ 2 \end{bmatrix}$ (B) $\begin{bmatrix} 6 & 1 \end{bmatrix}$ (C) $\begin{bmatrix} 1 & 2 \\ -1 & 0 \end{bmatrix}$ (D) $\begin{bmatrix} 1 & -5 \end{bmatrix}$

Solution The given transfer function can be written as

$$\frac{Y(s)}{U(s)} = \frac{s^{-1}+6s^{-2}}{1+5s^{-1}+6s^{-2}}$$

The CCF representation of this transfer function is given in Figure 16.16.

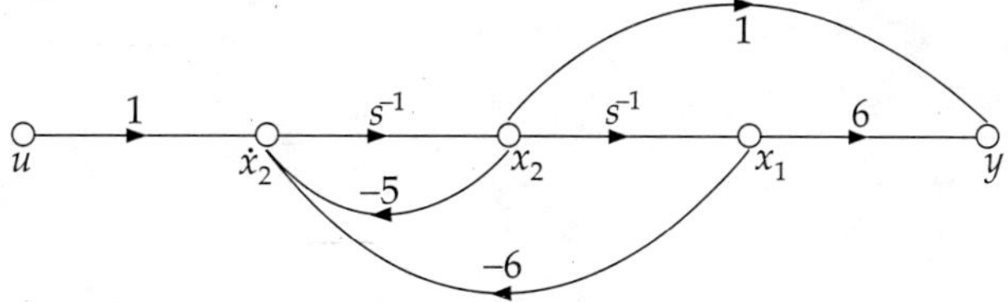

Figure 16.16: Example 16.17.

From the signal flow graph, we get the following equations:

$$\begin{aligned} \dot{x}_1 &= x_2 \\ \dot{x}_2 &= -6x_1 - 5x_2 + u \\ y &= 6x_1 + x_2 \end{aligned} \qquad \text{(i)}$$

The matrix representation of Eq. (i) is

$$\begin{bmatrix} \dot{x}_1 \\ \dot{x}_2 \end{bmatrix} = \begin{bmatrix} 0 & 1 \\ -6 & -5 \end{bmatrix} \begin{bmatrix} x_1 \\ x_2 \end{bmatrix} + \begin{bmatrix} 0 \\ 1 \end{bmatrix} u$$

$$y = \begin{bmatrix} 6 & 1 \end{bmatrix} \begin{bmatrix} x_1 \\ x_2 \end{bmatrix} \qquad \text{(ii)}$$

From Eq. (ii) we get $\mathbf{C} = \begin{bmatrix} 6 & 1 \end{bmatrix}$.
Ans. (B)

EXAMPLE 16.18. The state-space representation of a first-order system is given by

$$\dot{x} = -x + u$$

$$y = x$$

where x is the state variable, u is the control input and y is the controlled output. Let $= -Kx$ be the control law, where K is the controller gain. To place a closed-loop pole at -2, the value of K is ———.

Solution Given,

$$\dot{x} = -x + u = x(-K - 1) \qquad [\because u = -Kx]$$

$$y = x$$

For state space, by definition

$$\dot{x} = \mathbf{A}x \qquad [\because u \text{ is merged with } x]$$

Hence, $\mathbf{A} = -K - 1$. Now, closed-loop poles are roots of the characteristic equation and characteristic equation in state-space is given by [see Eq. (16.24)]

$$|s\mathbf{I} - \mathbf{A}| = 0$$

So,

$$|s - (-K - 1)| = 0$$

$\Rightarrow$

$$s = -K - 1$$

The question asks for a closed-loop pole at $s = -2$. So,

$$-K - 1 = -2$$

$\Rightarrow$

$$K = 1\,(\text{Ans.})$$

EXAMPLE 16.19. A linear time-invariant single-input single-output system has a state space model given by

$$\frac{d\mathbf{x}}{dt} = \mathbf{Fx} + \mathbf{G}u$$

$$y = \mathbf{Hx}$$

where

$$\mathbf{F} = \begin{bmatrix} 0 & 1 \\ -4 & -2 \end{bmatrix} \quad \mathbf{G} = \begin{bmatrix} 0 \\ 1 \end{bmatrix} \quad \mathbf{H} = \begin{bmatrix} 1 & 0 \end{bmatrix}$$

Here $\mathbf{x}$ is the state vector, u is the input and y is the output. The damping ratio of the system is

(A) 0.25 (B) 0.5 (C) 1 (D) 2

Solution The transfer function is calculated as follows:

$$[s\mathbf{I} - \mathbf{F}] = \begin{bmatrix} s & -1 \\ 4 & s+2 \end{bmatrix}$$

$$[s\mathbf{I} - \mathbf{F}]^{-1} = \frac{1}{s^2 + 2s + 4} \begin{bmatrix} s+2 & 1 \\ -4 & s \end{bmatrix}$$

$$\mathbf{G}(s) = \begin{bmatrix} 1 & 0 \end{bmatrix} \frac{1}{s^2 + 2s + 4} \begin{bmatrix} s+2 & 1 \\ -4 & s \end{bmatrix} \begin{bmatrix} 0 \\ 1 \end{bmatrix}$$

$$= \frac{1}{s^2 + 2s + 4} \begin{bmatrix} 1 & 0 \end{bmatrix} \begin{bmatrix} 1 \\ s \end{bmatrix} = \frac{1}{s^2 + 2s + 4}$$

Comparing the denominator of the transfer function to that of a standard second order system, we get

$$\omega_n^2 = 4 \quad \Rightarrow \quad \omega_n = 2$$

$$2\zeta\omega_n = 2 \quad \Rightarrow \quad \zeta = \frac{2}{2\omega_n} = 0.5$$

Ans. (B)

EXAMPLE 16.20. Considering the state representation given in Example 16.4, find $x_2(t)$, given that $u(t)$ is a unit step function and $x_2(0) = 0$.

Solution The state equation given in Example 16.4 is

$$\begin{bmatrix} \dot{x_1} \\ \dot{x_2} \end{bmatrix} = \begin{bmatrix} 1 & e^{-t} \\ 0 & -1 \end{bmatrix} \begin{bmatrix} x_1 \\ x_2 \end{bmatrix} + \begin{bmatrix} 0 \\ 1 \end{bmatrix} u(t)$$

From here, we get

$$\dot{x_2}(t) = -x_2(t) + u(t)$$

The Laplace transform of this equation yields

$$sX_2(s) - x_2(0) = -X_2(s) + U(s)$$

$$\Rightarrow \quad X_2(s)[s+1] = U(s) \qquad [\because x_2(0) = 0]$$

$$\Rightarrow \quad X_2(s) = \frac{1}{s(s+1)} \qquad \left[\because U(s) = \frac{1}{s}\right]$$

$$\Rightarrow \quad X_2(s) = \frac{1}{s} - \frac{1}{s+1}$$

$$\Rightarrow \quad x_2(t) = \mathscr{L}^{-1}\{X_2(s)\} = 1 - e^{-t}$$

EXAMPLE 16.21. Consider the state-space description of an LTI system with matrices

$$\mathbf{A} = \begin{bmatrix} 0 & 1 \\ -1 & -2 \end{bmatrix} \qquad \mathbf{B} = \begin{bmatrix} 0 \\ 1 \end{bmatrix} \qquad \mathbf{C} = \begin{bmatrix} 3 & -2 \end{bmatrix} \qquad \mathbf{D} = [1]$$

For the input $\sin(\omega t)$, $\omega > 0$, the value of ω for which the steady-state output of the system will be zero, is ———(round off to the nearest integer).

Solution From Eq. (16.23), the transfer function is given by

$$\mathbf{G}(s) = \mathbf{C}[s\mathbf{I} - \mathbf{A}]^{-1}\mathbf{B}$$

Now

$$[s\mathbf{I} - \mathbf{A}]^{-1} = \begin{bmatrix} s & -1 \\ 1 & s+2 \end{bmatrix}^{-1}$$

$$= \frac{1}{s^2 + 2s + 1} \begin{bmatrix} s+2 & 1 \\ -1 & s \end{bmatrix}$$

$$\mathbf{G}(s) = \frac{1}{s^2 + 2s + 1} \begin{bmatrix} 3 & -2 \end{bmatrix} \begin{bmatrix} s+2 & 1 \\ -1 & s \end{bmatrix} \begin{bmatrix} 0 \\ 1 \end{bmatrix} + 1$$

$$= \frac{1}{s^2 + 2s + 1} \begin{bmatrix} 3 & -2 \end{bmatrix} \begin{bmatrix} 1 \\ s \end{bmatrix} + 1$$

$$= \frac{3 - 2s}{s^2 + 2s + 1} + 1 = \frac{s^2 + 4}{s^2 + 2s + 1}$$

$$G(\jmath\omega) = \frac{4 - \omega^2}{1 - \omega^2 + 2\jmath\omega}$$

The output will be zero for

$$\omega^2 - 4 = 0$$

or,

$$\omega = 2\,\text{rad/s}$$

Ans. 2 rad/s

16.8 SOLVING STATE EQUATIONS

If the system is time-invariant, the state equation is given by

$$\dot{\mathbf{x}}(t) = \mathbf{A}\mathbf{x}(t) + \mathbf{B}\mathbf{u}(t) \tag{16.25}$$

In case the system is free-running with no application of control forces, $\mathbf{Bu}(\mathrm{t}) = 0$. Hence,

$$\dot{\mathbf{x}}(t) = \mathbf{A}\mathbf{x}(t) \tag{16.26}$$

where $\mathbf{A}$ is a constant matrix. Eqs. (16.25) and (16.26) are called *non-homogeneous* and *homogeneous* state equations, respectively. We will discuss methods of solving them in the reverse order. But before doing that, let us consider what is called the *state transition matrix* (STM) and how to calculate it because STM is necessary to solve both homogeneous and non-homogeneous equations.

State Transition Matrix (STM)

Let us consider Eq. (16.26). Its Laplace transform is

$$s\mathbf{X}(s) - \mathbf{X}(0) = \mathbf{A}\mathbf{X}(s)$$

$\Rightarrow$

$$(s\mathbf{I} - \mathbf{A})\mathbf{X}(s) = \mathbf{X}(0)$$

Therefore,

$$\mathbf{X}(s) = (s\mathbf{I} - \mathbf{A})^{-1}\mathbf{X}(0) \equiv \mathbf{\Phi}(s)\mathbf{X}(0) \tag{16.27}$$

where

$$\boxed{\mathbf{\Phi}(s) = (s\mathbf{I} - \mathbf{A})^{-1}} \tag{16.28}$$

$\mathbf{\Phi}(s)$ is called the *resolvent matrix*. Now,

$$\mathbf{\Phi}(s) = (s\mathbf{I} - \mathbf{A})^{-1} = \frac{1}{s}\left(\mathbf{I} - \frac{\mathbf{A}}{s}\right)^{-1} = \frac{1}{s}\left(\mathbf{I} + \frac{\mathbf{A}}{s} + \frac{\mathbf{A^2}}{s^2} + \cdots\right) = \frac{\mathbf{I}}{s} + \frac{\mathbf{A}}{s^2} + \frac{\mathbf{A}^2}{s^3} + \cdots \tag{16.29}$$

The inverse Laplace transform of Eq. (16.29) yields

$$\mathscr{L}^{-1}\{\mathbf{\Phi}(s)\} = \phi(t) = \mathbf{I} + \mathbf{A}t + \frac{\mathbf{A}^2t^2}{2!} + \frac{\mathbf{A}^3t^3}{3!} + \cdots = e^{\mathbf{A}t} \tag{16.30}$$

Taking inverse Laplace transform of Eq. (16.27) and substituting Eq. (16.30) therein, we get

$$\boxed{\mathbf{x}(t) = \phi(t)\mathbf{x}(0) = e^{\mathbf{A}t}\mathbf{x}(0)} \tag{16.31}$$

The matrix $\phi(t)$, which helps us know the system behaviour when it transits from $t = 0$ to $t = t$, is called the *state transition matrix* (STM).

Six properties of the STM, which are of importance to us, are listed in Table 16.2.

Table 16.2: Important properties of the STM

I	$\phi(0) = e^{\mathbf{\Lambda}0} = 1$
II	$\phi(t) = e^{\mathbf{\Lambda}t} = e^{-(-\mathbf{\Lambda}t)} = [\phi(-t)]^{-1}$
III	$\phi^{-1}(t) = (e^{\mathbf{\Lambda}t})^{-1} = \phi(-t)$
IV	$\phi(t_1 + t_2) = e^{\mathbf{\Lambda}t_1} \cdot e^{\mathbf{\Lambda}t_2} = \phi(t_1) \cdot \phi(t_2) = \phi(t_2) \cdot \phi(t_1)$
V	$[\phi(t)]^n = (e^{\mathbf{\Lambda}t})^n = e^{\mathbf{\Lambda}(nt)} = \phi(nt)$
VI	$\phi(t_2 - t_1)\phi(t_1 - t_0) = \phi(t_2 - t_0) = \phi(t_1 - t_0)\phi(t_2 - t_1)$

We will consider the following three methods for obtaining the STM:

1. Inverse Laplace transform method
2. Series summation method
3. $\mathbf{\Lambda}$-matrix method

1. Inverse Laplace transform method. We know that the resolvent matrix $\mathbf{\Phi}(s)$ is given by [see Eq. (16.28)]

$$\mathbf{\Phi}(s) = (s\mathbf{I} - \mathbf{A})^{-1}$$

Its inverse Laplace transform yields $\phi(t)$, i.e.

$$\boxed{\phi(t) = \mathscr{L}^{-1}\{\mathbf{\Phi}(s)\}}$$

Therefore, the STM can be obtained by an inverse Laplace transform of the resolvent matrix.

The following examples will make the procedure clear.

EXAMPLE 16.22. Find the STM for a system described by $\dot{\mathbf{x}}(t) = \mathbf{A}\mathbf{x}(t) + \mathbf{B}\mathbf{u}(t)$, where

$$\mathbf{A} = \begin{bmatrix} 1 & 0 \\ 1 & 1 \end{bmatrix} \quad \text{and} \quad \mathbf{B} = \begin{bmatrix} 1 \\ 1 \end{bmatrix}$$

Solution Here, $[s\mathbf{I} - \mathbf{A}] = \begin{bmatrix} s & 0 \\ 0 & s \end{bmatrix} - \begin{bmatrix} 1 & 0 \\ 1 & 1 \end{bmatrix} = \begin{bmatrix} s-1 & 0 \\ -1 & s-1 \end{bmatrix}$

Therefore,

$$\mathbf{\Phi}(s) = [s\mathbf{I} - \mathbf{A}]^{-1} = \frac{\begin{bmatrix} s-1 & 0 \\ 1 & s-1 \end{bmatrix}}{(s-1)^2} = \begin{bmatrix} \dfrac{1}{s-1} & 0 \\ \dfrac{1}{(s-1)^2} & \dfrac{1}{s-1} \end{bmatrix}.$$

We know that,

$$\mathscr{L}^{-1}\left\{\frac{1}{s-1}\right\} = e^t, \qquad \mathscr{L}^{-1}\left\{\frac{1}{(s-1)^2}\right\} = te^t$$

Thus,

$$\phi(t) = \mathscr{L}^{-1}\{\mathbf{\Phi}(s)\} = \begin{bmatrix} e^t & 0 \\ te^t & e^t \end{bmatrix}$$

EXAMPLE 16.23. The state model of a system is given by

$$\dot{\mathbf{x}}(t) = \begin{bmatrix} 0 & 1 \\ -2 & 3 \end{bmatrix} \mathbf{x}(t)$$

Find the STM of the system.

Solution Here,

$$[s\mathbf{I}-\mathbf{A}] = \begin{bmatrix} s & 0 \\ 0 & s \end{bmatrix} - \begin{bmatrix} 0 & 1 \\ -2 & 3 \end{bmatrix} = \begin{bmatrix} s & -1 \\ 2 & s-3 \end{bmatrix}$$

Therefore,

$$\mathbf{\Phi}(s) = [s\mathbf{I}-\mathbf{A}]^{-1} = \frac{\begin{bmatrix} s-3 & 1 \\ -2 & s \end{bmatrix}}{(s-1)(s-2)} = \begin{bmatrix} \dfrac{s-3}{(s-1)(s-2)} & \dfrac{1}{(s-1)(s-2)} \\ \dfrac{-2}{(s-1)(s-2)} & \dfrac{s}{(s-1)(s-2)} \end{bmatrix}$$

$$= \begin{bmatrix} \dfrac{2}{s-1} - \dfrac{1}{s-2} & -\dfrac{1}{s-1} + \dfrac{1}{s-2} \\ \dfrac{2}{s-1} - \dfrac{2}{s-2} & -\dfrac{1}{s-1} + \dfrac{2}{s-2} \end{bmatrix}$$

The inverse Laplace transform of the resolvent matrix $\mathbf{\Phi}(s)$ yields

$$\phi(t) = \begin{bmatrix} 2e^{t} - e^{2t} & -e^{t} + e^{2t} \\ 2e^{t} - 2e^{2t} & -e^{t} + 2e^{2t} \end{bmatrix}$$

EXAMPLE 16.24. A system is described by the following state and output equations:

$$\frac{dx_1(t)}{dt} = -3x_1(t) + x_2(t) + 2u(t)$$

$$\frac{dx_2(t)}{dt} = -2x_2(t) + u(t)$$

$$y(t) = x_1(t)$$

where $u(t)$ is the input and $y(t)$ is the output.

(a) The system transfer function is

(A) $\dfrac{s+2}{s^2+5s-6}$ (B) $\dfrac{s+3}{s^2+5s+6}$ (C) $\dfrac{2s+5}{s^2+5s+6}$ (D) $\dfrac{2s-5}{s^2+5s-6}$

(b) The state-transition matrix of the above system is

(A) $\begin{bmatrix} e^{-3t} & 0 \\ e^{-2t} + e^{-3t} & e^{-2t} \end{bmatrix}$ (B) $\begin{bmatrix} e^{-3t} & e^{-2t} - e^{-3t} \\ 0 & e^{-2t} \end{bmatrix}$

(C) $\begin{bmatrix} e^{-3t} & e^{-2t} + e^{-3t} \\ 0 & e^{-2t} \end{bmatrix}$ (D) $\begin{bmatrix} e^{3t} & e^{-2t} - e^{-3t} \\ 0 & e^{-2t} \end{bmatrix}$

Solution The given equations can be written in matrix forms as follows:

$$\begin{bmatrix} \dot{x}_1 \\ \dot{x}_2 \end{bmatrix} = \begin{bmatrix} -3 & 1 \\ 0 & -2 \end{bmatrix} \begin{bmatrix} x_1 \\ x_2 \end{bmatrix} + \begin{bmatrix} 2 \\ 1 \end{bmatrix} u(t)$$

$$y(t) = \begin{bmatrix} 1 & 0 \end{bmatrix} \begin{bmatrix} x_1 \\ x_2 \end{bmatrix}$$

Let us write the matrices as

$$\mathbf{A} = \begin{bmatrix} -3 & 1 \\ 0 & -2 \end{bmatrix} \qquad \mathbf{B} = \begin{bmatrix} 2 \\ 1 \end{bmatrix} \qquad \mathbf{C} = \begin{bmatrix} 1 & 0 \end{bmatrix}$$

(a) Then

$$[s\mathbf{I} - \mathbf{A}] = \begin{bmatrix} s+3 & -1 \\ 0 & s+2 \end{bmatrix}$$

$$[s\mathbf{I} - \mathbf{A}]^{-1} = \frac{1}{(s+2)(s+3)} \begin{bmatrix} s+2 & 1 \\ 0 & s+3 \end{bmatrix} \tag{i}$$

$$\Rightarrow \quad G(s) = \begin{bmatrix} 1 & 0 \end{bmatrix} \frac{1}{(s+2)(s+3)} \begin{bmatrix} s+2 & 1 \\ 0 & s+3 \end{bmatrix} \begin{bmatrix} 2 \\ 1 \end{bmatrix}$$

$$= \frac{1}{(s+2)(s+3)} \begin{bmatrix} 1 & 0 \end{bmatrix} \begin{bmatrix} 2(s+2)+1 \\ s+3 \end{bmatrix}$$

$$= \frac{2(s+2)+1}{(s+2)(s+3)} = \frac{2s+5}{s^2+5s+6}$$

Ans. (C)

(b) From Eq. (i), we have

$$\mathbf{\Phi}(s) = [s\mathbf{I} - \mathbf{A}]^{-1} = \frac{1}{(s+2)(s+3)} \begin{bmatrix} s+2 & 1 \\ 0 & s+3 \end{bmatrix}$$

$$= \begin{bmatrix} \dfrac{1}{s+3} & \dfrac{1}{s+2} - \dfrac{1}{s+3} \\ 0 & \dfrac{1}{s+2} \end{bmatrix}$$

$$\Rightarrow \quad \mathscr{L}^{-1}\{\mathbf{\Phi}(s)\} = \phi(t) = \begin{bmatrix} e^{-3t} & e^{-2t} - e^{-3t} \\ 0 & e^{-2t} \end{bmatrix}$$

Ans. (B)

2. Series summation method. We have seen from Eq. (16.30) that

$$\phi(t) = \mathbf{I} + \mathbf{A}t + \frac{\mathbf{A}^2 t^2}{2!} + \frac{\mathbf{A}^3 t^3}{3!} + \cdots$$

This series can be worked out to find $\phi(t)$. The method is particularly suitable for digital computation, as will be clear from the following discussion.

Let us put $\mathbf{P} = \mathbf{A}t$. Then,

$$\phi(t) = \mathbf{I} + \mathbf{P} + \frac{\mathbf{P}^2}{2!} + \frac{\mathbf{P}^3}{3!} + \cdots$$

$$= \mathbf{I} + \mathbf{P} + \frac{\mathbf{P}(\mathbf{P})}{2} + \frac{\mathbf{P}}{3}\left[\frac{\mathbf{P}(\mathbf{P})}{2}\right] + \cdots \tag{16.32}$$

Equation (16.32) shows that each term in the series contains the preceding term and a multiplier in a regular order. If we denote the terms in the series as T_1, T_2, ..., they are

$$T_1 = \mathbf{I}$$

$$T_2 = \mathbf{P}(\mathbf{I}) = \frac{\mathbf{P}}{2-1}(T_1)$$

$$T_3 = \frac{\mathbf{P}}{2}(\mathbf{P}) = \frac{\mathbf{P}}{3-1}(T_2)$$

$$\vdots$$

$$T_n = \frac{\mathbf{P}}{n-1}(T_{n-1})$$

The series may converge soon if terms contain exponential factors. However, for manual calculation of STM, it is not so suitable because we have to calculate it through Eq. (16.30). We consider a simple example.

EXAMPLE 16.25. Find $\phi(t)$ for the system given by the state equation

$$\dot{\mathbf{x}}(t) = \begin{bmatrix} -1 & 1 \\ 0 & -2 \end{bmatrix} \mathbf{x}(t) + \begin{bmatrix} 0 \\ 1 \end{bmatrix} \mathbf{u}(t)$$

through the series summation method.

Solution Given that

$$\mathbf{A} = \begin{bmatrix} -1 & 1 \\ 0 & -2 \end{bmatrix}$$

we have

$$\mathbf{A}^2 = \begin{bmatrix} -1 & 1 \\ 0 & -2 \end{bmatrix}\begin{bmatrix} -1 & 1 \\ 0 & -2 \end{bmatrix} = \begin{bmatrix} 1 & -3 \\ 0 & 4 \end{bmatrix}$$

$$\mathbf{A}^3 = \begin{bmatrix} -1 & 1 \\ 0 & -2 \end{bmatrix}\begin{bmatrix} 1 & -3 \\ 0 & 4 \end{bmatrix} = \begin{bmatrix} -1 & 7 \\ 0 & -8 \end{bmatrix}$$

Therefore,

$$\begin{aligned}
\phi(t) &= \begin{bmatrix} 1 & 0 \\ 0 & 1 \end{bmatrix} + t\begin{bmatrix} -1 & 1 \\ 0 & -2 \end{bmatrix} + \frac{t^2}{2}\begin{bmatrix} 1 & -3 \\ 0 & 4 \end{bmatrix} + \frac{t^3}{6}\begin{bmatrix} -1 & 7 \\ 0 & -8 \end{bmatrix} + \cdots \\
&= \begin{bmatrix} 1 & 0 \\ 0 & 1 \end{bmatrix} + \begin{bmatrix} -t & t \\ 0 & -2t \end{bmatrix} + \begin{bmatrix} \dfrac{t^2}{2} & -\dfrac{3t^2}{2} \\ 0 & 2t^2 \end{bmatrix} + \begin{bmatrix} -\dfrac{t^3}{6} & \dfrac{7t^3}{6} \\ 0 & -\dfrac{4t^3}{3} \end{bmatrix} + \cdots \\
&= \begin{bmatrix} 1 - t + \dfrac{t^2}{2} - \dfrac{t^3}{6} + \cdots & t - \dfrac{3}{2}t^2 + \dfrac{7}{6}t^3 - \cdots \\ 0 & 1 - 2t + 2t^2 - \dfrac{4}{3}t^3 + \cdots \end{bmatrix} \\
&= \begin{bmatrix} 1 - t + \dfrac{t^2}{2} - \dfrac{t^3}{6} + \cdots & \left\{\left(1 - t + \dfrac{t^2}{2} - \dfrac{t^3}{6} + \cdots\right) - \left(1 - 2t + \dfrac{(2t)^2}{2} - \dfrac{(2t)^3}{3!} + \cdots\right)\right\} \\ 0 & 1 - 2t + \dfrac{(2t)^2}{2} - \dfrac{(2t)^3}{3!} + \cdots \end{bmatrix} \\
&= \begin{bmatrix} e^t & e^t - e^{2t} \\ 0 & e^{2t} \end{bmatrix}
\end{aligned}$$

3. Λ-matrix method. The $\mathbf{\Lambda}$-matrix method is based on finding a nonsingular matrix $\mathbf{M}$, called the *modal matrix*, such that the transformation

$$\mathbf{x}(t) = \mathbf{M}\mathbf{z}(t) \tag{16.33}$$

transforms the linear system given by

$$\dot{\mathbf{x}}(t) = \mathbf{A}\mathbf{x}(t) + \mathbf{B}\mathbf{u}(t) \tag{16.34}$$

to

$$\dot{\mathbf{z}}(t) = \mathbf{\Lambda}\mathbf{z}(t) + \mathbf{\Gamma}\mathbf{u}(t)$$

We can verify that from the following manipulations. Let us substitute Eq. (16.33) into Eq. (16.34). Then we get

$$\mathbf{M}\dot{\mathbf{z}}(t) = \mathbf{A}\mathbf{M}\mathbf{z}(t) + \mathbf{B}\mathbf{u}(t) \tag{16.35}$$

Now we left multiply Eq. (16.35) by $\mathbf{M}^{-1}$ to get

$$\dot{\mathbf{z}}(t) = \mathbf{M}^{-1}\mathbf{A}\mathbf{M}\mathbf{z}(t) + \mathbf{M}^{-1}\mathbf{B}\mathbf{u}(t) \equiv \mathbf{\Lambda}\mathbf{z}(t) + \mathbf{\Gamma}\mathbf{u}(t) \tag{16.36}$$

where,

$$\mathbf{\Lambda} \equiv \mathbf{M}^{-1}\mathbf{A}\mathbf{M} \tag{16.37}$$

and

$$\mathbf{\Gamma} \equiv \mathbf{M}^{-1}\mathbf{B}$$

The transformation, given by the RHS of Eq. (16.37) is called *similarity transformation*, and the state equation given by Eq. (16.36) is known as the *canonical form*.We will presently see that the similarity transformation diagonalizes $\mathbf{A}$ having its eigenvalues[7] as its diagonal elements. That is,

$$\boxed{\mathbf{\Lambda} = \begin{bmatrix} \lambda_1 & 0 & 0 & 0 \\ 0 & \lambda_2 & 0 & 0 \\ \vdots & \vdots & \cdots & \vdots \\ 0 & 0 & 0 & \lambda_n \end{bmatrix}}$$

Now, the question is how to obtain the modal matrix $\mathbf{M}$.

Modal matrix. It can be shown that $\mathbf{M}$ can be formed from the eigenvectors[8] $\mathbf{p}_i$ of $\mathbf{A}$ as

$$\boxed{\mathbf{M} = \begin{bmatrix} \mathbf{p}_1 & \mathbf{p}_2 & \cdots & \mathbf{p}_n \end{bmatrix}}$$

where $\mathbf{p}_i$ is the eigenvector associated with the eigenvalue λ_i. Because then for $\mathbf{p}_1$,

$$\mathbf{A}\mathbf{p}_1 = \lambda_1\mathbf{p}_1$$

That is,

$$\begin{bmatrix} a_{11} & a_{12} & \cdots & a_{1n} \\ a_{21} & a_{22} & \cdots & a_{2n} \\ \vdots & \vdots & \ddots & \vdots \\ a_{n1} & a_{n2} & \cdots & a_{nn} \end{bmatrix} \begin{bmatrix} p_{11} \\ p_{12} \\ \vdots \\ p_{n1} \end{bmatrix} = \lambda_1 \begin{bmatrix} p_{11} \\ p_{12} \\ \vdots \\ p_{n1} \end{bmatrix} \tag{16.38}$$

Similarly,

$$\mathbf{A}\mathbf{p}_2 = \lambda_2\mathbf{p}_2$$

and so on.

Diagonalization of A by modal matrix. Interchanging sides of Eq. (16.38) and forming the $n \times n$ matrices, we get

$$\begin{bmatrix} \lambda_1 p_{11} & \lambda_2 p_{12} & \cdots & \lambda_n p_{1n} \\ \lambda_1 p_{21} & \lambda_2 p_{22} & \cdots & \lambda_n p_{2n} \\ \vdots & \vdots & \ddots & \vdots \\ \lambda_1 p_{n1} & \lambda_2 p_{n2} & \cdots & \lambda_n p_{nn} \end{bmatrix} \begin{bmatrix} p_{11} \\ p_{12} \\ \vdots \\ p_{n1} \end{bmatrix} = \begin{bmatrix} a_{11} & a_{12} & \cdots & a_{1n} \\ a_{21} & a_{22} & \cdots & a_{2n} \\ \vdots & \vdots & \ddots & \vdots \\ a_{n1} & a_{n2} & \cdots & a_{nn} \end{bmatrix} \begin{bmatrix} p_{11} & p_{12} & \cdots & p_{1n} \\ p_{21} & p_{22} & \cdots & p_{2n} \\ \vdots & \vdots & \ddots & \vdots \\ p_{n1} & p_{n2} & \cdots & p_{nn} \end{bmatrix}$$

or,

$$\begin{bmatrix} p_{11} & p_{12} & \cdots & p_{1n} \\ p_{21} & p_{22} & \cdots & p_{2n} \\ \vdots & \vdots & \ddots & \vdots \\ p_{n1} & p_{n2} & \cdots & p_{nn} \end{bmatrix} \begin{bmatrix} \lambda_1 & 0 & \cdots & 0 \\ 0 & \lambda_2 & \cdots & 0 \\ \vdots & \vdots & \ddots & \vdots \\ 0 & 0 & \cdots & \lambda_n \end{bmatrix} = \begin{bmatrix} a_{11} & a_{12} & \cdots & a_{1n} \\ a_{21} & a_{22} & \cdots & a_{2n} \\ \vdots & \vdots & \ddots & \vdots \\ a_{n1} & a_{n2} & \cdots & a_{nn} \end{bmatrix} \begin{bmatrix} p_{11} & p_{12} & \cdots & p_{1n} \\ p_{21} & p_{22} & \cdots & p_{2n} \\ \vdots & \vdots & \ddots & \vdots \\ p_{n1} & p_{n2} & \cdots & p_{nn} \end{bmatrix} \tag{16.39}$$

[7]See Section D.2 in Appendix *Matrix* at page 533.
[8]See Section D.2 in Appendix *Matrix* at page 533.

Equation (16.39) can be written in brief form as

$$\mathbf{M\Lambda} = \mathbf{AM} \tag{16.40}$$

Left multiplying both sides of Eq. (16.40) by $\mathbf{M}^{-1}$, we get

$$\mathbf{\Lambda} = \mathbf{M}^{-1}\mathbf{AM} \tag{16.41}$$

So, we see that the modal matrix $\mathbf{M}$, formed by the eigenvectors of $\mathbf{A}$, diagonalizes it by the similarity transformation.

Modal matrix for CCF-type A-matrix. However, if the matrix $\mathbf{A}$ is of the CCF-type (or phase-variable canonical form), it is known that the modal matrix may be the *Vandermonde matrix* given by

$$\begin{bmatrix} 1 & 1 & \cdots & 1 \\ \lambda_1 & \lambda_2 & \cdots & \lambda_n \\ \lambda_1^2 & \lambda_2^2 & \cdots & \lambda_n^2 \\ \vdots & \vdots & \ddots & \vdots \\ \lambda_1^{n-1} & \lambda_2^{n-1} & \cdots & \lambda_n^{n-1} \end{bmatrix} \tag{16.42}$$

where $\lambda_1,\ \lambda_2,\ \cdots \lambda_n$ are the eigenvalues of $\mathbf{A}$.

Obtaining the STM. The STM is obtained from the similarity transformation of $e^{\mathbf{\Lambda}t}$ by the modal matrix. That is,

$$\boxed{\phi(t) = e^{\mathbf{A}t} = \mathbf{M}e^{\mathbf{\Lambda}t}\mathbf{M}^{-1}}$$

Proof. We know from Eq. (16.41)

$$\mathbf{\Lambda} = \mathbf{M}^{-1}\mathbf{AM}$$

Therefore,

$$\begin{aligned} \mathbf{A} &= \mathbf{M\Lambda M}^{-1} \\ \mathbf{A}^2 &= (\mathbf{M\Lambda M}^{-1})(\mathbf{M\Lambda M}^{-1}) = \mathbf{M\Lambda}(\mathbf{M}^{-1}\mathbf{M})\mathbf{\Lambda M}^{-1} = \mathbf{M\Lambda}^2\mathbf{M}^{-1} \\ &\ \vdots \\ \mathbf{A}^n &= \mathbf{M\Lambda}^n\mathbf{M}^{-1} \end{aligned} \tag{16.43}$$

Thus,

$$\begin{aligned} \phi(t) = e^{\mathbf{A}t} &= \mathbf{I} + \mathbf{A}t + \frac{\mathbf{A}^2t^2}{2!} + \cdots \\ &= \mathbf{I} + \mathbf{M\Lambda M}^{-1}t + \frac{\mathbf{M\Lambda}^2\mathbf{M}^{-1}t^2}{2!} + \cdots && \text{[from Eq. (16.43)]} \\ &= \mathbf{I} + \mathbf{M}(\mathbf{\Lambda}t)\mathbf{M}^{-1} + \mathbf{M}\left[\frac{(\mathbf{\Lambda}t)^2}{2!}\right]\mathbf{M}^{-1} + \cdots \\ &= \mathbf{M}e^{\mathbf{\Lambda}t}\mathbf{M}^{-1} \end{aligned}$$

Let us work out a problem to make ourselves familiar with the procedure.

EXAMPLE 16.26. The state model of a system is given by

$$\dot{\mathbf{x}}(t) = \begin{bmatrix} 0 & 1 \\ -2 & 3 \end{bmatrix}\mathbf{x}(t)$$

Find eigenvalues, eigenvectors and the STM.

Solution Let

$$\mathbf{A} = \begin{bmatrix} 0 & 1 \\ -2 & 3 \end{bmatrix}$$

Then,

$$|\lambda\mathbf{I} - \mathbf{A}| = \left|\lambda\begin{bmatrix} 1 & 0 \\ 0 & 1 \end{bmatrix} - \begin{bmatrix} 0 & 1 \\ -2 & 3 \end{bmatrix}\right| = \begin{vmatrix} \lambda & -1 \\ 2 & \lambda - 3 \end{vmatrix} = \lambda^2 - 3\lambda + 2$$

Solving the equation

$$\lambda^2 - 3\lambda + 2 = 0$$

we get the eigenvalues of $\mathbf{A}$ which are: $\lambda_1 = 1$ and $\lambda_2 = 2$. If $\mathbf{p}_1$ is the eigenvector associated with the eigenvalue $\lambda_1 = 1$, it must satisfy the equation

$$[\lambda_1\mathbf{I} - \mathbf{A}]\mathbf{p}_1 = 0$$

That is,

$$1\begin{bmatrix} 1 & 0 \\ 0 & 1 \end{bmatrix} - \begin{bmatrix} 0 & 1 \\ -2 & 3 \end{bmatrix}\begin{bmatrix} p_{11} \\ p_{21} \end{bmatrix} = 0$$

$\Rightarrow$

$$\begin{bmatrix} 1 & -1 \\ 2 & -2 \end{bmatrix}\begin{bmatrix} p_{11} \\ p_{21} \end{bmatrix} = 0$$

This gives us the equations

$$p_{11} - p_{21} = 0$$
$$2p_{11} - 2p_{21} = 0$$

both of which yield $p_{11} = p_{21}$. Let us put both equal to 1, so that the eigenvector associated with the eigenvalue $\lambda_1 = 1$ is

$$\mathbf{p}_1 = \begin{bmatrix} 1 \\ 1 \end{bmatrix}$$

In a similar way, the matrix equation for the eigenvalue $\lambda_2 = 2$ is obtained as

$$\begin{bmatrix} 2 & -1 \\ 2 & -1 \end{bmatrix}\begin{bmatrix} p_{12} \\ p_{22} \end{bmatrix} = 0$$

which yields identical equations involving p_{12} and p_{22}. They give the relation $p_{22} = 2p_{12}$. Let us put $p_{12} = 1$. Then, $p_{22} = 2$. So, the eigenvector associated with the eigenvalue $\lambda_2 = 2$ is

$$\mathbf{p}_2 = \begin{bmatrix} 1 \\ 2 \end{bmatrix}$$

Therefore, the modal matrix and its inverse are

$$\mathbf{M} = [\mathbf{p}_1 \quad \mathbf{p}_2] = \begin{bmatrix} 1 & 1 \\ 1 & 2 \end{bmatrix}$$

$$\mathbf{M}^{-1} = \begin{bmatrix} 2 & -1 \\ -1 & 1 \end{bmatrix}$$

Note: 1. Since the matrix $\mathbf{A}$ is of the CCF-type, we could straight-away construct the modal matrix as the Vandermonde matrix

$$\mathbf{M} = \begin{bmatrix} 1 & 1 \\ \lambda_1 & \lambda_2 \end{bmatrix} = \begin{bmatrix} 1 & 1 \\ 1 & 2 \end{bmatrix}$$

without finding out eigenvectors. However, here we need to find out the eigenvectors because the problem specifically asks for them.

2. If $\mathbf{M}$ is correctly formed, it should diagonalize $\mathbf{A}$ through its similarity transformation with the eigenvalues as its diagonal elements. We observe that

$$\mathbf{M}^{-1}\mathbf{A}\mathbf{M} = \begin{bmatrix} 2 & -1 \\ -1 & 1 \end{bmatrix}\begin{bmatrix} 0 & 1 \\ -2 & 3 \end{bmatrix}\begin{bmatrix} 1 & 1 \\ 1 & 2 \end{bmatrix} = \begin{bmatrix} 2 & -1 \\ -1 & 1 \end{bmatrix}\begin{bmatrix} 1 & 2 \\ 1 & 4 \end{bmatrix} = \begin{bmatrix} 1 & 0 \\ 0 & 2 \end{bmatrix}$$

The result shows that the modal matrix is correct.

Now,

$$\mathbf{\Lambda} = \begin{bmatrix} 1 & 0 \\ 0 & 2 \end{bmatrix}, \qquad e^{\mathbf{\Lambda}t} = \begin{bmatrix} e^t & 0 \\ 0 & e^{2t} \end{bmatrix}$$

Therefore, the STM is given by

$$\phi(t) = \mathbf{M}e^{\mathbf{\Lambda}t}\mathbf{M}^{-1} = \begin{bmatrix} 1 & 1 \\ 1 & 2 \end{bmatrix}\begin{bmatrix} e^t & 0 \\ 0 & e^{2t} \end{bmatrix}\begin{bmatrix} 2 & -1 \\ -1 & 1 \end{bmatrix}$$

$$= \begin{bmatrix} 1 & 1 \\ 1 & 2 \end{bmatrix}\begin{bmatrix} 2e^t & -e^t \\ e^{-2t} & e^{2t} \end{bmatrix} = \begin{bmatrix} 2e^t - e^{2t} & -e^t + e^{2t} \\ 2e^t - 2e^{2t} & -e^t + 2e^{2t} \end{bmatrix}$$

Once $\phi(t)$ is found out, homogeneous and non-homogeneous equations can be easily solved.

EXAMPLE 16.27. The state transition matrix for the system $\dot{\mathbf{x}} = \mathbf{A}\mathbf{x}$ with initial state $\mathbf{x}(0)$ is

(A) $[s\mathbf{I} - \mathbf{A}]^{-1}$ (B) $e^{\mathbf{A}t}\mathbf{x}(0)$

(C) Laplace inverse of $[s\mathbf{I} - \mathbf{A}]^{-1}$ (D) Laplace inverse of $[s\mathbf{I} - \mathbf{A}]^{-1}\mathbf{x}(0)$

Ans. (C)

EXAMPLE 16.28. For the system

$$\dot{\mathbf{x}} = \begin{bmatrix} 2 & 0 \\ 0 & 4 \end{bmatrix}\mathbf{x} + \begin{bmatrix} 1 \\ 1 \end{bmatrix}u$$

$$\mathbf{y} = \begin{bmatrix} 4 & 0 \end{bmatrix}$$

with u as unit impulse and with zero initial state, the output, $\mathbf{y}$, becomes

(A) $2e^{2t}$ (B) $4e^{2t}$ (C) $2e^{4t}$ (D) $4e^{4t}$

Solution Given,

$$\mathbf{A} = \begin{bmatrix} 2 & 0 \\ 0 & 4 \end{bmatrix} \quad \mathbf{B} = \begin{bmatrix} 1 \\ 1 \end{bmatrix} \quad \mathbf{C} = \begin{bmatrix} 4 & 0 \end{bmatrix}$$

Also, $u = \delta(t)$, which implies $U(s) = \mathscr{L}\{\delta(t)\} = 1$. Therefore,

$$[s\mathbf{I} - \mathbf{A}] = \begin{bmatrix} s-2 & 0 \\ 0 & s-4 \end{bmatrix}$$

$$\mathbf{\Phi}(s) = [s\mathbf{I} - \mathbf{A}]^{-1} = \frac{1}{(s-2)(s-4)}\begin{bmatrix} s-4 & 0 \\ 0 & s-2 \end{bmatrix}$$

$$= \begin{bmatrix} \dfrac{1}{s-2} & 0 \\ 0 & \dfrac{1}{s-4} \end{bmatrix}$$

$$\mathbf{y}(s) = \mathbf{C}\mathbf{\Phi}(s)\mathbf{B} \qquad \text{[see Eq. (16.23), also } U(s) = 1]$$

$$= \begin{bmatrix} 4 & 0 \end{bmatrix}\begin{bmatrix} \dfrac{1}{s-2} & 0 \\ 0 & \dfrac{1}{s-4} \end{bmatrix}\begin{bmatrix} 1 \\ 1 \end{bmatrix} = \frac{4}{s-2}$$

Hence, $y(t) = 4e^{2t}$

Ans. (B)

EXAMPLE 16.29. Consider the following state-space representation of a linear time-invariant system:

$$\dot{\mathbf{x}}(t) = \begin{bmatrix} 1 & 0 \\ 0 & 2 \end{bmatrix} \mathbf{x}(t) \qquad \mathbf{y}(t) = \mathbf{C}^T \mathbf{x}(t)$$

$$\mathbf{C} = \begin{bmatrix} 1 \\ 1 \end{bmatrix}, \text{ and } \qquad \mathbf{x}(0) = \begin{bmatrix} 1 \\ 1 \end{bmatrix}$$

The value of $\mathbf{y}(t)$ for $t = \log_e 2$ is ———.

Solution Given, $\mathbf{A} = \begin{bmatrix} 1 & 0 \\ 0 & 2 \end{bmatrix}$. Therefore,

$$s\mathbf{I} - \mathbf{A} = \begin{bmatrix} s & 0 \\ 0 & s \end{bmatrix} - \begin{bmatrix} 1 & 0 \\ 0 & 2 \end{bmatrix} = \begin{bmatrix} s-1 & 0 \\ 0 & s-2 \end{bmatrix}$$

Therefore, $$\mathbf{\Phi}(s) = [s\mathbf{I} - \mathbf{A}]^{-1} = \frac{1}{(s-1)(s-2)} \begin{bmatrix} s-2 & 0 \\ 0 & s-1 \end{bmatrix} = \begin{bmatrix} \dfrac{1}{s-1} & 0 \\ 0 & \dfrac{1}{s-2} \end{bmatrix}$$

STM $$\phi(t) = \mathscr{L}^{-1}\{\Phi(s)\} = \begin{bmatrix} e^t & 0 \\ 0 & e^{2t} \end{bmatrix}$$

$$\mathbf{x}(t) = \phi(t)\mathbf{x}(0) = \begin{bmatrix} e^t & 0 \\ 0 & e^{2t} \end{bmatrix} \begin{bmatrix} 1 \\ 1 \end{bmatrix} = \begin{bmatrix} e^t \\ e^{2t} \end{bmatrix} \qquad \text{[see Eq. (16.31)]}$$

$$\mathbf{y}(t) = \begin{bmatrix} 1 & 1 \end{bmatrix} \begin{bmatrix} e^t \\ e^{2t} \end{bmatrix} = e^t + e^{2t}$$

Therefore, $y(t)\big|_{t=\ln 2} = 2 + 4 = 6$

Ans. 6

Homogeneous Equation

For a free-running system, $\mathbf{Bu}(t) = 0$. Hence the state equation becomes

$$\dot{\mathbf{x}}(t) = \mathbf{A}\mathbf{x}(t)$$

We know[9] that its solution is given by

$$\boxed{\mathbf{x}(t) = \phi(t)\mathbf{x}(0)} \tag{16.44}$$

where $\mathbf{x}(0)$ is the initial condition. So, to solve homogeneous equations, all we need to do is to find out the STM and multiply it with the given initial condition. Let us work out a few problems.

EXAMPLE 16.30. Given the homogeneous state equation

$$\dot{\mathbf{x}} = \begin{bmatrix} -3 & 1 \\ 0 & -2 \end{bmatrix} \mathbf{x}$$

the steady-state value of $\mathbf{x}_{\text{ss}} = \lim_{t \to \infty} \mathbf{x}(t)$, given the initial state value of $\mathbf{x}(0) = [10\ 10]^{\text{T}}$, is

(A) $\mathbf{x}_{\text{ss}} = \begin{bmatrix} 0 \\ 0 \end{bmatrix}$ (B) $\mathbf{x}_{\text{ss}} = \begin{bmatrix} -3 \\ -2 \end{bmatrix}$ (C) $\mathbf{x}_{\text{ss}} = \begin{bmatrix} -10 \\ 10 \end{bmatrix}$ (D) $\mathbf{x}_{\text{ss}} = \begin{bmatrix} \infty \\ \infty \end{bmatrix}$

[9] See Eq. (16.31) at page 446.

Solution Given,

$$\mathbf{A} = \begin{bmatrix} -3 & 1 \\ 0 & -2 \end{bmatrix}$$

We know that the solution of this equation is given by Eq. (16.44). So,

$$[s\mathbf{I} - \mathbf{A}] = \begin{bmatrix} s+3 & -1 \\ 0 & s+2 \end{bmatrix}$$

$$\mathbf{\Phi}(s) = [s\mathbf{I} - \mathbf{A}]^{-1} = \frac{1}{(s+2)(s+3)} \begin{bmatrix} s+2 & 1 \\ 0 & s+3 \end{bmatrix} = \begin{bmatrix} \dfrac{1}{s+3} & \dfrac{1}{(s+2)(s+3)} \\ 0 & \dfrac{1}{s+2} \end{bmatrix}$$

$$= \begin{bmatrix} \dfrac{1}{s+3} & \dfrac{1}{s+2} - \dfrac{1}{s+3} \\ 0 & \dfrac{1}{s+2} \end{bmatrix}$$

$$\phi(t) = \mathscr{L}^{-1} \begin{bmatrix} \dfrac{1}{s+3} & \dfrac{1}{s+2} - \dfrac{1}{s+3} \\ 0 & \dfrac{1}{s+2} \end{bmatrix} = \begin{bmatrix} e^{-3t} & e^{-2t} - e^{-3t} \\ 0 & e^{-2t} \end{bmatrix}$$

Therefore,

$$\mathbf{x}(t) = \phi(t)\mathbf{x}(0) = \begin{bmatrix} e^{-3t} & e^{-2t} - e^{-3t} \\ 0 & e^{-2t} \end{bmatrix} \begin{bmatrix} 10 \\ -10 \end{bmatrix} = \begin{bmatrix} 20e^{-3t} - 10e^{-2t} \\ -10e^{-2t} \end{bmatrix}$$

Hence,

$$\mathbf{x}_{ss} = \lim_{t\to\infty} \mathbf{x}(t) = \lim_{t\to\infty} \begin{bmatrix} 20e^{-3t} - 10e^{-2t} \\ -10e^{-2t} \end{bmatrix} = \begin{bmatrix} 0 \\ 0 \end{bmatrix}$$

Ans. (A)

EXAMPLE 16.31. Find $\mathbf{\Phi}(s)$, $\phi(t)$, $\phi^{-1}(t)$, and the solution to the system given by the state equation

$$\begin{bmatrix} \dot{x}_1 \\ \dot{x}_2 \end{bmatrix} = \begin{bmatrix} -1 & 1 \\ 0 & -2 \end{bmatrix} \begin{bmatrix} x_1 \\ x_2 \end{bmatrix} + \begin{bmatrix} 0 \\ 1 \end{bmatrix} u$$

with the initial state

$$\begin{bmatrix} x_1 \\ x_2 \end{bmatrix} = \begin{bmatrix} 1 \\ 0 \end{bmatrix}$$

Solution We note that although the state equation contains the term involving u, its value is not given. So, we need to solve it as a homogeneous equation. Here,

$$\mathbf{A} = \begin{bmatrix} -1 & 1 \\ 0 & -2 \end{bmatrix}$$

$$[s\mathbf{I} - \mathbf{A}] = \begin{bmatrix} s & 0 \\ 0 & s \end{bmatrix} - \begin{bmatrix} -1 & 1 \\ 0 & -2 \end{bmatrix} = \begin{bmatrix} s+1 & -1 \\ 0 & s+2 \end{bmatrix}$$

$$\mathbf{\Phi}(s) = [s\mathbf{I} - \mathbf{A}]^{-1} = \frac{\begin{bmatrix} s+2 & 1 \\ 0 & s+1 \end{bmatrix}}{(s+1)(s+2)}$$

$$= \begin{bmatrix} \dfrac{s+2}{(s+1)(s+2)} & \dfrac{1}{(s+1)(s+2)} \\ 0 & \dfrac{s+1}{(s+1)(s+2)} \end{bmatrix} = \begin{bmatrix} \dfrac{1}{s+1} & \dfrac{1}{(s+1)(s+2)} \\ 0 & \dfrac{1}{s+2} \end{bmatrix}$$

Now,

$$\mathscr{L}^{-1}\{\mathbf{\Phi}_{11}(s)\} = \mathscr{L}^{-1}\left\{\frac{1}{s+1}\right\} = e^{-t}$$

$$\mathscr{L}^{-1}\{\mathbf{\Phi}_{12}(s)\} = \mathscr{L}^{-1}\left\{\frac{1}{(s+1)(s+2)}\right\} = \mathscr{L}^{-1}\left\{\frac{1}{s+1} - \frac{1}{s+2}\right\} = e^{-t} - e^{-2t}$$

$$\mathscr{L}^{-1}\{\mathbf{\Phi}_{21}(s)\} = \mathscr{L}^{-1}\{0\} = 0$$

$$\mathscr{L}^{-1}\{\mathbf{\Phi}_{22}(s)\} = \mathscr{L}^{-1}\left\{\frac{1}{s+2}\right\} = e^{-2t}$$

Therefore,

$$\phi(t) = \begin{bmatrix} e^{-t} & e^{-t} - e^{-2t} \\ 0 & e^{-2t} \end{bmatrix}$$

$$\phi^{-1}(t) = \phi(-t) = \begin{bmatrix} e^{t} & e^{t} - e^{2t} \\ 0 & e^{2t} \end{bmatrix}$$

$$\mathbf{x}(t) = \phi(t)\mathbf{x}(0) = \begin{bmatrix} e^{-t} & e^{-t} - e^{-2t} \\ 0 & e^{-2t} \end{bmatrix}\begin{bmatrix} 1 \\ 0 \end{bmatrix} = \begin{bmatrix} e^{-t} \\ 0 \end{bmatrix}$$

EXAMPLE 16.32. The state model of a system is given by

$$\dot{\mathbf{x}}(t) = \begin{bmatrix} 0 & 1 \\ -2 & 3 \end{bmatrix}\mathbf{x}(t)$$

Find response of the system when

(a) $\mathbf{x}(0) = \begin{bmatrix} 1 \\ 1 \end{bmatrix}$

(b) $\mathbf{x}(0) = \begin{bmatrix} 1 \\ 2 \end{bmatrix}$

Solution We have worked out the STM for the system by the inverse Laplace transform method in Example 16.23 as well as by the $\mathbf{\Lambda}$-matrix method in Example 16.26. It is given by

$$\phi(t) = \begin{bmatrix} 2e^{t} - e^{2t} & -e^{t} + e^{2t} \\ 2e^{t} - 2e^{2t} & -e^{t} + 2e^{2t} \end{bmatrix}$$

Hence, the responses are

(a) $$\mathbf{x}(t) = \phi(t)\mathbf{x}(0) = \begin{bmatrix} 2e^{t} - e^{2t} & -e^{t} + e^{2t} \\ 2e^{t} - 2e^{2t} & -e^{t} + 2e^{2t} \end{bmatrix}\begin{bmatrix} 1 \\ 1 \end{bmatrix} = \begin{bmatrix} 2e^{t} - e^{2t} - e^{t} + e^{2t} \\ 2e^{t} - 2e^{2t} - e^{t} + 2e^{2t} \end{bmatrix} = \begin{bmatrix} e^{t} \\ e^{t} \end{bmatrix}$$

This means that $x_1(t) = e^{t}$, $x_2(t) = e^{t}$

(b) $$\mathbf{x}(t) = \phi(t)\mathbf{x}(0) = \begin{bmatrix} 2e^{t} - e^{2t} & -e^{t} + e^{2t} \\ 2e^{t} - 2e^{2t} & -e^{t} + 2e^{2t} \end{bmatrix}\begin{bmatrix} 1 \\ 2 \end{bmatrix} = \begin{bmatrix} 2e^{t} - e^{2t} - 2e^{t} + 2e^{2t} \\ 2e^{t} - 2e^{2t} - 2e^{t} + 4e^{2t} \end{bmatrix} = \begin{bmatrix} e^{2t} \\ 2e^{2t} \end{bmatrix}$$

This means that $x_1(t) = e^{2t}$, $x_2(t) = 2e^{2t}$

Non-homogeneous Equation

The Laplace transform of the non-homogeneous state equation, Eq. (16.25), is

$$
\begin{aligned}
& s\mathbf{X}(s) - \mathbf{X}(\mathbf{0}) = \mathbf{A}\mathbf{X}(s) + \mathbf{B}\mathbf{U}(s) \\
\Rightarrow\quad & (s\mathbf{I} - \mathbf{A})\mathbf{X}(s) = \mathbf{X}(0) + \mathbf{B}\mathbf{U}(s) \\
\Rightarrow\quad & \mathbf{X}(s) = (s\mathbf{I} - \mathbf{A})^{-1}\mathbf{X}(0) + (s\mathbf{I} - \mathbf{A})^{-1}\mathbf{B}\mathbf{U}(s) \qquad (16.45)
\end{aligned}
$$

The inverse Laplace transform of Eq. (16.45) yields

$$\mathbf{x}(t) = \mathscr{L}^{-1}\{(s\mathbf{I} - \mathbf{A})^{-1}\mathbf{X}(0)\} + \mathscr{L}^{-1}\{(s\mathbf{I} - \mathbf{A})^{-1}\mathbf{B}\mathbf{U}(s)\}$$

Therefore, the state vector can be calculated from the following equation:

$$\boxed{\mathbf{x}(t) = \underbrace{\phi(t)\mathbf{x}(0)}_{\text{Free response}} + \underbrace{\int_0^t \phi(t-\tau)\mathbf{B}\mathbf{u}(\tau)d\tau}_{\text{Forced response}}}$$

The output vector can be obtained from Eq. (16.3) as follows:

$$\boxed{\mathbf{y}(t) = \mathbf{C}\phi(t)\mathbf{x}(0) + \int_0^t \mathbf{C}\phi(t-\tau)\mathbf{B}\mathbf{u}(\tau)d\tau} \qquad (16.46)$$

The method will be clear from the following examples.

EXAMPLE 16.33. The state variable representation of a plant is given by

$$\dot{\mathbf{x}} = \mathbf{A}\mathbf{x} + \mathbf{B}\mathbf{u} \qquad \mathbf{y} = \mathbf{C}\mathbf{x}$$

where $\mathbf{x}$ is the state, $\mathbf{u}$ is the input and $\mathbf{y}$ is the output. Assuming zero initial conditions, the impulse response of a plant is given by

(A) $\exp(\mathbf{A}t)$
(B) $\int \exp[\mathbf{A}(t-\tau)]\mathbf{B}\mathbf{u}(\tau)d\tau$
(C) $\mathbf{C}\exp(\mathbf{A}t)\mathbf{B}$
(D) $\mathbf{C}\int \exp[\mathbf{A}(t-\tau)]\mathbf{B}\mathbf{u}(\tau)d\tau$

Solution The given data are:

(i) The initial conditions are zero.
(ii) The input is an impulse function.

The zero initial conditions make the first part of the output [see Eq. (16.46)] zero. The impulse input, which exists only at $t = 0$ with an area of 1, converts the integral in the second part of Eq. (16.46) to

$$\mathbf{C}\exp(\mathbf{A}t)\mathbf{B}$$

because

$$\phi(t) = \exp(\mathbf{A}t)$$

Ans. (C)

EXAMPLE 16.34. A system dynamics is given by

$$\dot{\mathbf{x}}(t) = \begin{bmatrix} -1 & 1 \\ 0 & -2 \end{bmatrix} \mathbf{x}(t) + \begin{bmatrix} 0 \\ 1 \end{bmatrix} \mathbf{u}(t)$$

Given, the initial vector

$$\mathbf{x}(0) = \begin{bmatrix} -1 \\ 0 \end{bmatrix}$$

and $\mathbf{u}(t)$ is a unit step function. Find the response of the system.

Solution We have already found (see Example 16.31 at page 456) that

$$\phi(t) = \begin{bmatrix} e^{-t} & e^{-t} - e^{-2t} \\ 0 & e^{-2t} \end{bmatrix}$$

Therefore,

$$\begin{aligned}
\mathbf{x}(t) &= \begin{bmatrix} e^{-t} & e^{-t} - e^{-2t} \\ 0 & e^{-2t} \end{bmatrix} \begin{bmatrix} -1 \\ 0 \end{bmatrix} + \int_0^t \begin{bmatrix} e^{-(t-\tau)} & e^{-(t-\tau)} - e^{-2(t-\tau)} \\ 0 & e^{-2(t-\tau)} \end{bmatrix} \begin{bmatrix} 0 \\ 1 \end{bmatrix} d\tau \quad [\because\ \mathbf{u}(t) = 1] \\
&= \begin{bmatrix} -e^{-t} \\ 0 \end{bmatrix} + \int_0^t \begin{bmatrix} e^{-(t-\tau)} - e^{-2(t-\tau)} \\ e^{-2(t-\tau)} \end{bmatrix} d\tau \\
&= \begin{bmatrix} -e^{-t} \\ 0 \end{bmatrix} + \begin{bmatrix} e^{-(t-\tau)} - \dfrac{e^{-2(t-\tau)}}{2} \\ \dfrac{e^{-2(t-\tau)}}{2} \end{bmatrix}_0^t \\
&= \begin{bmatrix} -e^{-t} \\ 0 \end{bmatrix} + \begin{bmatrix} \dfrac{1}{2} - e^{-t} - \dfrac{e^{-2t}}{2} \\ \dfrac{1}{2} - \dfrac{e^{-2t}}{2} \end{bmatrix} \\
&= \begin{bmatrix} \dfrac{1}{2} - 2e^{-t} - \dfrac{e^{-2t}}{2} \\ \dfrac{1}{2} - \dfrac{e^{-2t}}{2} \end{bmatrix}
\end{aligned}$$

which means that

$$x_1(t) = \frac{1}{2} - 2e^{-t} - \frac{e^{-2t}}{2} \qquad \text{and} \qquad x_2(t) = \frac{1}{2} - \frac{e^{-2t}}{2}$$

EXAMPLE 16.35. A first order matrix differential equation of a system is given as

$$\begin{bmatrix} \dot{x}_1 \\ \dot{x}_2 \end{bmatrix} = \begin{bmatrix} 0 & 1 \\ -2 & -3 \end{bmatrix} \begin{bmatrix} x_1 \\ x_2 \end{bmatrix} + \begin{bmatrix} 0 \\ 2 \end{bmatrix} u$$

$$y = [1 \quad 1] \begin{bmatrix} x_1 \\ x_2 \end{bmatrix}$$

(a) Find the transfer function of the system.

(b) Find the solution of the states and the output when the input is a unit step input and the initial condition of the states is $\begin{bmatrix} 0 \\ 1 \end{bmatrix}$.

Solution Given,

$$\mathbf{A} = \begin{bmatrix} 0 & 1 \\ -2 & -3 \end{bmatrix} \qquad \mathbf{B} = \begin{bmatrix} 0 \\ 2 \end{bmatrix} \qquad \mathbf{C} = [1 \;\; 1]$$

(a)
$$G(s) = [1 \;\; 1]\,[s\mathbf{I} - \mathbf{A}]^{-1} \begin{bmatrix} 0 \\ 2 \end{bmatrix}$$

$$[s\mathbf{I} - \mathbf{A}] = \begin{bmatrix} s & 0 \\ 0 & s \end{bmatrix} - \begin{bmatrix} 0 & 1 \\ -2 & -3 \end{bmatrix} = \begin{bmatrix} s & -1 \\ 2 & s+3 \end{bmatrix}$$

$\Rightarrow$
$$[s\mathbf{I} - \mathbf{A}]^{-1} = \frac{1}{s^2+3s+2} \begin{bmatrix} s+3 & 1 \\ -2 & s \end{bmatrix}$$

Therefore,
$$G(s) = \frac{1}{s^2+3s+2} [1 \;\; 1] \begin{bmatrix} s+3 & 1 \\ -2 & s \end{bmatrix} \begin{bmatrix} 0 \\ 2 \end{bmatrix}$$

$$= \frac{1}{s^2+3s+2} [1 \;\; 1] \begin{bmatrix} 2 \\ 2s \end{bmatrix} = \frac{2(s+1)}{s^2+3s+2}$$

(b)
$$\mathbf{\Phi}(s) = [s\mathbf{I} - \mathbf{A}]^{-1} = \frac{\begin{bmatrix} s+3 & 1 \\ -2 & s \end{bmatrix}}{(s+1)(s+2)} = \begin{bmatrix} \dfrac{s+3}{(s+1)(s+2)} & \dfrac{1}{(s+1)(s+2)} \\ \dfrac{-2}{(s+1)(s+2)} & \dfrac{s}{(s+1)(s+2)} \end{bmatrix}$$

$$= \begin{bmatrix} \dfrac{2}{s+1} - \dfrac{1}{s+2} & \dfrac{1}{s+1} - \dfrac{1}{s+2} \\ \dfrac{2}{s+2} - \dfrac{2}{s+1} & \dfrac{2}{s+2} - \dfrac{1}{s+1} \end{bmatrix}$$

$\Rightarrow$
$$\phi(t) = \begin{bmatrix} 2e^{-t} - e^{-2t} & e^{-t} - e^{-2t} \\ 2e^{-2t} - 2e^{-t} & 2e^{-2t} - e^{-t} \end{bmatrix}$$

Therefore,

$$\mathbf{x}(t) = \begin{bmatrix} 2e^{-t} - e^{-2t} & e^{-t} - e^{-2t} \\ 2e^{-2t} - 2e^{-t} & 2e^{-2t} - e^{-t} \end{bmatrix} \begin{bmatrix} 0 \\ 1 \end{bmatrix} + \int_0^t \begin{bmatrix} 2e^{-(t-\tau)} - e^{-2(t-\tau)} & e^{-(t-\tau)} - e^{-2(t-\tau)} \\ 2e^{-2(t-\tau)} - 2e^{-(t-\tau)} & 2e^{-2(t-\tau)} - e^{-(t-\tau)} \end{bmatrix} \begin{bmatrix} 0 \\ 2 \end{bmatrix} d\tau$$

$$= \begin{bmatrix} e^{-t} - e^{-2t} \\ 2e^{-2t} - e^{-t} \end{bmatrix} + \begin{bmatrix} 2\left\{ e^{-(t-\tau)} - \frac{e^{-2(t-\tau)}}{2} \right\} \\ 2\left\{ \frac{2e^{-2(t-\tau)}}{2} - e^{-(t-\tau)} \right\} \end{bmatrix}_0^t$$

$$= \begin{bmatrix} e^{-t} - e^{-2t} \\ 2e^{-2t} - e^{-t} \end{bmatrix} + \begin{bmatrix} 1 - 2e^{-t} + e^{-2t} \\ 2e^{-t} - 2e^{-2t} \end{bmatrix} = \begin{bmatrix} 1 - e^{-t} \\ e^{-t} \end{bmatrix}$$

EXAMPLE 16.36. The state variable formulation of a system is given as

$$\begin{bmatrix} \dot{x_1} \\ \dot{x_2} \end{bmatrix} = \begin{bmatrix} -2 & 0 \\ 0 & -1 \end{bmatrix} \begin{bmatrix} x_1 \\ x_2 \end{bmatrix} + \begin{bmatrix} 1 \\ 1 \end{bmatrix} u$$

$$\mathbf{y} = [1 \;\; 0] \begin{bmatrix} x_1 \\ x_2 \end{bmatrix}$$

$$x_1(0) = 0 \qquad x_2(0) = 0$$

The response $y(t)$ to a unit step input is

(A) $\frac{1}{2} - \frac{1}{2}e^{-2t}$ (B) $1 - \frac{1}{2}e^{-2t} - \frac{1}{2}e^{-t}$

(C) $e^{-2t} - e^{-t}$ (D) $1 - e^{-t}$

Solution Given,

$$\mathbf{A} = \begin{bmatrix} -2 & 0 \\ 0 & -1 \end{bmatrix} \qquad \mathbf{B} = \begin{bmatrix} 1 \\ 1 \end{bmatrix} \qquad \mathbf{C} = \begin{bmatrix} 1 & 0 \end{bmatrix}$$

We have, $$[s\mathbf{I} - \mathbf{A}] = \begin{bmatrix} s & 0 \\ 0 & s \end{bmatrix} - \begin{bmatrix} -2 & 0 \\ 0 & -1 \end{bmatrix} = \begin{bmatrix} s+2 & 0 \\ 0 & s+1 \end{bmatrix}$$

Therefore, $$[s\mathbf{I} - \mathbf{A}]^{-1} = \frac{1}{(s+2)(s+1)} \begin{bmatrix} s+1 & 0 \\ 0 & s+2 \end{bmatrix} = \begin{bmatrix} \dfrac{1}{s+2} & 0 \\ 0 & \dfrac{1}{s+1} \end{bmatrix}$$

STM $$\phi(t) = \mathscr{L}^{-1}\{[s\mathbf{I} - \mathbf{A}]^{-1}\} = \begin{bmatrix} e^{-2t} & 0 \\ 0 & e^{-t} \end{bmatrix}$$

Hence, $$\begin{aligned}
\mathbf{x}(t) &= \begin{bmatrix} e^{-2t} & 0 \\ 0 & e^{-t} \end{bmatrix} \begin{bmatrix} 1 \\ 1 \end{bmatrix} + \int_0^t \begin{bmatrix} e^{-2(t-\tau)} & 0 \\ 0 & e^{-(t-\tau)} \end{bmatrix} \begin{bmatrix} 1 \\ 1 \end{bmatrix} d\tau \\
&= \begin{bmatrix} e^{-2t} \\ e^{-t} \end{bmatrix} + \int_0^t \begin{bmatrix} e^{-2(t-\tau)} & 0 \\ 0 & e^{-(t-\tau)} \end{bmatrix} d\tau \\
&= \begin{bmatrix} e^{-2t} \\ e^{-t} \end{bmatrix} + \begin{bmatrix} \dfrac{e^{-2(t-\tau)}}{2} & 0 \\ 0 & e^{-(t-\tau)} \end{bmatrix}_0^t \\
&= \begin{bmatrix} e^{-2t} \\ e^{-t} \end{bmatrix} + \begin{bmatrix} \frac{1}{2} - \frac{1}{2}e^{-2t} \\ 1 - e^{-t} \end{bmatrix}
\end{aligned}$$

Therefore, $$\begin{aligned}
\mathbf{y}(t) &= \begin{bmatrix} 1 & 0 \end{bmatrix} \begin{bmatrix} e^{-2t} \\ e^{-t} \end{bmatrix} \begin{bmatrix} 0 \\ 0 \end{bmatrix} + \begin{bmatrix} 1 & 0 \end{bmatrix} \begin{bmatrix} \frac{1}{2} - \frac{1}{2}e^{-2t} \\ 1 - e^{-t} \end{bmatrix} \qquad \text{[see Eq. (16.46)]} \\
&= \frac{1}{2} - \frac{1}{2}e^{-2t}
\end{aligned}$$

Ans. (A)

EXAMPLE 16.37. A state variable system

$$\dot{\mathbf{x}}(t) = \begin{bmatrix} 0 & 1 \\ 0 & -3 \end{bmatrix} \mathbf{x}(t) + \begin{bmatrix} 1 \\ 0 \end{bmatrix} u(t)$$

with initial conditions $\mathbf{x}(0) = [-1 \quad 3]^{\mathrm{T}}$ and the unit step input $u(t)$ has

(a) The state transition matrix

(A) $\begin{bmatrix} 1 & \frac{1}{3}(1 - e^{-3t}) \\ 0 & e^{-3t} \end{bmatrix}$ (B) $\begin{bmatrix} 1 & \frac{1}{3}(e^{-t} - e^{-3t}) \\ 0 & e^{-t} \end{bmatrix}$

(C) $\begin{bmatrix} 1 & \frac{1}{3}(e^{-t} - e^{-3t}) \\ 0 & e^{-3t} \end{bmatrix}$ (D) $\begin{bmatrix} 1 & (1 - e^{-t}) \\ 0 & e^{-t} \end{bmatrix}$

(b) The state transition equation

(A) $\mathbf{x}(t) = \begin{bmatrix} t - e^{-t} \\ e^{-t} \end{bmatrix}$ (B) $\mathbf{x}(t) = \begin{bmatrix} t - e^{-t} \\ 3e^{-3t} \end{bmatrix}$

(C) $\mathbf{x}(t) = \begin{bmatrix} t - e^{-3t} \\ 3e^{-3t} \end{bmatrix}$ (D) $\mathbf{x}(t) = \begin{bmatrix} t - e^{-3t} \\ e^{-3t} \end{bmatrix}$

Solution

(a) Given

$$\mathbf{A} = \begin{bmatrix} 0 & 1 \\ 0 & -3 \end{bmatrix}$$

Then, the STM is calculated as follows

$$[s\mathbf{I} - \mathbf{A}] = \begin{bmatrix} s & -1 \\ 0 & s+3 \end{bmatrix}$$

$$[s\mathbf{I} - \mathbf{A}]^{-1} = \frac{1}{s(s+3)}\begin{bmatrix} s+3 & 1 \\ 0 & s \end{bmatrix} = \begin{bmatrix} \dfrac{1}{s} & \dfrac{1}{s(s+3)} \\ 0 & \dfrac{1}{s+3} \end{bmatrix} = \begin{bmatrix} \dfrac{1}{s} & \dfrac{1}{3}\left(\dfrac{1}{s} - \dfrac{1}{s+3}\right) \\ 0 & \dfrac{1}{s+3} \end{bmatrix}$$

Therefore, $\quad \phi(t) = \mathscr{L}^{-1}\{[s\mathbf{I} - \mathbf{A}]^{-1}\} = \begin{bmatrix} 1 & \frac{1}{3}(1 - e^{-3t}) \\ 0 & e^{-3t} \end{bmatrix}$

Ans. (A)

(b) The state transition equation is given by [see Eq. (16.8)]

$$\begin{aligned}
\mathbf{x}(t) &= \phi(t)\mathbf{x}(0) + \int_0^t \phi(t-\tau)\mathbf{B}\mathbf{u}(\tau)d\tau \\
&= \begin{bmatrix} 1 & \frac{1}{3}(1 - e^{-3t}) \\ 0 & e^{-3t} \end{bmatrix}\begin{bmatrix} -1 \\ 3 \end{bmatrix} + \int_0^t \begin{bmatrix} 1 & \frac{1}{3}[1 - e^{-3(t-\tau)}] \\ 0 & e^{-3(t-\tau)} \end{bmatrix}\begin{bmatrix} 1 \\ 0 \end{bmatrix}\begin{bmatrix} 1 \\ 1 \end{bmatrix} d\tau \\
&= \begin{bmatrix} -\cancel{1} + \cancel{1} - e^{-3t} \\ 0 + 3e^{3t} \end{bmatrix} + \int_0^t \begin{bmatrix} 1 \\ 0 \end{bmatrix} d\tau \\
&= \begin{bmatrix} -e^{-3t} \\ 3e^{-3t} \end{bmatrix} + \begin{bmatrix} t \\ 0 \end{bmatrix} \\
&= \begin{bmatrix} t - e^{-3t} \\ 3e^{-3t} \end{bmatrix}
\end{aligned}$$

Ans. (C)

EXAMPLE 16.38. (a) Find the state representation corresponding to the following transfer function:

$$\frac{\mathbf{Y}(s)}{\mathbf{U}(s)} = \frac{s+3}{s^2 + 2s - 3}$$

(b) Find the unit step response of the system using state transition matrix if the initial vector is given by

$$\mathbf{x}(0) = \begin{bmatrix} 1 \\ 0 \end{bmatrix}$$

Solution (a) The state representation, as we know, can be found by any of the following methods: (i) CCF, (ii) OCF, (iii) cascade decomposition and (iv) parallel decomposition. We do it by method (i). For this purpose, dividing the numerator and denominator by s^2, the transfer function becomes

$$\frac{\mathbf{Y}(s)}{\mathbf{U}(s)} = \frac{s^{-1} + 3s^{-2}}{1 + 2s^{-1} - 3s^{-2}}$$

The corresponding CCF signal flow graph is given in Figure 16.17.

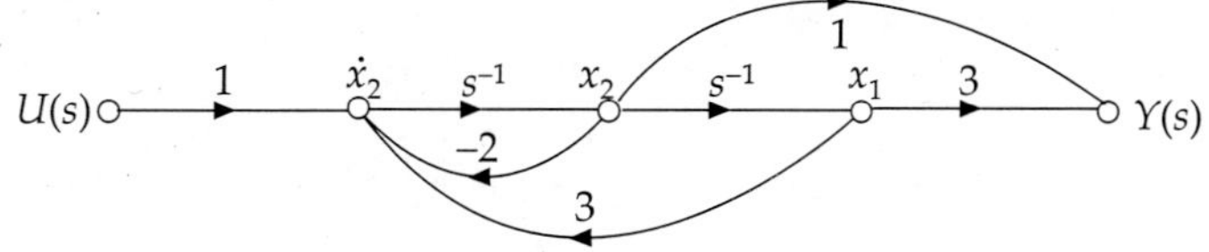

Figure 16.17: Signal flow graph (Example 16.38).

From the signal flow graph, we get

$$\begin{aligned}\dot{x}_1 &= x_2\\ \dot{x}_2 &= 3x_1 - 2x_2 + u\\ y &= 3x_1 + x_2\end{aligned}$$

Therefore, the state representation is

$$\begin{bmatrix}\dot{x}_1\\ \dot{x}_2\end{bmatrix} = \begin{bmatrix}0 & 1\\ 3 & -2\end{bmatrix}\begin{bmatrix}x_1\\ x_2\end{bmatrix} + \begin{bmatrix}0\\ 1\end{bmatrix}u$$

$$y = \begin{bmatrix}3 & 1\end{bmatrix}\begin{bmatrix}x_1\\ x_2\end{bmatrix}$$

(b)
$$[s\mathbf{I} - \mathbf{A}] = \begin{bmatrix}s & 0\\ 0 & s\end{bmatrix} - \begin{bmatrix}0 & 1\\ 3 & -2\end{bmatrix} = \begin{bmatrix}s & -1\\ -3 & s+2\end{bmatrix}$$

$$[s\mathbf{I} - \mathbf{A}]^{-1} = \frac{\begin{bmatrix}s+2 & 1\\ 3 & s\end{bmatrix}}{(s-1)(s+3)}$$

Therefore,

$$\mathbf{\Phi}(s) = [s\mathbf{I} - \mathbf{A}]^{-1} = \begin{bmatrix}\dfrac{s+2}{(s-1)(s+3)} & \dfrac{1}{(s-1)(s+3)}\\ \dfrac{3}{(s-1)(s+3)} & \dfrac{s}{(s-1)(s+3)}\end{bmatrix}$$

$$= \begin{bmatrix}\dfrac{3}{4}\cdot\dfrac{1}{s-1} + \dfrac{1}{4}\cdot\dfrac{1}{s+3} & \dfrac{1}{4}\cdot\dfrac{1}{s-1} - \dfrac{1}{4}\cdot\dfrac{1}{s+3}\\ \dfrac{3}{4}\cdot\dfrac{1}{s-1} + \dfrac{3}{4}\cdot\dfrac{1}{s+3} & \dfrac{1}{4}\cdot\dfrac{1}{s-1} + \dfrac{3}{4}\cdot\dfrac{1}{s+3}\end{bmatrix}$$

$$\phi(t) = \begin{bmatrix}\frac{3}{4}e^{t} + \frac{1}{4}e^{-3t} & \frac{1}{4}e^{t} - \frac{1}{4}e^{-3t}\\ \frac{3}{4}e^{t} - \frac{3}{4}e^{-3t} & \frac{1}{4}e^{t} + \frac{3}{4}e^{-3t}\end{bmatrix} = \frac{1}{4}\begin{bmatrix}3e^{t} + e^{-3t} & e^{t} - e^{-3t}\\ 3e^{t} - 3e^{-3t} & e^{t} + 3e^{-3t}\end{bmatrix}$$

Alternatively

The STM can be worked out by the $\mathbf{\Lambda}$-matrix method as follows. Here,

$$\mathbf{\Lambda} = \begin{bmatrix}0 & 1\\ 3 & -2\end{bmatrix}$$

The eigenvalues of the matrix can be found from the equation

$$|\lambda\mathbf{I} - \mathbf{A}| = 0$$

$$\Rightarrow \qquad \begin{bmatrix}\lambda & -1\\ -3 & \lambda+2\end{bmatrix} = 0$$

$$\Rightarrow \qquad \lambda^2 + 2\lambda - 3 = 0$$

Therefore, the eigenvalues are 1 and −3. Since the matrix $\mathbf{A}$ is of the CCF-type, the modal matrix is the Vandermonde matrix given by

$$\mathbf{M} = \begin{bmatrix} 1 & 1 \\ \lambda_1 & \lambda_2 \end{bmatrix} = \begin{bmatrix} 1 & 1 \\ 1 & -3 \end{bmatrix}$$

$$\mathbf{M}^{-1} = \frac{1}{4}\begin{bmatrix} 3 & 1 \\ 1 & -1 \end{bmatrix}$$

The diagonal $\mathbf{\Lambda}$-matrix is

$$\mathbf{\Lambda} = \begin{bmatrix} \lambda_1 & 0 \\ 0 & \lambda_2 \end{bmatrix} = \begin{bmatrix} 1 & 0 \\ 0 & -3 \end{bmatrix}$$

Therefore,

$$e^{\mathbf{\Lambda}t} = \begin{bmatrix} e^t & 0 \\ 0 & e^{-3t} \end{bmatrix}$$

Hence the STM is

$$\begin{aligned}
\phi(t) = \mathbf{M}e^{\mathbf{\Lambda}t}\mathbf{M}^{-1} &= \begin{bmatrix} 1 & 1 \\ 1 & -3 \end{bmatrix}\begin{bmatrix} e^t & 0 \\ 0 & e^{-3t} \end{bmatrix}\frac{1}{4}\begin{bmatrix} 3 & 1 \\ 1 & -1 \end{bmatrix} = \frac{1}{4}\begin{bmatrix} 1 & 1 \\ 1 & -3 \end{bmatrix}\begin{bmatrix} 3e^t & e^t \\ e^{-3t} & -e^{-3t} \end{bmatrix} \\
&= \frac{1}{4}\begin{bmatrix} 3e^t + e^{-3t} & e^t - e^{-3t} \\ 3e^t - 3e^{-3t} & e^t + 3e^{-3t} \end{bmatrix}
\end{aligned}$$

Thus, the unit step response is obtained as follows:

$$\begin{aligned}
\mathbf{x}(t) &= \frac{1}{4}\begin{bmatrix} 3e^t + e^{-3t} & e^t - e^{-3t} \\ 3e^t - 3e^{-3t} & e^t + 3e^{-3t} \end{bmatrix}\begin{bmatrix} 1 \\ 0 \end{bmatrix} + \int_0^t \frac{1}{4}\begin{bmatrix} 3e^{(t-\tau)} + e^{-3(t-\tau)} & e^{(t-\tau)} - e^{-3(t-\tau)} \\ 3e^{(t-\tau)} - 3e^{-3(t-\tau)} & e^{(t-\tau)} + 3e^{-3(t-\tau)} \end{bmatrix}\begin{bmatrix} 0 \\ 1 \end{bmatrix} d\tau \\
&= \frac{1}{4}\begin{bmatrix} 3e^t + e^{-3t} \\ 3e^t - 3e^{-3t} \end{bmatrix} + \frac{1}{4}\int_0^t \begin{bmatrix} e^{(t-\tau)} - e^{-3(t-\tau)} \\ e^{(t-\tau)} + 3e^{-3(t-\tau)} \end{bmatrix} \\
&= \frac{1}{4}\begin{bmatrix} 3e^t + e^{-3t} \\ 3e^t - 3e^{-3t} \end{bmatrix} + \frac{1}{4}\begin{bmatrix} e^{(t-\tau)} - \dfrac{e^{-3(t-\tau)}}{3} \\ e^{(t-\tau)} + e^{-3(t-\tau)} \end{bmatrix}_0^t \\
&= \frac{1}{4}\begin{bmatrix} 3e^t + e^{-3t} \\ 3e^t - 3e^{-3t} \end{bmatrix} + \frac{1}{4}\begin{bmatrix} \frac{2}{3} - e^t + \frac{1}{3}e^{-3t} \\ 2 - e^t - e^{-3t} \end{bmatrix} \\
&= \frac{1}{4}\begin{bmatrix} \frac{2}{3} + 2e^t + \frac{4}{3}e^{-3t} \\ 2 + 2e^t - 4e^{-3t} \end{bmatrix} \\
&= \begin{bmatrix} \frac{1}{6} - \frac{1}{2}e^t + \frac{1}{3}e^{-3t} \\ \frac{1}{2} + \frac{1}{2}e^t - e^{-3t} \end{bmatrix}
\end{aligned}$$

16.9 CONTROLLABILITY AND OBSERVABILITY

A system is said to be controllable if there exists some finite control vector $\mathbf{u}$ that will bring the system to any specified state $\mathbf{x}$ in the state space within a finite length of time $t = t_f$ from any state $\mathbf{x}(0)$ at $t = 0$.

Observability is the dual of controllability. Thus, the system is said to be observable at any time $t_f > 0$, a knowledge of the output $\mathbf{y}$ and the system matrices enables us to determine the initial state $\mathbf{x}(0)$.

According to these concepts, any system can be divided into the following four parts (see Figure 16.18):

A : Completely controllable and observable B : Controllable but unobservable
C : Uncontrollable but observable D : Uncontrollable and unobservable

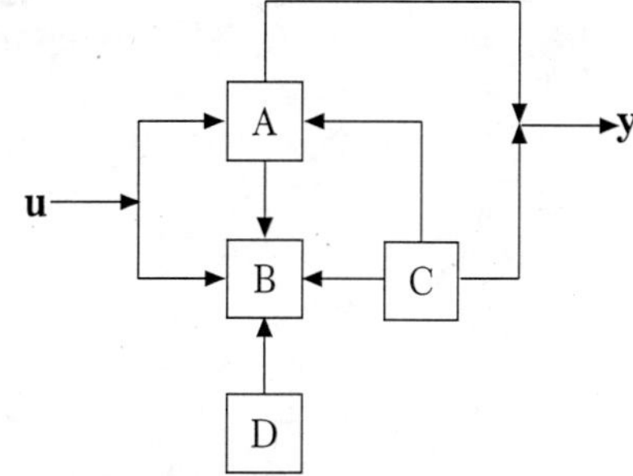

Figure 16.18: Four possible parts of a system.

Controllability Criterion

If the system is characterized by

$$\dot{\mathbf{x}} = \mathbf{Ax} + \mathbf{Bu}$$

then it is controllable *iff* the $n \times n$ matrix

$$\mathbf{G} = \left[\mathbf{B} : \mathbf{AB} : \mathbf{A^2B} : \cdots : \mathbf{A^{n-1}B}\right] \tag{16.47}$$

is of rank[10] n.

Observability Criterion

If the system is characterized by dynamical equations

$$\dot{\mathbf{x}} = \mathbf{Ax} + \mathbf{Bu}$$
$$\mathbf{y} = \mathbf{Cx} + \mathbf{Du}$$

then it is observable *iff* the $n \times n$ matrix

$$\mathbf{H} = \left[\mathbf{C}^{\mathrm{T}} : \mathbf{A}^{\mathrm{T}}\mathbf{C}^{\mathrm{T}} : \left(\mathbf{A}^{\mathrm{T}}\right)^2 \mathbf{C}^{\mathrm{T}} : \cdots : \left(\mathbf{A}^{\mathrm{T}}\right)^{n-1} \mathbf{C}^{\mathrm{T}}\right] \tag{16.48}$$

is of rank n.

EXAMPLE 16.39. Find the controllability and observability of the system

$$\begin{bmatrix} \dot{x}_1 \\ \dot{x}_2 \end{bmatrix} = \begin{bmatrix} -2 & 1 \\ 0 & -3 \end{bmatrix} \begin{bmatrix} x_1 \\ x_2 \end{bmatrix} + \begin{bmatrix} 4 \\ 1 \end{bmatrix} u$$
$$y = \begin{bmatrix} 1 & 0 \end{bmatrix} \begin{bmatrix} x_1 \\ x_2 \end{bmatrix}$$

Solution Here,

$$\mathbf{A} = \begin{bmatrix} -2 & 1 \\ 0 & -3 \end{bmatrix} \qquad \mathbf{B} = \begin{bmatrix} 4 \\ 1 \end{bmatrix} \qquad \mathbf{C} = \begin{bmatrix} 1 & 0 \end{bmatrix}$$

[10] For definition, see Section D.2 in Appendix D.

Therefore,

$$\mathbf{A}^{\mathrm{T}} = \begin{bmatrix} -2 & 0 \\ 1 & -3 \end{bmatrix} \qquad \mathbf{C}^{\mathrm{T}} = \begin{bmatrix} 1 \\ 0 \end{bmatrix}$$

To examine controllability of the system, we find that

$$\mathbf{AB} = \begin{bmatrix} -2 & 1 \\ 0 & -3 \end{bmatrix} \begin{bmatrix} 4 \\ 1 \end{bmatrix} = \begin{bmatrix} -8+1 \\ 0-3 \end{bmatrix} = \begin{bmatrix} -7 \\ -3 \end{bmatrix}$$

We need not calculate other terms of $\mathbf{G}$ of Eq. (16.47) which will be a 2×2 matrix. So,

$$\mathbf{G} = [\mathbf{B} : \mathbf{AB}] \begin{bmatrix} 4 & -7 \\ 1 & -3 \end{bmatrix}$$

$$\det \mathbf{G} = \begin{vmatrix} 4 & -7 \\ 1 & -3 \end{vmatrix} = -5$$

Since, $\mathbf{G}$ is non-singular and therefore is of rank 2, the system is controllable.

To examine observability of the system, we find

$$\mathbf{A}^{\mathrm{T}}\mathbf{C}^{\mathrm{T}} = \begin{bmatrix} -2 & 0 \\ 1 & -3 \end{bmatrix} \begin{bmatrix} 1 \\ 0 \end{bmatrix} = \begin{bmatrix} -2 \\ 1 \end{bmatrix}$$

We need not calculate further terms of $\mathbf{H}$ of Eq. (16.48) which will be a 2×2 matrix. So,

$$\mathbf{H} = [\mathbf{C}^{\mathrm{T}} : \mathbf{A}^{\mathrm{T}}\mathbf{C}^{\mathrm{T}}] = \begin{bmatrix} 1 & -2 \\ 0 & 1 \end{bmatrix}$$

$$\det \mathbf{H} = \begin{vmatrix} 1 & -2 \\ 0 & 1 \end{vmatrix} = 3$$

Since, $\mathbf{H}$ is non-singular, it is of rank 2. Therefore, the system is observable.

EXAMPLE 16.40. The second order dynamic system

$$\frac{d\mathbf{X}}{dt} = \mathbf{PX} + \mathbf{Q}u$$

$$\mathbf{y} = \mathbf{RX}$$

has the matrices as follows:

$$\mathbf{P} = \begin{bmatrix} -1 & 1 \\ 0 & -3 \end{bmatrix} \qquad \mathbf{Q} = \begin{bmatrix} 0 \\ 1 \end{bmatrix} \qquad \mathbf{R} = [0 \quad 1]$$

The system has the following controllability and observability properties:

(A) Controllable and observable (B) Not controllable but observable
(C) Controllable but not observable (D) Not controllable and not observable

Solution To check controllability, we observe

$$\mathbf{G} = [\mathbf{Q} \quad \mathbf{PQ}] = \begin{bmatrix} 0 & 1 \\ 1 & -3 \end{bmatrix}$$

$$\det \mathbf{G} = \begin{vmatrix} 0 & 1 \\ 1 & -3 \end{vmatrix} \neq 0$$

Since the matrix is non-singular and of rank = 2, therefore the system is controllable. To check observability, we note

$$\mathbf{H} = [\mathbf{R}^{\mathrm{T}} \quad \mathbf{R}^{\mathrm{T}}\mathbf{P}^{\mathrm{T}}] = \begin{bmatrix} 0 & 0 \\ 1 & -3 \end{bmatrix}$$

$$\det \mathbf{H} = \begin{vmatrix} 0 & 0 \\ 1 & -3 \end{vmatrix} = 0$$

Therefore, the system is not observable.
Ans. (C)

EXAMPLE 16.41. For the system

$$\dot{\mathbf{x}} = \begin{bmatrix} 2 & 3 \\ 0 & 5 \end{bmatrix} \mathbf{x} + \begin{bmatrix} 1 \\ 0 \end{bmatrix} u$$

which of the following statements is true?

(A) The system is controllable but unstable.
(B) The system is uncontrollable and unstable.
(C) The system is controllable and stable.
(D) The system is uncontrollable but stable.

Solution Given,

$$\mathbf{A} = \begin{bmatrix} 2 & 3 \\ 0 & 5 \end{bmatrix} \qquad \mathbf{B} = \begin{bmatrix} 1 \\ 0 \end{bmatrix}$$

To check controllability, we observe

$$\mathbf{AB} = \begin{bmatrix} 2 & 3 \\ 0 & 5 \end{bmatrix} [1] = \begin{bmatrix} 2 \\ 0 \end{bmatrix}$$

Therefore,

$$\mathbf{G} = [\mathbf{B} \quad \mathbf{AB}] = \begin{bmatrix} 1 & 2 \\ 0 & 0 \end{bmatrix}$$

This a singular matrix and its rank is not 2. So, the system is uncontrollable.

To check stability, we look at the characteristic equation [see Eq. (16.24)]

$$|s\mathbf{I} - \mathbf{A}| = 0$$

$$\Rightarrow \qquad \left| \begin{bmatrix} s & 0 \\ 0 & s \end{bmatrix} - \begin{bmatrix} 2 & 3 \\ 0 & 5 \end{bmatrix} \right| = 0$$

$$\Rightarrow \qquad \begin{vmatrix} s - 2 & -3 \\ 0 & s - 5 \end{vmatrix} = 0$$

$$\Rightarrow \qquad (s-2)(s-5) = 0$$

which points to the existence of two poles on the RHP. Hence, the system is unstable.
Ans. (B)

REVIEW QUESTIONS

16.1 Choose the correct answer

(i) A state variable approach converts an nth order system into

(A) n second order differential equations

(B) two differential equations
(C) n first order differential equations
(D) a lower order system

(ii) The matrix $\mathbf{A}$ in $\dot{\mathbf{x}} = \mathbf{A}\mathbf{x}$ for the system described by the differential equation $\ddot{y} + 2\dot{y} + 3y = 0$ is

(A) $\begin{bmatrix} 1 & 0 \\ -1 & -2 \end{bmatrix}$ (B) $\begin{bmatrix} 1 & 0 \\ -2 & -1 \end{bmatrix}$

(C) $\begin{bmatrix} 0 & 1 \\ -3 & -2 \end{bmatrix}$ (D) $\begin{bmatrix} 0 & 1 \\ -2 & 1 \end{bmatrix}$

(iii) If $\mathbf{\Lambda} = \begin{bmatrix} 1 & 0 \\ 0 & 2 \end{bmatrix}$, then $e^{\mathbf{\Lambda}t}$ will be

(A) $\begin{bmatrix} e^t & 0 \\ 0 & e^{2t} \end{bmatrix}$ (B) $\begin{bmatrix} e^t & 0 \\ 0 & 2e^t \end{bmatrix}$

(C) $\begin{bmatrix} 0 & e^t \\ e^{2t} & 0 \end{bmatrix}$ (D) $\begin{bmatrix} e^{2t} & 0 \\ 0 & e^t \end{bmatrix}$

(iv) If $\mathbf{A} = \begin{bmatrix} 0 & 1 \\ -2 & 3 \end{bmatrix}$, then eigenvalues of $\mathbf{A}$ are

(A) 0 and 3 (B) 1 and -2

(C) 1 and 2 (D) 1 and 4

(v) $\mathbf{A}$ is an $n \times n$ matrix. For the system to be controllable, the rank of the controllability matrix should be

(A) n (B) $> n$ (C) $\geq n$ (D) $\leq n$

16.2 Find the transfer function for the system given by

$$\begin{bmatrix} \dot{x}_1 \\ \dot{x}_2 \end{bmatrix} = \begin{bmatrix} 0 & 1 \\ -2 & -3 \end{bmatrix} \begin{bmatrix} x_1 \\ x_2 \end{bmatrix} + \begin{bmatrix} 0 \\ 1 \end{bmatrix} [u]$$

$$[y] = \begin{bmatrix} 1 & 0 \end{bmatrix} \begin{bmatrix} x_1 \\ x_2 \end{bmatrix}$$

16.3 A system is described by the state-space equation

$$\begin{bmatrix} \dot{x}_1 \\ \dot{x}_2 \end{bmatrix} = \begin{bmatrix} 0 & 1 \\ 0 & 0 \end{bmatrix} \begin{bmatrix} x_1 & x_2 \end{bmatrix} + \begin{bmatrix} 0 \\ 1 \end{bmatrix} \mathbf{u}$$

$$\mathbf{y} = \begin{bmatrix} 1 & 0 \end{bmatrix} \begin{bmatrix} x_1 & x_2 \end{bmatrix}$$

Find out the transfer function of the system.

16.4 Consider a state-variable model of a system

$$\begin{bmatrix} \dot{x}_1 \\ \dot{x}_2 \end{bmatrix} = \begin{bmatrix} 0 & 1 \\ -\alpha & -2\beta \end{bmatrix} \begin{bmatrix} x_1 \\ x_2 \end{bmatrix} + \begin{bmatrix} 0 \\ \alpha \end{bmatrix} r$$

$$y = \begin{bmatrix} 1 & 0 \end{bmatrix} \begin{bmatrix} x_1 \\ x_2 \end{bmatrix}$$

where y is the output, and r is the input. The damping ratio ζ and the undamped natural frequency ω_n (rad/s) of the system are given by

(A) $\zeta = \sqrt{\alpha};\ \omega_n = \dfrac{\beta}{\sqrt{\alpha}}$

(B) $\zeta = \dfrac{\sqrt{\alpha}}{\beta};\ \omega_n = \sqrt{\beta}$

(C) $\zeta = \sqrt{\beta};\ \omega_n = \sqrt{\alpha}$

(D) $\zeta = \dfrac{\beta}{\sqrt{\alpha}};\omega_n = \sqrt{\alpha}$

16.5 Express the following closed-loop transfer function in CCF and OCF and obtain corresponding state matrices.

$$\frac{Y(s)}{U(s)} = \frac{5s^2 + 2s + 6}{s^3 + 7s^2 + 11s + 8}$$

16.6 (a) Derive the expression for the transfer function from the state model

$$\dot{\mathbf{x}} = \mathbf{Ax} + \mathbf{Bu}$$

$$\mathbf{y} = \mathbf{Cx} + \mathbf{Du}$$

(b) For the electrical network shown in the following figure, determine the state model. Consider i_1, i_2 and V_c as state variables. The output variables are i_1 and i_2.

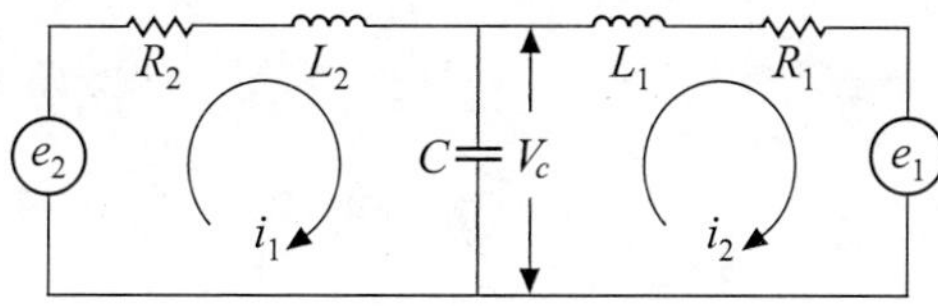

16.7 (a) The state equation of a control system is given by

$$\begin{bmatrix} \dot{x}_1 \\ \dot{x}_2 \end{bmatrix} = \begin{bmatrix} 0 & 1 \\ -2 & -3 \end{bmatrix} \begin{bmatrix} x_1 \\ x_2 \end{bmatrix}$$

Obtain the state transition matrix.

(b) Obtain the eigenvalues and eigenvectors for a system described by

$$\dot{\mathbf{x}} = \begin{bmatrix} 0 & 1 & 0 \\ 3 & 0 & 2 \\ -12 & -7 & -6 \end{bmatrix} \begin{bmatrix} x_1 \\ x_2 \\ x_3 \end{bmatrix} + \begin{bmatrix} 1 \\ 0 \\ 2 \end{bmatrix}$$

$$\mathbf{y} = \begin{bmatrix} 1 & 0 & 0 \end{bmatrix} \mathbf{x}$$

16.8 Obtain the state equation in matrix form for the network shown in the following figure.

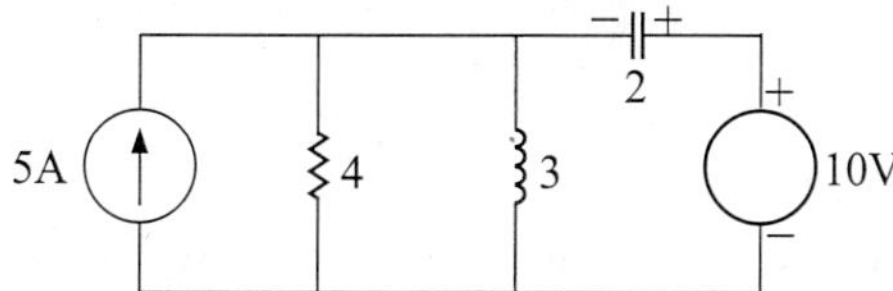

16.9 (a) Define state variable and explain its importance and use in mathematical modelling of system.

(b) Define state transition matrix and state its properties.

(c) Solve the following state equation:

$$\begin{bmatrix} \frac{d}{dt}x_1(t) \\ \frac{d}{dt}x_2(t) \end{bmatrix} = \begin{bmatrix} 0 & 1 \\ -2 & -3 \end{bmatrix} \begin{bmatrix} x_1(t) \\ x_2(t) \end{bmatrix} + \begin{bmatrix} 0 \\ 1 \end{bmatrix} r(t)$$

Assume $r(t) = 1$ for $t > 0$ and $r(t) = 0$ for $t = 0$.

16.10 (a) Find out a state-space model for the electrical network shown in the following figure. The system input is the voltage source $v_i(t)$ and the output is $v_o(t)$, the voltage across the capacitor C.

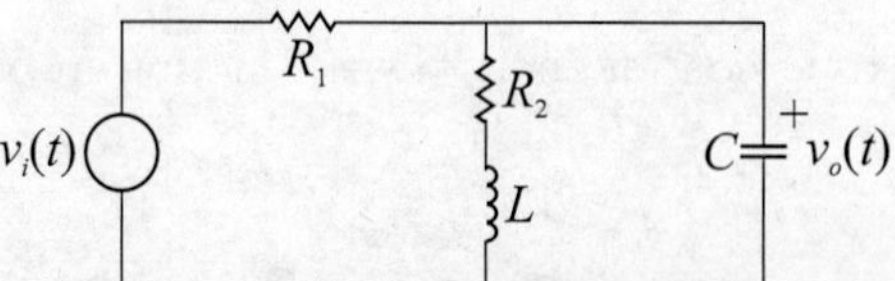

(b) Assuming zero initial condition, derive the transfer function $V_o(s)/V_i(s)$ from the state-space model.

16.11 A linear time-invariant system is described by the following state model:

$$\frac{dx_2}{dt} = -x_1 - x_2 + u$$

$$\frac{dx_1}{dt} = x_2$$

and

$$y = x_1$$

where, $u(t)$ is the system input, $y(t)$ is the system output and $[x1(t)x2(t)]^{\mathrm{T}}$ is the state vector.

(a) If $u(t)$ is a unit step, calculate the rise time, peak time, % peak overshoot and settling time (2% band) for the system.

(b) Write down the sinusoidal transfer function of the system and calculate the resonant frequency, resonant peak and -3 dB bandwidth.

16.12 The following figure shows the signal flow graph of a system.

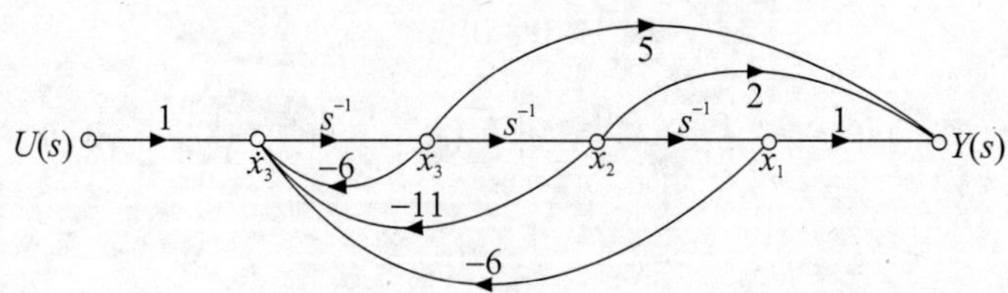

(a) Identify the forward paths and loops, and calculate the corresponding path gains and loop gains.

(b) Using Mason's gain formula obtain the transfer function $X_6(s)/X_1(s)$.

(c) If one of the poles of the system is at $s = -1$, obtain a state-space model of the system using canonical variables.

Answers to Selected Questions

16.1. (i) (B) (ii) (C) (iii) (A) (iv) (C) (v) (A)

16.2. $G(s) = \dfrac{1}{s^2+3s+2}$

16.3. $G(s) = \dfrac{1}{s^2}$

16.4. $G(s) = \dfrac{\alpha}{s^2+2\beta s+\alpha}$. *cf.* Example 16.19. Ans. (D)

16.5. CCF representation is

$$\begin{bmatrix}\dot{x}_1\\ \dot{x}_2\\ \dot{x}_3\end{bmatrix} = \begin{bmatrix}0 & 1 & 0\\ 0 & 0 & 1\\ -8 & -11 & -7\end{bmatrix}\begin{bmatrix}x_1\\ x_2\\ x_3\end{bmatrix} + \begin{bmatrix}0\\ 0\\ 1\end{bmatrix}[u] \qquad [y] = \begin{bmatrix}6 & 2 & 5\end{bmatrix}\begin{bmatrix}x_1\\ x_2\\ x_3\end{bmatrix} + [0][u]$$

OCF representation is

$$\begin{bmatrix}\dot{x}_1\\ \dot{x}_2\\ \dot{x}_3\end{bmatrix} = \begin{bmatrix}-7 & 1 & 0\\ -11 & 0 & 1\\ -8 & 01 & 0\end{bmatrix}\begin{bmatrix}x_1\\ x_2\\ x_3\end{bmatrix} + \begin{bmatrix}5\\ 2\\ 6\end{bmatrix}[u] \qquad [y] = \begin{bmatrix}1 & 0 & 0\end{bmatrix}\begin{bmatrix}x_1\\ x_2\\ x_3\end{bmatrix} + [0][u]$$

16.6. (b) $$\begin{bmatrix}\dot{i}_1\\ \dot{i}_2\\ \dot{V}_c\end{bmatrix} = \begin{bmatrix}-\dfrac{R_2}{L_2} & 0 & -\dfrac{1}{L_2}\\ 0 & -\dfrac{R_1}{L_1} & -\dfrac{1}{L_1}\\ -\dfrac{1}{C} & -\dfrac{1}{C} & 0\end{bmatrix}\begin{bmatrix}i_1\\ i_2\\ V_c\end{bmatrix} + \begin{bmatrix}0 & -\dfrac{1}{L_2}\\ -\dfrac{1}{L_1} & 0\\ 0 & 0\end{bmatrix} \qquad y = \begin{bmatrix}1 & 0 & 0\\ 0 & 1 & 0\\ 0 & 0 & 0\end{bmatrix}\begin{bmatrix}i_1\\ i_2\\ V_c\end{bmatrix}$$

16.7. (a) $\phi(t) = \begin{bmatrix}2e^{-t}-e^{-2t} & e^{-t}-e^{-2t}\\ 2e^{-2t}-2e^{-t} & 2e^{-2t}-e^{-t}\end{bmatrix}$

(b) For $\lambda_1 = -1,\ \mathbf{p}_1 = \begin{bmatrix}1\\ -1\\ -1\end{bmatrix}$, $\lambda_2 = -2,\ \mathbf{p}_2 = \begin{bmatrix}2\\ -4\\ 1\end{bmatrix}$ $\lambda_3 = -3,\ \mathbf{p}_3 = \begin{bmatrix}1\\ -3\\ 3\end{bmatrix}$

16.8. $$\begin{bmatrix}\dot{i}_L\\ \dot{V}_c\end{bmatrix} = \begin{bmatrix}0 & -\dfrac{1}{3}\\ -\dfrac{1}{2} & 0\end{bmatrix}\begin{bmatrix}i_L\\ V_c\end{bmatrix} + \begin{bmatrix}-\dfrac{10}{3}\\ \dfrac{5}{2}\end{bmatrix}u(t)$$

16.9. (c) $x(t) = \begin{bmatrix}-\dfrac{1}{2}+e^t-\dfrac{e^{2t}}{2}\\ e^t-e^{2t}\end{bmatrix}$ assuming $\mathbf{x}(0) = \begin{bmatrix}0\\ 0\end{bmatrix}$

16.10. (a) $$\begin{bmatrix}\dot{i}_L\\ \dot{v}_o\end{bmatrix} = \begin{bmatrix}-\dfrac{R_2}{L} & \dfrac{1}{L}\\ -\dfrac{1}{C} & -\dfrac{1}{R_1C}\end{bmatrix}\begin{bmatrix}i_L\\ v_a\end{bmatrix} + \begin{bmatrix}-\dfrac{2}{L}\\ \dfrac{1}{R_2C}\end{bmatrix}v_i \qquad y = \begin{bmatrix}0 & 1\end{bmatrix}\begin{bmatrix}i_L\\ v_a\end{bmatrix}$$

(b) $G(s) = \dfrac{sL+2R_1+R_2}{s^2R-1LC+s(R_1R_2C+L)+R_1+R_2}$

16.11. (a) $t_r = 2.42$ s, $t_p = 3.63$ s, $M_p = 12.31\%$, $t_s = 8$ s

(b) $G(\jmath\omega) = \dfrac{1}{-\omega^2+\jmath\omega+1}$, $\omega_r = 0.707$ rad/s, $M_r = 1.15$, Bw $= 1.27$ rad/s

16.12. (a) Forward paths and gains: $p_1 = s^{-3}$, $p_2 = 2s^{-2}$, $p_3 = 5s^{-1}$

Loops and gains: $L_1 = -6s^{-1}$, $L_2 = -11s^{-2}$, $L_3 = 6s^{-3}$

(b) $\dfrac{Y(s)}{U(s)} = \dfrac{5s^2 + 2s + 1}{s^3 + 6s^2 + 11s + 6}$ (c) $\begin{bmatrix} \dot{x}_1 \\ \dot{x}_2 \\ \dot{x}_3 \end{bmatrix} = \begin{bmatrix} 0 & 1 & 0 \\ 0 & 0 & 1 \\ -6 & -11 & -6 \end{bmatrix} \begin{bmatrix} x_1 \\ x_2 \\ x_3 \end{bmatrix} + \begin{bmatrix} 0 \\ 0 \\ 1 \end{bmatrix} u$

$$y = \begin{bmatrix} 1 & 2 & 5 \end{bmatrix} \begin{bmatrix} x_1 \\ x_2 \\ x_3 \end{bmatrix}$$

17

Discrete Time Systems

17.1 INTRODUCTION

So far we have considered analog systems with continuous control signals. With the advent of digital computers and a very powerful digital electronics, control signals are sent or the feedback is received in the form of discrete pulses. Systems incorporating digital monitoring and control are collectively called *discrete time systems*.

The numbers that enter a digital control system do so at some fixed time interval, T, called the *sampling period*. Consider the typical digital control system presented in Figure 17.1.

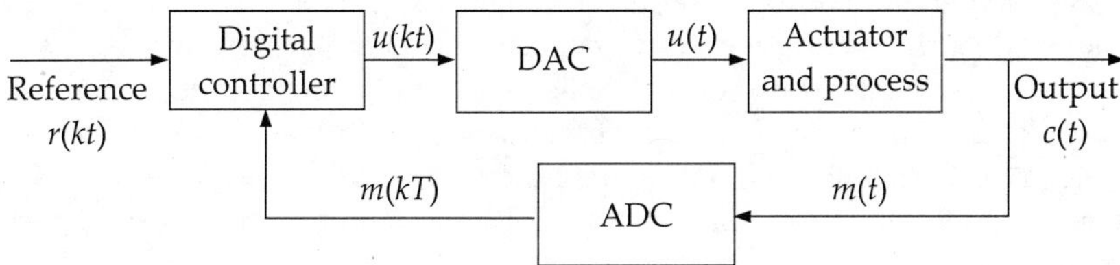

Figure 17.1: Typical digital control system.

Here, the reference input $r(kT)$ is a train of sampled values. The variables $r(kT)$, $u(kT)$ and $m(kT)$ are discrete signals in contrast to continuous signals $u(t)$, $c(t)$ and $m(t)$.

The sampling is done by a sampler which is basically a switch that becomes ON for an instant every T seconds (see Figure 17.2). So it can be thought of producing a train of impulse functions $\delta(t)$.

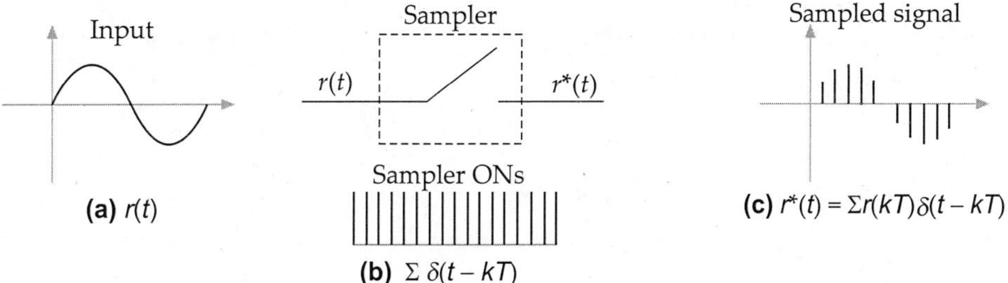

Figure 17.2: The sampling process.

When we sample a signal $r(t)$ to produce a sampled signal $r^*(t)$, we actually produce a train of impulses of amplitude $r(kT)$ starting at $t = 0$ and spaced T seconds apart. We will now discuss a mathematical tool, called the *z-transform*, to deal with such signals.

17.2 THE z-TRANSFORM

The Laplace transform converts linear differential equations to algebraic equations and thus makes their manipulation easy. The z-transform is a similar mathematical tool to deal with linear difference equations or digital data represented by sequences.

Let a discrete set of data be represented by a series of impulses, $\delta(t)$, of amplitudes $x(kT)$ where T is the time interval of sampling of data and $k = 0, 1, 2, \ldots \infty$. Then the equation for the entire series of impulses can be written as

$$\begin{aligned} x^*(t) &= x(0)\delta(t) + x(T)\delta(t-T) + x(2T)\delta(t-2T) + \cdots \\ &= \sum_{k=0}^{\infty} x(kT)\delta(t-kT) \qquad \text{for } t > 0 \end{aligned} \tag{17.1}$$

The Laplace transform of Eq. (17.1) yields

$$\mathscr{L}\{x^*(t)\} \equiv X^*(s) = \sum_{k=0}^{\infty} x(kT)e^{-ksT} \tag{17.2}$$

Equation (17.2) represents an infinite series involving the factor e^{-sT} and its powers. The transfer functions of continuous-data systems contain rational functions of s and are thus easy to manipulate. Here the appearance of e^{-sT} shows the difficulty of using Laplace transform for the general treatment of discrete-data systems because the transfer function relations will no longer be algebraic as they are in continuous-data systems. When the terms involving e^{-sT} appear in a transfer function, other than as a multiplying factor, it makes the inverse Laplace transform difficult.

Incidentally, we can make operations on infinite sequences a lot easier by transforming the irrational function $X^*(s)$ to a rational function, say $X(z)$, through a transformation from the complex s-plane to another complex z-plane by defining

$$z = e^{sT} \tag{17.3}$$

which leads to

$$s = \frac{1}{T}\ln z \tag{17.4}$$

From Eq. (17.3), we observe that the real and imaginary parts of the complex variable z are related to those of s by the following equations:

$$\begin{aligned} \Re\{z\} &= e^{\sigma T}\cos\omega T \\ \Im\{z\} &= e^{\sigma T}\sin\omega T \end{aligned}$$

since $s = \sigma + \jmath\omega$.

The relationship between z and s, enunciated by Eq. (17.3), is called the z-transformation. Substituting Eq. (17.3) in Eq. (17.2), we get

$$\boxed{X^*\left(s = \frac{1}{T}\ln z\right) \equiv X(z) = \sum_{k=0}^{\infty} x(kT)z^{-k}} \tag{17.5}$$

The z-transform is usually defined in either of the following two ways:

1. If values of $x(t)$ are defined as the sequence $x(kT)$, where $k = 0, 1, 2\cdots$, then they indicate the sampled values of $x(t)$ at time $t = kT$.

2. If the sequence of numbers of events of $x[n]$ for $n = 0,\ 1,\ 2\ldots$ is given, the z-transform of $x[n]$ is defined as

$$\mathcal{Z}\{x[n]\} = \sum_{n=0}^{\infty} x[n]z^{-n} \tag{17.6}$$

We note that in this case the sampling process and the sampling interval are not explicit. If the sequence is given in discrete form, $x[n]$, then it is written as

$$\boxed{X(z) = \sum_{n=0}^{\infty} x[n]z^{-n} \quad \text{or, shorthand:} \quad x[n] \overset{\mathcal{Z}}{\longleftrightarrow} X(z)}$$

Note:
1. Like the Laplace transform, the transformed quantity (here, X) is written in capital letter.
2. Like the Laplace plane is called the s-plane, we will call the complex plane z-plane.
3. In our discussion of the z-transform, we will use $x(kT)$ and $x[n]$ interchangeably.

Region of Convergence

Any time we consider a summation or integral with infinite limits, we must think about convergence. For example, the series

$$1 + \frac{1}{2} + \frac{1}{4} + \frac{1}{8} + \cdots = \frac{1}{1 - \frac{1}{2}} = 2$$

converges, whereas does the series

$$1 - 1 + 1 - 1 + 1 - \cdots$$

converge?

The region of convergence (ROC) for a given $x[n]$ is defined as the range of z for which the z-transform converges. Since the z-transform is a power series, it converges when $x[n]z^{-n}$ is absolutely summable. Stated differently,

$$\sum_{n=-\infty}^{\infty} \left|x[n]z^{-n}\right| < \infty$$

must be satisfied for convergence. Thus, written symbolically,

$$\text{ROC} = \left\{ z : \sum_{n=-\infty}^{\infty} \left|x[n]z^{-n}\right| < \infty \right\} \tag{17.7}$$

The z-transform $X(z)$ of $x[n]$ is unique when and only when the ROC is specified.

Properties of the ROC

1. The ROC of $X(z)$ consists of a ring in the z-plane centred around the origin (equivalent to a vertical strip in the s-plane).
2. The ROC does not contain any poles.
3. If $x[n]$ is finite, then the ROC is the entire z-plane, except possibly at $z = 0$ and/or $z = \infty$.
4. If $x[n]$ is a right-sided sequence , and if $|z| = r_o$ is in the ROC, then all *finite* values of z for which $|z| > r_o$ are also in the ROC [see Figure 17.3a].

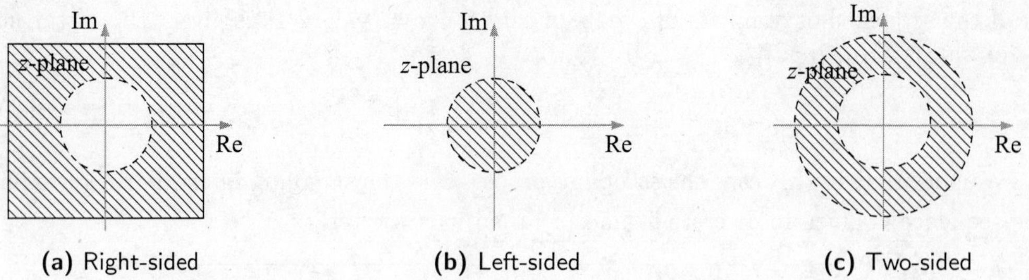

(a) Right-sided (b) Left-sided (c) Two-sided

Figure 17.3: ROCs of $X(z)$ for infinite duration sequences.

5. If $x[n]$ is a left-sided sequence, and if $|z| = r_o$ is in the ROC, then all *finite* values of z for which $0 < |z| < r_o$ are also in the ROC. However, if $x[n] = 0$ for all $n > 0$, then the ROC will include $z = 0$ [see Figure 17.3b].
6. If $x[n]$ is two-sided, and if $|z| = r_o$ is in the ROC, then the ROC consists of a ring in the z-plane including the circle $|z| = r_o$ [see Figure 17.3c].

We now work out ROC for z-transforms of a few typical sequences.

EXAMPLE 17.1. If $x[n] = \delta[n]$, find $X(z)$and ROC of the function defined as

$$\delta[k] = \begin{cases} 1 & \text{for } k = 0 \\ 0 & \text{for } k > 0 \end{cases}$$

Solution The function looks something like Figure 17.4a.

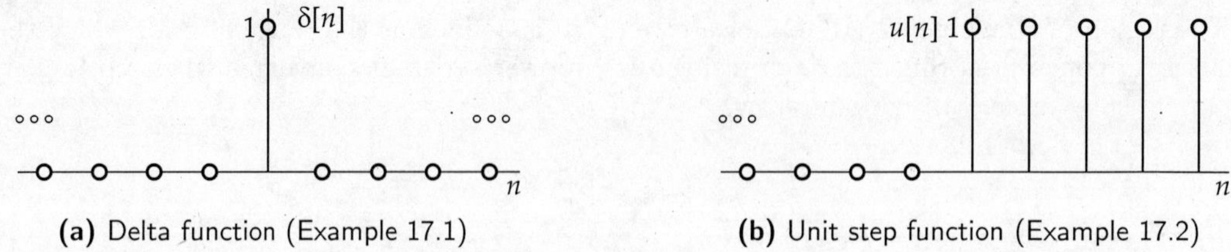

(a) Delta function (Example 17.1) (b) Unit step function (Example 17.2)

Figure 17.4: Discrete function plots.

Its z-transform is

$$X(z) = \sum_{k=0}^{\infty} \delta[k] z^{-k} = (1)z^{-0} + \sum_{k=1}^{\infty} (0) z^{-k} = 1$$

Clearly, this transform exists only when $|z| > 0$. This means that the entire z-plane is the ROC for this transform.

EXAMPLE 17.2. Determine the z-transform and ROC of the function given by

$$u[n] = \begin{cases} 1 & \text{for } k \geq 0 \\ 0 & \text{for } k < 0 \end{cases}$$

Solution This function looks something like Figure 17.4b. The z-transform can be obtained as follows:

$$\mathcal{Z}\{u[n]\} = \sum_{k=0}^{\infty} u[n]z^{-k} = \sum_{k=0}^{\infty} z^{-k} = 1 + \frac{1}{z} + \frac{1}{z^2} + \cdots$$

$$= \frac{1}{1-\frac{1}{z}} = \frac{z}{z-1} \qquad \text{ROC: } |z| > 1$$

The condition for the valid region of transform, i.e. $|z| > 1$, is the ROC here.

Note: The function is the unit step sequence.

EXAMPLE 17.3. Determine the z-transform and ROC for $a^{kT}, \quad k \geq 0$.

Solution The z-transform is found as follows:

$$\mathcal{Z}\{a^{kT}\} = \sum_{k=0}^{\infty} a^{kT}z^{-k} = \sum_{k=0}^{\infty} \left(a^T z^{-1}\right)^k = \frac{1}{1-a^T z^{-1}} \qquad \left|a^T z^{-1}\right| < 1$$

$$= \frac{z}{z-a^T} \qquad \text{ROC: } |z| > a^T$$

EXAMPLE 17.4. Find z-transform for

$$x[n] = 6, \underset{\uparrow}{3}, 0, \pi$$

where the $\uparrow$ indicates the time of origin, i.e. $n = 0$, of the sequence.

Solution

$X(z) = 6z + 3 + \pi z^{-2}$

ROC: For a finite duration signal, the entire z-plane except, possibly $z = 0$ and $z = \infty$

Note: For $k > 0$: $z^k = \infty$ for $z = \infty$, and $z^{-k} = \infty$ for $z = 0$; everywhere else the polynomials in z and z^{-2} are finite.

EXAMPLE 17.5. Find $X(z)$ when

(a) $x[n] = p^n u[n]$ and
(b) $x[n] = -p^n u[-n-1]$

where $p \neq 0$.

Solution (a) Here,

$$X(z) = \sum_{n=-\infty}^{\infty} x[n]z^{-n} = \sum_{n=0}^{\infty} p^n z^{-n} = \sum_{n=0}^{\infty} (pz^{-1})^n = 1 + \left(\frac{p}{z}\right) + \left(\frac{p}{z}\right)^2 + \left(\frac{p}{z}\right)^3 + \cdots = \frac{1}{1-pz^{-1}}$$

The sequence converges *iff* $|pz^{-1}| < 1$, i.e. if $|z| > |p|$. Thus,

$$p^n u[n] \overset{\mathcal{Z}}{\longleftrightarrow} \frac{1}{1-pz^{-1}} = \frac{z}{z-p} \qquad \text{for } |z| > |p| \tag{i}$$

Note: 1. If the sequence and the ROC are plotted, they will look like as shown Figure 17.5. This kind of sequence is called a **causal** sequence. A sequence is said to be causal when its output does not depend on any 'future' inputs. In more colourful terms, a sequence is causal if it does not 'laugh' before it is 'tickled'. In practical terms, a causal sequence is right-sided.

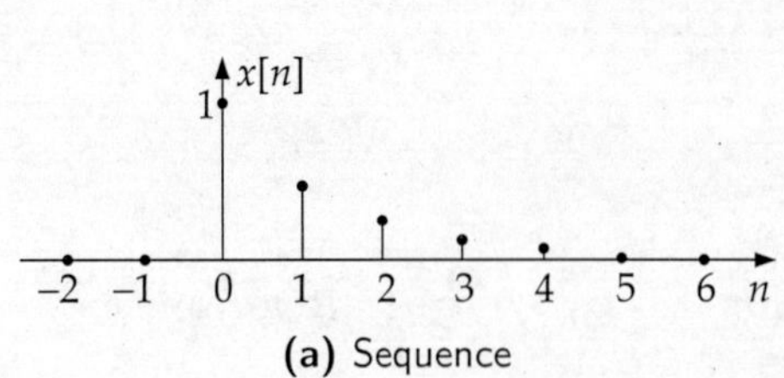

(a) Sequence

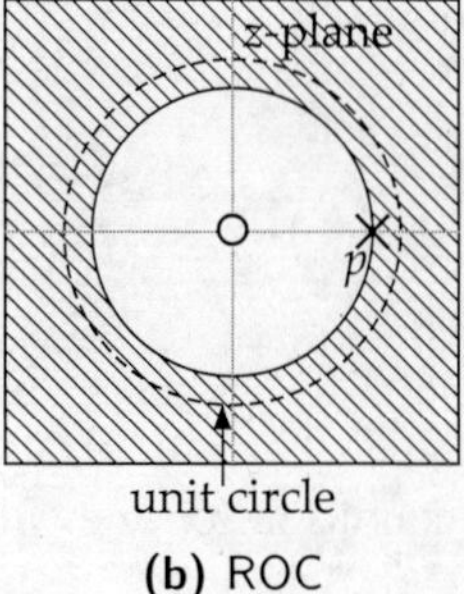

(b) ROC

Figure 17.5: Causal sequence and ROC.

2. The ROC here shades outside radius $|p|$. Smaller $|p|$ means faster decay means larger ROC. The ROC of a causal signal is the *exterior* of a circle of some radius r.
3. Important special case: $p = 1$ leaves the unit step function with

$$u[n] \overset{z}{\longleftrightarrow} U(z) = \frac{1}{1 - z^{-1}} \qquad |z| > 1$$

(b) In this case,

$$X(z) = \sum_{n=-\infty}^{-1} -p^n z^{-n} = -\sum_{k=1}^{\infty} (p^{-1}z)^k = -(p^{-1}z)\sum_{k=0}^{\infty} (p^{-1}z)^k = -(p^{-1}z)\frac{1}{1-(p^{-1}z)}$$
$$= \frac{1}{1 - pz^{-1}}$$

The sequence converges *iff* $|pz^{-1}| < 1$, i.e. if $|z| < |p|$. Thus,

$$-p^n u[-n-1] \overset{z}{\longleftrightarrow} \frac{1}{1 - pz^{-1}} \qquad \text{for } |z| < |p| \tag{ii}$$

Note: 1. If the sequence and ROC are plotted, they will look like what is shown Figure 17.6. This

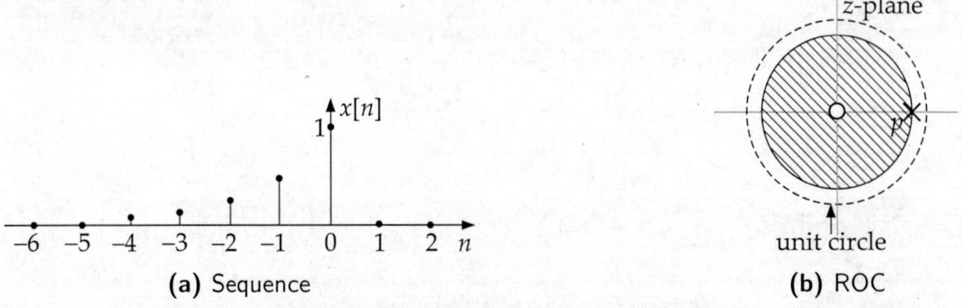

(a) Sequence (b) ROC

Figure 17.6: Anticausal sequence and ROC.

kind of sequence is called an *anticausal* sequence. In an anticausal sequence, the output anticipates the input one sample into the future. For example,

$$y(n) = x(n + 1)$$

where y is the output and x, the input, is an anticausal sequence. An anticausal sequence is, therefore, left-sided.

2. The ROC here shades inside radius $|p|$. The ROC of an anticausal signal is the *interior* of a circle of some radius r.
3. Equations (i) and (ii) have the same expression for $X(z)$. Thus, the *ROC is essential* for resolving this ambiguity!

EXAMPLE 17.6. The symbols, a and T, represent positive quantities, and $u(t)$ is the step function. Which of the following impulse responses is NOT the output of a causal linear time-invariant system?

(A) $e^{-a(t+T)}u(t)$ (B) $1+e^{-at}u(t)$ (C) $e^{-a(t-T)}u(t)$ (D) $e^{+at}u(t)$

Solution A linear time-invariant system is causal if the impulse response $h(t)=0$ where $t<0$. Options (A), (C) and (D) show that the corresponding impulse responses $h(t)=0$ where $t<0$. So, they are causal. However, for option (B), i.e. $h(t)=1+e^{-at}u(t)$, it is obvious that $h(t)=1$ for $t<0$. Hence, it is non-causal.
Ans. (B)

EXAMPLE 17.7. Let

$$X(z)=\frac{1}{1-z^{-3}}$$

be the z-transform of a causal signal $x[n]$. Then, the values of $x[2]$ and $x[3]$ are

(A) 0 and 0 (B) 0 and 1 (C) 1 and 0 (D) 1 and 1

Solution Given,

$$\begin{aligned}X(z)=\frac{1}{1-z^{-3}}&\equiv 1+z^{-3}+z^{-6}+z^{-9}+\ldots\\&=1+(0)z^{-1}+(0)z^{-2}+(1)z^{-3}+(0)z^{-4}+(0)z^{-5}+(1)z^{-6}+\ldots\\&\equiv x[0]+x[1]z^{-1}+x[2]z^{-2}+x[3]z^{-3}+\ldots\end{aligned}$$

Hence, $x[2]=0$ and $x[3]=1$.
Ans. (B)

EXAMPLE 17.8. Consider the discrete-time system shown in the following figure where the impulse response of $G(z)$ is $g(0)=0,\ g(1)=g(2)=1,\ g(3)=g(4)=\ldots=0$.

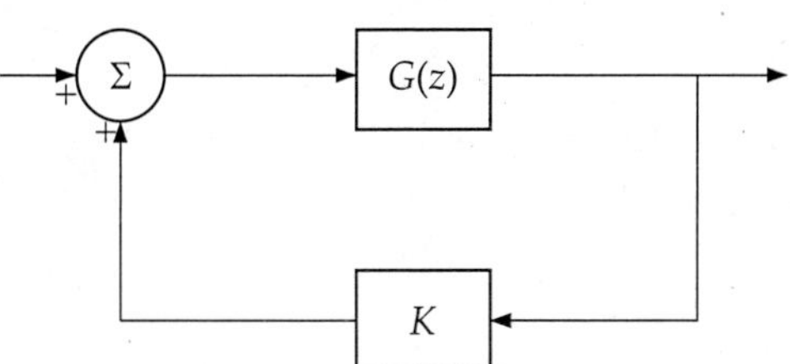

The system is stable for range of values of K

(A) $[-1,1/2]$ (B) $[-1,1]$ (C) $[-1/2,1]$ (D) $[-1/2,2]$

Solution The given condition is $g(0)=0,\ g(1)=g(2)=1,\ g(3)=g(4)=\ldots=0$. Which means,

$$g[n]=\delta[n-1]+\delta[n-2] \quad \text{[here, } \delta \text{ means Kronecker delta]} \tag{i}$$

From the diagram and given condition [Eq. (i)], we get the system response as

$$H(z) = \frac{G(z)}{1 - KG(z)}$$

From Eq. (i), $$g[n] = \delta[n-1] + \delta[n-2]$$

Therefore, $$G(z) = z^{-1} + z^{-2}$$

Thus, $$H(z) = \frac{(z^{-1} + z^{-2})}{1 - K(z^{-1} + z^{-2})} = \frac{z+1}{z^2 - Kz - K}$$

The characteristic equation is

$$z^2 - Kz - K \qquad \text{(ii)}$$

For the system to be stable, poles should lie inside the unit circle. This means that

$$|z| \leq 1$$

From Eq. (ii) $$z = \frac{K \pm \sqrt{K^2 + 4K}}{2} \leq 1$$

or, $$K \pm \sqrt{K^2 + 4K} \leq 2$$

or, $$\sqrt{K^2 + 4K} \leq 2 - K$$

or, $$\cancel{K^2} + 4K \leq 4 - 4K + \cancel{K^2}$$

or, $$8K \leq 4$$

or, $$K \leq \frac{1}{2}$$

Ans. (A)

EXAMPLE 17.9. The following discrete-time equations result from the numerical integration of the differential equations of an undamped simple harmonic oscillator with state variables x and y. The integration time step is h.

$$\frac{x_{k+1} - x_k}{h} = y_k$$

$$\frac{y_{k+1} - y_k}{h} = -x_k$$

For the discrete-time system, which one of the following statements is TRUE?

(A) The system is not stable for $h > 0$

(B) The system is stable for $h > \frac{1}{\pi}$

(C) The system is stable for $0 < h < \frac{1}{2\pi}$

(D) The system is stable for $\frac{1}{2\pi} < h < \frac{1}{\pi}$

Solution Since x and y are state variables and the integration step is h, we transform the given equations as

$$x_{k+1} = x_k + hy_k$$

$$y_{k+1} = -hx_k + y_k$$

In matrix form, $$\begin{bmatrix} \dot{x}_k \\ \dot{y}_k \end{bmatrix} = \begin{bmatrix} 1 & h \\ -h & 1 \end{bmatrix} \begin{bmatrix} x_k \\ y_k \end{bmatrix} \qquad \text{(i)}$$

Equation (i) can be written as

$$\dot{\mathbf{x}} = \mathbf{A}\mathbf{x}$$

where,
$$\mathbf{A} = \begin{bmatrix} 1 & h \\ -h & 1 \end{bmatrix}$$

For the transfer function in the z-domain, we calculate $[z\mathbf{I} - \mathbf{A}]$ as follows

$$z\mathbf{I} - \mathbf{A} = \begin{bmatrix} z-1 & -h \\ h & z-1 \end{bmatrix}$$

Therefore, characteristic equation
$$|z\mathbf{I} - \mathbf{A}| = (z-1)^2 + h^2 = 0$$

We note from the characteristic equation that poles lie at $z = 1 \pm \jmath h$ which are outside the unit circle for $h > 0$. Hence, the system is unstable for $h > 0$.
Ans. (A)

We will now discuss a few properties of the z-transform before we work out more examples.

Useful Properties of z-Transform

In the following discussions we will use $x[n]$ and $x(kT)$ interchangeably.

Linearity. If $x_1[n] \overset{\mathcal{Z}}{\longleftrightarrow} X_1(z)$ and $x_2[n] \overset{\mathcal{Z}}{\longleftrightarrow} X_2(z)$, then

$$x[n] = a_1 x_1[n] + a_2 x_2[n] \overset{\mathcal{Z}}{\longleftrightarrow} a_1 X_1(z) + a_2 X_2(z) \tag{17.8}$$

This means that the z-transform of the linear combination of two signals is the linear combination of the individual z-transforms.

EXAMPLE 17.10. Determine the z-transform of

$$x(t) = 2\delta(t) + \delta(t+T) + 4\delta(t-3T) - \delta(t-5T) \tag{i}$$

Solution From the linearity theorem, we have

$$\begin{aligned} X(z) &= \mathcal{Z}\{2\delta(t)\} + \mathcal{Z}\{\delta(t+T)\} + \mathcal{Z}\{4\delta(t-3T)\} - \mathcal{Z}\{\delta(t-5T)\} \\ &= 2 + z + 4z^{-3} - z^{-5} \\ &= z + 2 + 4z^{-3} - z^{-5} \end{aligned} \tag{ii}$$

EXAMPLE 17.11. Consider the discrete-time signal

$$x[n] = u[-n+5] - u[n+3]$$

where

$$u[n] = \begin{cases} 1 \text{ if } n \geq 0 \\ 0 \text{ if } n < 0 \end{cases}$$

The smallest n for which $x[n] = 0$ is ______

Solution The three signals, namely, $u[-n+5]$, $u[n+3]$ and $x[n]$ are shown in the following figure. According to the given condition, the first signal starts from $n = +5$ and extends to $n = -\infty$, whereas the second signal starts at $n = -3$ and extends to $n = +\infty$.

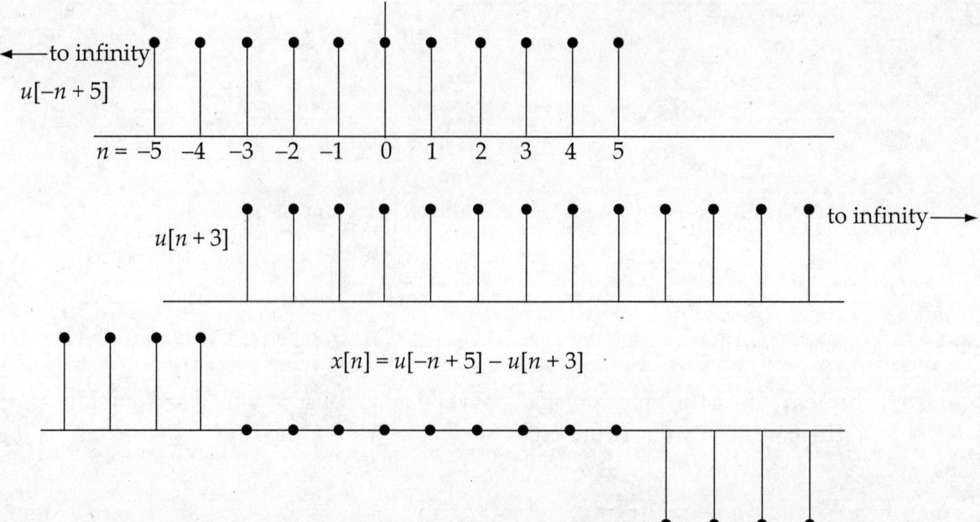

The third signal $x[n]$ is shown at the bottom. From the figure, it is clear that the smallest n for which $x[n] = 0$ is -3.

Note:
1. z-transforms of $\delta(t+T)$, $\delta(t-3T)$, $\delta(t-5T)$ will be discussed in the time shifting property below.
2. The ROC of the sum contains at least as much of the z-plane as the intersection of the ROCs.
3. In the sequence representation, Eq. (ii) is written as follows:

$$x[n] = 1,\ \underset{\uparrow}{2},\ 0,\ 0,\ 4,\ 0,\ -1$$

where the $\uparrow$ indicates the time of origin, i.e. $n = 0$, of the sequence. The sequence can be obtained by substituting $t = -T,\ 0,\ T,\ 3T$, and $5T$ in Eq. (i).

EXAMPLE 17.12. Find $x[n] \overset{\mathcal{Z}}{\longleftrightarrow} \left[\left(\frac{1}{2}\right)^n u[n] + \left(\frac{1}{4}\right)^n u[n]\right]$

Solution The z-transform of $\left(\frac{1}{2}\right)^n u[n]$ is

$$X_1(z) = \mathcal{Z}\left\{\left(\frac{1}{2}\right)^n u[n]\right\} = \sum_{-\infty}^{\infty}\left(\frac{1}{2}\right)^n z^{-n}$$

$$= \left(\frac{1}{2}\right)^0 z^{-0} + \left(\frac{1}{2}\right)^1 z^{-1} + \left(\frac{1}{2}\right)^2 z^{-2} + \cdots = 1 + \frac{1}{2z} + \frac{1}{4z^2} + \cdots$$

$$= \frac{1}{1-\frac{1}{2z}} = \frac{z}{z-\frac{1}{2}} \qquad \text{ROC: } |z| > \frac{1}{2} \tag{i}$$

Similarly, for

$$X_2(z) = \mathcal{Z}\left\{\left(\frac{1}{4}\right)^n u[n]\right\} = \frac{z}{z+\frac{1}{4}} \qquad \text{ROC: } |z| > \frac{-1}{4} \tag{ii}$$

Due to linearity,

$$X(z) = X_1(z) + X_2(z) = \frac{z}{z-\frac{1}{2}} + \frac{z}{z+\frac{1}{4}} = \frac{z\left(2z-\frac{1}{8}\right)}{\left(z-\frac{1}{2}\right)\left(z+\frac{1}{4}\right)} \tag{iii}$$

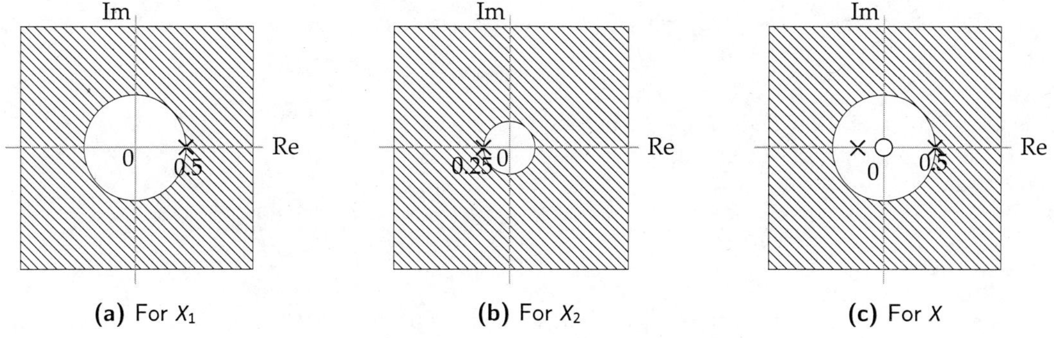

(a) For X_1 (b) For X_2 (c) For X

Figure 17.7: ROCs.

From Eq. (i), we find that there is a pole at $z = 0.5$ and a zero at $z = 0$. So, the ROC picture is as shown in Fig. 17.7a. Eq. (ii) shows a pole at $z = -0.25$ and a zero at $z = 0$ [see Fig. 17.7b]. However, Eq. (iii) indicates two poles at $z = -0.25$ and 0.5 and two zeros at $z = 0$ and 0.0625 [see Fig. 17.7c]. Since there cannot be poles within the ROC, the ROC is $|z| > 0.5$.

Time Shifting (or Real Translation). If $x[n] \xleftrightarrow{\mathcal{Z}} X(z)$ then $x[n-k] \xleftrightarrow{\mathcal{Z}} z^{-1}X(z)$.

Note: 1. The transform variable z^{-1} corresponds to a delay of one sampling period in the time domain. This is why z^{-1} is usually referred to as a *delay operator* in digital control systems.

2. This is why z, z^{-3}, z^{-5} were written in z-transforms of $\delta(t+T)$, $\delta(t-3T), \delta(t-5T)$ in Example 17.10.

Scaling. If $x[n] \xleftrightarrow{\mathcal{Z}} X(z)$ with ROC $= \{r_1 < |z| < r_2\}$, then $a^n x[n] \xleftrightarrow{\mathcal{Z}} X(az^{-1})$ with ROC $= \{|a|r_1 < |z| < |a|r_2\}$.

EXAMPLE 17.13. Using the scaling theorem, determine the z-transform of $x[n] = a^n \cos(\omega n)u[n]$.

Solution From Appendix *z-Transform Tables* at page 536 we see that

$$\mathcal{Z}\{\cos\omega\} = \frac{z^2 - z\cos\omega}{z^2 - 2z\cos\omega + 1} = \frac{1 - z^{-1}\cos\omega}{1 - 2z^{-1}\cos\omega + z^{-2}} \qquad |z| > 1$$

Substituting $z^{-1} = az^{-1}$ according to the scaling theorem, we get

$$\mathcal{Z}\{a^n \cos\omega\} = \frac{1 - az^{-1}\cos\omega}{1 - 2az^{-1}\cos\omega + a^2 z^{-2}} = \frac{z^2 - az\,\cos\omega}{z^2 - 2az\cos\omega + a^2} \qquad |z| > |a|$$

EXAMPLE 17.14. Determine the z-transform of $x[n] = 2^n u[n-2]$.

Solution We know,

$$\mathcal{Z}\{u[n]\} = \frac{z}{z-1} \qquad |z| > 1$$

By the time-shifting theorem

$$\mathcal{Z}\{u[n-2]\} = z^{-2} \cdot \frac{z}{z-1} = \frac{z^{-1}}{z-1} \qquad |z| > 1$$

Now using the scaling theorem, we get

$$\mathcal{Z}\{2^n u[n-2]\} = \frac{2z^{-1}}{\frac{z}{2} - 1} = \frac{4}{z(z-2)} \qquad |z| > 2$$

Time Reversal. If $x[n] \overset{\mathcal{Z}}{\longleftrightarrow} X(z)$ with ROC $=\{r_1 < |z| < r_2\}$, then $x[-n] \overset{\mathcal{Z}}{\longleftrightarrow} X(z^{-1})$ with ROC $= \{1/r_2 < |z| < 1/r_1$.

Real Convolution of Two Sequences. The convolution of two sequences $x_1(kT)$ and $x_2(kT)$, generating a sequence $x(kT)$ is defined as

$$x(kT) \equiv x_1(kT) \star x_2(kT) = \sum_{n=0}^{k} x_1[(k-n)T]x_2(nT) \tag{17.9}$$

The z-transform of the convolution of two sequences is the product of the individual z-transforms.

$$\mathcal{Z}\{x_1(kT) \star x_2(kT)\} = \mathcal{Z}\{x_1(kT)\}\mathcal{Z}\{x_2(kT)\} \equiv X_1(z)X_2(z)$$

EXAMPLE 17.15. If

$$x_1[n] = \{2,\ 2,\ 0,\ 1\}$$
$$x_2[n] = \{3,\ 3,\ 1,\ 1\}$$

find $x_1[n] \star x_2[n]$.

Solution Taking z-transform of $x_1[n]$ and $x_2[n]$, we get

$$X_1(z) = 2 + 2z^{-1} + z^{-3}$$
$$X_2(z) = 3 + 3z^{-1} + z^{-2} + z^{-3}$$

Therefore,

$$X_1(z)X_2(z) = 6 + 12z^{-1} + 8z^{-2} + 7z^{-3} + 5z^{-4} + z^{-5} + z^{-6} \tag{i}$$

Clearly Eq. (i) is the z-transform of

$$\{\underset{\uparrow}{6},\ 12,\ 8,\ 7,\ 5,\ 1,\ 1\} \equiv x_1[n] \star x_2[n]$$

<u>Alternative method</u>

We construct the convolution table as follows:

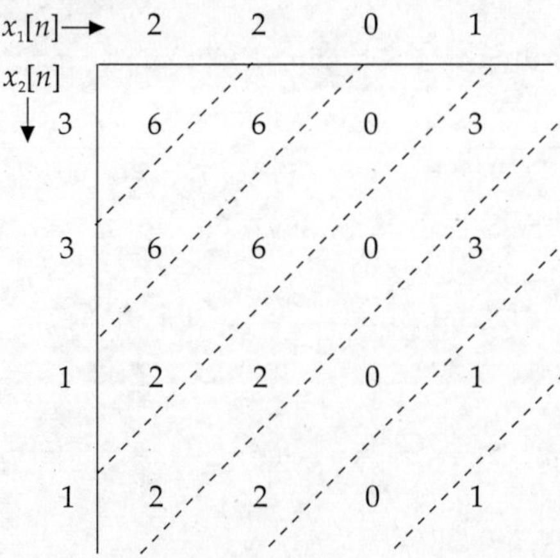

$x_2[n] \downarrow$ \ $x_1[n] \rightarrow$	2	2	0	1
3	6	6	0	3
3	6	6	0	3
1	2	2	0	1
1	2	2	0	1

We arrange the $x_1[n]$ sequence horizontally and the $x_2[n]$ sequence vertically. Then we multiply each element of the $x_1[n]$ sequence by an element of the $x_2[n]$ sequence and place the resulting sequence horizontally. The convolution $x_1[n] \star x_2[n]$ is obtained by adding elements of sequences along the slanted lines. Thus,

$$\begin{aligned} x_1[n] \star x_2[n] &= \{6,\ (6+6),\ (2+6+0),\ (2+2+0+3),\ (2+0+3),\ (0+1),\ 1\} \\ &= \{\underset{\uparrow}{6},\ 12,\ 8,\ 7,\ 5,\ 1,\ 1\} \end{aligned}$$

Note: Obviously, this method can be useful only for finite sequences.

EXAMPLE 17.16. The input signal shown below

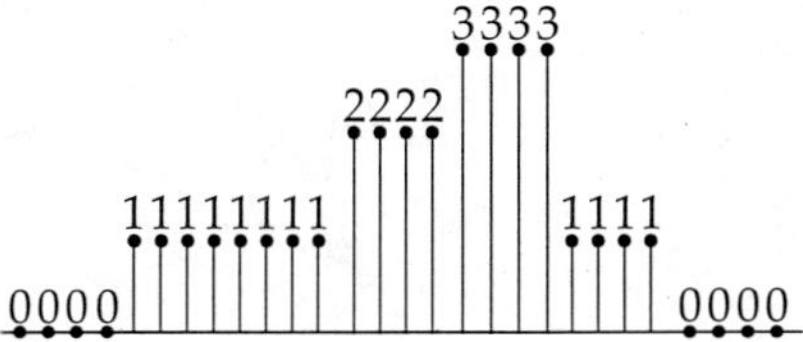

is passed through the filter with the following taps:

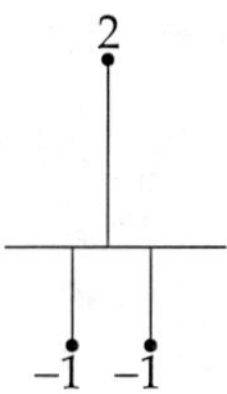

The number of non-zero output sample is ———

Solution We construct the convolution table as follows:

$h[n]$ \ $x[n]$	1	1	1	1	1	1	1	1	2	2	2	2	3	3	3	3	1	1	1	1
–1	–1	–1	–1	–1	–1	–1	–1	–1	–2	–2	–2	–2	–3	–3	–3	–3	–1	–1	–1	–1
2	2	2	2	2	2	2	2	2	4	4	4	4	6	6	6	6	2	2	2	2
–1	–1	–1	–1	–1	–1	–1	–1	–1	–2	–2	–2	–2	–3	–3	–3	–3	–1	–1	–1	–1

As in the previous example, from this, we get

$$y[n] = \{-1, 1, 0, 0, 0, 0, 0, 0, -1, 1, 0, 0, -1, 1, 0, 0, 2, -2, 0, 0, 1, -1\}$$

Therefore, the number of non-zero samples in $y[n] = 10$ (Ans.)

EXAMPLE 17.17. The input-output relationship of an LTI system is given below

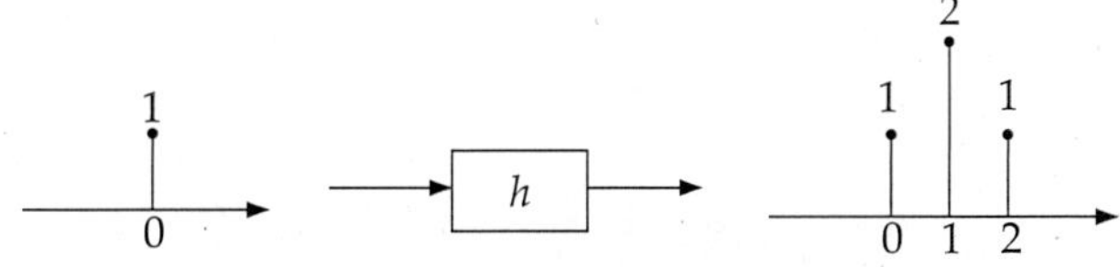

For an input $x[n]$ shown below,

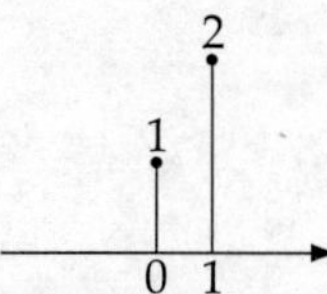

the peak value of the output when $x[n]$ is passed through h is

(A) 4 (B) 5 (C) 2 (D) 6

Solution The applied input is nothing but an impulse. So, the transfer function equals the output of the system. This means that

$$h[n] = \{1, 2, 1\} \equiv 1 + 2z^{-1} + z^{-2}$$

The given input is

$$x[n] = \{1, 2\} \equiv 1 + 2z^{-1}$$

Therefore, the output is

$$\begin{aligned} y[n] = h[n] \star x[n] &= 1 + 4z^{-1} + 5z^{-2} + 2z^{-3} \\ &= \{1, 4, 5, 2\} \end{aligned}$$

So, the peak value of $y[n] = 5$.
Ans. (B)

Alternatively

We construct the convolution table as follows:

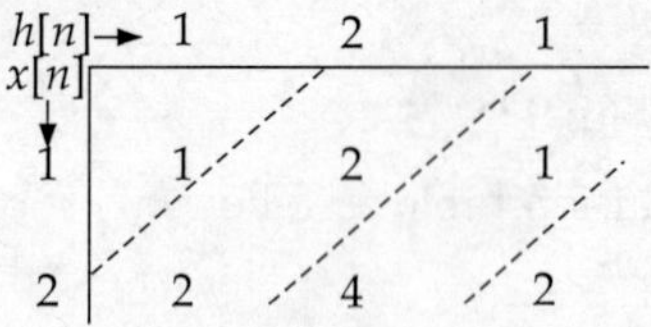

which gives

$$y[n] = \{1, 4, 5, 2\}$$

Ans. (B)

EXAMPLE 17.18. Let $x[n] = 0;\ n < -1,\ n > 0,\ x[-1] = -1,\ x[0] = 2$ be the input and

$$y[n] = 0;\ n < -1,\ n > 2,\ y[-1] = -1 = y[1],\ y[0] = 3,\ y[2] = -2$$

be the output of a discrete-time LTI system. The system impulse response $h[n]$ will be

(A) $h[n] = 0;\ n < 0,\ n > 2,\ h[0] = 1,\ h[1] = h[2] = -1$
(B) $h[n] = 0;\ n < -1,\ n > 1,\ h[-1] = 1,\ h[0] = h[1] = 2$
(C) $h[n] = 0;\ n < 0,\ n > 3,\ h[0] = -1,\ h[1] = 2,\ h[2] = 1$
(D) $h[n] = 0;\ n < -2,\ n > 1,\ h[-2] = h[1] = h[-1] = -h[0] = 3$

Solution The given input and output data indicate that the sequences are

$$\begin{aligned} x[n] =& \{-1, \underset{\uparrow}{2}\}, & -1 \le n \le 0 \\ y[n] =& \{-1, \underset{\uparrow}{3}, -1, -2\}, & -1 \le n \le 2 \end{aligned}$$

If $h[n]$ is the impulse response of the system, then output is

$$y[n] = h[n] \star x[n]$$

The length of the convolution $y[n]$ is -1 to 2, the length of $x[n]$ is -1 to 0. So, the length of $h[n]$ is 0 to 2.

$$\text{Let} \quad h[n] = \underset{\uparrow}{\{a}, b, c\}$$

The convolution is

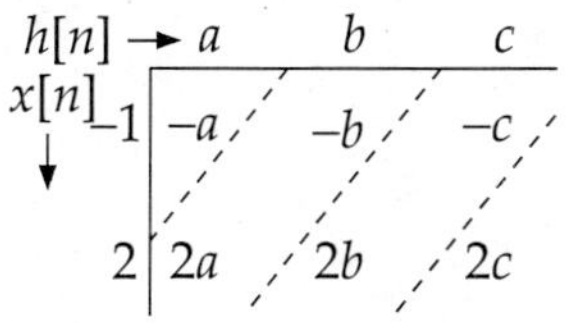

$x[n] \downarrow$ / $h[n] \rightarrow$	a	b	c
-1	$-a$	$-b$	$-c$
2	$2a$	$2b$	$2c$

Then,

$$y[n] = \{-a, \underset{\uparrow}{2a - b}, 2b - c, 2c\} \qquad \text{(i)}$$

$$\text{Given,} \quad y[n] = \{-1, \underset{\uparrow}{3}, -1, -2\} \qquad \text{(ii)}$$

Comparing corresponding terms of Eqs. (i) and (ii), we get

$$a = 1$$
$$2a - b = 3 \qquad \Rightarrow b = -1$$
$$2b - c = -1 \qquad \Rightarrow c = -1$$

Hence, impulse response is

$$h[n] = \{\underset{\uparrow}{1}, -1, -1\}$$

Ans. (A)

EXAMPLE 17.19. Two discrete-time linear time-invariant systems with impulse responses

$$h_1[n] = \delta[n-1] + \delta[n+1]$$
$$h_2[n] = \delta[n] + \delta[n-1]$$

are connected in cascade, where $\delta[n]$ is the Dirac delta function. The impulse response of the cascaded system is

(A) $\delta[n-2] + \delta[n+1]$ (B) $\delta[n-1]\delta[n] + \delta[n+1]\delta[n-1]$
(C) $\delta[n-2] + \delta[n-1] + \delta[n] + \delta[n+1]$ (D) $\delta[n]\delta[n-1] + \delta[n-2]\delta[n+1]$

Solution We know that for such systems

$$\begin{aligned} h[n] &= h_1[n] \star h_2[n] \\ &= \{\delta[n-1] + \delta[n+1]\} \star \{\delta[n] + \delta[n-1]\} \\ &= \delta[n-1] \star \delta[n] + \delta[n-1] \star \delta[n-1] + \delta[n+1] \star \delta[n] + \delta[n+1] \star \delta[n-1] \\ &= \delta[n-1] + \delta[n-1-1] + \delta[n+1] + \delta[n+1-1] \end{aligned}$$

$$(\because \ \delta[n-a] \star \delta[n-b] = \delta[n-a-b])$$

$$= \delta[n-2] + \delta[n-1] + \delta[n] + \delta[n-1] \qquad \text{(on rearranging terms)}$$

Ans. (B)

EXAMPLE 17.20. Suppose for input $x(t)$ a linear time-invariant system with impulse response $h(t)$ produces output $y(t)$, so that $x(t) \star h(t) = y(t)$. Further, if $|x(t)| \star |h(t)| = z(t)$, which of the following statements is true?

(A) For all $t \in$[1]$(-\infty, \infty)$, $z(t) \leq y(t)$
(B) For some but not all $t \in (-\infty, \infty)$, $z(t) \leq y(t)$
(C) For all $t \in (-\infty, \infty)$, $z(t) \geq y(t)$
(D) For some but not all $t \in (-\infty, \infty)$, $z(t) \geq y(t)$

Solution Given,

$$x(t) \star h(t) = y(t)$$
$$|x(t)| \star |h(t)| = z(t)$$

It is known that for any two signals of amplitudes x and h, $x \cdot h \leq |x| \cdot |h|$.
Hence, (C) is true.

Differentiation in z-domain. If $x[n] \overset{\mathcal{Z}}{\longleftrightarrow} X(z)$ then $n\,x[n] \overset{\mathcal{Z}}{\longleftrightarrow} -z\dfrac{d}{dz}X(z)$. The ROC is unchanged.

EXAMPLE 17.21. Find the z-transform for the unit ramp signal $n\,u[n]$.

Solution

We know $$U(z) = \frac{1}{1 - z^{-1}} \qquad \text{for } \{|z| > 1\}$$

Therefore, $$X(z) = -z\frac{d}{dz}U(z) = -z\frac{-z^{-2}}{\left(1 - z^{-1}\right)^2} = \frac{z^{-1}}{\left(1 - z^{-1}\right)^2} \qquad \text{for } \{|z| > 1\}$$

Initial Value Theorem. In case the z-transformed function $x[n]$ is causal, the initial value theorem helps us find the initial value $x[0]$ of the continuous-time function $x[n]$ without inverting $X(z)$. The theorem is stated as follows:

$$x[0] = \lim_{z \to \infty} X(z) \tag{17.10}$$

It follows from the definition

$$X(z) = x[0] + x[1]z^{-1} + x[2]z^{-2} + \cdots$$

EXAMPLE 17.22. Find the initial value of the signal $x[n]$ when its z-transform is given by

$$X(z) = \frac{2z^2 + 3z + 4}{z^2}$$

Solution Using Eq. (17.10), the required initial value is given by

$$x[0] = \lim_{z \to \infty} \frac{2z^2 + 3z + 4}{z^2} = \lim_{z \to \infty} \left(2 + \frac{3}{z} + \frac{4}{z^2}\right) = 2$$

[1] $f(x), x \in X$ means the range of f is defined as $f(X)$.

Final Value Theorem. The final value theorem helps us compute the final value of the causal function $x[n]$ at $n \to \infty$ without inverting the z-transformed function $X(z)$ to the t-space. The theorem states that

$$\lim_{n\to\infty} x[n] = \lim_{z\to 1} \left(1 - z^{-1}\right) X(z) \tag{17.11}$$

if the limit exists and the system is stable, i.e. all poles of $\left(1 - z^{-1}\right) X(z)$ are inside the unit circle $|z| = 1$ on the z-plane.

Note: The theorem is useful to obtain the steady state value of a discrete control system.

EXAMPLE 17.23. Find the value of $x[k]$ as $k \to \infty$ when

$$\mathcal{Z}\{x[k]\} \equiv X(z) = \frac{0.7z}{(z-1)(z^2 - 0.8z + 0.15)}$$

Solution We observe that

$$X(z) = \frac{0.7z}{(z-1)(z-0.3)(z-0.5)}$$

Therefore,

$$(1 - z^{-1})X(z) = \frac{0.7}{(z-0.3)(z-0.5)}$$

Since $(1 - z^{-1})X(z)$ does not have any pole on or outside the unit circle defined by $|z| = 1$ on the z-plane, we can apply the final value theorem defined by Eq. (17.11). Thus,

$$\lim_{k\to\infty} x[k] = \lim_{z\to 1} \frac{0.7}{(z-0.3)(z-0.5)} = 2$$

Having considered important properties of z-transform we now work out a few examples.

EXAMPLE 17.24. Find the z-transform of the following sequences

(a) $x[n] = \left(\frac{1}{2}\right)^{-n} u[n]$

(b) $x[n] = [(0.4)^n - 3(0.2)^n]u[n]$

(c) $x[n] = \begin{cases} (-1)^n & n = 2,\ 3,\ 4\cdots \\ 0 \quad n < 2 \end{cases}$

Solution

(a)
$$X(z) = \sum_{n=0}^{\infty} \left(\frac{1}{2}\right)^{-n} u[n]z^{-n} = \sum_{n=0}^{\infty} \left(\frac{1}{2z}\right)^{-n} = \frac{1}{1 - \dfrac{1}{2z}} = \frac{2z}{2z-1}$$

(b)
$$X(z) = \sum_{n=0}^{\infty} [(0.4)^n - 3(0.2)^n]\, u[n]z^{-n} = \sum_{n=0}^{\infty} \left[\left(\frac{0.4}{z}\right)^n - 3\left(\frac{0.2}{z}\right)^n\right]$$
$$= \frac{1}{1 - \dfrac{0.4}{z}} - 3\cdot\frac{1}{1 - \dfrac{0.2}{z}} = \frac{z}{z-0.4} - \frac{3z}{z-0.2}$$

(c)
$$X(z) = \sum_{n=2}^{\infty}(-1)^n z^{-n} = \sum_{n=0}^{\infty}\left(-\frac{1}{z}\right)^n - \left[\left(-\frac{1}{z}\right)^0 + \left(-\frac{1}{z}\right)^1\right]$$
$$= \sum_{n=0}^{\infty}\left(-\frac{1}{z}\right)^n + \left(1+\frac{1}{z}\right) = \frac{1}{1+\dfrac{1}{z}} + \frac{z+1}{z}$$
$$= \frac{z}{z+1} + \frac{z+1}{z} = \frac{2z^2+2z+1}{z(z+1)}$$

EXAMPLE 17.25. Find the z-transform of the sequence $x[n] = e^{-an}u[n]$ where a is a constant.

Solution

$$X(z) = \sum_{n=0}^{\infty} e^{-an} z^{-n} = \sum_{n=0}^{\infty}\left(\frac{1}{ze^a}\right)^n$$
$$= \frac{1}{1-\dfrac{1}{ze^a}} \qquad \left|\frac{1}{ze^a}\right| < 1$$
$$= \frac{z}{z-e^a} \qquad |z| > e^a$$

EXAMPLE 17.26. Find the right-handed (i.e. k ranging from 0 to ∞) z-transform $F(z)$ of the sequence $f(k)$, where

$$f(k) = \begin{cases} 1 & \text{for } k = 0 \text{ and even number} \\ -1 & \text{for } k = \text{odd number} \end{cases}$$

Solution By adding the two sequences, we get the following sequence:

$$f(k) = \sum_{k=0}^{\infty}(-1)^k$$

Therefore, the z-transform is given by

$$F(z) = \sum_{k=0}^{\infty}(-1)^k z^{-k} = \sum_{k=0}^{\infty}\left(-\frac{1}{z}\right)^k = \frac{1}{1+\dfrac{1}{z}} = \frac{z}{z+1}$$

EXAMPLE 17.27. Find the z-transform of the following sequence:

$$f(k) = \begin{cases} 2^k & k < 0 \\ \left(\dfrac{1}{2}\right)^k & k = 0 \text{ and even} \\ \left(\dfrac{1}{3}\right)^k & k = \text{ odd} \end{cases}$$

Solution The z-transformed function can be written as follows:

$$F(z) = \sum_{k=-\infty}^{-1} 2^k z^{-k} + \sum_{k=0,\ k \text{ even}}^{\infty}\left(\frac{1}{2}\right)^k z^{-k} + \sum_{k=\text{ odd}}^{\infty}\left(\frac{1}{3}\right)^k z^{-k}$$
$$= \sum_{l=1}^{\infty} 2^{-l} z^l + \sum_{m=0}^{\infty}\left(\frac{1}{2}\right)^{2m} z^{-2m} + \sum_{n=0}^{\infty}\left(\frac{1}{3}\right)^{2n+1} z^{-(2n+1)}$$
$$\equiv F_1(z) + F_2(z) + F_3(z)$$

Now, $$F_1(z) = \sum_{l=1}^{\infty}\left(\frac{z}{2}\right)^l = \frac{\frac{z}{2}}{1-\frac{z}{2}} = \frac{z}{2-z}$$

$$F_2(z) = \sum_{m=0}^{\infty}\left(\frac{1}{4z^2}\right)^m = \frac{1}{1-\frac{1}{4z^2}} = \frac{4z^2}{4z^2-1}$$

$$F_3(z) = \sum_{n=0}^{\infty}\left(\frac{1}{3z}\right)^{2n+1} = \frac{1}{3z} + \left(\frac{1}{3z}\right)^3 + \left(\frac{1}{3z}\right)^5 + \cdots = \frac{\frac{1}{3z}}{1-\left(\frac{1}{3z}\right)^2}.$$

$$= \frac{3z}{9z^2-1}$$

Therefore, $$F(z) = F_1(z) + F_2(z) + F_3(z)$$

$$= \frac{z}{2-z} + \frac{4z^2}{4z^2-1} + \frac{3z}{9z^2-1}$$

EXAMPLE 17.28. Find $F(z)$ if $f(s) = \dfrac{1}{s(s+a)}$

Solution Let $F(s) \equiv \dfrac{A}{s} + \dfrac{B}{s+a}$

Then, $$A = sF(s)|_{s=0} = \left.\frac{1}{s+a}\right|_{s=0} = \frac{1}{a}$$

$$B = (s+a)F(s)|_{s=-a} = \left.\frac{1}{s}\right|_{s=-a} = -\frac{1}{a}$$

Therefore,

$$F(z) = \frac{1}{a}\cdot\mathcal{Z}\left\{\frac{1}{s}\right\} - \frac{1}{a}\cdot\mathcal{Z}\left\{\frac{1}{s+a}\right\} = \frac{1}{a}\left[\frac{z}{z-1} - \frac{z}{z-e^{-aT}}\right] = \frac{1-e^{-aT}}{a}\cdot\frac{z}{(z-1)\left(z-e^{-aT}\right)}$$

EXAMPLE 17.29. Find the z-transform of the following function:

$$X(s) = \frac{1}{s^2+2s+2}$$

Solution Factoring the given function, we get

$$X(s) = \frac{1}{(s+1-j)(s+1+j)} \equiv \frac{A}{s+1-j} + \frac{B}{s+1+j}$$

Then, $$A = (s+1-j)X(s)|_{s=-1+j} = \left.\frac{1}{s+1+j}\right|_{s=-1+j} = \frac{1}{2j}$$

$$B = (s+1+j)X(s)|_{s=-1-j} = \left.\frac{1}{s+1-j}\right|_{s=-1-j} = -\frac{1}{2j}$$

Thus, $$X(s) = \frac{1}{2j}\left[\frac{1}{s+1-j} - \frac{1}{s+1+j}\right]$$

$$x(t) = \mathscr{L}^{-1}\{X(s)\} = \frac{1}{2j}\left[e^{-(1-j)T} - e^{-(1+j)T}\right]$$

Therefore,

$$X(z) = \frac{1}{2\jmath}\left[\frac{z}{z-e^{-(1-\jmath)T}} - \frac{z}{z-e^{-(1+\jmath)T}}\right]$$
$$= \frac{z}{2\jmath}\left[\frac{z-e^{-(1+\jmath)T} - z + e^{-(1-\jmath)T}}{\left\{z-e^{-(1-\jmath)T}\right\}\left\{z-e^{-(1+\jmath)T}\right\}}\right]$$

Now,
$$e^{-(1-\jmath)T} - e^{-(1+\jmath)T} = e^{-T}\left[e^{\jmath T} - e^{-\jmath T}\right]$$
$$= e^{-T}[\cos T + \jmath\sin T - \cos T + \jmath \sin T]$$
$$= 2\jmath\, e^{-T}\sin T$$

similarly
$$e^{-(1-\jmath)T} + e^{-(1+\jmath)T} = 2e^{-T}\cos T$$

Therefore,
$$X(z) = \frac{z}{e^T}\cdot\frac{\sin T}{z^2 - 2ze^{-T}\cos T + e^{-2T}}$$

Inverse z-transform

In Laplace transform, we can manipulate equations in the s-space and then revert to the t-space by the inverse Laplace transform. Similarly, in the z-transform we can revert to the t-space through the inverse z-transform, though with the difference that the reverted t-space is discrete here. This means that we can revert to the kT-space rather than continuous t-space through the inverse z-transform.

The inverse z-transform can be performed by a few methods. Of them, we will consider only two methods, namely

1. Partial fraction expansion method
2. Residue method

But, before discussing them, it is necessary to talk about poles and zeros of the pulse transfer function.

Poles and zeros in the z-plane. While performing z-transform or its inverse, we may encounter $X(z)$ of the form

$$X(z) = \frac{b_0 z^m + b_1 z^{m-1} + \cdots + b_m}{z^n + a_1 z^{n-1} + \cdots + a_n} \qquad (m \le n)$$
$$\equiv \frac{b_0(z-z_1)(z-z_2)\cdots(z-z_m)}{(z-p_1)(z-p_2)\cdots(z-p_m)}$$

where p_i's $(i = 1, 2, \ldots, n)$ and z_j's $(j = 1, 2, \ldots, m)$ are poles and zeros of $X(z)$, respectively.

The locations of poles and zeros in the z-plane determine the characteristics of $x[n]$, the sequence of values of numbers.

Sometimes, $X(z)$ is written in terms of z^{-1} as follows:

$$X(z) = \frac{b_0 z^{-(n-m)} + b_1 z^{-(n-m+1)} + \cdots + b_m z^{-n}}{1 + a_1 z^{-1} + a_2 z^{-2} + \cdots + a_n z^{-n}}$$

where z^{-1} is interpreted as the unit delay operator. However, in finding poles and zeros, one may miss zeros at the origin as may be evident from the following two ways of writing $X(z)$

$$X(z) = \frac{z^2+0.5z}{z^2+5z+3} = \frac{z(z+0.5z)}{(z+2)(z+3)} \tag{i}$$

$$X(z) = \frac{1+0.5z^{-1}}{1+5z^{-1}+3z^{-2}} = \frac{1+0.5z^{-1}}{\left(1+2z^{-1}\right)\left(1+3z^{-1}\right)} \tag{ii}$$

While poles at $z=-2$ and $z=-3$ are evident in both Eqs. (i) and (ii), the zero at $z=0$ is not explicit in Eq. (ii). So, for pole-zero plotting, the form represented by Eq. (i) is preferable.

A pole-zero plot is a graphic description of $X(z)$. We will use numbers to indicate number of poles or zeros at the same location.

By definition, the ROC will not contain any poles.

EXAMPLE 17.30. Given, $x[n] = n\,u[n]$. Make a pole-zero plot.

Solution

$$\mathcal{Z}\{x[n]\} = X(z) = \frac{z}{(z-1)^2} \qquad (|z|>1)$$

Here, $X(z)$ contains one zero at $z=0$ and two poles at $z=1$. The plot is shown in Figure 17.8.

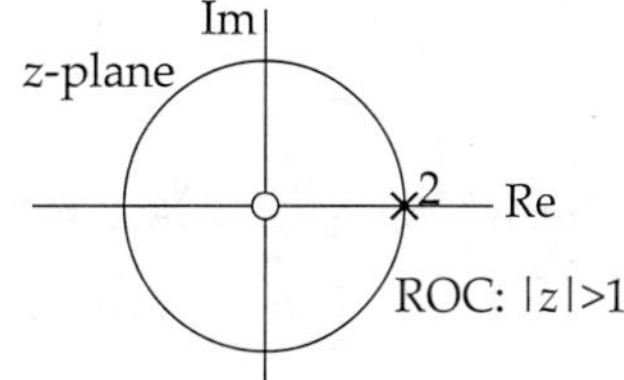

Figure 17.8: Pole-zero plot (Example 17.30).

We will now discuss the two methods of inverse z-transform with examples.

1. Partial Fraction Expansion Method. The procedure is almost the same as we do in the case of inverse Laplace transform except that here we break $F(z)/z$, rather than $F(z)$ itself, into partial fractions. The reason is that almost all z-transformed functions contain z in the numerator as is evident from Table E.1. So, we actually need to break $F(z)$ into partial fractions of the form

$$F(z) \equiv \frac{Az}{(z\text{-function})_1} + \frac{Bz}{(z\text{-function})_2} + \cdots \tag{17.12}$$

To do that, it is helpful to follow our standard procedure of finding coefficients A, B, ... if we write Eq. (17.12) as

$$\frac{F(z)}{z} \equiv \frac{A}{(z\text{-function})_1} + \frac{B}{(z\text{-function})_2} + \cdots$$

Let us work out a few examples.

EXAMPLE 17.31. Find the sequence corresponding to the z-transformed function

$$X(z) = \frac{z^3-4.8}{z(z-0.2)(z-0.4)}$$

Solution We rewrite the function as

$$\frac{X(z)}{z} = \frac{z^3 - 4.8}{z^2(z-0.2)(z-0.4)}$$

Let
$$\frac{X(z)}{z} \equiv \frac{A}{z^2} + \frac{B}{z} + \frac{C}{z-0.2} + \frac{D}{z-0.4} \tag{i}$$

Then,
$$A = \left.\frac{z^3 - 4.8}{(z-0.2)(z-0.4)}\right|_{z=0} = \frac{-4.8}{0.08} = -60$$

$$B = \frac{d}{dz}\left[\frac{z^3-4.8}{(z-0.2)(z-0.4)}\right]_{z=0}$$

$$= \left[\frac{3z^2}{(z-0.2)(z-0.4)} - \frac{(z^3-4.8)(2z-0.6)}{(z^2-0.6z+0.08)^2}\right]_{z=0} = 450$$

$$C = \left.\frac{z^3-4.8}{z^2(z-0.4)}\right|_{z=0.2} = 599$$

$$D = \left.\frac{z^3-4.8}{z^2(z-0.2)}\right|_{z=0.4} = -148$$

Therefore, from Eq. (i)

$$\frac{X(z)}{z} = -\frac{60}{z^2} + \frac{450}{z} + \frac{599}{z-0.2} - \frac{148}{z-0.4}$$

or,
$$X(z) = -\frac{60}{z} + 450 + \frac{599z}{z-0.2} - \frac{148z}{z-0.4} \tag{ii}$$

Equation (ii) yields on inversion

$$x[n] = -60\delta[n-1] + 450\delta[n] + [599(0.2)^n - 148(0.4)^n]u[n]$$

EXAMPLE 17.32. Given the z-transformed function

$$X(z) = \frac{(1-e^{-aT})z}{(z-1)(z-e^{-aT})}$$

Find out the sampled sequence.

Solution Dividing the function by z and writing it in the form of partial fractions, we have

$$\frac{X(z)}{z} = \frac{(1-e^{-aT})}{(z-1)(z-e^{-aT})} \equiv \frac{A}{z-1} + \frac{B}{z-e^{-aT}}$$

A and B are found out in the usual way as

$$A = \left.\frac{1-e^{-aT}}{z-e^{-aT}}\right|_{z=1} = 1$$

$$B = \left.\frac{1-e^{-aT}}{z-1}\right|_{z=e^{-aT}} = -1$$

Therefore,

$$\frac{X(z)}{z} = \frac{1}{z-1} - \frac{1}{z-e^{-aT}}$$

or,
$$X(z) = \frac{z}{z-1} - \frac{z}{z-e^{-aT}}$$

or,
$$x(kT) = u(kT) - e^{-akT} = 1 - e^{-akT}$$

Note: We repeat $x(kT) \equiv x(k) \equiv x[k]$.

EXAMPLE 17.33. Find $y(k)$ when

$$Y(z) = \frac{2z^2 - 1.5z}{z^2 - 1.5z + 0.5}$$

Solution Let

$$Y(z) = \frac{z(2z-1.5)}{(z-1)(z-0.5)} \equiv z\left[\frac{A}{z-1} + \frac{B}{z-0.5}\right]$$

Then

$$A = (z-1)\left.\frac{Y(z)}{z}\right|_{z=1} = \left.\frac{2z-1.5}{z-0.5}\right|_{z=1} = 1$$

$$B = (z-0.5)\left.\frac{Y(z)}{z}\right|_{z=0.5} = \left.\frac{2z-1.5}{z-1}\right|_{z=0.5} = 1$$

Therefore,

$$Y(z) = \frac{z}{z-1} + \frac{z}{z-0.5}$$

Its inverse z-transform yields

$$y(kT) = u(kT) + 0.5^{kT}$$
$$y[k] = 1 + 0.5^k \qquad k \geq 0$$

EXAMPLE 17.34. If $u(t)$ is the unit step and $\delta(t)$ is the unit impulse function, the inverse z-transform of

$$F(z) = \frac{1}{z+1} \qquad \text{for } k > 0$$

is

(A) $(-1)^k \delta(k)$ (B) $\delta(k) - (-1)^k$ (C) $(-1)^k u(k)$ (D) $u(k) - (-1)^k$

Solution

$$F(z) = \frac{1}{z+1} = \frac{z+1-z}{z+1} = 1 - \frac{z}{z-(-1)}$$

Therefore, $\mathcal{Z}^{-1}\{F(z)\} = \delta(k) - (-1)^k$ [see line 6 of Table E.1 at page 536]

Ans. (B)

2. Residue Method. The time sequence $x(kT)$ and the z-transformed function $X(z)$ are related by the formula

$$f(kT) = \frac{1}{2\pi j} \oint_{\Gamma} F(z)\, z^{k-1} dz$$

It involves a contour integration along the circular path Γ of radius $|z| = e^{cT}$ having its centre at the origin of the z-plane and the value of c is so chosen that all poles of $F(z)$ lie within the circle. It can be shown that

$$f(kT) = \sum \text{Residues of } F(z)\, z^{k-1} \text{ at the poles of } F(z)\, z^{k-1} \tag{17.13}$$

Equation (17.13) is utilized to find the inverse z-transform.

The residue to a repeated pole at $z = p$ of multiplicity n is given by

$$\rho = \frac{1}{(n-1)!} \lim_{z \to p} \left\{ \frac{\partial^{n-1}}{\partial z^{n-1}} \left[(z-p)^n F(z) z^{k-1} \right] \right\} \tag{17.14}$$

EXAMPLE 17.35. Find the inverse z-transform of

$$X(z) = \frac{z}{z - a^T}$$

by the residue method.

Solution The function has only one pole at $z = a^T$. Therefore,

$$x(kT) = z \cdot z^{k-1}\Big|_{z=a^T} = a^{kT}$$

EXAMPLE 17.36. Find the inverse z-transform of

$$F(z) = \frac{(1 - e^{-aT})z}{(z-1)(z - e^{-aT})}$$

by the residue method.

Solution The function has two poles at $z = 1$ and $z = e^{-aT}$. Therefore,

$$\begin{aligned} f(kT) &= \frac{(1 - e^{-aT})z \cdot z^{k-1}}{z - e^{-aT}}\Bigg|_{z=1} + \frac{(1 - e^{-aT})z \cdot z^{k-1}}{z - 1}\Bigg|_{z=e^{-aT}} \\ &= \frac{(1 - e^{-aT}) \cdot 1^{k-1}}{1 - e^{-aT}} + \frac{(1 - e^{-aT})e^{-akT}}{e^{-aT} - 1} \\ &= 1 - e^{-akT} \end{aligned}$$

EXAMPLE 17.37. Find by the residue method the continuous-time function corresponding to

$$F(z) = \frac{Tz}{(z-1)^2}$$

Solution We see that $F(z)$ has a pole at $z = 1$ with a multiplicity of 2. Therefore, from Eq. (17.14) we get

$$\begin{aligned} \rho &= \frac{1}{(2-1)!} \lim_{z \to 1} \left\{ \frac{\partial^{2-1}}{\partial z^{2-1}} \left[(z-1)^2 \cdot \frac{Tz}{(z-1)^2} z^{k-1} \right] \right\} \\ &= \lim_{z \to 1} \frac{\partial}{\partial z} (Tz^k) \\ &= kTz^{k-1}\Big|_{z=1} \\ &= kT \end{aligned}$$

Therefore, $f(kT) = kT$, or the continuous-time function is $f(t) = t$.

EXAMPLE 17.38. Using convolution, determine the inverse of the function given by

$$F(z) = \frac{z^2}{\left(z - \frac{1}{2}\right)\left(z + \frac{1}{4}\right)}$$

Solution The given function can be written as

$$\begin{aligned} F(z) &= \frac{1}{1 - \frac{1}{2}z^{-1}} \cdot \frac{1}{1 + \frac{1}{4}z^{-1}} \\ &\equiv F_1(z) \star F_2(z) \end{aligned}$$

Performing inverse z-transform of $F_1(z)$ and $F_2(z)$, we get

$$\mathcal{Z}^{-1}\{F_1(z)\} = \mathcal{Z}^{-1}\left\{\frac{1}{1-\frac{1}{2}z^{-1}}\right\} = \left(\frac{1}{2}\right)^{kT} u(kT) \equiv f_1(kT)$$

$$\mathcal{Z}^{-1}\{F_2(z)\} = \mathcal{Z}^{-1}\left\{\frac{1}{1+\frac{1}{4}z^{-1}}\right\} = \left(-\frac{1}{4}\right)^{kT} u(kT) \equiv f_2(kT)$$

Now, we know from Eq. (17.9) that

$$f(kT) = \sum_{n=0}^{k} f_1[(k-n)T]f_2(nT)$$

Therefore

$$\begin{aligned}
f(kT) &= \sum_{n=0}^{k}\left(\frac{1}{2}\right)^{(k-n)T}\left(-\frac{1}{4}\right)^{nT} u(kT) \\
&= \left(\frac{1}{2}\right)^{kT}\sum_{n=0}^{k}\left[\frac{(-1/4)}{1/2}\right]^{nT} u(kT) \\
&= \left(\frac{1}{2}\right)^{kT}\sum_{n=0}^{k}\left(-\frac{1}{2}\right)^{nT} u(kT) \\
&= \left(\frac{1}{2}\right)^{kT}\cdot\frac{1-\left(-\frac{1}{2}\right)^{(k+1)T}}{1-\left(-\frac{1}{2}\right)}\cdot u(kT) \qquad \left[\because \sum_{n=0}^{k} a^n = \frac{1-a^{k+1}}{1-a}\right] \\
&= \left(\frac{1}{2}\right)^{kT}\cdot\frac{2}{3}\left[1-\left(-\frac{1}{2}\right)\left(-\frac{1}{2}\right)^{kT}\right]u(kT) \\
&= \left[\frac{2}{3}\left(\frac{1}{2}\right)^{kT}+\frac{1}{3}\left(-\frac{1}{4}\right)^{kT}\right]u(kT)
\end{aligned}$$

A few z-transforms are given in Appendix E at page 536. The table helps perform both the z-transform and its inverse. We show here a typical example.

EXAMPLE 17.39. Find z-transform of the following sequences:

(a) $x[n] = 10\sin[0.25\pi n]u[n]$

(b) $x[n] = e^{-0.1n}\cos[0.25n]u[n]$

Solution

(a) From line 23 of the Table given in Appendix E, we have

$$\begin{aligned}
X(z) &= 10\mathcal{Z}\{\sin[0.25\pi n]u[n]\} \\
&= \frac{10z\sin(0.25\pi)}{z^2-2z\cos(0.25\pi)+1} = \frac{7.07z}{z^2-1.414z+1}
\end{aligned}$$

(b) From line 24 of the Table:

$$\begin{aligned}
X(z) &= \mathcal{Z}\{e^{-0.1n}\cos[0.25\pi n]u[n]\} \\
&= \frac{z\left[z-e^{-0.1}\cos(0.25\pi)\right]}{z^2-2e^{-0.1}z\cos(0.25\pi)+e^{-0.2}} = \frac{z(z-0.6397)}{z^2-1.2794z+0.8187}
\end{aligned}$$

EXAMPLE 17.40. Determine the inverse z-transform of $F(z)$ given by

$$F(z) = \frac{z^2 + 2z}{z^2 - 2z + 4}$$

Solution From the table given in Appendix E at page 536, we find (Serial Nos. 27 and 28)

$$\mathcal{Z}\{e^{-at} \sin \omega t\} = \frac{ze^{-aT} \sin \omega T}{z^2 - 2ze^{-aT} \cos \omega T + e^{-2aT}}$$

$$\mathcal{Z}\{e^{-at} \cos \omega t\} = \frac{z(z - e^{-aT} \cos \omega T)}{z^2 - 2ze^{-aT} \cos \omega T + e^{-2aT}}$$

If we put $e^{-at} = a^t$ and $e^{-aT} = a^T$, we get

$$\mathcal{Z}\{a^t \sin \omega t\} = \frac{za^T \sin \omega T}{z^2 - 2za^T \cos \omega T + a^{2T}} \tag{i}$$

$$\mathcal{Z}\{a^t \cos \omega t\} = \frac{z(z - a^T \cos \omega T)}{z^2 - 2za^T \cos \omega T + a^{2T}} \tag{ii}$$

The denominators of Eqs. (i) and (ii) are the same and they compare with the denominator of $F(z)$. Comparing the denominators, we get

$$a^{2T} = 4 \tag{iii}$$

$$a^T \cos \omega T = 1 \tag{iv}$$

Using Eqs. (iii) and (iv), we get

$$a^T = 2 \qquad \Rightarrow \quad a = 2^{\frac{1}{T}}$$

$$\cos \omega T = \frac{1}{2} \qquad \Rightarrow \quad \omega T = 60° = \frac{\pi}{3} \qquad \Rightarrow \quad \omega = \frac{\pi}{3T} \tag{v}$$

$$\sin \omega T = \sin \frac{\pi}{3} = \frac{\sqrt{3}}{2}$$

Now, let us write the numerator of $F(z)$ as sum of the numerators at the RHS of both Eqs. (i) and (ii) as follows:

$$z^2 + 2z \equiv z^2 - za^T \cos \omega T + A\,(za^T \sin \omega T) \tag{vi}$$

$$= z^2 - z(2)\left(\frac{1}{2}\right) + A\,(za^T \sin \omega T)$$

$$\Rightarrow \qquad 3z = A\,(za^T \sin \omega T)$$

$$\Rightarrow \qquad A = \frac{3}{a^T \sin \omega T} = \frac{3}{(2)\left(\frac{\sqrt{3}}{2}\right)} = \sqrt{3}$$

Substituting values of a^T, ω obtained in Eq. (v), we get from Eq. (vi)

$$z^2 + 2z = \left(z^2 - 2z \cos \frac{\pi}{3}\right) + \sqrt{3}\left(2z \sin \frac{\pi}{3}\right)$$

$$z^2 - 2z + 4 = z^2 - 2za^T \cos \omega T + a^{2T} = z^2 - 2z\left(2 \cos \frac{\pi}{3}\right) + 4$$

Thus, we have

$$F(z) = \frac{\left(z^2 - 2z\cos\frac{\pi}{3}\right) + \sqrt{3}\left(2z\sin\frac{\pi}{3}\right)}{z^2 - 2z\left(2\cos\frac{\pi}{3}\right) + 4} \quad \text{(vii)}$$

According to the table given in Appendix E, Eq (vii) gives us on inversion

$$f(t) = 2^{\frac{t}{T}}\left(\cos\frac{\pi t}{3T} + \sqrt{3}\sin\frac{\pi t}{T}\right)$$

17.3 DIFFERENCE EQUATIONS

We know that a continuous-time dynamic system is represented by a differential equation. But if it is a discrete-time dynamic system, then a *difference equation* is necessary to represent it.

A difference equation may be constructed as an approximation to samples drawn at regular intervals from an analog system described by a differential equation. Let us consider an example.

EXAMPLE 17.41. Construct a difference equation for the circuit given in Figure 17.9.

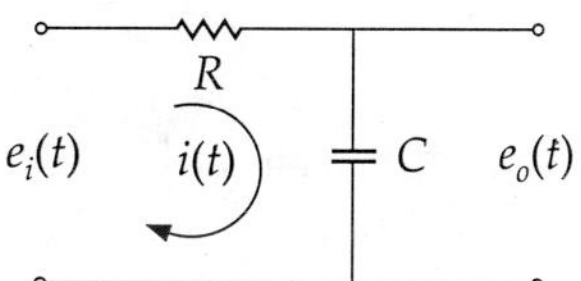

Figure 17.9: Circuit (Example 17.41).

Solution From the KVL we get

$$e_i(t) = Ri(t) + \frac{\int i(t)\,dt}{C} \quad \text{(i)}$$

$$e_o(t) = \frac{\int i(t)\,dt}{C} \quad \text{(ii)}$$

From Eq. (ii) we get

$$\frac{de_o(t)}{dt} = \frac{i(t)}{C} \quad \text{(iii)}$$

Substituting the value of $i(t)$ from Eq. (iii) in Eq. (i), we get

$$e_i(t) = RC\frac{de_o(t)}{dt} + e_o(t) \quad \text{(iv)}$$

We can write Eq. (iv) in the difference equation form as follows:

$$e_i(nT) = RC\frac{e_o[(n+1)T] - e_o(nT)}{T} + e_o(nT)$$

where T is the sampling interval and $n = 0,\ 1,\ 2, \cdots$.

Substituting $nT \equiv [n]$ and $(n+1)T \equiv [n+1]$, since this is a sequence, we have

$$\frac{T}{RC}e_i[n] = e_o[n+1] - e_o[n] + \frac{T}{RC}e_o[n]$$

$$\Rightarrow \quad e_o[n+1] = \left(1 - \frac{T}{RC}\right)e_o[n] + \frac{T}{RC}e_i[n]$$

$$\Rightarrow \quad e_o[n+1] = ae_o[n] + (1-a)e_i[n] \quad \text{(v)}$$

where $a = 1 - \dfrac{T}{RC}$.

Equation (v) is of the general form

$$y[n+1] = ay[n] + (1-a)u[n] \tag{17.15}$$

where y is the output and u, the input.

Note: 1. Where differential equations have *derivatives*, difference equations have *delays*. If the derivative is of higher order, the delay is also of higher order.
2. Figure 17.9 actually is a simple circuit for a low-pass filter. Therefore, Eq. (17.15) represents a difference equation for a low-pass filter.

From this example it is clear that, in general, a linear constant-coefficient differential equation of order n can be written as

$$a_0y[n] + a_1y[n-1] + \cdots + a_Ny[n-N] = b_0u[n] + b_1u[n-1] + \cdots + b_Lu[n-L]$$

$$\Rightarrow \quad \sum_{k=0}^{N} a_k[n-k] = \sum_{k=0}^{L} b_ku[n-k]$$

where we have L possible delays of input and N possible delays of output.

We can think that it is a process by which a sequence $u[n]$ is transformed into another sequence $y[n]$. If $u[n] = 0$ for all n, then this is called a *homogeneous difference equation.* A linear difference equation of order n is also called a *linear recurrence relation* of order n because it can be used to compute recursively each $y[n+1]$ from the preceding value $y[n]$.

EXAMPLE 17.42. A discrete system is represented by the difference equation

$$\begin{bmatrix} X_1(k+1) \\ X_2(k+1) \end{bmatrix} = \begin{bmatrix} a & a-1 \\ a+1 & a \end{bmatrix} \begin{bmatrix} X_1(k) \\ X_2(k) \end{bmatrix}$$

It has initial conditions $X_1(0) = 1$; $X_2(0) = 0$. The pole locations for the system for $a = 1$, are

(A) $1 \pm \jmath 0$ (B) $-1 \pm \jmath 0$ (C) $\pm 1 + \jmath 0$ (D) $0 \pm \jmath 1$

Solution

Characteristic equation $\quad |z\mathbf{I} - \mathbf{A}| = 0$

where $\quad \mathbf{A} = \begin{bmatrix} a & a-1 \\ a+1 & a \end{bmatrix}$

So, $\quad z\mathbf{I} - \mathbf{A} = \begin{bmatrix} z & 0 \\ 0 & z \end{bmatrix} - \begin{bmatrix} a & a-1 \\ a+1 & a \end{bmatrix}$

$$= \begin{bmatrix} z-a & -a+1 \\ -a-1 & z-a \end{bmatrix}$$

For $a = 1$, $\quad z\mathbf{I} - \mathbf{A} = \begin{bmatrix} z-1 & 0 \\ -2 & z-1 \end{bmatrix}$

Then, characteristic equation $\quad (z-1)^2 = 0$

which gives $\quad z = 1 \pm \jmath 0$

Ans. (A)

EXAMPLE 17.43. The z-transform of a signal $x[n]$ is given by $4z^{-3}+3z^{-1}+2-6z^2+2z^3$. It is applied to a system with a transfer function $H(z)=3z^{-1}-2$. Let the output be $y[n]$. Which of the following is true?

(A) $y[n]$ is non-causal with finite support
(B) $y[n]$ is causal with infinite support
(C) $y[n]=0;\ |n|>3$
(D) $\Re[Y(z)]_{z=e^{j\theta}}=-\Re[Y(z)]_{z=e^{j\theta}};\ \Im[Y(z)]_{z=e^{j\theta}}=-\Im[Y(z)]_{z=e^{j\theta}};\ \pi\le\theta<\pi$

Solution

Given, $X(z)=4z^{-3}+3z^{-1}+2-6z^2+2z^3$

TF $H(z)=3z^{-1}-2$

Therefore, output

$$\begin{aligned}Y(z)&=H(z)X(z)\\&=(3z^{-1}-2)(4z^{-3}+3z^{-1}+2-6z^2+2z^3)\\&=12z^{-4}-8z^{-3}+9z^{-2}-4-18z+18z^2-4z^3\end{aligned}$$

or,

$$\begin{aligned}y[n]=&\ 12\delta[n-4]-8\delta[n-3]+9\delta[n-2]-4\delta[n]-18\delta[n+1]+18\delta[n+2]\\&-4\delta[n-3]\end{aligned}$$

So, for $n<0$ $\quad y[n]\neq 0$

Hence, $y[n]$ is non-causal with finite support.
Ans. (A)

Calculating Responses from Difference Equations

As we know, responses can be static or dynamic.

Static Response

A static response means the steady-state value of the output when the input is constant. It can be calculated from the static version of the difference equation. The static version, in turn, is obtained by neglecting all time dependencies in the difference equation. For example, the output $y[n]$ is replaced with y_s where the subscript s indicates a static condition.

EXAMPLE 17.44. Calculate the static response for the low-pass filter of Example 17.41.

Solution The difference equation for the low-pass filter is given by

$$y[n+1]=ay[n]+(1-a)u[n]$$

In static condition, it assumes the following form:

$$y_s=(1-a)y_s+au$$

$\Rightarrow$

$$y_s(1-1+a)=au$$

Therefore,

$$y_s=u \qquad \text{(i)}$$

Equation (i) indicates that the output is equal to the input in static condition which is true for a low-pass filter.

Dynamic Response

A difference equation is, by itself, an algorithm for calculating responses in the form of time function. The following example shows that.

EXAMPLE 17.45. Calculate and the dynamic response for a low-pass RC filter when a step input of 1 V is applied to it. Assume, $R = 1\,\text{M}\Omega$ and $C = 2\,\mu\text{F}$.

Solution Let the sampling interval be 1 s. $RC = (1 \times 10^6)(2 \times 10^{-6}) = 2\,\text{s}$. Then from Eq. (v) of Example 17.41 we get

$$\begin{aligned} e_o[1] &= 0.5e_o[0] + 0.5 \times 1 = 0.5\,\text{V} \qquad [\because e_o(0) = 0] \\ e_o[2] &= 0.5e_o[1] + 0.5 = 0.75\,\text{V} \\ e_o[3] &= 0.5e_o[2] + 0.5 = 0.875\,\text{V} \\ &\vdots \end{aligned}$$

The response of first 10 s is plotted in Figure 17.10.

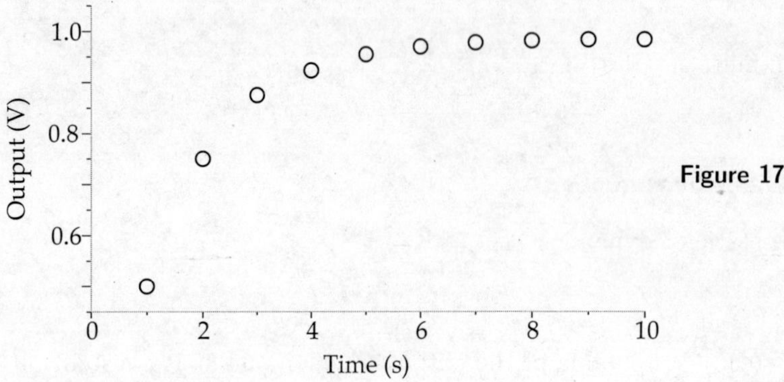

Figure 17.10: Response (Example 17.45).

Solution of Difference Equations through z-Transform

We have seen that Laplace transform helps us to solve *differential* equations. We will now see that the z-transform plays a similar role in solving *difference* equations. Let us consider the following examples.

EXAMPLE 17.46. The response of a low-pass filter is given by the difference equation [see Eq. (17.15)]

$$y[n+1] = ay[n] + (1-a)u[n] \tag{i}$$

Calculate the response by the z-transform method when a step input of 1 V is applied to it. Assume, $R = 1\,\text{M}\Omega$ and $C = 2\,\mu\text{F}$.

Solution Given, $R = 1\,\text{M}\Omega$ and $C = 2\,\mu\text{F}$. So, $a = 1 - \dfrac{T}{RC} = 0.5$ for $T = 1\,\text{s}$

Now, $\mathcal{Z}\{y[n+1]\} = zY(z) - zy(0)$ [see line 4 of Table E.2, Appendix E]

$\mathcal{Z}\{y[n]\} = Y(z)$ [line 5 of the same Table]

$\mathcal{Z}\{u[n]\} = \dfrac{z}{z-1}$ [line 3 of Table E.1, Appendix E]

Therefore, $zY(z) - zy(0) = aY(z) + (1-a)\dfrac{z}{z-1}$ [transforming Eq. (i)]

On rearranging, $Y(z)(z-a) = (1-a)\dfrac{z}{z-1}$ [$\because\ y(0) = 0$]

or,

$$\frac{Y(z)}{z} = (1-a)\frac{1}{(z-1)(z-a)}$$

$$\equiv (1-a)\left[\frac{A}{z-1} + \frac{B}{z-b}\right] \qquad \text{[say]}$$

$$A = \left.\frac{1}{z-a}\right|_{z=1} = \frac{1}{1-a}$$

$$B = \left.\frac{1}{z-1}\right|_{z=a} = -\frac{1}{1-a}$$

$$Y(z) = \frac{z}{z-1} - \frac{z}{z-a}$$

$\mathcal{Z}^{-1}\{Y(z)\} = y[n] = 1 - a^n u[n]$ [see line 6 of Table E.1, Appendix E]

Now, we calculate responses from this relation as follows:

$$y[0] = 1 - 1 = 0$$
$$y[1] = 1 - 0.5 = 0.5$$
$$y[2] = 1 - 0.5^2 = 0.75$$
$$y[3] = 1 - 0.5^3 = 0.875$$

They are the same as those calculated from the difference equation [see Example 17.45 at page 502].

EXAMPLE 17.47. The relation between the input $x[n]$ and the output $y[n]$ of a system is given by

$$y(n) - \frac{1}{2}y(n-1) = x(n)$$

Evaluate the output for the input

$$x[n] = \begin{cases} 1 & n \geq -2 \\ 0 & n < -2 \end{cases}$$

Solution Taking z-transform on both sides of the given difference equation, we get

$$Y(z) - \frac{1}{2}z^{-1}Y(z) = X(z)$$

$$\Rightarrow \qquad \left(1 - \frac{1}{2z}\right)Y(z) = X(z)$$

$$\Rightarrow \qquad \frac{Y(z)}{X(z)} = \frac{z}{z - \dfrac{1}{2}}$$

Now, the input is given by the sequence

$$x[n] = \sum_{n=-2}^{\infty} u[n]$$

$$\mathcal{Z}\{x[n]\} = X(z) = \mathcal{Z}\{u[n+2]\} = z^2U(z) - z^2u(0) - zu[1] \quad \text{[see Table E.2 of Appendix E]}$$

$$= z^2.\frac{z}{z-1} - \left[z^2 + z\right] = \frac{z}{z-1}$$

Therefore, the output corresponding to this input is

$$\frac{Y(z)}{z} = 2.\frac{z}{z-1} - 2.\frac{z}{z-\frac{1}{2}} \quad \text{[see Table E.1 of Appendix E]}$$

Therefore, $\mathcal{Z}^{-1}\{y[n]\} = 2u[n] - 2\left(\frac{1}{2}\right)^2 u[n] = 2\left[1 - \left(\frac{1}{2}\right)^n\right]u[n]$

EXAMPLE 17.48. Consider the difference equation

$$x(k+2) = x(k+1) + x(k)$$

with $x(0) = 0,\ x(1) = 1$.

(a) Derive an expression for the z-transform $X(z)$ of $[x(k)]$.

(b) Obtain an expression for $x(k)$ by inverting $X(z)$.

(c) Determine $\lim_{k\to\infty} \dfrac{x(k+1)}{x(k)}$.

Solution (a) Considering the z-transform of the three terms of the given difference equation, we get

$$\mathcal{Z}\{x(k+2)\} = z^2\left[X(z) - x(0) - x(1)z^{-1}\right]$$

$$= z^2X(z) - z \quad [\because x(0) = 0,\ x(1) = 1]$$

$$\mathcal{Z}\{x(k+1)\} = z[X(z) - x(0)] = zX(z) \quad \text{(i)}$$

$$\mathcal{Z}\{x(k)\} = X(z) \quad \text{(ii)}$$

Therefore,

$$z^2X(z) - z = zX(z) + X(z)$$

$$\Rightarrow \quad (z^2 - z - 1)X(z) = z$$

$$\Rightarrow \quad X(z) = \frac{z}{z^2 - z - 1} \quad \text{(iii)}$$

(b) From Eq. (iii), we get

$$X(z) = \frac{z}{z^2 - z - 1} = \frac{z}{\left(z - \frac{1+\sqrt{5}}{2}\right)\left(z - \frac{1-\sqrt{5}}{2}\right)} \quad \text{(iv)}$$

From Eq. (iv), we observe that the poles of multiplicity 1 are at $z = \dfrac{1+\sqrt{5}}{2}$ and $z = \dfrac{1-\sqrt{5}}{2}$. The corresponding residues are

$$\rho_1 = \left.\frac{z^n}{z - \frac{1-\sqrt{5}}{2}}\right|_{z=\frac{1+\sqrt{5}}{2}} = \frac{\left(\frac{1+\sqrt{5}}{2}\right)^n}{\sqrt{5}}$$

$$\rho_2 = \left.\frac{z^n}{z - \frac{1+\sqrt{5}}{2}}\right|_{z=\frac{1-\sqrt{5}}{2}} = \frac{\left(\frac{1-\sqrt{5}}{2}\right)^n}{-\sqrt{5}}$$

Therefore,

$$x(k) = \rho_1 + \rho_2 = \frac{1}{\sqrt{5}}\left[\left(\frac{1+\sqrt{5}}{2}\right)^k - \left(\frac{1-\sqrt{5}}{2}\right)^k\right]u(k)$$

(c) We know from Eqs. (i) and (ii) that

$$\mathcal{Z}\{x(k+1)\} = zX(z)$$
$$\mathcal{Z}\{x(k)\} = X(z)$$

Therefore,

$$\lim_{x\to\infty} \frac{x(k+1)}{x(k)} = \lim_{z\to 1} \frac{zX(z)}{X(z)} = 1$$

17.4 TRANSFER FUNCTIONS OF DISCRETE DATA SYSTEMS

The discrete data systems can be divided into the following five categories:

1. Simple digital system
2. Cascaded elements separated by a sampler
3. Cascaded elements not separated by a sampler
4. Zero order hold circuit
5. Closed-loop discrete data

We consider them in that order.

1. Simple Digital System

The input and output of a simple digital system consists of a train of signals as shown in Figure 17.11a.

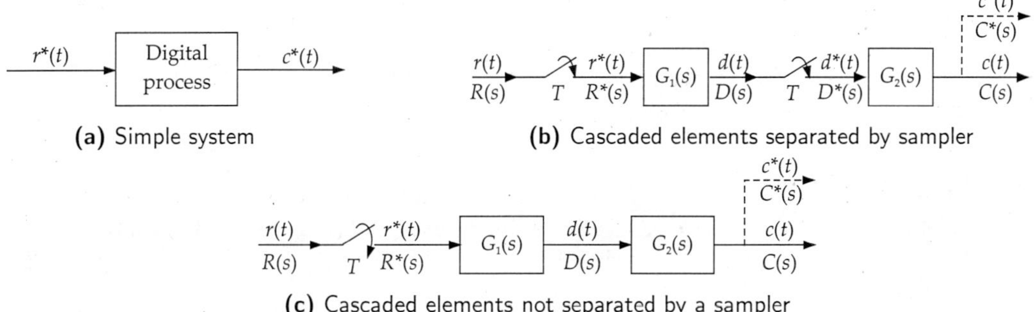

Figure 17.11: Digital systems.

The Laplace transformed equation of such a system can be written as

$$C^*(s) = G^*(s)\,R^*(s)$$

Its z-transformed form is

$$C(z) = G(z)\,R(z)$$

2. Cascaded Elements Separated by a Sampler

A digital system having cascaded linear elements separated by a sampler is shown in Figure 17.11b. For this system we have the following equations

$$\begin{aligned} D(s) &= G_1(s)R^*(s) \\ D^*(s) &= G_1^*(s)R^*(s) \\ C(s) &= G_2(s)D^*(s) = G_2(s)G_1^*(s)R^*(s) \end{aligned}$$

Therefore,

$$C^*(s) = G_2^*(s)G_1^*(s)R^*(s) \tag{17.16}$$

Equation (17.16) leads to

$$C(z) = G_2(z)G_1(z)R(z) \tag{17.17}$$

The import of Eq. (17.17) is that the z-transform of two linear elements separated by a sampler equals the product of z-transforms of two individual transfer functions.

3. Cascaded Elements Not Separated by a Sampler

The situation is as shown in Figure 17.11c. In this case the relevant equations are

$$\begin{aligned} C(s) &= G_1(s)G_2(s)R^*(s) \\ C^*(s) &= [G_1(s)G_2(s)]^*R^*(s) \end{aligned} \tag{17.18}$$

Let us rewrite Eq. (17.18) in a slightly different way to emphasize that the functions within the square bracket have to be multiplied first before they are sampled. So,

$$C^*(s) = [G_1G_2]^*(s)R^*(s)$$

which leads to

$$C(z) = [G_1G_2](z)R(z)$$

where $[G_1G_2](z)$ is the z-transform of the product of $G_1(s)$ and $G_2(s)$. It needs be kept in mind that

$$\begin{aligned} \left[G_1G_2\right]^*(s) &\neq G_1^*(s)\,G_2^*(s) \\ \left[G_1G_2\right](z) &\neq G_1(z)\,G_2(z) \end{aligned}$$

4. Zero Order Hold Circuit

A zero order hold circuit (often called ZOH) holds the input signal (Figure 17.12a) from a switch for a duration T, thus giving the output signal the shape of a unit step function of duration T (see Figure 17.12b).

In fact, it acts as an interface or filter which converts impulses into an analog signal. Its action can be visualized from the diagram of Figure 17.13.

The transfer function of such a circuit can be obtained as follows. We know that

$$u(t) = \begin{cases} 1 & t \geq 0 \\ 0 & t < 0 \end{cases}$$

$$u(t-T) = \begin{cases} 1 & (t-T) \geq 0 \\ 0 & (t-T) < 0 \end{cases}$$

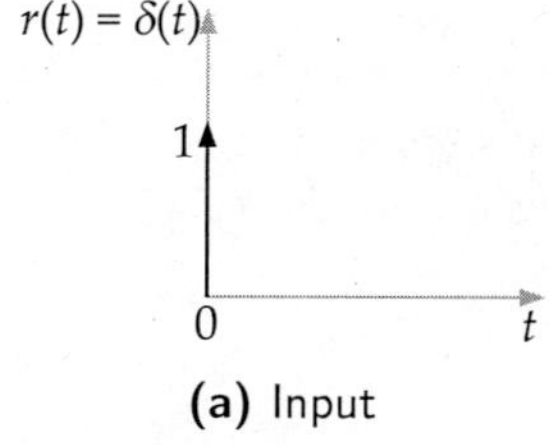

(a) Input

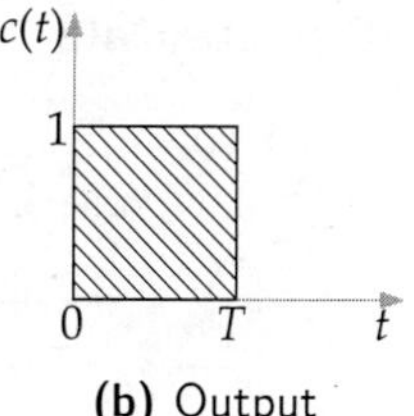

(b) Output

Figure 17.12: ZOH.

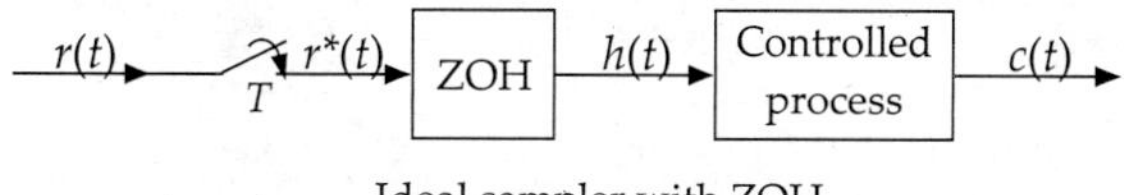

Ideal sampler with ZOH

Input signal, r(t), to ideal sampler

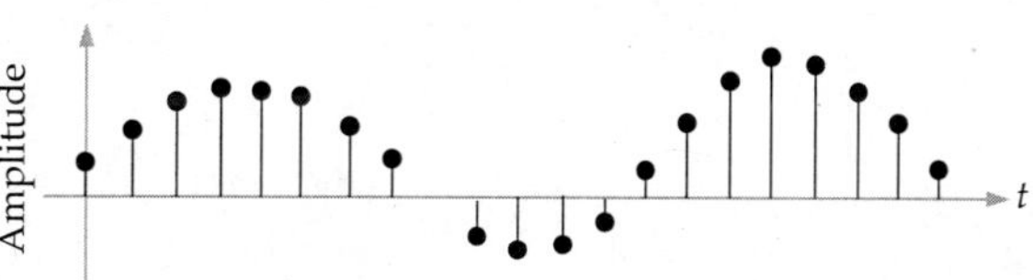

Output signal, r*(t), of ideal sampler

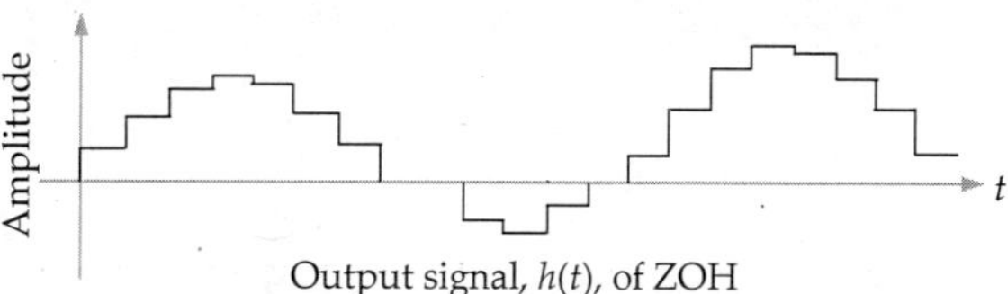

Output signal, h(t), of ZOH

Figure 17.13: Action of a ZOH.

Since the output lasts only for T seconds, it can be represented as

$$c(t) = u(t) - u(t - T) \tag{17.19}$$

Therefore, the Laplace transform of Eq. (17.19) is

$$\begin{aligned}\mathscr{L}\{c(t)\} \equiv C(s) &= \mathscr{L}\{u(t)\} - \mathscr{L}\{u(t-T)\} \\ &= \frac{1}{s} - \frac{e^{-sT}}{s} \qquad \text{[by time shifting}^2\text{property]}\end{aligned} \tag{17.20}$$

The input $r(t)$ is an impulse function and its Laplace transform is 1. Therefore,

$$\text{Transfer function, } G(s) = \frac{C(s)}{R(s)} = \frac{1 - e^{-sT}}{s}$$

[2] See Section A at page 516.

5. Closed-loop Discrete Data

A closed-loop digital control system contains a switching system over and above the other components of an analog control system. Such a system is shown in Figure 17.14.

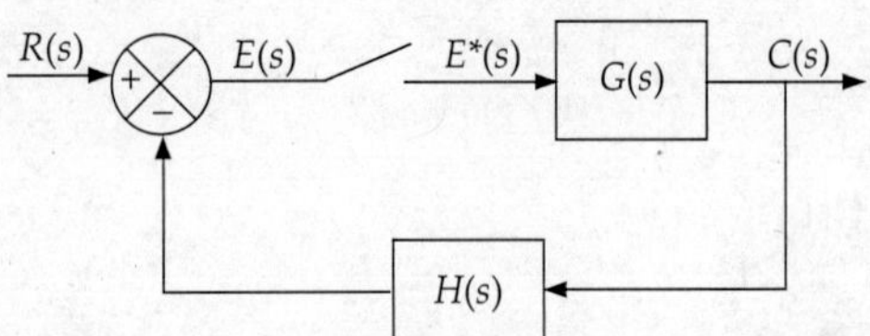

Figure 17.14: A closed-loop digital control system.

In a non-sampled system,

$$E(s) = R(s) - C(s)H(s)$$

$$C(s) = E(s)G(s)$$

Therefore,

$$E(s) = R(s) - E(s)G(s)H(s) \tag{17.21}$$

But, in a discrete time system, $E(s)$ is being sampled. Which means, all signals passing through the switch are being sampled. Therefore, for a discrete time system we can write Eq. (17.21) as

$$E^*(s) = R^*(s) - E^*(s)G^*(s)H^*(s)$$

or,

$$E^*(s)\left[1 + G^*(s)H^*(s)\right] = R^*(s) \tag{17.22}$$

Also,

$$C^*(s) = E^*(s)G^*(s) \tag{17.23}$$

Combining the results of Eqs. (17.22) and (17.23), we get

$$\frac{C^*(s)}{G^*(s)} = \frac{G^*(s)}{1 + G^*(s)H^*(s)}$$

In terms of z-transform,

$$\frac{C(z)}{R(z)} = \frac{G(z)}{1 + G(z)H(z)} \tag{17.24}$$

17.5 STABILITY OF DIGITAL CONTROL SYSTEMS

From Eq. (17.24) we observe that the characteristic equation of a digital control system is

$$1 + G(z)H(z) = 0 \tag{17.25}$$

The roots of Eq. (17.25) are the poles of the discrete time system.

We know that in an analog system, the characteristic equation is given by

$$1 + G(s)H(s) = 0$$

and we have seen that the poles of such a system must lie in the LHP of the s-plane to make the system stable. The discrete time system maps with the analog system through the relation

$$z = e^{sT}$$

For a complex analysis, as usual, we put $s = \jmath\omega$ and see

$$
\begin{aligned}
z = e^{\jmath\omega T} &= \cos\omega T + \jmath\sin\omega T \\
&\equiv \sqrt{\cos^2\omega T + \sin^2\omega T}\angle\tan^{-1}\frac{\sin\omega t}{\cos\omega T} \\
&= \sqrt{\cos^2\omega T + \sin^2\omega T}\angle\tan^{-1}\tan\omega T
\end{aligned}
$$

$$\Rightarrow \qquad z = |1|\angle\omega T$$

Thus, the LHP of the s-plane (Figure 17.15a) corresponds to a unit circle in the z-plane (Figure 17.15b).

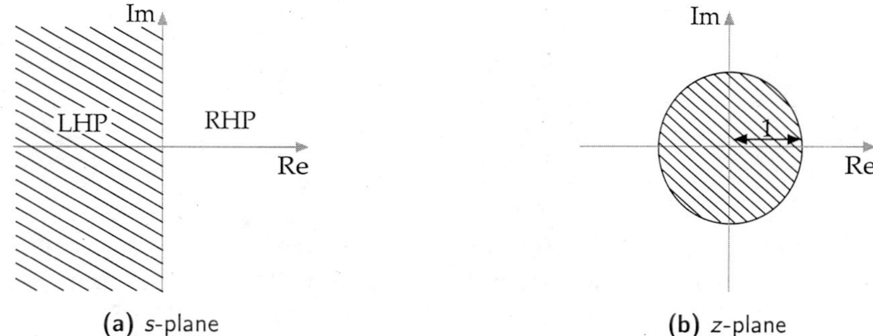

(a) s-plane (b) z-plane

Figure 17.15: Correspondence of LHP of s-plane and unit circle of z-plane.

The foregoing discussion can be summarized as follows:

1. The $\jmath\omega$-axis in the s-plane maps to the unit circle in the z-plane.
2. The right half of the s-plane maps to outside of the unit circle in the z-plane.
3. The left half of the s-plane maps to inside of the unit circle in the z-plane.

Therefore, a discrete time system is stable if the poles of the z-plane closed-loop transfer function lie inside a unit circle.

With this background of the z-transform, let us work out a few problems.

EXAMPLE 17.49. The transfer function of a system is given by

$$G(s) = \frac{10}{s(s+1)(s+3)}$$

If the system is sampled, find

(a) the z-transfer function of the system, and
(b) check its stability (T = sampling time).

Solution

(a) Let $G(s) \equiv \dfrac{A}{s} + \dfrac{B}{s+1} + \dfrac{C}{s+3}$. Then,

$$A = sG(s)|_{s=0} = \left.\frac{10}{(s+1)(s+3)}\right|_{s=0} = 3.33$$

$$B = (s+1)G(s)|_{s=-1} = \left.\frac{10}{s(s+3)}\right|_{s=-1} = -5$$

$$C = (s+3)G(s)|_{s=-3} = \left.\frac{10}{s(s+1)}\right|_{s=-3} = 1.67$$

Thus,

$$G(s) = \frac{3.33}{s} - \frac{5}{s+1} + \frac{1.67}{s+3}$$

Therefore,

$$G(z) = \frac{3.33z}{z-1} - \frac{5z}{z-e^{-T}} + \frac{1.67z}{z-e^{-3T}} \tag{i}$$

(b) It is apparent from Eq. (i) that the poles of $G(z)$ are at 1, e^{-T} and e^{-3T}. e^{-T} and $e^{-3T} < 1$ for $T > 0$. Therefore, they lie within the $|z| = 1$ circle. But the pole at 1 touches the unity circle. So, the system is marginally stable.

EXAMPLE 17.50. Find the transfer function for the arrangement given in Figure 17.16.

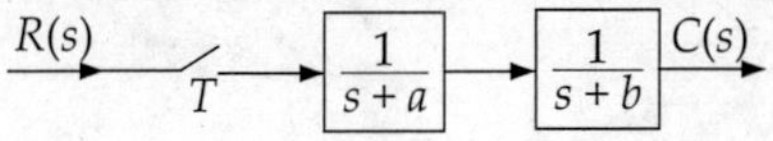

$R(s)$ —/ T → $\boxed{\dfrac{1}{s+a}}$ → $\boxed{\dfrac{1}{s+b}}$ → $C(s)$

Figure 17.16: Discrete time system (Example 17.50).

Solution Here, $G(s) = \dfrac{1}{(s+a)(s+b)} = \dfrac{1}{b-a}\left[\dfrac{1}{s+a} - \dfrac{1}{s+b}\right]$ Therefore,

$$g(t) \equiv \mathscr{L}^{-1}\{G(s)\} = \frac{1}{b-a}\left[e^{-at} - e^{-bt}\right]$$

For a discrete time system, $t = kT$ where $k = 0,\ 1,\ 2,\ \ldots, \infty$. Therefore,

$$g(kT) = \frac{1}{b-a}\left[e^{-akT} - e^{-bkT}\right]$$

Thus,

$$\begin{aligned} G(z) \equiv \mathcal{Z}\{g(kT)\} &= \frac{1}{b-a}\left[\mathcal{Z}\{e^{-akT}\} - \mathcal{Z}\{-e^{-bkT}\}\right] \\ &= \frac{1}{b-a}\left[\frac{z}{z-e^{-aT}} - \frac{z}{z-e^{-bT}}\right] \\ &= \frac{1}{b-a}\cdot\frac{z\left(e^{-aT} - e^{-bT}\right)}{\left(z-e^{-aT}\right)\left(z-e^{-bT}\right)} \end{aligned}$$

EXAMPLE 17.51. Find the transfer function for the arrangement given in Figure 17.17.

Figure 17.17: Discrete time system (Example 17.51).

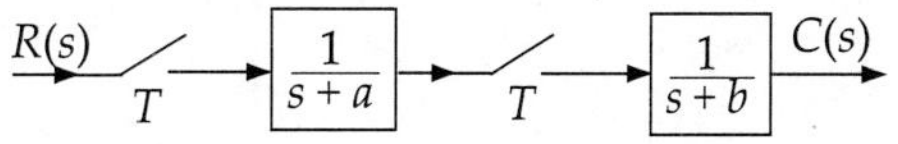

Solution The difference between the previous example and the present example is that here both the processes are made discrete separately. Hence,

$$G_1(s) = \frac{1}{s+a}$$

$$\Rightarrow \quad g_1(t) = e^{-at}$$

$$\Rightarrow \quad g_1(kT) = e^{-akT}$$

Therefore,

$$G_1(z) = \frac{z}{z - e^{-aT}} \tag{i}$$

Similarly,

$$G_2(z) = \frac{z}{z - e^{-bT}} \tag{ii}$$

Combining Eqs. (i) and (ii), we get the z-transform of the composite process as

$$G_1(z)G_2(z) = \frac{z^2}{(z - e^{-aT})(z - e^{-bT})}$$

EXAMPLE 17.52. Consider the open-loop sampled data system given in Figure 17.18.

Figure 17.18: Open-loop sampled data system (Example 17.52).

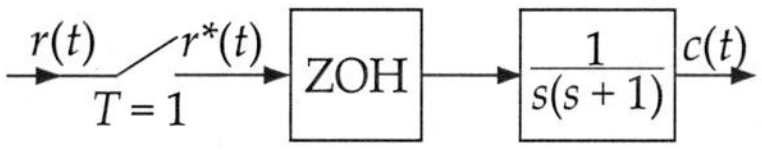

Find the following:

(a) The z-transfer function of the system.

(b) The response of the system to a unit impulse.

Solution

(a) Incorporating the transfer function of the ZOH [see Eq. (17.20)], the open loop transfer function of the system is given by

$$G(s) = \frac{1 - e^{-sT}}{s^2(s+1)}$$

Expanding it into partial fractions, we get

$$G(s) = (1 - e^{-sT})\left(\frac{1}{s^2} - \frac{1}{s} + \frac{1}{s+1}\right)$$

Therefore,

$$G(z) \equiv \mathcal{Z}\{G(s)\} = (1 - z^{-1})\mathcal{Z}\left\{\frac{1}{s^2} - \frac{1}{s} + \frac{1}{s+1}\right\}$$

$$= (1 - z^{-1})\left[\frac{Tz}{(z-1)^2} - \frac{z}{z-1} + \frac{z}{z - e^{-T}}\right]$$

$$= \frac{z-1}{z}\left[\frac{Tz(z - e^{-T}) - z(z-1)(z - e^{-T}) + z(z-1)^2}{(z-1)^2(z - e^{-T})}\right]$$

$$= \frac{(ze^{-T} - z + Tz) + (1 - e^{-T} - Te^{-T})}{(z-1)(z - e^{-T})}$$

(b) When $T = 1$, we have

$$G(z) = \frac{ze^{-1} + 1 - 2e^{-1}}{(z-1)(z-e^{-1})}$$
$$= \frac{0.3678z + 0.2644}{(z-1)(z-0.3678)}$$
$$= \frac{0.3678z + 0.2644}{z^2 - 1.3678z + 0.3678} \quad \text{(i)}$$

By long division of Eq. (i), we find

$$G(z) = 0.3678z^{-1} + 0.7675z^{-2} + 0.9145z^{-3} + \cdots \quad \text{(ii)}$$

Since, $R(z) = 1$ for an impulse input, $C(z) = G(z)$. So, we can write

$$C(z) \equiv C(kT) = C(0) + C(T)z^{-1} + C(2T)z^{-2} + C(3T)z^{-3} + \cdots \quad \text{(iii)}$$
$$= G(z)$$

Comparing the coefficients Eqs. (ii) and (iii), we obtain $C(kT)$ as follows:

$$C(0) = 0$$
$$C(T) = 0.3678$$
$$C(2T) = 0.7675$$
$$C(3T) = 0.9145$$

REVIEW QUESTIONS

17.1 (i) A cascade of 3 linear time-invariant systems is causal and unstable. From this, we conclude that

(A) each system in the cascade is individually causal and unstable
(B) at least one system is unstable and at least one system is causal
(C) at least one system is causal and all systems are unstable
(D) the majority are unstable and the majority are causal

(ii) An input signal $x(t) = 2 + \sin(100\pi t)$ is sampled with a sampling frequency of 400 Hz and applied to the system whose transfer function is represented by

$$\frac{Y(z)}{Z(z)} = \frac{1}{n}\left(\frac{1-z^{-N}}{1-z^{-1}}\right)$$

where, N represents the number of samples per cycle. The output $y(n)$ of the system under steady state is

(A) 0 (B) 1 (C) 2 (D) 5

(iii) The impulse response of a causal linear time-invariant system is given by $h(t)$. Now consider the following two statements:

Statement I—Principle of superposition holds
Statement II—$h(t) = 0$ for $t < 0$

Which one of the following statements is correct?

(A) Statement I is correct and Statement II is wrong
(B) Statement II is correct and Statement I is wrong
(C) Both Statement I and Statement II are wrong
(D) Both Statement I and Statement II are correct

(iv) $H(z)$ is a transfer function of a *real* system. When a signal $x[n] = \left(1 + \jmath\right)^n$ is the input to such a system, the otput is zero. Further, the region of convergence (ROC) of $\left(1 - \frac{1}{2}z^{-1}\right)H(z)$ is the entire z-plane (except $z = 0$). It can then be inferred that $H(z)$ can have a minimum of

(A) one pole and one zero (B) one pole and two zeros
(C) two poles and one zero (D) two poles and two zeros

(v) Given $X(z) = \dfrac{z}{(z-a)^2}$ with $|z| > a$, the residue of $X(z)z^{n-1}$ at $z = a$ will be

(A) a^{n-1} (B) a^n (C) na^n (D) na^{n-1}

17.2 (a) In brief, discuss z- and s- domain relationship.
(b) From this, describe stability criteria of a system in z-domain.
(c) In brief, discuss zero-order hold circuit.
(d) $x[n] = -a^n u[-n-1]$. Then calculate $x(z)$ and also its ROC.

17.3 (a) Find the z-transform of unit step.
(b) In brief discuss s-domain to z-domain mapping of a system.
(c) If the z-transfer function of a system is given, explain how you will comment about the stability of the system.

17.4 Explain sampling and hold. Obtain the unit-step response of the system shown in the following figure.

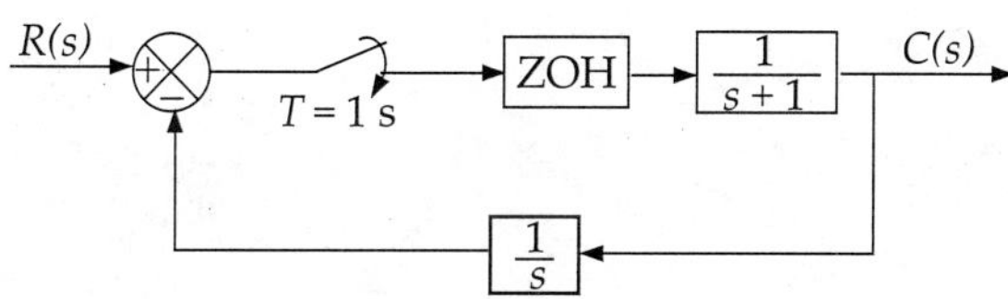

17.5 (a) What is a sampled data system?
(b) Express the output c(t) in the form of zero-order hold (ZOH) sampled data system given in the following figure.

e(t) T ZOH c(t)

17.6 (a) Find the z-transforms of
(i) $f(k) = ke^{-3k}$
(ii) $f(k) = k^2$
(b) Find $C(z)/R(z)$ for the sampled data system shown in the following figure.

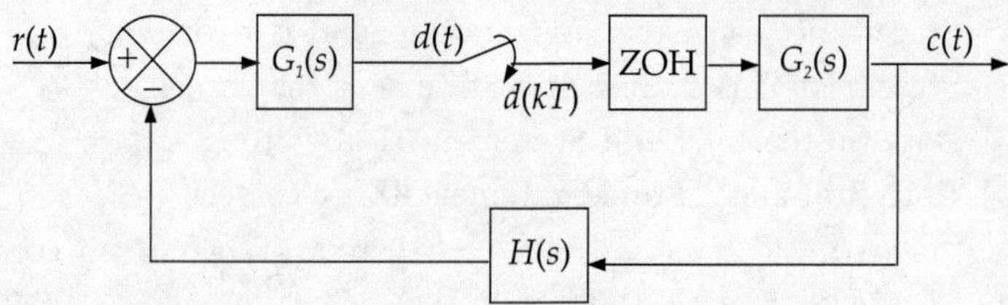

Answers to Selected Questions

17.1. (i) (B) (ii) (C) (iii) (D) (iv) (B) (v) (D)

17.2. (d) $\dfrac{z}{z-a}$ ROC: $|z| < a$ 17.4. $C(1) = 0,\ C(2) = 0.5335,\ C(3) = 0.4193,\ C(4) = 0.1332,$ $C(5) = -0.0495, C(6) = -0.0879,\ C(7) = -0.0508,\ C(8) = -0.0076,\ C(9) = 0.0127,$ $C(10) = 0.0128,\ C(11) = 0.0054$

17.6. (a)(i) $\dfrac{Tze^{-3T}}{\left(z-e^{-3T}\right)^2}$ (ii) $\dfrac{T^2z(z+1)}{(z-1)^3}$

A

Laplace Transform

The Laplace[1] transform is a very powerful tool for studying differential equations.

The Laplace transform helps us

1. Solve the homogeneous equation and the particular integral in one operation.
2. Convert differential equations to algebraic equations, solve such simultaneous equations with ease and then get the final solution through inverse Laplace transform.

Definition

If $f(t)$ is a function of t specified for $t > 0$, then Laplace transform of $f(t)$, denoted by $\mathscr{L} f(t)\}$, is defined as

$$\mathscr{L}\{f(t)\} \equiv F(s) = \int_0^\infty e^{-st} f(t)dt \tag{A.1}$$

The quantity s, which is often referred to as the Laplace operator, may be real or complex. It is more useful to consider s complex as $s = \sigma + \jmath\omega$ where ω is the frequency.

Actually, in mathematics the Laplace transform limits are from $-\infty$ to ∞. But here, in Eq. (A.1) we have chosen the limits from 0 to ∞. This may be called the *one-sided Laplace transform*. The one-sided Laplace transform implies that the initial conditions of the system at $t = 0$ were zero. Physically that is a perfectly acceptable situation; because, a system is not expected to produce a response before an input is applied at $t = 0$.

Important Properties of Laplace Transform

We now consider a few important properties of the Laplace transform. These will be of help to us while working out Laplace transforms of functions.

1. Multiplication by a constant

If $f(t)$ is a real function in the t-domain and $\mathscr{L}\{f(t)\} \equiv F(s)$ where $\mathscr{L}$ is the Laplace transform operator, then

$$\mathscr{L}\{kf(t)\} = kF(s) \tag{A.2}$$

2. Linearity

If c_1 and c_2 are constants and $\mathscr{L}\{f_1(t)\} \equiv F_1(s)$, $\mathscr{L}\{f_2(t)\} \equiv F_2(s)$, then

$$\begin{aligned}\mathscr{L}\{c_1 f_1(t) \pm c_2 f_2(t)\} &= c_1\mathscr{L}\{f_1(t)\} \pm c_2\mathscr{L}\{f_2(t)\}\\ &\equiv c_1F_1(s) \pm c_2F_2(s)\end{aligned}$$

Because of this property, $\mathscr{L}$ is called a linear operator.

[1]Pierre-Simon, marquis de Laplace (1749—1827) was a French mathematician and astronomer who contributed much to the development of mathematical astronomy and statistics.

EXAMPLE A.1. Find the Laplace transform of $3t^2 - 2\cos 2t + 4e^{-t}$.

Solution

$$\begin{aligned}\mathscr{L}\{3t^2 - 2\cos 2t + 4e^{-t}\} &= 3\mathscr{L}\{t^2\} - 2\mathscr{L}\{\cos 2t\} + 4\mathscr{L}\{e^{-t}\} \\ &= 3\left(\frac{2!}{s^3}\right) - 2\left(\frac{s}{s^2+4}\right) + 4\left(\frac{1}{s+1}\right) \\ &= \frac{6}{s^3} - \frac{2s}{s^2+4} + \frac{4}{s+1}\end{aligned}$$

For the Laplace transforms of different functions used here, see Appendix A.

3. Exponential shifting

If $\mathscr{L}\{f(t)\} \equiv F(s)$, then

$$\mathscr{L}\{e^{\mp at} f(t)\} = F(s \pm a) \tag{A.3}$$

EXAMPLE A.2. Find the Laplace transform of $e^{-t}\cos 2t$.

Solution Since $\mathscr{L}\{\cos 2t\} = \dfrac{s}{s^2+4}$

$$\mathscr{L}\{e^{-t}\cos 2t\} = \frac{s+1}{(s+1)^2+4} = \frac{s+1}{s^2+2s+5}$$

4. Time shifting

If $\mathscr{L}\{f(t)\} \equiv F(s)$ and

$$g(t) = f(t-a)u(t-a) \quad a \geq 0$$

then

$$\mathscr{L}\{g(t)\} = e^{-as}F(s) \tag{A.4}$$

EXAMPLE A.3. Find the Laplace transform of

$$g(t) = \begin{cases} (t-2)^3 & t > 2 \\ 0 & t < 2 \end{cases}$$

Solution Since $\mathscr{L}\{t^3\} = \dfrac{3!}{s^4} = \dfrac{6}{s^4}$,

$$\mathscr{L}\{g(t)\} = \frac{6e^{-2s}}{s^4}$$

5. Change of scale

If $\mathscr{L}\{f(t)\} \equiv F(s)$, then

$$\mathscr{L}\{f(at)\} = \frac{1}{a}F\left(\frac{s}{a}\right) \tag{A.5}$$

EXAMPLE A.4. Find the Laplace transform of $\sin 3t$.

Solution Since $\mathscr{L}\{\sin t\} = \dfrac{1}{s^2+1}$,

$$\mathscr{L}\{\sin 3t\} = \frac{1}{3} \cdot \frac{1}{\left(\frac{s}{3}\right)^2 + 1} = \frac{3}{s^2+9}$$

6. Differentiation

If $\mathscr{L}\{f(t)\} \equiv F(s)$, then

(a) $\mathscr{L}\{f'(t)\} = sF(s) - f(0)$ provided, $f(t)$ is continuous in $0 \le t \le N$ and exponential order for $t > N$, while $f'(t)$ is sectionally continuous for $0 \le t \le N$.
(b) $\mathscr{L}\{f''(t) = s^2F(s) - sf(0) - f''(0)$
(c) $\mathscr{L}\{f^{(n)}(t)\} = s^nF(s) - s^{n-1}f(0) - s^{n-2}f'(0) - \cdots sf^{(n-2)}(0) - f^{(n-1)}(0)$

EXAMPLE A.5. If $f(t) = \cos 3t$, then find the Laplace transform of $f'(t)$.

Solution $\mathscr{L}\{f(t)\} \equiv F(s) = \dfrac{s}{s^2+9}$. Therefore,

$$\begin{aligned}\mathscr{L}\{f'(t)\} &= sF(s) - f(0)\\ &= \frac{s^2}{s^2+9} - 1 = -\frac{9}{s^2+9}\end{aligned}$$

7. Integration

If $\mathscr{L}\{f(t)\} \equiv F(s)$, then

$$\mathscr{L}\left\{\int_0^t f(u)du\right\} = \frac{F(s)}{s} \tag{A.6}$$

EXAMPLE A.6. Find the Laplace transform of $\int_0^t \sin 2u du$.

Solution Since $\mathscr{L}\{\sin 2t\} = \dfrac{2}{s^2+4}$,

$$\mathscr{L}\left\{\int_0^t \sin 2u du\right\} = \frac{\dfrac{2}{(s^2+4)}}{s} = \frac{2}{s(s^2+4)}$$

8. Initial value theorem

If $\mathscr{L}\{f(t)\} \equiv F(s)$, then

$$\lim_{t\to 0} f(t) = \lim_{s\to\infty} sF(s) \tag{A.7}$$

EXAMPLE A.7. If

$$F(s) = \frac{1}{s\big(s^2+3s+2\big)} \tag{i}$$

then find the initial value of the corresponding $f(t)$.

Solution The required initial value is

$$\begin{aligned}\lim_{t\to 0} f(t) &= \lim_{s\to\infty} s \cdot \frac{1}{s\big(s^2+3s+2\big)}\\ &= \lim_{s\to\infty}\frac{1}{s^2+3s+2}\\ &= 0\end{aligned}$$

9. Final value theorem

If $\mathscr{L}\{f(t)\} \equiv F(s)$ and if $sF(s)$ is analytic on the imaginary axis and in the right half of the s-plane, then

$$\lim_{t\to\infty} f(t) = \lim_{s\to 0} sF(s) \tag{A.8}$$

EXAMPLE A.8. Find the final value of the corresponding $f(t)$ of Eq. (i) of Example A.7.

Solution The required final value is

$$\begin{aligned}\lim_{t\to\infty} f(t) &= \lim_{s\to 0} s \cdot \frac{1}{s(s^2+3s+2)} = \lim_{s\to 0} \frac{1}{s^2+3s+2} \\ &= \frac{1}{2}\end{aligned}$$

Note: 1. By virtue of the initial and final value theorems, we need not find out the t-space function $f(t)$ to figure out its initial and final values, if its Laplace transformed form is known.
2. The final value theorem is very helpful in evaluating the steady-state response of systems represented by their transfer functions. But the condition stated above must be kept in mind before applying it. Let us see what happens if the condition is not satisfied.

EXAMPLE A.9. If $F(s) = \dfrac{\alpha}{s^2+\alpha^2}$ find the steady-state value of $f(t)$ with the help of the final value theorem.

Solution If we apply the final value theorem to find the steady state value of the function, we get

$$\lim_{t\to\infty} f(t) = \lim_{s\to 0}\left(s \cdot \frac{\alpha}{s^2+\alpha^2}\right) = 0$$

Note: We know that $f(t) = \sin t$ which has no final value. The result obtained by applying the final value theorem is erroneous because the function $sF(s)$ has two poles on the imaginary axis and therefore, the final value theorem cannot be applied here.

Inverse Laplace Transform

We have already told that to get the real t-space solution from the Laplace transformed functions, we have to perform inverse Laplace transform. Stated mathematically,

$$f(t) = \mathscr{L}^{-1}\{F(s)\} \tag{A.9}$$

$$= \frac{1}{2\pi j}\int_{k-j\infty}^{k+j\infty} F(s)e^{st}ds \tag{A.10}$$

where k is a real constant which is greater than all the singularities of $F(s)$. However, for most of our tasks, the inversion can be performed with the help of the following inverse Laplace transform table.

Table A.1: Commonly Used Inverse Laplace Transforms

Serial No.	*Laplace transform* $F(s)$	*Inverse function* $f(t)$	*Serial No.*	*Laplace transform* $F(s)$	*Inverse function* $f(t)$
1	s	$\left.\frac{d}{dt}\delta(t)\right\|_{t=0}$	15	$\frac{s\sin\phi+\omega\cos\phi}{s^2+\omega^2}$	$\sin(\omega t+\phi)$
2	$e^{-as} \quad a\geq 0$	$\delta(t-a)$	16	$\frac{s\cos\phi-\omega\sin\phi}{s^2+\omega^2}$	$\cos(\omega t+\phi)$
3	$\frac{e^{-as}}{s}$	$u(t-a)$	17	$\frac{2\omega s}{(s^2+a^2)^2}$	$t\sin\omega t$
4	$\frac{e^{-as}-e^{-bs}}{s}$	$f(t)=\begin{cases}0 & \text{for } t<a\\ 1 & \text{for } a<t<b\\ 0 & \text{for } t>b\end{cases}$	18	$\tan^{-1}\left(\frac{\omega}{s}\right)$	$\frac{1}{t}\sin\omega t$
5	$\frac{n!}{s^{n+1}}$	$t^n \quad n=1,\ 2,\ 3\ldots$	19	$\frac{s}{s^2+\omega^2}$	$\cos\omega t$
6	$\frac{1}{s+a}$	e^{-at}	20	$\frac{s^2-\omega^2}{(s^2+\omega^2)^2}$	$t\cos\omega t$
7	$\frac{1}{(s+a)^2}$	te^{-at}	21	$\frac{1}{s(s^2+\omega^2)}$	$\frac{1-\cos\omega t}{\omega^2}$
8	$\frac{n!}{(s+a)^{n+1}}$	$t^n e^{-at} \quad n=1,\ 2,\ 3\ldots$	22	$\frac{1}{(s^2+a^2)^2}$	$\frac{\sin\omega t-\omega t\cos\omega t}{\omega^3}$
9	$\frac{1}{(s+a)^n}$	$\frac{t^{n-1}}{(n-1)!}e^{-at} \quad n=1,\ 2,\ 3\ldots$	23	$\frac{s^2}{(s^2+a^2)^2}$	$\frac{\sin\omega t+\omega t\cos\omega t}{2\omega}$
10	$\frac{a}{s(s+a)}$	$1-e^{-at}$	24	$\frac{1}{\sqrt{s}}$	$\frac{1}{\sqrt{\pi t}}$
11	$\frac{1}{(s+a)(s+b)}$	$\frac{e^{-at}-e^{-bt}}{b-a}$	25	$\frac{1}{\sqrt{s+\omega}}$	$\frac{1}{\sqrt{\pi t}}e^{-\omega t}$
12	$\frac{s}{(s+a)(s+b)}$	$\frac{be^{-bt}-ae^{-at}}{b-a}$	26	$\frac{1}{\sqrt{s^3}}$	$2\sqrt{\frac{t}{\pi}}$
13	$\frac{\omega}{s^2+\omega^2}$	$\sin\omega t$	27	$\int_s^\infty f(u)du$	$\frac{f(t)}{t}$
14	$\frac{b}{(s-a)^2+b^2}$	$e^{at}\sin(bt)$	28	$\frac{s-a}{(s-a)^2+b^2}$	$e^{at}\cos(bt)$

B

Rules of Partial Fractions

While expanding transfer functions by the partial fraction method, we may encounter any of the following two situations:

1. When all poles of $G(s)$ are simple
2. When some poles are of multiple order

B.1 WHEN ALL POLES ARE SIMPLE

The situation can be divided into two categories as follows:

1. When the denominator polynomial is of higher order
2. When the numerator polynomial is of higher order

We consider two examples to show the procedures.

When the denominator polynomial is of higher order

EXAMPLE B.1. Express the following function into partial fractions:

$$G(s) = \frac{4s+2}{s(s+1)(s+2)} \tag{i}$$

Solution We write Eq. (i), splitting denominator fractions, as

$$G(s) \equiv \frac{A}{s} + \frac{B}{s+1} + \frac{C}{s+2} \tag{ii}$$

where A, $B$and C are unknown constants. These constants can be evaluated as follows:

$$A = s \cdot G(s)|_{s=0} = s \cdot \left.\frac{4s+2}{s(s+1)(s+2)}\right|_{s=0} = \left.\frac{4s+2}{(s+1)(s+2)}\right|_{s=0} = 1$$

$$B = (s+1) \cdot G(s)|_{s=-1} = (s+1) \cdot \left.\frac{4s+2}{s(s+1)(s+2)}\right|_{s=-1} = \left.\frac{4s+2}{s(s+2)}\right|_{s=-1} = 2$$

$$C = (s+2) \cdot G(s)|_{s=-2} = (s+2) \cdot \left.\frac{4s+2}{s(s+1)(s+2)}\right|_{s=-2} = \left.\frac{4s+2}{s(s+1)}\right|_{s=-2} = -3$$

Substituting the values of A, B and C in Eq. (ii), we have

$$G(s) = \frac{1}{s} + \frac{2}{s+1} - \frac{3}{s+2}$$

When the numerator polynomial is of higher order

Note: This kind of transfer function is called *improper*[1]. We will not encounter such transfer function. The following is given only on academic interest.

[1]See Section 2.1 at page 11.

EXAMPLE B.2. Express the following function into appropriate partial fractions:

$$x(s) = \frac{4s^3 + s^2 + 8s + 4}{s^2 + 4}$$

Solution We note that the function cannot be solved using partial fractions as such because the numerator is of 3rd order and the denominator is only of 2nd order. Therefore long division can be used to reduce the order of the equation as follows:

$$\begin{array}{r|l} & 4s \quad +1 \\ \hline s^2+4 & 4s^3 + s^2 + 8s \; + 4 \\ & 4s^3 \qquad + 16s \\ \hline & \qquad s^2 - 8s \; + 4 \\ & \qquad s^2 \qquad + 4 \\ \hline & \qquad\quad - 8s \end{array}$$

The result helps us write a new function that has a reduced fraction as shown below

$$x(s) = 4s + 1 - \frac{8s}{s^2 + 4}$$

This fraction can be solved with partial fractions as

$$\frac{8s}{s^2 + 4} = \frac{\jmath\, 2s}{s + \jmath\, 2} - \frac{\jmath\, 2s}{s - \jmath\, 2}$$

B.2 WHEN SOME POLES ARE OF MULTIPLE ORDER

In this case, the constants of the multiple order pole C_n are obtained from the formula

$$C_n = \frac{1}{(n-1)!} \cdot \frac{d^{n-1}}{ds^{n-1}} \left[(s+a)^n G(s) \right] \tag{B.1}$$

The following example will demonstrate how to handle this situation.

EXAMPLE B.3. Find partial fractions for

$$G(s) = \frac{1}{s(s+1)^3(s+2)} \tag{i}$$

Solution We have to write Eq. (i) as

$$G(s) \equiv \frac{A}{s} + \frac{B}{s+2} + \frac{C_1}{(s+1)^3} + \frac{C_2}{(s+1)^2} + \frac{C_3}{s+1} \tag{ii}$$

A, B, C_1, C_2 and C_3 in Eq. (ii) are found out in the following way:

$$A = s \cdot G(s)\big|_{s=0} = \left.\frac{1}{(s+1)^3(s+2)}\right|_{s=0} = \frac{1}{2}$$

$$B = (s+2) \cdot G(s)\big|_{s=-2} = \left.\frac{1}{s(s+1)^3}\right|_{s=-2} = \frac{1}{2}$$

$$C_1 = (s+1)^3 \cdot G(s)\big|_{s=-1} = \left.\frac{1}{s(s+2)}\right|_{s=-1} = -1$$

$$C_2 = \left.\frac{d}{ds}\left[(s+1)^3 \cdot G(s)\right]\right|_{s=-1} = \left.\frac{d}{ds}\left[\frac{1}{s(s+2)}\right]\right|_{s=-1} = \left.\frac{-(2s+2)}{s^2(s+2)^2}\right|_{s=-1} = 0$$

$$C_3 = \left.\frac{1}{2!}\frac{d^2}{ds^2}\left[(s+1)^3 \cdot G(s)\right]\right|_{s=-1} = \left.\frac{1}{2!}\frac{d}{ds}\left[\frac{-(2s+2)}{s^2(s+2)^2}\right]\right|_{s=-1}$$

$$= \frac{1}{2}\left[\frac{-2}{s^2(s+2)^2} + \frac{2(2s+2)}{s^3(s+2)^2} + \frac{2(2s+2)}{s^2(s+2)^3}\right]_{s=-1} = -1$$

Thus,

$$G(s) = \frac{1}{2s} + \frac{1}{2(s+2)} - \frac{1}{(s+1)^3} - \frac{1}{s+1}$$

B.3 GENERAL METHOD FOR ALL SITUATIONS

We consider the foregoing examples to demonstrate the procedure.

EXAMPLE B.4. Find the partial fractions of

$$G(s) = \frac{4s+2}{s(s+1)(s+2)} \tag{i}$$

by the general method.

Solution We write it as

$$G(s) = \frac{A}{s} + \frac{B}{s+1} + \frac{C}{s+2}$$

$$= \frac{A(s+1)(s+2) + Bs(s+2) + Cs(s+1)}{s(s+1)(s+2)} \tag{ii}$$

Comparing the numerators of Eq. (i) and Eq. (ii), we get

$$A(s+1)(s+2) + Bs(s+2) + Cs(s+1) = 4s+2$$

or

$$(A+B+C)s^2 + (3A+2B+C)s + 2A = 4s+2 \tag{iii}$$

Comparing coefficients of equal powers of s on both sides of Eq. (iii), we have

$$A+B+C = 0 \tag{iv}$$

$$3A+2B+C = 4 \tag{v}$$

$$2A = 2 \tag{vi}$$

Solving Eqs. (iv) to (vi), we get

$$A = 1 \qquad B = 2 \qquad C = -3$$

EXAMPLE B.5. Find the partial fractions of

$$G(s) = \frac{1}{s(s+1)^3(s+2)} \tag{i}$$

by the general method.

Solution Writing it as

$$\begin{aligned} G(s) &= \frac{A}{s} + \frac{B}{s+2} + \frac{C_1}{(s+1)^3} + \frac{C_2}{(s+1)^2} + \frac{C_3}{s+1} \\ &= \frac{A(s+1)^3(s+2) + Bs(s+1)^3 + C_1 s(s+2) + C_2 s(s+1)(s+2) + C_3 s(s+1)^2(s+2)}{s(s+1)^3(s+2)} \end{aligned} \tag{ii}$$

Comparing the numerators of Eq. (i) and Eq. (ii), we get

$$\begin{aligned} (A+B+C_3)s^4 + (5A+3B+C_2+5C_3)s^3 + (9A+3B+C_1+3C_2+5C_3)s^2 \\ + (7A+B+2C_1+2C_2+2C_3)s + 2A = 1 \end{aligned} \tag{iii}$$

Comparing the coefficients of equal powers of s on both sides of Eq. (iii), we get

$$\begin{aligned} A+B+C_3 &= 0 \\ 5A+3B+C_2+5C_3 &= 0 \\ 9A+3B+C_1+3C_2+5C_3 &= 0 \\ 7A+B+2C_1+2C_2+2C_3 &= 0 \\ 2A &= 1 \end{aligned}$$

On solving the above simultaneous equations, we get the following five values for the constants

$$A = \frac{1}{2} \qquad B = \frac{1}{2} \qquad C_1 = -1 \qquad C_2 = 0 \qquad C_3 = -1$$

The transfer function of overdamped second order system[2] can also be handled in this way. The following example shows that.

EXAMPLE B.6. Find the partial fractions of

$$G(s) = \frac{K}{\dfrac{s^2}{\omega_n^2} + \dfrac{2\zeta s}{\omega_n} + 1} = \frac{K\omega_n^2}{s^2 + 2\zeta\omega_n s + \omega_n^2} \tag{i}$$

Solution

1. For $\zeta > 1$, the denominator of Eq. (i) can be factorised as

$$\begin{aligned} G(s) &= \frac{K\omega_n^2}{s\left(s+\zeta\omega_n - \omega_n\sqrt{\zeta^2-1}\right)\left(s+\zeta\omega_n + \omega_n\sqrt{\zeta^2-1}\right)} \\ &\equiv K\omega_n^2\left(\frac{A}{s} + \frac{B}{s+\zeta\omega_n - \omega_n\sqrt{\zeta^2-1}} + \frac{C}{s+\zeta\omega_n + \omega_n\sqrt{\zeta^2-1}}\right) \end{aligned} \tag{ii}$$

[2]See Section 6.3 at page 124.

where A, B and C are arbitrary coefficients to be determined.
We get the following condition to determine A, B and C:

$$A(s^2+2\zeta\omega_n s+\omega_n^2)+Bs(s+\zeta\omega_n+\omega_n\sqrt{\zeta^2-1})+Cs(s+\zeta\omega_n-\omega_n\sqrt{\zeta^2-1})=1$$

or $$(A+B+C)s^2+\left[(2A+B+C)\zeta\omega_n+(B-c)\omega_n\sqrt{\zeta^2-1}\right]s+A\omega_n^2=1 \tag{iii}$$

Comparing coefficients of different powers of s on both sides of Eq. (iii), we get

$$A\omega_n^2=1 \qquad \text{or,} \qquad A=\frac{1}{\omega_n^2}$$

$$A+B+C=0 \qquad \text{or,} \qquad B+C=-A=-\frac{1}{\omega_n^2} \tag{iv}$$

$$(2A+B+C)\zeta\omega_n+(B-C)\omega_n\sqrt{\zeta^2-1}=0 \tag{v}$$

Substituting $A+B+C=0$ in Eq. (v), we get

$$A\zeta+(B-C)\sqrt{\zeta^2-1}=0 \qquad \text{or,} \qquad B-C=-\frac{\zeta}{\omega_n^2\sqrt{\zeta^2-1}} \tag{vii}$$

From Eqs. (iv) and (vii), we get

$$B=\frac{\zeta+\sqrt{\zeta^2-1}}{2\omega_n^2\sqrt{\zeta^2-1}} \qquad C=\frac{\zeta-\sqrt{\zeta^2-1}}{2\omega_n^2\sqrt{\zeta^2-1}}$$

2. For $\zeta<1$, we have

$$C(s)=\frac{\omega_n^2}{s^2+2\zeta\omega_n s+\omega_n^2}\cdot R(s)=\frac{\omega_n^2}{s^2+2\zeta\omega_n s+\omega_n^2}\cdot\frac{1}{s} \tag{B.2}$$

$$=\frac{\omega_n^2}{s(s+\zeta\omega_n-\jmath\omega_n\sqrt{1-\zeta^2})(s+\zeta\omega_n+\jmath\omega_n\sqrt{1-\zeta^2})}$$

$$\equiv\frac{A}{s}+\frac{B}{(s+\zeta\omega_n-\jmath\omega_n\sqrt{1-\zeta^2})}+\frac{C}{(s+\zeta\omega_n+\jmath\omega_n\sqrt{1-\zeta^2})} \tag{B.3}$$

A, B, C can be obtained in the following way

$$A = sC(s)|_{s=0} = \left.\frac{\omega_n^2}{s(s+\zeta\omega_n - \jmath\omega_n\sqrt{1-\zeta^2})(s+\zeta\omega_n + \jmath\omega_n\sqrt{1-\zeta^2})}\right|_{s=0}$$

$$= \frac{\omega_n^2}{\zeta^2\omega_n^2 + \omega_n^2(1-\zeta^2)} = 1 \tag{B.4}$$

$$B = (s+\zeta\omega_n - \jmath\omega_n\sqrt{1-\zeta^2})C(s)\Big|_{s=-\zeta\omega_n+\jmath\omega_n\sqrt{1-\zeta^2}}$$

$$= \left.\frac{\omega_n^2}{s(s+\zeta\omega_n + \jmath\omega_n\sqrt{1-\zeta^2})}\right|_{s=-\zeta\omega_n+\jmath\omega_n\sqrt{1-\zeta^2}}$$

$$= \frac{1}{(-\zeta+\jmath\sqrt{1-\zeta^2})(2\jmath\sqrt{1-\zeta^2})}$$

$$= \frac{-\zeta-\jmath\sqrt{1-\zeta^2}}{(-\zeta-\jmath\sqrt{1-\zeta^2})(-\zeta+\jmath\sqrt{1-\zeta^2})(2\jmath\sqrt{1-\zeta^2})} = -\frac{\zeta}{2\jmath\sqrt{1-\zeta^2}} - \frac{1}{2} \tag{B.5}$$

$$C = (s+\zeta\omega_n + \jmath\omega_n\sqrt{1-\zeta^2})C(s)\Big|_{s=-\zeta\omega_n-\jmath\omega_n\sqrt{1-\zeta^2}}$$

$$= \left.\frac{\omega_n^2}{s(s+\zeta\omega_n - \jmath\omega_n\sqrt{1-\zeta^2})}\right|_{s=-\zeta\omega_n-\jmath\omega_n\sqrt{1-\zeta^2}}$$

$$= \frac{1}{(-\zeta-\jmath\sqrt{1-\zeta^2})(2\jmath\sqrt{1-\zeta^2})} = \frac{\zeta}{2\jmath\sqrt{1-\zeta^2}} - \frac{1}{2} \tag{B.6}$$

C

Frequency Domain Analysis

One of the most commonly used test functions for a circuit or system is the sine wave. The reason why sine waves are important involves Fourier[1] theory. As we know, any signal going into a circuit can be represented by a sum of sine and cosine waves of varying frequency and amplitude (often an infinite sum). As a simple example, consider the graphs shown in Figure C.1. The graph on the left [Figure C.1a] shows a ramp function. The graph in the middle [Figure C.1b] shows its Fourier components which is a combination of a constant plus sine and cosine waves of varying amplitude. The right graph [Figure C.1c] shows the sum of the waves shown in Figure C.1b and the original ramp function (continuous curve).

Evidently, even just one pair of sine and cosine waves is sufficient in this case to closely approximate the original function. Fourier's theorem states that any function (with some very minor restrictions), can be represented in this way.

This is why sine waves are important. Not because they are common, but because we can represent arbitrarily complex functions using only this rather simple function.

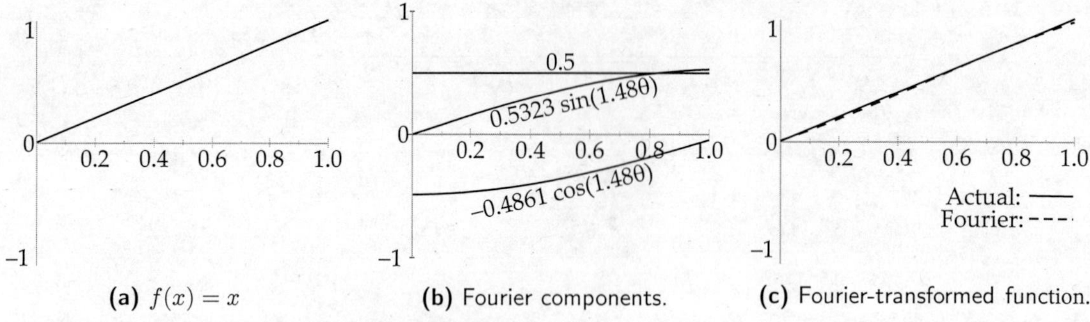

Figure C.1: Demonstration of Fourier transform: (a) a ramp function, (b) Fourier components — **0.5, 0.5323 sin(1.480)** and **–0.4861 cos(1.480)**—of the ramp function and (c) comparison of the Fourier-transformed representation and the original ramp function.

C.1 CONSEQUENCES OF SINUSOIDAL INPUT

The question is, what do we observe if a time-domain analysis of response to a sinusoidal input is made?

Suppose we give an input $x(t)$ to a system and the corresponding output is $y(t)$. If the Laplace transform of the input and output are $X(s)$ and $Y(s)$ respectively and the transfer function is $G(s)$, we know that they are related as

$$G(s) = \frac{Y(s)}{X(s)}$$

[1] Fourier (pronounced *Fooryay*), Jean Baptiste Joseph, Baron (1768–1830) was born in France. He was a mathematician and a physicist.

In general terms, the numerator and denominator can be expressed as product of m and n factors respectively [see, for example, Eq. (2.2)]. For the following study, we treat the numerator as one factor and write the transfer function as

$$G(s) = \frac{Y(s)}{X(s)} = \frac{N(s)}{(s+a_1)(s+a_2)\dots(s+s_n)} \tag{C.1}$$

The Laplace transform of Eq. (11.1) gives

$$X(s) = \frac{A\omega}{s^2+\omega^2}$$

So, we get from Eq. (C.1)

$$Y(s) = \frac{N(s)}{(s+a_1)(s+a_2)\dots(s+s_n)} \cdot \frac{A\omega}{s^2+\omega^2} \tag{C.2}$$

The right hand side of Eq. (C.2) can be expanded in terms of partial fractions as

$$Y(s) = \left(\frac{A_1}{s+a_1} + \frac{A_2}{s+a_2} + \dots \frac{A_n}{s+a_n}\right) + \frac{B_1}{s+\jmath\omega} + \frac{B_2}{s-\jmath\omega} \tag{C.3}$$

Inverse Laplace transform of Eq. (C.3) yields

$$y(t) = \underbrace{A_1e^{-a_1t} + A_2e^{-a_2t} + \dots + A_ne^{-a_nt}}_{\text{transients}} + \underbrace{B_1e^{-\jmath\omega t} + B_2e^{\jmath\omega t}}_{\text{steady state}} \tag{C.4}$$

As $t \to \infty$, the terms marked 'transients' in Eq. (C.4) will die out, but those marked 'steady state' will endure. So, we can write

$$y_{ss}(t) = B_1e^{-\jmath\omega t} + B_2e^{\jmath\omega t} \tag{C.5}$$

The coefficients can be found out as

$$B_1 = (s+\jmath\omega)G(s)\cdot\frac{A\omega}{s^2+\omega^2}\bigg|_{s=-\jmath\omega} = G(s)\cdot\frac{A\omega}{s-\jmath\omega}\bigg|_{s=-\jmath\omega} = -G(-\jmath\omega)\frac{A}{2\jmath} \tag{C.6}$$

$$B_2 = (s-\jmath\omega)G(s)\cdot\frac{A\omega}{s^2+\omega^2}\bigg| s=\jmath\omega = G(s)\cdot\frac{A\omega}{s+\jmath\omega}\bigg|_{s=\jmath\omega} = G(\jmath\omega)\frac{A}{2\jmath} \tag{C.7}$$

Since the complex quantity $G(\jmath\omega)$ can be written in the form of $G(\jmath\omega) = |G(\jmath\omega)|e^{\jmath\phi}$, using Eqs. (C.6) and (C.7) we have from Eq. (C.5)

$$y_{ss}(t) = A\,|G(\jmath\omega)|\left\{\frac{e^{\jmath(\omega t+\phi)}e^{-\jmath(\omega t+\phi)}}{2\jmath}\right\} = A\,|G(\jmath\omega)|\sin(\omega t+\phi) \tag{C.8}$$

Comparing the input [Eq. (11.1)] and the steady-state output [Eq. (C.8)], we observe the following:

1. The frequency of the output remains the same as the input.
2. The amplitude and phase of the output differ from those of the input.

The results are important and that they make the frequency-domain study much simpler.

C.2 SYSTEM OUTPUT WHEN INPUT AND TRANSFER FUNCTION ARE GIVEN

Given that sine waves are important, how can we analyse the response of a circuit or system to sinusoidal inputs?

The answer is that there are many ways of doing this, depending on the mathematical sophistication. Let us demonstrate a fairly basic technique that uses phasors[2].

[2]Description of phasors can be found in almost any circuit theory textbook.

Using complex impedances, it is possible to find the transfer function of a circuit. For example, we consider the circuit of Figure C.2a. Its transfer function can be obtained from the

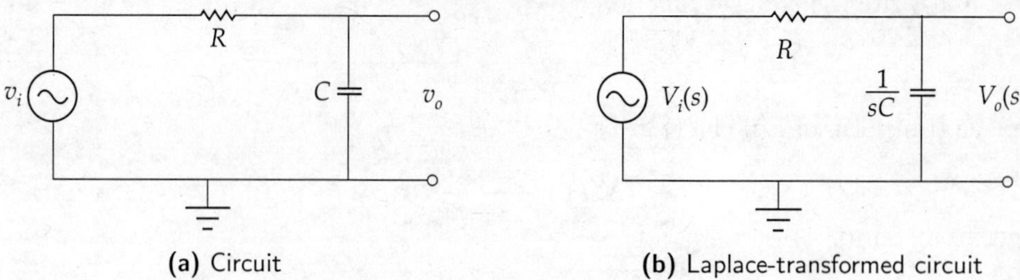

(a) Circuit (b) Laplace-transformed circuit

Figure C.2: The circuit and its Laplace transformed components.

Laplace transformed components [Figure C.2b] as,

$$V_o(s) = \frac{V_i(s)}{R + \frac{1}{sC}} \times \frac{1}{sC} = \frac{V_i(s)}{1 + sRC}$$

Therefore,

$$G(s) = \frac{V_o(s)}{V_i(s)} = \frac{1}{1 + sRC} \tag{C.9}$$

In the frequency domain $s \equiv \jmath\omega$.

Consider the case when $R = 1$ and $C = 0.1$. Then

$$G(s) = \frac{1}{1 + 0.1s} \equiv \frac{1}{1 + \jmath 0.1\omega}$$

Generally, we know the input V_i and want to find the output V_o. We can do this by simple multiplication

$$V_o = V_i \cdot G(s) = V_i \cdot \frac{1}{1 + \jmath 0.1\omega}$$

If we have a phasor representation for the input and the transfer function, the multiplication is simple (multiply magnitudes and add phases). Finding the output becomes easy. Let us have a look at a few examples.

EXAMPLE C.1. Find the output of the system given by Eq. (C.9) when the input is:

(A) $3\sin(30t)$
(B) $3\sin(30t + 40°)$
(C) $3\sin(10t + 40°)$
(D) $3\cos(10t)$

Solution (A) Here, the input is $V_i = 3\sin(30t)$ and the transfer function is

$$G(\jmath\omega) = \frac{1}{1 + \jmath 0.1\omega} = \frac{1}{1 + \jmath 3}$$

Therefore,

	Input	*Transfer characteristic*
Amplitude	$M_i = 3$	$M_t = \frac{1}{\sqrt{1^2 + 3^2}} = 0.316$
Phase	$\phi_i = 0°$	$\phi_t = \frac{1}{\tan^{-1}\frac{3}{1}} = -71.6°$

From the principle of phasor calculation,

$$\text{Output magnitude} = M_i \times M_t = 3 \times 0.316 = 0.95$$

$$\text{Output phase} = \phi_i - \phi_t = 0° - 71.6° = -71.6°$$

Thus,

$$V_o = 0.95\sin(30t - 71.6°)$$

So, while transmitting the signal, the circuit reduces the amplitude from 3 to 0.95 and changes the phase from 0° to −71.6°.

(B) Here, the input is $V_i = 3\sin(30t + 40°)$, transfer function remaining the same. We note that only the phase has changed from the previous case, the magnitude and frequency remaining the same.

Therefore,

	Input	*Transfer characteristic*
Phase	$\phi_i = 40°$	$\phi_t = \dfrac{1}{\tan^{-1}\dfrac{3}{1}} = -71.6°$

From the principle of phasor calculation,

$$\text{Output phase} = \phi_i - \phi_t = 40° - 71.6° = -31.6°$$

Thus,

$$V_o = 0.95\sin(30t - 31.6°)$$

(C) Here, the input is $V_i = 3\sin(10t + 40°)$, and the transfer function is

$$G(j\omega) = \frac{1}{1 + j0.1\omega} = \frac{1}{1 + j1}$$

Therefore,

	Input	*Transfer characteristic*
Amplitude	$M_i = 3$	$M_t = \dfrac{1}{\sqrt{1^2 + 1^2}} = 0.707$
Phase	$\phi_i = 40°$	$\phi_t = \dfrac{1}{\tan^{-1}\dfrac{1}{1}} = -45°$

From the principle of phasor calculation,

$$\text{Output magnitude} = M_i \times M_t = 3 \times 0.707 = 2.12$$

$$\text{Output phase} = \phi_i - \phi_t = 40° - 45° = -5°$$

Thus,

$$V_o = 2.12\sin(30t - 5°)$$

(D) Here, the input is $V_i = 3\cos(10t) = 3\sin(10t + 90°)$, the transfer function remains the same as in (c). We note that the transfer characteristics have remained unaltered, though the input phase has changed.

Therefore, from the principle of phasor calculation

$$\text{Output phase} = \phi_i - \phi_t = 90° - 45° = 45°$$

So,

$$V_o = 2.12\sin(30t + 45°)$$

D

Matrix

In the State Space Representation, we need to invert matrices pretty often. Here, we will show how to do that.

D.1 INVERSION

Standard Method

The standard way of inverting a matrix is to find the adjoint of the matrix and divide it by the determinant of the matrix. The adjoint is found by working out the minor and cofactor of the matrix.

Minor of a Matrix

Minor of a matrix, M_{ij}, is obtained by finding the determinant of the matrix after removing the ith row and jth column from it. Let

$$\mathbf{A} = \begin{bmatrix} a_{11} & a_{12} & a_{13} \\ a_{21} & a_{22} & a_{23} \\ a_{31} & a_{32} & a_{33} \end{bmatrix}$$

Then

$$M_{22} = \begin{vmatrix} a_{11} & \cancel{a_{12}} & a_{13} \\ \cancel{a_{21}} & \cancel{a_{22}} & \cancel{a_{23}} \\ a_{31} & \cancel{a_{32}} & a_{33} \end{vmatrix} = \begin{vmatrix} a_{11} & a_{13} \\ a_{31} & a_{33} \end{vmatrix} = a_{11}a_{33} - a_{13}a_{31}$$

which is obtained by removing the 2nd row and 2nd column from $\mathbf{A}$ and defining a determinant with the remaining elements, as shown.

Cofactor

The cofactor of the element a_{ij} of a matrix is defined as

$$\alpha_{ij} = (-1)^{i+j} M_{ij}$$

where M_{ij} is the minor of the matrix. Cofactor of a_{22} of matrix $\mathbf{A}$ above is

$$\begin{aligned} \alpha_{22} &= (-1)^{2+2}\big(a_{11}a_{33} - a_{13}a_{31}\big) \\ &= a_{11}a_{33} - a_{13}a_{31} \end{aligned}$$

EXAMPLE D.1. Find the cofactor of b_{32} of matrix $\mathbf{B}$ defined as follows:

$$\mathbf{B} = \begin{bmatrix} 1 & 0 & 2 \\ 3 & 1 & 4 \\ 0 & 1 & 5 \end{bmatrix}$$

Solution The minor is

$$M_{32} = \begin{vmatrix} 1 & 2 \\ 3 & 4 \end{vmatrix}$$

Therefore, the cofactor is

$$\alpha_{32} = (-1)^{3+2}(-2) = 2$$

Adjoint of a Matrix

The adjoint of a matrix is obtained after replacing elements of the matrix by their cofactors and then taking the transpose of the resulting cofactor matrix.

EXAMPLE D.2. Find the adjoint of the matrix **B** of Example D.1.

Solution The minors of the matrix **B** are

$$M_{11} = \begin{vmatrix} 1 & 4 \\ 1 & 5 \end{vmatrix} = 5 - 4 = 1 \quad M_{12} = \begin{vmatrix} 3 & 4 \\ 0 & 5 \end{vmatrix} = 15 - 0 = 15 \quad M_{13} = \begin{vmatrix} 3 & 1 \\ 0 & 1 \end{vmatrix} = 3 - 0 = 3$$

$$M_{21} = \begin{vmatrix} 0 & 2 \\ 1 & 5 \end{vmatrix} = 0 - 2 = -2 \quad M_{22} = \begin{vmatrix} 1 & 2 \\ 0 & 5 \end{vmatrix} = 5 - 0 = 5 \quad M_{23} = \begin{vmatrix} 1 & 0 \\ 0 & 1 \end{vmatrix} = 1 - 0 = 1$$

$$M_{31} = \begin{vmatrix} 0 & 2 \\ 1 & 4 \end{vmatrix} = 0 - 2 = -2 \quad M_{32} = \begin{vmatrix} 1 & 2 \\ 1 & 4 \end{vmatrix} = 4 - 6 = -2 \quad M_{33} = \begin{vmatrix} 1 & 0 \\ 3 & 1 \end{vmatrix} = 1 - 0 = 1$$

The cofactors are as follows:

$$\alpha_{11} = (-1)^{1+1}M_{11} = 1 \quad \alpha_{12} = (-1)^{1+2}M_{12} = -15 \quad \alpha_{13} = (-1)^{1+3}M_{13} = 3$$

$$\alpha_{21} = (-1)^{2+1}M_{21} = 2 \quad \alpha_{22} = (-1)^{2+2}M_{22} = 5 \quad \alpha_{23} = (-1)^{2+3}M_{23} = -1$$

$$\alpha_{31} = (-1)^{3+1}M_{31} = -2 \quad \alpha_{32} = (-1)^{3+2}M_{32} = 2 \quad \alpha_{33} = (-1)^{3+3}M_{33} = 1$$

Therefore

$$\text{Adj}\,(\mathbf{B}) = \begin{bmatrix} 1 & -15 & 3 \\ 2 & 5 & -1 \\ -2 & 2 & 1 \end{bmatrix}^{\text{T}} = \begin{bmatrix} 1 & 2 & -2 \\ -15 & 5 & 2 \\ 3 & -1 & 1 \end{bmatrix}$$

The determinant of **B** is $|\mathbf{B}| = 7$ The inverse of a matrix **B** is

$$\mathbf{B}^{-1} = \frac{\text{Adj}\,(\mathbf{B})}{|\mathbf{B}|} = \frac{\begin{bmatrix} 1 & 2 & -2 \\ -15 & 5 & 2 \\ 3 & -1 & 1 \end{bmatrix}}{7} = \begin{bmatrix} \frac{1}{7} & \frac{2}{7} & -\frac{2}{7} \\ -\frac{15}{7} & \frac{5}{7} & \frac{2}{7} \\ \frac{3}{7} & -\frac{1}{7} & \frac{1}{7} \end{bmatrix}$$

However, there is a simpler way of finding the inverse of a matrix as follows.

Simpler Method

If **A** is an $n \times n$ matrix, the $n \times 2n$ matrix that has entries of **A** on the left and of the identity matrix **I** on the right is constructed first.

$$\left[\begin{array}{cccc|cccc} a_{11} & a_{12} & \dots & a_{1n} & 1 & 0 & \dots & 0 \\ a_{21} & a_{22} & \dots & a_{2n} & 0 & 1 & \dots & 0 \\ \vdots & \vdots & \ddots & \vdots & \vdots & \vdots & \ddots & \vdots \\ a_{n1} & a_{n2} & \dots & a_{nn} & 0 & 0 & \dots & 1 \end{array}\right]$$

Then, elementary row operations are used on this large matrix so that the left side is changed to an identity matrix. The right side is automatically transformed into $\mathbf{A}^{-1}$.

EXAMPLE D.3. Find the inverse of the matrix

$$\begin{bmatrix} -3 & 2 & 3 \\ 0 & -1 & 3 \\ 1 & 0 & -2 \end{bmatrix}$$

Solution We begin with the 3×6 matrix whose left half is the given matrix and right half is an identity matrix

$$\left[\begin{array}{ccc|ccc} -3 & 2 & 3 & 1 & 0 & 0 \\ 0 & -1 & 3 & 0 & 1 & 0 \\ 1 & 0 & -2 & 0 & 0 & 1 \end{array}\right]$$

Next we transform the left hand side of this new matrix into an identity matrix by performing the following sequence of elementary operations (as indicated on the left; R_1, R_2 etc indicate rows) on the *entire* new matrix.

$$\xrightarrow{R_3 \leftrightarrow R_1} \left[\begin{array}{ccc|ccc} 1 & 0 & -2 & 0 & 0 & 1 \\ 0 & -1 & 3 & 0 & 1 & 0 \\ -3 & 2 & 3 & 1 & 0 & 0 \end{array}\right]$$

$$\xrightarrow{-R_2 \to R_2} \left[\begin{array}{ccc|ccc} 1 & 0 & -2 & 0 & 0 & 1 \\ 0 & 1 & -3 & 0 & -1 & 0 \\ -3 & 2 & 3 & 1 & 0 & 0 \end{array}\right]$$

$$\xrightarrow{R_3 + 3R_1 \to R_3} \left[\begin{array}{ccc|ccc} 1 & 0 & -2 & 0 & 0 & 1 \\ 0 & 1 & -3 & 0 & -1 & 0 \\ 0 & 2 & -3 & 1 & 0 & 3 \end{array}\right]$$

$$\xrightarrow{R_3 - 2R_2 \to R_3} \left[\begin{array}{ccc|ccc} 1 & 0 & -2 & 0 & 0 & 1 \\ 0 & 1 & -3 & 0 & -1 & 0 \\ 0 & 0 & 3 & 1 & 2 & 3 \end{array}\right]$$

$$\xrightarrow{\frac{1}{3}R_3 \to R_3} \left[\begin{array}{ccc|ccc} 1 & 0 & -2 & 0 & 0 & 1 \\ 0 & 1 & -3 & 0 & -1 & 0 \\ 0 & 0 & 1 & \frac{1}{3} & \frac{2}{3} & 1 \end{array}\right]$$

$$\xrightarrow{R_1 + 2R_3 \to R_1} \left[\begin{array}{ccc|ccc} 1 & 0 & 0 & \frac{2}{3} & \frac{4}{3} & 3 \\ 0 & 1 & -3 & 0 & -1 & 0 \\ 0 & 0 & 1 & \frac{1}{3} & \frac{2}{3} & 1 \end{array}\right]$$

$$\xrightarrow{R_2 + 3R_3 \to R_2} \left[\begin{array}{ccc|ccc} 1 & 0 & 0 & \frac{2}{3} & \frac{4}{3} & 3 \\ 0 & 1 & 0 & 1 & 1 & 3 \\ 0 & 0 & 1 & \frac{1}{3} & \frac{2}{3} & 1 \end{array}\right]$$

Therefore,

$$\mathbf{A}^{-1} = \begin{bmatrix} \frac{2}{3} & \frac{4}{3} & 3 \\ 1 & 1 & 3 \\ \frac{1}{3} & \frac{2}{3} & 1 \end{bmatrix}$$

We can verify it by observing

$$\mathbf{AA}^{-1} = \begin{bmatrix} -3 & 2 & 3 \\ 0 & -1 & 3 \\ 1 & 0 & -2 \end{bmatrix} \begin{bmatrix} \frac{2}{3} & \frac{4}{3} & 3 \\ 1 & 1 & 3 \\ \frac{1}{3} & \frac{2}{3} & 1 \end{bmatrix} = \begin{bmatrix} 1 & 0 & 0 \\ 0 & 1 & 0 \\ 0 & 0 & 1 \end{bmatrix} = \mathbf{I}$$

$$\mathbf{A}^{-1}\mathbf{A} = \begin{bmatrix} \frac{2}{3} & \frac{4}{3} & 3 \\ 1 & 1 & 3 \\ \frac{1}{3} & \frac{2}{3} & 1 \end{bmatrix} \begin{bmatrix} -3 & 2 & 3 \\ 0 & -1 & 3 \\ 1 & 0 & -2 \end{bmatrix} = \begin{bmatrix} 1 & 0 & 0 \\ 0 & 1 & 0 \\ 0 & 0 & 1 \end{bmatrix} = \mathbf{I}$$

Inverse of a 2 × 2 Matrix

The following rule provides a simple way for finding the inverse of a 2 × 2 matrix.

If $\mathbf{A} = \begin{bmatrix} a & b \\ c & d \end{bmatrix}$, then

$$\mathbf{A}^{-1} = \frac{1}{ad - bc} \begin{bmatrix} d & -b \\ -c & a \end{bmatrix}$$

If $ad - bc = 0$, then $\mathbf{A}$ has no inverse.

EXAMPLE D.4. Invert the matrix

$$\mathbf{A} = \begin{bmatrix} -2 & 1 \\ -3 & 0 \end{bmatrix}$$

Solution The inverse is

$$\mathbf{A}^{-1} = \frac{1}{(-2)(0) - (1)(-3)} \begin{bmatrix} 0 & -1 \\ 3 & -2 \end{bmatrix} = \frac{1}{3} \begin{bmatrix} 0 & -1 \\ 3 & -2 \end{bmatrix}$$

D.2 EIGENVALUES AND EIGENVECTORS

If $\mathbf{A}$ is an $n \times n$ matrix, $\mathbf{x}$ is a nonzero vector and λ is any scalar so that

$$\mathbf{A}\mathbf{x} = \lambda\mathbf{x} \tag{D.1}$$

then $\mathbf{x}$ is called an eigenvector and λ, an eigenvalue of $\mathbf{A}$. Eigenvalues and eigenvectors are not independent of each other. Hence, it is better said that $\mathbf{x}$ is the eigenvector associated with or corresponding to λ and vice versa.

Suppose

$$\mathbf{A} = \begin{bmatrix} -4 & 2 \\ 3 & -5 \end{bmatrix}$$

then

$$\mathbf{x} = \begin{bmatrix} 1 \\ 1 \end{bmatrix}$$

is an eigenvector associated with the eigenvalue -2 because

$$\begin{bmatrix} -4 & 2 \\ 3 & -5 \end{bmatrix} \begin{bmatrix} 1 \\ 1 \end{bmatrix} = -2 \begin{bmatrix} 1 \\ 1 \end{bmatrix}$$

conforming to the definition given by Eq. (D.1).

Eigenvalues

To find eigenvalues of a matrix $\mathbf{A}$, we have to solve the characteristic equation given by

$$\Delta\,(\lambda\mathbf{I} - \mathbf{A}) = 0 \tag{D.2}$$

EXAMPLE D.5. Find the eigenvalues of the matrix

$$\mathbf{A} = \begin{bmatrix} -4 & 2 \\ 3 & -5 \end{bmatrix}$$

Solution For the matrix **A** defined above, the characteristic function is

$$\Delta(\lambda\mathbf{I} - \mathbf{A}) = \begin{vmatrix} \lambda + 4 & -2 \\ -3 & \lambda + 5 \end{vmatrix}$$

The characteristic equation $\Delta(\lambda\mathbf{I} - \mathbf{A}) = 0$ gives

$$(\lambda + 4)(\lambda + 5) - 6 = 0$$

or

$$\lambda^2 + 9\lambda + 14 = 0$$

yielding

$$\lambda_1 = -2, \qquad \lambda_2 = -7$$

Hence, the eigenvalues of **A** are -2 and -7.

Eigenvectors

To find eigenvectors, we have to proceed as shown in the following example.

EXAMPLE D.6. Find the eigenvectors of the matrix

$$\mathbf{A} = \begin{bmatrix} -4 & 2 \\ 3 & -5 \end{bmatrix}$$

Solution We know from Example D.5 that the eigenvalues of the matrix are -7 and -2. Suppose $\mathbf{c}_1$ is the *column eigenvector* associated with the eigenvalue $\lambda_1 = -7$. Then $\mathbf{c}_1$ must satisfy

$$(\lambda_1\mathbf{I} - \mathbf{A}) = 0 \tag{i}$$

or

$$\begin{vmatrix} \lambda + 4 & -2 \\ -3 & \lambda + 5 \end{vmatrix} \begin{bmatrix} c_1 \\ c_2 \end{bmatrix} = 0 \tag{ii}$$

Substituting $\lambda_1 = -7$, we get from Eq. (ii)

$$\begin{aligned} -3c_1 - 2c_2 &= 0 \\ -3c_1 - 2c_2 &= 0 \end{aligned} \tag{iii}$$

Both of the equations in Eq. (iii) give the relation $3c_1 = -2c_2$. Let us put $c_1 = 2$ which gives $c_2 = -3$. Then the eigenvector corresponding to the eigenvalue -7 is

$$\mathbf{c}_1 = \begin{bmatrix} 2 \\ -3 \end{bmatrix} \tag{iv}$$

In a similar way, for the eigenvalue -2, we get equations

$$2c_1 - 2c_2 = 0$$

$$-3c_1 + 3c_2 = 0$$

leading to the relation $c_1 = c_2$. Let us put $c_1 = c_2 = 1$. Then the column eigenvector of **A** corresponding to the eigenvalue -2 is

$$\mathbf{c}_2 = \begin{bmatrix} 1 \\ 1 \end{bmatrix} \tag{v}$$

Hence, the eigenvector matrix is given by

$$\mathbf{P} = \begin{bmatrix} \mathbf{c}_1 & \mathbf{c}_2 \end{bmatrix} = \begin{bmatrix} 2 & 1 \\ -3 & 1 \end{bmatrix}$$

The eigenvectors that we presented in Eqs. (iv) and (v) are often called *right vectors*[1].

The matrix $\mathbf{A}$ also possesses *row* or *left* eigenvectors which differ from the column vectors. These are found out as follows:
For $\lambda = -7$

$$\begin{bmatrix} r_1 & r_2 \end{bmatrix} \begin{bmatrix} -3 & -2 \\ -3 & -2 \end{bmatrix} = 0$$

This yields the relation

$$-3r_1 - 3r_2 = 0$$

$\Rightarrow$ $$r_1 = -r_2$$

Therefore, $$\mathbf{r_1} = \begin{bmatrix} 1 & -1 \end{bmatrix} \quad \text{(vi)}$$

Similarly, for $\lambda = -2$

$$\begin{bmatrix} r_1 & r_2 \end{bmatrix} \begin{bmatrix} 2 & -2 \\ 3 & 3 \end{bmatrix} = 0$$

$\Rightarrow$ $$2r_1 - 3r_2 = 0$$

$\Rightarrow$ $$2r_1 = 3r_2$$

Therefore, $$\mathbf{r_2} = \begin{bmatrix} 3 & 2 \end{bmatrix} \quad \text{(vii)}$$

Rank

A matrix $\mathbf{A}$ is said to be of rank r if there exists an $r \times r$ submatrix of $\mathbf{A}$ which is non-singular and all other $q \times q$ (where $q \geq r+1$) submatrices are singular. Let us consider the 4×4 matrix

$$\mathbf{A} = \begin{bmatrix} 1 & 2 & 3 & 4 \\ 0 & 1 & -1 & 0 \\ 1 & 0 & 1 & 2 \\ 1 & 1 & 0 & 2 \end{bmatrix}$$

Here, det $\mathbf{A} = 0$ but the determinant of the submatrix consisting of first three columns by first three rows is not zero. Hence, the rank of $\mathbf{A}$ is 3.

[1]The prefix eigen- is adopted from the German word eigen (cognate with the English word own) for 'right', 'characteristic', 'own'.

E

z-Transform Tables

Table E.1: Commonly used z-transforms

Line No.	*Time function* $x[n],\ n > 0$	*z-transform* $X(z)$	*Region of Convergence*
1	$x[n]$	$\sum_{n=0}^{\infty} x[n]z^{-n}$	$\|z\| > 0$
2	$\delta[n]$	1	
3	$au[n]$	$\frac{az}{z-1}$	$\|z\| > 1$
4	$nu[n]$	$\frac{z}{(z-1)^2}$	$\|z\| > 1$
5	$n^2u[n]$	$\frac{z(z+1)}{(z-1)^3}$	$\|z\| > 1$
6	$a^n u[n]$	$\frac{z}{z-a}$	$\|z\| > a$
7	$e^{-na}u[n]$	$1 + \frac{z}{z-e^{-a}}$	$\|z\| > e^{-a}$
8	$na^n u[n]$	$\frac{az}{(z-a)^2}\ n$	$\|z\| > a$
9	$\sin[an]u[n]$	$\frac{z\sin(a)}{z^2 - 2z\cos(a) + 1}$	$\|z\| > 1$
10	$\cos[an]u[n]$	$\frac{z[z-\cos(a)]}{z^2 - 2z\cos(a) + 1}$	$\|z\| > 1$
11	$a^n \sin[bn]u[n]$	$\frac{[a\sin(b)]z}{z^2 - [2a\cos(b)]z + a^2}$	$\|z\| > \|a\|$
12	$a^n \cos b[n]u[n]$	$\frac{z[z - a\cos(b)]}{z^2 - [2a\cos(b)]z + a^2}$	$\|z\| > \|a\|$
13	$e^{-a[n]} \sin b[n]u[n]$	$\frac{[e^{-a\sin(b)}]z}{z^2 - [2e^{-a}\cos(b)]z + e^{-2a}}$	$\|z\| > e^{-a}$
14	$e^{-a[n]} \cos b[n]u[n]$	$\frac{z[z - e^{-a}\cos(b)]}{z^2 - [2e^{-a}\cos(b)] + e^{-2a}}$	$\|z\| > e^{-a}$

Table E.2: z-transforms of $x[n+m]$ and $x[n\text{-}m]$

Line	*Discrete function*	*z-transform*
1	$x[n+4]$	$z^4X(z) - z^4x(0) - z^3x(1) - z^2x(2)$
2	$x[n+3]$	$z^3X(z) - z^3x(0) - z^2x(1) - zx(2)$
3	$x[n+2]$	$z^2X(z) - z^2x(0) - zx(1)$
4	$x[n+1]$	$zX(z) - zx(0)$
5	$x[n]$	$X(z)$
6	$x[n-1]$	$z^{-1}X(z)$
7	$x[n-2]$	$z^{-2}X(z)$
8	$x[n-3]$	$z^{-3}X(z)$
9	$x[n-4]$	$z^{-4}X(z)$

Index